Somatosensory Feedback for Neuroprosthetics

Somatosensory Feedback for Neuroprosthetics

Edited by

Burak Güçlü
Institute of Biomedical Engineering, Boğaziçi University, Istanbul, Turkey

ELSEVIER

ACADEMIC PRESS
An imprint of Elsevier

Academic Press is an imprint of Elsevier
125 London Wall, London EC2Y 5AS, United Kingdom
525 B Street, Suite 1650, San Diego, CA 92101, United States
50 Hampshire Street, 5th Floor, Cambridge, MA 02139, United States
The Boulevard, Langford Lane, Kidlington, Oxford OX5 1GB, United Kingdom

British Library Cataloguing-in-Publication Data
A catalogue record for this book is available from the British Library

Library of Congress Cataloging-in-Publication Data
A catalog record for this book is available from the Library of Congress

ISBN: 978-0-12-822828-9

For Information on all Academic Press publications
visit our website at https://www.elsevier.com/books-and-journals

Publisher: Mara Conner
Acquisitions Editor: Sonnini Yura
Editorial Project Manager: Isabella Silva
Production Project Manager: Poulouse Joseph
Cover Designer: Victoria Pearson

Typeset by MPS Limited, Chennai, India

Dedication

There is nothing in the real world which is merely an inert fact. Every reality is there for feeling: it promotes feeling; and it is felt.

—*Process and Reality: An Essay in Cosmology*
by Alfred North Whitehead (1929)

Dedicated to my parents and all teachers who taught me how to learn.

Contents

List of contributors

Dilek Betul Arslan Institute of Biomedical Engineering, Boğaziçi University, Istanbul, Turkey

Marion Badi Bertarelli Foundation Chair in Translational Neuroengineering, Center for Neuroprosthetics and Institute of Bioengineering, Ecole Polytechnique Federale de Lausanne, Lausanne, Switzerland

Rebecca Baldi The BioRobotics Institute, Scuola Superiore Sant'Anna, Pisa, Italy; Department of Excellence in Robotics and AI, Scuola Superiore Sant'Anna, Pisa, Italy

Sliman J. Bensmaia Department of Organismal Biology and Anatomy, University of Chicago, Chicago, IL, United States

David A. Bjånes Department of Electrical and Computer Engineering, University of Washington, Seattle, WA, United States; Center for Neurotechnology, University of Washington, Seattle, WA, United States

Kyle P. Blum Department of Physiology, Feinberg School of Medicine, Northwestern University, Chicago, IL, United States

Simon Borgognon Department of Neuroscience and Movement Science, Platform of Translational Neurosciences, University of Fribourg, Fribourg, Switzerland; Center for Neuroprosthetics and Brain Mind Institute, School of Life Sciences, Swiss Federal Institute of Technology (EPFL), Geneva, Switzerland

David J. Caldwell Center for Neurotechnology, University of Washington, Seattle, WA, United States; Department of Bioengineering, University of Washington, Seattle, WA, United States; Medical Scientist Training Program, University of Washington, Seattle, WA, United States

Thierri Callier Department of Organismal Biology and Anatomy, University of Chicago, Chicago, IL, United States

Leonardo Cappello The BioRobotics Institute, Scuola Superiore Sant'Anna, Pisa, Italy; Department of Excellence in Robotics and AI, Scuola Superiore Sant'Anna, Pisa, Italy

Raeed H. Chowdhury Department of Bioengineering, Swanson School of Engineering, University of Pittsburgh, Pittsburgh, PA, United States

Andrea Cimolato NEAR Lab, Department of Electronics, Information Science, and Bioengineering (DEIB), Politecnico di Milano, Milan, Italy; Rehab Technologies, Istituto Italiano di Tecnologia (IIT), Genova, Italy

Christian Cipriani The BioRobotics Institute, Scuola Superiore Sant'Anna, Pisa, Italy; Department of Excellence in Robotics and AI, Scuola Superiore Sant'Anna, Pisa, Italy

Jeneva A. Cronin Center for Neurotechnology, University of Washington, Seattle, WA, United States; Department of Bioengineering, University of Washington, Seattle, WA, United States

Jaume del Valle Institute of Neurosciences and Department of Cell Biology, Physiology and Immunology, Universitat Autònoma de Barcelona, and Centro de Investigación Biomédica en Red sobre Enfermedades Neurodegenerativas (CIBERNED), Bellaterra, Spain

İsmail Devecioğlu Department of Biomedical Engineering, Çorlu Faculty of Engineering, Tekirdağ Namık Kemal University, Tekirdağ, Turkey

Jakob Dideriksen Department of Health Science and Technology, Aalborg University, Aalborg, Denmark

Strahinja Dosen Department of Health Science and Technology, Aalborg University, Aalborg, Denmark

Sigrid Dupan Edinburgh Neuroprosthetics Laboratory, School of Informatics, Edinburgh University, Edinburgh, United Kingdom

Ezgi Tuna Erdogan Department of Physiology, School of Medicine, Koç Universitesi, Istanbul, Turkey

Leonard Frederik (Engels) The BioRobotics Institute, Scuola Superiore Sant'Anna, Pisa, Italy; Department of Excellence in Robotics and AI, Scuola Superiore Sant'Anna, Pisa, Italy

Burak Güçlü Institute of Biomedical Engineering, Boğaziçi University, İstanbul, Turkey

Jacqueline S. Hebert Division of Physical Medicine & Rehabilitation, Department of Medicine, Faculty of Medicine & Dentistry, University of Alberta 5-005 Katz Group Centre Edmonton, Alberta, AB, Canada

Francesco Iberite The BioRobotics Institute, Scuola Superiore Sant'Anna, Pisa, Italy; Department of Excellence in Robotics and AI, Scuola Superiore Sant'Anna, Pisa, Italy

Leen Jabban Department of Electronic and Electrical Engineering, University of Bath, Bath, United Kingdom

Sacit Karamursel Department of Physiology, School of Medicine, Koç Universitesi, Istanbul, Turkey

Natalija Katic Institute Mihajlo Pupin, Belgrade, Serbia; School of Electrical Engineering, University of Belgrade, Belgrade, Serbia

Oğuzhan Kırtaş Department of Health Science and Technology, Aalborg University, Aalborg, Denmark

Lila H. Levinson Center for Neurotechnology, University of Washington, Seattle, WA, United States; Graduate Program in Neuroscience, University of Washington, Seattle, WA, United States

Paul D. Marasco Department of Biomedical Engineering, Lerner Research Institute, Cleveland Clinic, Cleveland, OH, United States

Alberto Mazzoni The BioRobotics Institute, Scuola Superiore Sant'Anna, Pisa, Italy; Department of Excellence in Robotics and AI, Scuola Superiore Sant'Anna, Pisa, Italy

Vincent Mendez Bertarelli Foundation Chair in Translational Neuroengineering, Center for Neuroprosthetics and Institute of Bioengineering, Ecole Polytechnique Federale de Lausanne, Lausanne, Switzerland

Daniel R. Merrill Dan Merrill Consulting, LLC, Salt Lake City, UT, United States

Benjamin W. Metcalfe Department of Electronic and Electrical Engineering, University of Bath, Bath, United Kingdom

Silvestro Micera Bertarelli Foundation Chair in Translational Neuroengineering, Center for Neuroprosthetics and Institute of Bioengineering, Ecole Polytechnique Federale de Lausanne, Lausanne, Switzerland; The BioRobotics Institute, Scuola Superiore Sant'Anna, Pisa, Italy; Department of Excellence in Robotics and AI, Scuola Superiore Sant'Anna, Pisa, Italy

Lee E. Miller Department of Physiology, Feinberg School of Medicine, Northwestern University, Chicago, IL, United States; Department of Biomedical Engineering, McCormick School of Engineering, Northwestern University, Evanston, IL, United States; Department of Physical Medicine and Rehabilitation, Feinberg School of Medicine, Northwestern University, Chicago, IL, United States; Shirley Ryan AbilityLab, Chicago, IL, United States

Chet T. Moritz Department of Electrical and Computer Engineering, University of Washington, Seattle, WA, United States; Center for Neurotechnology, University of Washington, Seattle, WA, United States; Department of Rehabilitation Medicine, University of Washington, Seattle, WA, United States; Department of Physiology & Biophysics, University of Washington, Seattle, WA, United States

Senol Mutlu Department of Electrical and Electronics Engineering, Boğaziçi University, Istanbul, Turkey

Xavier Navarro Institute of Neurosciences and Department of Cell Biology, Physiology and Immunology, Universitat Autònoma de Barcelona, and Centro de Investigación Biomédica en Red sobre Enfermedades Neurodegenerativas (CIBERNED), Bellaterra, Spain

Kianoush Nazarpour Edinburgh Neuroprosthetics Laboratory, School of Informatics, Edinburgh University, Edinburgh, United Kingdom

Sevgi Öztürk Institute of Biomedical Engineering, Boğaziçi University, İstanbul, Turkey

Esin Ozturk-Isik Institute of Biomedical Engineering, Boğaziçi University, Istanbul, Turkey

Joseph E. O'Doherty Neuralink Corp., San Francisco, CA, United States

Rajesh P.N. Rao Center for Neurotechnology, University of Washington, Seattle, WA, United States; Paul G. Allen School of Computer Science & Engineering, University of Washington, Seattle, WA, United States

Greta Preatoni Neuroengineering Laboratory, Department of Health Sciences and Technology, Institute for Robotics and Intelligent Systems, ETH Zürich, Zürich, Switzerland

Stanisa Raspopovic Neuroengineering Laboratory, Department of Health Sciences and Technology, Institute for Robotics and Intelligent Systems, ETH Zürich, Zürich, Switzerland

Bruno Rodríguez-Meana Institute of Neurosciences and Department of Cell Biology, Physiology and Immunology, Universitat Autònoma de Barcelona, and Centro de Investigación Biomédica en Red sobre Enfermedades Neurodegenerativas (CIBERNED), Bellaterra, Spain

Evren Samur Department of Mechanical Engineering, Boğaziçi University, Istanbul, Turkey

Solaiman Shokur Bertarelli Foundation Chair in Translational Neuroengineering, Center for Neuroprosthetics and Institute of Bioengineering, Ecole Polytechnique Federale de Lausanne, Lausanne, Switzerland; The BioRobotics Institute and Department of Excellence in Robotics and AI, Scuola Superiore Sant'Anna, Pisa, Italy

Joseph Sombeck Department of Biomedical Engineering, McCormick School of Engineering, Northwestern University, Evanston, IL, United States

Giacomo Valle Neuroengineering Laboratory, Department of Health Sciences and Technology, Institute for Robotics and Intelligent Systems, ETH Zürich, Zürich, Switzerland

Christopher Versteeg Department of Biomedical Engineering, McCormick School of Engineering, Northwestern University, Evanston, IL, United States

Buse Buz Yalug Institute of Biomedical Engineering, Boğaziçi University, Istanbul, Turkey

Preface

It is my great pleasure to present this book that focuses on particular neuroprostheses for compensating the loss of somatosensory function due to amputation or various injuries, such as those caused by trauma or neurodegenerative diseases. The somatosensory system works in tandem with the motor system, but so far, the majority of prosthetics research and commercial efforts have concentrated on accommodating movement deficits. With the development of state-of-the-art neuroprostheses in the last few decades, it has become evident that somatosensory input (mainly as touch and proprioception) is essential for motor control, manipulating objects, and embodiment, in addition to its primary role in sensory perception. Therefore designing prosthetic and robotic devices that can provide the "feel" of touch and limb position/movement to the user is one of the major challenges of biomedical and neural engineering today.

There have been exciting studies published in scientific journals regarding somatosensory feedback and its use in bidirectional neuroprostheses, but these are quite spread out across the literature. Several excellent books on neuroprostheses do exist, but they have wider coverage and less information about somatosensory feedback. This book aims to collect the relevant information from a wide spectrum of sources, which include noninvasive versus invasive approaches, peripheral versus central (i.e., cortical) feedback, animal models, and human patients. Such a collection, written by experts in the field, is a unique reference and presents the latest advances concisely. I hope it will be useful for students, researchers, and also engineers from the industry.

The book starts with chapters reviewing the basic anatomy, physiology, and psychophysics of the somatosensory system, sensorimotor control, and instrumentation (including basic electrochemistry) in Part I. Most of the remaining chapters are grouped based on different approaches to somatosensory feedback in neuroprostheses. These chapters are dedicated to noninvasive methods (Part II), invasive peripheral nerve stimulation (Part III), and invasive cortical stimulation (Part IV). The last part of the book (Part V) contains chapters related to future technologies (novel neural interfaces, miniaturized and wireless electronics, signal processing, soft robotics, artificial intelligence), and a chapter on safety and regulations.

I would like to thank all the chapter contributors who had to work during an extraordinary time in human history (i.e., COVID-19 lockdowns). Without their support and patience for the delays in my editorial duties, this book would not be possible. I am also indebted to Sonnini Ruiz Yura and Isabella Conti Silva from the Elsevier office in Brazil for giving me the initial boost of motivation for engaging in this kind of work. In addition, Isabella had to fine-tune my course many times as things slowed down and the line between home and work blurred; I appreciate her patience. Although we had to drop a few chapters on the way, the outcome is quite similar to what we planned at the beginning, for the material to be covered. I also thank my former student and now highly regarded colleague, Dr. İsmail Devecioğlu, for his help at the proofing stage, and the production staff of Elsevier in India.

My research for this work was supported by TÜBİTAK Grant 117F481 within European Union's FLAG-ERA JTC 2017 project GRAFIN (PI: Xavier Navarro). Prof. Navarro has been influential in our collaboration and the dissemination activities as part of the consortium he leads. On behalf of our team at the Tactile Research Laboratory in the Institute of Biomedical Engineering, I send my sincere thanks to him. We also benefited from the synergy established by iNavigate (H2020-MSCA-RISE-2019, Grant agreement ID: 873178) project (PI: Tansu Celikel). Many researchers at Boğaziçi University now have new opportunities to learn more about neurotechnology and neural engineering. As you read this book, probably newer papers will already have been published. There are always things evolving very fast, and it is becoming ever more difficult to keep up with digitalization and shared scientific knowledge. Each chapter presents a wealth of references, and I am grateful to all the authors for their hard and valuable work. I take the blame for anything missed and also hope the chapters are enjoyable to read for everyone interested in somatosensory feedback.

Burak Güçlü
Boğaziçi University, Istanbul, Turkey

Part I

Background and fundamentals

Chapter 1

Introduction to somatosensory neuroprostheses

Burak Güçlü
Institute of Biomedical Engineering, Boğaziçi University, İstanbul, Turkey

ABSTRACT
Although simple prostheses and orthoses date back to antiquity, there has been an explosion of research and development for neuroprostheses in recent decades. This is largely due to advances in basic neuroscience, neuroengineering, miniaturization and fabrication of electromechanical systems, development of novel algorithms for signal processing, and increased synergy among all partners, from the laboratory to the clinical setting and commercialization. This chapter covers a brief history and overview of the field, emphasizing the need for somatosensory function in motor neuroprostheses. Basic components of somatosensation and current approaches for providing somatosensory feedback information are introduced.

Keywords: Somatosensory feedback; prosthesis; touch; proprioception; haptics; electrotactile; vibrotactile; nerve stimulation; cortical stimulation

1.1 Scope and history of neuroprostheses

Prosthesis is the ancient Greek word (πρόσθεσις) for addition, attachment, and application. It is still used in this context (e.g., arithmetic addition) as well as in its medical meaning. Its standard technical meaning in many languages refers to a device for replacing a missing body part. Neuroprosthesis, in contrast, is a very recent word, emphasizing the direct or indirect communication of a prosthetic device with the nervous system. The meaning can be extended to include devices which communicate with muscles (e.g., myoelectric prostheses), since muscles are indeed anatomically and functionally connected to the nervous system. Therefore the wide definition of neuroprosthesis may include all active devices, which record neural and muscle signals to detect physiological states and/or which stimulate associated tissues, for partially restoring function in affected or replaced organs. For example, although a simple electrocardiograph (ECG) machine would not

Somatosensory Feedback for Neuroprosthetics. DOI: https://doi.org/10.1016/B978-0-12-822828-9.00022-8

qualify for that definition, an automated external defibrillator may, because it is used to restore the normal cardiac rhythm. Nevertheless, the most common use for the term neuroprosthesis refers to *implantable* devices for direct therapeutic intervention, or sensory/motor restoration and rehabilitation. A bidirectional hand neuroprosthesis which helps an amputee move a robotic hand with his/her neural signals, and which also provides somatosensory feedback (SF), typically requires a state-of-the-art implant for direct neural communication. Similarly, electroceuticals, the novel therapeutic devices which target specific organs by electrical stimulation, have recently gained significant attention regarding certain diseases and medical conditions. A vagus nerve stimulator for epilepsy is a good example. The 21st century has already seen the significant impact of neuroengineering, neurobionics, and bioelectronic medicine. It is remarkable, however, that such innovative ideas have old roots.

The earliest prostheses are thought to date back to ancient Egypt and Iran (c.2800–3000 BCE). An ocular prosthesis was discovered in the excavations of the Burnt City (Shahr-e Sukhteh, Iran); the world's oldest prosthesis known was in the left eye socket of a woman's remains (Moghadasi, 2014). There have been many accounts of wooden or metal limbs and hooks throughout history, but the oldest wooden splints and a prosthetic limb part, that is, a toe, were found in Egyptian mummies from c.1000 to 2000 BCE (Nerlich et al., 2000). Other notable historical prostheses are the artificial limb of Capri (c. 300 BCE), the iron "hand" of Roman general Marcus Sergius (c. 200 BCE), prosthetic limbs designed by the amputation surgeon Páre (16th century), the Hanger limb during the American Civil War, body-powered upper limb prosthesis by Baliff (early 1800s), and leg prosthetics with sockets by Dubois Parmelee (1863) from New York (McNamee, 2014; Spires et al., 2014)

Within the realm of the definition given above, the earliest neuroprostheses (e.g., from the 1950s) with implantable parts may be considered as the cardiac pacemakers for regulating the heart rhythm, which is naturally under neural control. Historically, the use of electricity for medical purposes, for example, to alleviate pain, may have its beginnings in ancient Egypt, Babylon, and Greece. Small electric fish supposedly provided relief by numbing the sensation of pain. As a matter of fact, Roman physician Largus (1st century CE) prescribed torpedo fish for headache and gout (Tsoucalas et al., 2014). However, as is widely known, the link between biology and electricity was first established by the work of Galvani, Volta, and others (18th and 19th centuries); whereas du Bois-Reymond is considered as the father of electrophysiology by recording action potentials in 1843 (Finkelstein, 2015). In the 20th century, electrophysiology and neuroscience developed at an incredible pace due to advances in engineering and technology, and established the basic scientific background for neuroprostheses. For example, the fully implantable Elmqvist-Senning (1958) pacemaker would not have been possible without the invention of the transistor (1947). Fig. 1.1 shows the timeline of some important milestones achieved in neuroprosthetics research.

The first myoelectric prosthesis was created by a German student, Reinhold Reiter, in 1948; however, it was Alexander Kobrinski's "Russian Hand" (1960) which first gained clinical use and was sold in Britain and Canada (Zuo & Olson, 2014). By the 1970s, it was understood that neuroprostheses critically depend on four pillars of research: biomaterials, neural stimulation, neural recording, and electronics (Donaldson & Brindley, 2016). Advances in microelectronics, materials and hermetic packaging, electrochemistry and battery technology, microelectromechanical systems and electrode technology, clinical and surgical practice, and rehabilitation methods all contributed significantly to the development of neuroprostheses into a multi-billion-dollar industry. One could also add the impetus of digitalization, novel signal processing techniques, various design and simulation software platforms, communication technology, and connectivity in the success of modern devices.

Pioneers of sensory neuroprostheses were the early cochlear implant (1961) by House and Doyles (Mudry & Mills, 2013), the 80-channel visual prosthesis for cortical surface stimulation (Brindley & Lewin, 1968), and perhaps, the auditory prosthesis based on cortical surface stimulation (Dobelle et al., 1973). Wilder Penfield's seminal work in the 1930s may be considered as the first step for the development of a cortical somatosensory neuroprosthesis in humans (Catani, 2017); since then, there have been numerous studies which have applied electrical stimulation to somatosensory areas in epileptic patients as part of long-term presurgical investigations (e.g., see Libet, 1982; Woolsey et al., 1979). Some of these indeed utilized implanted electrodes (Johnson et al., 2013; Richer et al., 1993). However, the feasibility of an implantable somatosensory neuroprosthesis for the brain was established very recently, using cortical surface (Cronin et al., 2016; Hiremath et al., 2017) and intracortical (Flesher et al., 2016) stimulation. Peripheral nerve stimulation for SF in prostheses was applied as early as the 1970s (Clippinger et al., 1974). Its success was unequivocally established by Kenneth Horch and co-workers (Dhillon & Horch, 2005); and this area of neuroprostheses advanced rapidly in the subsequent years. The late 2000s also included exciting results from the clinical trials of retinal neuroprostheses (Humayun et al., 2009; for review, see Bloch et al., 2019).

While a sensory neuroprosthesis aims to convey sensory information to the nervous system, a motor neuroprosthesis reads out motor-related signals to activate external devices, or sometimes to target dysfunctional muscles, voluntarily. The devices which control robotic limbs, computers, or other machines by signals recorded from the brain are typically called brain–computer interfaces (BCIs). The earliest human BCI was noninvasive and based on electroencephalogram (EEG) signals to control an object on a

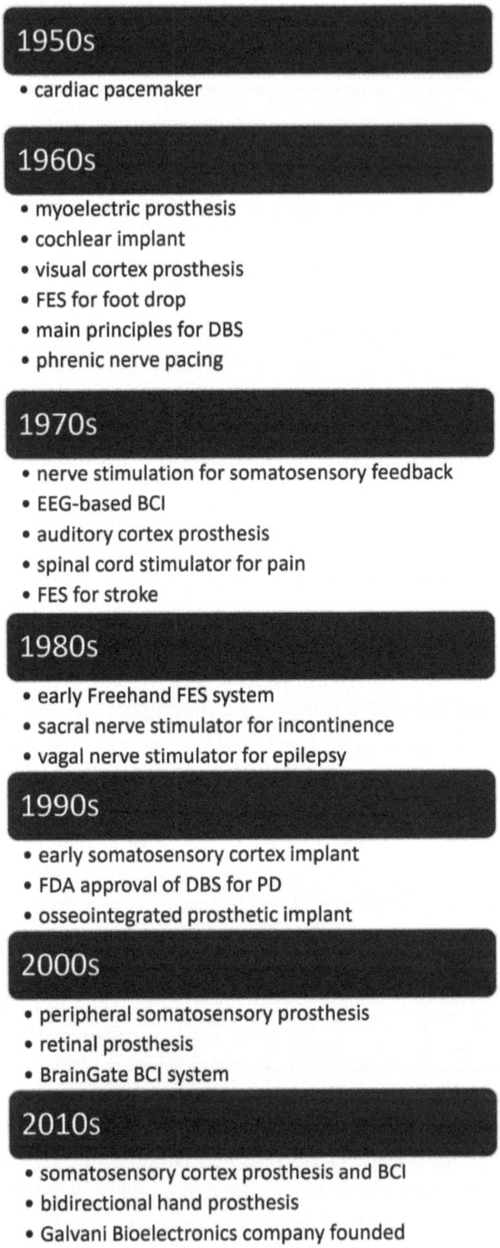

FIGURE 1.1 Timeline for important milestones in human neuroprosthetics research. Basic and applied research with experimental animals mostly predate these achievements and are essential before clinical trials. See text for the abbreviations.

computer display (Vidal, 1977). Since then, there have been tremendous efforts to make reliable and safe BCI systems, especially for severe conditions such as amyotrophic lateral sclerosis (ALS) and tetraplegia. Invasive BCIs can provide more neural information for control and hold great promise as demonstrated initially by the famous BrainGate system (Hochberg et al., 2006). Today, it is possible to communicate with the brain (Quick et al., 2020) or the peripheral nerves and muscles (e.g., see Mastinu et al., 2020) bidirectionally to partially compensate for lost sensorimotor functions.

Functional electrical stimulation (FES) has a long history; and devices to treat foot drop are among the early examples of neuroprostheses (Liberson et al., 1961; Moe & Post, 1962). FES is basically a neuromuscular stimulation technique which allows some control over paralyzed muscles for desired periods (e.g., during standing, grasping, walking, urination) in permanent use, and also as a temporary, short-term therapy for restoration of voluntary function. As a matter of fact, significant know-how on FES has accumulated for patients suffering from stroke and spinal cord injuries (SCIs). Through the work of research groups such as Kralj and co-workers (e.g., see Kralj et al., 1980), and those at the Cleveland FES Center (e.g., see Keith et al., 1989), the technology is quite mature now and may be preferred over BCI-driven robotics in less severe cases. For example, in the implementation of the Freehand system (Kilgore, 2015a, 2015b), the stimulator is controlled by implanted sensors for measuring joint angle or myoelectric signals.

Electroceuticals (a.k.a. bioelectronic medicine, neuromodulation) is a novel approach to treat chronic diseases, including psychiatric conditions (e.g., obsessive-compulsive disorder, depression), Parkinson disease (PD), epilepsy, pain, metabolic diseases, and inflammatory disorders. Its goal is, by electrical, magnetic, or chemical stimulation, to restore healthy patterns of activity in dysfunctional neural circuits which affect innervated organs (Peeples, 2019). The exact mechanisms of several electroceutical therapies are currently not fully explained; however, the positive clinical outcomes attract scientific and commercial investments, which are also used to elucidate the basic neuroscientific and clinical mechanisms behind those diseases. It is particularly enlightening to study the historical development of deep-brain stimulation (DBS) for PD (Gardner, 2013). Although the main parameters for treatment were known in the 1960s, medical, technological, economic, regulatory, and sociocultural factors prevented DBS from being adopted until its approval by the Food and Drug Administration (FDA) in 1997.

This section focused on some historical landmarks regarding only devices applied to humans. The term SF is used here as the means for providing artificial tactile and proprioceptive information to the nervous system, by a somatosensory neuroprosthesis. Artificial sensor data have also been used "internally," in short closed-loop designs, by some prosthetic devices to achieve better movement and/or grasp control automatically, not necessarily for applying SF to the user. For additional information on specific topics related to neuroprostheses, excellent review articles (e.g., Bensmaia & Miller, 2014; Bensmaia et al., 2020; Wolf et al., 2019), monographs (Rao, 2013; Sanchez, 2016), and other edited books (Chapin & Moxon, 2001; Dornhege et al., 2007; Horch & Dhillon, 2004; Kilgore, 2015a; Shepherd, 2016) are available, as well as the remaining chapters of the current book *Somatosensory Feedback for Neuroprosthetics.*

1.2 Classification of neuroprostheses

Neuroprostheses can be classified based on various criteria, such as their purpose (e.g., visual sensation), medical indication (e.g., for epilepsy), location of use (e.g., peripheral nerve), invasiveness (e.g., intracortical implant), or directionality (e.g., unidirectional myoelectric control). Two examples are presented here to highlight the possible classification criteria. There is a growing trend in the research for visual prostheses. For example, Argus II system is an epiretinal device with both FDA and European Commission (EC) approval for retinitis pigmentosa (RP). It is, therefore, helpful to study the multiple approaches to visual prostheses and identify them in a classification tree. Another helpful example is prosthetic/robotic hand systems. This is also a very active area for research, and significant advances have occurred during the last two decades. There are quite a few commercial hand prostheses with myoelectric control (e.g., Bebionic, i-Limb Quantum, Michelangelo, Motion Control, TASKA, Vincent Evolution), and novel bidirectional prostheses are in development (Ciancio et al., 2016). Since the neuroprosthetics field is highly dynamic, the classification trees given in Fig. 1.2 should be considered as suggestive.

Within neuroprosthetics research, sensory neuroprostheses have gained an important place due to the success and widespread use of cochlear implants. Many people's lives have been significantly improved by this technology. Currently, there is an optimism for retinal implants too, which, nevertheless, still need further development (Mirochnik & Pezaris, 2019). Fig. 1.2A shows a classification subtree associated with a certain group of visual neuroprostheses. Some visual diseases are also listed. Several approaches may be indicated for a particular disease; and several diseases or medical conditions may be compensated with a given device. For example, RP is one of the most studied diseases for visual neuroprosthetics. The next level of the classification hierarchy shows the general location of the neuroprosthetic implant [e.g., eye, optic nerve, lateral geniculate nucleus (LGN), or visual cortex]. Finally, there are three main types of implants which can be used in the eye. Epiretinal implants are placed on the inner surface of the retina for stimulating ganglion cells; and they connect to an external camera which is controlled by head movement. Subretinal implants usually have an integrated imaging sensor which enables a more natural gaze. These are placed on the outer layer of the retina for stimulating adjacent bipolar cells. Suprachoroidal

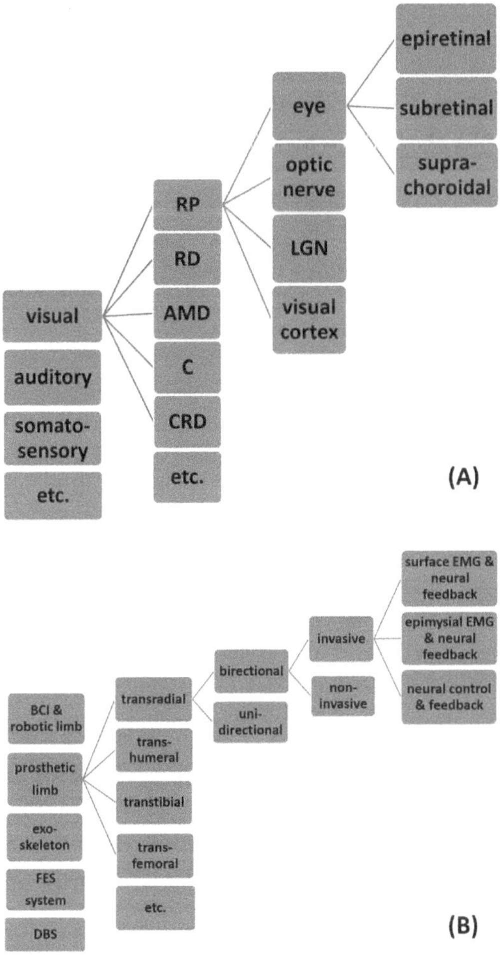

FIGURE 1.2 Examples for classification of neuroprostheses. (A) Among sensory neuroprostheses, a classification subtree for a certain group of visual prostheses is presented. The second column represents medical indications. The third column is for the general location of the implant; and the final column gives the specific type of implant with various advantages and disadvantages including invasiveness. (B) A classification subtree for invasive bidirectional hand prostheses in development. Prosthetic limbs are used to provide movement after amputation (examples listed in the second column) or for congenital deficiencies. Bidirectional ones also utilize SF; and invasive versions provide better control and sensory experience. The final column shows various approaches to obtain biological control signals, with peripheral nerve stimulation for the sensory feedback. *AMD*, age-related macular degeneration; *C*, choroidermia; *CRD*, cone–rod dystrophy; *RD*, retinal degeneration; *RP*, retinal pigmentosa.

implants are placed between the choroid and sclera; they also require an external camera. The implant location and invasiveness typically go hand in hand. However, invasiveness not only implies the difficulty of the initial surgery, it should also involve the risk of complications and long-term stability. For example, the suprachoroidal location is away from the more delicate retina, but requires higher electrical currents which may be unsuitable in the long term, as well as decreased resolution due to current spread. Implants in the visual pathway of the central nervous system (CNS) may be applicable for many types of blindness. Indeed, three cortical visual prosthesis systems (Orion, CORTIVIS, NeuroPace) are currently undergoing clinical trials (Mirochnik & Pezaris, 2019). Regarding directionality, visual neuroprostheses are unidirectional because of the one-way communication in today's systems. However, a future bionic eye may replace the entire eyeball and provide eye movements, and thus become bidirectional.

Neuroprostheses for providing and facilitating movement encompass a huge corpus of literature, and successful clinical examples. At one end, DBS has a miraculous outcome to alleviate the motor symptoms of PD and other movement disorders. Similarly, FES systems can activate paralyzed muscles for on-demand daily tasks. BCI-controlled exoskeletons and robotic limbs hold great promise for patients with SCI or neurodegenerative diseases such as ALS. Nevertheless, designing prosthetic limbs for amputations or congenital deficiencies is also quite challenging, especially due to the difficulty of the embodiment of the external device. Embodiment requires SF. Most advanced ones, currently in use, are for the upper extremity and provide proportional control by using myoelectric signals. Those utilizing invasive SF have not been commercialized yet, but the electronic-Osseoanchored Prostheses for the Rehabilitation of Amputees (e-OPRA) by Integrum AB (Sweden) is currently in clinical trials. In this device, the prosthetic limb is directly connected to the bone, and thus, it has excellent mechanical stability. Myoelectric signals are collected by epimysial electrodes implanted in the muscles. Sensory feedback is provided by stimulation of the peripheral nerves by implanted cuff electrodes. Mastinu et al. (2020) describe the results from this system tested on three transhumeral amputees. Along similar lines, Fig. 1.2B shows a conceptual classification regarding invasive, bidirectional, transradial prostheses. The last column shows three approaches possible with the current technology. Traditional surface electromyography (EMG) (e.g., see Raspopovic et al., 2014) or implanted epimysial electrodes can be used for recording myoelectric signals. In either case, however, the SF is by invasive neural electrodes [e.g., transversal intrafascicular multichannel electrodes (TIMEs) or cuff electrodes]. The third approach would involve invasive neural electrodes for movement control also (e.g., see Dhillon & Horch, 2005).

1.3 Basic components of the somatosensory system

Somatosensation is critical for the survival of an organism, and includes the modalities of touch, thermoreception, nociception, and proprioception. It is subserved by the afferent axons in the somatic part of the peripheral nervous system (PNS), which also comprises the efferent axons, that is, motor nerve fibers. Somatic nerves are integrated within the body, and besides the required sensorimotor function for many tasks, they also supply closed-loop signals to the CNS for sensing stimuli and body ownership. This was appreciated long ago by Aristotle: "Without touch it is impossible for an animal to exist... the loss of this one sense alone must bring death" (from *de Anima:* 435b4–5, cited in Smith, 2000). Mechanosensitivity is very ancient from an evolutionary point of view; prokaryotes can detect osmotic forces by mechanosensitive channels. The unicellular *Paramecium* displays escape and avoidance reactions to contact stimuli. The evolution of the nervous system brought an enormous number of possibilities for animals, which have elaborate ways to interact with the external world. Humans, however, excel in the use of somatosensation, especially that originating from their highly developed hands. The functionality of the hand in human culture goes well beyond its use for basic survival skills, for example, think about social gestures, manual labor, fine tool use, performance of music, art, sports, and various forms of recreation. Two other dramatic examples are the Tadoma method of communication by deaf–blind individuals and massage therapy, which clearly has rehabilitative and psychophysiological effects.

The neurons supplying the somatosensory nerves of the limbs and the trunk have their soma in the dorsal root ganglia (DRG). They are derived from the neural crest, and are created very early, by about the 4th week of gestational age in humans. Cutaneous innervation starts around 8 weeks, while the hair follicles are innervated at 22–24 weeks in the fetus. On the other hand, at around 6–7 weeks of gestation, the human embryo may display movement in response to cutaneous stimulation (Rees et al., 2010). Severe neurological disorders may result if this development is disrupted due to external or genetic factors. Therefore, unlike those with congenital blindness or deafness, there are very few people born with widespread loss of somatosensation, which usually accompanies movement disorders. Toxicity due to drugs and inherited diseases such as Friedreich ataxia cause peripheral loss of large sensory fibers. Autoimmune neuropathies (e.g., Guillain-Barré syndrome), metabolic diseases (e.g., diabetes), vascular diseases, infections, tumors, and trauma are common health conditions which affect somatosensory nerves. On the other hand, trauma, stroke, and neurodegenerative diseases damage CNS neurons and impact somatosensory processing in the spinal cord or the brain, while studies on some psychiatric disorders such as autism and obsessive-compulsive disorder (e.g., see Güçlü et al., 2015) imply modified somatosensory circuits. A major group of cases which are pertinent to neuroprostheses, and which have anatomically restricted, but complete loss of somatosensation (and motor function), are amputations and congenital limb deficiencies.

Large fiber peripheral neuronopathy, which was originally described as a post-infective and nonprogressive condition, has been helpful in understanding the functional implications of selective tactile and proprioceptive loss (Cole, 2008). In the acute and severe version, all large cell bodies of the DRG are destroyed in days; nociception, thermoreception, and peripheral motor nerve function are conserved. One patient was described as completely unable to move after deafferentation; he could not feel the bed and needed 2 years of rehabilitation to recover motor abilities (given the lack of proprioception) with visual feedback and great mental effort. However, that mental effort was essential at all times, even after the rehabilitation. Movement is not automatic when proprioception does not exist; it requires constant thinking and concentration. A common cold could keep him from moving an arm to get a cup, because of the difficulty in planning this simple task. Even after 30 years of living without SF, the fact that movement *always* requires effortful thinking is emotionally very disturbing. Since he does not have discriminative touch, he cannot pick up small things, and grasping objects is problematic. Active tactile exploration is impossible.

The somatosensory system consists of encapsulated receptors, receptor nerve endings, afferent peripheral nerves, pathways within the spinal cord and the brain, and the associated processing centers in the CNS. The following subsections present a summary of these components. Peripheral nerves and the somatosensory cortex are typical interfacial targets for neuroprostheses. Regarding the neural mechanisms of somatosensation, in particular for the hand, one can also refer to the comprehensive book by Mountcastle (2005).

1.3.1 Somatosensory receptors and afferent nerves

Cutaneous nerves consist of myelinated Aα, Aβ, Aδ, and unmyelinated C fibers, associated with mechanoreceptors, thermoreceptors, and nociceptors in the skin. For proprioceptive afferents, a different terminology is used: myelinated I, II, III, and unmyelinated IV fibers. All four groups may exist in muscle nerves, while joint nerves have much less of group I (Willis & Coggeshall, 2004). Proprioceptors include muscle spindle terminals, Golgi tendon organs, and joint receptors.

Cutaneous mechanoreceptors vary depending on the type of skin, that is, glabrous or hairy. In glabrous skin (Fig. 1.3A), Meissner corpuscles are located in the dermal papillae, and line up at both sides of the intermediate ridges (Güçlü et al., 2003). They are innervated by fast-adapting type I (FAI) nerve fibers (Lindblom, 1965). There are no Meissner corpuscles in hairy skin (Fig. 1.3B). Merkel cells are epithelial cells at the epidermal junction, and they participate in light touch sensation by chemical and mechanical interaction with associated nerve endings (Maricich et al., 2009; Press et al., 2010). The combined end organ is called the Merkel cell–neurite complex and is innervated by slowly adapting type I (SAI) fibers (Iggo & Muir, 1969). Merkel cells are mostly found at the tips of epidermal ridges in the

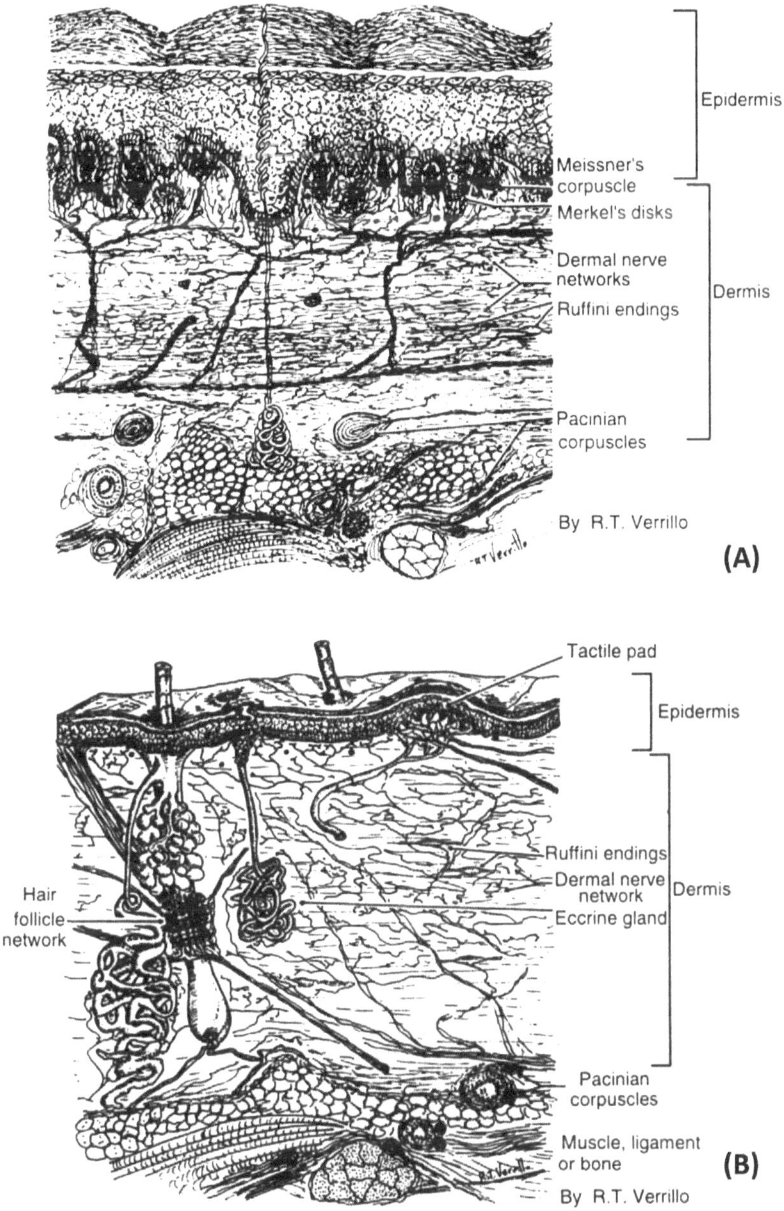

FIGURE 1.3 Mechanoreceptors and important structures of (A) glabrous and (B) hairy skin. *Reproduced with permission from Greenspan, J. D., & Bolanowski, S. J. (1996). The psychophysics of tactile perception and its peripheral physiological basis. In L. Kruger (Ed.),* Pain and touch: Handbook of perception and cognition *(2nd ed.) (pp. 25–103). Academic Press. Artist: R. T. Verrillo.*

glabrous skin (Güçlü et al., 2008), and especially in the touch domes between and near the hair follicles of hairy skin. Pacinian corpuscles are located deeper in the dermis and subcutaneous tissue of both glabrous and hairy skin (Fig. 1.3A and B). They have multilayered capsules, and are the largest mechanoreceptors in the skin. Each Pacinian corpuscle is innervated by only one fast-adapting type II (FAII) fiber (Bolanowski & Zwislocki, 1984). Ruffini endings are fusiform shaped and often associated with slowly adapting type II (SAII) fibers. They are located in the dermis of hairy skin, mostly around the fingernails; a smaller number can be found in the glabrous skin (Paré et al., 2003).

In hairy skin, there are several other types of mechanoreceptive fibers, which are associated with hair follicles (Lechner & Lewin, 2013). There are three main types of hair in mammals: guard, awl/auchene, and zigzag hairs. In addition, sinus hairs of some animals are particularly specialized for tactile sensing, and restricted to certain locations (e.g., vibrissae). Nerve fibers innervating the hair follicles terminate in circumferential arrays or lanceolate endings (Fig. 1.3B). Large Aβ fibers innervating guard and awl/auchene hairs are rapidly (i.e., fast) adapting (G1 and G2 receptors). Aδ fibers (i.e., D-hair receptor) and low-threshold mechanoreceptive C (i.e., C-tactile) fibers innervate awl/auchene and zigzag hairs. D-hair receptors are rapidly adapting; and C-tactile fibers have intermediate adaptation properties. The molecular, anatomical, and physiological correlates of mechanosensation via human hair follicles are somewhat incomplete, given the fact that human hair is quite distinct (much lower in number and of terminal, vellus, intermediate types) from that of other mammals. Field receptors, described in cats and monkeys, only respond when large numbers of hairs are bent and have mostly rapid adaptation (Willis & Coggeshall, 2004). C-tactile fibers respond best when hairy skin is lightly stroked with a speed of around 3 cm/s; and they are thought to mediate pleasant/social touch sensation, also in humans (Olausson et al., 2010).

Thermoreceptors can be grouped into two major classes, one which responds to increases and another to decreases in skin temperature. They are subserved by thinly myelinated (Aδ) fibers in cool receptors and C fibers in warm receptors. These two classes cover a skin temperature range of 5°C−45°C (Gardner & Johnson, 2013). Above these limits, heat and cold nociceptors signal the excessive, harmful temperatures with pain. Nociceptors are diverse; some respond to only mechanical damage (Aδ mechanical nociceptors), some to noxious mechanical stimuli and high temperatures (Aδ mechanoheat nociceptors). Most C nociceptors are polymodal; they respond to extreme mechanical and heat/cold stimuli, and also to chemical stimuli. These cause a slow, burning pain. Some C fibers are purely chemoreceptors; they respond to irritating chemicals. Muscles and joints also contain nociceptive fibers which are usually classified as groups III and IV (for a more in-depth review, see Willis & Coggeshall, 2004).

Among the proprioceptors, muscle mechanoreceptors consist of spindle primary and secondary endings, Golgi tendon organs, and Paciniform corpuscles in the fasciae. Each spindle has a primary ending supplied by a large myelinated fiber (group Ia); and this fiber responds mostly to dynamic stretch. The sensitivity of the dynamic response depends on dynamic γ motor neuron activity. There may be one or more secondary spindle endings (of fiber group II). They mainly respond to static stretch, which has sensitivity also dependent on another subpopulation of static γ motor neurons. Due to the unloading effect, during a twitch, muscle spindle afferents decrease their firing, whereas Golgi tendon organs increase firing. Golgi tendon organs are located within the muscle tendons and aponeuroses; they have a structure similar to Ruffini endings. They are innervated by large myelinated fibers (group Ib). They signal high levels of muscle tension, but can also be activated at low threshold by the contraction of nearby muscles (Willis & Coggeshall, 2004). Other proprioceptors are the mechanoreceptors in the joints. Encapsulated mechanoreceptors such as Ruffini endings and Paciniform corpuscles have been found in some joints. These typically signal flexion and extension to large angles, and mechanical transients, respectively.

1.3.2 Central pathways and cortical areas

During active tactile exploration of objects, somatosensory and motor systems are highly involved and interact in complex ways. Here, the passive flow of somatosensory information is described briefly in the associated pathways. It is important to note that slow versus rapid adaptation properties of mechanoreceptive fibers, especially, are often considered as critical to shaping the population response of afferents. However, this terminology refers to ramp-and-hold stimuli; as a matter of fact, all mechanoreceptive fibers respond to skin vibration (Talbot et al., 1968). Since any mechanical stimulus can be considered as a sum of spatiotemporal sine waves, it is more useful to describe intensity and frequency characteristics of responses to vibrotactile stimuli (e.g., see Güçlü & Bolanowski, 2003a, 2003b, 2004a). Even during "static" indentations of the skin while interacting with objects, Pacinian fibers may fire due to tiny fluctuations of contact force. They are very sensitive to transients, and have spike thresholds down to 0.01 μm at 200–300 Hz when studied in isolation (Bolanowski & Zwislocki, 1984). In addition to varying response properties, different types of cutaneous mechanoreceptive fibers have different innervation densities and receptive fields. For example, SAI and FAI afferents in the glabrous skin are very dense at the fingertips, and they have small punctate receptive fields. On the other hand, since the associated mechanoreceptors are deeper in the skin, FAII and SAII afferents have large receptive fields with blurry borders, and their densities are much lower.

The somatosensory afferents of the limbs and the trunk, which send discriminative touch and proprioceptive information, branch upon entering the spinal cord. The main axons ascend in the ipsilateral dorsal columns and first synapse in the dorsal column nuclei (DCN) (Fig. 1.4) of the medulla. The second-order neurons send axons that cross the midline within the medial lemniscus tract and project to the contralateral thalamus. On the other hand, the afferents which convey crude touch, temperature, and pain information synapse in the dorsal horn of the spinal cord. The second-order neurons of this system cross the midline at the spinal cord level and ascend in the anterolateral column, that is, anterolateral spinothalamic tract, projecting to the contralateral thalamus. The medial lemniscus fibers terminate in the ventroposterior nucleus (VP) of the thalamus, in which the proprioceptive

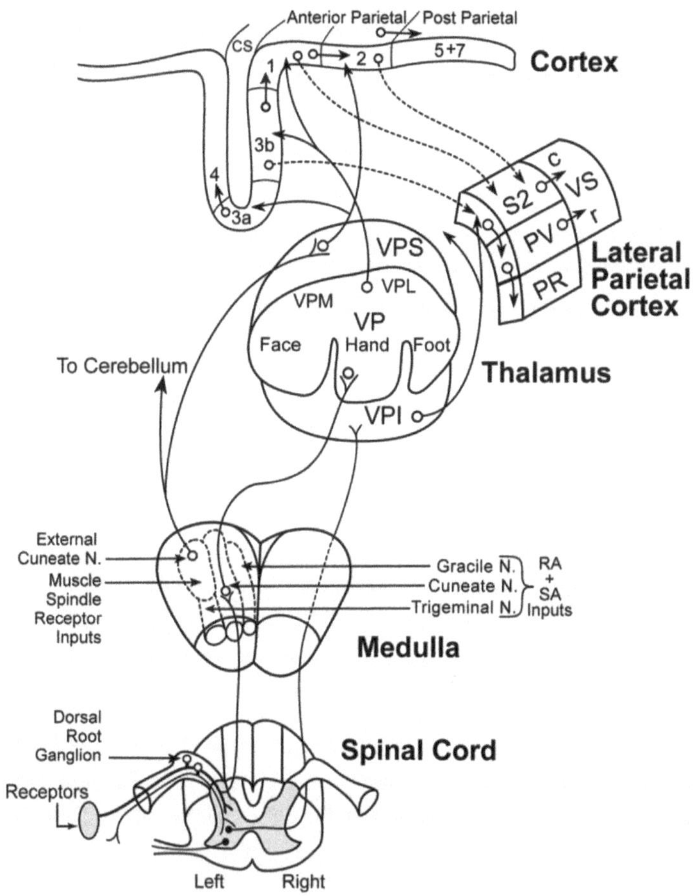

FIGURE 1.4 Summary of somatosensory pathways. *Reproduced with permission from Kaas, J. H. (2012). Somatosensory system. In J. K. Mai, & G. Paxinos (Eds.),* The human nervous system *(3rd ed.) (pp. 1074–1109). Academic Press. https://doi.org/10.1016/B978-0-12-374236-0.10030-6.*

information is represented more superiorly (VPS). The spinothalamic fibers terminate inferior (VPI) to the VP proper, and also in the ventromedial posterior nucleus (VMpo) (Kaas, 2012). The dorsal-column pathway maintains a detailed somatotopy in its trajectory, whereas the anterolateral pathway has less defined borders.

Within the primary somatosensory cortex (S1), areas 3b and 1 have parallel somatotopic maps of cutaneous inputs (largely from VP), whereas areas 3a and 2 receive direct proprioceptive inputs (from VPS), and are also organized somatotopically. S1 makes up the anterior parietal cortex. The secondary somatosensory cortex (S2) is in the lateral parietal cortex, and it receives inputs from VPI of thalamus and from S1 (Fig. 1.4). Areas 3b, 1, and 2 represent a hierarchical chain of somatosensory processing. Area 3a is closely associated with the primary motor cortex (area 4). Areas 5 and 7 in the posterior parietal cortex participate in the sensory guidance of motor actions; as such, there areas are connected to premotor and motor areas (Kaas, 2012).

Besides the main ascending information described above, there is evidence of additional more complex pathways, such as second-order neurons in ipsilateral dorsal columns (e.g., postsynaptic dorsal column pathway), the spinocervical pathway which is reduced in primates, and variations regarding upper versus lower limbs (Kaas, 2012). For example, gracile and cuneate tracts (from the lower and upper body and limbs, respectively) of the dorsal column pathway contain different information at higher spinal cord levels. At lower levels, the gracile tract includes tactile and proprioceptive afferents; however, most of the proprioceptive afferents terminate in Clarke's nucleus (a.k.a. posterior thoracic nucleus) which projects to nucleus Z (just rostral to gracile nucleus, and projecting to contralateral thalamus) via the dorsolateral tract and to cerebellum via the spinocerebellar tract. Therefore, at higher levels, the gracile tract includes mostly tactile afferents which terminate in the gracile nucleus. Within the cuneate tract, the tactile inputs terminate in the cuneate nucleus, and the proprioceptive inputs terminate in both the cuneate and the external cuneate nucleus, the latter of which projects to the VPS and cerebellum (Fig. 1.4). By also including the descending pathways not described here, the DCN complex (consisting of several associated structures, as well as the special zones of cuneate and the gracile nuclei) functions as an important sensorimotor integration and distribution center. Loutit et al. (2021) present in-depth coverage of the DCN complex with all its inputs and outputs.

1.3.3 Psychophysical processing and perception

How the brain combines the modalities of somatosensation is not fully understood currently. This integration is rather important, because it gives the richness of sensations in daily life. For example, to appreciate dampness on a surface, mechanical aspect of touch, thermoreception, and proprioception are all involved, respectively, to sense the smoothness, heat transfer, and frictionless movement across it. On the other hand, the most critical tasks require the mechanical sense of touch for fine object and texture discrimination, and proprioception for sensorimotor control. The psychophysical four-channel theory of touch gives a reliable framework for understanding the detection of threshold-level vibrotactile stimuli (i.e., those having sinusoidal waveforms) which activate cutaneous mechanoreceptive fibers (Bolanowski et al., 1988). The theory is mostly applicable to glabrous skin on which stimulus control is better achieved due to the lack of hair follicles and their associated fibers. Although the idea of psychophysical channels is relatively well established in sensory neuroscience (e.g., see Campbell & Robson, 1968), there seems to be some confusion regarding the interpretation of the four-channel theory as it appears in some recent, mostly haptics engineering, literature.

Psychophysical channels are abstract information-processing units in the CNS that link physical stimuli with psychological responses used for understanding perception. In this black-box approach, the physical parameters of the stimulus are very accurately controlled (e.g., at submicrometer scale for touch) and the responses are measured probabilistically because of the properties of physiological components, the nature of decision-making, and various sources of noise in the overall system. In the four-channel theory, test stimuli consist of very weak sinusoidal mechanical vibrations with slow rise/fall times, and they are applied with various contactor sizes on a pre-indented skin site. The biomechanical properties of the skin are mostly linear at this regime and transients are avoided by windowing the signals. There are range limits for the amplitude and frequency of the signals to prevent decoupling of the contactor probe from the skin (Cohen et al., 1999). This allows for a faithful representation of the stimulus waveform (e.g., mechanical displacement) on the skin surface.

The four-channel theory only predicts average sensitivities of tactile channels (i.e., detection thresholds) as a function of frequency, and does not study channel interactions at suprathreshold levels except regarding psychophysical masking phenomena. With the restrictions given above, the theory shows that the four channels mostly act "independently" for the detection of weak stimuli. In other words, for a set of given experimental conditions (such as test frequency, masking frequency and level, test and masking skin sites, test and masking contactor sizes), only one channel will be responsible for the threshold-level detection of the vibrotactile test stimulus, regardless of the existence of the other channels. This absolute threshold refers to the amplitude or level of the test stimulus which yields a predefined probability [e.g., 75% in a two-interval forced-choice task (2IFC)] of detection.

The four psychophysical channels were based on the anatomical, receptive-field, and spike-response properties of cutaneous myelinated mechanoreceptive fibers, that is, FAI, FAII, SAI, and SAII fibers. They were originally called non-Pacinian I (NP I), Pacinian (P), NP III, and NP II for historical reasons. In current usage, they are also called RA (rapidly adapting), PC (Pacinian corpuscle), SA I, and SA II channels, respectively (Gescheider et al., 2009). However, psychophysical channels do not need to have very clear-cut anatomical and physiological substrates. For example, although the psychophysical PC channel predominantly gets input from Pacinian (FAII) fibers (Greenspan & Bolanowski, 1996), convergence from multiple afferent inputs begins as early in the ascending pathway as the DCN (Ferrington et al., 1987), and cortical responses are highly nonlinear with submodality and even cross-modality convergence (Macgillis et al., 1983). The modern view for the classification of cortical neurons is based on function and not submodality composition (Saal & Bensmaia, 2014). Therefore the channel inputs cannot entirely be tied to one class of afferents, although, within the conditions used to characterize the channel, one class may be more dominant and its name is used in the nomenclature. If very low-level, high-frequency (say 250 Hz) stimuli are applied with a large contactor (Fig. 1.5A), it will be mostly the FAII fibers that are activated, and thus the PC channel. As a matter of fact, the detection threshold of the PC channel is around 0.1 μm at 250 Hz (−20 dB in Fig. 1.5A); that is to say, humans can detect this stimulus in approximately 75% of 2IFC trials. If the same stimulus is applied at higher levels, it will be detected more often (as characterized by the psychometric function), and then, at some level, eventually in 100% of the trials. It is expected that many more fibers (including those from the other mechanoreceptive classes) will be activated with these higher level stimuli at 250 Hz. However, apparently another tactile channel is not recruited until 3.2 μm (10 dB in Fig. 1.5A); after this point, the theory posits that both PC and SA II channels are activated. It is important to reemphasize that, between 0.1−3.2 μm, the four-channel theory does not claim that only Pacinian fibers are activated. It merely states that only the *psychophysical* PC channel is activated at a suprathreshold level. Cortical neurons contributing to the processing in the PC channel may indeed be supplied by information from other afferent classes; nevertheless, Pacinian fibers would display vigorous activity within that range (Bolanowski & Zwislocki, 1984) and dominate the percept. Due to the probabilistic nature of detection and other psychophysical channel properties (e.g., spatial and temporal summation), it appears that only the PC channel can be characterized distinctly at this latter condition.

Consider another example: a vibrotactile stimulus at frequency 10 Hz and with amplitude 5 dB (1.8 μm). Fig. 1.5A shows that this stimulus is below the threshold of all four channels. However, it can indeed activate some mechanoreceptors just under the contactor probe, and even be psychophysically detected with lower probability than 75%. If the stimulus level is increased to 15 dB, it becomes just as strong to be detected about 75% by the RA channel, and probably with much less probability by the SA I

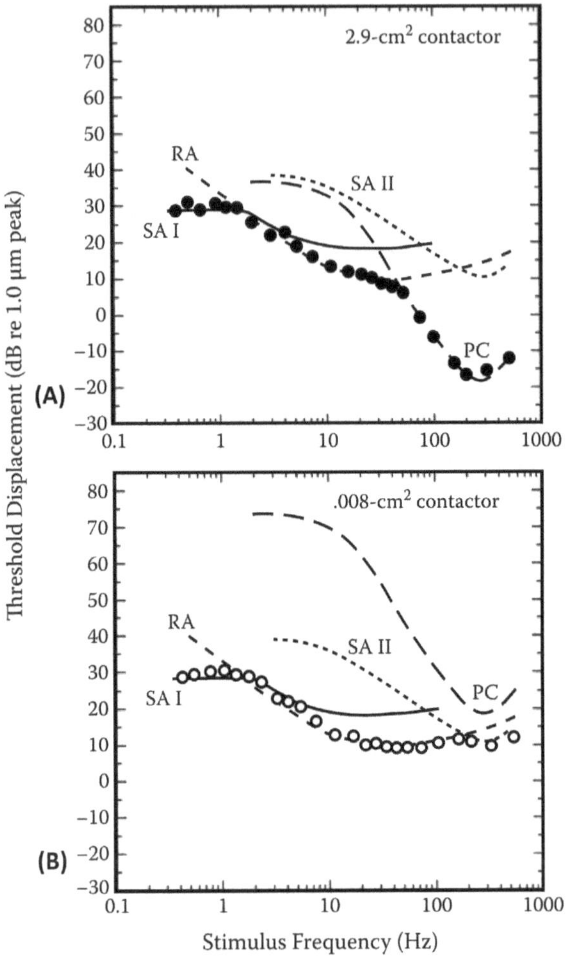

FIGURE 1.5 Sensitivity curves of tactile psychophysical channels measured with a large (A) and a small (B) contactor. *Reproduced with permission from Gescheider, G. A., Wright, J. H., & Verrillo, R. T. (2009).* Information-processing Channels in the tactile sensory system: A psychophysical and physiological analysis. *Psychology Press. https://doi.org/10.4324/9780203890004.*

channel. We cannot for sure know whether the SA I channel activation is above chance level (50% detection in 2IFC task), since the curves in Fig. 1.5 are averages across human participants. Each participant will have a different set of sensitivity curves. There is typically 5 dB intrasubject and 10 dB intersubject variation in this type of measurement. If the 10-Hz stimulus is at 30 dB (31.6 µm), it is considered "operationally" to be suprathreshold for both RA and SA I channels, and subthreshold for PC and SA II channels, but a more in-depth analysis must consider the actual detection probabilities.

Nevertheless, the four-channel theory gives a useful interpretation, as described by the examples above, to guide further experiments and computational modeling of such probabilities (Güçlü & Bolanowski, 2004b, 2005b; Güçlü et al., 2005; Güçlü, 2007).

How can the sensitivity curves of psychophysical channels with higher thresholds be derived? In Fig. 1.5A, the thresholds of RA and SA II channels are much higher than those of the PC channel at frequencies >100 Hz. However, because of less spatial summation in the PC channel, its sensitivity curve will be much more elevated if the vibrotactile stimuli are applied with a small contactor (Fig. 1.5B). Therefore the detection thresholds will be accounted by the RA or SA II channels at higher frequencies (> 50 Hz). In summary, if a large contactor (2.9 cm^2) is used, the detection thresholds will be mediated from low to high frequencies, in order, by SA I, RA, and PC channels (filled circles in Fig. 1.5A). On the other hand, with a small contactor (0.008 cm^2), they will be mediated by SA I, RA, and SA II channels as such (open circles in Fig. 1.5B). The better, systematic way to investigate "hidden" sensitivity curves of tactile channels is by adaptation or masking. A psychophysical channel can be selectively masked, which causes its sensitivity curve to shift upwards (Güçlü & Bolanowski, 2005a; Güçlü & Öztek, 2007) similar to that of the PC channel in Fig. 1.5B. However, by changing the masking stimulus level, the amount of masking can be controlled. The existence of four tactile channels was discovered by this method. In brief, the plateau regions in the masking function (i.e., constant masking effect) as the masking intensity gets higher, suggest detection by a channel different to the one which is being masked. Information processing by tactile psychophysical channels is lucidly explained by Gescheider et al. (2009), as well as their characteristic properties such as temporal and spatial summation.

There are many other psychophysical measurements which would be of interest for somatosensory neuroprostheses, such as discrimination and magnitude estimation, and vibrotactile stimuli applied passively at a single site convey limited information compared to those experienced during active tactile perception. Recognition of objects is better with haptic exploration (Heller & Myers, 1983), which includes both tactile and proprioceptive inputs, and is possibly also influenced by motor signals. On the other hand, active and passive touch yield similar results regarding texture perception (Lamb, 1983), pattern recognition (Vega-Bermudez et al., 1991), and roughness estimation (Verrillo et al., 1999) if the stimuli affect a small area on the skin, which does not necessitate large exploratory movements. However, it is important to note that vibrotactile sensitivity is reduced during movement via a gating effect, especially at high speeds (Yıldız et al., 2015). Therefore sensorimotor integration in a neuroprosthesis ought to be designed quite differently depending on the target task. There is yet another mode of tactile perception, that is, affective touch, which is mediated by C-tactile fibers and is considered to be important for social communication (Cascio et al., 2019).

Additional information on the perceptual and cognitive aspects of somatosensation can be found in various monographs and edited books (e.g., see Grunwald, 2008; Jones & Lederman, 2006; Nelson, 2002).

1.4 Overview of somatosensory neuroprostheses

The need for SF can be clearly observed in individuals with neurological problems affecting the somatosensory system, and especially in amputees who currently use prostheses without adequate sensory feedback. In these cases, the patient has to rely on visual, auditory, and incidental (i.e., mechanical sounds of a prosthesis) cues, and possibly some proprioceptive inputs from residual muscles. These feedback sources are essentially incomplete (e.g., no texture information) and place a high cognitive load, yielding a slow, clumsy movement. There have been numerous studies reporting the effects of similar deficits (e.g., see Augurelle et al., 2003; Sainburg et al., 1993) and various methods exist for providing artificial SF (Bensmaia & Miller, 2014; Bensmaia et al., 2020; Svensson et al., 2017; Weber et al., 2012). Inclusion of SF for tactile and proprioceptive information improves movement performance and grasp-force accuracy in motor prostheses; the user may also benefit from better embodiment and reduced phantom limb pain (PLP). The principles and application of somatosensory neuroprostheses have been demonstrated in experimental animals and humans. The field is very active with ongoing research and development. Only some representative examples are presented next, because other chapters of the current book *Somatosensory Feedback for Neuroprosthetics* go into more detail. Fig. 1.6 shows three main approaches for SF: noninvasive through skin stimulation, invasive through peripheral nerve stimulation, and invasive through brain (i.e., cortical) stimulation. In each case, artificial sensor data from the prosthetic limb, regarding movement and contact-force information, are converted to SF signals. The first two are more suitable for amputees, and the third can be used in severe neurological conditions such as ALS and tetraplegia due to SCI.

Ideally, SF should be modality-matched and spatiotemporally congruent. For example, if a vibrotactile stimulus is applied on the fingertip of a robotic hand, the SF signals need to activate tactile channels concurrently in the user and at the same somatotopic neural representation of the fingertip. These requirements are almost impossible to satisfy all at the same time due to biological and technical limitations. Consider a sensorimotor neuroprosthesis for a person with SCI; cortical surface stimulation for SF requires an initial surgery for the implant, but is a feasible approach due to damage to the ascending/descending pathways of the spinal cord. As the robotic arm/hand controlled by the person makes contact with an object, electrical current pulses can be concurrently applied to the somatotopic representation in the cortex. However, the current state of knowledge and technology does not

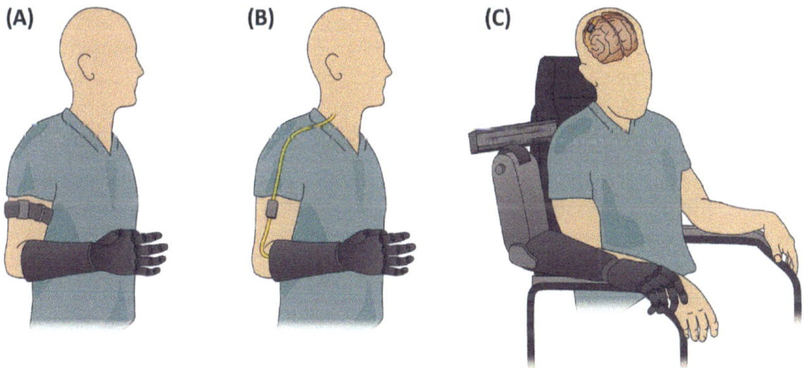

FIGURE 1.6 Three main approaches for somatosensory feedback (SF) in neuroprostheses. (A) SF is provided noninvasively by mechanical or electrical stimulation of the skin (shown here with actuators or electrodes placed in an armband) for an amputee. (B) An invasive neuroelectric interface is placed on/in the peripheral nerve to provide SF by electrical current pulses for an amputee. (C) Although the biological limbs are available, they cannot be used; sensorimotor function is disabled due to damage to ascending/descending pathways. An invasive neuroelectric interface is implanted on/in the brain cortex to provide SF by electrical current pulses. *Reproduced with permission from Bensmaia, S. J., Tyler, D. J., & Micera, S. (2020). Restoration of sensory information via bionic hands.* Nature Biomedical Engineering, *Nov 23, online ahead of print. https://doi.org/10.1038/s41551-020-00630-8.*

reproduce the natural activation pattern in the brain, and thus the stimulation often feels artificial (buzz, tingling, etc.). In this example, the modality is somewhat altered, and the spatial extent of the artificial sensation (i.e., projected field) may not be as crisp as that obtained in peripheral intraneural stimulation because of the converging inputs at this level, as evidenced by the cortical magnification of receptive fields, and because of the electrical current spread. On the other hand, slow forces applied to the myoelectric hand of an amputee can be mimicked by the noninvasive mechanotactile stimulation of an intact skin site. Although the original modality and temporal congruency are achieved in this type of SF, there is no somatotopic match, since the biological limb does not exist. However, in some cases, stimulation of the stump may evoke sensations referred to the phantom limb.

1.4.1 Noninvasive methods for feedback

Noninvasive methods for SF consist of vibrotactile and mechanotactile modalities which apply mechanical vibrations on the skin surface (either by fast vibrations or slowly changing indentations, respectively), and electrotactile modality which applies small electrical currents to excite the nerves transcutaneously (Fig. 1.6A). Transcranial (noninvasive) electrical stimulation was also shown to have modulatory effects on somatosensory processing (Saito et al., 2019), but it has not been used as SF in prosthetic applications.

Current noninvasive methods for SF usually have a disadvantage because of the somatotopic mismatch, but the strategy called "targeted reinnervation" can partially compensate for that at the price of a complex surgical intervention (Hebert et al., 2014). Although there is both modality and somatotopic mismatch with electrotactile stimulation, better skin coupling (by monitoring the electrical impedance) and small electrode size are desirable factors for neuroprosthetics. Mechanotactile actuators are quite bulky in that respect, but can be modality-matched to indentation or force. Vibrotactile SF is usually applied by small vibration motors or voice coils, which have limited output characteristics, unlike laboratory-grade shakers. Therefore vibrotactile SF signals do not usually match with the stimulus waveforms impinging on a robotic limb, despite the fact that both are mechanical. Skin-coupling issues for vibrotactile SF feedback can be remedied by better device design and psychophysical calibration.

1.4.1.1 Vibrotactile stimulation

A simple object manipulation task, for example, pick and lift, consists of a series of actions separated by temporally discrete tactile events (Johansson & Flanagan, 2009). Analyzing the somatosensory input at these transition events yields information quite adequate to guide further movement. Cipriani et al. (2014) showed that healthy humans can integrate vibrotactile SF for controlling a robotic hand. This so-called discrete event-driven sensory feedback control (DESC) policy was later applied for transradial amputees with conventional myoelectric prostheses (Clemente et al., 2016). A modified box-and-blocks test, that is, the virtual eggs test, was used to assess the performance of the amputees based on picking and repositioning "fragile" objects. When SF was on, the number of "broken" blocks was reduced and there was also a significant decreasing trend over the course of test weeks. Additionally, the block transfer rate slightly increased, somewhat faster than the SF-off condition. This study suggests that amputees can quickly learn to utilize vibrotactile SF for a given task. To benefit from vibrotactile SF optimally, it was recently shown that a psychophysical calibration procedure should be adopted with respect to detection thresholds and sensation magnitudes at different frequencies and for each stimulation site (Karakuş & Güçlü, 2020). By establishing a careful psychophysical model for each prosthesis user, sequential vibrotactile patterns, signaling manipulated object and movement information, can be recognized at decent recall and precision rates without the contribution of additional cues or the knowledge of physical rules.

1.4.1.2 Electrotactile stimulation

Electrotactile SF improves grasp-force control in amputees using myoelectric prostheses (Štrbac et al., 2017). Nine amputees were tested in multiple sessions which included open-loop (i.e., no SF) and feedback control to achieve four target grasp forces. Force levels were mapped to separate electrode channels by spatial coding. The performance was specifically assessed in several blocks (a baseline, two feedback, and a validation block) for each day. In a given daily session, electrotactile SF reduced the generated force errors significantly compared to the baseline, and interestingly, the performance in the open-loop final validation block stayed similar to the SF blocks, which imply short-term learning in force control. Furthermore, the performance in all blocks improved across days. The trends were similar for the dispersion of the errors and for the number of saturations in the generated force, thus precision was also improved. Considering that only three of the nine amputees had some experience with myoelectric prostheses, this study suggested that SF by electrotactile stimulation can be easily integrated with EMG-based control.

1.4.2 Invasive methods for feedback

The common forms of invasive SF are peripheral nerve stimulation (Fig. 1.6B) and cortical stimulation (Fig. 1.6C). The latter may be via penetrating electrodes, that is, intracortical microstimulation (ICMS), or surface electrodes which are placed epi- or subdurally. Surface electrodes are similar to electrocorticographic (ECoG) electrodes, but are currently developed for enabling safe, accurate, and long-term electrical stimulation; they may also be smaller than the traditional ones for higher spatial resolution (e.g., μECoG). Similarly, there are various neuroelectric interfaces available for peripheral nerve stimulation (del Valle & Navarro, 2013). Recently, epidural spinal cord stimulation has also shown promise for neuroprosthetic applications (Chandrasekaran et al., 2020), with the added benefit of clinically approved technology which has been used to treat chronic pain for quite a long time. Additionally, spinal cord or dorsal root stimulation would be clinically relevant if the amputation level of a limb is high, such that the desired part of the peripheral nerve is not available. However, here, the focus will be on peripheral nerve stimulation and brain, that is, specifically cortical, stimulation for SF. The somatosensory pathways can also be stimulated at several supraspinal levels, but cortical stimulation is by far more suitable regarding the surgery, electrode technology, and the availability of a clear somatotopic map.

Invasive neural interfaces have the typical trade-off regarding selectivity and invasiveness. That is to say, they have better selectivity, and thus can provide more accurate and precise SF as they become more invasive. For example, in theory, intraneural electrodes can provide better stimulation selectivity than extraneural electrodes. A similar statement can be made for ICMS versus cortical surface stimulation. However, as the invasiveness increases, the trauma and tissue response worsen, which hinder both the short-term and long-term performance of SF. A major part of research in neuroprosthetics is therefore targeted toward novel, biocompatible neural interfaces for long-term applications.

1.4.2.1 Peripheral nerve stimulation

One of the first demonstrations of a bidirectional hand prosthesis, which uses EMG signals for control and peripheral nerve stimulation for SF, was made by Raspopovic et al. (2014). Transversal intrafascicular multichannel electrodes (TIMEs) were placed in median and ulnar nerve fascicles. The amputee participant could distinguish three force levels and use SF to modulate grasp force without visual or auditory cues in a myoelectric hand. Furthermore, he could select the appropriate grasp (accuracy: 97%) based on the placement of an object on the hand surface. Object stiffness (soft, medium, hard) and object shape (cylindrical, spherical) tests also yielded quite high recognition rates. While this study enabled artificial tactile sensation in the phantom hand digits, Wendelken et al. (2017) showed both artificial cutaneous sensation and proprioception using Utah slanted electrode arrays (USEAs). Closed-loop control was achieved with neural decoding and SF based on a virtual object. The artificial sensations were very rich: flexion/extension, adduction/abduction of fingers, vibration, pressure, tingling, and stinging on various phantom locations consistent with the intact hand innervation. On the other hand, biomimetic SF, that is, one that mimics the mechanoreceptive afferent firing rates, enables faster identification of objects (George et al., 2019). In general, this type of SF improved the manipulation of fragile objects, and grip precision using a DEKA LUKE arm controlled via intramuscular EMG.

As mentioned earlier, the e-OPRA system is one of the most promising approaches and is also close to commercialization. Mastinu et al. (2020) studied the performance of this system with three transhumeral amputees. SF was provided during a pick-and-lift task with an instrumented object, which is able to record grip and load forces (Fig. 1.7). Different neuromodulation methods were used: continuous amplitude modulation, DESC, and hybrid (continuous + DESC). Hybrid SF yielded the best temporal correlation between grip and load forces, which is an indicator for motor coordination. The results also showed that SF is more important under uncertainty, which was introduced by object weight changes; again, hybrid SF gave the lowest load-phase durations (i.e., faster lifting). With the added benefit of osseointegration for stability, this latest bidirectional technology includes high-quality epimysial EMG recordings for control and cuff electrodes for the electrical stimulation of median and ulnar nerves. Therefore the entire system has been described as a neuromusculoskeletal prosthesis.

1.4.2.2 Brain cortex stimulation

After the success of nonhuman primate experiments, direct cortical stimulation for SF was tested on humans. As expected, ICMS, applied within the

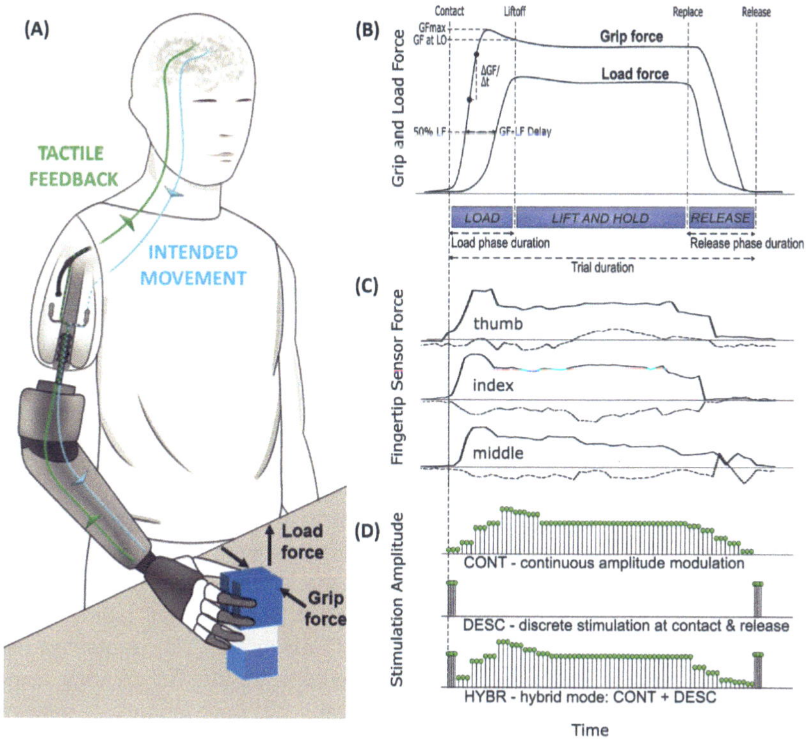

FIGURE 1.7 A bidirectional neuromusculoskeletal prosthesis with an osseointegrated implant (e-OPRA system). (A) Intended movement is decoded from EMG signals recorded by epimysial electrodes, and tactile feedback is provided by peripheral nerve stimulation via cuff electrodes during closed-loop control. The amputees performed a pick-and-lift task with an instrumented object, which is able to record grip and load forces. (B) Time course of grip and load forces on an object. Motor performance metrics were derived from such data. (C) Tactile neural feedback was based on force sensor data from the prosthetic fingertips. (D) Nerve stimulation for sensory feedback was tested with three neuromodulation methods. CONT: the electrical-current pulse amplitudes were continuously modulated proportional to fingertip force-sensor data. DESC: stimulation was applied at discrete events, that is, contact and release. HYBR: a hybrid method was used combining CONT and DESC modes. *Reproduced with permission from Mastinu, E., Engels, L. F., Clemente, F., Dione, M., Sassu, P., Aszmann, O., Brånemark, R., Håkansson, B., Controzzi, M., Wessberg, J., Cipriani, C., & Ortiz-Catalan, M. (2020). Neural feedback strategies to improve grasping coordination in neuromusculoskeletal prostheses.* Scientific Reports, 10*(1), 11793. https://doi.org/10.1038/s41598-020-67985-5.*

hand area of the S1 cortex of a person with SCI, produced artificial tactile sensations originating from somatotopically corresponding locations on the hand (Flesher et al., 2016). These sensations were stable for months; additionally, they felt quasi-naturalistic (e.g., touch, pressure, warmth, but also some electrical) and not painful. Current pulse amplitudes could be scaled for yielding varying sensation magnitudes within safety limits. Armenta

Salas et al. (2018) showed that proprioceptive sensations can also be evoked by ICMS; the participant, who had SCI for 1.5 years, reported directional (e.g., right, forward, upward) movement sensations as well as cutaneous sensations. However, these were dependent on stimulus parameters (amplitude and frequency) and particular electrode channels. The utility of SF by ICMS in a closed-loop control task is currently undergoing rigorous testing. For example, Quick et al. (2020) recently presented promising results which show that grasp-force accuracy can be improved in a virtual environment. This study is highly significant for future neuroprostheses suitable for SCIs, because it is one of the first demonstrations of a bidirectional, cortical BCI with neural spike decoding *and* ICMS for closed-loop control.

Similar advances have been made with recording and stimulation systems using cortical surface electrodes, which are less invasive, but less selective, than intracortical electrodes. Task-specific SF by cortical surface stimulation was found to be feasible by Cronin et al. (2016). In this study, participants wore data gloves and attempted to adjust their hand apertures by using information provided by SF. Only one of three participants achieved a satisfactory performance in this task. Since these participants were patients hospitalized for epilepsy monitoring after neurosurgery, the conditions were not ideal for attending to the task, and two became fatigued. On the other hand, cortical surface stimulation by a high-density ECoG electrode grid, implanted in a patient with severe arm paralysis, yielded reliable results for creating somatotopic artificial sensations (Hiremath et al., 2017). Stimulus parameters could modulate the magnitude and quality of sensations (such as electrical buzz, tingling, vibration, arm movement, vertigo). The participant could also discriminate stimulation through different electrodes quite well. Lee et al. (2018) reported results for artificial somatosensation by stimulation through a mini-ECoG grid. Almost all (out of nine) participants had very high accuracy in behavioral tasks, in which they had to discriminate left/right targets and identify orientations of targets.

1.5 Multidisciplinary approach and future directions

The 21st century has brought significant achievements for neuroengineering and neuroprostheses, given the fundamental knowledge acquired in neuroscience. However, there remains a tremendous way to go for the neuroprostheses with SF to become widely available. The major reason is the requirement of a multidisciplinary approach and large consortiums to reach the ambitious goals introduced in this chapter. In addition to neuroscience and neuroengineering, neuroprostheses depend on the research and developments regarding the miniaturization and fabrication of electromechanical systems, novel algorithms for signal processing, and perhaps more importantly, the neural interfaces, that is, electrodes and their assemblies, which directly communicate with the neuromuscular system. As discussed earlier, noninvasive alternatives exist, but direct neural interfaces increase signal quality for recording and enable a more accurate and precise sensory feedback. New developments on conductive materials, such as conductive polymers and graphene, appear to offer advantages over conventional noble metal electrodes, and may be the sources for the next-generation interfacing devices (Kostarelos et al., 2017; Won et al., 2018). Biocompatibility is the key factor for the long-term success of a neural interface.

Given the difficulty of producing a long-term viable and biocompatible neuroprosthesis, only some devices have reached wide commercial use, for example, spinal cord stimulators for chronic pain, cochlear implants, vagus stimulators for epilepsy, and DBS devices for PD. However, with the large investments in bioelectronic medicine (e.g., see Galvani Bioelectronics), many other types of devices (already approved or in development) are expected to be in use in the following decades. Government agencies have also been bolstering the innovation (e.g., see The BRAIN Initiative). The synergy among multidisciplinary partners has decreased the time for an idea to move from the laboratory to the clinical setting, and finally to commercialization. As the Fourth Industrial Revolution is in effect, the future neuroprostheses will probably be part of the Internet of Things and will also heavily incorporate artificial intelligence algorithms (Schwab, 2017).

Nevertheless, there are several points to consider regarding the technical and biological limitations of somatosensory neuroprostheses. These may be critical in the development of new devices or further improvement of the available designs. It seems that, with the current neuromodulation techniques, it is possible to adjust the intensity and location of the artificial sensation, but how about the perceptual quality? It has been argued that it may not be possible to achieve a "natural" sensation per se, but the unnatural sensation is still useful in closed-loop control. This is quite acceptable (as in cochlear implants), since it would increase the person's involvement with the activities of daily living and his/her independence. Most of the studies have not reported uncomfortable sensations, if the stimulation is within certain limits, but some have also indicated that PLP may decrease, which may be the most desirable function of neuromodulation for amputees. As a matter of fact, a recent study with the e-OPRA system showed that the users (2–6 years) experienced positive psychosocial effects at home and in their professional lives because of the reliability of the device and the cessation of PLP when SF was provided (Middleton & Ortiz-Catalan, 2020). However, the users' subjective reports stated that SF was of limited benefit during motor control, perhaps because they used the devices initially without SF for quite a long time. Future studies may also focus on separating cutaneous sensations from proprioceptive sensations, since that would make a big difference regarding the use of SF. Current electronic sensors are highly capable of measuring movement-based or contact-based information as the prosthetic device interacts with the environment. Feedback of independent somatic modalities may increase both the functionality of the device and embodiment. A future neuroprosthesis may also signal temperature and pain measured by artificial skin (Ho et al., 2016).

Due to low functionality, discomfort, and difficulty of use, conventional upper-limb prostheses are rejected by many users, for example, 28% in a study by Biddiss and Chau (2007). About 85% of prosthesis rejecters listed lack of SF as important among other factors. Given the great financial cost and effort in developing medical devices, regulatory agencies now require incorporating patient preferences into the total product lifecycles (Benz & Civillico, 2017). This is especially important, because the preferences may differ for the particular clinical condition. For example, tetraplegic patients wish some sort of arm and hand function (48.7%), and sensation is very secondary (6.1%) (Tsu et al., 2015). Of course, they may not know that limb function would be considerably improved if SF was efficiently included and if embodiment was also achieved. In any case, a human-centered design approach leads to safe and effective systems which minimize the mismatch between humans and machines (Contreras-Vidal et al., 2015).

Embodiment can be loosely defined as the sense of ownership for the artificial limb. The conventional method to quantify it has been by using questionnaires and behavioral or psychophysical measurements of perceptual illusions, for example, the rubber hand illusion (RHI). Guterstam et al. (2019) have recently shown the first data on the electrophysiological correlates of body ownership. Fig. 1.8 shows the setup, stimulation conditions, and questionnaire results for the RHI. The participant only sees a rubber hand. If the tactile stimulus applied to the rubber hand is spatiotemporally congruent with the same stimulus applied to the hidden biological hand, the illusion develops after about 5–10 trials. For a control, the stimuli can be made asynchronous or the rubber hand can be rotated. The questionnaire results had significant illusion effects only in the statements related to RHI. In Guterstam et al. (2019), cortical surface potentials were also measured during the development of RHI (Fig. 1.8A). Cortical activity due to RHI was mostly concentrated in the high-gamma band over the intraparietal sulcus and premotor cortex. However, the activity of the S1 cortex was not modulated with synchronous tactile stimulation as compared to control conditions. This technique shows the electrophysiological signature of visuotactile integration in related brain areas and may be a good marker for embodiment.

An important limitation of SF is adaptation due to repetitive stimulation (Bensmaia et al., 2020). Prolonged electrical stimulation may induce local changes related to the excitability of the neural tissue, for example, as in peripheral nerve stimulation, and also more widespread changes regarding network excitability. For example, ICMS also excites inhibitory neurons leading to significant depression of activity. Understanding these mechanisms can help to develop biomimetic neuromodulation paradigms. However, for rigorous testing of many hypotheses, it may also be worthwhile to

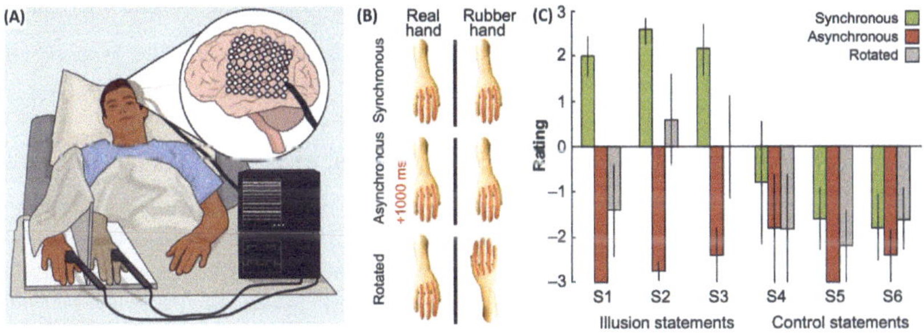

FIGURE 1.8 Rubber hand illusion (RHI) studied during ECoG recording. (A) The setup for inducing the RHI by applying spatiotemporally congruent tactile stimuli to a visible rubber hand and the hidden biological hand. ECoG recordings show the development of activity due to RHI in brain areas responsible for visuotactile integration. (B) Three test conditions: synchronous stimulation, asynchronous stimulation (temporal control), and rotated rubber hand (limb control). (C) Questionnaire results show significant effects of RHI only in the statements related to the illusion. *Reproduced with permission from Guterstam, A., Collins, K. L., Cronin, J. A., Zeberg, H., Darvas, F., Weaver, K. E., Ojemann, J. G., & Ehrsson, H. H. (2019). Direct electrophysiological correlates of body ownership in human cerebral cortex.* Cerebral Cortex, *29(3), 1328–1341. https://doi.org/10.1093/cercor/bhy285.*

consider rodent, especially rat, models in addition to nonhuman primates. Although human-like neuroprosthetic applications will be very limited with rats, and training may take a considerable time, many phenomena can still be demonstrated (Beygi et al., 2016; Devecioğlu & Güçlü, 2017; Öztürk et al., 2019). Furthermore, biocompatibility and technical properties of novel interfaces can be tested effectively and with more subjects. An interesting recent study combines FES with BCI and shows significant recovery in stroke patients due to neuroplasticity occurring with contingent activation of afferent and efferent pathways (Biasiucci et al., 2018). The anatomical and physiological substrates of this recovery can be investigated in much more detail with a rat model.

Neuroprosthetics research has a great impact on society as a whole; it has implications far beyond the improvement of patients' daily lives, such as reintegration into education, workforce, and culture. Responsible research and innovation practices would ensure that all partners engage in open activities such that the outcomes meet the needs, expectations, and values of society. Ethics may be thought of as one of the most important components of the complex interaction between scientific research and society. From a clinical point of view, autonomy, nonmaleficence, beneficence, and justice need to be followed (Lane et al., 2016). Glannon (2016) argues that, depending on the recovery potential in each application, there may indeed also be an ethical obligation to conduct related neuroprosthetics research. It seems that the current benefits of sensorimotor neuroprostheses can outweigh the risks for the most severe neurological cases. However, this still does not make a device suitable for use; there is an arduous path governed by regulatory and commercial issues (Parker, 2016). The performance of a device is assessed in multiple dimensions, such as quality-adjusted life years, mean time before failure, habilitation, ergonomics, cosmetics, and total cost including running and maintenance costs. Along these dimensions, criteria are set for a given device toward commercialization. If the minimal functional performance of the device is close to those criteria, the device is likely to be practically used in the near future. It is encouraging to see that neuroprostheses with somatosensory function have apparently passed a critical threshold.

Acknowledgments

This work was supported by TÜBİTAK Grant 117F481 within the European Union's FLAG-ERA JTC 2017 project GRAFIN. The author also thanks Xavier Navarro for helpful comments on the manuscript.

References

Armenta Salas, M., Bashford, L., Kellis, S., Jafari, M., Jo, H., Kramer, D., Shanfield, K., Pejsa, K., Lee, B., Liu, C. Y., & Andersen, R. A. (2018). Proprioceptive and cutaneous sensations in humans elicited by intracortical microstimulation. *Elife*, *7*, e32904. Available from https://doi.org/10.7554/eLife.32904.

Augurelle, A. S., Smith, A. M., Lejeune, T., & Thonnard, J. L. (2003). Importance of cutaneous feedback in maintaining a secure grip during manipulation of hand-held objects. *Journal of Neurophysiology*, *89*(2), 665–671. Available from https://doi.org/10.1152/jn.00249.2002.

Bensmaia, S. J., & Miller, L. E. (2014). Restoring sensorimotor function through intracortical interfaces: Progress and looming challenges. *Nature Reviews Neuroscience*, *15*(5), 313–325. Available from https://doi.org/10.1038/nrn3724.

Bensmaia, S. J., Tyler, D. J., & Micera, S. (2020). Restoration of sensory information via bionic hands. *Nature Biomedical Engineering*. Available from https://doi.org/10.1038/s41551-020-00630-8, *Nov 23, online ahead of print*.

Benz, H. L., & Civillico, E. F. (2017). Neuroprosthetics and the science of patient input. *Experimental Neurology*, *287*(Pt 4), 486–491. Available from https://doi.org/10.1016/j.expneurol.2016.07.017.

Beygi, M., Mutlu, S., & Güçlü, B. (2016). A microfabricated strain gauge array on polymer substrate for tactile neuroprostheses in rats. *Journal of Micromechanics and Microengineering*, *26*(8), 084006. Available from https://doi.org/10.1088/0960-1317/26/8/084006.

Biasiucci, A., Leeb, R., Iturrate, I., Perdikis, S., Al-Khodairy, A., Corbet, T., Schnider, A., Schmidlin, T., Zhang, H., Bassolino, M., Viceic, D., Vuadens, P., Guggisberg, A. G., & Millán, J. D. R. (2018). Brain-actuated functional electrical stimulation elicits lasting arm motor recovery after stroke. *Nature Communication*, *9*(1), 2421. Available from https://doi.org/10.1038/s41467-018-04673-z.

Biddiss, E., & Chau, T. (2007). Upper-limb prosthetics: Critical factors in device abandonment. *Americal Journal of Physical Medicine & Rehabilitation*, *86*(12), 977–987. Available from https://doi.org/10.1097/PHM.0b013e3181587f6c.

Bloch, E., Luo, Y., & da Cruz, L. (2019). Advances in retinal prosthesis systems. *Therapeutic Advances in Ophthalmology*, *11*, 1–16. Available from https://doi.org/10.1177/2515841418817501.

Bolanowski, S. J., Jr., Gescheider, G. A., Verrillo, R. T., & Checkosky, C. M. (1988). Four channels mediate the mechanical aspects of touch. *Journal of the Acoustical Society of America*, *84*(5), 1680–1694. Available from https://doi.org/10.1121/1.397184.

Bolanowski, S. J., Jr., & Zwislocki, J. J. (1984). Intensity and frequency characteristics of Pacinian corpuscles. I. Action potentials. *Journal of Neurophysiology*, *51*(4), 793–811. Available from https://doi.org/10.1152/jn.1984.51.4.793.

Brindley, G. S., & Lewin, W. S. (1968). The sensations produced by electrical stimulation of the visual cortex. *Journal of Physiology*, *196*(2), 479–493. Available from https://doi.org/10.1113/jphysiol.1968.sp008519.

Campbell, F. W., & Robson, J. G. (1968). Application of Fourier analysis to the visibility of gratings. *Journal of Physiology*, *197*(3), 551–566. Available from https://doi.org/10.1113/jphysiol.1968.sp008574.

Catani, M. (2017). A little man of some importance. *Brain*, *140*(11), 3055–3061. Available from https://doi.org/10.1093/brain/awx270.

Cascio, C. J., Moore, D., & McGlone, F. (2019). Social touch and human development. *Developmental Cognitive Neuroscience*, *35*, 5–11. Available from https://doi.org/10.1016/j.dcn.2018.04.009.

Chandrasekaran, S., Nanivadekar, A. C., McKernan, G., Helm, E. R., Boninger, M. L., Collinger, J. L., Gaunt, R. A., & Fisher, L. E. (2020). Sensory restoration by epidural stimulation of the lateral spinal cord in upper-limb amputees. *Elife*, *9*, e54349. Available from https://doi.org/10.7554/eLife.54349.

Chapin, J. K., & Moxon, K. A. (Eds.), (2001). *Neural prostheses for restoration of sensory and motor function*. CRC Press.

Ciancio, A. L., Cordella, F., Barone, R., Romeo, R. A., Bellingegni, A. D., Sacchetti, R., Davalli, A., Di Pino, G., Ranieri, F., Di Lazzaro, V., Guglielmelli, E., & Zollo, L. (2016). Control of prosthetic hands via the peripheral nervous system. *Frontiers in Neuroscience*, *10*, 116. Available from https://doi.org/10.3389/fnins.2016.00116.

Cipriani, C., Segil, J. L., Clemente, F., ff. Weir, R. F., & Edin, B. (2014). Humans can integrate feedback of discrete events in their sensorimotor control of a robotic hand. *Experimental Brain Research*, *232*(11), 3421–3429. Available from https://doi.org/10.1007/s00221-014-4024-8.

Clemente, F., D'Alonzo, M., Controzzi, M., Edin, B. B., & Cipriani, C. (2016). Non-invasive, temporally discrete feedback of object contact and release improves grasp control of closed-loop myoelectric transradial prostheses. *IEEE Transactions on Neural Systems and Rehabilitation Eng*, *24*(12), 1314–1322. Available from https://doi.org/10.1109/TNSRE.2015.2500586.

Clippinger, F. W., Avery, R., & Titus, B. R. (1974). A sensory feedback system for an upper-limb amputation prosthesis. *Bulletin of Prosthetics Research*, *10–22*, 247–258.

Cohen, J. C., Makous, J. C., & Bolanowski, S. J. (1999). Under which conditions do the skin and probe decouple during sinusoidal vibrations? *Experimental Brain Research*, *129*(2), 211–217. Available from https://doi.org/10.1007/s002210050891.

Cole, J. (2008). The neuroscience and phenomenology of sensory loss. In M. Grunwald (Ed.), *Human haptic perception: Basics and applications* (pp. 295–301). Birkhäuser.

Contreras-Vidal, J. L., Kilicarslan, A., Huang, H. (H., & Grossman, R. G. (2015). Human-centered design of wearable neuroprostheses and exoskeletons. *AI Magazine*, *36*(4), 12–22. Available from https://doi.org/10.1609/aimag.v36i4.2613.

Cronin, J. A., Wu, J., Collins, K. L., Sarma, D., Rao, R. P. N., Ojemann, J. G., & Olson, J. D. (2016). Task-specific somatosensory feedback via cortical stimulation in humans. *IEEE Transactions on Haptics*, *9*(4), 515–522. Available from https://doi.org/10.1109/TOH.2016.2591952.

del Valle, J., & Navarro, X. (2013). Interfaces with the peripheral nerve for the control of neuroprostheses. *International Review of Neurobiology*, *109*, 63–83. Available from https://doi.org/10.1016/B978-0-12-420045-6.00002-X.

Devecioğlu, İ., & Güçlü, B. (2017). Psychophysical correspondence between vibrotactile intensity and introcortical microstimulation for tactile neuroprostheses in rats. *Journal of Neural Engineering*, *14*(1), 016010. Available from https://doi.org/10.1088/1741-2552/14/1/016010.

Dhillon, G. S., & Horch, K. W. (2005). Direct neural sensory feedback and control of a prosthetic arm. *IEEE Transactions on Neural Systems and Rehabilitation Engineering*, *13*(4), 468–472. Available from https://doi.org/10.1109/TNSRE.2005.856072.

Dobelle, W. H., Stensaas, S. S., Mladejoversusky, M. G., & Smith, J. B. (1973). A prosthesis for the deaf based on cortical stimulation. *Annals of Otology, Rhinology, and Laryngology*, *82* (4), 445–463. Available from https://doi.org/10.1177/000348947308200404.

Donaldson, N., & Brindley, G. S. (2016). The historical foundations of bionics. In R. K. Shepherd (Ed.), *Neurobionics: The biomedical engineering of neural prostheses* (pp. 3–37). Wiley Blackwell. Available from https://doi.org/10.1002/9781118816028.ch1.

Dornhege, G., Millán, J., del, R., Hinterberger, T., McFarland, D. J., & Müller, K.-R. (Eds.), (2007). *Toward brain-computer interfacing*. The MIT Press.

Ferrington, D. G., Horniblow, S., & Rowe, M. J. (1987). Temporal patterning in the responses of gracile and cuneate neurones in the cat to cutaneous vibration. *Journal of Physiology*, *386*, 277–291. Available from https://doi.org/10.1113/jphysiol.1987.sp016534.

Finkelstein, G. (2015). Mechanical neuroscience: Emil du Bois-Reymond's innovations in theory and practice. *Frontiers in System Neuroscience*, *9*(133), 1–4. Available from https://doi.org/10.3389/fnsys.2015.00133.

Flesher, S. N., Collinger, J. L., Foldes, S. T., Weiss, J. M., Downey, J. E., Tyler-Kabara, E. C., Bensmaia, S. J., Schwartz, A. B., Boninger, M. L., & Gaunt, R. A. (2016). Intracortical microstimulation of human somatosensory cortex. *Science Translational Medicine*, *8*. Available from https://doi.org/10.1126/scitranslmed.aaf8083, 361ra141.

Gardner, E. P., & Johnson, K. E. (2013). The somatosensory system: Receptors and central pathways. In E. R. Kandel, J. H. Schwartz, T. M. Jessell, S. A. Siegelbaum, & A. J. Hudspeth (Eds.), *Principles of neural science* (5rd (ed.), pp. 475–497). McGraw-Hill Medical.

Gardner, J. (2013). A history of deep brain stimulation: Technological innovation and the role of clinical assessment tools. *Social Studies of Science*, *43*(5), 707–728. Available from https://doi.org/10.1177/0306312713483678.

George, J. A., Kluger, D. T., Davis, T. S., Wendelken, S. M., Okorokova, E. V., He, Q., Duncan, C. C., Hutchinson, D. T., Thumser, Z. C., Beckler, D. T., Marasco, P. D., Bensmaia, S. J., & Clark, G. A. (2019). Biomimetic sensory feedback through peripheral nerve stimulation improves dexterous use of a bionic hand. *Science Robotics*, *4*(32), eaax2352. Available from https://doi.org/10.1126/scirobotics.aax2352.

Gescheider, G. A., Wright, J. H., & Verrillo, R. T. (2009). *Information-processing channels in the tactile sensory system: A psychophysical and physiological analysis*. Psychology Press. Available from https://doi.org/10.4324/9780203890004.

Glannon, W. (2016). Ethical issues in neuroprosthetics. *Journal of Neural Engineering*, *13*(2), 021002. Available from https://doi.org/10.1088/1741-2560/13/2/021002.

Greenspan, J. D., & Bolanowski, S. J. (1996). The psychophysics of tactile perception and its peripheral physiological basis. In L. Kruger (Ed.), *Pain and touch: Handbook of perception and cognition* (2nd ed., pp. 25–103). Academic Press.

Grunwald, M. (Ed.), (2008). *Human haptic perception: Basics and applications*. Birkhäuser.

Güçlü, B. (2007). Deviation from Weber's law in the non-Pacinian I tactile channel: A psychophysical and simulation study of intensity discrimination. *Neural Computing*, *19*(10), 2638–2664. Available from https://doi.org/10.1162/neco.2007.19.10.2638.

Güçlü, B., & Bolanowski, S. J. (2003a). Distribution of the intensity-characteristic parameters of cat rapidly adapting mechanoreceptive fibers. *Somatosensory & Motor Research*, *20*(2), 149–155. Available from https://doi.org/10.1080/0899022031000105181.

Güçlü, B., & Bolanowski, S. J. (2003b). Frequency responses of cat rapidly adapting mechanoreceptive fibers. *Somatosensory & Motor Research*, *20*(3–4), 249–263. Available from https://doi.org/10.1080/08990220310001622988.

Güçlü, B., & Bolanowski, S. J. (2004a). Tristate Markov model for the firing statistics of rapidly-adapting mechanoreceptive fibers. *Journal of Computational Neuroscience*, *17*(2), 107–126. Available from https://doi.org/10.1023/B:JCNS.0000037680.56375.85.

Güçlü, B., & Bolanowski, S. J. (2004b). Probability of stimulus detection in a model population of rapidly adapting fibers. *Neural Computing*, *16*(1), 39–58. Available from https://doi.org/10.1162/08997660460733985.

Güçlü, B., & Bolanowski, S. J. (2005a). Vibrotactile thresholds of the Non-Pacinian I channel: I. Methodological issues. *Somatosensory & Motor Research*, *22*(1–2), 49–56. Available from https://doi.org/10.1080/08990220500084719.

Güçlü, B., & Bolanowski, S. J. (2005b). Vibrotactile thresholds of the Non-Pacinian I channel: II. Predicting the effects of contactor location on the phalanx. *Somatosensory & Motor Research*, *22*(1–2), 57–68. Available from https://doi.org/10.1080/08990220512331387971.

Güçlü, B., & Öztek, Ç. (2007). Tactile sensitivity of children: Effects of frequency, masking, and the non-Pacinian I psychophysical channel. *Journal of Experimental Child Psychology*, *98*(2), 113–130. Available from https://doi.org/10.1016/j.jecp.2007.05.003.

Güçlü, B., Bolanowski, S. J., & Pawson, L. (2003). End-to-end linkage (EEL) clustering algorithm: A study on the distribution of Meissner corpuscles in the skin. *Journal of Computational Neuroscience*, *15*(1), 19–28. Available from https://doi.org/10.1023/a:1024466617694.

Güçlü, B., Gescheider, G. A., Bolanowski, S. J., & İstefanopulos, Y. (2005). Population-response model for vibrotactile spatial summation. *Somatosensory & Motor Research*, *22*(4), 239–253. Available from https://doi.org/10.1080/08990220500262075.

Güçlü, B., Mahoney, G. K., Pawson, L. J., Pack, A. K., Smith, R. L., & Bolanowski, S. J. (2008). Localization of Merkel cells in the monkey skin: An anatomical model. *Somatosensory & Motor Research*, *25*(2), 123–138. Available from https://doi.org/10.1080/08990220802131234.

Güçlü, B., Tanıdır, C., Çanayaz, E., Güner, B., İpek Toz, H., Üneri, Ö. Ş., & Tommerdahl, M. (2015). Tactile processing in children and adolescents with obsessive-compulsive disorder. *Somatosensory & Motor Research*, *32*(3), 163–171. Available from https://doi.org/10.3109/08990220.2015.1023950.

Guterstam, A., Collins, K. L., Cronin, J. A., Zeberg, H., Darvas, F., Weaver, K. E., Ojemann, J. G., & Ehrsson, H. H. (2019). Direct electrophysiological correlates of body ownership in human cerebral cortex. *Cerebral Cortex*, *29*(3), 1328–1341. Available from https://doi.org/10.1093/cercor/bhy285.

Hebert, J. S., Olson, J. L., Morhart, M. J., Dawson, M. R., Marasco, P. D., Kuiken, T. A., & Chan, K. M. (2014). Novel targeted sensory reinnervation technique to restore functional hand sensation after transhumeral amputation. *IEEE Transactions on Neural Systems and Rehabiliation Engineering*, *22*(4), 765–773. Available from https://doi.org/10.1109/TNSRE.2013.2294907.

Heller, M. A., & Myers, D. S. (1983). Active and passive tactual recognition of form. *Journal of General Psychology*, *108*(2nd Half), 225–229. Available from https://doi.org/10.1080/00221309.1983.9711496.

Hiremath, S. V., Tyler-Kabara, E. C., Wheeler, J. J., Moran, D. W., Gaunt, R. A., Collinger, J. L., Foldes, S. T., Weber, D. J., Chen, W., Boninger, M. L., & Wang, W. (2017). Human perception of electrical stimulation on the surface of somatosensory cortex. *PLoS One*, *12* (5), e0176020. Available from https://doi.org/10.1371/journal.pone.0176020.

Ho, D. H., Sun, Q., Kim, S. Y., Han, J. T., Kim, D. H., & Cho, J. H. (2016). Stretchable and multimodal all graphene electronic skin. *Advanced Mater*, *28*(13), 2601–2608. Available from https://doi.org/10.1002/adma.201505739.

Hochberg, L. R., Serruya, M. D., Friehs, G. M., Mukand, J. A., Saleh, M., Caplan, A. H., Branner, A., Chen, D., Penn, R. D., & Donoghue, J. P. (2006). Neuronal ensemble control of prosthetic devices by a human with tetraplegia. *Nature*, *442*(7099), 164–171. Available from https://doi.org/10.1038/nature04970.

Horch, K. W., & Dhillon, G. S. (Eds.), (2004). *Neuroprosthetics: Theory and practice*. World Scientific Publishing.

Humayun, M. S., Dorn, J. D., Ahuja, A. K., Caspi, A., Filley, E., Dagnelie, G., Salzmann, J., Santos, A., Duncan, J., daCruz, L., Mohand-Said, S., Eliott, D., McMahon, M. J., & Greenberg, R. J. (2009). Preliminary 6-month results from the Argus II epiretinal prosthesis feasibility study. *Annual International Conference on IEEE Engineering in Medicine and Biological Society*, *2009*, 4566–4568. Available from https://doi.org/10.1109/IEMBS.2009.5332695.

Iggo, A., & Muir, A. R. (1969). The structure and function of a slowly adapting touch corpuscle in hairy skin. *Journal of Physiology*, *200*(3), 763–796. Available from https://doi.org/10.1113/jphysiol.1969.sp008721.

Johansson, R. S., & Flanagan, J. R. (2009). Coding and use of tactile signals from the fingertips in object manipulation tasks. *Nature Reviews Neuroscience*, *10*(5), 345–359. Available from https://doi.org/10.1038/nrn2621.

Johnson, L. A., Wander, J. D., Sarma, D., Su, D. K., Fetz, E. E., & Ojemann, J. G. (2013). Direct electrical stimulation of somatosensory cortex in humans using electrocorticography electrodes: A qualitative and quantitative report. *Journal of Neural Engineering*, *10*(3), 036021. Available from https://doi.org/10.1088/1741-2560/10/3/036021.

Jones, L. A., & Lederman, S. J. (2006). *Human hand function*. Oxford University Press.

Kaas, J. H. (2012). Somatosensory system. In J. K. Mai, & G. Paxinos (Eds.), *The human nervous system* (3rd ed., pp. 1074–1109). Academic Press. Available from https://doi.org/10.1016/B978-0-12-374236-0.10030-6.

Karakuş, İ., & Güçlü, B. (2020). Psychophysical principles of discrete event-driven vibrotactile feedback for prostheses. *Somatosensory & Motor Research*, *37*(3), 186–203. Available from https://doi.org/10.1080/08990220.2020.1769055.

Keith, M. W., Peckham, P. H., Thrope, G. B., Stroh, K. C., Smith, B., Buckett, J. R., Kilgore, K. L., & Jatich, J. W. (1989). Implantable functional neuromuscular stimulation in the tetraplegic hand. *Journal of Hand Surgery (American Volume)*, *14*(3), 524–530. Available from https://doi.org/10.1016/s0363-5023(89)80017-6.

Kilgore, K. (Ed.), (2015a). *Implantable neuroprostheses for restoring function*. Woodhead Publishing.

Kilgore, K. L. (2015b). Hand grasp and reach in spinal cord injury. In K. Kilgore (Ed.), *Implantable neuroprostheses for restoring function* (pp. 209–235). Woodhead Publishing. Available from https://doi.org/10.1016/B978-1-78242-101-6.00009-4.

Kostarelos, K., Vincent, M., Hebert, C., & Garrido, J. A. (2017). Graphene in the design and engineering of next generation neural interfaces. *Advanced Materials*, *29*(42), 1700909. Available from https://doi.org/10.1002/adma.201700909.

Kralj, A., Bajd, T., & Turk, R. (1980). Electrical stimulation providing functional use of paraplegic patient muscles. *Medical Progress Technology*, *7*(1), 3–9.

Lamb, G. D. (1983). Tactile discrimination of textured surfaces: Psychophysical performance measurements in humans. *Journal of Physiology*, *338*, 551–565. Available from https://doi.org/10.1113/jphysiol.1983.sp014689.

Lane, F. J., Nitsch, K. P., & Scherer, M. (2016). Ethical considerations in the development of neural prostheses. In R. K. Shepherd (Ed.), *Neurobionics: The biomedical engineering of neural prostheses* (pp. 3–37). Wiley Blackwell. Available from https://doi.org/10.1002/9781118816028.ch11.

Lechner, S. G., & Lewin, G. R. (2013). Hairy sensation. *Physiology (Bethesda)*, *28*(3), 142–150. Available from https://doi.org/10.1152/physiol.00059.2012.

Lee, B., Kramer, D., Armenta Salas, M., Kellis, S., Brown, D., Dobreva, T., Klaes, C., Heck, C., Liu, C., & Andersen, R. A. (2018). Engineering artificial somatosensation through cortical stimulation in humans. *Frontiers in Systems Neuroscience*, *12*, 24. Available from https://doi.org/10.3389/fnsys.2018.00024.

Liberson, W. T., Holmquest, H. J., Scot, D., & Dow, M. (1961). Functional electrotherapy: Stimulation of the peroneal nerve synchronized with the swing phase of the gait of hemiplegic patients. *Archives of Physical Medicine and Rehabilitation*, *42*, 101–105.

Libet, B. (1982). Brain stimulation in the study of neuronal functions for conscious sensory experiences. *Human Neurobiology*, *1*(4), 235–242.

Lindblom, U. (1965). Properties of touch receptors in distal glabrous skin of the monkey. *Journal of Neurophysiology*, *28*(5), 966–985. Available from https://doi.org/10.1152/jn.1965.28.5.966.

Loutit, A. J., Vickery, R. M., & Potas, J. R. (2021). Functional organization and connectivity of the dorsal column nuclei complex reveals a sensorimotor integration and distribution hub. *Journal of Comparative Neurology*, *529*(1), 187–220. Available from https://doi.org/10.1002/cne.24942.

Macgillis, M., Kang, R., Herman, D., & Zarzecki, P. (1983). Interactions among convergent inputs to somatosensory cortex neurons. *Brain Research*, *276*(2), 329–332. Available from https://doi.org/10.1016/0006-8993(83)90741-2.

Maricich, S. M., Wellnitz, S. A., Nelson, A. M., Lesniak, D. R., Gerling, G. J., Lumpkin, E. A., & Zoghbi, H. Y. (2009). Merkel cells are essential for light-touch responses. *Science*, *324*(5934), 1580–1582. Available from https://doi.org/10.1126/science.1172890.

Mastinu, E., Engels, L. F., Clemente, F., Dione, M., Sassu, P., Aszmann, O., Brånemark, R., Håkansson, B., Controzzi, M., Wessberg, J., Cipriani, C., & Ortiz-Catalan, M. (2020). Neural feedback strategies to improve grasping coordination in neuromusculoskeletal prostheses. *Scientific Reports*, *10*(1), 11793. Available from https://doi.org/10.1038/s41598-020-67985-5.

McNamee, S. (2014). History of amputation: From the past to the present. In D. Murphy (Ed.), *Fundamentals of amputation care and prosthetics* (pp. 1–6). Demos Medical Publishing.

Middleton, A., & Ortiz-Catalan, M. (2020). Neuromusculoskeletal arm prostheses: Personal and social implications of living with an intimately integrated bionic arm. *Frontiers in Neurorobotics*, *14*, 39. Available from https://doi.org/10.3389/fnbot.2020.00039.

Mirochnik, R. B., & Pezaris, J. S. (2019). Contemporary approaches to visual prostheses. *Military Medical Research*, *6*(1), 19. Available from https://doi.org/10.1186/s40779-019-0206-9.

Moe, J. H., & Post, H. W. (1962). Functional electrical stimulation for ambulation in hemiplegia. *Journal of Lancet*, *82*, 285–288.

Moghadasi, A. N. (2014). Artificial eye in Burnt City and theoretical understanding of how vision works. *Iranian Journal of Public Health*, *43*(11), 1595–1596.

Mountcastle, V. B. (2005). *The sensory hand: Neural mechanisms of somatic sensation*. Harvard University Press.

Mudry, A., & Mills, M. (2013). The early history of the cochlear implant: A retrospective. *JAMA Otolaryngology Head Neck Surgery*, *139*(5), 446–453. Available from https://doi.org/10.1001/jamaoto.2013.293.

Nelson, R. J. (Ed.), (2002). *The somatosensory system: Deciphering the brain's own body image*. CRC Press.

Nerlich, A. G., Zink, A., Szeimies, U., & Hagedorn, H. G. (2000). Ancient Egyptian prosthesis of the big toe. *Lancet*, *356*(9248), 2176–2179. Available from https://doi.org/10.1016/S0140-6736(00)03507-8.

Olausson, H., Wessberg, J., Morrison, I., McGlone, F., & Vallbo, Å. (2010). The neurophysiology of unmyelinated tactile afferents. *Neuroscience & Biobehavioral Reviews*, *34*(2), 185–191. Available from https://doi.org/10.1016/j.neubiorev.2008.09.011.

Öztürk, S., Devecioğlu, İ., Beygi, M., Atasoy, A., Mutlu, Ş., Özkan, M., & Güçlü, B. (2019). Real-time performance of a tactile neuroprosthesis on awake behaving rats. *IEEE Transactions on Neural Systems and Rehabilitation Engineering*, *27*(5), 1053–1062. Available from https://doi.org/10.1109/TNSRE.2019.2910320.

Paré, M., Behets, C., & Cornu, O. (2003). Paucity of presumptive Ruffini corpuscles in the index finger pad of humans. *Journal of Comparative Neurology*, *456*(3), 260–266. Available from https://doi.org/10.1002/cne.10519.

Parker, J. L. (2016). Taking a device to market: Regulatory and commercial issues. In R. K. Shepherd (Ed.), *Neurobionics: The biomedical engineering of neural prostheses* (pp. 3–37). Wiley Blackwell. Available from https://doi.org/10.1002/9781118816028.ch10.

Peeples, L. (2019). Core concept: The rise of bioelectric medicine sparks interest among researchers, patients, and industry. *Proceedings of the National Academy Science of USA*, *116*(49), 24379–24382. Available from https://doi.org/10.1073/pnas.1919040116.

Press, D., Mutlu, S., & Güçlü, B. (2010). Evidence of fast serotonin transmission in frog slowly adapting type 1 responses. *Somatosensory & Motor Research*, *27*(4), 174–185. Available from https://doi.org/10.3109/08990220.2010.516670.

Quick, K. M., Weiss, J. M., Clemente, F., Gaunt, R. A., & Collinger, J. L. (2020). Intracortical microstimulation feedback improves grasp force accuracy in a human using a brain-computer interface. *Annual International Conference on IEEE Engineering in Medicine and Biological Society*, *2020*, 3355–3358. Available from https://doi.org/10.1109/EMBC44109.2020.9175926.

Rao, R. P. N. (2013). *Brain-computer interfacing: An introduction*. Cambridge University Press.

Raspopovic, S., Capogrosso, M., Petrini, F. M., Bonizzato, M., Rigosa, J., Di Pino, G., Carpaneto, J., Controzzi, M., Boretius, T., Fernandez, E., Granata, G., Oddo, C. M., Citi, L., Ciancio, A. L., Cipriani, C., Carrozza, M. C., Jensen, W., Guglielmelli, E., Stieglitz, T., . . . Micera, S. (2014). Restoring natural sensory feedback in real-time bidirectional hand prostheses. *Science Translational Medicine*, *6*(222), 222ra19. Available from https://doi.org/10.1126/scitranslmed.3006820.

Rees, S., Walker, D., & Jennings, E. (2010). Development of the somatosensory system. In H. Lagercrantz, M. A. Hanson, L. R. Ment, & D. M. Peebles (Eds.), *The newborn brain: Neuroscience and clinical applications* (2nd ed., pp. 129–145). Cambridge University Press.

Richer, F., Martinez, M., Robert, M., Bouvier, G., & Saint-Hilaire, J.-M. (1993). Stimulation of human somatosensory cortex: Tactile and body displacement perceptions in medial regions. *Experimental in Brain Research*, *93*(1), 173–176. Available from https://doi.org/10.1007/BF00227792.

Saal, H. P., & Bensmaia, S. J. (2014). Touch is a team effort: Interplay of submodalities in cutaneous sensibility. *Trends Neuroscience*, *37*(12), 689–697. Available from https://doi.org/10.1016/j.tins.2014.08.012.

Sainburg, R. L., Poizner, H., & Ghez, C. (1993). Loss of proprioception produces deficits in interjoint coordination. *Journal of Neurophysiology*, *70*(5), 2136–2147. Available from https://doi.org/10.1152/jn.1993.70.5.2136.

Saito, K., Otsuru, N., Inukai, Y., Miyaguchi, S., Yokota, H., Kojima, S., Sasaki, R., & Onishi, H. (2019). Comparison of transcranial electrical stimulation regimens for effects on inhibitory circuit activity in primary somatosensory cortex and tactile spatial discrimination performance. *Behavioural Brain Research*, *375*, 112168. Available from https://doi.org/10.1016/j.bbr.2019.112168.

Sanchez, J. C. (2016). *Neuroprosthetics: Principles and applications*. CRC Press.

Schwab, K. (2017). *The fourth industrial revolution*. Portfolio Penguin.

Shepherd, R. K. (Ed.), (2016). *Neurobionics: The biomedical engineering of neural prostheses*. Wiley Blackwell.

Smith, C. U. M. (2000). *Biology of sensory systems*. John Wiley & Sons.

Spires, M. C., Kelly, B. M., & Davis, A. J. (Eds.), (2014). *Prosthetic restoration and rehabilitation of the upper and lower extremity*. Demos Medical Publishing.

Štrbac, M., Isaković, M., Belić, M., Popović, I., Simanić, I., Farina, D., Keller, T., & Došen, S. (2017). Short- and long-term learning of feedforward control of a myoelectric prosthesis with sensory feedback by amputees. *IEEE Transactions on Neural Systems and Rehabilitation Engineering*, *25*(11), 2133–2145. Available from https://doi.org/10.1109/TNSRE.2017.2712287.

Svensson, P., Wijk, U., Björkman, A., & Antfolk, C. (2017). A review of invasive and non-invasive sensory feedback in upper limb prostheses. *Expert Review of Medical Devices*, *14*(6), 439–447. Available from https://doi.org/10.1080/17434440.2017.1332989.

Talbot, W. H., Darian-Smith, I., Kornhuber, H. H., & Mountcastle, V. B. (1968). The sense of flutter-vibration: Comparison of the human capacity with response patterns of mechanoreceptive afferents from the monkey hand. *Journal of Neurophysiology*, *31*(2), 301–334. Available from https://doi.org/10.1152/jn.1968.31.2.301.

Tsoucalas, G., Karamanou, M., Lymperi, M., Gennimata, V., & Androutsos, G. (2014). The "torpedo" effect in medicine. *International Maritime Health*, *65*(2), 65–67. Available from https://doi.org/10.5603/IMH.2014.0015.

Tsu, A. P., Burish, M. J., GodLove, J., & Ganguly, K. (2015). Cortical neuroprosthetics from a clinical perspective. *Neurobiology of Diseases*, *83*, 154–160. Available from https://doi.org/10.1016/j.nbd.2015.07.015.

Vega-Bermudez, F., Johnson, K. O., & Hsiao, S. S. (1991). Human tactile pattern recognition: Active vs passive touch, velocity effects, and patterns of confusion. *Journal of Neurophysiology*, *65*(3), 531–546. Available from https://doi.org/10.1152/jn.1991.65.3.531.

Verrillo, R. T., Bolanowski, S. J., & McGlone, F. P. (1999). Subjective magnitude of tactile roughness. *Somatosensory & Motor Research*, *16*(4), 352–360. Available from https://doi.org/10.1080/08990229970401.

Vidal, J. J. (1977). Real-time detection of brain events in EEG. *Proceedings of the IEEE, 65*(5), 633–641. Available from https://doi.org/10.1109/PROC.1977.10542.

Weber, D. J., Friesen, R., & Miller, L. E. (2012). Interfacing the somatosensory system to restore touch and proprioception: Essential considerations. *Journal of Motor Behavior, 44*(6), 403–418. Available from https://doi.org/10.1080/00222895.2012.735283.

Wendelken, S., Page, D. M., Davis, T., Wark, H. A. C., Kluger, D. T., Duncan, C., Warren, D. J., Hutchinson, D. T., & Clark, G. A. (2017). Restoration of motor control and proprioceptive and cutaneous sensation in humans with prior upper-limb amputation via multiple Utah Slanted Electrode Arrays (USEAs) implanted in residual peripheral arm nerves. *Journal of Neuroengineering and Rehabilitation, 14*(1), 121. Available from https://doi.org/10.1186/s12984-017-0320-4.

Willis, W. D., Jr., & Coggeshall, R. E. (2004). (3rd ed.). *Sensory mechanisms of the spinal cord: Primary afferent neurons and the spinal dorsal horn*, (Vol. 1). Kluwer Academic/Plenum Publishers.

Wolf, E. J., Cruz, T. H., Emondi, A. A., Langhals, N. B., Naufel, S., Peng, G. C. Y., Schulz, B. W., & Wolfson, M. (2019). Advanced technologies for intuitive control and sensation of prosthetics. *Biomedical Engineering Letters, 10*(1), 119–128. Available from https://doi.org/10.1007/s13534-019-00127-7.

Won, S. M., Song, E., Zhao, J., Li, J., Rivnay, J., & Rogers, J. A. (2018). Recent advances in materials, devices, and systems for neural interfaces. *Advanced Materials, 30*(30), e1800534. Available from https://doi.org/10.1002/adma.201800534.

Woolsey, C. N., Erickson, T. C., & Gilson, W. E. (1979). Localization in somatic sensory and motor areas of human cerebral cortex as determined by direct recording of evoked potentials and electrical stimulation. *Journal of Neurosurgery, 51*(4), 476–506. Available from https://doi.org/10.3171/jns.1979.51.4.0476.

Yıldız, M. Z., Toker, İ., Özkan, F. B., & Güçlü, B. (2015). Effects of passive and active movement on vibrotactile detection thresholds of the Pacinian channel and forward masking. *Somatosensory & Motor Research, 32*(4), 262–272. Available from https://doi.org/10.3109/08990220.2015.1091771.

Zuo, K. J., & Olson, J. L. (2014). The evolution of functional hand replacement: From iron prostheses to hand transplantation. *Plastics and Surgery (Oakv), 22*(1), 44–51.

Chapter 2

Proprioception: a sense to facilitate action

Kyle P. Blum[1], Christopher Versteeg[2], Joseph Sombeck[2], Raeed H. Chowdhury[3] and Lee E. Miller[1,2,4,5]

[1]Department of Physiology, Feinberg School of Medicine, Northwestern University, Chicago, IL, United States, [2]Department of Biomedical Engineering, McCormick School of Engineering, Northwestern University, Evanston, IL, United States, [3]Department of Bioengineering, Swanson School of Engineering, University of Pittsburgh, Pittsburgh, PA, United States, [4]Department of Physical Medicine and Rehabilitation, Feinberg School of Medicine, Northwestern University, Chicago, IL, United States, [5]Shirley Ryan AbilityLab, Chicago, IL, United States

ABSTRACT

Proprioception is commonly described as the sense of body position and movement. However, an underappreciated aspect of this "hidden sixth sense" is how intimately it is tied to the systems that execute movement. Loss of proprioception devastates our ability to move; without the ability to replace it, a motor brain-computer interface cannot be expected to function well. Proprioceptive receptors, located mostly within the muscles and joints, signal limb state that is the consequence of descending motor commands to the muscles and their interaction with the effects of external forces. As proprioceptive afferents ascend the neuroaxis, they contribute to reflexes and muscle coordination before reaching the brainstem and thalamus, subject to descending modulatory influence from both the brainstem and sensorimotor cortex at all levels. Within the somatosensory cortex, area 3a, with its largely muscle-receptor inputs, shares many common features with the adjacent motor cortex and is certainly intimately involved in movement execution. The higher level representation that results from combination with cutaneous inputs in area 2 likely provides critical limb state information to the posterior parietal cortex for use in movement planning. Proprioceptive signals are also involved in internal model formation and error calculation in the cerebellum, where deficits share many features in common with those experienced following deafferentation. The challenge of developing a proprioceptive brain interface is certainly a technical one, but also one of reaching a deeper understanding of all the systems—and their interconnectedness—outlined in this chapter.

Keywords: Proprioceptive system; movement; muscle receptor; sensorimotor control; cerebellum; dorsal column

Somatosensory Feedback for Neuroprosthetics. DOI: https://doi.org/10.1016/B978-0-12-822828-9.00017-4

2.1 Introduction

Proprioception is commonly defined as the awareness of the position and movement of the body. However, its further designation as "the silent sixth sense" suggests something hidden. Proprioception for subconscious motor control arises largely from the same sensors as does conscious proprioception, but its function is to facilitate the control of the simplest spinal reflexes as well as to plan and execute the most dexterous volitional actions. At every level, these proprioceptive systems are well integrated with the motor systems they serve. Brodmann's area 3a, a primary cortical area receiving inputs mainly from muscle receptors, is immediately adjacent to the primary motor cortex (area 4), to which it supplies inputs. The close relation between the two areas led Edward Jones to comment in a 1978 paper, "The connectivity of area 3a, as conventionally identified, is such that it is probably best regarded not as an entity, but as a part of area 4" (Jones et al., 1978). In addition to its afferent inputs, 3a is a source of descending projections to the brainstem and spinal cord, likely regulating the function of reflexes and even setting the gain of the peripheral sensors themselves. It is with this perspective that we aim to view proprioception for motor control: a sense that is virtually inseparable from its motor counterpart.

The proprioceptive system differs from the other sensory systems as a result of biomechanics. There are around 10,000 spindle receptors in the human arm (Banks, 2005; Scott & Loeb, 1994) and roughly that number of cutaneous receptors in the palm (Saal et al., 2017; Vallbo & Johansson, 1984). However, the practical dimensionality of the spindles is much lower than that of touch. Remarkably, it was claimed that humans can perceive the stimulation of even a single rapidly adapting cutaneous afferent (Macefield et al., 1990); if that is the case, the dimensionality of the tactile system would be nearly as high as the number of receptors. In the intact limb, spindles are largely activated in concert by whole-muscle length change. Experimentally, there is no realistic way to present stimuli that activate arbitrary combinations of muscle receptors. Studies of proprioception, in awake animals at least, are largely limited to the combinations of inputs that can be evoked by natural movements (although for an heroic effort using magnetically activated, implanted steel slugs, see Wolpaw & Colburn, 1978).

Despite its apparent low dimensionality, the proprioceptive system is crucial to our ability to move. Most of us make rapid reaching movements effortlessly—movements that are straight and accurate—even with our eyes closed. The critical role of proprioception for motor control is clearly evident when patients lacking proprioception attempt to reach. Their movements are typically highly inaccurate and variable, often starting in an incorrect direction and drifting at termination (Ghez et al., 1990; Fig. 2.1A). Large-fiber sensory neuropathies can cause a loss of touch and proprioception with minimal effects on motor fibers. These patients typically struggle to perform activities of daily living (e.g., dressing, drinking from a cup) and are

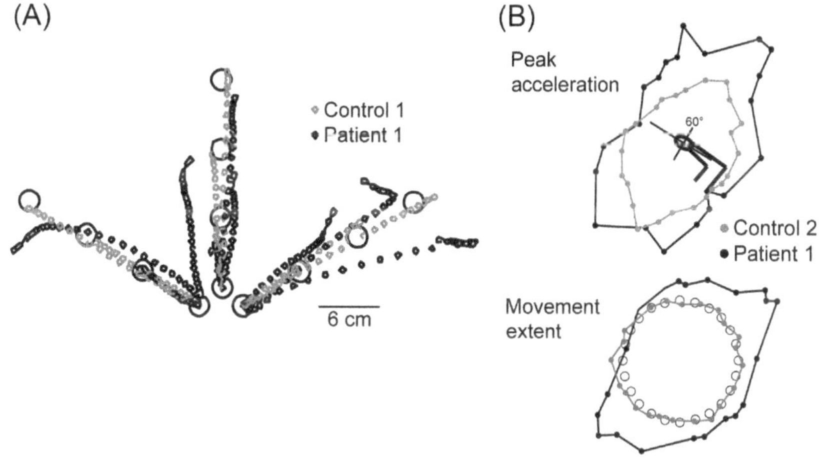

FIGURE 2.1 Loss of proprioception severely disrupts reaching. (A) Reaches made by a control subject (*gray*) and a deafferented patient (*black*) to targets (*large black circles*) in different directions and at different distances. (B) Peak acceleration (*top*) and movement extent (*bottom*) for reaches made by a control subject (*gray*) and deafferented patient (*black*). The length of the lines extending from the stick figure arm drawn in the center of the upper panel illustrates the magnitude of the impedance of the arm for hand movements in the corresponding direction. The large overshoots along the 60 degrees (low-impedance) axis imply that deafferented patients do not account for the impedance anisotropy of the arm. *Adapted from Ghez, C., Gordon, J., Ghilardi, M. F., Christakos, C. N., & Cooper, S. E. (1990). Roles of proprioceptive input in the programming of arm trajectories.* Cold Spring Harbor Symposia on Quantitative Biology, 55, *837–847.* *<https://doi.org/10.1101/SQB.1990.055.01.079>.*

sometimes wheelchair-bound. One notable exception, Ian Waterman, partially overcame the loss of almost all touch and proprioception that developed suddenly following a fever. After intensely practicing movements of different trajectories and speeds for years, he was able to perform effective, but slow, reaching movements and walk slowly, albeit awkwardly (McNeill et al., 2010). In the absence of visual feedback, he is unable to make accurate movements and struggles even to stand.

The slow, inaccurate movements made by patients without proprioception are likely explained by the loss of rapid feedback provided by proprioception at both the spinal and supraspinal levels. Their movements typically accelerate more rapidly in directions requiring largely forearm motion compared to those involving the upper limb (Fig. 2.1B). Further, it is as though even their initial movement plan does not account for the posture-dependent inertial anisotropy of the arm (Ghez et al., 1995). More generally, deafferented patients struggle to achieve normal joint coordination. Planar movements, such as those made when slicing bread, require synchronous rotation of the shoulder and elbow joints. Deafferented patients make extremely nonplanar movements in these attempts (Sainburg et al., 1993). Deafferented patients struggle to quickly reverse their movements, instead moving in a direction that is not directly opposite of the outward movement (Sainburg et al., 1995). As in the case of the reaching, the difficulty coordinating joint rotation may be due to an inability to plan for the interaction forces between limb segments. While visual feedback during planning decreases their endpoint error and visual feedback during movement further decreases error, these patients still make larger errors compared to controls, indicating that vision cannot fully replace proprioception.

Much like the motor deficits seen with loss of somatosensation, patients with cerebellar damage also struggle to make rapidly alternating movements and to account for interaction forces between limb segments (Bastian et al., 1996). Likewise, disrupting cerebellar thalamus (VL) slows the adaptation of movements to novel environments, an ability that is likely to depend on a comparison of predicted and actual sensory feedback (Chen et al., 2006). The similarity between proprioceptive loss and cerebellar damage leads naturally to the idea that many proprioceptive deficits may essentially be the result of deafferenting the cerebellum, which Sherrington described as, "the head-ganglion of the proprioceptive system" (Sherrington, 1906).

On the other hand, the effect of somatosensory cortical lesions is surprisingly limited. Somatosensory cortex appears to be critical for skill learning but not the performance of learned skills (Pavlides et al., 1993; Sakamoto et al., 1989; Vidoni et al., 2010). S1 is also needed to adapt learned behaviors. Optogenetically silencing S1 in mice prevents adaptation of perturbed reaching movements (Mathis et al., 2017). Similarly, disrupting somatosensory cortex with transcranial magnetic stimulation after reach adaptation reduces consolidation of the adapted behavior (Kumar et al., 2019). The observations are all consistent with the idea that S1 may communicate with the cerebellum to update an internal model of the limb.

The primary goal of this chapter is to describe the characteristics of proprioception that are fundamental to the control of movement and the effort to develop neuroprosthetic devices. In this chapter, we first discuss the receptors contributing to proprioception and the diversity in their form and function. We then examine the coding properties of neurons in the proprioceptive afferent pathways to the cerebral cortex and the descending control of reflexes. Finally, we consider the role of the cerebellum in processing these proprioceptive signals.

2.2 Sensors contributing to proprioception

Proprioception arises from sensory receptors within skeletal muscles, joints, and skin whose signals are combined with descending motor signals to provide a high-level sense of body configuration and to inform the control of movement. Skin stretch near a joint causes a perception like that of the corresponding movement, suggesting that cutaneous receptors also contribute to the conscious perception of movement (Proske & Gandevia, 2012). Less research has been done to determine their contribution to motor control.

Although joint receptor afferents usually respond only near the limits of flexion or extension (Burgess & Clark, 1969; Rossi & Grigg, 1982), contractions of muscles acting around a particular joint will activate joint capsule afferents, even in the absence of joint angle change, implying that joint afferents are also responsive to loads on the joint (Grigg & Greenspan, 1977). In the remainder of this section, we will focus primarily on the muscle receptors (muscle spindles and Golgi tendon organs) that contribute to both conscious proprioception and the subconscious planning and control of movement.

2.2.1 Muscle spindles

Muscle spindles are elongated fusiform sensors situated in parallel with the skeletal muscles and widely considered the primary mechanoreceptors supplying proprioception. Each consists of "intrafusal" muscle fibers surrounding the sensory element, innervated by motor and sensory axons, respectively (Fig. 2.2). Contraction of the intrafusal muscles changes the

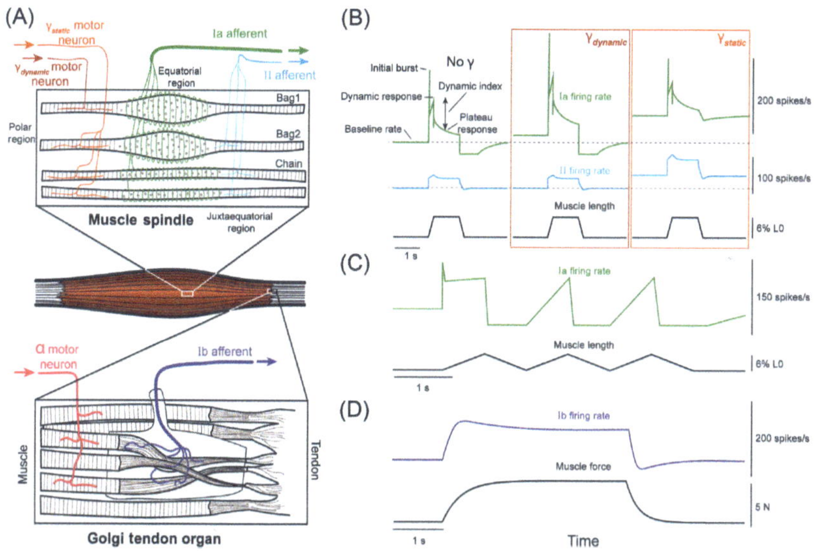

FIGURE 2.2 Anatomy and coding properties of muscle proprioceptors. (A) General anatomy of a muscle spindle (*top panel*) and Golgi tendon organ (*bottom panel*). Mammalian muscle spindles comprise three types of intrafusal fibers (bag1, bag2, and chain) as well as four types of motor and sensory neurons (static gamma motor neurons—*orange*, dynamic gamma motor neurons—*red*, group Ia sensory neurons—*green*, and group II sensory neurons—*cyan*). (B) Muscle spindle coding properties during passive stretch and the effect of constant static and dynamic gamma drive during stretch. Stereotypical groups Ia (*green*) and II (*cyan*) firing rate responses to a ramp-hold-release muscle stretch (*black*) are shown in the left column. (C) History dependence of the Ia afferent response to repeated stretch. When the muscle is repeatedly subjected to the same stretch, the dynamic response is reduced, and the initial burst disappears altogether. (D) Stereotypical Golgi tendon organ Ib afferent response to an increase in muscle force. The firing rate increases dramatically when force begins to rise, then settles to a plateau at steady state.

response characteristics of the sensory endings. In mammalian muscle spindles, there are two main types of intrafusal muscle fibers: nuclear bag and nuclear chain fibers (Fig. 2.2A; Ruffini, 1898). The bag and chain fibers, respectively, are constructed with slow and fast myosin (Thornell et al., 2015) and contribute dynamic and static components of spindle afferent responses (Banks et al., 1997).

Classically, muscle spindle afferents have been described by their sensitivity to muscle length and length changes, properties which have been extensively studied in reduced anesthetized or decerebrate preparations (Banks et al., 1997; Boyd, 1962, 1976; Carrasco et al., 2017; Honeycutt et al., 2012; Houk et al., 1981; Matthews, 1933, 1963; Proske & Gregory, 1977; Vincent et al., 2017). When a resting muscle is stretched, the Ia afferents respond with a burst of spikes within a few milliseconds, which scales linearly with acceleration (Blum et al., 2017; Schäfer & Schäfer, 1969). Following the initial burst is a response with a fractional power relationship to velocity (Houk et al., 1981; Matthews, 1963; Prochazka & Gorassini, 1998a,b). Finally, there is a maintained firing rate related to the length (Fig. 2.2B). The "dynamic index" compares the firing frequency at the end of a ramp to that during hold (Fig. 2.2B; Jansen & Matthews, 1962). The dynamic index of Ia afferents is significantly greater than that of group II afferents, which closely resemble muscle length (Houk et al., 1981; Matthews, 1963; Prochazka & Gorassini, 1998a,b). Human microneurography studies have echoed these observations, including the initial burst and dynamic response to ramp-and-hold joint angle changes (Cordo et al., 2002; Edin & Vallbo, 1990; Vallbo, 1974).

The sensitivity of muscle spindle afferents is influenced by motor axons projecting to the intrafusal muscle fibers (Crowe & Matthews, 1964a,b; Jansen & Matthews, 1962). Activation of dynamic fusimotor axons leads to increased sensitivity of primary endings to length change (Fig. 2.2B), whereas static fusimotor activation leads to increased static stretch sensitivity of both primary and secondary muscle spindle afferents. In the absence of increased fusimotor drive, muscle spindles would fall silent as the extrafusal muscle is shortened. However, if the fusimotor effects are great enough to overcome the effect of the shortening, the firing of the spindle afferent may actually increase (Blum et al., 2020; Prochazka et al., 1979). In cats, voluntary muscle shortening speeds greater than $\sim$0.2 muscle lengths per second tended to dominate the afferent response, whereas slower movements were more influenced by fusimotor activation (Blum et al., 2020; Prochazka et al., 1979). In the extreme of isometric muscle contraction, Ia afferent firing rates in the contracting muscle increase proportionally to joint torque through a combination of alpha and fusimotor coactivation (Vallbo, 1970). In short, external loading on the limb and central drive from alpha motor and fusimotor neurons all play a role in shaping the muscle spindle afferent firing rates.

The Ia afferent passive stretch response also depends on the history of movement (Fig. 2.2C). When a muscle is repeatedly stretched and shortened, the afferent responses diminish and the initial burst at stretch onset disappears altogether (Blum et al., 2019; Haftel et al., 2004; Proske et al., 1992, 1993). An analogous phenomenon is seen in force responses of isolated single extrafusal muscle fibers (Campbell & Lakie, 1998; Campbell & Moss, 2000, 2002), due at least in part to the dynamics of cross bridge cycling (Nichols & Cope, 2004; Proske et al., 1993). A recent modeling study demonstrated that many of the nonlinear length and history-dependent responses in Ia afferents can alternatively be explained by a linear combination of intrafusal force and its first time-derivative (Blum et al., 2017, 2020; Lin et al., 2019).

2.2.2 Golgi tendon organs

The Golgi tendon organ (GTO) is another type of muscle-bound mechanoreceptor that contributes to proprioception (Fig. 2.2A). In contrast to spindles, GTOs are situated at the junction between muscle fibers and tendons and are directly sensitive to tension. GTOs are innervated by Ib sensory axons that are compressed by collagen fibers when the muscle fibers contract (Fig. 2.2D; Crago et al., 1982; Jami, 1992; Stuart et al., 1972). A single motor unit may stimulate multiple GTOs, and a single GTO may be activated by 25–50 in-series muscle fibers from as many as 10–20 fast and slow motor units (Jami, 1992; Prochazka & Ellaway, 2012). As a result, the receptive field of a given GTO is distributed throughout a muscle and its firing reflects a selective average of the tension across motor units (Horcholle-Bossavit et al., 1990). Perhaps for this reason, most of the Ib afferents in a given muscle fire throughout approximately the same force range. Another consequence of this mechanical arrangement is that contraction of muscle fibers that are not in series with a given GTO can unload it from passive forces and paradoxically cause it to pause its firing (Houk & Henneman, 1967). This complicated mechanical relationship between individual motor units and GTOs means that the force resulting from individual motor unit stimulation is not well represented by individual Ib afferents. However, an ensemble average across multiple Ib afferents is typically well correlated with whole-muscle force (Horcholle-Bossavit et al., 1990).

For many years, Ib sensitivity to passive stretch was thought to be very low and unlikely to make any meaningful contribution to proprioception (Jami, 1992; Jansen & Rudjord, 1964). However, later experiments revealed that GTOs in the gastrocnemius muscle of cats can fire strongly when stretched passively within the muscle's physiological range, especially following an eccentric contraction, a result of the high number of intact muscle cross-bridges under those conditions (Gregory et al., 2002). A more recent study also showed GTOs from triceps surae of mice are roughly as sensitive as some group II muscle spindle afferents to passive stretch (Vincent et al., 2017). In summary, individual GTOs likely signal localized in-series force, regardless of whether this force is developed actively or passively.

2.3 Proprioceptive coding along the cerebral cortical pathway

Although there have been many fewer studies of the physiology of proprioceptive neurons than there have been of the motor system, both research fields have used similar experimental and analytical approaches. In order to study areas representing the proximal arm, animals are typically trained to make voluntary movements and to accept passive movements of their limbs. The resulting neural responses are analyzed with respect to the timing of firing rate modulation (their "dynamics," or temporal modulation) as well as their spatial tuning, the dependence on the joint(s) being rotated, or the direction of hand movement. In the following sections, we discuss what is known about proprioceptive coding in the dorsal column medial lemniscal pathway through to the parietal cortex (Fig. 2.3).

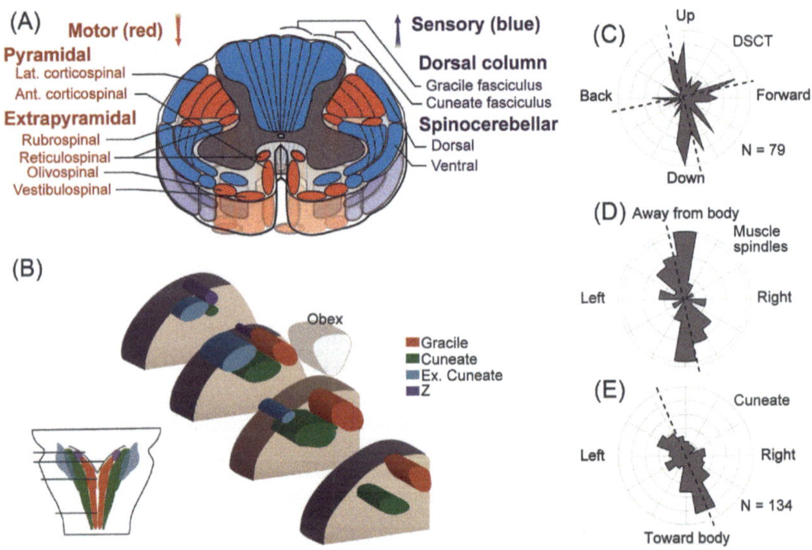

FIGURE 2.3 Dorsal column system anatomy and coding properties. (A) Diagram of the major ascending and descending tracts in the spinal cord. Afferent pathways are shown in blue, efferent pathways in red. (B) Diagram of dorsal column nuclei in the medulla. Note the variation in size of the nuclei along the rostrocaudal axis. (C) Distribution of hind limb preferred directions (PDs) in sagittal plane for neurons recorded in cat DSCT (N = 79 neurons). (D) Horizontal-plane PDs of 1000 simulated spindles in the muscles of a monkey's arm. The majority of PDs fall along the axis toward and away from the body. (E) CN neuron PDs (N = 134 neurons) from three monkeys during planar reaching with the right arm. *(Figure 2.3A) Adapted from Wikipedia common license. (Figure 2.3C) Redrawn from Bosco, G., & Poppele, R.E. (2001). Proprioception from a spinocerebellar perspective.* Physiological Reviews, 81*(2), 539–568. <https://doi.org/10.1152/physrev.2001.81.2.539>.*

2.3.1 Dorsal column pathway

Proprioceptor cell bodies with axons projecting to their receptor organs in the upper and lower limbs are collected in the dorsal root ganglia (DRG) located just lateral to the spinal cord. Graded potentials from the receptors are transformed into action potentials in the distal portions of the afferents and travel through pseudo-unipolar axons to the soma within the DRG and then to the spinal cord. Proprioceptive and cutaneous (as well as nociceptive and thermoceptive) afferents from the proximal upper limb make modality-specific projections into the spinal cord through the cervical 5, 6, and 7 DRG (Fig. 2.3A).

The dorsal column pathways differ for the lower and upper limbs. The gracile fasciculus contains fibers originating from the lower limb, while the cuneate fasciculus contains fibers from the upper limb. These tracts contain both first-order sensory afferents and oligosynaptic projections from spinal interneurons. The axons synapse onto the gracile (GN) and cuneate nuclei (CN), respectively part of the dorsal column nuclei (DCN) which extend rostrocaudally along the dorsal surface of the medulla (Fig. 2.3B). While CN has both proprioceptive and cutaneous neurons, GN is predominantly cutaneous (Loutit et al., 2020). The proprioceptive signals from the lower limb reach the thalamus and cortex via nucleus Z, just rostral to GN (Berkley et al., 1986). The external cuneate nucleus receives afferent input from proprioceptors of the upper limb and projects to the cerebellum.

The magnitude of muscle length changes as the hand (or foot) moves in different directions is highly nonuniform, the greatest changes occurring for movements along the axis connecting the shoulder and hand. The sensitivity of a neuron to movement direction can be quantified by a tuning curve and "preferred direction" (PD), the movement direction associated with the highest firing rates. The strongly bimodal distribution of PDs for neurons in the dorsal spinocerebellar tract (DCST) during paw movements in the cat is a signature of their biomechanical origin (Fig. 2.3C). Equally bimodal is the distribution of PDs of a simulated population of muscle spindles, computed from the kinematics of arm movement when a monkey reaches in the horizontal plane (Fig. 2.3D). Even in CN, the distribution PDs during planar reaches is strongly bimodal (Fig. 2.3E), much like that of the simulated spindles.

At first blush, this spindle-like distribution of PDs suggests that there may be rather limited convergence from multiple muscles onto single neurons in CN. In the literature, however, the evidence is mixed. Stimulation of peripheral nerves in monkeys under propofol anesthesia suggests high convergence; 87% of CN neurons responded to stimulation of more than one peripheral nerve (Witham & Baker, 2011). However, earlier experiments found that a much smaller proportion of neurons responded to stretch of more than one muscle in barbiturate-anesthetized monkeys (Hummelsheim & Wiesendanger, 1985). In an effort to address this apparent contradiction, we recorded the response of CN neurons to muscle vibration, which strongly activates muscle spindle Ia afferents. Vibration activated some neurons, but only rarely were we able to activate a given CN neuron from more than one muscle.

A possible explanation for these disparate observations is hinted at by the apparent role of CN in the development and response to injury. When the dorsal roots are cut, the dorsal column nuclei undergo a functional remapping in which neurons that previously represented the deafferented segment begin to represent adjacent body regions. This effect may account for much of the observed shifts in cortical somatotopy following loss of afferent input (Kambi et al., 2014). Similarly, recent evidence in the rat suggests that muscle afferent projections to CN are numerous early in development, before extensive pruning during the period of invasion by corticocuneate fibers (Fisher & Clowry, 2009). While these results required active synapse formation, other studies found a remapping of the input to DCN neurons within tens of minutes after lidocaine injection into the peripheral receptive fields or cold block of the lumbar cord (Dostrovsky et al., 1976; Pettit & Schwark, 1993). The high convergence seen during electrical stimulation studies may result from widely branched, but functionally weak synapses recruited by the strong, synchronous inputs.

2.3.2 Thalamic proprioceptive encoding

As the medial lemniscus leaves the dorsal column nuclei, it decussates before synapsing in the ventroposterior complex of the thalamus. The nomenclature used to delineate proprioceptive regions of the thalamus is inconsistent across species and even research groups [see Bota et al. (2019) for a review]. In rats, a complex proprioceptive somatotopy exists in the most rostral regions of the ventral posterolateral nucleus (rVPL) (Francis et al., 2008). In monkeys, classic studies describe a region rostral and dorsal within the thalamus, termed the ventroposterior superior nucleus or sometimes ventroposterior oralis that projects to cortical area 3a (Bota et al., 2019; Krubitzer et al., 2004). Other thalamic "motor" nuclei (ventrolateral; VL and ventral intermediate nucleus; Vim) also receive inputs from the cerebellum, carrying signals partially derived from proprioceptive mossy fiber inputs [for a review, see Shadmehr (2020)]. To our knowledge, no recordings have been made of the primate proprioceptive thalamus during reaching. Proprioceptive recordings from monkey VPL during head rotations have been reported, containing a mixture of proprioceptive and vestibular modalities (Dale & Cullen, 2017).

2.3.3 Somatosensory cortex

The somatosensory thalamus projects broadly to the primary somatosensory cortices as well as association areas in the posterior parietal cortex (S1; Fig. 2.4A). VPL projects roughly equally to 3a, 3b, and area 1, VPS primarily to area 2 and area 5, while 3a is the most common cortical target for VL

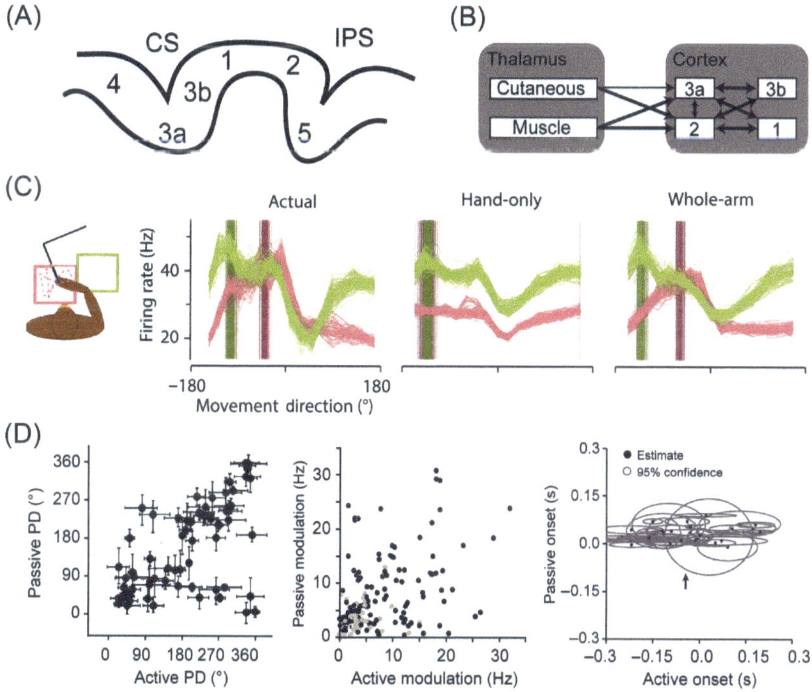

FIGURE 2.4 Anatomy and coding properties of proprioceptive primary somatosensory cortices. (A) Anatomical schematic of primary motor cortex (area 4), primary somatosensory cortex (areas 3a, 3b, 1 and 2), posterior parietal cortex (area 5), and the central (CS) and intraparietal (IPS) sulci. (B) Arrows represent projections from the thalamus and from areas within the primary somatosensory cortex to *proprioceptive* areas (3a, 2) of primary somatosensory cortex. Arrow thickness roughly represents the strength of projections. (C) The actual tuning curve (*left*) of an example area 2 neuron for planar movements in different directions to randomly placed targets in each of two workspaces (colors). Each trace represents one cross-validation fold and vertical lines represent preferred directions (PD). The tuning curves for this neuron were predicted by models using only hand kinematics (*middle*) and hand plus elbow kinematics (*right*). Note that the latter whole-arm model better predicts the actual PD shift across workspaces. (D) Comparison of PDs (*left*), modulation depth (*middle*), and onset latency (*right*) for different area 2 neurons during active reaches and passive perturbations applied to the hand. Error bars on the left plot represent 95% confidence intervals on the PDs. Gray symbols (*middle plot*) represent neurons that were not sinusoidally tuned for either active or passive movements. Onset for neurons that responded in both active and passive conditions. Gray ellipses trace the 95% confidence intervals. Arrow indicates time of the earliest observed average EMG onset. *(Figure 2.4C) Adapted from Chowdhury, R. H., Glaser, J. I., & Miller, L. E. (2020). Area 2 of primary somatosensory cortex encodes kinematics of the whole arm.* ELife, 9, *e48198. <https://doi.org/10.7554/eLife.48198>. (Figure 2.4D) Adapted from London, B. M., & Miller, L. E. (2013). Responses of somatosensory area 2 neurons to actively and passively generated limb movements.* Journal of Neurophysiology, 109*(6), 1505–1513. <https://doi.org/10.1152/jn.00372.2012>.*

(Padberg et al., 2009). Area 3a, near the bottom of the central sulcus in the macaque monkey, receives afferent input that originates mostly from muscle spindles and joint receptors, with some minor input from cutaneous receptors (Heath et al., 1976; Kim et al., 2015; Krubitzer et al., 2004; Yamada et al., 2016). Area 2, located on the postcentral gyrus caudal of area 1 (Huffman & Krubitzer, 2001; Padberg et al., 2018; Fig. 2.4A), receives projections from cortical areas 3 and 1, as well as from both muscle and cutaneous thalamic inputs. Because of the complexity of its multimodal inputs, some do not consider area 2 to be a part of S1. Areas 3a and 2 receive inputs not only from the somatosensory thalamus, but also from thalamic motor nuclei, including VL (Krubitzer et al., 2004; Padberg et al., 2009). In fact, most of the thalamic input to area 3a may be from the motor thalamus (Huffman & Krubitzer, 2001; Padberg et al., 2009).

Although proprioception arises chiefly from muscle receptors, most people would be hard pressed to say how their individual muscles were changing length during a movement. When reaching toward an object, we generally focus on the motion of the hand, rather than that of individual joints, let alone muscles. Likewise, following passive displacement of the arm in the dark, we have a more accurate perception of hand position than of joint angles (Fuentes & Bastian, 2010). How and where this transformation occurs from discrete muscle receptors to the perception of hand movement in space remains a persistent question.

Early experiments exploring somatosensation, much like other sensory areas of the brain, sought to characterize the neural activity resulting from well-controlled, independent sensory stimuli. This approach, in the form of controlled single-muscle stretches (Hore et al., 1976; Lucier et al., 1975; McIntyre et al., 1984; Schwarz et al., 1973), cutaneous receptive field mapping (Kaas et al., 1979; Pons et al., 1985; Yau et al., 2013), and peripheral nerve stimulation (Heath et al., 1976; Phillips et al., 1971; Yamada et al., 2016), revealed the topography and modalities of different portions of S1. While the cutaneous areas of S1 have a remarkably detailed body map, the proprioceptive areas are much less detailed. There is a rough medial to lateral progression of body part representation in areas 3a and 2, with feet and legs medial, face most lateral, and hand in the middle (Krubitzer et al., 2004; Seelke et al., 2011), but there is little evidence for a finer-grained somatotopy.

As an alternative to the reductionist approach that seeks to activate receptors independently, other investigations of proprioceptive cortex adopted methods first applied to motor cortex (Georgopoulos et al., 1982). As in M1, many somatosensory neurons have broad, sinusoidal tuning curves (London & Miller, 2013; Prud'homme & Kalaska, 1994) like that illustrated in Fig. 2.4C. Across movement directions, this example neuron responded most during movements in the −160 degrees direction and least to movements roughly in the opposite direction. If tuning with respect to movement direction is a fundamental cortical property, we might expect neurons with similar preferred directions to cluster together. In fact, pairs of area 2 neurons recorded on a given electrode are somewhat more likely than neurons recorded on different electrodes to respond to movements of the same joint and have a similar PDs (Gardner & Costanzo, 1981; Weber et al., 2011). However, the effect is small, and there does not appear to be any longer range topography across area 2 related to the direction of hand movement.

Given the likely role of area 2 in limb movement execution, a representation that includes only the direction of hand movement would not provide adequate limb state information to plan and execute reaching movements. Consistent with this idea, a recent study showed that area 2 neurons represent the kinematics of the whole arm rather than just those of hand movement (Chowdhury et al., 2020). A model incorporating information from the whole arm better predicted neuronal figure rates as well as tuning shifts across workspaces than a model incorporating only information from the hand (Fig. 2.4C). That model was purely kinematic, but we know that neural firing rates in both areas 3a and 2 correlate with joint torques as well as limb kinematics (Jennings et al., 1983). This information, related to the dynamics of the arm and its substantial inertial and segment-coupling torques, is also critical for movement planning and execution, as illustrated by the movement deficits that occur in the absence of proprioceptive input (Sainburg et al., 1993; Sainburg et al., 1995).

Perhaps as a consequence of coding both limb kinetic and kinematic information, somatosensory neurons respond differently to active and passive movements (Jennings et al., 1983; Prud'homme & Kalaska, 1994; Tanji, 1975; Wise & Tanji, 1981; Yumiya, Kubota, & Asanuma, 1974). Although the PDs for area 2 neurons appear to be similar for active and passive movements (London & Miller, 2013), other tuning properties, like modulation depth and onset time, differ (Fig. 2.4D). These results are also consistent with the idea that the area receives efference copy signals from motor areas during active movement (Chowdhury et al., 2020; London & Miller, 2013; Nelson, 1987).

Rather than being a systematic map of the input space, it may be the case that proprioceptive cortex is optimized to supply control variables needed by the motor system. The strong inputs from motor thalamus, the presence of corticomotor neuronal cells that likely modify reflex gains (Rathelot & Strick, 2009), and the interconnections with M1 (Huffman & Krubitzer, 2001; Padberg et al., 2018) all lend credence to the idea that 3a in particular operates to optimize motor execution. On the other hand, neurons in area 2 combine cutaneous and muscle information and have more complex receptive fields than in areas 3a, 3b, and 1 (Rincon-Gonzalez et al., 2011), likely encoding movement at a higher level than does area 3a. We speculate that arm area 2, with its strong projections into area 5 may allow the arm posture and the location of encountered objects to be mapped with respect to the body, much like the role of its hand-area counterpart in stereognosis (*below*). Importantly, the strong force representation present in both areas 3a and 2, is lost in area 5 (Kalaska, 1988). This purely kinematic representation is ideal for the subsequent combination with vision that occurs in area 7 (Bolognini & Maravita, 2007).

Often, the goal at the end of a reach is to grasp an object. Many psychophysical experiments have demonstrated close coupling between reaching and grasping. This coupling is mediated through forward models of the limb (Oztop & Kawato, 2009; Takemura et al., 2015), as well as the visual system during perturbations of object location during grasp (Gentilucci et al., 1992). However, reaching and grasping information also seems to be coupled in primary somatosensory cortex—roughly half of recorded tactile hand neurons responded both during movement toward an object and during contact (Rincon-Gonzalez et al., 2011).

The need to couple a well-timed grasp to the end of a reach means that the planning of the hand preshaping during transport will be indirectly affected by the limb's changing inertia and the complex, velocity-related coupling forces between limb segments. However, the role of proprioception in interacting with a grasped object is very different. The hand itself is a highly dexterous manipulator, complete with ~17,000 touch receptors distributed across its surface, a unique, deformable sensor surface optimized for manipulating objects (Johansson & Vallbo, 1979; Yau et al., 2016). The conjunction of tactile and proprioceptive inputs governing sensation from the hand is at least as critical as it is for the arm. The "cutaneous rabbit" effect, the inaccurate perception of the location of a cutaneous stimulus caused by rapidly tapping the skin at multiple locations, provides a striking example. When evoked in the fingertips, this illusory sensation is dependent on the posture of the hand (Warren et al., 2011).

This combination of tactile and proprioceptive inputs may be particularly important for stereognosis, the ability to recognize the form of an object in the absence of visual input. This ability requires information about an object's multiple points of contact with the skin, combined with that of the relative locations of those contact points on different fingers and the palm (Yau et al., 2016). Area 2, with its combination of proprioceptive and tactile inputs, seems ideally suited for performing stereognosis (Hsiao, 2008). Roughly 60% of hand area 2 neurons have receptive fields spanning multiple digits or other areas that are typically stimulated together. Many of these neurons seem to represent the posture of the hand as they lack velocity sensitivity (Gardner & Costanzo, 1981; Gardner, 1988; Goodman et al., 2019). It should also be noted however, that some degree of interaction between tactile and proprioceptive modalities is present in all four of the classic regions of S1 (Kim et al., 2015).

2.4 Somato-motor connections and control of proprioceptive feedback

2.4.1 Spinal reflexes

Simple proprioceptive reflex functions are as diverse as postural stabilization (e.g., the stretch reflex at the knee or ankle), trauma avoidance (e.g., the flexor withdrawal reflex), and the rapid modulation of the mechanical properties of individual muscles. The fastest and most elemental role of the feedback from muscle spindles may be to regulate muscle stiffness and to linearize its force output (Nichols & Houk, 1976). When an active muscle is stretched in the absence of reflex action, its force response is highly nonlinear due to the high stiffness of cross-bridges that can act only across a short range before being broken (Banks et al., 1997; Campbell & Lakie, 1998; Morgan, 1977). Instead, when reflexes are intact, these cross-bridges rapidly reform and the force response becomes much more spring-like.

This linearizing reflex function is rather distant from the typical clinical picture of reflexes. In one of the most basic neurological tests, originating in the mid-19th century, a clinician uses a hammer to strike and thereby stretch a tendon. The stretch activates muscle spindle primary afferents, which form a monosynaptic excitatory connection with alpha motor neurons innervating the stretched muscle (Fig. 2.5). The timing and vigor of the resulting muscle contraction can be used to diagnose basic nerve conduction deficits. In this role, the monosynaptic stretch reflex acts as a fast, negative feedback servo mechanism, counteracting length changes within 20–40 ms.

This clinical use of the monosynaptic stretch reflex, while an effective diagnostic tool, does not nearly capture the complexity of reflexes in natural behavior. The monosynaptic spinal stretch reflex is complemented by a disynaptic

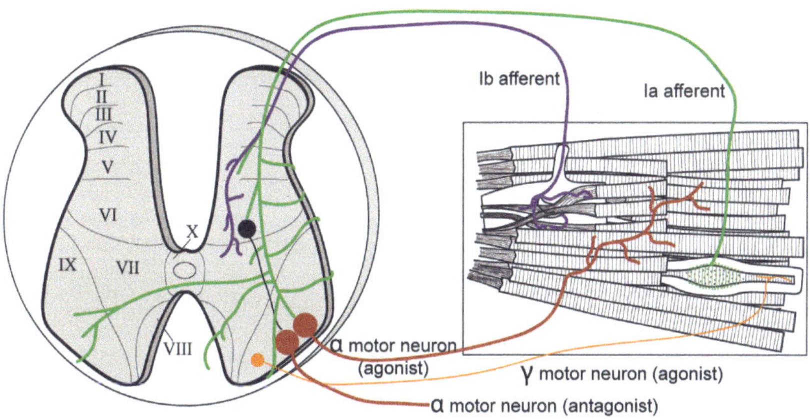

FIGURE 2.5 Spinal projections of proprioceptive afferents and the monosynaptic stretch reflex. When a muscle spindle is stretched, the afferent signals travel to the spinal cord via Ia axons (*green* path). Upon entering the dorsal horn, these afferents project to multiple laminae and adjacent white matter, demonstrating the widespread effect of this proprioceptive feedback. Some projections to lamina IX form synapses with alpha motor neurons, which activate the spindle-bearing muscle and complete the monosynaptic reflex arc. Antagonist alpha motor neurons can be inhibited through an inhibitory interneuron (*black*). Gamma motor neurons (*yellow*) are also present in lamina IX, although their spinal control is not well-understood. Ib afferents also project to multiple spinal lamina, although to a lesser extent than Ia afferents, and can have complex effects on muscle coordination.

pathway that inhibits the muscles acting in opposition to the stretched muscle(s) (Fig. 2.5). In this pathway, the stretch-evoked sensory signals synapse on a spinal interneuron which inhibits the alpha motor neurons of the antagonist(s). The excitatory monosynaptic and inhibitory disynaptic pathways of the spinal stretch reflex exemplify one fundamental function of proprioceptive reflexes: to coordinate multiple muscles in response to external perturbations.

Other so-called heterogenic reflexes involve input from receptors of a variety of modalities, and influence muscles beyond just synergists (Nichols, 2018). One simple demonstration of the breadth of proprioceptive spinal connections within the spinal cord can be seen in Fig. 2.5. Muscle spindle Ia afferents project to ipsilateral lamina V–IX and adjacent white matter, in addition to some contralateral laminae IX projections (Brown & Fyffe, 1978). Groups Ib and II afferents also project to multiple ipsilateral lamina and can thus affect the actions of many muscles through spinal interneurons. Just one example is the force feedback from GTOs within one muscle, which can inhibit or excite other muscles, in addition to their more basic autogenic inhibitory effects (Laporte & Lloyd, 1952). Functional muscle groups can be formed from such connections, such as the inhibitory force feedback that originates from proximal muscles and acts on distal ones (Ross & Nichols, 2009). It is believed that these complex proprioceptive spinal pathways promote coordination between joints and regulate limb mechanics in the face of external perturbations (Nichols, 2018).

2.4.2 Longer latency reflexes and sensorimotor connections

There are three partially overlapping EMG responses when a muscle is rapidly stretched, known as the short-, medium-, and long-latency reflexes (Fig. 2.6A). The short-latency reflexes arise primarily from the monosynaptic reflex discussed above. The neural pathways for the medium- and long-latency reflexes are more varied and less completely understood but are thought to represent

FIGURE 2.6 Concept of nested feedback loops and descending modulation. (A) Diagram of short-, medium-, and long-latency reflex loops. Proprioceptive perturbations result in muscle responses at different latencies. Short-latency reflexes are mediated by spinal circuits, while medium- and long-latency reflexes are likely also mediated by higher level brain structures. Further, the gains on these reflex loops can be modulated by descending commands. (B) Example responses in cervical spinal inter neurons to stimulation of deep radial and superficial radial nerves (muscle and cutaneous nerves, respectively) while the monkey performs a wrist flexion task. Vertical lines indicate stimulation time. Neurons receiving input from muscle proprioceptors responded more strongly during the active condition than the rest condition, while cutaneous units had smaller responses. (C) Response gain pooled across neurons, segregated by movement epoch. In general, muscle responses were potentiated while cutaneous responses were attenuated. *(Figure 2.6B,C) Adapted from Azim, E., & Seki, K. (2019). Gain control in the sensorimotor system.* Current Opinion in Physiology, 8, *177–187. <https://doi.org/10.1016/j.cophys.2019.03.005>.*

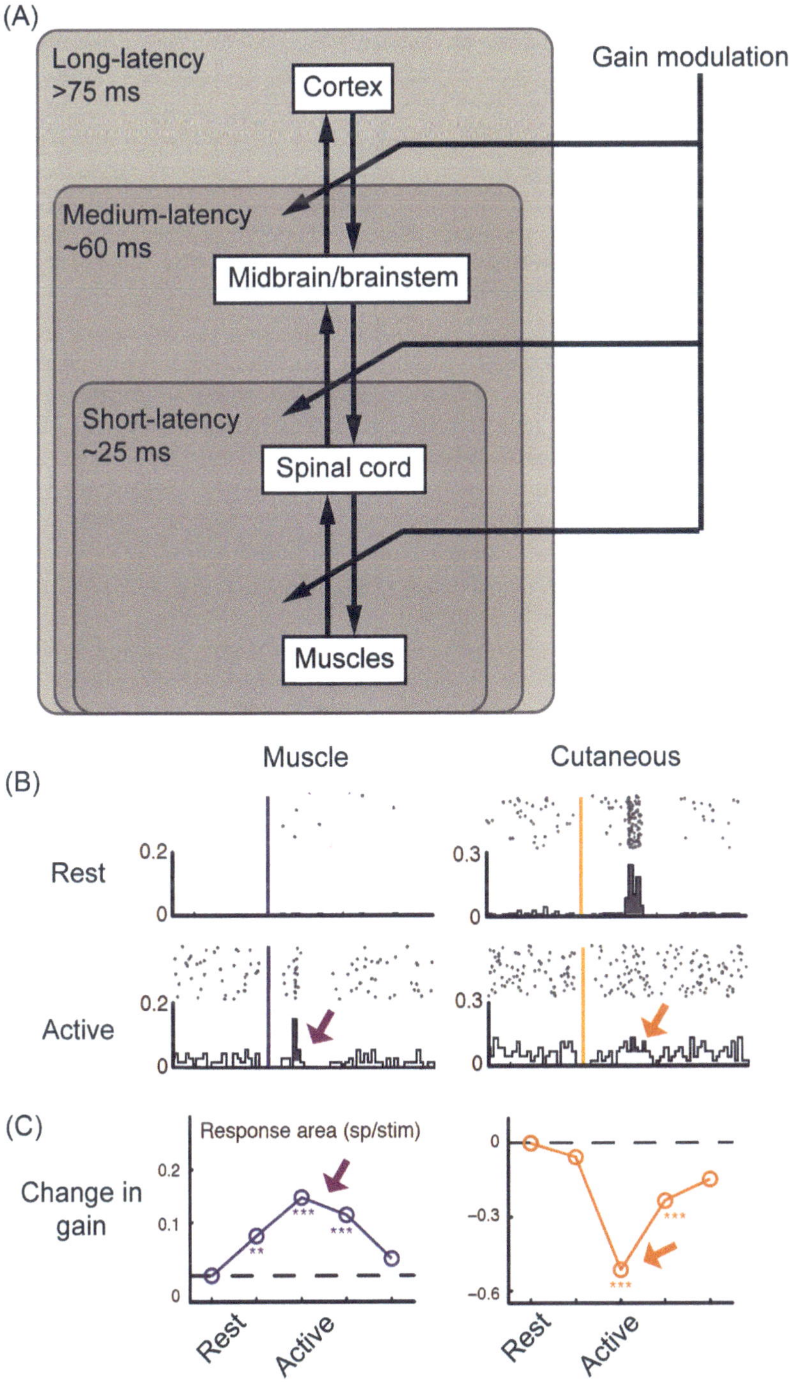
(A)
Long-latency
>75 ms
Cortex
Gain modulation
Medium-latency
~60 ms
Midbrain/brainstem
Short-latency
~25 ms
Spinal cord
Muscles
(B)
Muscle
Cutaneous
Rest
Active
0.2
0
0.3
(C)
Change in gain
Response area (sp/stim)
0.2
0.1
0
−0.3
−0.6
Rest
Active

slower spinal pathways and paths through the brainstem and cortex where additional processing and top-down modulation may occur to create a more flexible and context-dependent response. Factors such as conduction velocity, axon length, and processing complexity all contribute to reflex latency. Therefore, precise categorization of these reflexes into groups by latency alone is imprecise.

It is likely that the soleus medium-latency reflex is a spinal reflex mediated by group II afferents. When the soleus muscle is stretched during locomotion (Grey et al., 2001), short- and medium-latency bursts of muscle activity occur at around 60 and 85 ms, respectively (Toft et al., 1989). The medium-latency response depends less on stretch velocity than the short-latency response (consistent with a group II origin) and is consistent with the slower conduction velocity of the smaller group II axons (Schieppati & Nardone, 1997). When pharmacological agents are used to suppress the oligosynaptic group II afferent pathway, only the medium-latency responses are significantly reduced (Grey et al., 2001).

The longest latency reflexes could arise from multisynaptic transcortical or brainstem pathways, which presumably allows for their greater sophistication (Cheney & Fetz, 1984; Pruszynski et al., 2011; Wolpaw, 1980). Indeed, motor cortical responses to muscle stretches during arm movement tend to have two phases: one at short latency (20–25 ms) that depends solely on the perturbation and another at slightly longer latency (40–50 ms) that depends on the goal of the task; responses to perturbations during reaching have been shown to depend on whether there are obstacles to reach around (Evarts & Tanji, 1976; Nashed et al., 2012; Omrani et al., 2014). Furthermore, these motor cortical responses can integrate information from several muscles to create multijoint responses to perturbations, allowing for control that is both rapid and sophisticated (Pruszynski et al., 2011). These reflexes can even coordinate responses across arms during a bimanual task (Omrani et al., 2013) and use the brain's internal inverse model to account for the arm's dynamics (Kurtzer et al., 2008).

Long-latency EMG responses in the legs can serve to coordinate postural responses based on task-level goals. When stance is perturbed, such as when a bus accelerates and the standing passengers are not anticipating the sudden movement, the proprioceptive system shapes the long-latency activity of multiple muscles to stabilize posture in a manner that takes into account the direction and intensity of the perturbation (Safavynia & Ting, 2013; Welch & Ting, 2008, 2009). An early acceleration-related response is likely governed by Ia muscle spindle afferents, as it disappears after a large afferent fiber lesion with pyridoxine (Lockhart & Ting, 2007).

2.4.3 Top-down modulation of proprioceptive signals

The circuits that govern these reflex loops receive descending inputs which alter their behavior in different contexts. There are two primary mechanisms thought to perform this function: alterations of motor excitability (which is outside the scope of this chapter) or modulations of sensory sensitivity (or gain, Fig. 2.6A). In this section, we describe two such mechanisms, one, receptor-level modulation through altered fusimotor drive to muscle spindles, the other, presynaptic inhibition or direct excitation of relay neurons along the afferent pathway.

2.4.3.1 Control of the fusimotor system

As is the case for alpha motoneurons, fusimotor neurons are subject to a combination of spinal, brainstem, and cortical control. The simplest spinal reflexes involving fusimotor neurons arise from the group II afferents of homonymous muscles, which can have either inhibitory or excitatory actions (Appelberg et al., 1983). Dimitriou showed that reciprocal inhibition between antagonistic muscle pairs exists for the fusimotor, as well as alpha motor neurons. More complex spinal reflexes involving contralateral limbs have also been described (Johansson et al., 1987). The brainstem has neurons that control spinal fusimotor neurons directly. In a series of papers (Durbaba et al., 2001, 2006; Taylor et al., 1998, 1999, 2000, 2006), Taylor and colleagues described an area dorsal to the mesencephalic locomotor region in the cat, where stimulation evoked widespread activation of dynamic and static fusimotor axons.

Sensorimotor cortex is also a likely candidate for fusimotor control. In the baboon, M1 projects to fusimotor neurons via the pyramidal tract (Grigg & Preston, 1971). In humans, transcutaneous stimulation of the motor cortex was used to activate afferents recorded by microneurography, with latency consistent with beta or gamma motor neuron activation (Rothwell et al., 1990). In addition to M1, projections to the cord from all regions of S1 were demonstrated in cynomolgus monkeys using anterograde autoradiographic tracing (Coulter & Jones, 1977). Those from 3a extend ventrolaterally into the spinal motor nucleus, overlapping with and extending medially and dorsally from the M1 corticospinal terminations. More recently, using transsynaptic viral retrograde tracing, Rathelot and Strick found that ~15% of corticomotor neuronal cells originated in area 3a (Rathelot & Strick, 2006, 2009).

Perhaps the most prominent, and certainly the simplest model of the control of these descending projections is that of alpha–gamma coactivation. In this model, gamma motor axons fire concomitantly with the alpha motor axons innervating the extrafusal fibers of the same muscle, essentially to keep the spindle sensory region taut and responsive during muscle shortening. The strongest evidence of this idea is the existence of motor neurons which project to both extrafusal and intrafusal fibers within the same muscle, called beta motor or skeletofusimotor neurons. More than 30% of motor axons in the alpha range of conduction velocities (> 85 m/s) are projections of beta motor neurons (Manuel & Zytnicki, 2011). Mammals have gamma motor neurons that can be controlled independent of alpha motor neurons in addition to the "hard-wired" alpha-linked drive from beta motor neurons.

One piece of countervailing evidence to strict alpha–gamma coactivation comes from locomoting cats, in which static fusimotor neurons are modulated slowly throughout the step cycle, in phase with muscle shortening, despite marked modulation in the parent muscle EMG (Taylor et al., 2006). Observations like this have motivated other models, including the "fusimotor set," which hypothesizes that fusimotor drive is set to different tonic levels depending on the task (Prochazka et al., 1985). As task novelty or difficulty increases, this model predicts a shift from predominantly static fusimotor drive to more dynamic drive (Prochazka & Ellaway, 2012; Prochazka et al., 1985). More recent experiments in which muscle spindles were recorded from both upper and lower limbs in humans also revealed signs of independent gamma motor control. In the lower limb, feedback from a bit over half of group Ia afferents (but not group II) was enhanced when subjects were instructed simply to attend to the trajectory of passively imposed ankle movements (Hospod et al., 2007).

2.4.3.2 Neural sensory gain modulation

In addition to altering muscle spindle gain through the fusimotor system, reflexes can be potentiated or attenuated through control of neurons along the afferent pathway. A prime example is found in the spinal stretch reflex. To generate efficient movements, it is important that this reflex does not activate when a muscle is stretched by a voluntary activation of an antagonist. To this end, presynaptic inhibition is exerted upon the Ia afferent terminals synapsing on the reflexively activated motor neurons (Meunier & Pierrot-Deseilligny, 1989). This inhibits the activation of the muscle by its monosynaptic reflex pathway, but not by other sources of input. This is one simple example of how a central motor command can exert a modulatory influence on sensory gain, damping the feedback loop to allow controlled movement. More complex spinal-circuit conditioning has been revealed in the mouse using genetic manipulations. Ablation of spinal interneurons that form inhibitory presynaptic connections leads to motor oscillations reminiscent of an underdamped sensorimotor loop (Fink et al., 2014). These spinal interneurons receive both ascending and descending inputs, as well as inputs from local spinal interneurons, allowing a complex context-dependence in the regulation of sensory gain.

In addition to mediating flexible reflex activity, another possible effect of gain modulation may be to attenuate distracting signals. Limb movements generate sensory reafference (sensation resulting from the movement), much of which is not functionally relevant. The sensation of a shirt sliding across your skin, for instance, likely does not give much useful information about the reach, and therefore it may be better to reduce the intensity of those signals. Ghez and Pisa found that tactile signals recorded from neurons in the medial lemniscus of a cat are attenuated during stepping, potentially to prevent the afferent barrage from interfering with the execution of centrally planned, ballistic movements (Ghez & Pisa, 1972). Analogous observations have been made in the psychophysics literature and in somatosensory evoked potentials (Cohen & Starr, 1987; Rushton et al., 1981; Schmidt et al., 1990; Yıldız et al., 2015).

While evidence for attenuation of tactile responses by descending inputs is well described in the literature, there is little such evidence on the proprioceptive system. Several observations make the prospect of substantial attenuation of proprioceptive signals unlikely. First, single neurons in cortical area 2 encode actively and passively generated reaching movements with similar strength (Fig. 2.4D) (London & Miller, 2013; Prud'homme & Kalaska, 1994). Second, precise sensory cancellation is known to occur in the cerebellum (see Section 2.5; Brooks et al., 2015; Herzfeld et al., 2018). It seems unlikely that this information would be substantially attenuated, only to be selectively canceled by a forward model. Finally, studies demonstrate mixed potentiation and attenuation of afferent information as a function of descending drive from sensorimotor cortex (Leiras et al., 2010; Palmeri et al., 1999). Many of these studies also suggest that the sign of this modulation may depend on the modality and relative locations of the afferent fields. For example, electrical stimulation of arm M1 tended to enhance the responsiveness of shoulder-receiving neurons in the cuneate nucleus, but suppress hand-related neurons. These findings suggest a more nuanced effect, in which some receptors are potentiated and others are attenuated in order to reinforce or filter out information to improve perceptual acuity or for use in reflex pathways.

Experiments in the lab of Dr Seki, demonstrated a striking example of these differential effects in spinal cervical interneurons. During a wrist flexion task, gain of tactile inputs to a cervical spinal interneuron was reduced during movement relative to rest (Fig. 2.6B, right column), while gain to a proprioceptive neuron increased (Fig. 2.6B, left column). Across neurons, proprioceptive inputs were more potent during reaching, while tactile inputs were attenuated (Fig. 2.6C). A key task in understanding how the brain utilizes proprioceptive feedback is to find general principles that explain how and for what purpose sensory gain is modulated across contexts and modalities.

2.5 Cerebellar involvement in proprioception

As described in Section 2.1, aspects of the deficits due to cerebellar damage share a strong resemblance to complete loss of proprioceptive inputs, highlighting how crucial the cerebellum is in processing proprioceptive information needed to plan and guide movement. In this section, we first discuss the anatomy of the proprioceptive inputs to the cerebellum, then highlight studies that demonstrate the role of the cerebellum in movement planning and error correction.

2.5.1 Cerebellar afferent pathway

There are four fiber tracts supplying sensory information from the spinal cord to both the cerebellar cortex and nuclei: two from the upper limbs (the cuneo and rostral spinocerebellar tracts) and two from the lower limbs (dorsal and ventral tracts). The cuneo and dorsal tracts run in the dorsal aspect of the cord and carry largely unprocessed sensory information from muscle spindles and GTOs. The rostral and ventral tracts arise mainly from laminae 5–7 in the ventral horn (Sengul & Watson, 2012). The ventral spinocerebellar tract in particular, is thought to be driven strongly by information from spinal motor circuits, including those making up spinal pattern generators, though they are influenced by signals from the periphery as well (Arshavsky et al., 1978). Less is known of its forelimb analog, the rostral spinocerebellar tract.

To understand how proprioceptive signals are carried to the cerebellum, we might look to one of the more well studied spinocerebellar tracts: the dorsal tract carrying lower limb signals (DSCT). Neurons in the DSCT are only one or two synapses away from the muscle receptors, but despite this proximity, have complex responses to stimuli, appearing to respond inconsistently to the motion of individual joints, depending on the state of the rest of the limb (Osborn & Poppele, 1992). This complexity can be accounted for by a wide convergence of muscle inputs, which results in a coding scheme that represents the state of the whole limb (Bosco & Poppele, 2000; Bosco et al., 1996; Chowdhury et al., 2017). This work suggests that DSCT may combine peripheral inputs in many different ways, with each neuron representing the whole limb in a slightly different way. Downstream areas, like the cerebellum, could then build from these simple linear combinations of inputs to those needed to generate, coordinate, and adapt movement.

2.5.2 Sensorimotor adaptation

A key characteristic of our motor system is its ability to adapt to changing environments. Proprioception plays an important role in enabling this flexibility. Consider how a tennis player might adapt to a light-weight racquet. Switching to a lighter-than-normal racquet would initially result in a faster-than-expected swing, requiring the player to adapt the descending motor commands to bring their swing back to normal. If then they pick up a normal racquet, their swing would briefly be slower than normal, a so-called "after-effect" of having adapted to the lighter racquet (Fig. 2.7A, right section, red line). Laboratory experiments often study this kind of adaptation by having subjects reach to targets in the presence of an imposed force field (Shadmehr & Mussa-Ivaldi, 1994; Fig. 2.7A). In both these laboratory tasks and the tennis racquet swing, motor adaptation is dependent on the discrepancy between the expected and actual sensory feedback, otherwise known as the sensory prediction error (Miall & Wolpert, 1996; Tseng et al., 2007). The motor

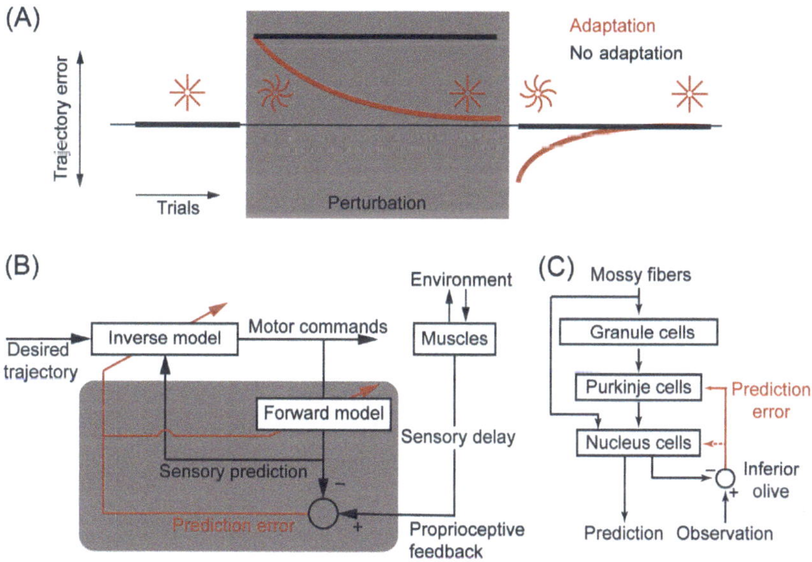

FIGURE 2.7 The use of proprioception and internal models for motor adaptation and sensory cancellation. (A) Example of reach adaptation during a force field perturbation. Example trajectories show evolving adaptation (*red trace*). At perturbation onset, trajectories become curved, leading to a high trajectory error. Through adaptation, trajectories straighten, but at perturbation offset, they deviate in the opposite direction. Black trace shows what happens without updates to internal model. (B) Block diagram of internal model framework. Inverse model computes motor commands from a desired trajectory. The forward model receives a copy of motor commands to predict sensory consequences of movement, which might be used for surrogate feedback in place of actual, later feedback. Models are updated by computing the error between predicted feedback and actual feedback. (C) Possible implementation of forward model updating (*gray box* in A) in the cerebellum. Other models exist with other sites acting as the comparator. *(Figure 2.7C) Adapted from Shadmehr, R. (2020). Population coding in the cerebellum: A machine learning perspective.* Journal of Neurophysiology, 124*(6), 2022–2051. <https://doi.org/10.1152/jn.00449.2020>.*

system uses this error signal to adapt a set of internal models of the body to the new environment, which are then used to generate the required motor commands for feedforward control (Shadmehr & Mussa-Ivaldi, 1994; Wolpert et al., 1995), as well as for feedback, reflexive motor control (Kurtzer et al., 2008; Maeda et al., 2018).

In the internal model framework, the sensory prediction error provides a mechanism to determine whether a motor action is "going according to plan." For the tennis player, the plan is the desired trajectory of the swing. An "inverse model" of the body converts this desired trajectory into motor commands, which, when sent to the muscles, result in the desired swing. At the same time, the motor command is sent to a "forward model," which predicts the sensory consequences of the anticipated swing, faster than the actual feedback can be returned to the nervous system (Fig. 2.7A; Shadmehr

& Krakauer, 2008). In the case that the swing goes awry (perhaps due to an unexpectedly light racquet), the actual feedback will not match the predicted feedback. Calculating the sensory prediction error is then a matter of comparing the output of the forward model to the incoming sensory information. A nonzero prediction error means that one or both of the models was incorrect and needs to be updated (Fig. 2.7A, red lines).

The cerebellum is strongly implicated in this process for the visual, vestibular, and proprioceptive systems (Brooks et al., 2015; Gilbert & Thach, 1977; Herzfeld et al., 2018; Ito, 2013). In addition to observations in patients (Holmes, 1939), cerebellar-mediated adaptation has been demonstrated in monkeys for saccadic eye movements (Optican & Robinson, 1980), and arm movements (Gilbert & Thach, 1977). Perhaps the most compelling example of sensory error calculation in the proprioceptive system was obtained by recording from the cerebellum of a monkey learning to compensate for an abruptly added inertial load on its head movements (Brooks et al., 2015). Firing rates of neurons recorded from the rostral fastigial nucleus (rFN) of the cerebellum reflect the sensory prediction error as the monkey adapted to the unexpected load (Fig. 2.7A, *red lines*).

While a limb forward model has not been definitively identified, Purkinje cells (PCs) in the cerebellar vermis display many characteristics that we might expect of an adaptive forward model (Fig. 2.7C; Shadmehr, 2020). The sensory prediction (i.e., position of a saccade target on the retina) is encoded in the simple spike firing rate transmitted to neurons in the deep cerebellar nuclei. Neurons in the inferior olive, to which the deep nuclei project, encode sensory prediction errors, though evidence suggests that this prediction error may be inherited from the cerebellar nuclei rather than computed in the olive. The olive sends an error signal back to the PCs via climbing fibers, resulting in complex spikes which drive long-term depression at the synapses between parallel fibers and PCs (Herzfeld et al., 2018; Ito & Kano, 1982). Thus, errors in prediction change the simple spike rates of PCs, thereby altering the predicted consequence of a motor output. These plastic changes are thought to underlie adaptation of the forward model and altered sensory predictions for a given motor plan.

2.6 Summary

Proprioception, despite its common description as a sensation of body position and movement, is also so intimately tied to the systems governing motor control that the two can scarcely be considered distinct at any level. As the proprioceptive afferents reach the spinal cord, collateral axons project to many spinal laminae, shaping muscle coordination and reflex actions. The afferent proprioceptive feedback is subject to additional processing at multiple stages, including the brainstem and thalamus, before reaching the brain. These areas all receive top-down modulation from motor areas, more closely entwining the two systems, and ultimately converge in somatosensory cortical areas 3a and 2, where proprioceptive inputs from the thalamus are further processed by modality convergence as well as projections from motor cortices that shape their coding properties. In parallel with the pathway ascending to the cerebrum, proprioceptive signals from the thalamus also reach the cerebellum, where they are involved in internal model formation and error calculation, critical for movement planning, execution, and adaptation.

Considering the interconnectedness of the proprioceptive and motor systems, it is no surprise that the loss of proprioceptive feedback leads to profound motor deficits. The challenge of developing proprioceptive prosthetic devices should therefore not be underestimated. It is only through a deeper understanding of the systems outlined in this chapter that a suitable replacement of proprioception can be developed.

References

Appelberg, B., Hulliger, M., Johansson, H., & Sojka, P. (1983). Actions on gamma-motoneurones elicited by electrical stimulation of group II muscle afferent fibres in the hind limb of the cat. *The Journal of Physiology*, *335*(1), 255–273.

Arshavsky, Y. I., Gelfand, I. M., Orlovsky, G. N., & Pavlova, G. A. (1978). Messages conveyed by spinocerebellar pathways during scratching in the cat. II. Activity of neurons of the ventral spinocerebellar tract. *Brain Research*, *151*(3), 493–506. Available from https://doi.org/10.1016/0006-8993(78)91082-X.

Azim, E., & Seki, K. (2019). Gain control in the sensorimotor system. *Current Opinion in Physiology*, *8*, 177–187. Available from https://doi.org/10.1016/j.cophys.2019.03.005.

Banks, R. W. (2005). *The muscle spindle. Peripheral neuropathy*. Elsevier.

Banks, R. W., Hulliger, M., Scheepstra, K. A., & Otten, E. (1997). Pacemaker activity in a sensory ending with multiple encoding sites: The cat muscle spindle primary ending. *Journal of Physiology*, *498*(Pt 1), 177–199. Available from https://doi.org/10.1111/(ISSN)1469-7793.

Bastian, A. J., Martin, T. A., Keating, J. G., & Thach, W. T. (1996). Cerebellar ataxia: Abnormal control of interaction torques across multiple joints. *Journal of Neurophysiology*, *76*(1), 492–509. Available from https://doi.org/10.1152/jn.1996.76.1.492.

Berkley, K. J., Budell, R. J., Blomqvist, A., & Bull, M. (1986). *Output systems of the dorsal column nuclei in the cat, . Brain research reviews* (Vol. 11). Elsevier. Available from https://doi.org/10.1016/0165-0173(86)90012-3.

Blum, K. P., Campbell, K. S., Horslen, B. C., Nardelli, P., Housley, S. N., Cope, T. C., & Ting, L. H. (2020). Diverse and complex muscle spindle afferent firing properties emerge from multiscale muscle mechanics. *Elife*, *9*, e55177.

Blum, K. P., Lamotte D'Incamps, B., Zytnicki, D., & Ting, L. H. (2017). Force encoding in muscle spindles during stretch of passive muscle. *PLOS Computational Biology*, *13*(9), e1005767. Available from https://doi.org/10.1371/journal.pcbi.1005767.

Blum, K. P., Nardelli, P., Cope, T. C., & Ting, L. H. (2019). Elastic tissue forces mask muscle fiber forces underlying muscle spindle Ia afferent firing rates in stretch of relaxed rat muscle. *Journal of Experimental Biology*, *222*(15). Available from https://doi.org/10.1242/jeb.196287, jeb196287.

Bologna, N., & Maravita, A. (2007). Proprioceptive alignment of visual and somatosensory maps in the posterior parietal cortex. *Current Biology*, *17*(21), 1890–1895. Available from https://doi.org/10.1016/j.cub.2007.09.057.

Bosco, G., & Poppele, R. E. (2000). Reference frames for spinal proprioception: Kinematics based or kinetics based? *Journal of Neurophysiology*, *83*(5), 2946–2955. Available from https://doi.org/10.1152/jn.2000.83.5.2946.

Bosco, G., & Poppele, R. E. (2001). Proprioception from a spinocerebellar perspective. *Physiological Reviews*, *81*(2), 539–568. Available from https://doi.org/10.1152/physrev.2001.81.2.539.

Bosco, G., Rankin, A., & Poppele, R. (1996). Representation of passive hindlimb postures in cat spinocerebellar activity. *Journal of Neurophysiology*, *76*(2), 715–726. Available from https://doi.org/10.1152/jn.1996.76.2.715.

Bota, M., Ten Donkelaar, H. J., Mai, J. K., & Majtanik, M. (2019). Toward a common terminology for the thalamus. *Frontiers in Neuroanatomy*, *1*, 114. Available from https://doi.org/10.3389/fnana.2018.00114.

Boyd, I. A. (1962). The structure and innervation of the nuclear bag muscle fibre system and the nuclear chain muscle fibre system in mammalian muscle spindles. *Philosophical Transactions of the Royal Society of London. Series B, Biological Sciences*, *245*(720), 92.

Boyd, I. A. (1976). The response of fast and slow nuclear bag fibres and nuclear chain fibres in isolated cat muscle spindles to fusimotor stimulation, and the effect of intrafusal contraction on the sensory endings. *Quarterly Journal of Experimental Physiology and Cognate Medical Sciences*, *61*(3), 203–254.

Brooks, J. X., Carriot, J., & Cullen, K. E. (2015). Learning to expect the unexpected: Rapid updating in primate cerebellum during voluntary self-motion. *Nature Neuroscience*, *18*(9), 1310–1317. Available from https://doi.org/10.1038/nn.4077.

Brown, A. G., & Fyffe, R. E. (1978). The morphology of group Ia afferent fibre collaterals in the spinal cord of the cat. *The Journal of Physiology*, *274*(1), 111–127. Available from https://doi.org/10.1113/jphysiol.1978.sp012137.

Burgess, P. R., & Clark, F. J. (1969). Characteristics of knee joint receptors in the cat. *The Journal of Physiology*, *203*(2), 317–335. Available from https://doi.org/10.1113/jphysiol.1969.sp008866.

Campbell, K. S., & Lakie, M. (1998). A cross-bridge mechanism can explain the thixotropic short-range elastic component of relaxed frog skeletal muscle. *Journal of Physiology*, *510*(Pt 3), 941–962.

Campbell, K. S., & Moss, R. L. (2000). A thixotropic effect in contracting rabbit psoas muscle: Prior movement reduces the initial tension response to stretch. *Journal of Physiology*. Available from https://doi.org/10.1111/j.1469-7793.2000.00531.x.

Campbell, K. S., & Moss, R. L. (2002). History-dependent mechanical properties of permeabilized rat soleus muscle fibers. *Biophysical Journal*, *82*(2), 929–943. Available from https://doi.org/10.1016/S0006-3495(02)75454-4.

Carrasco, D. I., Vincent, J. A., & Cope, T. C. (2017). Distribution of TTX-sensitive voltage-gated sodium channels in primary sensory endings of mammalian muscle spindles. *Journal of Neurophysiology*, *117*(4), 1690–1701. Available from https://doi.org/10.1152/jn.00889.2016.

Chen, H., Hua, S. E., Smith, M. A., Lenz, F. A., & Shadmehr, R. (2006). Effects of human cerebellar thalamus disruption on adaptive control of reaching. *Cerebral Cortex*, *16*(10), 1462–1473. Available from https://doi.org/10.1093/cercor/bhj087.

Cheney, P. D., & Fetz, E. E. (1984). Corticomotoneuronal cells contribute to long-latency stretch reflexes in the rhesus monkey. *The Journal of Physiology*, *349*(1), 249–272. Available from https://doi.org/10.1113/jphysiol.1984.sp015155.

Chowdhury, R. H., Glaser, J. I., & Miller, L. E. (2020). Area 2 of primary somatosensory cortex encodes kinematics of the whole arm. *ELife*, *9*, e48198. Available from https://doi.org/10.7554/eLife.48198.

Chowdhury, R. H., Tresch, M. C., & Miller, L. E. (2017). Musculoskeletal geometry accounts for apparent extrinsic representation of paw position in dorsal spinocerebellar tract. *Journal of Neurophysiology*, *118*(1), 234–242. Available from https://doi.org/10.1152/jn.00695.2016.

Cohen, L. G., & Starr, A. (1987). Localization, timing and specificity of gating of somatosensory evoked potentials during active movement in man. *Brain*, *110 (Pt 2)*, 451–467. Available from https://doi.org/10.1093/brain/110.2.451, 3567532.

Cordo, P. J., Flores-Vieira, C., Verschueren, S. M. P., Inglis, J. T., & Gurfinkel, V. (2002). Position sensitivity of human muscle spindles: Single afferent and population representations. *Journal of Neurophysiology*, *87*(3), 1186–1195. Available from https://doi.org/10.1152/jn.00393.2001.

Coulter, J., & Jones, E. (1977). Differential distribution of corticospinal projections from individual cytoarchitectonic fields in the monkey. *Brain Research*, *129*(2), 135–140.

Crago, P. E., Houk, J. C., & Rymer, W. Z. (1982). Sampling of total muscle force by tendon organs. *Journal of Neurophysiology*, *47*(6), 1069–1083. Available from https://doi.org/10.1152/jn.1982.47.6.1069.

Crowe, A., & Matthews, P. B. C. (1964a). Further studies of static and dynamic fusimotor fibres. *The Journal of Physiology*, *174*(1), 132–151. Available from https://doi.org/10.1113/jphysiol.1964.sp007477.

Crowe, A., & Matthews, P. B. C. (1964b). The effects of stimulation of static and dynamic fusimotor fibres on the response to stretching of the primary endings of muscle spindles. *The Journal of Physiology*, *174*(1), 109–131. Available from https://doi.org/10.1113/jphysiol.1964.sp007476.

Dale, A., & Cullen, K. E. (2017). The ventral posterior lateral thalamus preferentially encodes externally applied versus active movement: Implications for self-motion perception. *Cerebral Cortex*, *29*(1), 305–318. Available from https://doi.org/10.1093/cercor/bhx325.

Dostrovsky, J. O., Millar, J., & Wall, P. D. (1976). The immediate shift of afferent drive of dorsal column nucleus cells following deafferentation: A comparison of acute and chronic deafferentation in gracile nucleus and spinal cord. *Experimental Neurology*, *52*(3), 480–495. Available from https://doi.org/10.1016/0014-4886(76)90219-3.

Durbaba, R., Taylor, A., Ellaway, P. H., & Rawlinson, S. (2001). Modulation of primary afferent discharge by dynamic and static gamma motor axons in cat muscle spindles in relation to the intrafusal fibre types activated. *The Journal of Physiology*, *532*(2), 563–574.

Durbaba, R., Taylor, A., Ellaway, P. H., & Rawlinson, S. (2006). Classification of longissimus lumborum muscle spindle afferents in the anaesthetized cat. *The Journal of Physiology*, *571* (Pt 2), 489–498. Available from https://doi.org/10.1113/jphysiol.2005.102731.

Edin, B. B., & Vallbo, A. B. (1990). Muscle afferent responses to isometric contractions and relaxations in humans. *Journal of Neurophysiology*, *63*(6), 1307–1313. Available from https://doi.org/10.1152/jn.1990.63.6.1307.

Evarts, E. V., & Tanji, J. (1976). Reflex and intended responses in motor cortex pyramidal tract neurons of monkey. *Journal of Neurophysiology*, *39*(5), 1069–1080. Available from https://doi.org/10.1152/jn.1976.39.5.1069.

Fink, A. J. P., Croce, K. R., Huang, Z. J., Abbott, L. F., Jessell, T. M., & Azim, E. (2014). Presynaptic inhibition of spinal sensory feedback ensures smooth movement. *Nature*, *509* (7498), 43–48. Available from https://doi.org/10.1038/nature13276.

Fisher, T., & Clowry, G. J. (2009). Elimination of muscle afferent boutons from the cuneate nucleus of the rat medulla during development. *Neuroscience*, *161*(3), 787–793. Available from https://doi.org/10.1016/j.neuroscience.2009.04.009.

Francis, J. T., Xu, S., & Chapin, J. K. (2008). Proprioceptive and cutaneous representations in the rat ventral posterolateral thalamus. *Journal of Neurophysiology*, *99*, 2291–2304. Available from https://doi.org/10.1152/jn.01206.2007.

Fuentes, C. T., & Bastian, A. J. (2010). Where is your arm? Variations in proprioception across space and tasks. *Journal of Neurophysiology*, *103*(1), 164–171.

Gardner, E. P. (1988). Somatosensory cortical mechanisms of feature detection in tactile and kinesthetic discrimination. *Canadian Journal of Physiology and Pharmacology*, *66*(4), 439–454.

Gardner, E. P., & Costanzo, R. M. (1981). Properties of kinesthetic neurons in somatosensory cortex of awake monkeys. *Brain Research*, *214*(2), 301–319.

Gentilucci, M., Chieffi, S., Scarpa, M., & Castiello, U. (1992). Temporal coupling between transport and grasp components during prehension movements: Of visual perturbation. *Behavioural Brain Research*, *47*(1), 71–82. Available from https://doi.org/10.1016/S0166-4328(05)80253-0.

Georgopoulos, A. P., Kalaska, J. F., Caminiti, R., & Massey, J. T. (1982). On the relations between the direction of two-dimensional arm movements and cell discharge in primate motor cortex. *Journal of Neuroscience*, *2*(11), 1527–1537.

Ghez, C., & Pisa, M. (1972). Inhibition of afferent transmission in cuneate nucleus during voluntary movement in the cat. *Brain Research*, *40*(1), 145–155.

Ghez, C., Gordon, J., & Ghilardi, M. F. (1995). Impairments of reaching movements in patients without proprioception. II. Effects of visual information on accuracy. *Journal of Neurophysiology*, *73*(1), 361–372. Available from https://doi.org/10.1152/jn.1995.73.1.361.

Ghez, C., Gordon, J., Ghilardi, M. F., Christakos, C. N., & Cooper, S. E. (1990). Roles of proprioceptive input in the programming of arm trajectories. *Cold Spring Harbor Symposia on Quantitative Biology*, *55*, 837–847. Available from https://doi.org/10.1101/SQB.1990.055.01.079.

Gilbert, P. F. C., & Thach, W. T. (1977). Purkinje cell activity during motor learning. *Brain Research*, *128*(2), 309–328. Available from https://doi.org/10.1016/0006-8993(77)90997-0.

Goodman, J. M., Tabot, G. A., Lee, A. S., Suresh, A. K., Rajan, A. T., Hatsopoulos, N. G., & Bensmaia, S. (2019). Postural representations of the hand in the primate sensorimotor cortex. *Neuron*, *104*(5), 1000–1009, e7.

Gregory, J. E., Brockett, C. L., Morgan, D. L., Whitehead, N. P., & Proske, U. (2002). Effect of eccentric muscle contractions on Golgi tendon organ responses to passive and active tension in the cat. *The Journal of Physiology*, *538*(Pt 1), 209–218. Available from https://doi.org/10.1013/jphysiol.2001.012785.

Grey, M. J., Ladouceur, M., Andersen, J. B., Nielsen, J. B., & Sinkjær, T. (2001). Group II muscle afferents probably contribute to the medium latency soleus stretch reflex during walking in humans. *The Journal of Physiology*, *534*(Pt 3), 925–933. Available from https://doi.org/10.1111/j.1469-7793.2001.00925.x.

Grigg, P., & Greenspan, B. J. (1977). Response of primate joint afferent neurons to mechanical stimulation of knee joint. *Journal of Neurophysiology.*, *40*(1), 1–8. Available from https://doi.org/10.1152/jn.1977.40.1.1.

Grigg, P., & Preston, J. B. (1971). Baboon flexor and extensor fusimotor neurons and their modulation by motor cortex. *Journal of Neurophysiology*, *34*(3), 428–436.

Haftel, V. K., Bichler, E. K., Nichols, T. R., Pinter, M. J., & Cope, T. C. (2004). Movement reduces the dynamic response of muscle spindle afferents and motoneuron synaptic potentials in rat. *Journal of Neurophysiology*, *91*(5), 2164–2171. Available from https://doi.org/10.1152/jn.01147.2003.

Heath, C. J., Hore, J., & Phillips, C. G. (1976). Inputs from low threshold muscle and cutaneous afferents of hand and forearm to areas 3a and 3b of baboon's cerebral cortex. *The Journal of Physiology*, *257*(1), 199–227. Available from https://doi.org/10.1113/jphysiol.1976.sp011364.

Herzfeld, D. J., Kojima, Y., Soetedjo, R., & Shadmehr, R. (2018). Encoding of error and learning to correct that error by the Purkinje cells of the cerebellum. *Nature Neuroscience*, *21*(5), 736–743. Available from https://doi.org/10.1038/s41593-018-0136-y.

Holmes, G. (1939). The cerebellum of man. *Brain*, *62*(1), 1–30. Available from https://doi.org/10.1093/brain/62.1.1.

Honeycutt, C. F., Nardelli, P., Cope, T. C., & Nichols, T. R. (2012). Muscle spindle responses to horizontal support surface perturbation in the anesthetized cat: Insights into the role of autogenic feedback in whole body postural control. *Journal of Neurophysiology*, *108*(5), 1253–1261.

Horcholle-Bossavit, G., Jami, L., Petit, J., Vejsada, R., & Zytnicki, D. (1990). Ensemble discharge from Golgi tendon organs of cat peroneus tertius muscle. *Journal of Neurophysiology*, *64*(3), 813–821.

Hore, J., Preston, J. B., & Cheney, P. D. (1976). Responses of cortical neurons (areas 3a and 4) to ramp stretch of hindlimb muscles in the baboon. *Journal of Neurophysiology.*, *39*(3), 484–500. Available from https://doi.org/10.1152/jn.1976.39.3.484.

Hospod, V., Aimonetti, J. M., Roll, J. P., & Ribot-Ciscar, E. (2007). Changes in human muscle spindle sensitivity during a proprioceptive attention task. *Journal of Neuroscience*, *27*(19), 5172–5178. Available from https://doi.org/10.1523/JNEUROSCI.0572-07.2007.

Houk, J., & Henneman, E. (1967). Responses of Golgi tendon organs to active contractions of the soleus muscle of the cat. *Journal of Neurophysiology*, *30*(3), 466–481. Available from https://doi.org/10.1152/jn.1967.30.3.466.

Houk, J. C., Rymer, W. Z., & Crago, P. E. (1981). Dependence of dynamic response of spindle receptors on muscle length and velocity. *Journal of Neurophysiology*, *46*(1), 143–166. Available from https://doi.org/10.1152/jn.1981.46.1.143.

Hsiao, S. (2008). Central mechanisms of tactile shape perception. *Current Opinion in Neurobiology*, *18*(4), 418–424. Available from https://doi.org/10.1016/j.conb.2008.09.001.

Huffman, K. J., & Krubitzer, L. (2001). Area 3a: Topographic organization and cortical connections in marmoset monkeys. *Cerebral Cortex (New York, N.Y.: 1991)*, *11*(9), 849–867.

Hummelsheim, H., & Wiesendanger, M. (1985). Neuronal responses of medullary relay cells to controlled stretches of forearm muscles in the monkey. *Neuroscience*, *16*(4), 989–996.

Ito, M. (2013). Error detection and representation in the olivo-cerebellar system. *Frontiers in Neural Circuits*, *7*(Jan), 1. Available from https://doi.org/10.3389/fncir.2013.00001.

Ito, M., & Kano, M. (1982). Long-lasting depression of parallel fiber-Purkinje cell transmission induced by conjunctive stimulation of parallel fibers and climbing fibers in the cerebellar cortex. *Neuroscience Letters*, *33*(3), 253–258. Available from https://doi.org/10.1016/0304-3940(82)90380-9.

Jami, L. (1992). Golgi tendon organs in mammalian skeletal muscle: Functional properties and central actions. *Physiological Reviews*, *72*(3), 623–666.

Jansen, J. K. S., & Matthews, P. B. C. (1962). The effects of fusimotor activity on the static responsiveness of primary and secondary endings of muscle spindles in the decerebrate cat.

Acta Physiologica Scandinavica, 55(4), 376–386. Available from https://doi.org/10.1111/j.1748-1716.1962.tb02451.x.

Jansen, J. K. S., & Rudjord, T. (1964). On the silent period and golgi tendon organs of the soleus muscle of the cat. *Acta Physiologica Scandinavica, 62*(4), 364–379. Available from https://doi.org/10.1111/j.1748-1716.1964.tb10435.x.

Jennings, V. A., Lamour, Y., Solis, H., & Fromm, C. (1983). Somatosensory cortex activity related to position and force. *Journal of Neurophysiology, 49*(5), 1216–1229. Available from https://doi.org/10.1152/jn.1983.49.5.1216.

Johansson, H., Sjölander, P., Sojka, P., & Wadell, I. (1987). Fusimotor reflexes to antagonistic muscles simultaneously assessed by multi-afferent recordings from muscle spindle afferents. *Brain Research, 435*(1–2), 337–342.

Johansson, R. S., & Vallbo, A. (1979). Tactile sensibility in the human hand: Relative and absolute densities of four types of mechanoreceptive units in glabrous skin. *The Journal of Physiology., 286*(1), 283–300.

Jones, E. G., Coulter, J. D., & Hendry, S. H. C. (1978). Intracortical connectivity of architectonic fields in the somatic sensory, motor and parietal cortex of monkeys. *Journal of Comparative Neurology, 181*(2), 291–347. Available from https://doi.org/10.1002/cne.901810206.

Kaas, J., Nelson, R., Sur, M., Lin, C., & Merzenich, M. (1979). Multiple representations of the body within the primary somatosensory cortex of primates. *Science, 204*(4392), 521–523. Available from https://doi.org/10.1126/science.107591.

Kalaska, J. F. (1988). The representation of arm movements in postcentral and parietal cortex. *Canadian Journal of Physiology and Pharmacology*. Available from https://doi.org/10.1139/y88-075.

Kambi, N., Halder, P., Rajan, R., Arora, V., Chand, P., Arora, M., & Jain, N. (2014). Large-scale reorganization of the somatosensory cortex following spinal cord injuries is due to brainstem plasticity. *Nature Communications, 5*(1), 1–10. Available from https://doi.org/10.1038/ncomms4602.

Kim, S. S., Gomez-Ramirez, M., Thakur, P. H., & Hsiao, S. S. (2015). Multimodal interactions between proprioceptive and cutaneous signals in primary somatosensory cortex. *Neuron, 86*(2), 555–566.

Krubitzer, L., Huffman, K. J., Disbrow, E., & Recanzone, G. (2004). Organization of area 3a in macaque monkeys: Contributions to the cortical phenotype. *The Journal of Comparative Neurology, 471*(1), 97–111. Available from https://doi.org/10.1002/cne.20025.

Kumar, N., Manning, T. F., & Ostry, D. J. (2019). Somatosensory cortex participates in the consolidation of human motor memory. *PLOS Biology, 17*(10), e3000469. Available from https://doi.org/10.1371/journal.pbio.3000469.

Kurtzer, I. L., Pruszynski, J. A., & Scott, S. H. (2008). Long-latency reflexes of the human arm reflect an internal model of limb dynamics. *Current Biology, 18*(6), 449–453. Available from https://doi.org/10.1016/j.cub.2008.02.053.

Laporte, Y., & Lloyd, D. P. C. (1952). Nature and significance of the reflex connections established by large afferent fibers of muscular origin. *American Journal of Physiology-Legacy Content, 169*(3), 609–621. Available from https://doi.org/10.1152/ajplegacy.1952.169.3.609.

Leiras, R., Velo, P., Martín-Cora, F., & Canedo, A. (2010). Processing afferent proprioceptive information at the main cuneate nucleus of anesthetized cats. *Journal of Neuroscience, 30*(46).

Lin, D. C., McGowan, C. P., Blum, K. P., & Ting, L. H. (2019). Yank: The time derivative of force is an important biomechanical variable in sensorimotor systems. *The Journal of Experimental Biology, 222*(18), jeb180414. Available from https://doi.org/10.1242/jeb.180414.

Lockhart, D. B., & Ting, L. H. (2007). Optimal sensorimotor transformations for balance. *Nature Neuroscience*, *10*(10), 1329–1336. Available from https://doi.org/10.1038/nn1986.

London, B. M., & Miller, L. E. (2013). Responses of somatosensory area 2 neurons to actively and passively generated limb movements. *Journal of Neurophysiology*, *109*(6), 1505–1513. Available from https://doi.org/10.1152/jn.00372.2012.

Loutit, A. J., Vickery, R. M., & Potas, J. R. (2020). Functional organization and connectivity of the dorsal column nuclei complex reveals a sensorimotor integration and distribution hub. *Journal of Comparative Neurology*. Available from https://doi.org/10.1002/cne.24942.

Lucier, G. E., Rüegg, D. C., & Wiesendanger, M. (1975). Responses of neurones in motor cortex and in area 3A to controlled stretches of forelimb muscles in cebus monkeys. *The Journal of Physiology*, *251*(3), 833–853.

Macefield, G., Gandevia, S. C., & Burke, D. (1990). Perceptual responses to microstimulation of single afferents innervating joints, muscles and skin of the human hand. *The Journal of Physiology*, *429*(1), 113–129. Available from https://doi.org/10.1113/jphysiol.1990.sp018247.

Maeda, R. S., Cluff, T., Gribble, P. L., & Pruszynski, J. A. (2018). Feedforward and Feedback Control Share an Internal Model of the Arm's Dynamics. *The Journal of Neuroscience*, *38*(49), 10505–10514. Available from https://doi.org/10.1523/JNEUROSCI.1709-18.2018, 30355628.

Manuel, M., & Zytnicki, D. (2011). Alpha, beta, and gamma motoneurons: Functional diversity in the motor system's final pathway. *Journal of Integrative Neuroscience*, *10*(03), 243–276. Available from https://doi.org/10.1142/S0219635211002786.

Mathis, M. W., Mathis, A., & Uchida, N. (2017). Somatosensory cortex plays an essential role in forelimb motor adaptation in mice. *Neuron*, *93*(6), 1493–1503. Available from https://doi.org/10.1016/j.neuron.2017.02.049, e6.

Matthews, B. H. C. (1933). Nerve endings in mammalian muscle. *The Journal of Physiology*, *78*(1), 1–53. Available from https://doi.org/10.1113/jphysiol.1933.sp002984.

Matthews, P. B. C. (1963). The response of de-efferented muscle spindle receptors to stretching at different velocities. *The Journal of Physiology*, *168*(3), 660–678. Available from https://doi.org/10.1113/jphysiol.1963.sp007214.

McIntyre, A. K., Proske, U., & Rawson, J. A. (1984). Cortical projection of afferent information from tendon organs in the cat. *The Journal of Physiology*, *354*, 395–406.

McNeill, D., Quaeghebeur, L., & Duncan, S. (2010). IW—"The man who lost his body". In D. Schmicking, & S. Gallagher (Eds.), *Handbook of phenomenology and cognitive science* (pp. 519–543). Netherlands: Springer. Available from https://doi.org/10.1007/978-90-481-2646-0_27.

Meunier, S., & Pierrot-Deseilligny, E. (1989). Gating of the afferent volley of the monosynaptic stretch reflex during movement in man. *The Journal of Physiology*, *419*(1), 753–763. Available from https://doi.org/10.1113/jphysiol.1989.sp017896.

Miall, R. C., & Wolpert, D. M. (1996). Forward models for physiological motor control. *Neural Networks*, *9*(8), 1265–1279. Available from https://doi.org/10.1016/S0893-6080(96)00035-4.

Morgan, D. L. (1977). Separation of active and passive components of short-range stiffness of muscle. *American Journal of Physiology-Cell Physiology*, *232*(1), 45–49.

Nashed, J. Y., Crevecoeur, F., & Scott, S. H. (2012). Influence of the behavioral goal and environmental obstacles on rapid feedback responses. *Journal of Neurophysiology*, *108*(4), 999–1009. Available from https://doi.org/10.1152/jn.01089.2011.

Nelson, R. J. (1987). Activity of monkey primary somatosensory cortical neurons changes prior to active movement. *Brain Research*, *406*(1–2), 402–407. Available from https://doi.org/10.1016/0006-8993(87)90815-8.

Nichols, T. R. (2018). Distributed force feedback in the spinal cord and the regulation of limb mechanics. *Journal of Neurophysiology*, *119*(3), 1186–1200. Available from https://doi.org/10.1152/jn.00216.2017.

Nichols, T. R., & Cope, T. C. (2004). Cross-bridge mechanisms underlying the history-dependent properties of muscle spindles and stretch reflexes. *Canadian Journal of Physiology and Pharmacology*, *82*(8–9), 569–576. Available from https://doi.org/10.1139/y04-074.

Nichols, T. R., & Houk, J. C. (1976). Improvement in linearity and regulation of stiffness that results from actions of stretch reflex. *Journal of Neurophysiology*, *39*(1), 119–142. Available from https://doi.org/10.1152/jn.1976.39.1.119.

Omrani, M., Diedrichsen, J., & Scott, S. H. (2013). Rapid feedback corrections during a bimanual postural task. *Journal of Neurophysiology*, *109*(1), 147–161. Available from https://doi.org/10.1152/jn.00669.2011.

Omrani, M., Pruszynski, J. A., Murnaghan, C. D., & Scott, S. H. (2014). Perturbation-evoked responses in primary motor cortex are modulated by behavioral context. *Journal of Neurophysiology*, *112*(11), 2985–3000. Available from https://doi.org/10.1152/jn.00270.2014.

Optican, L. M., & Robinson, D. A. (1980). Cerebellar-dependent adaptive control of primate saccadic system. *Journal of Neurophysiology*, *44*(6), 1058–1076. Available from https://doi.org/10.1152/jn.1980.44.6.1058.

Osborn, C. E., & Poppele, R. E. (1992). Parallel distributed network characteristics of the DSCT. *Journal of Neurophysiology.*, *68*(4), 1100–1112. Available from https://doi.org/10.1152/jn.1992.68.4.1100.

Oztop, E., & Kawato, M. (2009). Models for the control of grasping. In D. A. Nowak, & J. Hermsdorfer (Eds.), *Sensorimotor control of grasping* (pp. 110–124). Cambridge University Press. Available from https://doi.org/10.1017/CBO9780511581267.010.

Padberg, J., Cerkevich, C., Engle, J., Rajan, A. T., Recanzone, G., Kaas, J., & Krubitzer, L. (2009). Thalamocortical connections of parietal somatosensory cortical fields in macaque monkeys are highly divergent and convergent. *Cerebral Cortex*, *19*(9), 2038–2064. Available from https://doi.org/10.1093/cercor/bhn229.

Padberg, J., Cooke, D. F., Cerkevich, C. M., Kaas, J. H., & Krubitzer, L. (2018). Cortical connections of area 2 and posterior parietal area 5 in macaque monkeys. *The Journal of Comparative Neurology*, *3*(15). Available from https://doi.org/10.1002/cne.24453, 299–20.

Palmeri, A., Bellomo, M., Giuffrida, R., & Sapienza, S. (1999). Motor cortex modulation of exteroceptive information at bulbar and thalamic lemniscal relays in the cat. *Neuroscience*, *88*(1), 135–150.

Pavlides, C., Miyashita, E., & Asanuma, H. (1993). Projection from the sensory to the motor cortex is important in learning motor skills in the monkey. *Journal of Neurophysiology.*, *70* (2), 733–741. Available from https://doi.org/10.1152/jn.1993.70.2.733.

Pettit, M. J., & Schwark, H. D. (1993). Receptive field reorganization in dorsal column nuclei during temporary denervation. *Science*, *262*(5142), 2054–2056. Available from https://doi.org/10.1126/science.8266104.

Phillips, C. G., Powell, T. P. S., & Wiesendanger, M. (1971). Projection from low-threshold muscle afferents of hand and forearm to area 3a of baboon's cortex. *The Journal of Physiology*, *217*(2), 419–446. Available from https://doi.org/10.1113/jphysiol.1971.sp009579.

Pons, T. P., Garraghty, P. E., Cusick, C. G., & Kaas, J. H. (1985). The somatotopic organization of area 2 in macaque monkeys. *Journal of Comparative Neurology*, *241*(4), 445–466. Available from https://doi.org/10.1002/cne.902410405x.

Prochazka, A., & Ellaway, P. (2012). Sensory systems in the control of movement. In R. Terjung (Ed.), *Comprehensive physiology* (p. c100086). John Wiley & Sons, Inc.. Available from https://doi.org/10.1002/cphy.c100086.

Prochazka, A., & Gorassini, M. (1998a). Models of ensemble firing of muscle spindle afferents recorded during normal locomotion in cats. *Journal of Physiology*..

Prochazka, A., & Gorassini, M. (1998b). Ensemble firing of muscle afferents recorded during normal locomotion in cats. *The Journal of Physiology*, *507*(1), 293–304.

Prochazka, A., Hulliger, M., Zangger, P., & Appenteng, K. (1985). 'Fusimotor set': New evidence for α-independent control of γ-motoneurones during movement in the awake cat. *Brain Research*, *339*(1), 136–140. Available from https://doi.org/10.1016/0006-8993(85)90632-8.

Prochazka, A., Stephens, J. A., & Wand, P. (1979). Muscle spindle discharge in normal and obstructed movements. *The Journal of Physiology*, *287*(1), 57–66. Available from https://doi.org/10.1113/jphysiol.1979.sp012645.

Proske, U., & Gandevia, S. C. (2012). The proprioceptive senses: Their roles in signaling body shape, body position and movement, and muscle force. *Physiological Reviews*, *92*(4), 1651–1697. Available from https://doi.org/10.1152/physrev.00048.2011.

Proske, U., & Gregory, J. E. (1977). The time-course of recovery of the initial burst of primary endings of muscle spindles. *Brain Research*, *121*(2), 358–361. Available from https://doi.org/10.1016/0006-8993(77)90159-7.

Proske, U., Morgan, D. L., & Gregory, J. E. (1992). Muscle history dependence of responses to stretch of primary and secondary endings of cat soleus muscle spindles. *Journal of Physiology*, *445*, 81–95.

Proske, U., Morgan, D. L., & Gregory, J. E. (1993). Thixotropy in skeletal muscle and in muscle spindles: A review. *Progress in Neurobiology*, *41*(6), 705–721.

Prud'homme, M. J., & Kalaska, J. F. (1994). Proprioceptive activity in primate primary somatosensory cortex during active arm reaching movements. *Journal of Neurophysiology*, *72*(5), 2280–2301. Available from https://doi.org/10.1152/jn.1994.72.5.2280.

Pruszynski, J. A., Kurtzer, I., Nashed, J. Y., Omrani, M., Brouwer, B., & Scott, S. H. (2011). Primary motor cortex underlies multi-joint integration for fast feedback control. *Nature*, *478*(7369), 387–390. Available from https://doi.org/10.1038/nature10436.

Rathelot, J. A., & Strick, P. L. (2006). Muscle representation in the macaque motor cortex: An anatomical perspective. *Proceedings of the National Academy of Sciences*, *103*(21), 8257–8262. Available from https://doi.org/10.1073/pnas.0602933103.

Rathelot, J.-A., & Strick, P. L. (2009). Subdivisions of primary motor cortex based on cortico-motoneuronal cells. *Proceedings of the National Academy of Sciences*, *106*(3), 918–923. Available from https://doi.org/10.1073/pnas.0808362106.

Rincon-Gonzalez, L., Warren, J. P., Meller, D. M., & Tillery, S. H. (2011). Haptic interaction of touch and proprioception: Implications for neuroprosthetics. *IEEE Transactions on Neural Systems and Rehabilitation Engineering*, *19*(5), 490–500.

Ross, K. T., & Nichols, T. R. (2009). Heterogenic feedback between hindlimb extensors in the spontaneously locomoting premammillary cat. *Journal of Neurophysiology*, *101*(1), 184–197. Available from https://doi.org/10.1152/jn.90338.2008.

Rossi, A., & Grigg, P. (1982). Characteristics of hip joint mechanoreceptors in the cat. *Journal of Neurophysiology*, *47*(6), 1029–1042. Available from https://doi.org/10.1152/jn.1982.47.6.1029.

Rothwell, J. C., Gandevia, S. C., & Burke, D. (1990). Activation of fusimotor neurones by motor cortical stimulation in human subjects. *The Journal of Physiology*, *431*(1), 743–756. Available from https://doi.org/10.1113/jphysiol.1990.sp018357.

Ruffini, A. (1898). On the minute anatomy of the neuromuscular spindles of the cat, and on their physiological significance. *Journal of Physiology*, *23*(3), 190–208, 3.

Rushton, D. N., Rothwell, J. C., & Craggs, M. D. (1981). Gating of somatosensory evoked potentials during different kinds of movement in man. *Brain*, *104*, 465–491. Available from https://academic.oup.com/brain/article-abstract/104/3/465/296781.

Saal, H. P., Delhaye, B. P., Rayhaun, B. C., & Bensmaia, S. J. (2017). Simulating tactile signals from the whole hand with millisecond precision. *Proceedings of the National Academy of Sciences*, *114*(28), E5693–E5702. Available from https://doi.org/10.1073/pnas.1704856114.

Safavynia, S. A., & Ting, L. H. (2013). Sensorimotor feedback based on task-relevant error robustly predicts temporal recruitment and multidirectional tuning of muscle synergies. *Journal of Neurophysiology*, *109*(1), 31–45. Available from https://doi.org/10.1152/jn.00684.2012.

Sainburg, R. L., Ghilardi, M. F., Poizner, H., & Ghez, C. (1995). Control of limb dynamics in normal subjects and patients without proprioception. *Journal of Neurophysiology*, *73*(2), 820–835. Available from https://doi.org/10.1152/jn.1995.73.2.820.

Sainburg, R. L., Poizner, H., & Ghez, C. (1993). Loss of proprioception produces deficits in interjoint coordination. *Journal of Neurophysiology*, *70*(5), 2136–2147. Available from https://doi.org/10.1152/jn.1993.70.5.2136.

Sakamoto, T., Arissian, K., & Asanuma, H. (1989). Functional role of the sensory cortex in learning motor skills in cats. *Brain Research*, *503*(2), 258–264. Available from https://doi.org/10.1016/0006-8993(89)91672-7.

Schäfer, S.-S., & Schäfer, S. (1969). Die Eigenschaften einer primären Muskelspindelafferenz bei rampenförmiger Dehnung und ihre mathematische Beschreibung. *Pflügers Archiv*, *310*(3), 206–228. Available from https://doi.org/10.1007/BF00587210.

Schieppati, M., & Nardone, A. (1997). Medium-latency stretch reflexes of foot and leg muscles analysed by cooling the lower limb in standing humans. *The Journal of Physiology*, *503*(3), 691–698. Available from https://doi.org/10.1111/j.1469-7793.1997.691bg.x.

Schmidt, R. F., Schady, W. J. L., & Torebjörk, H. E. (1990). Gating of tactile input from the hand—I. Effects of finger movement. *Experimental Brain Research*, *79*(1), 97–102. Available from https://doi.org/10.1007/BF00228877.

Schwarz, D., Deecke, L., & Fredrickson, J. (1973). Cortical projection of group I muscle afferents to areas 2, 3a, and the vestibular field in the rhesus monkey. *Experimental Brain Research*, *17*(5), 516–526.

Scott, S. H., & Loeb, G. E. (1994). The computation of position sense from spindles in mono- and multiarticular muscles. *The Journal of Neuroscience*, *14*(12), 7529–7540.

Seelke, A. M. H., Padberg, J. J., Disbrow, E., Purnell, S. M., Recanzone, G., & Krubitzer, L. (2011). Topographic maps within brodmann's area 5 of macaque monkeys. *Cerebral Cortex*, *22*(8), 1834–1850. Available from https://doi.org/10.1093/cercor/bhr257.

Sengul, G., & Watson, C. (2012). *Spinal cord: Connections. The human nervous system*. Academic Press.

Shadmehr, R. (2020). Population coding in the cerebellum: A machine learning perspective. *Journal of Neurophysiology*, *124*(6), 2022–2051. Available from https://doi.org/10.1152/jn.00449.2020.

Shadmehr, R., & Krakauer, J. W. (2008). A computational neuroanatomy for motor control. In. *Experimental Brain Research, 185*. Available from https://doi.org/10.1007/s00221-008-1280-5.

Shadmehr, R., & Mussa-Ivaldi, F. A. (1994). Adaptive representation of dynamics during learning of a motor task. *Journal of Neuroscience, 14*(5), 3208–3224. Available from https://doi.org/10.1523/JNEUROSCI.14-05-03208.1994.

Sherrington, C. (1906). *The integrative action of the nervous system*. Yale University Press.

Stuart, D. G., Mosher, C. G., Gerlach, R. L., & Reinking, R. M. (1972). Mechanical arrangement and transducing properties of Golgi tendon organs. *Experimental Brain Research, 14*(3), 274–292. Available from https://doi.org/10.1007/BF00816163.

Takemura, N., Fukui, T., & Inui, T. (2015). A computational model for aperture control in reach-to-grasp movement based on predictive variability. *Frontiers in Computational Neuroscience, 9*, 9. Available from https://doi.org/10.3389/fncom.2015.00143.

Tanji, J. (1975). Activity of neurons in cortical area 3a during maintenance of steady postures by the monkey. *Brain Research, 88*(3), 549–553.

Taylor, A., Durbaba, R., Ellaway, P. H., & Rawlinson, S. (2000). Patterns of fusimotor activity during locomotion in the decerebrate cat deduced from recordings from hindlimb muscle spindles. *The Journal of Physiology, 522*(3), 515–532.

Taylor, A., Durbaba, R., Ellaway, P. H., & Rawlinson, S. (2006). Static and dynamic γ-motor output to ankle flexor muscles during locomotion in the decerebrate cat. *The Journal of Physiology, 571*(3), 711–723.

Taylor, A., Ellaway, P. H., & Durbaba, R. (1998). Physiological signs of the activation of bag2 and chain intrafusal muscle fibers of gastrocnemius muscle spindles in the cat. *Journal of Neurophysiology, 80*(1), 130–142.

Taylor, A., Ellaway, P. H., & Durbaba, R. (1999). Why are there three types of intrafusal muscle fibers? *Progress in Brain Research, 123*, 121–131.

Thornell, L.-E., Carlsson, L., Eriksson, P.-O., Liu, J.-X., Österlund, C., Stål, P., & Pedrosa-Domellöf, F. (2015). Fibre typing of intrafusal fibres. *Journal of Anatomy, 227*(2), 136–156. Available from https://doi.org/10.1111/joa.12338.

Toft, E., Sinkjaer, T., & Espersen, G. T. (1989). Quantitation of the stretch reflex. *Acta Neurologica Scandinavica, 79*(5), 384–390. Available from https://doi.org/10.1111/j.1600-0404.1989.tb03805.x.

Tseng, Y., Diedrichsen, J., Krakauer, J. W., Shadmehr, R., & Bastian, A. J. (2007). Sensory prediction errors drive cerebellum-dependent adaptation of reaching. *Journal of Neurophysiology, 98*(1), 54–62. Available from https://doi.org/10.1152/jn.00266.2007.

Vallbo, Å. B. (1970). Discharge patterns in human muscle spindle afferents during isometric voluntary contractions. *Acta Physiologica Scandinavica, 80*(4), 552–566. Available from https://doi.org/10.1111/j.1748-1716.1970.tb04823.x.

Vallbo, A. B. (1974). Human muscle spindle discharge during isometric voluntary contractions. Amplitude relations between spindle frequency and torque. *Acta Physiologica Scandinavica, 90*(2), 319–336.

Vallbo, A. B., & Johansson, R. S. (1984). Properties of cutaneous mechanoreceptors in the human hand related to touch sensation. *Human Neurobiology, 3*(1), 3–14.

Vidoni, E. D., Acerra, N. E., Dao, E., Meehan, S. K., & Boyd, L. A. (2010). Role of the primary somatosensory cortex in motor learning: An rTMS study. *Neurobiology of Learning and Memory, 93*(4), 532–539. Available from https://doi.org/10.1016/j.nlm.2010.01.011.

Vincent, J. A., Gabriel, H. M., Deardorff, A. S., Nardelli, P., Fyffe, R. E. W., Burkholder, T., & Cope, T. C. (2017). Muscle proprioceptors in adult rat: Mechanosensory signaling and synapse distribution in spinal cord. *Journal of Neurophysiology, 118*(5), 2687–2701. Available from https://doi.org/10.1152/jn.00497.2017.

Warren, J. P., Santello, M., & Tillery, S. I. H. (2011). Effects of fusion between tactile and proprioceptive inputs on tactile perception. *PLOS One, 6*(3), e18073. Available from https://doi.org/10.1371/journal.pone.0018073.

Weber, D. J., London, B. M., Hokanson, J. A., Ayers, C. A., Gaunt, R. A., Torres, R. R., Zaaimi, B., & Miller, L. E. (2011). Limb-state information encoded by peripheral and central somatosensory neurons: Implications for an afferent interface. *IEEE Transactions on Neural Systems and Rehabilitation Engineering, 19*(5), 501–513. Available from https://doi.org/10.1109/TNSRE.2011.2163145.

Welch, T. D. J., & Ting, L. H. (2008). A feedback model reproduces muscle activity during human postural responses to support-surface translations. *Journal of Neurophysiology, 99*(2), 1032–1038. Available from https://doi.org/10.1152/jn.01110.2007.

Welch, T. D. J., & Ting, L. H. (2009). A feedback model explains the differential scaling of human postural responses to perturbation acceleration and velocity. *Journal of Neurophysiology, 101*(6), 3294–3309. Available from https://doi.org/10.1152/jn.90775.2008.

Wise, S. P., & Tanji, J. (1981). Neuronal responses in sensorimotor cortex to ramp displacements and maintained positions imposed on hindlimb of the unanesthetized monkey. *Journal of Neurophysiology, 45*(3), 482–500. Available from https://doi.org/10.1152/jn.1981.45.3.482.

Witham, C. L., & Baker, S. N. (2011). Modulation and transmission of peripheral inputs in monkey cuneate and external cuneate nuclei. *Journal of Neurophysiology, 106*(5), 2764–2775. Available from https://doi.org/10.1152/jn.00449.2011, 21865437.

Wolpaw, J. R. (1980). Correlations between task-related activity and responses to perturbation in primate sensorimotor cortex. *Journal of Neurophysiology, 44*(6), 1122–1138. Available from https://doi.org/10.1152/jn.1980.44.6.1122.

Wolpaw, J. R., & Colburn, T. R. (1978). Electromagnetic stretch of individual muscles in behaving primates. *Brain Research, 141*(1), 193–196. Available from https://doi.org/10.1016/0006-8993(78)90630-3.

Wolpert, D. M., Ghahramani, Z., & Jordan, M. I. (1995). An internal model for sensorimotor integration. *Science, 269*(5232), 1880–1882. Available from https://doi.org/10.1126/science.7569931.

Yamada, H., Yaguchi, H., Tomatsu, S., Takei, T., Oya, T., & Seki, K. (2016). Representation of afferent signals from forearm muscle and cutaneous nerves in the primary somatosensory cortex of the macaque monkey. *PLOS One, 11*(10). Available from https://doi.org/10.1371/journal.pone.0163948, e0163948-16.

Yau, J. M., Connor, C. E., & Hsiao, S. S. (2013). Representation of tactile curvature in macaque somatosensory area 2. *Journal of Neurophysiology, 109*(12), 2999–3012. Available from https://doi.org/10.1152/jn.00804.2012.

Yau, J. M., Kim, S. S., Thakur, P. H., & Bensmaia, S. J. (2016). Feeling form: The neural basis of haptic shape perception. *Journal of Neurophysiology., 115*(2), 631–642. Available from https://doi.org/10.1152/jn.00598.2015.

Yıldız, M. Z., Toker, İ., Özkan, F. B., & Güçlü, B. (2015). Effects of passive and active movement on vibrotactile detection thresholds of the Pacinian channel and forward masking. *Somatosensory & Motor Research, 32*(4), 262–272. Available from https://doi.org/10.3109/08990220.2015.1091771.

Yumiya, H., Kubota, K., & Asanuma, H. (1974). Activities of neurons in area 3a of the cerebral cortex during voluntary movements in the monkey. *Brain Research, 78*(2), 169–177. Available from https://doi.org/10.1016/0006-8993(74)90544-7.

Chapter 3

Electrodes and instrumentation for neurostimulation

Daniel R. Merrill

Dan Merrill Consulting, LLC, Salt Lake City, UT, United States

ABSTRACT

The requirements for electrical stimulation and recording of excitable tissue are listed and analyzed within the overarching context of design for safety and efficacy. The physical basis for electrical stimulation and recording is presented, and charge transfer mechanisms are presented and contrasted. Electrical models of the electrode–tissue interface for stimulation and recording are given. Principles of stimulation of excitable tissue are described, and the mechanisms of damage to tissue and the electrode are reviewed. Design compromises for safety and efficacy are discussed, commonly used and emerging electrode materials are reviewed, and the requirements and design considerations for instrumentation are discussed. Finally, the concept of impedance is properly defined and how it is commonly used in bioengineering is discussed.

Keywords: Electrode; stimulation; recording; safety; efficacy; charge transfer; electrode model; impedance

3.1 Two fundamental requirements

Any medical device, drug, or biologic must meet two fundamental requirements: (1) safety and (2) efficacy/effectiveness. These requirements often act in opposition; for example, as the dose of injected charge from a neurostimulator increases, the effectiveness of action potential generation may increase, but this occurs at the expense of decreased safety if tissue is damaged due to the generation of toxic electrochemical products or biologic effects of overstimulation. For designers of neurostimulation systems it is a routine exercise to determine appropriate compromises which provide both safety and efficacy. This chapter will be presented from the perspective of designing to meet these two requirements.

Somatosensory Feedback for Neuroprosthetics. DOI: https://doi.org/10.1016/B978-0-12-822828-9.00002-2

3.2 Recording and stimulating

The function of any instrumentation system can broadly be classified as either sensing (observing) or controlling; sometimes colloquially called "peek and poke." In neural engineering we refer to recording and stimulation, respectively.

Electrodes form the terminal devices in a neuroprosthetic system, that is they are the system element which interfaces between the instrumentation electronics and tissue. For recording purposes, electrodes are used to detect the electric potential at some point on or in the body. The potential is usually recorded as a dynamic signal, that is, we are interested in how the electric potential changes in time. It can also be measured statically on an absolute or thermodynamic scale, although this properly requires a stable reference electrode (RE) and is less common. For stimulation purposes, electrodes are used to change the electric field within tissue to excite or inhibit neural firing. Recording and stimulating can be combined in a neuroprosthetic system to enable closed-loop control.

3.3 Requirements for efficacy and safety of a stimulating device

Table 3.1 lists the primary points of consideration regarding design of a neural stimulating system. This is meant as a high-level list and brief introduction. Individual topics will be expanded upon in the following sections. The major points below are divided into the topics of (1) efficacy/effectiveness, (2) safety, (3) performance, and (4) reliability.

Working definitions for these terms are as follows. *Efficacy* (from "efficacious") and *effectiveness* (from "effective") refer to producing the intended objective or results. Efficacy is determined from a randomized controlled trial (RCT) with a prescribed sample set of subjects. Effectiveness is determined from within the general population of patients. It is a goal of clinical study design to have a subject sample set which well represents the general population so that efficacy is close to effectiveness, but this ideal is rarely met. The underlying reasons are an important subject, beyond the scope of this chapter. *Performance* is a subset of effectiveness and refers to meeting quantitative functional specifications, in particular as claimed in the device labeling. *Safety* refers to avoiding unacceptable levels of damage to the subject or patient, or to the system. Some threshold for damage may be acceptable as determined on a case-by-case basis. The term "safe" should be used cautiously from a regulatory perspective. *Reliability* refers to maintaining requirements for effectiveness and safety over the intended lifetime of a system.

TABLE 3.1 Primary considerations in the design of a neural stimulation system, based on (1) efficacy/effectiveness, (2) safety, (3) performance, and (4) reliability. The terms "sensitive" and "specific," properly defined for a recording system, are abstracted here.

Efficacy/effectiveness
E1. The system is effective. This requires that sufficient charge can be injected with the chosen electrode material and surface area to elicit action potentials or other desired physiological response
Safety
S1. The system delivers a small enough charge so that the electrode overpotential does not cause excessive Faradaic reactions. The appropriate metrics generally associated with an electrode are the reversible charge storage capacity (CSC) and the reversible charge injection capacity (CIC)
S1a. Faradaic reactions do not occur at levels that damage tissue (generally but not always from Faradaic reduction processes). The level of reaction product that is tolerated may be higher for acute stimulation than chronic stimulation
S1b. Faradaic corrosion reactions do not occur at levels that will cause premature failure of the electrode (generally from Faradaic oxidation processes). This again depends greatly on the intended duration of use. During acute stimulation, corrosion is rarely a concern, whereas a device that is intended for a 30-year implant must have a very low corrosion rate
S2. Tissue is not damaged by intrinsic biological processes due to overstimulation
S3. Tissue is not damaged from excessive mechanical forces
S4. The passive (unstimulated) device materials in contact with tissue are biocompatible, defined bidirectionally:
S4a. The tissue is not damaged by the device. The device does not induce a toxic or necrotic response, nor an excessive foreign body or immune response
S4b. The device is not damaged by the tissue
S5. Mechanical moduli of the system are compatible with tissue
S6. The implanted device is minimally invasive. As a minimum, invasiveness is justified by the benefit
S7. Requirements for safety are met as specified in standards including IEC 60601-1 and its collateral and particular standards, and ISO 14708-1 and -3
S8. No galvanic cells are formed by dissimilar metals which drive consequential currents
Performance
P1. All specifications as claimed in the device labeling are met
P2. Electrodes are relatively low impedance, thus decreasing load on the source

(*Continued*)

TABLE 3.1 (Continued)

P3. The overall system is "sensitive," broadly defined as able to affect physiological response with minimum charge injection
P4. The overall system is "specific," broadly defined as able to effect physiological response in a specific region while not affecting unintended regions. The placement and protocol facilitate minimum charge injection so the desired response occurs without inducing undesired responses
Reliability
R1. The system is mechanically acceptable for the application
R1a. Implanted devices tolerate the insertion procedure, for example, the material does not buckle if it passes through the meninges
R1b. If a device is to be used chronically, it is flexible and fatigue-resistant to withstand movement between the device and tissue following implantation
R2. The material characteristics of the system are acceptably stable for the duration of the implant
R2a. Mechanical properties remain intact given the intended tissue, surgical procedure, and duration of use
R2b. Electrode electrical impedance is stable
R2c. Conducting and insulating properties of the leads remain intact
R3. Materials are corrosion resistant
R4. The packaging maintains hermeticity to prevent water ingress and damage to the electronics

3.4 Electrical model of stimulation: the electrode–tissue interface

This section is adapted from a previously published chapter (Merrill, 2010).

3.4.1 Physical basis of the electrode–tissue interface

When a metal electrode is placed inside a physiological medium such as extracellular fluid (ECF), an interface is formed between the two phases. In the metal electrode phase and in attached electrical circuits, charge is carried by electrons. In the physiological medium, or in more general electrochemical terms the electrolyte, charge is carried by ions, including sodium, potassium, and chloride in the ECF. The central process that occurs at the electrode–electrolyte interface is a transduction of charge carriers from electrons in the metal electrode to ions in the electrolyte.

In the simplest system, two electrodes are placed in an electrolyte and electrical current may pass between the electrodes through the electrolyte.

One of the two electrodes is termed a working electrode (WE), and the second is termed a counter electrode (CE). The WE is defined as the electrode that one is interested in studying, with the CE being necessary to complete the circuit for charge conduction. In electrochemistry experiments it is common to use a third electrode termed the reference electrode (RE), which defines a reference for electrical potential measurements.

A change in electrical potential occurs upon crossing from one conducting phase to another (from the metal electrode to the electrolyte) at the interface itself, in a very narrow interphase region (at most a few hundred angstroms in width). The basis for this is described in more detail in Section 3.4.5. The change or gradient in electrical potential corresponds to an electric field, measured in volts/meter, at the interface. This gradient exists even in the equilibrium condition when there is no current flow. Electrochemical reactions may occur in this interphase region if the electrical potential profile is forced away from the equilibrium condition. In the absence of current, the electrical potential is constant throughout the electrolyte beyond the narrow interphase region. During current flow, a potential gradient exists in the electrolyte, generally many orders of magnitude smaller than at the interface.

There are two primary mechanisms of charge transfer at the electrode–electrolyte interface, as illustrated in Fig. 3.1. One is a non-Faradaic reaction, where no electrons are transferred between the electrode and electrolyte. Non-Faradaic reactions include redistribution of charged chemical species in the electrolyte. The second mechanism is a Faradaic reaction, in which electrons are transferred between the electrode and electrolyte,

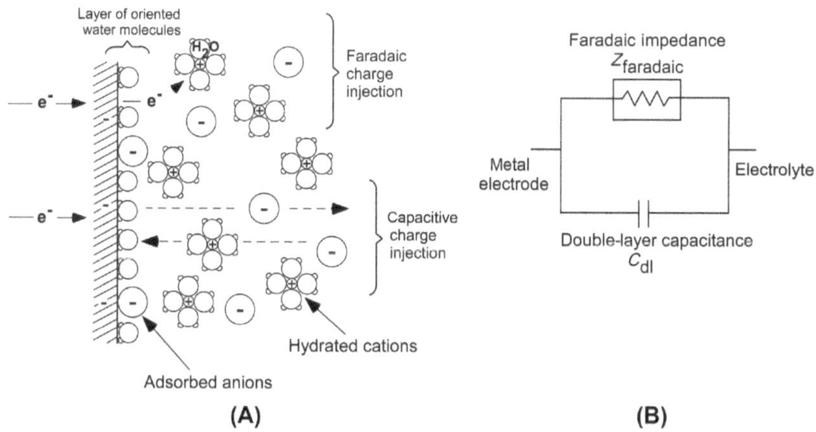

FIGURE 3.1 The electrode–electrolyte interface, illustrating Faradaic charge transfer (top) and capacitive redistribution of charge (bottom) as the electrode is driven negative. (A) Physical representation; (B) two-element electrical circuit model for mechanisms of charge transfer at the interface.

resulting in the reduction or oxidation of chemical species in the electrolyte. Faradaic reactions are further divided into reversible and nonreversible Faradaic reactions, which are detailed in Section 3.4.4. Reversible Faradaic reactions include those where the products either remain bound to the electrode surface, or do not diffuse far away from the electrode. In an irreversible Faradaic reaction, the products diffuse away from the electrode.

3.4.2 Capacitive/non-Faradaic charge transfer

If only non-Faradaic redistribution of charge occurs, the electrode–electrolyte interface may be modeled as a simple electrical capacitor called the double-layer capacitor C_{dl}. This capacitor is formed due to several physical phenomena (Chapman, 1913; Grahame, 1947; Guoy, 1910; Stern, 1924; von Helmholtz, 1853). First, when a metal electrode is placed in an electrolyte, charge redistribution occurs as metal ions in the electrolyte combine with the electrode. This involves a transient transfer of electrons between the two phases, resulting in a plane of charge at the surface of the metal electrode, opposed by a plane of opposite charge, as counterions, in the electrolyte. The excess charge on the electrode surface, symbolized by q^M or σ^M, takes the form of an excess or deficiency of electrons and is present on a very thin layer (< 0.1 Å thick) at the surface. In the electrolyte counterions take the form of excess cations or anions, symbolized by q^S. If q^M is an excess of electrons, then q^S is an excess of cations, and if q^M is a deficiency of electrons, then q^S is an excess of anions; that is, net electroneutrality is maintained and $q^M = -q^S$. A second reason for formation of the double layer is that some chemical species such as halide anions may specifically adsorb to the solid electrode, acting to separate charge. A third reason is that polar molecules such as water may have a preferential orientation at the interface, and the net orientation of polar molecules separates charge.

If the net charge on the metal electrode is forced to vary (as occurs with charge injection during stimulation), a redistribution of charge occurs in the solution. Consider two metal electrodes immersed in an electrolytic salt solution. A voltage source is applied across the two electrodes so that one electrode is driven to a relatively negative potential and the other to a relatively positive potential. At the electrode interface that is driven negative, the metal electrode has an excess of negative charge (Fig. 3.1). This will attract positive charge (cations) in solution toward the electrode and repel negative charge (anions). In the interfacial region, there will be net electroneutrality, because the negative charge excess on the electrode surface will equal the positive charge in solution near the interface. The bulk solution will also have net electroneutrality. At the second electrode (not shown in Fig. 3.1) the opposite processes occur, that is, the repulsion of anions by the negative electrode is countered by attraction of anions at the positive electrode. If the total amount of charge delivered is sufficiently small, only charge redistribution occurs, there is no transfer of electrons across the interface, and the interface is well modeled as a simple capacitor. *If the polarity of the applied voltage source is then reversed, the direction of current is reversed, the charge redistribution is reversed, and charge that was injected from the electrode into the electrolyte and stored by the capacitor may be recovered.*

3.4.3 Faradaic charge transfer and the electrical model of the electrode–electrolyte interface

Charge may also be injected from the electrode to the electrolyte by Faradaic processes of reduction and oxidation, whereby electrons are transferred between the two phases. Reduction, which requires the addition of an electron, occurs at the electrode that is driven negative, while oxidation, requiring the removal of an electron, occurs at the electrode that is driven positive. Faradaic charge injection results in the creation of chemical species, which may either go into the solution or remain bound to the electrode surface. *Unlike the capacitive charge injection mechanism, if these Faradaic reaction products diffuse sufficiently far away from the electrode, they cannot be recovered upon reversing the direction of current.* Fig. 3.1B illustrates a simple electrical circuit model of the electrode–electrolyte interface, consisting of two elements (Bard & Faulkner, 1980; Gileadi et al., 1975; Randles, 1947). C_{dl} is the double-layer capacitance, representing the ability of the electrode to cause charge flow in the electrolyte without electron transfer. $Z_{faradaic}$ is the Faradaic impedance, representing the Faradaic processes of reduction and oxidation where electron transfer occurs between the electrode and electrolyte. One may generally think of the capacitance as representing charge storage, and the Faradaic impedance as representing charge dissipation.

The following are examples of Faradaic electrode reactions. Cathodic processes, defined as those where a reduction of species in the electrolyte occurs as electrons are transferred from the electrode to the electrolyte, include such reactions as:

$$2\,H_2O + 2\,e^- \rightarrow H_2\uparrow + 2\,OH^- \quad \text{reduction of water} \tag{3.1}$$

$$Fe^{3+} + e^- \leftarrow\rightarrow Fe^{2+} \quad \text{simple electron transfer} \tag{3.2}$$

$$Cu^{2+} + 2\,e^- \leftarrow\rightarrow Cu \quad \text{metal deposition} \tag{3.3}$$

$$PtO + 2\,H^+ + 2\,e^- \leftarrow\rightarrow Pt + H_2O \quad \text{oxide formation and reduction} \tag{3.4}$$

$$IrO + 2\,H^+ + 2\,e^- \leftarrow\rightarrow Ir + H_2O \quad \text{oxide formation and reduction} \tag{3.5a}$$

$$IrO_2 + 4\,H^+ + 4\,e^- \leftarrow\rightarrow Ir + 2\,H_2O \quad \text{oxide formation and reduction} \tag{3.5b}$$

$$2\,IrO_2 + 2\,H^+ + 2\,e^- \leftarrow\rightarrow Ir_2O_3 + H_2O \quad \text{oxide formation and reduction} \tag{3.5c}$$

$$Pt + H^+ + e^- \leftarrow\rightarrow Pt - H \quad \text{hydrogen atom plating} \tag{3.6}$$

$$\begin{gathered} M^{(n+1)+}(OH)_{(n+1)} + H^+ + e^- \leftarrow\rightarrow M^{n+}(OH)_n + H_2O \\ \text{valency changes within an oxide} \end{gathered} \tag{3.7}$$

$$Ag^+ + e^- \leftarrow\rightarrow Ag \quad \text{reduction of silver ions} \tag{3.8a}$$

$$AgCl \leftarrow\rightarrow Ag^+ + Cl^- \quad \text{dissolution of silver chloride} \tag{3.8b}$$

Anodic processes, defined as those where oxidation of species in the electrolyte occurs as electrons are transferred to the electrode, include:

$$2\,H_2O \rightarrow O_2\uparrow + 4\,H^+ + 4\,e^- \quad \text{oxidation of water} \tag{3.9}$$

$$Pt + 4\,Cl^- \rightarrow [PtCl_4]^{2-} + 2\,e^- \quad \text{corrosion} \tag{3.10}$$

$$2\,Cl^- \rightarrow Cl_2\uparrow + 2\,e^- \quad \text{gas evolution} \tag{3.11}$$

$$Fe \rightarrow Fe^{2+} + 2\,e^- \rightarrow \text{anodic dissolution} \tag{3.12}$$

$$2\,Ag + 2\,OH^- \leftarrow\rightarrow Ag_2O + H_2O + 2\,e^- \quad \text{oxide formation} \tag{3.13}$$

Reaction 3.1 is the irreversible reduction of water (which is typically abundant as a solvent at 55.5 M), forming hydrogen gas and hydroxyl ions. The formation of hydroxyl raises the pH of the solution. Reversible reactions where species remain bound or close to the electrode surface are demonstrated by reactions (3.2)–(3.8). In reaction (3.2), the electrolyte consists of ferric and ferrous ions. By driving the metal electrode to more negative potentials, electrons are transferred to the ferric ions, forming ferrous ions. In reaction (3.3), a copper metal electrode is immersed in a solution of cuprous ions. The cuprous ions in the solution are reduced, building up the copper electrode. Reactions (3.4) and (3.5a)–(3.5c) are the reversible formation and reduction of an oxide layer on platinum and iridium, respectively. Reaction (3.6) is reversible adsorption of hydrogen onto a platinum surface, responsible for the so-called pseudocapacity of platinum. Reaction (3.7) is the general form of reversible valency changes that occur in a multilayer oxide film of iridium, ruthenium, or rhodium, with associated proton or hydroxyl ion transfer (Dautremont-Smith, 1982; Frazer & Woods, 1979; Gottesfeld, 1980; Rand & Woods, 1974). Reactions (3.8a) and (3.8b) are the reversible reactions of a silver chloride electrode driven cathodically. Silver ions in solution are reduced to solid silver on the electrode [reaction (3.8a)]. To maintain the solubility constant $K_S \equiv (a_{Ag+})(a_{Cl-})$, where a is the ionic activity, as silver ions in solution are reduced the AgCl salt covering the electrode dissolves to form silver and chloride ions in solution [reaction (3.8b)]. In reaction (3.9), water molecules are irreversibly oxidized, forming oxygen gas and hydrogen ions, and thus lowering the pH. Reaction (3.10) is the corrosion of a platinum electrode in a chloride-containing media. In reaction (3.11), chloride ions in solution are oxidized, forming chlorine gas. In reaction (3.12), an iron metal electrode is dissolved, forming ferrous ions that go into solution. Reaction (3.13) represents a reversible oxide formation on a silver electrode. As electrons are removed from the silver metal, Ag + ions are formed. These Ag + ions then combine with hydroxyl (OH^-) ions from the solution, forming an oxide layer (Ag_2O) on the surface of the silver electrode. Note the transfer of charge that occurs. As electrons are transferred to the electrode and then the external electrical circuit, the silver electrode is oxidized (Ag → Ag +). Because hydroxyl ions associate with the silver ions, the silver oxide is electroneutral. However, since hydroxyl has been removed from the solution, there is a net movement of negative charge from the electrolyte (loss of hydroxyl) to the electrode (electrons transferred to the electrode and then to the electrical circuit). The loss of hydroxyl lowers the solution pH.

3.4.4 Reversible and irreversible Faradaic reactions

There are two limiting cases that may define the net rate of a Faradaic reaction (Bard & Faulkner, 1980; Delahay, 1965; Pletcher & Walsh, 1990). At one extreme, the reaction rate is under kinetic control; at the other extreme, the reaction rate is under mass transport control. For a given metal electrode and electrolyte, there is an electrical potential called the equilibrium potential where no net current passes between the two phases. At potentials sufficiently close to equilibrium, the reaction rate is under kinetic control. Under kinetic control the rate of electron transfer at the interface is determined by the electrode potential, and is *not* limited by the rate at which reactant is delivered to the electrode surface (the reaction site). When the electrode potential is sufficiently different from equilibrium, the reaction rate is under mass transport control. In this case, all reactant that is delivered to the surface reacts immediately and the reaction rate is limited by the rate of reactant delivery to the surface.

Faradaic reactions are divided into reversible and irreversible reactions (Bard & Faulkner, 1980). The degree of reversibility depends on the relative rates of kinetics (electron transfer at the interface) and mass transport. A Faradaic reaction with very fast kinetics relative to the rate of mass transport is reversible. With fast kinetics, large currents occur with small potential excursions away from equilibrium. Since the electrochemical product does not move away from the surface extremely quickly (relative to the kinetic rate), there is an effective storage of charge near the electrode surface, and *if the direction of current is reversed then some product that has been recently formed may be reversed back into its initial (reactant) form.*

Fast Kinetics Relative to Mass Transport → Charge Storage Capacity = Reversible

In a Faradaic reaction with slow kinetics, large potential excursions away from equilibrium are required for significant currents to flow. In such a reaction the potential must be forced far from equilibrium before the mass transport rate limits the net reaction rate. In the lengthy time frame imposed by the slow electron transfer kinetics, chemical reactant is able to diffuse to the surface to support the kinetic rate, and product diffuses away quickly relative to the kinetic rate. Because product diffuses away, there is no effective storage of charge near the electrode surface, in contrast to reversible reactions. *If the direction of current is reversed, product will not be reversed back into its initial (reactant) form, since it has diffused away within the slow time frame of the reaction kinetics.* Irreversible products may include species that are soluble in the electrolyte [e.g., reaction (3.12)], precipitate in the electrolyte, or evolve as a gas [e.g., reactions (3.1), (3.9), and (3.11)]. Irreversible Faradaic reactions result in a net change in the chemical environment, potentially creating chemical species that are damaging to tissue or the electrode. Thus as a general principle, an objective of electrical stimulation design is to avoid irreversible Faradaic reactions.

Slow Kinetics → No Charge Storage Capacity (Product Diffuses Away) = Irreversible

In certain Faradaic reactions, the product remains bound to the electrode surface. Examples include hydrogen atom plating on platinum [reaction (3.6)] and oxide formation [reaction (3.13)]. These can be considered a logical extreme of slow mass transport. Since the product remains next to the electrode, such reactions are a basis for reversible charge injection.

3.4.5 The origin of electrode potentials and the three-electrode electrical model

Electrochemical potential is a parameter that defines the driving force for all chemical processes, and is the sum of a chemical potential term and an electrical potential term (Silbey & Alberty, 2001). It is defined as

$$\overline{\mu}_i^{\beta} \equiv \mu_i^{\beta} + z_i\, e\, \phi^{\beta} \tag{3.14}$$

where $\overline{\mu}_i^{\beta}$ is the electrochemical potential of particle i in phase β, μ_i^{β} is the chemical potential of particle i in phase β, and ϕ^{β} is the inner potential of the particle in phase β (the electrical potential in the bulk).

Two phases in contact are defined to be in electrochemical equilibrium when the electrochemical potential of any given chemical species is the same in each phase. If the electrochemical potentials of some species are unequal, there is a driving force for the net transfer of such species between the phases. For a metal electrode and a solution of metal ions in contact to be in equilibrium, the electrochemical potential of an electron must be the same in each phase. When two isolated phases are brought into contact, electron transfer may occur if the electrochemical potentials are unequal. Consider immersing a metal electrode into an electrolyte with a reduction/oxidation (redox) couple, for example, ferric and ferrous ions (Fe^{3+} and Fe^{2+}). Assume that while in isolation the metal electrode has a higher chemical potential for electrons than the redox couple. Upon bringing the electrode into contact with the electrolyte, the chemical potential gradient will transfer electrons from the metal to the redox couple, driving the reaction $Fe^{3+} + e^- \rightarrow Fe^{2+}$ to the right as ferric ions are reduced to ferrous ions. Upon transferring electrons, an electrical potential difference develops between the phases that repels further transfer. Equilibrium is reached when the electrostatic force equals the driving force due to a difference in chemical potentials for an electron. At equilibrium there is no net electron transfer, and a distinctive difference in inner potentials $\Delta\phi$ exists between the two phases (the inner potential ϕ is the electrical potential inside the bulk of the phase). The difference in inner potentials between a metal phase and solution phase in contact, $\Delta\phi_{\text{metal-solution}}$, defines the electrode interfacial potential.

It is an experimental limitation that a single interfacial potential cannot be measured. Whenever a measuring instrument is introduced, a new interface is created, and one is unable to separate the effects of the two interfaces. It is tempting to wonder why one cannot simply place one voltmeter probe on a metal electrode and a second voltmeter probe into the electrolyte and measure an electrode potential, as shown in Fig. 3.2. The electrode potential of interest is $\Delta\phi_{\text{electrode-electrolyte}}$. By introducing the measuring device (a metal voltmeter probe) into the electrolyte solution, a new interface is created with its own difference in potentials $\Delta\phi_{\text{electrolyte-probe}}$. It is impossible to separate the components $\Delta\phi_{\text{electrode-electrolyte}}$ and $\Delta\phi_{\text{electrolyte-probe}}$ from the measured potential. Note that if the other voltmeter probe (touching the metal electrode, not shown in Fig. 3.2) consists of a different material than the electrode, a third interface is formed, with a third difference in inner potentials.

Evaluation must be of a complete electrochemical cell, which is considered as two electrodes separated by an electrolyte. Practically, potentials are measured as complete cell potentials between two electrodes, either from the

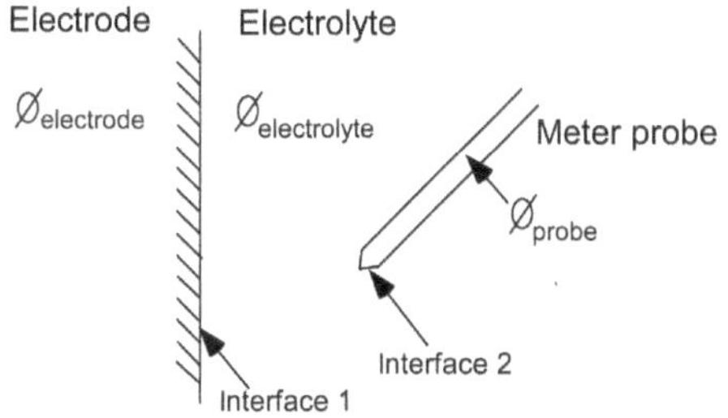

FIGURE 3.2 Voltmeter probe in an electrolyte. Introduction of the metal probe creates a second electrode–electrolyte interface.

WE to the CE or from the WE to an RE. A cell potential is the sum of two interfacial potentials (electrode$_1$ to electrolyte plus electrolyte to electrode$_2$), as well as any potential difference occurring across the electrolyte as current flows. In the absence of current, the cell potential between the WE and second (counter or reference) electrode is called the open-circuit potential (OCP), and is the sum of two equilibrium interfacial potentials from the WE to the electrolyte and from the electrolyte to the second electrode.

The term "electrode potential" is not defined consistently in the literature. Some authors define electrode potential as the potential between an electrode and a RE, and others define it as the (immeasurable) interfacial potential. For clarity and accuracy, when the term "electrode potential" is used it should be specified what this potential is with respect to, for example, the electrolyte, a RE, or another electrode.

Consider the electron transfer reaction between a metal electrode and a reduction/oxidation (redox) couple O and R in solution,

$$O + n\,e^{-} \leftarrow \rightarrow R \tag{3.15}$$

where O is the oxidized species of the couple, R is the reduced species, and n is the number of electrons transferred.

If the concentrations of both O and R in solution are equal, then the electrical potential of the redox couple equilibrates at $E^{\ominus\prime}$, defined as the formal potential. More generally, if the concentrations of O and R are unequal, the equilibrium potential or Nernst potential E_{eq} may be calculated by the Nernst equation (Bard & Faulkner, 1980; Silbey & Alberty, 2001):

$$E_{eq} = E^{\ominus\prime} + (RT/nF)\ln\{[O]/[R]\} \tag{3.16}$$

where [O] and [R] are concentrations in the bulk solution, R is the gas constant ~ 8.314 J/mole-°K, T is the absolute temperature, and F is Faraday's constant ~ 96,485 C/mole of electrons.

The Nernst Eq. (3.16) relates the equilibrium electrode potential E_{eq} (the electrical potential of the WE with respect to any convenient RE) to the bulk solution concentrations [O] and [R] when the system is in equilibrium. As the bulk concentration [O] increases or the bulk concentration [R] decreases, the equilibrium potential becomes more positive.

In a system containing only one redox couple that has fairly fast kinetics, the measured OCP is determined by the equilibriumWE potential of the redox couple. If the kinetics of the redox couple are slow, the OCP (an empirical parameter) may not quickly attain the equilibrium potential after a perturbation, and if other contaminating redox couples (affectionately known as "dirt") are present that affect the equilibrium state, the measured OCP does not readily correlate with any single redox equilibrium potential.

If one begins with a system that is in equilibrium and then forces the potential of an electrode away from its equilibrium value, for example by

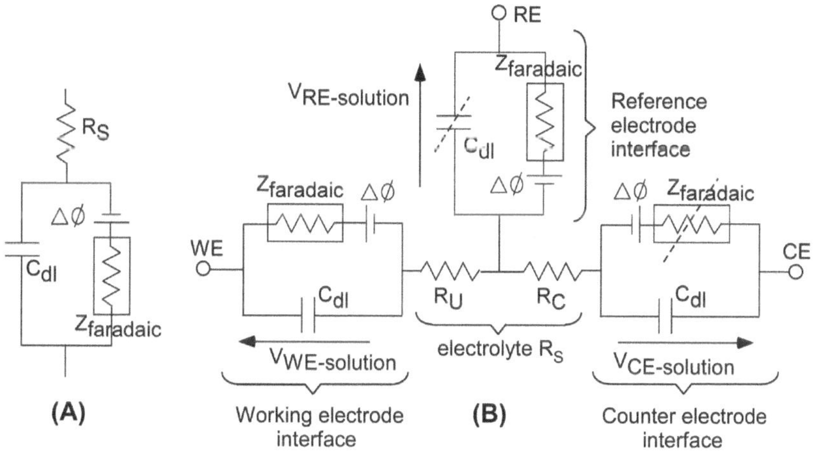

FIGURE 3.3 Electrical circuit models: (A) Single electrode–electrolyte interface; (B) three-electrode system. External access to the system is at three points labeled "WE," "CE," and "RE." If the counter electrode has a large surface area, it may be considered as strictly a capacitance as shown. A reference electrode with very low-valued Faradaic impedance will maintain the interfacial potential $V_{\text{RE-solution}}$ constant.

connecting a current source between the WE and CE, the electrode is said to become polarized. Polarization is measured by the overpotential η (eta), which is the difference between an electrode's potential and its equilibrium potential (both measured with respect to some RE):

$$\eta \equiv E - E_{\text{eq}} \tag{3.17}$$

The electrode interface model of Fig. 3.1B demonstrates the mechanisms of charge injection from an electrode; however, it neglects the equilibrium interfacial potential $\Delta\phi$ that exists across the interface at equilibrium. This is modeled as shown in Fig. 3.3A. In addition to the electrode interface, the solution resistance R_S (alternatively referred to as the access resistance R_A or the ohmic resistance R_Ω) that exists between two electrodes in solution is modeled.

An electrical potential difference, or voltage, is always defined between two points in space. During electrical stimulation, the potentials of both the working and CE may vary with respect to some third reference point. A third electrode whose potential does not change over time, known as the RE, may be used for making potential measurements. Potentials of the WE and CE may then be given with respect to the RE. An electrical circuit model of a three-electrode system, including the WE, CE, and RE immersed into an electrolyte, is shown in Fig. 3.3B.

As current is passed between the working and CE through the electrolytic solution in a two-electrode system, the interfacial potentials $V_{\text{WE-solution}}$ and $V_{\text{CE-solution}}$ will vary from their equilibrium values, that is, there are overpotentials associated with both interfaces (Fig. 3.4A). Also, as current flows

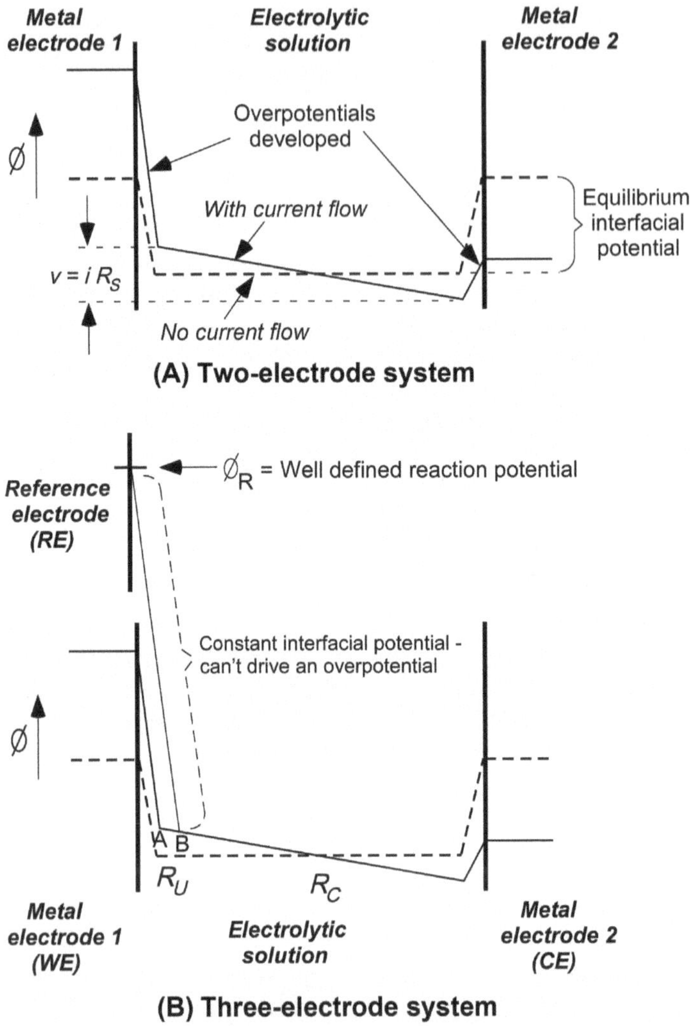

FIGURE 3.4 **Electrical potential profiles.** (A) Two-electrode system. In the absence of current, two equilibrium interfacial potentials exist, and the cell potential measured between the two electrodes is the difference between these equilibrium potentials. As shown, the equilibrium potentials are the same (as would be the case if the same metal was used for both electrodes) and the cell potential would be zero. Upon passing current, overpotentials develop at both interfaces (one interfacial potential becomes greater, one smaller). The net change in measured cell potential is due to three sources: the voltage drop in solution $i\ R_S$ and two overpotentials η_1 and η_2. (B) Three-electrode system. The measured potential is between the working electrode and reference electrode. Since no substantial overpotential can be developed at the reference electrode, any change in measured potential upon passing current is due to two sources: the overpotential at the working electrode–solution interface, and the solution drop $i\ R_U$, where the uncorrected resistance R_U is the solution resistance between the WE interface and RE interface.

there is a voltage drop across the resistive solution equal to the product of current and solution resistance: $v = i\ R_S$. Thus if current flows and there is a change in the measured $V_{\text{WE-CE}}$, this change may have contributions from three sources: (1) an overpotential at the WE as the interfacial potential $V_{\text{WE-solution}}$ varies, (2) an overpotential at the CE as the interfacial potential $V_{\text{CE-solution}}$ varies, and (3) the voltage drop $i\ R_S$ in solution. In the two-electrode system, one may only measure the cell potential $V_{\text{WE-CE}}$, and the individual components of the two overpotentials and $i\ R_S$ cannot be resolved. A third (reference) electrode may be used for potential measurements, although it is not required to pass current for stimulation; a two-electrode system (working and CE) is sufficient for stimulation. An ideal RE has a Faradaic reaction with very fast kinetics, which appears in the electrical model as a very low resistance for the Faradaic impedance Z_{faradaic}. In this case, no significant overpotential occurs at the RE during current flow, and the interfacial potential $V_{\text{RE-solution}}$ is considered constant. Examples of common RE are the reversible hydrogen electrode (RHE), the saturated calomel electrode (SCE), and the silver–silver chloride electrode (Ives & Janz, 1961). In the three-electrode system, if current flows through the working and CE and a change is noted in the measured potential $V_{\text{WE-RE}}$, this change may be from either of two sources: (1) an overpotential at the WE as the interfacial potential $V_{\text{WE-solution}}$ varies, and (2) the voltage drop $i\ R_S$ in solution. Unlike the two-electrode system, only one overpotential contributes to the measured potential change since the RE does not develop any overpotential. Furthermore, the overpotential at the WE can be estimated using the process of correction. This involves estimating the value of the solution resistance between the WE interface and the RE interface, called the uncorrected solution resistance R_U, and multiplying R_U by the measured current. This product $V_{\text{corr}} = i\ R_{U\ \text{estimated}}$ is then subtracted from the measured $V_{\text{WE-RE}}$ to yield the two interfacial potentials $V_{\text{WE-solution}}$ and $V_{\text{RE-solution}}$. Since $V_{\text{RE-solution}}$ is constant, any change in $V_{\text{WE-RE}}$ is attributed to an overpotential at the WE interface. Fig. 3.4 illustrates the electrical potential profiles of a two-electrode system and a three-electrode system, under conditions of no current flow and with current flow.

3.4.6 Faradaic processes: quantitative description

Eq. (3.18), the current-overpotential equation (Bard & Faulkner, 1980), relates the net current density through an electrode going into a Faradaic reaction to the overpotential, and defines the full characteristics of the Faradaic impedance.

$$i_{\text{net}} = i_0 \left\{ \frac{[\text{O}](0,t)}{[\text{O}]_\infty} \exp(-\alpha_c n f\ \eta) - \frac{[\text{R}](0,t)}{[\text{R}]_\infty} \exp(+(1-\alpha_c) n f\ \eta) \right\} \quad (3.18)$$

where i_{net} is the net Faradaic current density across the electrode−electrolyte interface, i_0 is the exchange current density, [O](0,t) and [R](0,t) are concentrations at the electrode surface ($x = 0$) as a function of time, $[O]_\infty$ and $[R]_\infty$ are bulk concentrations, α_c is the cathodic transfer coefficient and equals ~ 0.5, n is the number of moles of electrons per mole of reactant oxidized [Eq. (3.15)], $f \equiv F/R\,T$, F is Faraday's constant ~ 96,485 C/mole of electrons, R is the gas constant ~ 8.314 J/mole-°K, and T is the absolute temperature.

This equation relates the net current of a Faradaic reaction to three factors of interest: (1) the exchange current density i_0, which is a measure of the kinetic rate of the reaction, (2) an exponential function of the overpotential, and (3) the concentration of reactant at the electrode interface. The exponential dependence of Faradaic current on overpotential indicates that for a sufficiently small overpotential, there is very little Faradaic current, that is, for small potential excursions away from equilibrium, current flows primarily through the capacitive branch of Fig. 3.1, charging the electrode capacitance, not through the Faradaic branch. As more charge is delivered through an electrode interface, the electrode capacitance continues to charge, the overpotential increases, and the Faradaic current [proportional to exp (η)] begins to be a significant fraction of the total injected current. For substantial cathodic overpotentials the left term of Eq. (3.18) dominates; for substantial anodic overpotentials the right term dominates.

The two exponential terms represent the reduction and oxidation rates, respectively. The net current is the sum of the reduction and oxidation currents, as shown in Fig. 3.5. At the equilibrium potential E_{eq}, when $\eta = 0$, the rates are equal and opposite and may be relatively small (compared to when driven away from equilibrium), and the net current is zero. As the electrode potential moves away from equilibrium, one or the other term will begin to dominate.

A large value for i_0 represents a reaction with rapid electron exchange between the electrode and electrolyte (called the heterogeneous reaction); a small value for i_0 represents a reaction with slow electron transfer in the heterogeneous reaction. The values for exchange current density i_0 may range over several orders of magnitude, for example, from 10^{-12} to 10^{+1} A/cm^2. In the example shown in Fig. 3.5, i_0 is 0.1 of i_L, the limiting current. For a kinetically fast system with a large exchange current density, such as $i_0 = 10^{-3}$ A/cm^2, no significant overpotential may be achieved before a large current ensues. As the exchange current density decreases, the electrode must go to higher overpotentials (further from the equilibrium value of $\eta = 0$) before a given current is noted. For a finite detection level of current (a real instrument), a reaction with low exchange current density will not manifest until relatively high overpotentials are achieved.

If currents are low or if the electrolytic solution is well stirred, so that the surface concentrations [O](0,t) and [R](0,t) are essentially equal to the bulk concentrations, then Eq. (3.18) reduces to

$$i_{net} = i_0\left\{\exp(-\alpha_c\, n f\, \eta) - \exp(+(1 - \alpha_c) n f\, \eta)\right\} \tag{3.19}$$

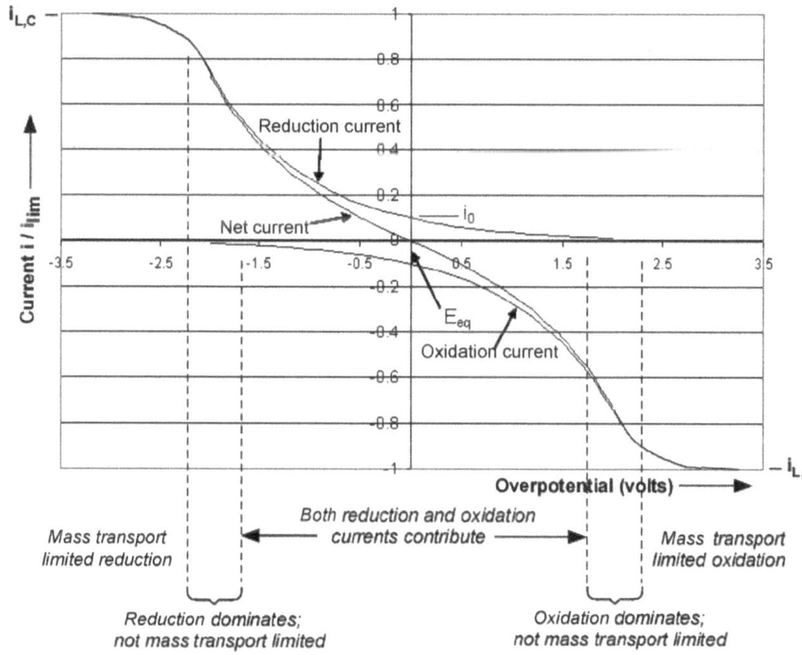

FIGURE 3.5 Net current versus overpotential, oxidation, and reduction curves. Three characteristic regions are displayed: (1) near equilibrium, both reduction and oxidation currents contribute. As the overpotential increases, either reduction or oxidation dominate, (2) initially in the absence of mass transport limitation, and then (3) with mass transport limitation.

This is the Butler–Volmer equation, which describes the current–overpotential relationship when mass transfer effects are negligible. This may be a useful approximation of (3.18) when the current is less than 10% of the limiting current.

As η increases away from zero, one of the two terms of the current–overpotential relationship (representing either reduction or oxidation) will dominate:

$$\text{for negative overpotentials} \quad i_{\text{net}} = i_0 \exp(-\alpha_c n f \eta) \tag{3.20a}$$

$$\text{for positive overpotentials} \quad i_{\text{net}} = i_0\{-\exp(+(1-\alpha_c)n f \eta)\} \tag{3.20b}$$

Near equilibrium, the surface concentrations of O and R are approximately equal to the bulk concentrations. As more charge is delivered and the overpotential continues to increase (in either direction), the surface concentration of reactant may decrease. The Faradaic current will then begin to level off, corresponding to the current becoming limited by mass transport of reactant, not electron transfer kinetics. At the limiting currents $i_{L,c}$ (cathodic, for negative overpotentials) or $i_{L,a}$ (anodic, for positive overpotentials), the reactant concentration at the electrode surface approaches zero, and the terms $[O](0,t)/[O]_\infty$ or $[R](0,t)/[R]_\infty$ counteract the exponential terms in Eq. (3.18), dominating the solution for net reaction rate. At the limiting currents, the slope of the reactant concentration gradient between the electrode surface and the bulk electrolyte determines the rate of reactant delivery and thus the current. At overpotentials where mass transport limitation effects occur (but prior to $i_{L,c}$ or $i_{L,a}$), Faradaic current takes the form

$$\text{for negative overpotentials} \quad i_{\text{net}} = \frac{[O](0.t)}{[O]_\infty} i_0 \exp(-\alpha_c n f \eta) \qquad (3.21a)$$

$$\text{for positive overpotentials} \quad i_{\text{net}} = -\frac{[R](0.t)}{[R]_\infty} i_0 \exp(+(1-\alpha_c) n f \eta) \qquad (3.21b)$$

Eqs. (3.19)–(3.21) are illustrated as three regions on the current–overpotential plot, shown in Fig. 3.5.

The mass transport limited currents $i_{L,c}$ and $i_{L,a}$ are given by Eqs. (3.22a) and (3.22b).

$$i_{L,c} = n\,F\,A\,k_{d,O}[O]_\infty \qquad (3.22a)$$

$$i_{L,a} = -\,n\,F\,A\,k_{d,R}[R]_\infty \qquad (3.22b)$$

where A is the electrode area and k_d is the mass transport rate, given by $k_d = D/\delta$, where D is the diffusion coefficient and δ is the diffusion layer thickness.

For very small overpotentials, the Butler–Volmer Eq. (3.19) can be approximated by

$$i_{\text{net}} = i_0(-\,n f \eta) \qquad (3.23)$$

since $e^x \sim 1 + x$ for small x. Thus at small overpotentials, the current is a linear function of overpotential. The ratio $-\,\eta/i$ is called the charge transfer resistance R_{ct}, given by

$$R_{\text{ct}} = R\,T/n\,F\,i_0 \qquad (3.24)$$

A small value for R_{ct} corresponds to a kinetically fast reaction.

As current is passed between a WE and CE through an electrolyte, both the working and counter electrode's potentials move away from their equilibrium values, with one moving positive of its equilibrium value and the other moving negative of its equilibrium value. Total capacitance is proportional to area, with capacitance C_{dl} = (capacitance/area) × area. Capacitance/area is an intrinsic material property. Capacitance is defined as the ability to store charge, and is given by

$$C_{\text{dl}} \equiv \mathrm{d}q/\mathrm{d}V \qquad (3.25)$$

where q = charge and V = the electrode potential with respect to some RE.

Thus an electrode with a relatively large area and total capacitance (as is often the case for a CE) can store a large amount of charge (dq) with a small overpotential (dv). During stimulation, the use of a large CE keeps the potential of the CE fairly constant during charge injection (near its equilibrium value), and there is little Faradaic current [Eq. (3.18)]. Significant overpotentials may be realized at a small WE. A typical reason for using a small electrode area is to achieve high spatial resolution during recording or stimulation. It is common to neglect the CE in analysis, and while this is often a fair assumption it is not always the case.

3.4.7 Charge injection during electrical stimulation: interaction of capacitive and Faradaic mechanisms

As illustrated in Fig. 3.1, there are two primary mechanisms of charge injection from a metal electrode into an electrolyte. The first consists of charging and discharging the double-layer capacitance, causing a redistribution of charge in the electrolyte but no electron transfer from the electrode to the electrolyte. C_{dl} for a metal in aqueous solution has values on the order of 10–20 μF/cm^2 of real area (geometric area multiplied by the roughness factor). For a small enough total injected charge, all charge injection is by charging and discharging of the double layer. Above some injected charge density, a second mechanism occurs consisting of Faradaic reactions where electrons are transferred between the electrode and electrolyte, thus changing the chemical composition in the electrolyte by reduction or oxidation reactions. Fig. 3.1 illustrates a single Faradaic impedance representing the electron transfer reaction $O + n\,e^- \leftarrow \rightarrow R$. Generally there may be more than one Faradaic reaction possible, which is modeled by several branches of $Z_{faradaic}$ (one for each reaction), all in parallel with the double-layer capacitance. The current–overpotential Eq. (3.18) and Fick's first and second laws for diffusion give the complete description of processes occurring for any Faradaic reaction.

In addition to the double-layer capacitance, some metals have the property of pseudocapacity (Gileadi et al., 1975), where a Faradaic electron transfer occurs, but because the product remains bound to the electrode surface the reactant may be recovered (the reaction may be reversed) if the direction of current is reversed. Although electron transfer occurs, in terms of the electrical model of Fig. 3.1 the pseudocapacitance is better modeled as a capacitor since it is a charge storage (not dissipative) process. Platinum is commonly used for stimulating electrodes as it has a pseudocapacity [by reaction (3.6)] of 210 μC/cm^2 real area (Rand & Woods, 1971), or equivalently 294 μC/cm^2 geometric area using a roughness factor of 1.4. The relationship between capacitance and stored charge is given by Eq. (3.28).

A 1 volt potential excursion applied to a double-layer capacitance of 20 μF/cm^2 yields 20 μC/cm^2 stored charge, which is an order of magnitude lower than the total charge storage available from platinum pseudocapacitance.

It is a general principle when designing electrical stimulation systems that one should avoid onset of irreversible Faradaic processes which may potentially create damaging chemical species, and keep the injected charge at a low enough level where it may be accommodated strictly by reversible charge injection processes. Unfortunately, this is not always possible, because a larger injected charge may be required to cause the desired effect (e.g., initiating action potentials). Reversible processes include (1) charging and discharging of the double-layer capacitance, (2) reversible Faradaic processes involving products that remain bound to the surface such as plating of hydrogen atoms on platinum [reaction (3.6)] or the reversible formation and reduction of a surface oxide [reactions (3.4) and (3.5)], and (3) reversible Faradaic processes where the solution phase product remains near the electrode due to mass transport limitations.

When the exchange current density is very low, Faradaic currents are not observed until significant overpotentials occur, and a relatively large total charge can be injected through the capacitive mechanism before Faradaic reactions commence. When the exchange current density is high, little injected charge is accommodated by capacitive charge, and significant Faradaic reactions commence with only small overpotentials. The desirable paradigm for a stimulating electrode is to use either capacitive charge injection or charge injection through reversible Faradaic processes (such as reversible oxide formation), thus minimizing irreversible Faradaic reactions that lead to either electrode or tissue damage.

The net current passed by an electrode, modeled as shown in Fig. 3.1, is the sum of currents through the two parallel branches. The total current through the electrode is given by

$$i_{\mathrm{total}} = i_C + i_f \tag{3.26}$$

where i_C is the current through the capacitance and i_f is the current through Faradaic processes.

The current through Faradaic processes is given by the current–overpotential Eq. (3.18). The current through the capacitance is given by Eq. (3.27).

$$i_C = C_{\mathrm{dl}}\,\mathrm{d}v/\mathrm{d}t = C_{\mathrm{dl}}\,\mathrm{d}\eta/\mathrm{d}t \tag{3.27}$$

The capacitive current depends upon the rate of potential change, but not the magnitude of the potential. The Faradaic current, by contrast, is exponentially dependent upon the overpotential, or departure from the equilibrium potential. Thus as an electrode is driven away from its equilibrium potential, essentially all charge initially flows through the capacitive branch since the overpotential is small near equilibrium. As the overpotential increases, the Faradaic branch begins to conduct a relatively larger fraction of the injected current. When the overpotential becomes great enough, the Faradaic impedance becomes sufficiently small (equivalently, the Faradaic admittance becomes large) that the Faradaic current equals the injected current. At this point the Faradaic process of reduction or oxidation conducts all injected charge, and the potential of the electrode does not change, corresponding to the capacitor not charging any further.

In terms of charge going into the different processes, the charge on the double-layer capacitance is proportional to the voltage across the capacitance:

$$q_C - C_{\mathrm{dl}} \Delta V \tag{3.28}$$

thus if the electrode potential does not change in time, neither does the stored charge. The charge into Faradaic processes however continues to flow for any nonzero overpotential. The Faradaic charge is the integration of Faradaic current over time, which by Eq. (3.18) is proportional to an exponential of the overpotential integrated over time:

$$q_f = \int i_f \, \mathrm{d}t \propto \int \exp(\eta) \, \mathrm{d}t \tag{3.29}$$

The charge delivered into Faradaic reactions is directly proportional to the mass of Faradaic reaction product formed, which may be potentially damaging to the tissue or the electrode.

Consider the flow of charge across an electrode interface during a single pulse, as illustrated in Fig. 3.6. Two "pipes" are shown next to the two charge injection elements, representing the available admittance to charge. Consider what happens as a 1 mA current step is applied to an electrode interface that is initially at its equilibrium potential. The overpotential at $t = 0^+$ is zero, so the Faradaic current $= \exp(\eta)$ is essentially zero (Fig. 3.6A). All current immediately after application of the current step is through the double-layer capacitance. The voltage immediately after current onset shows a simple linear charging of the capacitor (the *iR* drop of solution resistance is corrected out of these figures). As charge continues to flow through the double-layer capacitance, it charges up and an overpotential develops. At some point the Faradaic current proportional to $\exp(\eta)$ will happen to equal half of the injected current (Fig. 3.6B). Once the Faradaic current becomes a substantial fraction of the injected current, a visible inflection of the voltage waveform will be noted since a relatively smaller fraction of the injected charge per time is charging the double layer. Note in Fig. 3.6B that as the Faradaic admittance (the pipe) opens up, the capacitance pipe is closing. This is because the source is constant current, so as more current flows through the Faradaic impedance, less is available to

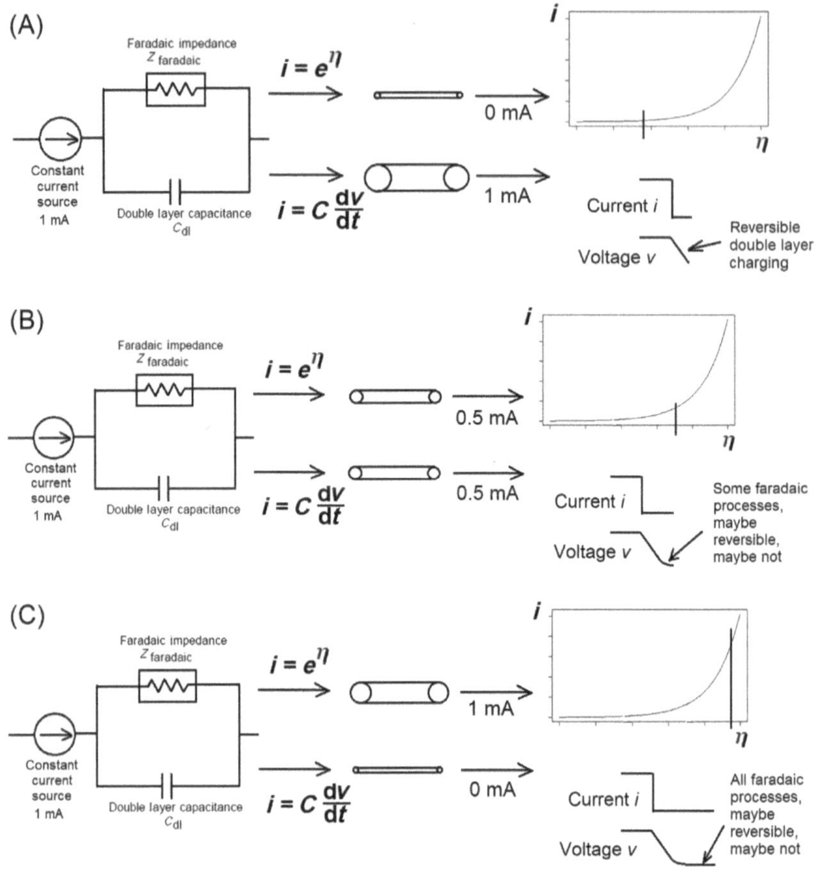

FIGURE 3.6 A walk-through charge injection during a single pulse. The second (counter) electrode and solution resistance are not shown. (A) At $t = 0+$ after initiation of 1 mA current step. The interface is near equilibrium potential and essentially all charge goes through capacitive charging. (B) Partially through the pulse, whereupon the Faradaic admittance is equal to the capacitive admittance. (C) Enough charge has been injected to charge the double-layer capacitance such that the Faradaic current [proportional to $\exp(\eta)$] exactly equals the injected current, thus no current flows through the double-layer capacitance and the interfacial potential remains constant.

charge the capacitor, resulting in a smaller dv/dt. If current continues to flow long enough, ultimately the capacitance will charge up to a point where the Faradaic current proportional to $\exp(\eta)$ equals the injected current. Then no further current flows through the capacitance, all current is accommodated by the Faradaic mechanism, and the electrode voltage remains constant (Fig. 3.6C).

If the injected current were not 1 mA but instead some large value (say 1 A), which cannot be accommodated by the Faradaic process in Fig. 3.6

due to a limitation in mass transport (the pipe cannot open far enough), current will continue to charge the double-layer capacitance and the overpotential will continue to increase until another Faradaic reaction with a lower exchange current density (thus more irreversible) than the first may start. In the case where this second reaction is the reduction of water, the reaction will not become mass transport limited (water at 55.5 molar will support substantial current), and an electrode potential will be reached where the non-mass transport limited reduction of water accepts all further injected charge. The "water window" is a potential range that is defined by the reduction of water in the negative direction, forming hydrogen, and the oxidation of water in the positive direction, forming oxygen. Because water does not become mass transport limited in an aqueous solution, the potentials where water is reduced and oxidized form lower and upper limits, respectively, for electrode potentials that may be attained, and any electrode driven to large enough potentials in water will evolve either hydrogen gas or oxygen gas. Upon reaching either of these limits, all further charge injection is accommodated by the reduction or oxidation of water.

3.4.8 Common waveforms used in neural stimulation

The current controlled method is commonly used for electrical stimulation of excitable tissue. This typically takes the form of rectangular pulses. In monophasic pulsing, a constant current is passed for a period of time (on the order of tens to hundreds of microseconds), then the external stimulator circuit is open-circuited (it is effectively electrically removed from the electrodes) until the next pulse. In biphasic pulsing, a constant current is passed in one direction, then the direction of current is reversed, and then the circuit is open-circuited until the next pulse.

In biphasic pulsing the first phase, or stimulating phase, is used to elicit the desired physiological effect such as initiation of an action potential, and the second phase, or reversal phase, is used to reverse electrochemical processes occurring during the stimulating phase. It is common to use a cathodic pulse as the stimulating phase (the WE is driven negative with respect to its prepulse potential), followed by an anodic reversal phase (the WE is driven positive), although anodic pulsing may also be used for stimulation (discussed in Section 3.5). Fig. 3.7 illustrates definitions of key parameters in pulsing. The frequency of stimulation is the inverse of the period, or time between pulses. The interpulse interval (IPI) is the time between pulses. Fig. 3.7B illustrates charge balanced biphasic pulsing, where the charge in the stimulation phase equals the charge in the reversal phase. Fig. 3.7C illustrates charge imbalanced biphasic pulsing where there are two phases, but the reversal phase has less charge than the stimulating phase. Fig. 3.7D illustrates the use of an interphase delay, where an open circuit is introduced between the stimulating and reversal phases.

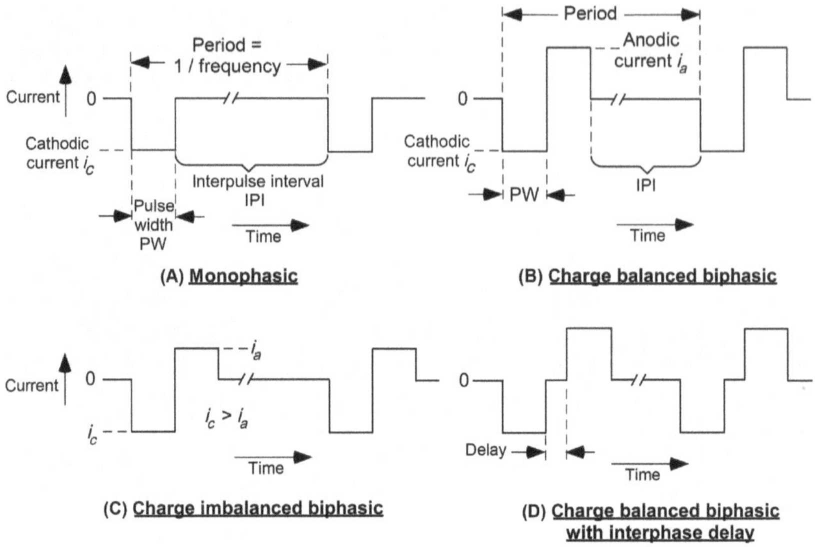

FIGURE 3.7 Common pulse types and parameters.

3.4.9 Pulse train response and ratcheting

Based on the simple electrical model of Fig. 3.1, one may predict characteristics of the potential waveforms resulting from monophasic pulsing, charge balanced biphasic pulsing, and charge imbalanced biphasic pulsing. Consider what occurs when an electrode, starting from the OCP, is pulsed with a single cathodic pulse and then left open circuit (illustrated in Fig. 3.8A, pulse 1). Upon pulsing the electrode charges, with injected charge being stored reversibly on the double-layer capacitance, causing the electrode potential to move negative. As the potential continues to move negative, charge begins to be delivered into Faradaic currents (whose magnitude is an exponential function of the overpotential). After the end of the pulse (i.e., during the IPI) when the external circuit is opened, charge on the double-layer capacitance continues to discharge through Faradaic reactions. This causes the electrode potential to move positive, and as the electrode discharges and the overpotential decreases, the Faradaic current decreases, resulting in an exponential discharge of the electrode. Given a sufficiently long time, the electrode potential will approach the OCP. However, if the electrode is pulsed with a train whose period is short with respect to the time constant for discharge (as may occur with neural stimulation, with a period of perhaps 20 ms), that is, if a second cathodic pulse arrives before the electrode has completely discharged, then the potential at the start of the second pulse (the prepulse potential) is more negative than the prepulse potential of the first pulse (the OCP). Because the potential during the second pulse begins at a more

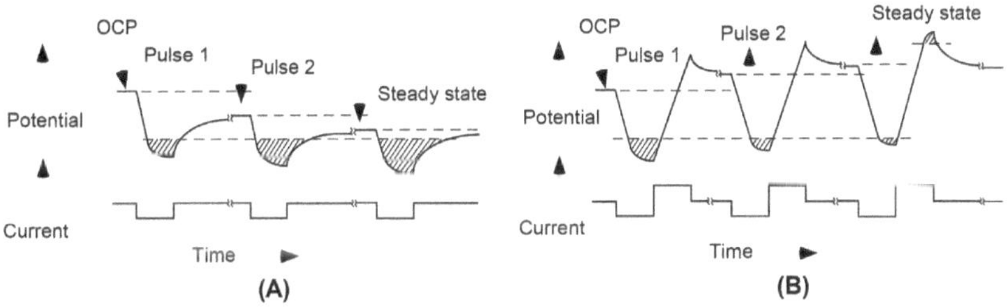

FIGURE 3.8 Electrode potentials in response to monophasic and charge-balanced biphasic pulse trains. (A) Ratcheting of potential during monophasic cathodic pulsing. The prepulse potential of successive pulses moves negative until all injected charge goes into irreversible processes. (B) Ratcheting during charge-balanced cathodic first biphasic pulsing. The prepulse potential of successive pulses moves positive until the same amount of charge is lost irreversibly during the cathodic and anodic phases. Shaded areas represent periods of irreversible reactions.

negative potential than the first, a smaller fraction of the injected charge goes into reversible charging of the double-layer capacitance. The Faradaic reactions begin accepting significant charge at an earlier time than in the first pulse, and there is more charge delivered to irreversible reactions during the second pulse than during the first as the overall potential range traversed is more negative during the second pulse (Fig. 3.8A, pulse 2). Upon going to open circuit after the second pulse, the electrode discharges through Faradaic reactions. Because the potential at the end of the second current pulse is more negative than the potential at the end of the first current pulse, the potential range during discharge between pulses 2 and 3 is also more negative than between pulses 1 and 2, and likewise the prepulse potential of pulse 3 is more negative than the prepulse potential of pulse 2. This "ratcheting" of the electrode potential continues until the following condition is met:

$$\text{Unrecoverable charge } (Q_{\text{ur}})\text{per pulse} = \text{Injected charge}\left(Q_{\text{inj}}\right)\text{per pulse} \quad (3.30)$$

that is, all injected charge goes into irreversible Faradaic reactions that occur either during the pulse or during the open-circuit IPI period. Charge delivered into irreversible processes is defined as unrecoverable charge Q_{ur}. Once condition (3.30) is met, the pulsing is in the steady state, and the potential excursions repeat themselves with each pulse cycle.

Next consider the electrode response when a charge-balanced stimulation protocol is used; cathodic then anodic, followed by open circuit. The electrode begins from OCP. Upon applying the first cathodic pulse, the double layer reversibly charges, and then the electrode may begin to transfer charge into Faradaic reactions as the potential moves negative. The anodic pulse then actively causes the electrode potential to move back positive (illustrated in Fig. 3.8B, pulse 1). Unlike the exponential decay during the monophasic pulsing, the electrode potential now changes according to the anodic current and the double-layer capacitance, and there is reversal of charge from the double layer. Because not all of the injected charge during the cathodic phase went into charging of the double layer, only some fraction of the charge injected during the cathodic phase would be required in the anodic phase to bring the electrode potential back to the prepulse value. However, since the anodic phase is balanced with the cathodic phase, the electrode potential at the end of the anodic phase of pulse 1 continues positive of the prepulse potential of pulse 1 (the OCP). During the anodic phase and during the open circuit following the anodic phase, if the potential becomes sufficiently positive, Faradaic oxidation reactions such as electrode corrosion may occur. During the open-circuit period, the electrode discharges exponentially through Faradaic oxidation reactions back toward the OCP, becoming negative with time. By the beginning of pulse 2 the potential is still positive of the prepulse potential for pulse 1 (the OCP). Thus as long as any charge is lost irreversibly during the cathodic phase, the potential at the end of the charge-balanced anodic phase will be positive of the prepulse potential, and a ratcheting effect is seen. Unlike the monophasic case, the ratcheting of the electrode prepulse potential is now in a positive direction. Steady state occurs when one of the two following conditions is met:

1. *There are no irreversible Faradaic reactions during either the cathodic or anodic phases, and the electrode simply charges and then discharges the double layer (the potential waveform appears as a sawtooth)*:

$$Q_{\text{ur cathodic}} = Q_{\text{ur anodic}} = 0 \tag{3.31}$$

or

2. *The same amount of charge is lost irreversibly during the cathodic phase and during the combined anodic phase and IPI*:

$$Q_{\text{ur cathodic}} = Q_{\text{ur anodic+IPI}} \neq 0 \tag{3.32}$$

If irreversible processes do occur, for cathodic first charge balanced biphasic pulsing, the electrode potential will move positive of the OCP, and during steady-state continuous pulsing there is an equal amount of unrecoverable charge delivered into irreversible reduction and oxidation processes.

It is important to note that charge balance does not equate to chemical balance. As illustrated in Fig. 3.8B, a charge-balanced waveform may result in two distinct Faradaic reactions, one being a reduction reaction occurring during the cathodic phase and a second being an oxidation reaction occurring during the anodic phase and IPI. Generally, reduction reactions may create species damaging to tissue, such as reactive oxygen species, and oxidation reactions cause damage to the electrode (corrosion), but these generalizations are not universal.

Finally, consider the electrode response when a charge-imbalanced stimulation protocol is used (not illustrated). The electrode begins from OCP. The response to the first cathodic pulse is the same as with the monophasic or charge-balanced biphasic waveforms. The anodic phase then causes the electrode potential to move back to positive, but since there is less charge in the anodic phase than the cathodic one, the electrode potential does not move as far positive as it did with the charge-balanced biphasic waveform. The potential at the end of the anodic phase will be closer to the OCP than during charge-balanced pulsing. The maximum positive potential will be less when using the charge-imbalanced waveform than when using the charge-balanced waveform. This has the advantage that charge delivered into anodic Faradaic processes, such as metal corrosion, is reduced with respect to charge-balanced stimulation. The prepulse potential will move under factors as explained for the monophasic and charge-balanced biphasic waveforms until the following condition is met:

The net imbalance in injected charge is equal to the net difference in unrecoverable charge between the cathodic phase and the combined anodic phase and IPI:

$$\left(Q_{\text{inj cathodic}} - Q_{\text{inj anodic}}\right) \equiv Q_{\text{imbal}} = \left(Q_{\text{ur cathodic}} - Q_{\text{ur anodic+IPI}}\right) \tag{3.33}$$

During charge-imbalanced stimulation the shift in prepulse potential may be either positive or negative of the OCP, depending on the amount of imbalance. Thus charge-imbalanced waveforms provide a unique opportunity to titrate the potential range over which an electrode is exposed.

Based on these considerations, monophasic pulsing causes the greatest shift of the electrode potential during pulsing away from the equilibrium potential, thus causing the most accumulation of unrecoverable charge (corresponding to products of irreversible Faradaic reactions) of the three protocol types (monophasic, charge-balanced biphasic, charge-imbalanced biphasic). Furthermore, since during monophasic pulsing the electrode potential is not brought back toward the equilibrium potential by an anodic phase, there is accumulation of unrecoverable charge during the open-circuit IPI.

3.4.10 Electrochemical reversal

The purpose of the reversal phase during biphasic stimulation is to reverse the electrochemical processes that occurred during the stimulating phase, minimizing unrecoverable charge. A reversible process is one where the reactants are reformed from the products upon reversing the direction of current. Upon delivering current in the stimulation phase and then reversing the direction of current, charge on the electrode capacitance will discharge, returning the electrode potential toward its prepulse value. If only double-layer charging occurs, then upon passing an amount of charge in the reversal phase equal to the charge delivered in the stimulation phase (a charge-balanced protocol), the electrode potential will return precisely to its prepulse potential by the end of the reversal phase and the potential curve will be a simple sawtooth as shown in Fig. 3.9A (corrected for solution

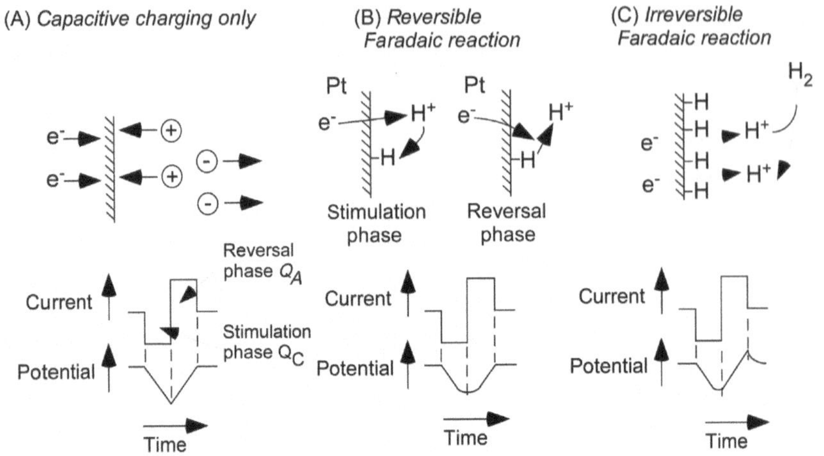

FIGURE 3.9 Electrochemical processes and potential waveforms during charge-balanced stimulation: (A) capacitive charging only, (B) reversible hydrogen plating, (C) irreversible hydrogen evolution.

resistance). If reversible Faradaic reactions occur during the stimulation phase, then charge in the reversal phase may go into reversing these reactions. Fig. 3.9B illustrates an example reversible Faradaic process, in this case charging of the pseudocapacitance (reduction of protons and plating of monatomic hydrogen onto the metal electrode surface) as may occur on platinum. During reversal the plated hydrogen is oxidized back to protons. Because the electrochemical process occurring during the reversal phase is the exact opposite of that occurring during the stimulation phase, there is zero net accumulation of electrochemical species. Reversible Faradaic reactions include adsorption processes, as in Fig. 3.9B, as well as processes where the solution phase product remains near the electrode due to mass transport limitations. If irreversible Faradaic reactions occur, upon passing current in the reverse direction, reversal of electrochemical product does not occur since the product is no longer available for reversal (it has diffused away). An example shown in Fig. 3.9C is the formation of hydrogen gas after a monolayer of hydrogen atoms has been adsorbed onto the platinum surface, that is, all possible sites on the platinum surface are occupied. In the case where a Faradaic reaction occurs during the stimulation phase, the potential waveform during the stimulation phase is not linear, but displays a slope inflection as Faradaic processes consume charge (this is charge that does *not* charge the capacitance C_{dl}, thus does not change the potential across C_{dl} which is identical to the potential across the electrode interface). If irreversible Faradaic reactions occur, then when an equal amount of charge is passed in the reversal phase, the electrode potential goes positive of the prepulse potential. To return the electrode potential exactly to its prepulse value would require that the charge in the reversal phase be equal to only the amount of charge that went onto the capacitance during the stimulation phase (a charge-imbalanced waveform).

The use of biphasic stimulation (either charge balanced or charge imbalanced) moves the electrode potential out of the most negative ranges immediately after stimulation. In comparison (as shown in Fig. 3.8), the monophasic stimulation protocol allows the electrode potential to remain relatively negative during the IPI, and during this time Faradaic reduction reactions may continue. In the presence of oxygen, these reactions may include reduction of oxygen and formation of reactive oxygen species, which have been implicated in tissue damage (Bergamini et al., 2004; Halliwell, 1992; Hemnani & Parihar, 1998; Imlay, 2003; Stohs, 1995). The charge-imbalanced waveform has the added advantage that the electrode potential at the end of the anodic pulse is less positive than with charge-balanced biphasic pulsing, thus less charge goes into irreversible oxidation reactions such as corrosion. Charge-imbalanced biphasic waveforms provide a method to reduce unrecoverable charge in the cathodic direction (with respect to monophasic stimulation) and in the anodic direction (with respect to charge-balanced biphasic stimulation), thus they are an attractive solution to minimizing damage to either the stimulated tissue or the metal electrode.

A key parameter describing any candidate material for charge injection is the reversible charge storage capacity, or simply charge storage capacity (RCSC or CSC) (Robblee & Rose, 1990). This is the total amount of charge that can be stored in reversible processes including charging the double-layer capacitance, adsorption processes, and processes where the solution phase product remains near the electrode due to mass transport limitations. A closely associated parameter is the reversible charge injection capacity (CIC), which is that fraction of the CSC which can effectively be exploited for stimulation of excitable tissue. We generally desire materials with a large CIC, as this supports a high charge to be injected (providing efficacy) without onset of deleterious reactions (providing safety). As an example of why the CIC may be lower than CSC, consider platinum, a common material for stimulation. Brummer and Turner (Brummer & Turner, 1975, 1977a, 1997b, 1977c) have reported on the electrochemical processes of charge injection using a platinum electrode. They found three processes that could store charge reversibly, including charging of the double-layer capacitance, hydrogen atom plating and oxidation (pseudocapacity), and reversible oxide formation and reduction on the electrode surface. In artificial cerebrospinal fluid, they reported that 300–350 $\mu C/cm^2$ (real area) could be stored reversibly by these processes [equivalently 420–490 $\mu C/cm^2$ (geometric area) using a typical roughness factor of 1.4]. This is a maximum RCSC under optimum conditions, including relatively long pulse widths, and optimum conditions are seldom used. Rose and Robblee (1990) reported on the charge injection limits for a platinum electrode using 200 μs charge-balanced biphasic pulses, and determined the charge injection limit to be 50–100 μC/cm^2 (geometric) using anodic first pulses, and 100–150 μC/cm^2 (geometric) using cathodic first pulses. These values are considerably lower than the theoretical values determined by Brummer and Turner, since the electrode potential at the beginning of a pulse begins somewhere intermediate to oxygen and hydrogen evolution, and not all of the three reversible processes accommodate charge during the stimulating pulse.

Cyclic voltammetry (CV) is a popular method for characterizing an electrode. In its simplest form, a WE is forced to follow a sawtooth-shaped voltage waveform with respect to some RE. A CE carries the current required to maintain the commanded voltage. This procedure is implemented with a potentiostat, and the data are displayed as current versus voltage as a cyclic voltammogram.

If a sawtooth voltage is applied across a capacitor, the i–v plot is a simple rectangle with peak amplitude of $i = C\, dv/dt$. Thus when the voltage applied to a WE remains within a range where all charge flows through the double-layer capacitance, the CV is a rectangle and C_{dl} can easily be calculated. As the applied potential increases, once overpotentials are reached where substantial Faradaic currents flow, these currents are observed as peaks on the plot of current versus $V_{WE\text{-}RE}$. If a Faradaic process is reversible there will be two peaks, one above the zero current axis and one below the zero current axis, representing the oxidation and reduction reactions of the reversible process, respectively. The area under the curve of any peak has units of charge. For a truly reversible process, the integrated areas under the oxidation and reduction peaks will be equal and opposite. If a Faradaic process is not reversible, there will be no corresponding peak of equal area.

Fig. 3.10 is an example CV of platinum in phosphate-buffered saline in the absence of oxygen. Reversible processes are shown in green and include double-layer charging/discharging (a rectangle), oxide formation and reduction, and hydrogen atom plating (accounting for the pseudocapacity of platinum). Irreversible processes are shown in red and include water breakdown to form hydrogen gas and water breakdown to form oxygen.

Because the area under the curve of a CV has units of charge, one can estimate the total CSC of a material by integrating the total reversible processes in a CV, including both the double-layer charging (observed as a rectangular region) and reversible Faradaic processes (observed as opposing peaks).

Returning to the difference between maximum CSC under optimum conditions and realistic CIC under neural stimulation conditions (420–490 μC/cm^2 vs 100–150 μC/cm^2 for platinum), at issue is the prepulse potential that takes on a value intermediate in the CV (far from hydrogen or oxygen evolution), perhaps in the double-layer charging region of Fig. 3.10. In this case, a cathodic sweep (similarly for anodic) only exploits some fraction of the available CSC before hitting hydrogen (or oxygen) evolution. It has been shown that by

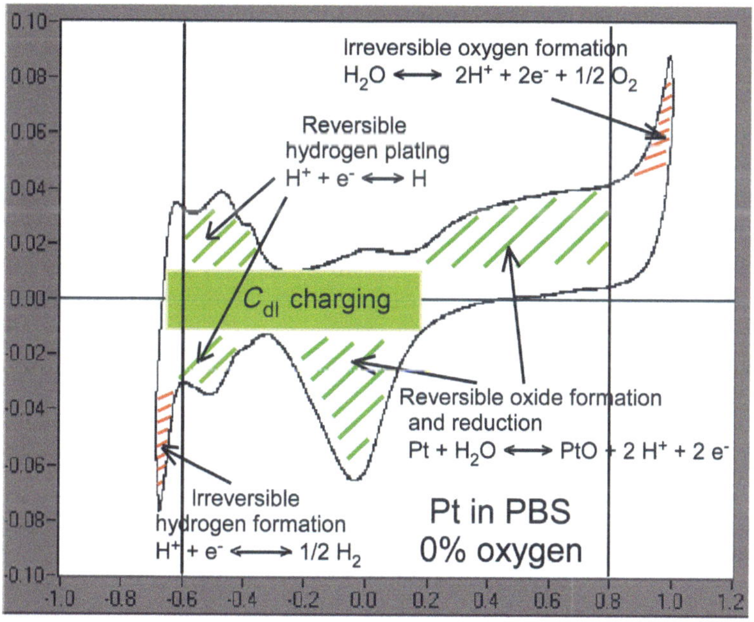

FIGURE 3.10 Slow cyclic voltammogram of platinum in phosphate-buffered saline, 0% oxygen. Voltage limits are −700 mV to +1.0 V versus saturated calomel electrode (SCE). Sweep rate is 50 mV/s. *y* axis is in μA, *x* axis is in volts versus SCE. Reversible processes are shown in green, irreversible in red.

applying a positive bias of 0.4−0.8 V versus Ag/AgCl during the IPI, the CIC of iridium oxide is at least tripled (Beebe & Rose, 1988; Cogan et al., 2006) since the bias causes the pulse to begin at a more positive potential, allowing more reversible reduction to occur prior to hydrogen evolution. This is an attractive method for increasing CIC; however, one must carefully place the bias to avoid invoking unacceptable net DC current during the IPI. Methods for achieving this are discussed by Troyk et al. (2004).

3.5 Introduction to extracellular stimulation of excitable tissue

3.5.1 Cathodic and anodic stimulation

The goal of electrical stimulation of excitable tissue is to control, by activation or inhibition, the initiation and propagation of action potentials and in turn the release of neurotransmitters and thus nervous system signaling. The initiation of action potentials requires the artificial depolarization of some portion of the neural membrane to threshold. In the process of extracellular stimulation, the extracellular region is driven to relatively more negative potentials, equivalent to driving the intracellular compartment of a cell to relatively more positive potentials. Charge is transferred across the membrane due to both passive (capacitive and resistive) membrane properties as well as through active ion channels (Hille, 1984). The process of physiological action potential generation is well reviewed in the literature (Kandel et al., 2000), and models have been proposed (Chiu et al., 1979; Sweeney et al., 1987) for mammalian myelinated axons in terms of the parameters "*m*" and "*h*" as defined by Hodgkin and Huxley (1952a–d) in their studies of the squid giant axon.

The mechanisms underlying electrical excitation of nerve have been reviewed elsewhere (Durand, 1995; McNeal, 1976; Merrill et al., 2005a; Mortimer, 1990; Ranck, 1981). In brief, in the simplest case of stimulation, a monopolar electrode (a single current-carrying conductor) is placed in the vicinity of excitable tissue. Current passes from the electrode, through the ECF surrounding the tissue of interest, and ultimately to a distant CE. For a current I flowing through the monopolar electrode located a distance r away from a segment of excitable tissue, and uniform conductivity in the fluid of σ, the extracellular potential V_e at the tissue is

$$V_e = \frac{\mathrm{I}}{4\,\pi\,\sigma\,r} \tag{3.34}$$

The electric field generated by a monopolar electrode will interact with an axon membrane (this may be generalized to any excitable tissue). During cathodic stimulation the negative charge of the WE causes a redistribution of charge across the axon membrane, with negative charge on the outside of the membrane underneath the cathode, and movement of positive charge intracellularly from the distant axon to the region under the electrode, causing depolarization of the membrane near the cathode. Associated with the membrane depolarization near the electrode is hyperpolarization of the membrane at a distance away from the electrode (Fig. 3.11A).

During cathodic stimulation, the hyperpolarized regions of the axon distant from the cathode may suppress an action potential that has been initiated near the electrode, known as "anodic surround block." This effect is observed at higher current levels than the threshold values required for initiation of action potentials with cathodic stimulation.

If the electrode is instead driven as an anode (to more positive potentials), hyperpolarization occurs under the anode, and depolarization occurs at a distance away from the anode (Fig. 3.11B). Action potentials may be initiated at the regions distant from the anode electrode where depolarization occurs, known as virtual cathodes. The depolarization that occurs with such anodic stimulation is less than that of the depolarization with cathodic stimulation; thus cathodic stimulation requires less current to bring an axon to threshold.

It is notable that when in the subthreshold domain (when ion channels have minimal conductance), the axon can be modeled as a closed volume;

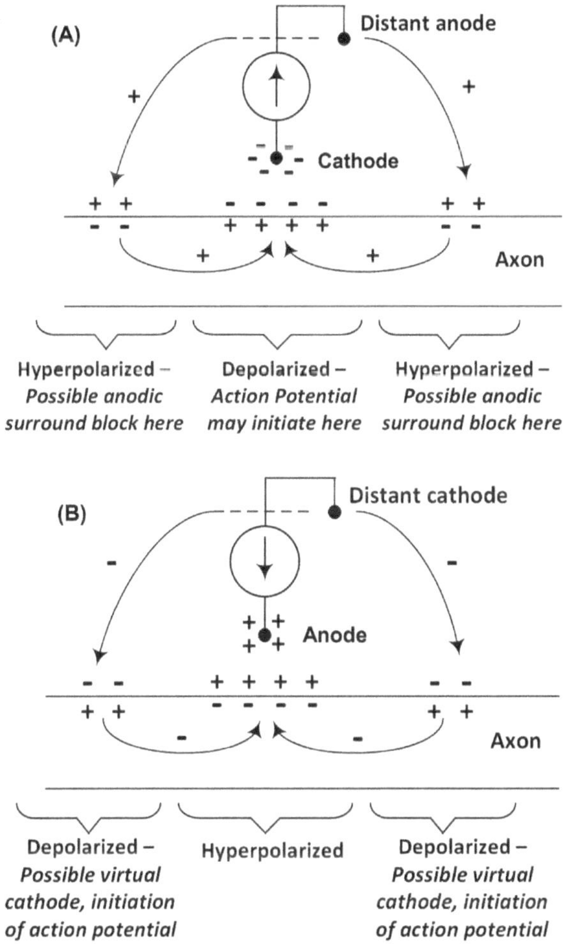

FIGURE 3.11 Fundamentals of (A) cathodic stimulation and (B) anodic stimulation.

thus to maintain charge neutrality, if depolarization occurs in one region hyperpolarization must occur in another region and vice versa. This is observed in Fig. 3.11, where cathodic stimulation causes local depolarization and distant hyperpolarization, and the opposite occurs during anodic stimulation.

3.5.2 Exploiting the voltage-gated sodium channel

Fig. 3.12 illustrates the dynamics of the voltage-gated sodium channel (VGSC) which underlies the physiological process of action potential propagation. A simple model of the channel contains two gates, known as the activation gate and inactivation gate. For sodium to pass from the extracellular

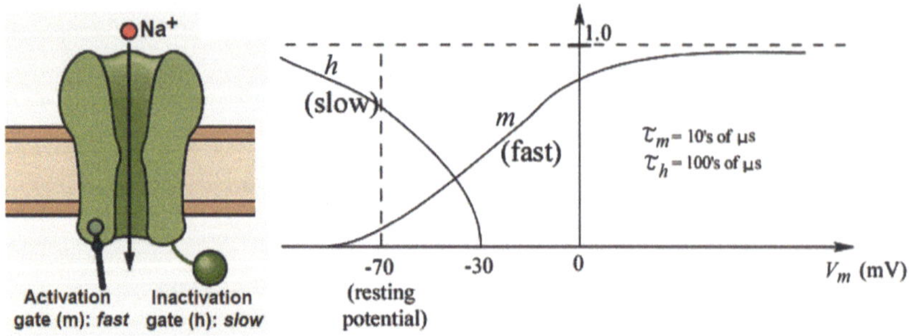

FIGURE 3.12 Dynamics of the voltage-gated sodium channel.

to intracellular space and cause depolarization, both gates must at some point be open simultaneously. The chart in Fig. 3.12 shows the relationship of transmembrane voltage V_m and the probability of a gate being open. The parameter "*m*" is the probability that an activation gate will be open at a particular potential (or alternatively, the fraction of activation gates open in a population), and the parameter "*h*" is the probability that an inactivation gate will be open at a particular potential. At the resting transmembrane potential, about 20% of activation gates are open and 70% of inactivation gates are open. Importantly, the activation gates respond quicker to changes in potential than inactivation gates, with time constants on the order of tens of microseconds for activation gates and hundreds of microseconds for inactivation gates. When a local region of membrane is depolarized (the potential moves to the right on the chart), activation gates open up within tens of microseconds, allowing more sodium to flow in. Then hundreds of microseconds to a millisecond later, the inactivation gates begin to close, thus temporally limiting sodium influx. In their 1952 seminal studies of action potential initiation and propagation in squid giant axon, Hodgkin and Huxley (1952a–d) demonstrated that the conductance of sodium through a VGSC is proportional to m^3h:

$$g_{Na} = \overline{g}_{Na}\, m^3\, h \tag{3.35}$$

where g_{Na} is the sodium channel conductance and $\overline{g}_{Na}$ is the maximum channel conductance.

Professor Tom Mortimer, a pioneer in our field, has been known to assert that Fig. 3.12 provides all a neural engineer needs to know. While this may be a stretch, it underscores the value in understanding and exploiting VGSC dynamics. A few examples will illustrate this.

In mammalian axon, hyperpolarizing (moving to the left in Fig. 3.12) with a pulse that is long relative to the sodium inactivation gate time constant will remove the normal partial inactivation (*h* will be higher than normal). If the hyperpolarizing current is then abruptly terminated (as with a rectangular pulse), the sodium activation gate conductance increases back to the rest value relatively quickly, but the opening of the slower inactivation gate remains high for a period of time; thus the net sodium conductance is briefly higher than normal and an action potential may be initiated. This phenomenon, known as anodic break, may be observed with either cathodic or anodic stimulation, since both cause some region of hyperpolarization in the axon. Fang and Mortimer (1991) have shown that anodic break may be prevented by using stimulating waveforms with slowly decaying phases such as an exponential decay instead of abrupt terminations.

Long-duration subthreshold stimuli can produce the phenomenon of accommodation. A cathodic pulse that produces subthreshold depolarization (i.e., does not trigger an action potential) will lower the h value (the inactivation gates close), thus increasing the current required for recruitment. This is not a problem with pulses that are shorter than the time constant of sodium channel inactivation, but can be problematic with prolonged pulses.

Accommodation can also be exploited to gain selectivity. Selectivity is the ability to activate one population of neurons without activating a neighboring population. Fiber diameter selectivity is the ability to activate fibers within a certain range of diameters. Fibers with larger diameters experience greater changes in the transmembrane potential due to electrical excitation (Rattay, 1989). Using conventional electrical stimulation waveforms with relatively brief pulses, the largest diameter fibers are activated at the lowest stimulus amplitude. In motor nerves, activating large-diameter fibers first corresponds to activating the largest motor units first and coarse functionality. This recruitment order is opposite to the physiological case where the smallest motor units are recruited first, providing fine control. To address this, one can apply a long duration (long relative to the inactivation gate time constant) subthreshold depolarizing pulse, followed by a higher current stimulating pulse. The subthreshold pulse causes accommodation, which preferentially affects the large-diameter fibers; then smaller diameter fibers are more susceptible to the follow-up stimulating pulse. Electrical stimulation protocols have also been developed (McIntyre & Grill, 2002) for triggering of action potentials in specific cell types (e.g., interneurons) and structures (e.g., nerve terminals).

3.5.3 Quantifying action potential initiation

Cable theory was originally described in 1855 by William Thomson (later Lord Kelvin) to describe signal propagation along the trans-Atlantic telegraph cable. It has more recently been applied to axons (Rall, 1977). As shown in Fig. 3.13A, the passive properties of the axonal membrane are modeled with a transmembrane capacitance C_m in parallel with transmembrane resistance r_m, plus intraaxonal resistance r_i and extra-axonal resistance r_e. E_m is the resting transmembrane voltage. The parallel branches of C_m and

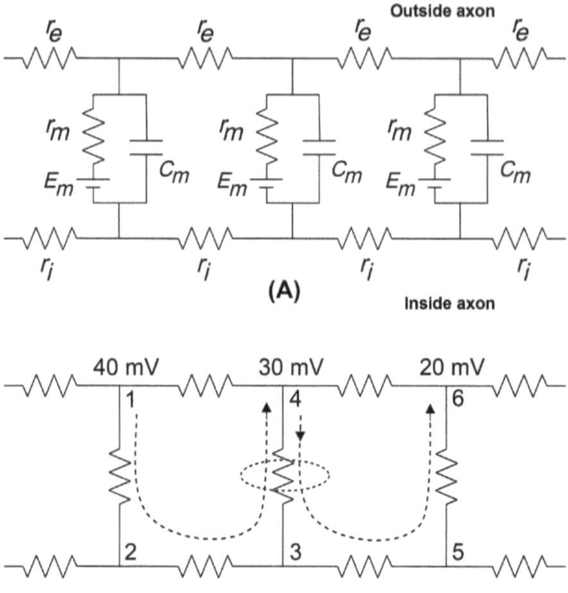

FIGURE 3.13 Passive electrical model of axon used for cable equation. (A) Passive elements of the axon; (B) generalized resistive ladder network showing no change in transmembrane voltage with a constant E field along the axon length.

r_m can represent the nodes of Ranvier in myelinated axon; alternatively, the model can represent unmyelinated axon if the lumped parameter model shown is appropriately extended to a distributed parameter model.

Define V_e as the extra-axonal voltage due to some source, V as the deviation of the transmembrane voltage away from resting value, and λ as the space constant and τ as the time constant for the axon. The distance along the length of the axon is given by x. The cable equation as applied to axons then states

$$\lambda^2 \frac{\partial^2 V}{\partial x^2} - \tau_m \frac{\partial V}{\partial t} - V = -\lambda^2 \frac{\partial^2 Ve}{\partial x^2} \tag{3.36}$$

This describes the deviation of the transmembrane voltage away from the resting value (therefore we can determine whether the threshold is reached) in terms of the second spatial derivative of the extra-axonal voltage along the length of the axon. This second spatial derivative $\partial^2 V_e/\partial x^2$ is called the activating function. It is important to note that the activating function must exceed some value as measured along the length of the axon. As shown in Fig. 3.14A, if electrodes are placed parallel to the axis of an axon, the activating function can be nonzero and stimulation may occur; however, if two electrodes are placed transversely, as shown in Fig. 3.14B with the

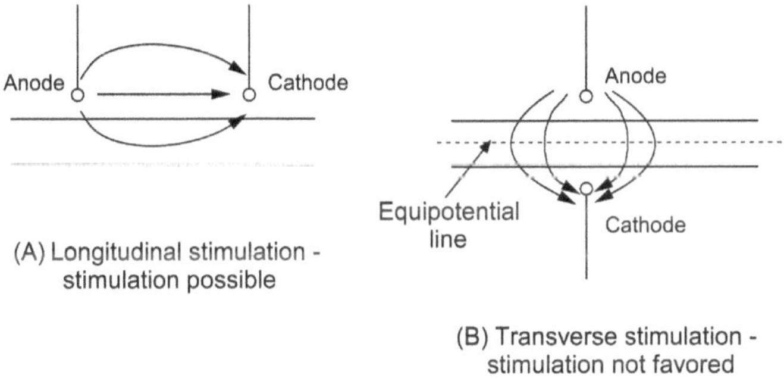

FIGURE 3.14 Transverse versus longitudinal stimulation.

equipotential line along the axon, the activating function is zero, and no stimulation occurs. Stimulation is thus vitally dependent on relative electrode orientation.

Providing proper electrode orientation is insufficient to guarantee stimulation. Consider Fig. 3.13B, where the passive axon model is simplified to its resistive elements, and a constant electric field is placed along the axon length. Two loops of current are shown, one from electrical node 1 through nodes 2, 3, and 4, and a second loop from electrical node 4 through nodes 3, 5, and 6. With a constant applied E field (dV_e/dx) along nodes 1, 4, and 6, the two loops of current are equal in value, so there is zero net current through the middle transmembrane resistance (shown by the dashed ellipse) and zero change in transmembrane voltage. It now appears that stimulation is quite challenging! One must achieve a sufficiently large second spatial derivative of extra-axonal voltage (or first derivative of the E field) along the length of the membrane. This is favored by having abrupt edges, and in fact maintaining a constant E field everywhere would be difficult. There will be edges near the electrodes, near nonhomogeneous structures, etc. The important point is that abrupt spatial changes in the E field promote stimulation.

The cable equation with its activating function provides a mechanistic description of the requirements for stimulation. More commonly, we empirically derive the current or charge required to elicit stimulation for a particular electrode arrangement, and express this using a strength-duration or charge-duration curve, respectively.

The relationship between the strength (current) of an applied constant current pulse required to initiate an action potential and the duration of the pulse, known as the strength–duration curve, is shown in Fig. 3.15A. The threshold current I_{th} decreases with increasing pulse width. At very long pulse widths, the current is a minimum, called the rheobase current I_{rh}. The following relationship has been derived experimentally to quantify the strength–duration curve (Lapicque, 1907):

$$I_{th} = \frac{I_{rh}}{1 - \exp(-W/\tau_m)} \tag{3.37}$$

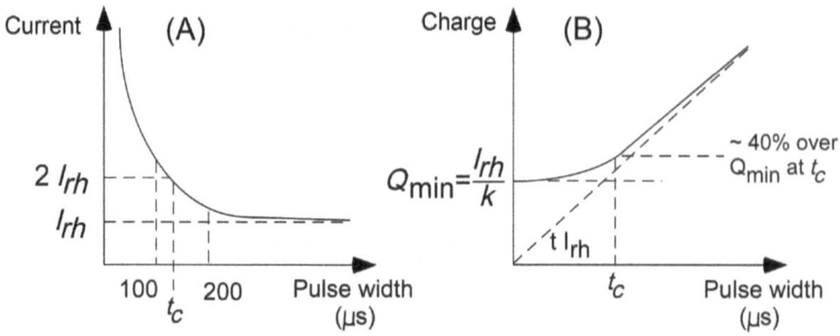

FIGURE 3.15 Strength–duration and charge–duration curves for initiating an action potential. Rheobase current I_{rh} is the current required when using an infinitely long pulse width. Chronaxie time t_c is the pulse width corresponding to two times the rheobase current.

where I_{th} is the current required to reach the threshold, I_{rh} is the rheobase current, W is the pulse width, and τ_m is the membrane time constant.

The qualitative nature of the strength–duration curve shown is representative of typical excitable tissue. The quantitative aspects, for example, the rheobase current, depend upon factors such as the distance between the neuron population of interest and the electrode, and are determined empirically. Fig. 3.15B illustrates the charge–duration curve, which plots the threshold charge $Q_{th} = I_{th}W$ versus pulse width. At longer pulse widths, the required charge to elicit an action potential increases. This occurs for two reasons: (1) charge is redistributed through the length of the axon and does not all participate in changing the transmembrane potential at the site of injection, and (2) at long pulse widths, accommodation occurs (inactivation gates close, see Section 3.5.2). The minimum charge Q_{min} occurs as the pulse width approaches zero. In practice, the Q_{th} is near Q_{min} when narrow pulses are used (tens of microseconds).

It is generally best to keep the pulse width narrow (and charge minimized) to minimize any electrochemical reactions occurring on the electrode surface. The narrowness of a pulse is often limited by the amount of current that can be delivered by a stimulator, especially if it is battery operated. Furthermore, some kinds of stimulation, such as selective activation of certain axons of a nerve, require pulses longer than tens of microseconds.

3.5.4 Bipolar configurations; voltage-controlled stimulation

In a so-called monopolar configuration, one electrode is placed in the vicinity of excitable tissue, current passes from the electrode through the ECF surrounding the tissue of interest, and ultimately to a distant CE. In fact, there must always be at least two electrodes to carry current. The term "monopolar" means only one electrode has a substantial effect on the electric field near the target tissue, with any other electrodes being sufficiently distant to be inconsequential to the field near the target tissue. In a bipolar configuration, there are two electrodes sufficiently near the target tissue to influence the electric field. Bipolar and other electrode configurations have complex voltage and current patterns and will not be discussed further here. Durand (1995) reviewed solutions for electrical potential profiles of various systems.

There is an unfortunate tendency to mistake "monopolar/bipolar" and "monophasic/biphasic" by the less experienced in the field. Monopolar/bipolar refers to the number of electrodes; monophasic/biphasic refers to the waveform (Fig. 3.7).

During current-controlled stimulation, the current is constant throughout the period of the pulse; thus the electric field (the gradient of V_e) at any point in tissue is constant during the pulse. During voltage-controlled stimulation, whereby the voltage between a WE and CE is commanded, current is not constant throughout the period of the pulse and the extracellular electric field in tissue generally decreases during the duration of the pulse as the double-layer capacitance charges (by Kirchoff's voltage law).

3.6 Mechanisms of damage

An improperly designed electrical stimulation system may cause damage to the tissue or damage to the electrode itself.

Damage to an electrode can occur in the form of corrosion if the electrode is driven anodically such that the electrode potential exceeds a value where significant metal oxidation occurs. An example of such a reaction is the corrosion of platinum in a chloride-containing medium such as ECF, reaction (3.10). Corrosion is an irreversible Faradaic process. It may be due to dissolution where the electrochemical product goes into solution, or the product may form an outer solid layer as a passivation film that cannot be recovered. Charge-balanced waveforms (Fig. 3.7B) are more likely to reach potentials where corrosion may occur during the anodic reversal phase and the open-circuit IPI than are monophasic waveforms (Fig. 3.8). The charge-imbalanced waveform (Fig. 3.7C) has advantages both in preventing tissue damage due to sustained negative potentials during the IPI, and in preventing corrosion by reducing the maximum positive potential during the anodic reversal phase (Section 3.4.9).

While evidence supports the value of charge-imbalanced protocols (Scheiner & Mortimer, 1990), charge-balanced protocols remain the industry standard. Technical implementation of charge imbalance is straightforward. Perhaps the greatest barrier to novel protocols is the timeline of regulatory processes. As discussed in a later chapter, regulatory clearance or approval is favored by claiming equivalence to prior technologies; thus the time-to-market imperative motivates industry (especially small companies) to take incremental steps to achieve access to the market.

Two major classes of mechanisms for stimulation-induced tissue damage are:

1. Damage may be caused by intrinsic biological processes as excitable tissue is overstimulated. This is called the mass action theory, and occurs due to the induced hyperactivity of many neurons firing or neurons firing for an extended period of time, thus changing the local environment. Mass action mechanisms include depletion of oxygen or glucose, or changes in ionic concentrations both intracellularly and extracellularly, for example, an increase in extracellular potassium. In the CNS, excessive release of excitatory neurotransmitters such as glutamate may cause excitotoxicity.
2. The second mechanism for tissue damage is the creation of toxic electrochemical reaction products at the electrode surface during stimulation at a rate greater than that which can be tolerated by the physiological system.

3.6.1 Tissue damage from intrinsic biological processes

Several studies from the Huntington Medical Research Institute (Agnew et al., 1990, 1993; McCreery et al., 1988) have concluded that neural injury from prolonged electrical stimulation is linked to neuronal activity produced by the stimulation, consistent with the mass action hypothesis. In Agnew's 1993 study of chronic stimulation of cat brain, the authors state "The most useful hypothesis may be that the activation of excitatory amino acid receptors does not 'cause' neural injury, but merely lowers its threshold in response to a given stressor by magnifying the destructive effect of the stressor, and that the manner and degree to which this occurs depends upon the physical and pharmacological composition of the particular neural substrate, as well as the stimulus parameters." Such an interpretation indicates how the mass action and electrochemical theories of tissue damage are not exclusive.

McCreery et al. (1990) have shown that both charge per phase and charge density are important factors in determining neuronal damage to cat cerebral cortex pulsed continuously for 7 hours. In terms of the mass action theory of damage, charge per phase determines the total volume within which neurons are excited, and the charge density determines the proportion of neurons close to an electrode that are excited; thus both factors determine the total change in the extracellular environment. The McCreery data show that as the charge per phase increases, the charge density for safe stimulation decreases. When the total charge is small (as with a microelectrode) a relatively large charge density might safely be used. Shannon (1992) reprocessed the McCreery data and developed an expression for the maximum safe level for stimulation, given by

$$\log(Q/A) = k - \log(Q) \tag{3.38}$$

where Q is the charge per phase (μC/phase), Q/A is charge density per phase (μC/cm^2/phase), and $1.5 < k < 2.0$, fit to the empirical data.

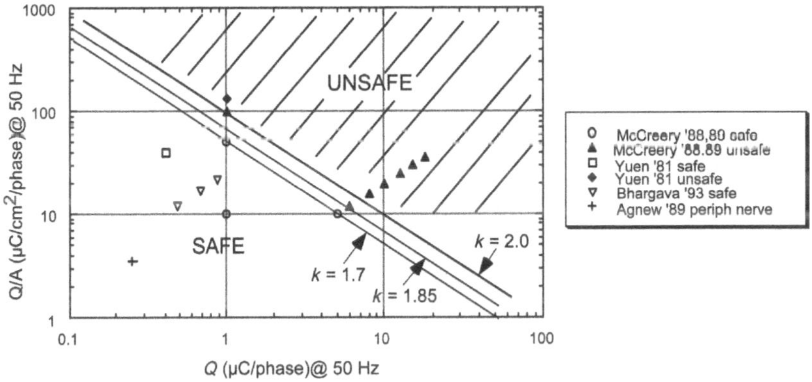

FIGURE 3.16 The Shannon plot: charge (Q) versus charge density (Q/A) for safe stimulation. A microelectrode with relatively small total charge per phase might safely stimulate using a large charge density, whereas a large surface area electrode with greater total charge per phase must use a lower charge density.

Fig. 3.16 illustrates the charge versus charge density relationship of Eq. (3.38) using k values of 1.7, 1.85, and 2.0, with histological data from the 1990 McCreery study using cat parietal cortex as well as data on cat parietal cortex from Yuen, Agnew, Bullara, Jacques, and McCreery (1981), cat peroneal nerve (Agnew et al., 1989), and cat sacral anterior roots (Bhargava, 1993). Above the threshold for damage, experimental data demonstrate tissue damage, and below the threshold line, experimental data indicate no damage.

Studies of penetrating microelectrodes (geometric surface areas $\leq$ 2000 μm^2) in cat cerebral cortex and cochlear nucleus demonstrated the efficacy of stimulation at about 1 nC/phase, and damage occurring at 4 nC/phase but not at 2 nC/phase (McCreery et al., 1994, 2000; McCreery et al., 2002), suggesting a fairly narrow therapeutic window. Microelectrodes behave as point sources from the perspective of tissue. Under the point source idealization, tissue damage is correlated with charge per phase but not charge density. As discussed in Cogan et al. (2016) for microelectrodes the Shannon relationship between charge and charge density should be replaced with a single charge per phase limit independent of charge density. Experimental data with microelectrodes (McCreery et al., 1992, 1997) suggest that this limit is about 4 nC/phase.

3.6.2 Tissue damage from electrochemical reaction products

Supporting the concept that damage is due to electrochemical reaction products is the seminal work by Lilly et al. (1952), which demonstrated that loss of electrical excitability and tissue damage occur when the cerebral cortex of monkey is stimulated using monophasic current pulses. Later, Lilly et al. (1955) showed that biphasic stimulation caused no loss of excitability or tissue damage after 15 weeks of stimulation for 4–5 hours/day. Lilly interpreted these results as due to movement of charged particles such as proteins out of physiological position. The concept that monophasic is a more damaging form of stimulation than charge-balanced biphasic was confirmed by Mortimer et al. (1970), who reported that breakdown of the blood–brain barrier during stimulation of the surface of cat cerebral cortex occurs when monophasic pulses were used at power densities greater than 0.003 W/in^2 (0.5 mW/cm^2), but does not occur with charge-balanced biphasic pulses until a power density of 0.05 W/in^2 (8 mW/cm^2) is exceeded. Pudenz, Bullara, Dru, et al. (1975) and Pudenz, Bullara, Jacques, et al. (1975) further showed that monophasic stimulation of the cat cerebral cortex causes vasoconstriction, thrombosis in venules and arterioles, and blood–brain barrier breakdown within 30 seconds of stimulation when used at levels required for a sensorimotor response; however charge-balanced biphasic stimulation could be used for up to 36 hours continuously without tissue damage if the charge per phase was below 0.45 μC (4.5 $\mu C/cm^2$).

Also supporting the hypothesis that damage is due to electrochemical products are observations of cat muscle that suggest some nonzero level of reaction product can be tolerated (Mortimer et al., 1980; Scheiner & Mortimer, 1990). Scheiner and Mortimer (1990) demonstrated that charge-imbalanced biphasic stimulation allows greater cathodic charge densities than monophasic prior to the onset of tissue damage as reactions occurring during the cathodic phase are reversed by the anodic phase, and also that greater cathodic charge densities can be used than with the charge-balanced waveform prior to electrode corrosion since the anodic phase is no longer constrained to be equal to the cathodic phase, thus the electrode potential reaches less positive values during the anodic phase and IPI.

In 1975, Brummer and Turner gave an alternative explanation to Lilly's for why biphasic pulses were less damaging than monophasic. They proposed that two principles should be followed to achieve electrochemically safe conditions during tissue stimulation:

(1) Perfect symmetry of the electrochemical processes in the two half-waves of the pulses should be sought. This implies that we do not generate any electrolysis products in solution. One approach to achieve this would appear to involve the use of perfectly charge-balanced waveforms of controlled magnitude. (2) The aim should be to inject charge via non-Faradaic or surface-Faradaic processes, to avoid injecting any possibly toxic materials into the body.

Their model for safe stimulation interprets the charge-balanced waveform in electrochemical terms. Any process occurring during the first (stimulating) phase, whether it is charging of the electrode or a reversible Faradaic process, is reversed during the second (reversal) phase, with no net charge delivered. The observation that monophasic stimulation causes greater tissue damage than biphasic stimulation at the same amplitude, pulse width, and frequency is explained by the fact that during monophasic stimulation, all injected charge results in the generation of electrochemical reaction products [Fig. 3.8A; Eq. (3.30)].

Reversible processes include charging and discharging of the double-layer capacitance and surface-bound reversible Faradaic processes such as reactions (3.3)–(3.8), and (3.13). Reversible reactions often involve the production or consumption of hydrogen or hydroxyl ions as the charge counter-ion. This causes a change in the pH of the solution immediately adjacent to the electrode surface. Ballestrasse et al. (1985) gave a mathematical description of these pH changes, and determined that the pH may range from 4 to 10 near a 1 μm diameter electrode during biphasic current pulses, but this change extended for only a few microns. Irreversible processes include Faradaic reactions where the product does not remain near the electrode surface, such as reactions (3.1) and (3.9)–(3.12).

Free radicals are known to cause damage to myelin, the lipid cell membrane, and DNA of cells. A likely candidate for a mechanism of neural tissue damage due to electrochemical products is peroxidation of the myelin by free radicals produced on the electrode surface. Several researchers (Buettner, 1993; Chan et al., 1982; Chia et al., 1983; Griot et al., 1990; Konat & Wiggins, 1985; Sevanian, 1988) have demonstrated the great susceptibility of myelin to free radical damage. Damage occurs as fatty acyl chains move apart and the myelin goes from a crystalline (ordered) state to a liquid (disordered) state.

Morton et al. (1994) have shown that oxygen reduction occurs on a gold electrode in phosphate-buffered saline under typical neural-stimulating conditions. Oxygen reduction reactions that may occur during the cathodic stimulating phase include reactions that generate free radicals such as superoxide and hydroxyl, and hydrogen peroxide, collectively known as reactive oxygen species. These species may have multiple deleterious effects on tissue (Bergamini et al., 2004; Halliwell, 1992; Hemnani & Parihar, 1998; Imlay, 2003; Stohs, 1995). As free radicals are produced they may interfere with chemical signaling pathways that maintain proper perfusion of nervous tissue. Nitric oxide has been identified as the endothelium-derived relaxing factor, the primary vasodilator (Furchgott, 1988; Ignarro et al., 1988; Umans & Levi, 1995). Nitric oxide is also known to prevent platelet aggregation and adhesion (Azuma et al., 1986; Moncada et al., 1991; Radomski et al., 1987). Beckman et al. (1990) have shown that the superoxide radical reacts with nitric oxide to form the peroxynitrite radical. Oxygen-derived free radicals from the electrode may reduce the nitric oxide concentration and diminish its ability as the principal vasodilator and as an inhibitor of platelet aggregation. Superoxide depresses vascular smooth muscle relaxation by inactivating nitric oxide, as reviewed by Rubanyi (1988).

An electrochemical product may accumulate to detrimental concentrations if the rate of Faradaic reaction, given by the current–overpotential relationship of Eq. (3.18), exceeds the rate for which the physiological system can tolerate the product. For most reaction products of interest there is some sufficiently low concentration near the electrode that can be tolerated over the long term. This level for a tolerable reaction may be determined by the capacity of an intrinsic buffering system. For example, changes in pH are buffered by several systems including the bicarbonate buffer system, the phosphate buffer system, and intracellular proteins. The superoxide radical, a product of the reduction of oxygen, is converted by superoxide dismutase and cytochrome c to hydrogen peroxide and oxygen. The diffusion rate of a toxic product must be considered, as it may be the case that high concentrations only exist very near the site of generation (the electrode surface).

In addition to mass action and electrochemical factors, mechanical forces may be responsible for tissue damage. McCreery, Agnew, et al. (1992) reviewed the damage from electrical stimulation of peripheral nerve. They concluded that damage may be from mechanical constriction of the nerve as well as neuronal hyperactivity and irreversible reactions at the electrode.

3.6.3 Multiple contributing factors

Fig. 3.17 illustrates the interplay of intrinsic biological processes (mass action) and Faradaic product generation in stimulation-induced damage of tissue and the electrode. Charge density and charge per phase, the two factors on the Shannon plot, can be related to mass action. Charge density determines the proportion of neurons close to an electrode that are excited, and charge per phase determines the total volume within which neurons are excited. Charge and charge density can alternatively be useful in quantifying Faradaic damage. A given Faradaic process is likely to occur if a sufficient overpotential (with units of voltage) is attained. Because capacitance per area is an intrinsic material property, charge density is sufficient to

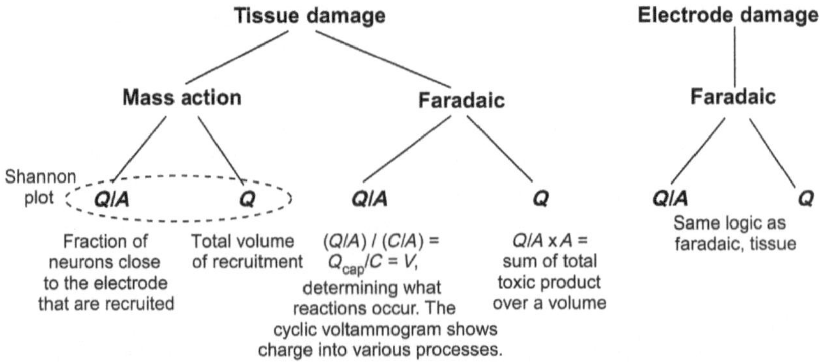

FIGURE 3.17 Mechanisms of stimulation-induced damage: mass action and Faradaic (electrochemical).

determine the overpotential and possible Faradaic processes. Electrode damage does not have an equivalent to mass action. The logic for charge density being sufficient to determine overpotential and therefore possible reactions is maintained for the electrode—if a sufficient charge density is injected anodically, the electrode will attain an electrical potential where damaging oxidative reactions occur.

While it would be convenient if one or two factors determined whether tissue damage will occur, this is generally not the case. The causes of tissue damage are multidimensional, including intrinsic biological processes (mass action), electrochemical product creation, and mechanical factors. The presence and extent of damage depend on many factors such as waveform specification (charge, charge density, pulse width, frequency, duty cycle, etc.) as well as species and tissue type. Theoretical understanding and empirical data allow us to make good estimates of likelihood for damage, but ultimately bench studies followed by preclinical (animal) studies prior to clinical use remain the gold standard to demonstrate safety.

There are a handful of useful tools for estimating likelihood of damage. The Shannon plot (Fig. 3.16) is commonly taken as a starting point for mass action effects. Recall that the Shannon relationship was derived from data on cat cerebral cortex pulsed continuously for 7 hours. Thus for extended continuous stimulation of CNS tissue, the Shannon plot is often a good model. As one moves away from those parameters, for example, by using a low-duty cycle stimulation, the Shannon plot becomes less appropriate. A strict reliance on the Shannon plot considers only charge and charge density as factors in damage; in fact, other factors may be important. A stimulation protocol with a given charge per phase, charge density, and pulse width may be safe at low frequency but cause excitotoxicity at higher frequencies. Lastly, and as previously mentioned, for microelectrodes the Shannon relationship might be replaced with a single charge per phase limit of about 4 nC/phase.

While the Shannon relationship may be a good indicator of mass action effects, we begin with RCSC, a material characteristic, to estimate damage from electrochemical products. The CSC quantifies the total amount of charge that may be injected through capacitive charging/discharging and reversible Faradaic processes before the onset of irreversible Faradaic reactions, which change the environment with possible deleterious effects.

3.7 Design compromises for efficacy and safety

The RCSC or CSC of an electrode is the total amount of charge that may be stored reversibly, including storage in the double-layer capacitance, pseudocapacitance, or any reversible Faradaic reaction (Robblee & Rose, 1990). The slow cyclic voltammogram (SCV) for a material displays the current density into various electrochemical processes as a function of the electrode potential as the potential is slowly cycled (Fig. 3.10). At any point in time, the current going into a particular process is determined by the potential as well as by the reactant concentration, as given by Eq. (3.18). The water window is defined as the potential region between the oxidation of water to form oxygen and the reduction of water to form hydrogen (processes shown in red shading, Fig. 3.10). For common noble metal electrodes, including platinum and iridium oxide, the water window is often taken to be between −0.6 and +0.8 V versus Ag/AgCl.

A common parameter for characterizing stimulating electrodes is the cathodal charge storage capacity (CSC_c). This is defined as the integrated area under the SCV curve, defined inside the water window potential range, and negative of zero current. This represents the maximum charge that may be reversibly injected during a cathodic pulse under optimal conditions.

It is important to note that potentially deleterious processes might occur at potentials within the water window, including irreversible oxygen reduction (Merrill, 2002; Merrill et al., 2005b; Morton et al., 1994). Therefore the common belief that any electrode potential within the water window is safe should be viewed with caution.

The reversible CIC is that subset of the CSC that is available for injection during pulsing. In electrical stimulation of excitable tissue, it is desirable to have a large CIC so that a large amount of charge may be injected (thus being efficacious for stimulation) prior to the onset of irreversible Faradaic reactions (which may be deleterious to the tissue being stimulated or to the electrode itself). The CIC depends upon the material used for the electrode, the size and shape of the electrode, the electrolyte composition, and parameters of the electrical stimulation waveform.

A stimulating system must be both effective and safe. The effectiveness of stimulation means the ability to elicit the desired physiological response, which can include initiation or suppression of action potentials. Safety has two primary aspects: first, the tissue being stimulated must not be damaged, and second, the stimulating electrode itself must not be damaged as in corrosion. An electrode implanted into a human as a prosthesis may need to meet these requirements for decades. In animal experimentation, damage to the tissue or the electrode can seriously complicate or invalidate the interpretation of results.

Effectiveness requires that the charge injected must exceed some threshold (Fig. 3.15). However, as the charge per phase increases, the overpotential of the electrode increases, as does the fraction of the current going into Faradaic reactions (which may be damaging to tissue or the electrode if the reaction is irreversible) (Fig. 3.6). Judicious design of stimulation protocols involves acceptable compromises between stimulation effectiveness, requiring a sufficiently high charge per phase, and safety, requiring a sufficiently low charge per phase, thus preventing the electrode from reaching potentials where deleterious Faradaic reactions occur at an intolerable rate. The overpotential an electrode reaches, and thus Faradaic reactions that can occur, depends on several factors in addition to the charge per phase, including (1) waveform type (Figs. 3.7), (2) stimulation frequency, (3) electrode material (a high charge-storage capacity allows large charge storage prior to reaching overpotentials where irreversible Faradaic reactions occur), (4) electrode geometric area and roughness (determining real area) and therefore total capacitance, and (5) train effects (Section 3.4.9). Increasing either the stimulus phase pulse width or the reversal phase pulse width of a charge-balanced stimulation protocol has the effect of increasing unrecoverable charge into irreversible reactions. Any factor which either drives the electrode potential into a range where irreversible reactions occur (such as a long stimulus phase pulse width) or fails to quickly reverse the electrode potential out of this range (such as a long reversal phase pulse width) will allow accumulation of unrecoverable charge. The overpotential an electrode must be driven to before any given current will be achieved is highly dependent on the kinetics of the system, characterized by the exchange current density i_0. For a system with a large exchange current density (e.g., $i_0 = 10^{-3}$ A/cm^2), no significant overpotential may be achieved before a large Faradaic current ensues [Eq. (3.18)]. When i_0 is many orders of magnitude smaller (e.g., $i_0 = 10^{-9}$ A/cm^2), a large overpotential must be applied before there is substantial Faradaic current. When i_0 is very low, a large total charge can be injected through the capacitive mechanism before significant Faradaic reactions commence.

This is the generally desirable paradigm for a stimulating electrode, minimizing Faradaic reactions that lead to either tissue damage or electrode damage. The fundamental design criteria for an electrochemically safe stimulation protocol can be stated as:

The electrode potential must be kept within a potential window where irreversible Faradaic reactions do not occur at levels that are intolerable to the physiological system or the electrode.

If irreversible Faradaic reactions do occur, one must ensure that they can be tolerated (e.g., that physiological buffering systems can accommodate any toxic products) or that their detrimental effects are low in magnitude (e.g., that corrosion occurs at a very slow rate, and the electrode will remain intact for its design lifetime).

Fig. 3.18 summarizes key features of various stimulation waveform types. The cathodic monophasic waveform (Fig. 3.18A) consists of pulses of current passed in one direction, with an open-circuit condition during the IPI. At no time does current pass in the opposite direction. Commonly the WE is pulsed cathodically for stimulation of tissue (as shown), although anodic stimulation may also be used (Section 3.5.1). Of the waveforms illustrated in Fig. 3.18, the monophasic is the most effective for stimulation. However, monophasic pulses are not used in long-term stimulation where tissue damage is to be avoided. Greater negative overpotentials are reached during monophasic pulsing than with biphasic pulsing (Fig. 3.8). Furthermore, the

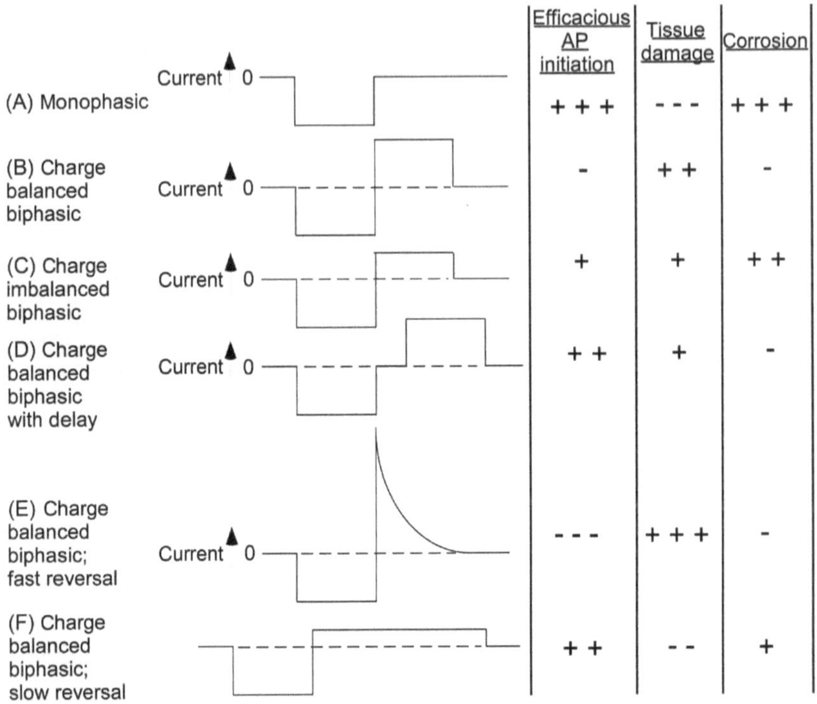

FIGURE 3.18 Comparison of stimulating waveforms. Six archetypical waveforms are rated for relative merit in efficacy and safety. "+ + +" = best (most efficacious, least damaging to tissue or the electrode), "– – –" = worst.

electrode potential during the IPI of cathodic monophasic pulsing remains relatively negative as the charged electrode capacitance slowly discharges through Faradaic reactions, allowing reduction reactions which may be deleterious to tissue to proceed throughout the entire period of stimulation. Biphasic waveforms are illustrated in Fig. 3.18B–F. The first (stimulating) phase elicits the desired physiological effect such as initiation of an action potential, and the second (reversal) phase is used to reverse the direction of electrochemical processes occurring during the stimulating phase. If all processes of charge injection during the stimulating phase are reversible, then the reversal phase will prevent net changes in the chemical environment of the electrode, as desired. The charge-balanced biphasic waveform (Fig. 3.18B) is widely used to prevent tissue damage. It should be noted that charge balance does not necessarily equate to electrochemical balance. As given by Eq. (3.32), during certain instances of stimulation there are irreversible Faradaic reactions during the cathodic phase (e.g., oxygen reduction), then different irreversible reactions during the anodic phase (e.g., electrode corrosion) that are not the reverse of the cathodic Faradaic reactions. Such electrochemical imbalance leads to a potential waveform as illustrated in Fig. 3.8B, where the potential at the end of the anodic phase is positive of the prepulse potential, allowing irreversible reactions such as electrode corrosion to occur. The charge-imbalanced waveform, illustrated in Fig. 3.18C, may be used to reduce the most positive potentials during the anodic phase with respect to the charge-balanced waveform, and prevent electrode corrosion (Scheiner & Mortimer, 1990). Ideally, the charge in the reversal phase is equal to the charge going into reversible processes during the stimulation phase, in which case the electrode potential returns to its prepulse value at the end of the reversal phase.

In addition to electrode corrosion, a second concern with the charge-balanced biphasic waveform is that the reversal phase not only reverses electrochemical processes of the stimulation phase, but may also reverse some of the desired physiological effect of the stimulation phase, that is, it may suppress an action potential that would otherwise be induced by a monophasic waveform. This effect causes an increased threshold for biphasic stimulation relative to monophasic. Gorman and Mortimer (1983) have shown that by introducing an open-circuit interphase delay between the stimulating and reversal phases, the threshold for biphasic stimulation is similar to that for monophasic. This is illustrated in Fig. 3.18D. Although the introduction of an interphase delay improves the threshold, it also allows the electrode potential to remain relatively negative during the delay period. A delay of 100 μs is typically sufficient to prevent the suppressing effect of the reversal phase, and may be a short enough period that deleterious Faradaic reaction products do not accumulate to an unacceptable level.

As illustrated in Fig. 3.18E and F, the more rapidly charge is injected during the anodic reversal phase, the more quickly the electrode potential is brought out of the most negative range, and thus the less likely that tissue damage will occur. A high current reversal phase however, means more of a suppressing effect on action potential initiation, and also means the electrode potential will move positive during the reversal phase, thus risking electrode corrosion.

3.8 Requirements for efficacy and safety of a recording device

Table 3.2 lists the primary points of consideration regarding the design of a neural recording system. This is meant as a high-level list and brief introduction. Individual topics will be expanded upon in the following sections. The major points below are divided into topics of (1) efficacy/effectiveness, (2) safety, (3) performance, and (4) reliability.

The safety requirements for a recording system are in many ways simpler than for a stimulating system, due to the passive nature of the terminal devices (electrodes). In a stimulating system the actively driven electrodes

TABLE 3.2 Primary considerations in design of a neural recording system, based on (1) efficacy/effectiveness, (2) safety, (3) performance, and (4) reliability.

Efficacy/effectiveness
E1. The system is effective. The combined electrodes, lead, and electronics are capable of acquiring and resolving signals of interest within the real-world environment
Safety
S1. Tissue is not damaged from excessive mechanical forces
S2. The passive (unstimulated) device materials in contact with tissue are biocompatible, defined bidirectionally:
S2a. The tissue is not damaged by the device. The device does not induce a toxic or necrotic response, nor an excessive foreign body or immune response
S2b. The device is not damaged by the tissue
S3. Mechanical moduli of the system are compatible with tissue
S4. The implanted device is minimally invasive. As a minimum, invasiveness is justified by the benefit
S5. Requirements for safety are met as specified in standards including IEC 60601-1 and its collateral and particular standards, and ISO 14708-1
S6. No galvanic cells are formed by dissimilar metals which drive consequential currents
Performance
P1. All specifications as claimed in the device labeling are met
P2. Electrodes are sufficiently low impedance for the intended application, mitigating excessive noise and shunt losses
P3. The overall system is "sensitive," defined as able to detect the intended signals, potentially of low amplitude
P4. The overall system is "specific," defined as able to resolve signals of interest in the presence of other confounding signals, that is, to reject signals not of interest
P5. The overall system has sufficiently low noise figure over the bandwidth of interest
Reliability
R1. The system is mechanically acceptable for the application
R1a. Implanted devices tolerate the insertion procedure, for example, the material does not buckle if it passes through the meninges
R1b. If a device is to be used chronically, it is flexible and fatigue-resistant to withstand movement between the device and tissue following implantation

(*Continued*)

TABLE 3.2 (Continued)

R2. The material characteristics of the system are acceptably stable for the duration of the implant
R2a. Mechanical properties remain intact given the intended tissue, surgical procedure, and duration of use
R2b. Electrode electrical impedance is stable
R2c. Conducting and insulating properties of the leads remain intact
R3. Materials are corrosion resistant
R4. The packaging maintains hermeticity to prevent water ingress and damage to the electronics

develop overpotentials which may drive damaging Faradaic reactions. This does not occur in a recording system (see Section 3.9 and Figs. 3.4 and 3.19 for a comparison of electrical potential profiles). The biocompatibility (Section 3.10) of the passive electrodes and entire implanted device must be considered. One must also ensure that there are no substantial galvanic cells set up by dissimilar metals, sufficient to drive unacceptable currents.

Many of the requirements listed in Table 3.2 must be evaluated at the level of the complete system. For example, "sensitivity," which is the ability to detect potentially low-level signals of interest, and "specificity," the ability to reject signals not of interest, depend on the construction of individual electrodes (surface area, material, impedance, etc.), the relative placement of multiple electrodes, and upstream signal processing as implemented in hardware and software/firmware. Section 3.11 discusses further instrumentation issues.

3.9 Electrical model of the recording electrode

As discussed in Section 3.4.5, when a current or voltage source is applied across electrodes in tissue the interfacial potentials at each electrode deviate from their equilibrium values, that is, overpotentials develop (Fig. 3.19A). If the overpotentials reach sufficient values, Faradaic reactions may commence. The electrical potential profiles are quite different in the recording situation (Fig. 3.19B). When an electric field is imposed by an external source, for example, firing neurons, the electron energy levels in the bulk of the metal electrode are raised or lowered, as are the energy levels in the tissue. Unlike the stimulation case, this does not change the relative energy levels between the metal electrode and the electrolyte at the interface—they are raised and lowered in unison; thus no overpotentials develop to drive Faradaic reactions. While current may flow in tissue due to the imposed voltage, for a

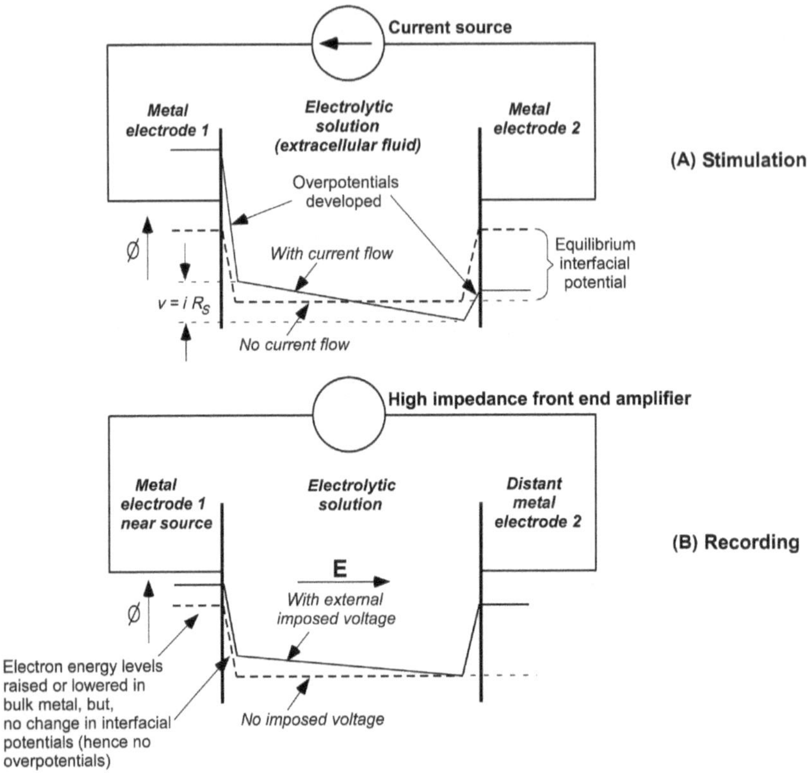

FIGURE 3.19 **Comparison of electrical models for stimulation and recording.** (A) During stimulation, two electrodes are actively driven by a source and the interfacial potentials change at both electrodes (thus developing overpotentials). (B) During recording, an electric field is imposed by an external source (perhaps a firing neuron). The E field changes the energy levels in the nearby electrode but does not change the interfacial potential. The potential gradient in tissue can cause current flow in tissue, although there should be minimal current into the high-input impedance front-end amplifier.

properly designed high-input impedance front-end amplifier there should be very little current flow through the electrodes and into the front end.

3.10 Materials used for stimulating and recording electrodes

Tables 3.1 and 3.2 list the requirements for neuroengineering systems. Among these, materials requirements for electrodes include the following. (1) The passive material is biocompatible, so it does not induce a toxic or necrotic response in the adjacent tissue, nor an excessive foreign body or immune response. (2) The material is mechanically acceptable for the application. (3) No galvanic cells are formed by dissimilar metals which drive consequential currents. Additionally, for an actively driven (stimulating) system the following apply. (4) Sufficient charge is injected with the chosen material and electrode area to elicit the desired physiological response. (5) Faradaic reactions do not occur at levels that are toxic to the surrounding tissue, and (6) Faradaic corrosion reactions do not occur at levels that will cause premature failure of the electrode. The extent of charge which may be injected prior to the onset of irreversible Faradaic reactions is characterized by the reversible charge storage capacity (CSC) and reversible CIC (Sections 3.4.10, 3.7).

Dymond et al. (1970) tested the toxicity of several metals implanted into cat cerebral cortex for 2 months. Materials were deemed toxic if the reaction to the implanted metal was significantly greater than the reaction to a puncture made from the same metal that was immediately withdrawn. Stensaas and Stensaas (1978) reported on the biocompatibility of several materials implanted passively into rabbit cerebral cortex. Materials were classified into one of three categories depending upon changes occurring at the implant–cortex interface. (1) *Nonreactive*. For these materials, little or no gliosis occurred, and normal CNS tissue with synapses was observed within 5 μm of the interface. (2) *Reactive*. Multinucleate giant cells and a thin layer (10 μm) of connective tissue surrounded the implant. Outside of this was a zone of astrocytosis. Normal CNS tissue was observed within 50 μm of the implant. (3) *Toxic*. These materials are separated from the cortical tissue by a capsule of cellular connective tissue and a surrounding zone of astrocytosis. Loeb et al. (1977) studied the histological response to materials used by the microelectronics industry implanted chronically in the subdural space of cats, and found reactions to be quite dependent on specific material formulations and surface preparations. Biocompatibility results from these investigators and others are summarized in Table 3.3.

Table 3.4 lists the CSC, CIC, corrosion resistance, and mechanical characteristics of several common and emerging electrode materials.

The first intracortical electrodes consisted of single-site conductive microelectrodes made of material stiff enough to penetrate the meninges, either an insulated metallic wire such as platinum, gold, tungsten, iridium, or stainless steel, or a glass micropipette filled with conductive electrolyte. Because of their small size and high impedance, microelectrodes only acquire signals from sources in very close proximity to the electrode, and useful recordings seldom last more than a few hours. Advances in microelectronics technology have allowed the development of multiple-site electrodes built onto a single substrate, using silicon micromachining and planar photolithographic technologies. Early silicon-based electrode arrays include the Utah Electrode Array (UEA) (Campbell et al., 1991; Rousche & Normann, 1998) and the planar "Michigan electrode" (Kipke et al., 2003).

Commonly used electrode materials include platinum, platinum–iridium, iridium oxide, gold, stainless steel, tungsten, and titanium alloys. The RCSC of bare iridium is similar to that of platinum; however, when a surface oxide

TABLE 3.3 Biocompatibility of common materials.

	Classification by Dymond et al. (1970)	Classification by Stensaas and Stensaas (1978)	Other references
Conductors			
Aluminum		Nonreactive	
Cobalt		Toxic	
Copper		Toxic	Toxic (Babb & Kupfer, 1984; Fisher et al., 1961; Sawyer & Srinivasan, 1974)
Gold	Nontoxic	Nonreactive	
Gold–nickel–chromium	Nontoxic		
Gold–palladium–rhodium	Nontoxic		
Iron		Toxic	
Molybdenum		Reactive	
Nickel–chromium (Nichrome)		Reactive	Nontoxic (Babb & Kupfer, 1984)
Nickel–chromium–molybdenum	Nontoxic		
Nickel–titanium (Nitinol)			Biocompatible (Bogdanski et al., 2002; Ryhanen et al., 1998)
Polypyrrole and poly(3,4-ethylenedioxythiophene)			Biocompatible (Cui et al., 2003; Stritesky et al., 2018)
Platinum	Nontoxic	Nonreactive	Biocompatible (Chouard & Pialoux, 1995; Majji et al., 1999)

Platinum–iridium	Nontoxic		Biocompatible (Niparko et al., 1989)
Pt-Ir, high surface area			Biocompatible (Dalrymple et al., 2020)
Platinum–nickel	Nontoxic		
Platinum–rhodium	Nontoxic		
Platinum–tungsten	Nontoxic		
Platinized platinum (Pt black)	Nontoxic		
Rhenium	Nontoxic		
Silver	Toxic	Toxic	Toxic (Babb & Kupfer, 1984; Fisher et al., 1961; Sawyer & Srinivasan, 1974)
Stainless steel	Nontoxic		Nontoxic (Babb & Kupfer, 1984)
Tantalum		Reactive	
Titanium			Biocompatible (Cui et al., 2003)
Tungsten		Nonreactive	
Insulators			
Amorphous silicon carbide (a-SiC)			Biocompatible (Cogan et al., 2003; Frewin et al., 2011)
Alumina ceramic		Nonreactive	Biocompatible (Chouard & Pialoux, 1995)
Araldite (epoxy plastic resin)		Reactive	
Polyethylene		Nonreactive	
Polyimide			Biocompatible (Stieglitz et al., 1999)

(*Continued*)

TABLE 3.3 (Continued)

	Classification by Dymond et al. (1970)	Classification by Stensaas and Stensaas (1978)	Other references
Polypropylene		Nonreactive	
Silastic RTV		Toxic	
Silicon dioxide (Pyrex)		Reactive	
Teflon TFE (high purity)		Nonreactive	
Teflon TFE (shrinkable)		Reactive	
Titanium dioxide		Reactive	
Semiconductors			
Germanium		Toxic	
Silicon		Nonreactive	Biocompatible (Hoogerwerf & Wise, 1994; Kristensen et al., 2001; Schmidt et al., 1993)
Assemblies			
Gold–silicon dioxide passivated microcircuit		Reactive	

TABLE 3.4 Characteristics of common and emerging electrode materials.

Electrode material	Reversible charge storage capacity (μC/cm²)	Reversible charge injection capacity (μC/cm²)	Corrosion resistance	Mechanical characteristics
Platinum	300–350 (r) (Brummer & Turner, 1977c); 550 (Cogan, 2008)		Resistant; greatly increased resistance with protein adsorption	Relatively soft
Anodic first, charge balanced, 200 μs		50–100 (g) (Rose & Robblee, 1990)		
Cathodic first, charge balanced, 200 μs		100–150 (g) (Rose & Robblee, 1990)		
Platinum/iridium alloys	Similar to Pt			Stronger than Pt
Platinum/iridium, high roughness factor	6000 (pre), 2500 (p-i) (Dalrymple et al., 2020)	55 (pre), 60 (p-i) (Dalrymple et al., 2020)		
Iridium	Similar to Pt		Resistant	Stronger than Pt
Iridium oxide			Highly resistant (Agnew et al., 1986)	Oxide is moderately acherent (Meyer et al., 2001)
AIROF		2200 (g, AF) (Beebe & Rose, 1988; Kelliher & Rose, 1989); 1200 (g, CF) (Beebe & Rose, 1988; Kelliher & Rose, 1989); 3500 (g, CF, AB) (Agnew et al., 1986; Beebe & Rose, 1988; Kelliher & Rose, 1989); 3300 (CF, AB) (Cogan et al., 2006); 9600 (AF, AB) (Cogan et al., 2006)		

(*Continued*)

TABLE 3.4 (Continued)

Electrode material	Reversible charge storage capacity ($\mu C/cm^2$)	Reversible charge injection capacity ($\mu C/cm^2$)	Corrosion resistance	Mechanical characteristics
EIROF	>25,000 (Meyer et al., 2001)	1200 (Meyer et al., 2001); >17,000 (Cogan, 2018)		
SIROF	2800 (Cogan, 2008)			
316 LVM stainless steel	40–50 (g)		Resistant in passive region; rapid breakdown in transpassive region	Strong and flexible
Tantalum/tantalum pentoxide	700 (g) (Guyton & Hambrecht, 1974); 200 (g) (Rose et al., 1985)		Resistant (Johnson et al., 1977)	
Titanium nitride	250 (Cogan, 2008)	870 (Weiland et al., 2002)		
Poly(3,4-ethylenedioxythiophene) (PEDOT)		2920 (Venkatraman et al., 2011)		Soft polymer
Carbon nanotubes		1000–1600 (Wang et al., 2006)		

AB, anodic bias; AF, anodic first biphasic pulsing; CF, cathodic first biphasic pulsing; g, geometric area; pre, preimplant; p-i, 5 weeks postimplant; r, real area.

is present on iridium it has greatly increased CSC over platinum. This occurs via reversible conversion between Ir^{3+} and Ir^{4+} states within the oxide. By using an anodic bias, cathodic charge densities of 3.5 mC/cm^2 (geometric) have been demonstrated in vivo (Agnew et al., 1986). Meyer et al. (2001) reported on a method to electrodeposit iridium oxide films onto substrates of gold, platinum, platinum–iridium, and 316LVM stainless steel, achieving a CSC of > 25 mC/cm^2.

Platinum is a relatively soft material and may not be mechanically acceptable for all stimulation applications. Platinum is often alloyed with iridium to increase the mechanical strength. Alloys of platinum with 10%–30% iridium have similar CSC to pure platinum (Robblee et al., 1983). Iridium is a much harder metal than platinum, with mechanical properties that make it suitable as an intracortical electrode.

The stainless steels (303, 316, and 316LVM) and cobalt–nickel–chromium–molybdenum alloy MP35N are protected from corrosion by a thin passivation layer that forms a barrier to reaction. For stainless steel this layer consists of iron oxides, iron hydroxides, and chromium oxides. Charge is injected by reversible oxidation and reduction of the passivation layers. If the electrode potential becomes too positive (the transpassive region), breakdown of the passivation layer and irreversible metal dissolution may occur (Greatbatch & Chardack, 1968; Loucks et al., 1959; White & Gross, 1974), leading to failure of the electrode. A charge imbalance has been shown to allow increased charge injection without electrode corrosion (McHardy et al., 1977; Scheiner & Mortimer, 1990). Titanium and cobalt–chromium alloys are also protected from corrosion by a surface oxide passivation layer, and demonstrate better corrosion resistance than does stainless steel (Gotman, 1997). 316LVM stainless steel has good mechanical properties and has been used for intramuscular electrodes. The CSC of 316LVM is only 40–50 $\mu C/cm^2$ (geometric), potentially necessitating large surface area electrodes.

Increasing the roughness factor (real area/geometric area) of an electrode increases the geometric CSC and CIC (from the tissue perspective). This has been exploited for titanium nitride, providing a CIC of 870 $\mu C/cm^2$ (Weiland et al., 2002), and for platinum–iridium. Dalrymple et al. (2020) demonstrated early in vivo results with high surface area Pt-Ir of about 6 mC/cm^2 preimplant and 2.5 mC/cm^2 after 5 weeks stimulation in rat cochlea.

Inherently conducting polymers (ICPs) including polypyrrole and poly (3,4-ethylenedioxythiophene) (PEDOT) have been investigated for use as electrodes. PEDOT is attractive as a chronic neural electrode for several reasons: (1) It is soft and flexible. (2) It provides a large, distributed surface area for contact with neural tissue. (3) It has been shown to extend beyond the glial scar. (4) Coating an electrode with PEDOT decreases the impedance by up to two orders of magnitude. (5) PEDOT offers high CSC for stimulation. (6) It demonstrates electrochemically stability. (7) It can incorporate bioactive molecules.

Richardson-Burns et al. (2007) demonstrated electrochemical polymerization of PEDOT in vivo using gold wire in mouse brain slice. A fuzzy cloud of conductive filaments was formed. Electrical impedance decreased by one to two orders of magnitude, attributed to the high surface area. Venkatraman et al. (2011) performed in vitro comparisons of PEDOT electrodes to Pt-Ir and iridium oxide at zero volts bias, and in vivo comparisons to Pt-Ir. PEDOT electrodes had a 15-fold increase in CSC relative to Pt-Ir and iridium oxide. In vivo experiments in rat cortex demonstrated improved signal-to-noise ratio and increased CIC relative to Pt-Ir over 2 weeks of implantation. PEDOT showed long-term stability using voltage cycling between −0.6 and +0.8 V, 24 hours continuous biphasic pulsing at 3 mC/cm^2, and accelerated lifetime testing for 4 weeks at 67°C.

Growth factors, cell adhesion peptides, and extracellular matrix proteins to enhance neural growth and binding to the electrode sites may be incorporated into ICP including polypyrrole and PEDOT (Cui et al., 2001; Cui & Martin, 2003; Kim et al., 2007). Electrodes coated with PEDOT doped with multiwalled carbon nanotubes demonstrated significant improvement in recording stability compared to PEDOT/PSS recording sites (Kozai et al., 2016).

For further discussion of electrode materials considerations, refer to Merrill (2014).

3.11 Instrumentation

This chapter concludes with a discussion of the "upstream" instrumentation which drives stimulation electrodes and acquires and processes information from recording electrodes. Essential considerations are presented although design details are beyond the scope of this work.

3.11.1 Stimulation parameters of interest

Table 3.5 lists several key parameters to be specified for a stimulation system. This assumes charge-balanced biphasic stimulation (vs monophasic, which is unsafe for chronic stimulation).

The energy source type, that is, current or voltage control, should be specified. Current-controlled stimulation has the advantage that current is constant throughout the pulse period; whereas during voltage-controlled stimulation current is not constant and the extracellular voltage generally decreases during the pulse as the electrode interfaces charge up. Minimum and maximum current or voltage and step size resolution should be specified. For current control, the maximum source voltage (called compliance voltage) should be stated. If the product of commanded current and load impedance

TABLE 3.5 Key specifications for a stimulation system.

1. Source type: current or voltage
2. Current maximum
3. Voltage maximum (compliance voltage for a current source)
4. Current or voltage resolution
5. Pulse width min and max, resolution
6. Frequency min and max, resolution
7. Maximum charge per phase
8. Programmability for arbitrary train definition
9. Interphase delay
10. Number of independent channels
11. Channel configuration: monopolar versus bipolar
12. Charge balance: mechanism and extent of balance
13. Symmetry of charge balance
14. Output impedance
15. Compliance with safety standards

exceeds the compliance voltage, the commanded current will not be met. Similarly, a voltage control system will fail if the load impedance is too small and maximum current is exceeded. The minimum, maximum, and resolution of pulse width and frequency should be specified.

Maximum charge per pulse (not necessarily the product of maximum current and pulse width) should be specified. For safety reasons, active measures may be instituted to limit this.

The extent of pulsing protocol flexibility should be given. Many stimulators will provide choice of pulse amplitude, width, and frequency only. Advanced systems may allow train durations, ramp up and down times, etc. The ability to insert an interphase delay should be specified.

The number of independent channels should be stated, as well as channel configuration. Monopolar stimulation provides each specified channel as an independent sink or source, all with respect to a distant CE (perhaps the stimulator metal case). Bipolar stimulation allows current flow between two specified electrodes.

The process for charge balance should be stated. This may be through passive capacitive discharge following the stimulation phase, or it may be active with the advantage of faster charge reversal. The claimed maximum charge imbalance should be stated. Symmetry of phases should be stated, that is, whether charge balance is by symmetric phases, by asymmetric (e.g., 2:1, 4:1, etc.), or by exponential decay with a capacitor.

Compliance with safety standards should be specified, including IEC 60601-1 and its collaterals (60601-1-X) and particulars (60601-2-X), ISO 14708-1 and -3, electromagnetic compatibility standards such as IEC 60601-1-2, and biocompatibility by ISO 10993-1.

3.11.2 Recording architecture and parameters of interest

A typical recording front end comprises a chain of electrode → low-noise amplifier (LNA) → filtering and further amplification → sampling → digitization. The first active element, the LNA, acquires the low-level signal. The necessity for low noise drives design creativity in these elements. The next stage is a second level of gain with analog filtering. The amplifier may be preceded by a high-pass filter to minimize noise and followed by low-pass filtering. The analog signal is then sampled at a specified rate and digitized by an analog-to-digital converter (ADC) with specified bit resolution. Further processing is in the digital domain.

Table 3.6 lists several key parameters to be specified for a recording system.

Specifications for an LNA include the maximum linear input signal (routinely a few mV), maximum input voltage without damage to the LNA, input

TABLE 3.6 Key specifications for a recording system.

1. Maximum linear input signal
2. Maximum input voltage without device damage
3. Input impedance
4. Low frequency gain
5. 3 dB bandwidth
6. Noise (over a specified frequency range)
7. Common mode rejection ratio
8. Sensitivity
9. Analog-to-digital converter (ADC) sample rate
10. ADC bit depth
11. Number of channels
12. Channel configuration: single-ended versus differential
13. Crosstalk between channels
14. Compliance with safety standards

impedance, low frequency gain, 3 dB bandwidth, noise, and common mode rejection ratio (CMRR). Input impedance is often given as a resistance (on the order of Gigaohms) in parallel with a capacitance (a few pF). Noise and CMRR are discussed in Sections 3.11.3 and 3.11.4.

Sensitivity is defined as the minimum signal level that can be detected with acceptable quality and is dependent on the overall system design.

ADC resolution is defined by the number of output bits. An ADC with M bits can represent 2^M values; for example, a 16-bit ADC represents $2^{16} = 65{,}536$ values. ADC resolution can also be given in volts. If V_{FS} is the full-scale voltage measurement range, then the analog input change required to toggle an output bit is given by $V_{FS} / 2^M$.

Recording channels may be differential or single-ended. A differential or floating channel is not referenced to ground; rather it is derived between two electrodes (analogous to a bipolar stimulation system). This has the advantage of noise rejection, since noise is acquired by both electrodes and can be filtered by upstream common mode rejection electronics. A single-ended recording measures the voltage between an electrode and ground.

Crosstalk is the amount of signal that is inadvertently coupled between channels and is given as a percent of coupled signal or as dB of rejection. An issue with microelectrode design is that as electrode sizes are reduced, capacitive coupling may increase, leading to increased electrical crosstalk (Najafi et al., 1990).

As with stimulation systems, compliance with safety standards should be specified.

For a combined system where the ability to stimulate and record is intended to be near-simultaneous, the switching time between modes should be specified. The methods for artifact rejection (blanking out the large stimulation signal from an immediate poststim recording) may be implemented in both hardware and software.

3.11.3 Noise

Noise can be defined as variations in an acquired signal due to numerous sources other than the intended signal of interest. These noise sources interfere with interpretation of the signal of interest. Many noise sources are random with a Gaussian distribution of instantaneous amplitudes versus time. A common measure of noise amplitude is the root mean square (rms) value.

When considering noise in the time domain, it is important to note the bandwidth over which the noise is observed. Some noise sources have a limited bandwidth, but most require filtering to restrict the noise bandwidth.

Noise from external sources (power supplies, lights, etc.) as 50 or 60 Hz and harmonics can usually be minimized with proper grounding, shielding, and filtering techniques.

Common noise sources include: (1) Thermal (also known as Johnson) noise is due to random movement of thermally excited charge carriers in a conductor such as a resistor. Higher temperature and higher resistance cause higher thermal noise. (2) Shot noise occurs as carriers cross a potential barrier, for example, a p–n junction, causing the current to consist of random current pulses. (3) Flicker or 1/f noise occurs in most electronics. "1/f" means the noise amplitude decreases with increasing frequency.

Uncorrelated random noise sources add in quadrature. If V_1, V_2, and V_3 are the rms values of three noise sources, the total rms noise of the three sources summed is $V_T = [\, V_1^2 + V_2^2 + V_3^2]^{½}$.

This has the effect that the largest individual source will dominate the total noise.

The intrinsic noise of an amplifier is specified as referred-to-input (RTI). This is the value of noise that, when applied to the input of an ideal noiseless amplifier, generates the same output noise as the actual circuit. If an amplifier with a gain of 50 has 100 μV noise at the output, the RTI noise is 2 μV. Noise values should always be specified over a certain frequency range.

As the surface area of an electrode decreases, impedance increases. Johnson noise increases proportional to impedance (Hassibi et al., 2004). Since uncorrelated noise sources add in quadrature and the largest noise source dominates, the Johnson noise is a small fraction of total noise until some electrode impedance threshold is crossed, after which Johnson noise dominates. In addition to Johnson noise, one must consider shunt losses (Ludwig et al., 2011), defined as the loss of signal to ground. As with Johnson noise, shunt losses increase proportional to impedance (Robinson, 1968).

Another source of noise, called quantizing noise, is from digitization in an ADC. Quantization is the approximation of a signal by an integer multiple of a quantity d, the quantizing step. For a 16-bit ADC with full-scale range of ± 10 V, the quantizing step $d = 20\ \text{V}/2^{12} = 305\ \mu\text{V}$. The quantization error is the rounding error due to finite input voltage changes required to toggle the output digitized value.

3.11.4 Common mode rejection

For a differential amplifier with output voltage V_O, inverting input voltage V_1, and noninverting voltage V_2, the differential gain is $A_D \equiv V_O/(V_2 - V_1)$, and the common mode gain is $A_C \equiv V_O/V_c$, where V_c is the input voltage when $V_1 = V_2$ (the inputs are shorted). The CMRR is the ratio of differential gain to common mode gain:

$$\text{CMRR} = A_D/A_C \tag{3.39}$$

Ideally CMRR is infinite, corresponding to zero common mode gain. A typical op-amp has a CMRR of 80–100 dB.

3.11.5 Loading and impedance

It is a design goal that an electrode, whether stimulating or recording, be low impedance relative to the connected instrumentation. For stimulation, a high-impedance electrode presents a large load. If the stimulator is a current source, "large load" means large voltage drop, possibly reaching the compliance voltage. If the stimulator is a voltage source, "large load" means decreased current and effectiveness. For recording, we again prefer a low-impedance electrode relative to the instrumentation input impedance. High-impedance recording electrodes load the amplifier, and increase noise (Hassibi et al., 2004). The load on a first-stage amplifier consists of the electrode, any encapsulation tissue, and extracellular impedance between electrodes. For an amplifier with gigaohm input impedance, the load is generally much smaller. It has been shown that under worst-case conditions of encapsulation by cells forming tight junctions, the impedance of microelectrodes only tripled compared to controls in saline (Merrill & Tresco, 2005). While increased tissue impedance may not adversely load the amplifier, it may affect the time constant determined by distributed load impedance and lead capacitance.

Changes in tissue impedance over time are routine. Upon surgical trauma the impedance around an electrode might drop for minutes to days as bleeding and edema occur, to be followed by an increase and then stabilization in impedance as encapsulation of the electrode occurs.

The term "electrode impedance" is nuanced, and characterization techniques often do not reflect this. Impedance is defined as a complex ratio of voltage to current for a sinusoid at a particular frequency. This is how electrodes are usually quantified. Impedance generally changes with the frequency of measurement. It is common for an electrode impedance to be reported at one frequency, often 1 kHz. A more thorough description of impedance would be a Bode plot of impedance magnitude and phase versus frequency. Stimulation of excitable tissue is generally performed with pulses, not sinusoids. A pulse can be thought of as a summation of sinusoids (as a Fourier decomposition). A sharp edge on a pulse corresponds to a high-frequency component.

Consider applying a pulse with a sharp edge to a capacitor. Since the voltage across a capacitance cannot change instantaneously, the derivative dv/di (loosely, but not exactly impedance) approaches zero. If this concept is applied to an electrode interface which is modeled as a double-layer capacitance in parallel with one or more Faradaic impedances, at the pulse onset it appears that the impedance is very small. However, since impedance is reported for a sinusoid at a particular frequency, and stimulation occurs with pulses, the utility of impedance for a pulse is only indirect.

An electrode with a small area has a small total capacitance ($C = \varepsilon A/d$) so the impedance ($Z = 1/j\omega C$) is large. One generally cares about how much charge can be injected prior to the onset of Faradaic reactions. Increasing the capacitance of an electrode increases the CSC. Since $\Delta v = Q/C = i\ t/C$, increasing capacitance minimizes the voltage excursion across the electrode interface and prevents Faradaic reactions.

Consider an electrode with impedance measured as 100 kΩ at 1 kHz. It is desired to inject 100 μA as a pulse. It is incorrect to say that the electrode interface drops $v = i\ Z = (10^{-4}\ \text{A})\ (10^{5}\ \Omega) = 10\ \text{V}$. The electrode interface does not appear the same way to a pulse as it does to a sinusoid at 1 kHz. But all hope is not lost. As the area of an electrode decreases, the total capacitance decreases, and two observables are correlated: (1) the impedance increases and (2) the CSC decreases. An increased impedance does mean a reduced ability to inject total charge, but the style of calculation as above is inaccurate. The total safe injectable charge is not only a function of current, but also pulse width, since $Q = i\ t$.

Since calculations of total voltage across an electrode interface (e.g., to determine whether compliance voltage is reached) cannot be made by multiplying the impedance at a certain frequency by the injected current using a pulse, how should this be done? Consider driving a 100 μA pulse across an electrode with a double-layer capacitance of 10 nF. The magnitude of impedance for this capacitance at $\omega = 1$ kHz is $(1/\omega\ C) = 100\ \text{k}\Omega$. It would be incorrect to say that the electrode drops $v = i\ Z = (100\ \mu\text{A})(100\ \text{k}\Omega) = 10\ \text{V}$. Instead, the capacitor charges up with time, with voltage $v = Q/C = i\ t/C = (10^{-4})\ t/(10^{-8}) = 10^{4}\ t$. To hit 10 V with 100 μA would require 1 ms, a rather long pulse.

The difficulties described here arise from improperly applying steady-state analysis techniques to a transient case. A complete solution can be derived by representing impedance in the complex s-domain, allowing for the correct steady-state and transient solutions.

References

Agnew, W. F., McCreery, D. B., Yuen, T. G. H., & Bullara, L. A. (1989). Histologic and physiologic evaluation of electrically stimulated peripheral nerve: Considerations for the selection of parameters. *Annals of Biomedical Engineering*, *17*, 39–60.

Agnew, W. F., McCreery, D. B., Yuen, T. G. H., & Bullara, L. A. (1990). Local anesthetic protects against electrically-induced neural damage in peripheral nerve. *Journal of Biomedical Engineering*, *12*, 301–308.

Agnew, W. F., McCreery, D. B., Yuen, T. G. H., & Bullara, L. A. (1993). MK-801 protects against neuronal injury induced by electrical stimulation. *Neuroscience*, *52*(1), 45–53.

Agnew, W. F., Yuen, T. G. H., McCreery, D. B., & Bullara, L. A. (1986). Histopathologic evaluation of prolonged intracortical electrical stimulation. *Experimental Neurology*, *92*, 162–185.

Azuma, H., Ishikawa, M., & Sekizaki, S. (1986). Endothelium-dependent inhibition of platelet aggregation. *British Journal of Pharmacology*, *88*, 411–415.

Babb, T. L., & Kupfer, W. (1984). Phagocytic and metabolic reactions to chronically implanted metal brain electrodes. *Experimental Neurology*, *86*(2), 171–182.

Ballestrasse, C. L., Ruggeri, R. T., & Beck, T. R. (1985). Calculations of the pH changes produced in body tissue by a spherical stimulation electrode. *Annals of Biomedical Engineering*, *13*, 405–424.

Bard, A. J., & Faulkner, L. R. (1980). *Electrochemical methods*. New York: John Wiley and Sons.

Beckman, J. S., Beckman, T. W., Chen, J., Marshall, P. A., & Freeman, B. A. (1990). Apparent hydroxyl radical production by peroxynitrite: Implications for endothelial injury from nitric oxide and superoxide. *Proceedings of the National Academy of Sciences of the United States of America*, *87*, 1620–1624.

Beebe, X., & Rose, T. L. (1988). Charge injection limits of activated iridium oxide electrodes with 0.2 msec pulses in bicarbonate buffered saline. *IEEE Transactions on Bio-Medical Engineering*, *35*, 494–495.

Bergamini, C. M., Gambetti, S., Dondi, A., & Cervellati, C. (2004). Oxygen, reactive oxygen species and tissue damage. *Current Pharmaceutical Design*, *10*(14), 1611–1626.

Bhargava, A. (1993). *Long-term effects of quasi-trapezoidal pulses on the structure and function of sacral anterior roots*. (M.S. thesis), Case Western Reserve University, Department of Biomedical Engineering, Cleveland, OH.

Bogdanski, D., Koller, M., Muller, D., Muhr, G., Bram, M., Buchkremer, H. P., Stover, D., Choi, J., & Epple, M. (2002). Easy assessment of the biocompatibility of Ni-Ti alloys by in vitro cell culture experiments on a functionally graded Ni-NiTi-Ti material. *Biomaterials*, *22*(23), 4549–4555.

Brummer, S. B., & Turner, M. J. (1975). Electrical stimulation of the nervous system: The principle of safe charge injection with noble metal electrodes. *Bioelectrochemistry and Bioenergetics (Lausanne, Switzerland)*, *2*, 13–25.

Brummer, S. B., & Turner, M. J. (1977a). Electrochemical considerations for safe electrical stimulation of the nervous system with platinum electrodes. *IEEE Transactions on Bio-Medical Engineering*, *24*, 59–63.

Brummer, S. B., & Turner, M. J. (1977b). Electrical stimulation with Pt electrodes. I. A method for determination of 'real' electrode areas. *IEEE Transactions on Bio-Medical Engineering*, *24*, 436–439.

Brummer, S. B., & Turner, M. J. (1977c). Electrical stimulation with Pt electrodes. II. Estimation of maximum surface redox (theoretical non-gassing) limits. *IEEE Transactions on Bio-Medical Engineering*, *24*, 440–443.

Buettner, G. R. (1993). The pecking order of free radicals and antioxidants: Lipid peroxidation, alpha-tocopherol, and ascorbate. *Archives of Biochemistry and Biophysics*, *300*(2), 535–543.

Campbell, P. K., Jones, K. E., Huber, R. J., Horch, K. W., & Normann, R. A. (1991). A silicon-based 3-dimensional neural interface: Manufacturing processes for an intracortical electrode array. *IEEE Transactions on Bio-Medical Engineering*, *38*(8), 758–768.

Chan, P. H., Yurko, M., & Fishman, R. (1982). Phospholipid degradation and cellular edema induced by free radicals in brain slice cortical slices. *Journal of Neurochemistry*, *38*, 525–531.

Chapman, D. L. (1913). A contribution to the theory of electrocapillarity. *Philosophical Magazine*, *25*, 475–481.

Chia, L. S., Thompson, J. E., & Moscarello, M. A. (1983). Disorder in human myelin induced by superoxide radical: An in vitro investigation. *Biochemical and Biophysical Research Communication*, *117*(1), 141–146.

Chiu, S. Y., Ritchie, J. M., Rogart, R. B., & Stagg, D. (1979). A quantitative description of membrane current in rabbit myelinated nerve. *The Journal of Physiology*, *292*, 149–166.

Chouard, C. H., & Pialoux, P. (1995). Biocompatibility of cochlear implants. *Bulletin de l'Academie Nationale de Medecine*, *179*(3), 549–555.

Cogan, S. F. (2008). Neural stimulation and recording electrodes. *Annual Review of Biomedical Engineering*, *10*, 275–309.

Cogan, S.F. (2018). Personal communication (5/16/18).

Cogan, S. F., Edell, D. J., Guzelian, A. A., et al. (2003). Plasma-enhanced chemical vapor deposited silicon carbide as an implantable dielectric coating. *Journal of Biomedical Materials Research. Part A*, *67*(3), 856–867.

Cogan, S. F., Ludwig, K. A., Welle, C. G., & Takmakov, P. (2016). Tissue damage thresholds during therapeutic electrical stimulation. *Journal of Neural Engineering*, *13*, 1–13.

Cogan, S. F., Troyk, P. R., Ehrlich, J., Plante, T. D., & Detlefsen, D. E. (2006). Potential-biased, asymmetric waveforms for charge-injection with activated iridium oxide (AIROF) neural stimulation electrodes. *IEEE Transactions on Bio-Medical Engineering*, *53*, 327–332.

Cui, X., Wiler, J., Dzaman, M., et al. (2003). In vivo studies of polypyrrole/peptide coated neural probes. *Biomaterials*, *24*, 777–787.

Cui, X. Y., Lee, V. A., Raphael, Y., Wiler, J. A., Hetke, J. F., Anderson, D. J., et al. (2001). Surface modification of neural recording electrodes with conducting polymer/ biomolecule blends. *Journal of Biomedical Materials Research*, *56*(2), 261–272.

Cui, X. Y., & Martin, D. C. (2003). Fuzzy gold electrodes for lowering impedance and improving adhesion with electrodeposited conducting polymer films. *Sensors and Actuators A: Physical*, *103*, 384–394.

Dalrymple, A., Huynh, M., Nayagam, B., et al. (2020). Electrochemical and biological characterization of thin-film platinum-iridium alloy electrode coatings: A chronic in vivo study. *Journal of Neural Engineering*, *17*, 1–16.

Dautremont-Smith, W. C. (1982). Transition metal oxide electrochromic materials and displays: A review. Part 2. Oxides with anodic coloration. *Displays*, *3*(3), 67–80.

Delahay, P. (1965). *Double layer and electrode kinetics*. New York: Interscience Publishers.

Durand, D. (1995). In J. D. Bronzino (Ed.), *Biomedical engineering handbook*. Boca Raton, FL: CRC Press.

Dymond, A. M., Kaechele, L. E., Jurist, J. M., & Crandall, P. H. (1970). Brain tissue reaction to some chronically implanted metals. *Journal of Neurosurgery*, *33*, 574–580.

Fang, Z., & Mortimer, J. T. (1991). Selective activation of small motor axons by quasitrapezoidal current pulses. *IEEE Transactions on Bio-Medical Engineering*, *38*(2), 168–174.

Fisher, G., Sayre, G. P., & Bickford, R. C. (1961). Histological changes in the cat's brain after introduction of metallic and plastic-coated wire. In D. E. Sheer (Ed.), *Electrical stimulation of the brain*. Austin, TX: University of Texas Press. (pp. 55-50).

Frazer, E. J., & Woods, R. (1979). The oxygen evolution reaction on cycled iridium electrodes. *Journal of Electroanalytical Chemistry*, *102*, 127–130.

Frewin, C.L., Locke, C., Saddow, S.E., & Weeber, E.J. (2011). Single-crystal cubic silicon carbide: An in vivo biocompatible semiconductor for brain machine interface devices. In *Proceedings of the thirty-third international conference on IEEE-EMBS* (pp. 2957–2960).

Furchgott, R. F. (1988). *Studies on relaxation of rabbit aorta by sodium nitrite: The basis for the proposal that the acid-activatable inhibitory factor from retractor penis is inorganic nitrite and the endothelium-derived relaxing factor is nitric oxide, vasodilatation. Vascular smooth muscle, peptides, autonomic nerves and endothelium* (pp. 401–414). New York: Raven Press.

Gileadi, E., Kirowa-Eisner, E., & Penciner, J. (1975). *Interfacial electrochemistry: An experimental approach*. Reading, MA: Addison-Wesley.

Gorman, P. H., & Mortimer, J. T. (1983). The effect of stimulus parameters on the recruitment characteristics of direct nerve stimulation. *IEEE Transactions on Bio-Medical Engineering*, *30*, 407–414.

Gotman, I. (1997). Characteristics of metals used in implants. *Journal of Endourology / Endourological Society*, *11*(6), 383–389.

Gottesfeld, S. (1980). The anodic rhodium oxide film: A two-color electrochromic system. *Journal of the Electrochemical Society*, *127*, 272–277.

Grahame, D. C. (1947). The electrical double layer and the theory of electrocapillarity. *Chemical Reviews*, *41*, 441–501.

Greatbatch, W., & Chardack, W. M. (1968). Myocardial and endocardiac electrodes for chronic implantation. *Annals of the New York Academy of Sciences*, *148*, 234–251.

Griot, C., Vandevelde, R. A., Peterhans, E., & Stocker, R. (1990). Selective degeneration of oligiodendrocytes mediated by reactive oxygen species. *Free Radical Research Communication*, *11*(4,5), 181–193.

Guoy, G. (1910). Constitution of the electric charge at the surface of an electrolyte. *Journal de Physique*, *9*, 457–467.

Guyton, D. L., & Hambrecht, F. T. (1974). Theory and design of capacitor electrodes for chronic stimulation. *Medical & Biological Engineering*, *12*(5), 613–620.

Halliwell, B. (1992). Reactive oxygen species and the central nervous system. *Journal of Neurochemistry*, *59*(5), 1609–1623.

Hassibi, A., Navid, R., Dutton, R. W., & Lee, T. H. (2004). Comprehensive study of noise processes in electrode electrolyte interfaces. *Journal of Applied Physics*, *96*, 1074–1082.

Hemnani, T., & Parihar, M. S. (1998). Reactive oxygen species and oxidative DNA damage. *Indian Journal of Physiology and Pharmacology*, *42*(4), 440–452.

Hille, B. (1984). *Ionic channels of excitable membranes*. Sunderland, MA: Sinauer Associates.

Hodgkin, A. L., & Huxley, A. F. (1952a). Currents carried by sodium and potassium ions through the membrane of the giant axon of Loligo. *The Journal of Physiology*, *116*, 449–472.

Hodgkin, A. L., & Huxley, A. F. (1952b). The Components of membrane conductance in the giant axon of logigo. *The Journal of Physiology*, *116*, 473–496.

Hodgkin, A. L., & Huxley, A. F. (1952c). The dual effect of membrane potential on sodium conductance in the giant axon of logigo. *The Journal of Physiology*, *116*, 497–506.

Hodgkin, A. L., & Huxley, A. F. (1952d). A quantitative description of the membrane current and its application to conduction and excitation in nerve. *The Journal of Physiology*, *117*, 500–544.

Hoogerwerf, A. C., & Wise, K. D. (1994). A three-dimensional microelectrode array for chronic neural recording. *IEEE Transactions on Bio-Medical Engineering*, *41*, 1136–1146.

Ignarro, L. J., Byrns, R. E., & Wood, K. S. (1988). *Biochemical and pharmacological properties of endothelium-derived relaxing factor and its similarity to nitric oxide radical. Vascular smooth muscle, peptides, autonomic nerves and endothelium* (pp. 427–436). New York: Raven Press.

Imlay, J. A. (2003). Pathways of oxidative damage. *Annual Review of Microbiology*, *57*, 395–418.

Ives, D. J. G., & Janz, G. J. (1961). *Reference electrodes: Theory and practice*. New York: Academic Press.

Johnson, P. F., Bernstein, J. J., Hunter, G., Dawson, W. W., & Hench, L. L. (1977). An in vitro and in vivo analysis of anodized tantalum capacitive electrodes: Corrosion response, physiology and histology. *Journal of Biomedical Materials Research*, *11*(5), 637–656.

Kandel, E. R., Schwartz, J. H., & Jessell, T. M. (2000). *Principles of neural science* (4th ed.). New York: McGraw Hill.

Kelliher, E.M. & Rose, T.L. (1989). Evaluation of charge injection properties of thin film redox materials for use as neural stimulation electrodes. In *Materials research society symposium proceedings, vol. 110*, (pp. 23–7).

Kim, D. H., Sequerah, C., Hendricks, J. L., et al. (2007). Effect of immobilized nerve growth factor (NGF) on conductive polymers electrical properties and cellular response. *Advanced Functional Materials, 17*, 79–86.

Kipke, D. R., Vetter, R. J., Williams, J. C., & Hetke, J. F. (2003). Silicon-substrate intracortical microelectrode arrays for long-term recording of neuronal spike activity in cerebral cortex. *IEEE Transactions on Neural Systems and Rehabilitation Engineering: A Publication of the IEEE Engineering in Medicine and Biology Society, 11*(2), 151–155.

Konat, G., & Wiggins, R. C. (1985). Effect of reactive oxygen species on myelin membrane proteins. *Journal of Neurochemistry, 45*, 1113–1118.

Kozai, T. D. Y., Catt, K., Du, Z., et al. (2016). Chronic *in vivo* evaluation of PEDOT/CNT for stable neural recordings. *IEEE Transactions on Bio-Medical Engineering, 63*(1), 111–119.

Kristensen, B. W., Noraberg, J., Thiebaud, P., Koudelka-Hep, M., & Zimmer, J. (2001). Biocompatibility of silicon-based arrays of electrodes coupled to organotypic hippocampal brain slice cultures. *Brain Research, 896*, 1–17.

Lapicque, L. (1907). Recherches quantitatives sur l'excitation electrique des nerfs traites comme une polarization. *The Journal of Physiology (Paris), 9*, 622–635.

Lilly, J. C., Austin, G. M., & Chambers, W. W. (1952). Threshold movements produced by excitation of cerebral cortex and efferent fibers with some parametric regions of rectangular current pulses (cats and monkeys). *Journal of Neurophysiology, 15*, 319–341.

Lilly, J. C., Hughes, J. R., Alvord, E. C., & Garkin, T. W. (1955). Brief noninjurious electric waveforms for stimulation of the brain. *Science (New York, N.Y.), 121*, 468–469.

Loeb, G. E., Walker, A. E., Vematsu, S., & Konigsmark, B. W. (1977). Histological reaction to various conductive and dielectric films chronically implanted in the subdural space. *Journal of Biomedical Materials Research, 11*(2), 195–210.

Loucks, R. B., Weinberg, H., & Smith, M. (1959). The erosion of electrodes by small currents. *Electroencephalography and Clinical Neurophysiology, 11*, 823–826.

Ludwig, K. A., Langhals, N. B., Joseph, M. D., et al. (2011). Poly(3,4-ethylenedioxythiophene) (PEDOT) polymer coatings facilitate smaller neural recording electrodes. *Journal of Neural Engineering, 8*(1). Available from https://doi.org/10.1088/1741-2560/8/1/014001.

Majji, A. B., Humayun, M. S., Weiland, J. D., Suzuki, S., D'Anna, S. A., & deJuan, E. (1999). Long-term histological and electrophysiological results of an inactive epiretinal electrode array implantation in dogs. *Investigative Ophthalmology & Visual Science, 40*(9), 2073–2081.

McCreery, D. B., Agnew, W. F., & Bullara, L. A. (2002). The effects of prolonged intracortical microstimulation on the excitability of pyramidal tract neurons in the cat. *Annals of Biomedical Engineering, 30*, 107–119.

McCreery, D. B., Agnew, W. F., Yuen, T. G. H., & Bullara, L. A. (1988). Comparison of neural damage induced by electrical stimulation with faradaic and capacitor electrodes. *Annals of Biomedical Engineering, 16*, 463–481.

McCreery, D. B., Agnew, W. F., Yuen, T. G. H., & Bullara, L. A. (1990). Charge density and charge per phase as cofactors in neural injury induced by electrical stimulation. *IEEE Transactions on Bio-Medical Engineering, 37*(10), 996–1001.

McCreery, D. B., Agnew, W. F., Yuen, T. G. H., & Bullara, L. A. (1992). Damage in peripheral nerve from continuous electrical stimulation: Comparison of two stimulus waveforms. *Medical & Biological Engineering & Computing*, *30*(1), 109–114.

McCreery, D. B., Yuen, T. G. H., Agnew, W. F., & Bullara, L. A. (1992). Stimulation with chronically implanted microelectrodes in the cochlear nucleus of the cat: Histologic and physiologic effects. *Hearing Research*, *62*, 42–56.

McCreery, D. B., Yuen, T. G., Agnew, W. F., & Bullara, L. A. (1994). Stimulus parameters affecting tissue injury during microstimulation in the cochlear nucleus of the cat. *Hearing Research*, *77*, 105–115.

McCreery, D. B., Yuen, T. G., & Bullara, L. A. (2000). Chronic microstimulation in the feline ventral cochlear nucleus: Physiologic and histologic effects. *Hearing Research*, *149*, 223–238.

McCreery, D. B., Yuen, T. G. H., Agnew, W. F., & Bullara, L. A. (1997). A characterization of the effects on neuronal excitability due to prolonged microstimulation with chronically implanted microelectrodes. *IEEE Transactions on Bio-Medical Engineering*, *44*, 931–939.

McHardy, J., Geller, D., & Brummer, S. B. (1977). An approach to corrosion control during electrical stimulation. *Annals of Biomedical Engineering*, *5*, 144–149.

McIntyre, C. C., & Grill, W. M. (2002). Extracellular stimulation of central neurons: Influence of stimulus waveform and frequency on neuronal output. *Journal of Neurophysiology*, *88*(4), 1592–1604.

McNeal, D. R. (1976). Analysis of a model for excitation of myelinated nerve. *IEEE Transactions on Bio-Medical Engineering*, *23*(4), 329–337.

Merrill, D.R. (2002). *Electrochemical processes occurring on gold in sulfuric acid under neural stimulation conditions*. (Ph.D. thesis), Department of Biomedical Engineering, Case Western Reserve University, Cleveland,OH.

Merrill, D. R. (2010). The electrochemistry of charge injection at the electrode/electrolyte interface. In D. D. Zhou, & E. Greenbaum (Eds.), *Implantable neural prostheses* (vol. 2, pp. 85–138). New York: Springer.

Merrill, D. R. (2014). Materials considerations of clinical implantable neuroengineering devices. *Current Opinion in Solid State and Materials Science*, *18*, 329–336.

Merrill, D. R., Bikson, M., & Jefferys, J. G. R. (2005a). Electrical stimulation of excitable tissue: Design of efficacious and safe protocols. *Journal of Neuroscience Methods*, *141*, 171–198.

Merrill, D. R., Stefan, I. C., Scherson, D. A., & Mortimer, J. T. (2005b). Electrochemistry of gold in aqueous sulfuric acid solutions under neural stimulation conditions. *Journal of the Electrochemical Society*, *152*(7), E212–E221.

Merrill, D. R., & Tresco, P. A. (2005). Impedance characterization of microarray recording electrodes *in vitro*. *IEEE Transactions on Bio-Medical Engineering*, *52*(11), 1960–1965.

Meyer, R. D., Cogan, S. F., Nguyen, T. H., & Rauh, R. D. (2001). Electrodeposited iridium oxide for neural stimulation and recording electrodes. *IEEE Transactions on Neural Systems and Rehabilitation Engineering*, *9*(1), 2–11.

Moncada, S., Palmer, R. M. J., & Higgs, E. A. (1991). Nitric oxide: Physiology, pathology and pharmacology. *Pharmacological Reviews*, *43*(2), 109–142.

Mortimer, J. T. (1990). In W. F. Agnew, & D. B. McCreery (Eds.), *Neural prostheses: Fundamental studies*. Englewood Cliffs, NJ: Prentice-Hall.

Mortimer, J. T., Kaufman, D., & Roessmann, U. (1980). Intramuscular electrical stimulation: Tissue damage. *Annals of Biomedical Engineering*, *8*, 235–244.

Mortimer, J. T., Shealy, C. N., & Wheeler, C. (1970). Experimental nondestructive electrical stimulation of the brain and spinal cord. *Journal of Neurosurgery*, *32*(5), 553–559.

Morton, S. L., Daroux, M. L., & Mortimer, J. T. (1994). The role of oxygen reduction in electrical stimulation of neural tissue. *Journal of the Electrochemical Society*, *141*, 122–130.

Najafi, K., Ji, J., & Wise, K. D. (1990). Scaling limitations of silicon multichannel recording probes. *IEEE Transactions on Bio-Medical Engineering*, *37*, 1–11.

Niparko, J. K., Altschuler, R. A., Xue, X. L., Wiler, J. A., & Anderson, D. J. (1989). Surgical implantation and biocompatibility of central nervous system auditory prostheses. *The Annals of Otology, Rhinology, and Laryngology*, *98*(12 Pt 1), 965–970.

Pletcher, D., & Walsh, F. C. (1990). *Industrial electrochemistry* (2nd ed.). London: Chapman and Hall.

Pudenz, R. H., Bullara, L. A., Dru, D., & Talalla, A. (1975). Electrical stimulation of the brain II: Effects on the blood-brain barrier. *Surgical Neurology*, *4*, 265–270.

Pudenz, R. H., Bullara, L. A., Jacques, P., & Hambrecht, F. T. (1975). Electrical stimulation of the brain III: The neural damage model. *Surgical Neurology*, *4*, 389–400.

Radomski, M. W., Palmer, R. M. J., & Moncada, S. (1987). Endogenous nitric oxide inhibits human platelet adhesion to vascular endothelium. *Lancet*, *2*, 1057–1058.

Rall, W. (1977). *Handbook of physiology – The nervous system I.* American Physiological Society, *vol. 1/Part 1*.

Ranck, J. B. (1981). In M. M. Patterson, & R. P. Kesner (Eds.), *Electrical stimulation research techniques*. New York: Academic Press.

Rand, D. A. J., & Woods, R. (1971). The nature of adsorbed oxygen on rhodium, palladium, and gold electrodes. *Journal of Electroanalytical Chemistry and Interfacial Electrochemistry*, *31*, 29–38.

Rand, D. A. J., & Woods, R. (1974). Cyclic voltammetric studies on iridium electrodes in sulfuric acid solutions. Nature of oxygen layer and metal dissolution. *Journal of Electroanalytical Chemistry and Interfacial Electrochemistry*, *55*, 375–381.

Randles, J. E. B. (1947). Rapid electrode reactions. *Discussions of the Faraday Society*, *1*, 11–19.

Rattay, F. (1989). Analysis of models for extracellular fiber stimulation. *IEEE Transactions on Bio-Medical Engineering*, *36*, 676–682.

Richardson-Burns, S. M., Hendricks, J. L., & Martin, D. C. (2007). Electrochemical polymerization of conducting polymers in living neural tissue. *Journal of Neural Engineering*, *4*(2), L6–L13.

Robblee, L. S., & Rose, T. L. (1990). Electrochemical guidelines for selection of protocols and electrode materials for neural stimulation. In W. F. Agnew, & D. B. McCreery (Eds.), *Neural prostheses: Fundamental studies* (pp. 25–66). Englewood Cliffs, NJ: Prentice-Hall.

Robblee, R. S., Lefko, J. L., & Brummer, S. B. (1983). Activated iridium: An electrode suitable for reversible charge injection in saline solution. *Journal of the Electrochemical Society*, *130*, 731–733.

Robinson, D. A. (1968). The electrical properties of metal microelectrodes. *Proceedings of the IEEE*, *56*, 1065.

Rose, T. L., Kelliher, E. M., & Robblee, L. S. (1985). Assessment of capacitor electrodes for intracortical neural stimulation. *Journal of Neuroscience Methods*, *12*(3), 181–193.

Rose, T. L., & Robblee, L. S. (1990). Electrical stimulation with Pt electrodes. VIII. Electrochemically safe charge injection limits with 0.2 ms pulses. *IEEE Transactions on Bio-Medical Engineering*, *37*(11), 1118–1120.

Rousche, P. J., & Normann, R. A. (1998). Chronic recording capability of the Utah intracortical electrode array in cat sensory cortex. *Journal of Neuroscience Methods*, *82*(1), 1–15.

Rubanyi, G. M. (1988). Vascular effects of oxygen derived free radicals. *Free Radical Biology and Medicine*, *4*, 107–120.

Ryhanen, J., Kallioinen, M., Tuukkanen, J., Junila, J., Niemela, E., Sandvik, P., & Serlo, W. (1998). In vivo biocompatibility evaluation of nickel-titanium shape memory alloy: Muscle and perineural tissue responses and encapsule membrane thickness. *Journal of Biomedical Materials Research, 41*(3), 481–488.

Sawyer, P.N. & Srinivasan, S. (1974). CD Ray, C.D. (Ed.), *Medical engineering* (pp. 1099–1110). Chicago.

Scheiner, A., & Mortimer, J. T. (1990). Imbalanced biphasic electrical stimulation: Muscle tissue damage. *Annals of Biomedical Engineering, 18*, 407–425.

Schmidt, S., Horch, K., & Normann, R. (1993). Biocompatibility of silicon-based electrode arrays implanted into feline cortical tissue. *Journal of Biomedical Materials Research, 27* (11), 1393–1399.

Sevanian, A. (1988). Lipid peroxidation, membrane damage, and phospholipase A2 action. In *CRC reviews, cellular antioxidant defense mechanisms, vol. II* (pp.77–95).

Shannon, R. V. (1992). A model of safe levels for electrical stimulation. *IEEE Transactions on Bio-Medical Engineering, 39*(4), 424–426.

Silbey, R. J., & Alberty, R. A. (2001). *Physical chemistry* (3rd ed.). New York: John Wiley and Sons.

Stensaas, S. S., & Stensaas, L. J. (1978). Histopathological evaluation of materials implanted in the cerebral cortex. *Acta Neuropathologica, 41*, 145–155.

Stern, O. (1924). Zur theorie der elektrolytischen doppelschicht. *Zeitschrift für Elektrochemie, 30*, 508–516.

Stieglitz, T., & Meyer, J. U. (1999). Implantable Microsystems. Polyimide-based neuroprostheses for interfacing nerves. *Medical Device Technologies, 10*(6), 28–30.

Stohs, S. J. (1995). The role of free radicals in toxicity and disease. *Journal of Basic and Clinical Physiology and Pharmacology, 6*(3-4), 205–228.

Stritesky, S., Markova, A., Vitecek, J., et al. (2018). Printing inks of electroactive polymer PEDOT:PSS: The study of biocompatibility, stability, and electrical properties. *Journal of Biomedical Materials Research. Part A, 106*(4), 1121–1128.

Sweeney, J.D., Durand, D., & Mortimer, J.T. (1987). Modeling of mammalian myelinated nerve for functional neuromuscular stimulation. In *Proceedings of the ninth international conference on IEEE-EMBS* (pp. 1577–1578).

Troyk, P.R., Detlefsen, D.E., & Cogan, S.F., et al. (2004). 'Safe' charge-injection waveforms for iridium oxide (AIROF) microelectrodes. In *Proceedings of the conference on IEEE-EMBS* (pp. 4141–4144).

Umans, J., & Levi, R. (1995). Nitric oxide in the regulation of blood flow and arterial pressure. *Annual Review of Physiology, 57*, 771–790.

Venkatraman, S., Hendricks, J., King, Z. A., Sereno, A. J., Richardson-Burns, S., Martin, D., et al. (2011). In vitro and in vivo evaluation of PEDOT microelectrodes for neural stimulation and recording. *IEEE Transactions on Neural Systems and Rehabilitation Engineering, 19*(3), 307–316.

von Helmholtz, H. L. F. (1853). Ueber einige gesetze der vertheilung elektrischer strome in korperlichen leitern mit anwendung auf die thierisch-elektrischen versuche. *Annalen der Physik, 89*, 211–233.

Wang, K., Fishman, H. A., Dai, H., & Harris, J. S. (2006). Neural stimulation with a carbon nanotubule microeletrode array. *Nano Letters, 6*(9), 2043–2048.

Weiland, J. D., Anderson, D. J., & Humayun, M. S. (2002). In vitro electrical properties for iridium oxide versus titanium nitride stimulating electrodes. *IEEE Transactions on Bio-Medical Engineering, 49*(12 Pt 2), 1574–1579.

White, R. L., & Gross, T. J. (1974). An evaluation of the resistance to electrolysis of metals for use in biostimulation probes. *IEEE Transactions on Bio-Medical Engineering*, *21*, 487–490.

Yuen, T. G. H., Agnew, W. F., Bullara, L. A., Jacques, S., & McCreery, D. B. (1981). Histological evaluation of neural damage from electrical stimulation: Considerations for the selection of parameters for clinical application. *Neurosurgery*, *9*(3), 292–299.

Chapter 4

Stimulus interaction in transcutaneous electrical stimulation

Sigrid Dupan[1], Leen Jabban[2], Benjamin W. Metcalfe[2] and Kianoush Nazarpour[1]
[1]*Edinburgh Neuroprosthetics Laboratory, School of Informatics, Edinburgh University, Edinburgh, United Kingdom,* [2]*Department of Electronic and Electrical Engineering, University of Bath, Bath, United Kingdom*

ABSTRACT
This chapter gives an overview of how our understanding of the nervous system and human motor control has shaped strategies to deliver functional sensory feedback for prosthetic control. We show how event-related feedback takes into account the naturally occurring spiking patterns of afferent sensory fibers. Recent studies employing this method show that it can lead to a range of benefits. However, it is not yet clear if the chosen stimulation paradigms are optimal in the context of human motor control, and the physiology of sensory feedback. At the end of the chapter, we focus on a study investigating how changing temporal patterns between identical stimuli can change their perception, thereby increasing the amount of information we can provide for prosthetic users. The results also show that stimulus interaction occurs on different time scales due to both neural behavior and human perception limitations.

Keywords: Prosthetic control; sensory feedback; motor control; event-related feedback; transcutaneous electrical stimulation; stimulus interaction; sensory discrimination

4.1 Introduction

The human hand is a diverse and complex tool that allows people to interact with the world around them. We not only use our hands to manipulate objects and communicate our intentions through gestures, but also to receive information about the world around us through sensory feedback. A complex neurophysiological network enables detailed temporal and spatial resolution in this feedback, and high levels of movement dexterity. This network encompasses the cerebral cortex, cerebellum, brainstem, spinal cord, peripheral nerves,

Somatosensory Feedback for Neuroprosthetics. DOI: https://doi.org/10.1016/B978-0-12-822828-9.00014-9

skeletal muscles, and sensory receptors. Pathologies where any part of this neurophysiological network is affected, such as stroke, dystonia, spinal cord injury, or limb loss, can therefore have a large impact on people's daily life.

Limb difference has several causes, and most can be categorized as either congenital or acquired. Upper limb prostheses allow people with limb difference to regain some of the lost function, and generally fit into three categories: passive, body-powered, or myoelectric prostheses. Passive prostheses focus on cosmetic improvement, mimicking the appearance of a hand, but provide limited functionality. Body-powered prostheses allow functional movement of the prosthesis, where movements of other body parts are relayed to the prosthesis with a harness. Myoelectric prosthesis allow for battery-powered movement that is controlled based on the electrical activity of residual muscles. Despite the wide range of available prostheses, only one commercially available device, the Vincent Evolution, was designed to include the controlled supply of force feedback (Vincent Systems GmbH, 2020).

Research into restoring sensory feedback for people with limb difference has been on-going for at least half a century (Childress, 1985), with an increased interest in the last two decades. Recent research has shown that providing sensory feedback can lead to a range of advantages for prosthetic users, such as improved hand control (D'Anna et al., 2019; George et al., 2019; Tan et al., 2014; Valle et al., 2018), increased prosthesis embodiment (D'Anna et al., 2019; George et al., 2019; Marasco et al., 2011), and a reduction in phantom limb pain (George et al., 2019; Rossini et al., 2010). Most devices, even if they were not specifically designed to provide sensory feedback, do generate external information through what is called "incidental feedback" (Mann & Reimers, 1970). Examples include the force on a harness, the sound of a myoelectric hand, or the forces on a prosthetic socket. In this sense, most prosthetic devices already provide some type of implicit feedback, even if it is not optimized.

In this chapter, we ask how we can provide better feedback for prosthetic users. We will highlight a study investigating how changing temporal patterns between identical stimuli can change their perception, thereby increasing the amount of information provided. However, before we go into details of stimulation patterns, we need to understand if users actually want sensory feedback, and why sensory feedback is important when performing movements. Then, we will give an overview of how sensory information is received and processed. Finally, we will investigate how this information can be incorporated into sensory feedback strategies, and which steps need to be taken to optimize the feedback.

4.2 User opinions on sensory feedback

Despite the range of devices available for people with upper limb difference, abandonment rates remain problematic. Studies report varying numbers of abandonment, ranging between 6%–100% for passive devices, 26%–87% for body-powered prostheses, and 0%–75% for myoelectric prostheses (Atkins et al., 1996; Biddiss & Chau, 2007; Smail, Neal, Wilkins, & Packham, 2020). While these numbers are very high, 68% of nonusers also stated that they were willing to reconsider using a prosthesis if improvements were made at a reasonable cost (Biddiss et al., 2007). People who report that they use their prostheses also demonstrate a range of wear/use time (Chadwell et al., 2018). These findings have led to an increase in surveys investigating the reasons for abandonment, and the needs perceived by people with limb difference. Identified priorities include comfort (Biddiss et al., 2007; Jang et al., 2011), appearance (Jang et al., 2011; Kyberd & Hill, 2011), and improved function (Biddiss et al., 2007; Engdahl et al., 2015; Kyberd & Hill, 2011; Luchetti et al., 2015; Østlie et al., 2012). However, Kumar et al. (2019) identified that the analysis of user needs and abandonment literature show contradictions, where devices that meet the identified requirements are not necessarily accepted. Therefore targeted prosthetic training may increase the actual use of the prosthesis and reduce abandonment (Østlie et al., 2011). Another study shows that not all users are inclined to accept new technologies, with users who are younger, have acquired limb loss, and have a unilateral limb difference being more interested in adopting new technologies (Engdahl et al., 2017).

The user needs surveys indicate that receiving sensory feedback is not necessarily seen as the main priority by people with limb difference. A survey including 1575 participants from 1996 reported that reducing the required visual attention through sensory feedback received a fifth place out of 10 possible priorities for transradial body-powered prostheses, and the third place out of 17 possible priorities for transradial myoelectric prostheses. Wrist rotation of the terminal device and the introduction of simultaneous movements of two joints were the top two priorities for both groups (Atkins et al., 1996). A 2007 survey asking for consumer design priorities similarly found sensory feedback as the eighth and fourth priorities out of 10 for body-powered and myoelectric prostheses, respectively (Biddiss et al., 2007). One user in this survey pointed out that they also desired proprioceptive sensory feedback as "[s]ome way to keep tabs on where it is" (Biddiss et al., 2007). When these same authors studied the decision not to wear a prosthesis, they found that 85% of those who abandoned their prosthesis agreed with the statement that they received more sensory feedback without their prosthesis. Forty-four percent of frequent wearers also agreed with this statement, indicating that a prosthesis can actually be an impediment to sensory feedback (Biddiss & Chau, 2007).

More recently, surveys have investigated user preferences related to sensory feedback in greater depth. Two surveys by Lewis et al. (2012) and Stephens-Fripp et al. (2020) show that respondents are most interested in receiving information related to the grip strength. In the 2012 paper by

Lewis et al. (2012), which had 108 respondents, proprioceptive information on the movement and position of the arm followed in reported importance. Information about events related to object handling, such as first contact and end of contact were the fourth and fifth most important, respectively. The temperature and surface texture of the object were reported as the least important. The same survey also queried the type of sensory feedback respondents would prefer. People had a clear preference for modalities that relay their information through changes in temperature, vibration, electric stimulation, and pressure. Visual and acoustic cues to relate sensory information were rejected as preferred options by the respondents (Lewis et al., 2012). The preference to temperature might be related to the fact that respondents also reported that their residual limb was least sensitive to temperature. Unfortunately, this lower sensitivity, together with the fact that transferring information through temperature would be slow and consume more energy than the other options, makes temperature less feasible as an option to transfer sensory information. The rejection of visual and acoustic cues is interesting, as existing devices provide implicit feedback using both these mechanisms. However, the survey question did not specify if the preference included current incidental feedback. As such, it is impossible to determine if the respondents want to reduce incidental feedback, or if they only reject additional cues.

4.3 The role of sensory feedback in motor control

The previous section showed that prosthetic users are interested in receiving targeted sensory feedback, even if it is not their main priority. So why is sensory feedback important, and why do researchers focus on implementing it if users have greater priorities, such as wrist rotation or the combined movement of two degrees of freedom (Atkins et al., 1996)? To answer this question, we need to investigate how sensory feedback influences the movements we make.

The common objective of human movement is to achieve a specific goal, such as picking up an object, forming specific gestures, or communicating in sign language. As we use a combination of muscles, joints, and limbs to create the movement, there are multiple realizations of different muscle activations to achieve this goal. However it has been found that repeated movements show consistent patterns, for example, in their velocity profile (Morasso, 1981), while allowing some variability in the execution of the movement if it does not stop us from completing our task (Scott, 2004; Todorov & Jordan, 2002). This suggests that the sensorimotor system has a means of determining what movement is optimal to reach its goal. Understanding how our brain chooses this optimal solution, and what factors influence it, could help in creating prosthetic solutions that are more easily integrated with our sensorimotor system. In other words, could we design

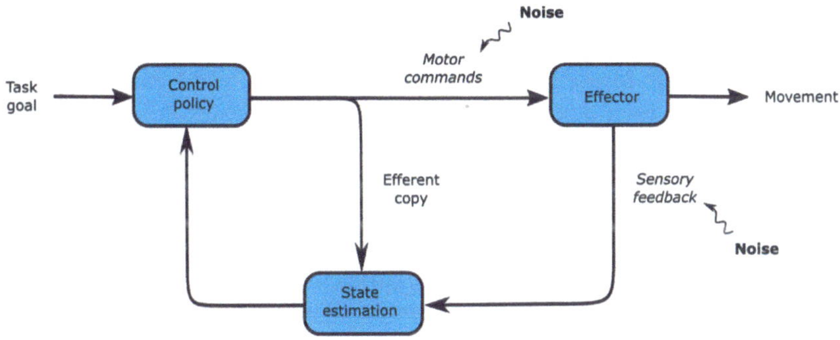

FIGURE 4.1 Simplified overview of optimal feedback control framework (Scott, 2004). *Based on Scott, S. H. (2004). Optimal feedback control and the neural basis of volitional motor control.* Nature Reviews Neuroscience, *5(7), 532–546. Credit: Springer Nature License No. 4986440284121.*

prosthetic feedback so that it is more easily integrated in the existing sensorimotor system and thereby improve control.

Optimal feedback control, a theory introduced by Todorov and Jordan (2002), is a computational framework that accounts for both repeatability in task completion and variability of movement details (Fig. 4.1). The system is of importance when attempting to model artificial movement, such as prosthetic control, within the sensorimotor system. Optimal feedback control not only states that the task goal is determining the motor commands that are sent to the effector, but also that sensory information is taken into account when updating these commands (Todorov & Jordan, 2002). It is therefore a prime example of closed-loop control. In short, a control policy determines which motor commands are sent to the effector. Sensory feedback from the receptors is relayed back to compare with the expected outcome of the motor command, herein called the efferent copy. The difference between the predicted and actual feedback enables the system to update the estimated state, and therefore to update its motor commands. To account for the variability, the framework points out that both the motor command and the sensory feedback are noisy. In this section, we will look at this framework, and its consequences for feedback in prosthetic control. Those who want a more in-depth overview of the role of sensory feedback in prosthetics in the context of human motor control can read the recent review by Sensinger and Dosen (2020).

4.3.1 Control policy

We mentioned previously that any movement can be completed in numerous ways. Accordingly, once the goal of a movement is determined, a decision has to be made regarding how to complete the task. At this moment, the known factors are the state of the musculoskeletal system, such as the position and speed of our hand, and the environment, such as the object we want to grab. The control policy dictates which factors the motor system takes into account to decide how to complete the task, and how the different factors relate to each other. Mathematically, this translates into a cost function, which determines the costs and rewards associated with all possible movements, and the weighting of each of these (Diedrichsen et al., 2010; Todorov & Jordan, 2002). The movement that minimizes the associated costs and maximizes its rewards will be executed.

When completing a movement, there are multiple factors that might be considered in the cost function, and these may favor different behaviors. As optimal feedback control is a computational model based on human behavior, it enables the comparison of different iterations of the model to identify which factors determine the movements we make. For example, we know people want movements to be executed within a certain time frame, while also wanting to minimize the associated energy expenditure. Other factors that might be considered in the cost function are the accuracy and variability of the movement. After comparison of these, and many other models, to behavioral data, these four factors—speed, effort, accuracy, and variability—are commonly accepted to be the main components of the cost functions in recent implementations of optimal feedback control models (Sensinger & Dosen, 2020; Shadmehr & Mussa-Ivaldi, 2012). These can therefore also be expected to be the main drivers for the control policy in prosthetic control. However, testing which factors make up the control policy in prosthetic control is difficult, as the chosen assessment for prosthetic control might bias the optimal feedback control model. Many established prosthetic assessment procedures, such as the Southampton Hand Assessment Procedure (SHAP), box and blocks test, and clothes pin test, focus on the time people need to complete a certain task (a range of daily activities, moving blocks, or relocating clothes pins, respectively). As such, participants will be motivated to prioritize the speed of task completion over other factors such as the effort, and therefore speed might be weighted higher in the cost function than it would be in movements prosthetic users perform in daily life (Liu & Todorov, 2007).

4.3.2 Efferent copy

Once the control policy has determined the best possible movement based on the cost function, our motor system needs to decide what motor command is associated with this movement. This is made possible by taking the inverse of our "internal model," a set of mappings between all possible motor commands and their expected outcome (Kawato, 1999). This internal model represents the prior knowledge of our motor system, and plays an important role in both feedforward control and closed-loop feedback control.

Internal models enable us to achieve feedforward movement, such as ballistic movements, and movements where no sensory feedback is present. In this case, the movements will be determined solely by the internal model, and will represent the motor system's best guess at how to achieve our goal.

However, these movements might result in small errors, which would be repeated indefinitely if the system is not able to update itself. Feedforward control is also important when initiating a movement, even if feedback is present, as sensory feedback is delayed. This delay is determined by the pathway the feedback takes, and can differ significantly between different feedback modalities. For example, visual feedback is around 10 times slower than tactile feedback (Chen & Thompson, 1995; Johansson & Flanagan, 2009; Shimada, Qi, & Hiraki, 2010).

In closed-loop feedback control, an efferent copy is sent out along with the motor command. This efferent copy represents the expected outcome of this specific motor command based on the internal model, and acts as a baseline to compare our sensory feedback with (Scott, 2004; Wolpert, Ghahramani, & Jordan, 1995). When discrepancies between the sensory feedback and expected outcomes occur, the internal model is updated. How much the internal model is updated will depend on the uncertainty of both the internal model and the incoming sensory information (Orbán & Wolpert, 2011). For example, when first using a prosthesis, the internal model related to controlling the prosthesis will be very uncertain, while the information received through visual feedback will have a high certainty. This updating of the internal model represents motor learning, whereby a new task is learned until the internal model is stable, and movements can therefore be performed largely in a feedforward manner.

Comparing the sensory feedback and the efferent copy also allows the motor system to update the current state of our musculoskeletal system and environment if the commands had an outcome that was different than expected. This in turn allows the motor commands to be updated so that the movement goal can be completed. The role of sensory feedback in adjusting movements while in progress highlights the importance of fast sensory feedback. The longer it takes for the feedback to be integrated and compared with the efferent copy, the bigger the error in a movement can become before it is adjusted. This speed is important in prosthetic control, as visual feedback has a relatively large delay of around 250 milliseconds (Chen & Thompson, 1995; Shimada et al., 2010). As a reference, tactile stimuli only take 14–28 milliseconds to reach the brainstem (Johansson & Flanagan, 2009). As such, errors detected by visual feedback are corrected late in a movement, and can make prosthetic control cumbersome or even unstable.

4.3.3 Signal noise

The optimal feedback control theory acknowledges that the system contains noise in both the motor commands and the sensory signals (Scott, 2004). Here, we will focus on noise in the sensory system, and how the associated uncertainty influences sensory integration. When performing a movement, we commonly receive sensory information from multiple sources. When lifting a coffee cup, we not only see what we are doing, but also receive tactile and proprioceptive feedback. Similarly, someone using a prosthesis will see their movement, and might feel a difference in force distribution of the socket on their residual limb when lifting up the cup.

To create the best possible estimation of the current state, the information of different sensory information sources is integrated. However, all these sources are associated with a level of uncertainty, which should be taken into account during the integration. This strategy, known as Bayesian integration, means that if one source is less certain, the estimated state should be biased toward the information from the more reliable source (Knill & Pouget, 2004). Experiments have shown that humans weight sensory information based on the uncertainty of the signals (Knill & Pouget, 2004).

4.3.4 Implications

This overview of the role of sensory feedback in motor control has shown the importance of sensory feedback, both in motor learning and adapting movements while in progress. However, the existence of feedback delays and noise indicate that different types of feedback will have different impacts on motor control. Currently, in the context of prosthetic control, information from a range of incidental sources is integrated to estimate the state of the arm, the prosthesis, and the environment. Some of the sources are visual feedback, the force on a harness, the sound of a myoelectric hand, and the forces on the prosthetic socket.

The fact that prosthetic users already have multiple sensory feedback sources does not necessarily produce satisfactory results. In terms of uncertainty, visual feedback is a reliable source. However, the long time delay implies that adaptation during movement is slow. Additionally, users indicate that they want to reduce visual attention (Atkins et al., 1996), and visual and auditory cues were rejected as preferred sensory feedback modalities (Lewis et al., 2012). The fact that 85% of those who abandoned their prosthesis and 44% of users agree with the statement that they receive more sensory feedback without their prosthesis (Biddiss & Chau, 2007) implies that the incidental feedback is not valued as a source of sensory feedback compared with the tactile information provided by the residual limb.

At the start of this section, we posed the question "is it possible to design prosthetic feedback so that it is more easily integrated in the existing sensorimotor system?". We hope it is now clear that, to increase functionality, a sensory feedback modality should have as short a delay as possible, and should confer reliable and meaningful information. Vibrotactile and electrical stimulation therefore may be better modalities, as they can provide faster feedback. However, the certainty of this information might be low when first presented to users as it will take time to be integrated into the internal model.

4.4 Physiology of sensory feedback

To restore sensory information for prosthetic users, the prosthesis should reliably inform the user about its state and its interaction with the environment. In the case of prosthetic users, the receptors that normally pick up this information, mechanoreceptors, are missing. Therefore understanding how these receptors work and how the signals are integrated can help in substituting their information.

4.4.1 Mechanoreceptors

When our fingers come into contact with an object, for example, when typing on a keyboard, their skin deforms. Different mechanoreceptors and their associated nerve fibers in the skin will react differently to this deformation. Therefore they are categorized based on the main stimuli (i.e., adequate stimuli) they react to, their depth within the skin, and their rate of adaptation. Mechanoreceptors mostly responding to static pressure, Merkel cells (SA1), and skin stretch, Ruffini endings (SA2), are slow adapting. The receptors mostly reacting to lateral motion and low-frequency vibrations, Meissner corpuscles (FA1), and high-frequency vibrations, Pacinian corpuscles (FA2), are fast adapting. The letter nomenclature typically refers to the associated nerve fibers which innervate these mechanoreceptors.

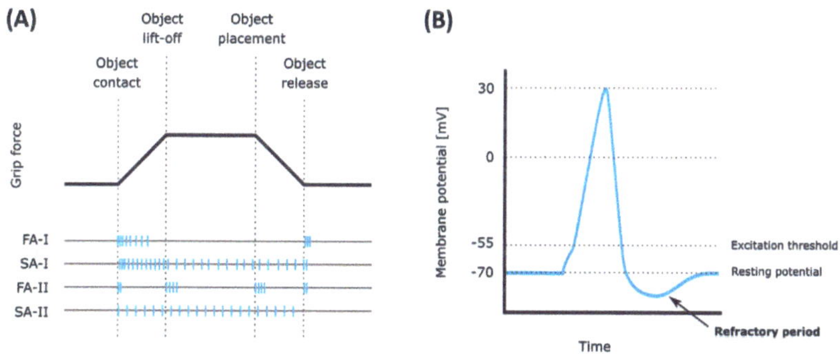

FIGURE 4.2 Physiology of sensory feedback. (A) Characteristic firing of different mechanoreceptive fibers to events during object manipulation. (B) illustration of neuronal membrane potential when an action potential is triggered. During the relative refractory period, approximately coinciding with the time in which the membrane hyperpolarizes below the resting threshold, new action potentials are more difficult to generate. *Figure (A): Based on Johansson, R. S., & Flanagan, J. R. (2009). Coding and use of tactile signals from the fingertips in object manipulation tasks.* Nature Reviews Neuroscience, *10(5), 345–359. and Johansson, R. S., & Westling, G. (1991). Afferent signals during manipulative tasks in humans. In O. Franzen & J. Westman (Eds.),* Information Processing in the Somatosensory System. *Macmillan Press. Credit: Springer Nature License No. 4986440614258.*

Microneurography measurements, where the spiking activity of a single peripheral nerve fiber is recorded, show that the fast- and slow-adapting fibers have distinct spiking patterns during object manipulation (Fig. 4.2A) (Johansson & Westling, 1991). Fast-adapting fibers have no lasting response, and show peaks of spike rates at the start and end of a mechanical stimulus. Slow-adapting receptors show a peak rate in spiking activity at the start of the mechanical stimulus, after which the spiking frequency (i.e., firing rate) slows but continues when a constant mechanical stimulus is present (Mcglone & Reilly, 2010). As a result, mechanoreceptive afferent nerves show distinct spike trains for particular events during object manipulation, such as object contact, lift-off, table contact, and object release (Johansson, 1992). Different types of fibers will mostly react to different events, for example, FA1 fibers will react to object contact and release, but not much to lift-off or placing the object back on a surface, while other fibers (SA1 and SA2) will show sustained spiking throughout the whole manipulation task (Johansson, 1992). Intracortical measurements from the hand representation in the primary somatosensory cortex of Rhesus macaques show that these event-related spiking patterns in afferent peripheral nerves are somewhat reflected in cortical neurons, where the population responses to object contact and release are much higher than those during the sustained pressure provided by the object (Callier et al., 2019). Analysis of the cortical spiking patterns showed that the increase in the overall activity is the effect of both changes in the firing rates of individual neurons, and an increase in the amount of neurons recruited in the population firing (Callier et al., 2019).

While we have only presented the main stimuli used for classifying mechanoreceptive fibers, they all respond to a range of stimuli. In other words, all would respond to complex spatiotemporal vibration patterns, but with different sensitivities to frequency components which make up the complex pattern. Meissner afferents, for example, also react to vibrations, although they react to lower frequency vibrations better than Pacinian afferents. As a result, Meissner afferents detect slip between the skin and an object, while Pacinian afferents can pick up vibrations transmitted by objects we hold in our hands (Mcglone & Reilly, 2010). Most dynamic skin deformations activate all types of mechanoreceptors and their afferents, and this information is integrated into our tactile perception. It is estimated that the tips of the fingers contain 241 mechanoreceptive fibers per cm^2, while the palm contains 58 fibers/cm^2. When all these are added up, the hand is innervated by an estimated 17,000 mechanoreceptive fibers (Johansson & Vallbo, 1979). This amounts to around 15% of the total tactile afferent fibers (Corniani & Saal, 2020), while the hands only account for around 2% of the total body surface area (Rhodes et al., 2013).

The sheer volume of mechanoreceptive fibers and their complex spiking patterns in reaction to different stimuli highlight the complexity of substituting sensory feedback, especially for the hand. Current technology cannot approximate the complex spatial and temporal resolution of mechanoreceptors and their associated nerve fibers. Therefore feedback strategies should focus on other aspects such as fast and slow adaptation, and the fact that neurons react more to changes in conditions rather than to sustained states.

4.4.2 Stimulus interaction

So far, we have seen that sensory information is encoded and propagated through specific series of spikes, also called action potentials. These action potentials are created if a neuron receives input from other neurons that is strong enough, and is generated by ions flowing in and out of the cell. The characteristic pattern in the membrane potential during an action potential is illustrated in Fig. 4.2B. Action potentials are the mechanism by which nerves propagate information over long distances, which is important when we want to use sensory information to update our movements. Therefore an understanding of how action potentials are formed might tell us how we can create artificial feedback that utilizes nerve properties optimally.

At rest, a typical neuron is negatively charged, with the inside of the cell about 70 mV below the outside. Incoming action potentials from other cells can change this polarization, and when the excitation threshold of approximately −55 mV is reached, an action potential is triggered. After the peak of around +30 mV is reached, the neuron repolarizes back, and then hyperpolarizes below the resting potential. Coinciding approximately with the depolarization/repolarization phases, a neuron is at an absolute refractory period and cannot generate a new action potential (Dayan & Abbott, 2001). After this, action potential generation is possible but difficult during the relative refractory period. Reported values for the refractory period vary slightly, but are generally below 2 milliseconds for the absolute refractory period and between 2–5 milliseconds for the relative refractory period (Wesselink, Holsheimer, & Boom, 1999), although some mention that this might last "up to tens of milliseconds after a spike" (Dayan & Abbott, 2001). These biophysical properties vary across different neuron types, and depend mainly on ion channel dynamics.

The refractory period sets a limit of how fast effective stimuli can follow each other for a given neuron or along the nerve fiber. While the neural system will usually be able to distinguish two stimuli with an interstimulus interval surpassing that limit, this may not be enough for the person on the receiving end to perceive two sequential stimuli because psychophysical processes result from complex excitatory and inhibitory interactions. Previous studies show that healthy participants require a pause of around 60–70 milliseconds between consecutive stimuli to perceive them as separate events

(Fiorio et al., 2003; Fiorio et al., 2008; Pastor et al., 2004). A psychophysical study with interstimulus intervals ranging from 5 to 140 milliseconds showed that for participants to perceive the two stimuli separately 100% of the time, an interstimulus interval of at least 100 milliseconds was required (Pastor et al., 2004). These sensory results are corroborated by findings from the motor system, where neurons show a silent period in the range of 110 milliseconds after transcranial magnetic stimulation (Fuhr et al., 1991).

Sensory feedback for prosthetics will most likely consist of a sequence of stimuli, and their temporal pattern will determine how the feedback is perceived. Stimuli should take into account the all-or-none nature of action potentials, and their temporal profile. In other words, stimuli should be large enough that they exceed the excitation threshold (suprathreshold), and they should be spaced far enough apart to avoid the refractory period. Increasing the interstimulus interval will only lead to the perception of two separate stimuli once the temporal discrimination threshold is reached. Changing the stimulus frequency of stimuli with a constant amplitude between these two limits can be used to create the perception of a change in amplitude as stimulus amplitude and frequency interact with each other (Macfie & Thomson, 1981; Van Doren, 1997). These physiological characteristics of sensory feedback enable the creation of many different perceptions with the use of a limited amount of stimuli, and could be exploited to create distinct sensory events when using a prosthesis.

4.5 Event-related feedback in upper-limb prosthetics

Understanding of the temporal spiking patterns of peripheral nerves and cortical neurons has led to a series of studies linking prosthetic feedback to events, mirroring the firing of mechanoreceptive fibers during movement. In 1993 Johansson and Edin proposed the concept of "sensory discrete event-driven control" (Johansson & Edin, 1993). They argued that providing discrete sensory events related to changes in movement phases might help the completion of the movement goal (Johansson & Edin, 1993), as the absence of these sensory events slows down movement transitions, and impairs corrections when errors are detected (Johansson & Bäckström, 1992; Johansson & Westling, 1991). One example of such a movement phase is the "preload phase," when the grip is established, after which the force is increased during the "loading phase" until it can overcome the vertical forces. The discrete sensory events associated with the end of these phases are object contact, and the start of movement. They also point out that identical sensory events can relay very different information depending on the phase of the task or the context (Johansson & Edin, 1993). Since then, the feasibility and effect of this feedback strategy have been studied for both lower and upper limb prosthetics.

Event-related feedback, whereby vibrotactile patterns were associated with symmetrical walking, led to an improved temporal gait symmetry in patients after only 2 hours of training (Crea et al., 2017). Event-related feedback for upper limb prosthetics led to faster reaction times (Aboseria et al., 2018), finer force control in grasping events (Clemente et al., et al., 2016; Engels et al., 2019; George et al., 2019), and better identification of object size (George et al., 2019) when compared to continuous feedback. These improvements were independent of the feedback modality, as both vibrotactile (Aboseria et al., 2018; Clemente et al., 2016) and electrical stimulation (George et al., 2019) led to improvements. Algorithms approximating the neural responses to events were also used to directly control a prosthesis. In this case, algorithms aiming to allow compliant grasping and prevent slip transformed pressure information from the prosthesis fingertips into neural-like responses, which were sent to the prosthesis to make fine force adjustments (Osborn et al., 2016). This led to a decrease in broken objects and objects slipping during grasping (Osborn et al., 2016).

Event-related feedback is delivered by approximating neural responses, as their complexity does not allow for an exact replication. Two main strategies have been adopted so far, where the characteristics of the stimuli are either independent (Aboseria et al., 2018; Cipriani et al., 2014; Clemente et al., 2016) or related to the magnitude of the event (George et al., 2019; Okorokova et al., 2018; Saal et al., 2017). A series of vibrotactile studies showed that approximating sensory stimuli with a fixed pattern can transfer the necessary information. A study with limb-intact participants showed that event-related feedback enabled them to control a robotic hand, and that delaying the feedback led to participants delaying subsequent phases of the task (Cipriani et al., 2014). In this study, stimulus amplitude, frequency, and duration were fixed, and the same was seen for contact, lift-off, placement, and release of the object. To increase the information transferred to the participant, three vibrotactile units were used. Two of these were fixed to the participant's thumb and index finger, and provided information of contact and release. The third unit was placed on the ring finger and conveyed the information related to object lift-off and placement (Cipriani et al., 2014). The same strategy, but now with only two vibrotactile units in an arm-cuff which translated contact and release, was used during a study where five transradial amputees used the device for 1 month at home (Clemente et al., 2016). The study showed that vibrotactile stimulation led to fewer broken blocks during a modified box and blocks test, and that more training led to a decrease in broken blocks, and an increase in speed (Clemente et al., 2016). A different study with limb-intact participants used this fixed stimulus to translate the occurrence of slipping events by sending a vibration burst when a decrease in force was detected (Aboseria et al., 2018). Comparison to visual and constant pressure feedback demonstrated that the discrete feedback outperformed both of the other modalities consistently over multiple days of training, showing faster reaction times, lower crush rate, and higher success rate (Aboseria et al., 2018).

The level of detail in event-related feedback can be increased by modifying the sensory stimulus based on the characteristics of the sensory event. Saal et al. (2017) created a computational model that simulates the population response of afferent fibers to any spatiotemporal stimulus applied to the skin. This model was used as a basis for an event-related encoding model, where firing rates were estimated based on the indentation depth, rate, and acceleration of the skin deformation (Okorokova et al., 2018). Later, different intraneural stimulation paradigms that depended on all three of these parameters, or only the absolute sensor values and rate of change of the contact sensor signals were tested (George et al., 2019). This last study included one participant with one implanted Utah Slanted Electrode Array in their median nerve, and one in their ulnar nerve. The study showed greater precision in grip force and better handling of fragile objects when sensory feedback was present. Event-related feedback also allowed the participant to identify object compliance and size faster than when feedback was either constant, or linearly related to the absolute sensory values (George et al., 2019).

4.6 Optimizing event-related feedback strategies

The results of the studies described in the previous section show a clear benefit of event-related feedback. However, it is not yet clear if the chosen stimulation paradigms are optimal in the context of human motor control, and the physiology of sensory feedback. In this section, we try to answer the following questions. Do these feedback paradigms lead to a change in our internal models? Do frequency and stimulus duration change perception? What should be the timing in the stimulation paradigms to avoid—or exploit—stimulus interaction?

4.6.1 Testing the internal model

Early studies in able-bodied participants controlling a robotic hand showed that delays in the feedback stimuli resulted in correlated delays in the subsequent movement phases (Cipriani et al., 2014). This suggests that the feedback was incorporated in the participants' internal models. However, a later study probing the strength of the internal model showed that this explanation may not be so straightforward (Engels et al., 2019). The strength of the internal model was determined through a set of metrics looking into the adaptation rate based on the error, the size of the error that leads to adaptation, and the participant's trust in the sensory information he/she receives. In this study, participants used a prosthesis mounted on a custom brace, and received one of four possible types of feedback: visual feedback only, visual and auditory feedback, visual and vibrotactile feedback, or a combination of all three types of feedback. Visual and auditory feedback were provided continuously, while vibrotactile feedback was event-related. A range of psychophysical and object-handling tasks showed that both types of feedback increased the functional outcomes, but that this is not necessarily related to an increase in strength of the internal model. The addition of auditory feedback to the control loop led to a stronger internal model with lower uncertainty, while the addition of only vibrotactile feedback did not significantly decrease the uncertainty. The combination of all three types of feedback resulted in more uncertainty compared to the visual and auditory feedback, showing that this may have had an adverse effect (Engels et al., 2019).

These results show that while additional sensory feedback might lead to functional outcomes, it does not necessarily improve the internal model. These results could be related to the novel aspect of artificial event-related feedback, which might cause this modality to be associated with a lower certainty compared to visual and auditory feedback. Therefore it is important for future studies to test the internal model development, and to do this over longer periods of time. The work of Engels et al. (2019) provides a range of different psychophysical metrics that can probe the internal model, with metrics that test the internal model uncertainty, the sensorimotor threshold, the sensory uncertainty, and the adaptation rate.

4.6.2 Effect of stimulation pattern

Multiple studies have investigated the effect of stimulation characteristics such as pulse width, amplitude, and frequency in the context of event-related feedback. A vibrotactile study sought to increase the amount of transferred information by changing frequency and magnitude settings on two vibrotactile actuators, thereby allowing to encode both object type/force and movement type. They found that participants were better at discriminating magnitude than frequency settings (Karakuş & Güçlü, 2020). The participants subsequently learned to associate 15 possible events with specific stimulation patterns, with each electrode having three settings for magnitude and frequency (off—low—high). Afterwards, the recollection of the participants was tested by presenting them with sequences of two or three events, each separated by 300 milliseconds. The participant recall rate was much higher than chance, but did not reach 100% as similar sequences were often confused (Karakuş & Güçlü, 2020). This suggests that people can learn to recall distinct stimulation patterns, and can associate them with specific task events, even if they might need more training to become proficient at this task.

Two limb-different participants with chronically implanted peripheral nerve electrodes participated in a range of psychophysical tasks to understand the influence of pulse width and frequency on their perception (Graczyk et al., 2016). The study found that both pulse width and frequency had systematic effects on perceived tactile intensity, and that a combination of the two factors could predict the magnitude of artificial tactile percepts (Graczyk et al., 2016). A later study compared the experience of intraneural microstimulation in a limb-different participant with transcutaneous electrical stimulation to their reinnervated skin on their residual limb, and the experience of transcutaneous stimulation on the hand of limb-intact participants (George et al., 2020). They found that intensity discrimination is similar across all stimulation conditions, and that low stimulation frequencies lead to better intensity discrimination performance than high stimulation frequencies (George et al., 2020). These results suggest that the outcomes of future studies investigating the effect of transcutaneous electrical stimulation patterns done with limb-intact participants can inform about useful prosthetic feedback strategies. The ability to test stimulation characteristics with limb-intact participants might speed up research, as recruiting a sufficient amount of limb-different participants to achieve statistical power can take a long time.

4.6.3 Testing stimulus interaction

The studies investigating optimization of stimulation patterns for event-related feedback so far have all investigated the effects of entire stimulus trains. As such, they do not incorporate knowledge of stimulus interaction at the level of single stimulation pulses. Therefore we performed a study to investigate how sensory perception can be modified by modulating the interval between transcutaneous electrical stimuli. We introduced a two-alternative force choice task, whereby two temporal factors were varied to probe both the refractory period and the discrimination of consecutive stimuli. Catch trials, with both stimuli identical, allowed us to study if the participants had a bias toward either the first or second stimulus.

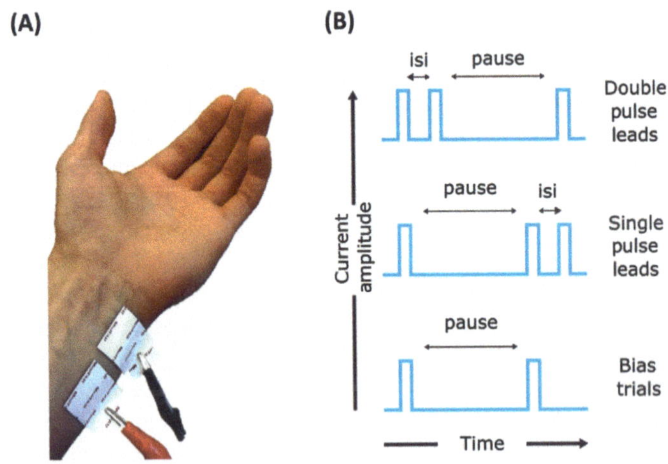

FIGURE 4.3 Methodology: (A) electrode placement; (B) different stimuli included in the experiment. *Courtesy: Authors.*

4.6.3.1 Methods

Ten participants took part in the experiment. The study was approved by the local ethics committee at Newcastle University (19-NAZ-043) and all participants provided written informed consent prior to participating in the study.

Transcutaneous electrical nerve stimulation was delivered through a pair of surface electrodes (MedTAB, Medgraphics Ltd., UK) placed proximal to the left wrist (Fig. 4.3A). The electrodes were positioned to target the ulnar nerve, with an interelectrode distance of 20 mm. Unipolar, anodic current pulses with a fixed with of 200 μs were delivered using a DS7A Constant Current High Voltage Stimulator (Digitimer Ltd, UK). Participants used their right hand to indicate which of the stimuli they received were perceived as having a higher amplitude.

Individual sensory thresholds were determined for all participants to account for variations in skin impedance. The threshold was determined using a staircase procedure, and was defined as the lowest stimulus amplitude, where participants perceived at least 5 out of 10 stimuli. During the experiment, participants received pulses with an amplitude of 120% sensory threshold.

The experiment consisted of a two-alternative forced choice task, where two stimuli were presented to the participants sequentially, and they had to report which of these they perceived as having the higher amplitude. The pairs of stimuli consisted of a single and double pulse, or in the case of the bias trials, two identical single pulses (see Fig. 4.3B). The interstimulus interval (isi) of the double pulse, and the pause between the two stimuli, were varied. The order of single and double pulses was balanced and randomized over the experiment.

The study included three isi (6, 8, and 10 milliseconds) to test the influence of the refractory period, and three pauses (250, 350, and 450 milliseconds) to investigate stimulus interaction. All parameters of the study (isi, pause, and type of pulse leading) were balanced, resulting in a $3 \times 3 \times 2$ design. Participants completed 6 blocks of 42 trials, where each block included six bias trials (two repetitions of three pause durations). Overall, each condition was repeated 12 times.

Participant perception was tested by analyzing the percentage of trials in which the double pulse was perceived as the one with the higher amplitude. Type A order effect, the preference of participants for either the first or second stimulus, was analyzed by determining the percentage of bias trials in which the first stimulus was perceived as having a higher amplitude. A type B order effect is present when the difference between two distinct stimuli is divergent based on which stimulus is leading. This was analyzed by studying whether there is a difference in amplitude perception when the double pulse either leads or lags.

Normality analysis (Shapiro−Wilk normality test) showed that not all data were normally distributed. Therefore nonparametric statistical analysis

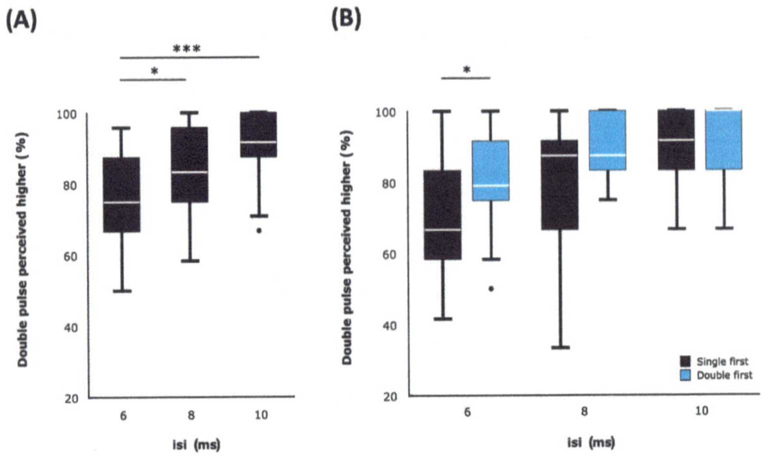

FIGURE 4.4 Analysis of the effects of the refractory period for transcutaneous stimulation. (A) Identification of the double pulse as the one with the higher amplitude is lower for an isi of 6 milliseconds than for 8 or 10 milliseconds; (B) there is a type B order effect, that is, influence of stimulus order on the effects of the refractory period, when the double pulse has an isi of 6 milliseconds, which disappears for higher isi. *Courtesy: Authors.*

was used throughout the study. When the Friedman test showed significant differences at the group level, Mann–Whitney U-tests with Bonferroni corrections were applied as the post hoc analysis.

4.6.3.2 Results

The length of the refractory period can be estimated by comparing how the double pulse is perceived for the different isi. If the isi is sufficiently long, the perceived amplitude of the double pulse will be higher than that of the single pulse. This distinction would not be clear for stimuli where the second pulse of the double pulse falls within the refractory period. A Friedman test showed that isi significantly influenced the perception of double stimuli ($P = 0.0008$), irrespective of the pause duration. Shorter isi were associated with lower percentages of double pulse perceived as having the higher amplitude (Fig. 4.4A). Post hoc comparison of the influence of isi on sensory perception showed that an isi of 6 milliseconds led to a lower identification of the double pulse as stronger (76.11 ± 10.53%) than compared to an isi of 8 milliseconds (84.86 ± 9.18%; $P = 0.046$) or 10 milliseconds (90.83 ± 7.09%; $P = 0.0005$). There was no statistical difference between the trials with an isi of 8 and 10 milliseconds ($P = 0.363$).

A second approach to investigate the refractory period is to look at the type B order effect. This analysis, presented in Fig. 4.4B, studies whether the identification of the double pulse as stronger is different when this stimulus is presented first or second. A Friedman test showed that there is an overall

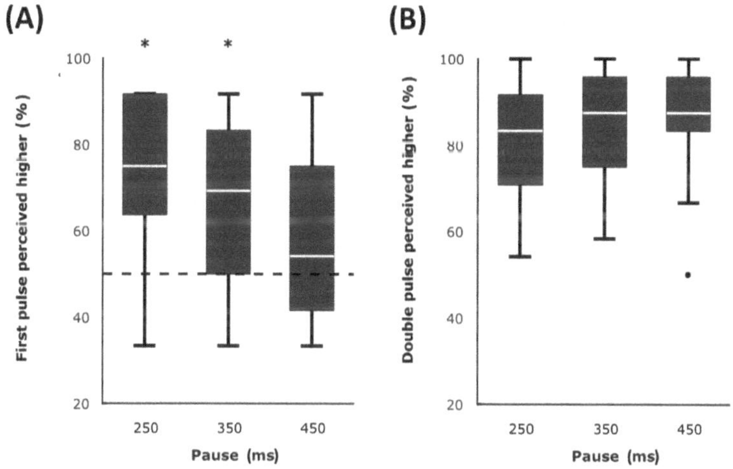

FIGURE 4.5 Influence of pause duration on stimulus interaction. (A) Type A order effect analysis on the bias trials shows that pauses of 250 and 350 milliseconds show a bias toward the first stimulus. (B) Effect of the pause duration on the perceived amplitude difference between single and double pulses. *Courtesy: Authors.*

difference in how the double pulse is perceived based on the order of stimuli ($P = 0.0087$), with the double pulse perceived as higher more often when it is leading ($87.31 \pm 12.13\%$) compared to when it is lagging ($80.56 \pm 16.76\%$). Post hoc analysis showed that this difference only existed for an isi of 6 milliseconds ($P = 0.043$), but not for 8 and 10 milliseconds ($P = 0.595$ and $P = 0.763$, respectively).

All participants reported that they perceived both stimuli in all trials, thereby confirming that the pause between the stimuli exceeds the temporal discrimination threshold. To test if there is an interaction between the stimuli beyond this threshold, we studied both the type A order effect, and the difference in perception of single and double pulses for the different pauses. The type A order effect tests if there is a bias toward the first or second pulse when they are identical by testing if the amplitude of first pulse is identified as highest more than 50% of the time. Analysis showed an overall bias toward the first stimulus as participants identified it as highest $64.24 \pm 19.80\%$ of the time ($P = 0.0157$). Fig. 4.5A shows the influence of the pause duration on the type A order effect, where a bias is present for pauses of 250 and 350 milliseconds ($P = 0.016$ and $P = 0.047$, respectively), but not for 450 milliseconds ($P = 0.481$).

Analysis of the difference in perception between single and double pulses based on the pause duration—irrespective of the isi—showed an increase in identification of the double pulse as stronger as pause time increases (identification of $80.56 \pm 9.39\%$, $84.86 \pm 9.88\%$, and $86.39 \pm 6.95\%$ for pauses of 250,

350, and 450 milliseconds, respectively; see Fig. 4.5B). However, a Friedman test showed that this difference did not reach significance ($P = 0.053$).

4.6.3.3 Implications for prosthetic control

This study showed that sensory perception of transcutaneous pulses can be modified through changes in the temporal profile of successive stimuli, even when they contain basic patterns such as single and double pulses. The fact that participants were able to distinguish the single and double pulse based on their perceived amplitude shows that increasing the interstimulus interval between pulses above the effective refractory period leads to a change in amplitude perception. Our results suggest that distinct sensory perceptions can be created by modulating the timing of identical stimuli, thereby increasing the amount of information that can be encoded for prosthetic feedback.

Participants reported that they perceived the stimuli as separate "mouse clicks" on their wrist, indicating that the pause durations in this study were long enough to allow temporal stimulus discrimination. This was expected, as previous studies on discrimination of peripheral sensory stimuli and the silent period after transcranial magnetic stimulation suggest timings below 100 milliseconds would lead to a lack of temporal discrimination (Fuhr et al., 1991; Pastor et al., 2004). However, while participants were able to distinguish the two stimuli, we saw interaction between them in the trials with 250 and 350 milliseconds pause, as we found a type A order effect for these trials. This suggests that such stimuli used for event-related feedback in prosthetic control should have an interstimulus interval of at least 450 milliseconds.

Previous studies have reported a total refractory period of between 2 and 5 milliseconds (Wesselink et al., 1999), while others stated that full recovery of the neurons might take up to tens of milliseconds (Dayan & Abbott, 2001). Here, we show a significant difference for trials with an isi of 6 milliseconds compared to those with 8 or 10 milliseconds, where shorter isi led to a lower identification of the double pulse as having a higher amplitude. This suggests that the associated neural networks may take at least 8 milliseconds to recover from the first pulse, which is in accordance with previous rodent work from our lab (Brunton et al., 2019). Although individual mechanoreceptive afferents may be driven up to 1000 spikes per second in acute in vivo electrophysiology of Merkel cell–neurite complexes in cats, higher order excitatory and inhibitory processes seem to make fast information transfer unlikely (Gottschaldt & Vahle-Hinz, 1981). Our results show that increasing the pulse frequency to increase the amplitude sensation of a stimulus has an upper limit for transcutaneous stimulation. Providing intervals between pulses lower than 8 milliseconds might not have the desired effect on sensory perception, and should therefore be avoided in prosthetic feedback.

4.7 Conclusion

This chapter has given an overview of the importance of providing sensory feedback in prosthetics, both from the perspective of user preference, and to improve prosthetic control. We have given an overview of the important physiological aspects of sensory feedback, and how these can be incorporated into prosthetic feedback strategies by adopting an event-related feedback strategy. We highlighted some of the efforts to optimize the stimulation parameters for event-related feedback, and how these are linked to our understanding of human motor control and neural physiology. Finally, we explored one optimization study more in depth, and showed how the temporal modulation of transcutaneous electrical pulses can create distinctly perceived stimulus amplitudes. Exploration of interstimulus intervals between double pulses and pauses between different stimuli identified lower boundaries for both that should be adopted when designing event-related feedback strategies with transcutaneous electrical stimulation.

References

Aboseria, M., Clemente, F., Engels, L. F., & Cipriani, C. (2018). Discrete vibro-tactile feedback prevents object slippage in hand prostheses more intuitively than other modalities. *IEEE Transactions on Neural Systems and Rehabilitation Engineering*, *26*(8), 1577–1584. Available from https://doi.org/10.1109/TNSRE.2018.2851617.

Atkins, D. J., Heard, D. C. Y., & Donovan, W. H. (1996). Epidemiologic overview of individuals with upper-limb loss and their reported research priorities. *Journal of Prosthetics and Orthotics*, *8*(1), 2–11. Available from https://doi.org/10.1097/00008526-199600810-00003.

Biddiss, E., Beaton, D., & Chau, T. (2007). Consumer design priorities for upper limb prosthetics. *Disability and Rehabilitation: Assistive Technology*, *2*(6), 346–357. Available from https://doi.org/10.1080/17483100701714733.

Biddiss, E., & Chau, T. (2007). Upper-limb prosthetics: Critical factors in device abandonment. *American Journal of Physical Medicine and Rehabilitation*, *86*(12), 977–987. Available from https://doi.org/10.1097/PHM.0b013e3181587f6c.

Brunton, E. K., Silveira, C., Rosenberg, J., Schiefer, M. A., Riddell, J., & Nazarpour, K. (2019). Temporal modulation of the response of sensory fibers to paired-pulse stimulation. *IEEE Transactions on Neural Systems and Rehabilitation Engineering: A Publication of the IEEE Engineering in Medicine and Biology Society*, *27*(9), 1676–1683. Available from https://doi.org/10.1109/TNSRE.2019.2935813.

Callier, T., Suresh, A. K., & Bensmaia, S. J. (2019). Neural coding of contact events in somatosensory cortex. *Cerebral Cortex*, *29*(11), 4613–4627. Available from https://doi.org/10.1093/cercor/bhy337.

Chadwell, A., Kenney, L., Granat, M. H., Thies, S., Head, J., Galpin, A., Baker, R., & Kulkarni, J. (2018). Upper limb activity in myoelectric prosthesis users is biased towards the intact limb and appears unrelated to goal-directed task performance. *Scientific Reports*, *8*(1), 1–12. Available from https://doi.org/10.1038/s41598-018-29503-6.

Chen, C., & Thompson, R. F. (1995). Temporal specificity of long-term depression in parallel fiber–Purkinje synapses in rat cerebellar slice. *Learning & Memory (Cold Spring Harbor, N.Y.)*, *2*(3–4), 185–198. Available from https://doi.org/10.1101/lm.2.3-4.185.

Childress, D. S. (1985). Historical aspects of powered limb prostheses. *Clinical Prosthetics & Orthotics*, *9*(1), 2–13.

Cipriani, C., Segil, J. L., Clemente, F., Richard, R. F., & Edin, B. (2014). Humans can integrate feedback of discrete events in their sensorimotor control of a robotic hand. *Experimental Brain Research*, *232*(11), 3421–3429. Available from https://doi.org/10.1007/s00221-014-4024-8.

Clemente, F., D'Alonzo, M., Controzzi, M., Edin, B. B., & Cipriani, C. (2016). Non-invasive, temporally discrete feedback of object contact and release improves grasp control of closed-loop myoelectric transradial prostheses. *IEEE Transactions on Neural Systems and Rehabilitation Engineering*, *24*(12), 1314–1322. Available from https://doi.org/10.1109/TNSRE.2015.2500586.

Corniani, G., & Saal, H. P. (2020). Tactile innervation densities across the whole body. *Journal of Neurophysiology*.

Crea, S., Edin, B. B., Knaepen, K., Meeusen, R., & Vitiello, N. (2017). Time-discrete vibrotactile feedback contributes to improved gait symmetry in patients with lower limb amputations: Case series. *Physical Therapy*, *97*(2), 198–207.

Anna, E., Valle, G., Mazzoni, A., Strauss, I., Iberite, F., Patton, J., Petrini, F. M., Raspopovic, S., Granata, G., Di Iorio, R., Controzzi, M., Cipriani, C., Stieglitz, T., Rossini, P. M., & Micera, S. (2019). A closed-loop hand prosthesis with simultaneous intraneural tactile and position feedback. *Science Robotics*, *4*(27). Available from https://doi.org/10.1126/scirobotics.aau8892.

Dayan, P., & Abbott, L. F. (2001). *Theoretical neuroscience: Computational and mathematical modeling of neural systems*. MIT Press.

Diedrichsen, J., Shadmehr, R., & Ivry, R. B. (2010). The coordination of movement: Optimal feedback control and beyond. *Trends in Cognitive Sciences*, *14*(1), 31–39. Available from https://doi.org/10.1016/j.tics.2009.11.004.

Engdahl, S. M., Chestek, C. A., Kelly, B., Davis, A., & Gates, D. H. (2017). Factors associated with interest in novel interfaces for upper limb prosthesis control. *PLoS One*, *12*(8), 1–17. Available from https://doi.org/10.1371/journal.pone.0182482.

Engdahl, S. M., Christie, B. P., Kelly, B., Davis, A., Chestek, C. A., & Gates, D. H. (2015). Surveying the interest of individuals with upper limb loss in novel prosthetic control techniques. *Journal of NeuroEngineering and Rehabilitation*, *12*(1), 1–11. Available from https://doi.org/10.1186/s12984-015-0044-2.

Engels, L. F., Shehata, A. W., Scheme, E. J., Sensinger, J. W., & Cipriani, C. (2019). When less is more—Discrete tactile feedback dominates continuous audio biofeedback in the integrated percept while controlling a myoelectric prosthetic hand. *Frontiers in Neuroscience*, *13*, 1–10. Available from https://doi.org/10.3389/fnins.2019.00578.

Fiorio, M., Tinazzi, M., Bertolasi, L., & Aglioti, S. M. (2003). Temporal processing of visuotactile and tactile stimuli in writer's cramp. *Annals of Neurology*, *53*(5), 630–635. Available from https://doi.org/10.1002/ana.10525.

Fiorio, M., Tinazzi, M., Scontrini, A., Stanzani, C., Gambarin, M., Fiaschi, A., Moretto, G., Fabbrini, G., & Berardelli, A. (2008). Tactile temporal discrimination in patients with blepharospasm. *Journal of Neurology, Neurosurgery and Psychiatry*, *79*(7), 796–798. Available from https://doi.org/10.1136/jnnp.2007.131524.

Fuhr, P., Agostino, R., & Hallett, M. (1991). Spinal motor neuron excitability during the silent period after cortical stimulation. *Electroencephalography and Clinical Neurophysiology/Evoked Potentials*, *81*(4), 257–262. Available from https://doi.org/10.1016/0168-5597(91)90011-L.

George, J.A., Brinton, M.R., Colgan, P.C., Colvin, G.K., Bensmaia, S.J., & Clark, G.A. (2020). Intensity discriminability of electrocutaneous and intraneural stimulation pulse frequency in intact individuals and amputees. *ArXiv Preprint:2001.08808*. http://arxiv.org/abs/2001.08808

George, J. A., Kluger, D. T., Davis, T. S., Wendelken, S. M., Okorokova, E. V., He, Q., Duncan, C. C., Hutchinson, D. T., Thumser, Z. C., Beckler, D. T., Marasco, P. D., Bensmaia, S. J., & Clark, G. A. (2019). Biomimetic sensory feedback through peripheral nerve stimulation improves dexterous use of a bionic hand. *Science Robotics*, *4*(32), 1–12. Available from https://doi.org/10.1126/scirobotics.aax2352.

Gottschaldt, K. M., & Vahle-Hinz, C. (1981). Merkel cell receptors: Structure and transducer function. *Science*, *214*(4517), 183–186. Available from https://doi.org/10.1126/science.7280690.

Graczyk, E. L., Schiefer, M. A., Saal, H. P., Delhaye, B. P., Bensmaia, S. J., & Tyler, D. J. (2016). The neural basis of perceived intensity in natural and artificial touch. *Science Translational Medicine*, *8*(362), 1–11. Available from https://doi.org/10.1126/scitranslmed.aaf5187.

Jang, C. H., Yang, H. S., Yang, H. E., Lee, S. Y., Kwon, J. W., Yun, B. D., Choi, J. Y., Kim, S. N., & Jeong, H. W. (2011). A survey on activities of daily living and occupations of upper extremity amputees. *Annals of Rehabilitation Medicine*, *35*(6), 907. Available from https://doi.org/10.5535/arm.2011.35.6.907.

Johansson, R. S. (1992). Sensory-motor coordination during grasping and manipulative actions. *Current Opinion in Neurobiology*, *2*, 815–823. Available from http://www.sciencedirect.com/science/article/pii/095943889290139C%5Cnpapers3://publication/uuid/A43C6CB5-10C0-45D2-B97C-DA8A7DB15DF6.

Johansson, R. S., & Bäckström, L. (1992). Somatosensory control of precision grip during unpredictable pulling loads. III. Impairments during digital anesthesia. *Experimental Brain Research*, *89*(1), 204–213.

Johansson, R. S., & Edin, B. B. (1993). Predictive feed-forward sensory control during grasping and manipulation in man. *Biomedical Research*, *14*(Suppl. 4), 95–106.

Johansson, R. S., & Flanagan, J. R. (2009). Coding and use of tactile signals from the fingertips in object manipulation tasks. *Nature Reviews Neuroscience*, *10*(5), 345–359. Available from https://doi.org/10.1038/nrn2621.

Johansson, R. S., & Vallbo, A. B. (1979). Tactile sensibility in the human hand: Relative and absolute densities of four types of mechanoreceptive units in glabrous skin. *J Physiol*, *286*, 283–300.

Johansson, R. S., & Westling, G. (1991). Afferent signals during manipulative tasks in humans. In O. Franzen, & J. Westman (Eds.), *Information processing in the somatosensory system*. Macmillan Press.

Karakuş, İ., & Güçlü, B. (2020). Psychophysical principles of discrete event-driven vibrotactile feedback for prostheses. *Somatosensory & Motor Research*, *37*(3), 1–18. Available from https://doi.org/10.1080/08990220.2020.1769055.

Kawato, M. (1999). Internal models for motor control and trajectory planning. *Current Opinion in Neurobiology*, *9*(6), 718–727.

Knill, D. C., & Pouget, A. (2004). The Bayesian brain: The role of uncertainty in neural coding and computation. *Trends in Neurosciences*, *27*(12), 712–719. Available from https://doi.org/10.1016/j.tins.2004.10.007.

Kumar, D. K., Jelfs, B., Sui, X., & Arjunan, S. P. (2019). Prosthetic hand control: A multidisciplinary review to identify strengths, shortcomings, and the future. *Biomedical Signal Processing and Control*, *53*, 101588. Available from https://doi.org/10.1016/j.bspc.2019.101588.

Kyberd, P. J., & Hill, W. (2011). Survey of upper limb prosthesis users in Sweden, the United Kingdom and Canada. *Prosthetics and Orthotics International*, *35*(2), 234–241. Available from https://doi.org/10.1177/0309364611409099.

Lewis, S., Russold, M.F., Dietl, H., & Kaniusas, E. (2012). User demands for sensory feedback in upper extremity prostheses. *Proceedings of the MeMeA 2012 – 2012 IEEE symposium on medical measurements and applications* (pp. 188–191). https://doi.org/10.1109/MeMeA.2012.6226669

Liu, D., & Todorov, E. (2007). Evidence for the flexible sensorimotor strategies predicted by optimal feedback control. *Journal of Neuroscience*, *27*(35), 9354–9368. Available from https://doi.org/10.1523/JNEUROSCI.1110-06.2007.

Luchetti, M., Cutti, A. G., Verni, G., Sacchetti, R., & Rossi, N. (2015). Impact of Michelangelo prosthetic hand: Findings from a crossover longitudinal study. *Journal of Rehabilitation Research and Development*, *52*(5), 605–618. Available from https://doi.org/10.1682/JRRD.2014.11.0283.

Macfie, H. J. H., & Thomson, D. M. H. (1981). Perception of two-component electrocutaneous stimuli. *Perception & Psychophysics*, *30*(5), 473–482. Available from https://doi.org/10.3758/BF03204844.

Mann, R. W., & Reimers, S. D. (1970). Kinesthetic sensing for the EMG controlled "Boston arm. *IEEE Transactions on Man-Machine Systems*, *11*(1), 110–115.

Marasco, P. D., Kim, K., Colgate, J. E., Peshkin, M. A., & Kuiken, T. A. (2011). Robotic touch shifts perception of embodiment to a prosthesis in targeted reinnervation amputees. *Brain*, *134*(3), 747–758. Available from https://doi.org/10.1093/brain/awq361.

Mcglone, F., & Reilly, D. (2010). The cutaneous sensory system. *Neuroscience and Biobehavioral Reviews*, *34*, 148–159. Available from https://doi.org/10.1016/j.neubiorev.2009.08.004.

Morasso, P. (1981). Spatial control of arm movements. *Experimental Brain Research*, *42*, 223–227.

Okorokova, E. V., He, Q., & Bensmaia, S. J. (2018). Biomimetic encoding model for restoring touch in bionic hands through a nerve interface. *Journal of Neural Engineering*, *15*(6). Available from https://doi.org/10.1088/1741-2552/aae398.

Orbán, G., & Wolpert, D. M. (2011). Representations of uncertainty in sensorimotor control. *Current Opinion in Neurobiology*, *21*(4), 629–635. Available from https://doi.org/10.1016/j.conb.2011.05.026.

Osborn, L., Kaliki, R. R., Soares, A. B., & Thakor, N. V. (2016). Neuromimetic event-based detection for closed-loop tactile feedback control of upper limb prostheses. *IEEE Transactions on Haptics*, *9*(2), 196–206. Available from https://doi.org/10.1109/TOH.2016.2564965.

Østlie, K., Lesjø, I. M., Franklin, R. J., Garfelt, B., Skjeldal, O. H., & Magnus, P. (2012). Prosthesis rejection in acquired major upper-limb amputees: A population-based survey. *Disability and Rehabilitation: Assistive Technology*, *7*(4), 294–303. Available from https://doi.org/10.3109/17483107.2011.635405.

Østlie, K., Skjeldal, O. H., Garfelt, B., & Magnus, P. (2011). Adult acquired major upper limb amputation in Norway: Prevalence, demographic features and amputation specific features. A population-based survey. *Disability and Rehabilitation*, *33*(17–18), 1636–1649. Available from https://doi.org/10.3109/09638288.2010.541973.

Pastor, M. A., Day, B. L., Macaluso, E., Friston, K. J., & Frackowiak, R. S. J. (2004). The functional neuroanatomy of temporal discrimination. *Journal of Neuroscience*, *24*(10), 2585–2591. Available from https://doi.org/10.1523/JNEUROSCI.4210-03.2004.

Rhodes, J., Clay, C., & Phillips, M. (2013). The surface area of the hand and the palm for estimating percentage of total body surface area: Results of a meta-analysis. *British Journal of Dermatology*, *169*(1), 76–84. Available from https://doi.org/10.1111/bjd.12290.

Rossini, P. M., Micera, S., Benvenuto, A., Carpaneto, J., Cavallo, G., Citi, L., Cipriani, C., Denaro, L., Denaro, V., Di Pino, G., Ferreri, F., Guglielmelli, E., Hoffmann, K. P., Raspopovic, S., Rigosa, J., Rossini, L., Tombini, M., & Dario, P. (2010). Double nerve intraneural interface implant on a human amputee for robotic hand control. *Clinical Neurophysiology*, *121*(5), 777–783. Available from https://doi.org/10.1016/j.clinph.2010.01.001.

Saal, H. P., Delhaye, B. P., Rayhaun, B. C., & Bensmaia, S. J. (2017). Simulating tactile signals from the whole hand with millisecond precision. *Proceedings of the National Academy of Sciences of the United States of America*, *114*(28), E5693–E5702. Available from https://doi.org/10.1073/pnas.1704856114.

Scott, S. H. (2004). Optimal feedback control and the neural basis of volitional motor control. *Nature Reviews Neuroscience*, *5*(7), 532–546. Available from https://doi.org/10.1038/nrn1427.

Sensinger, J. W., & Dosen, S. (2020). A review of sensory feedback in upper-limb prostheses from the perspective of human motor control. *Frontiers in Neuroscience*, *14*(June), 1–24. Available from https://doi.org/10.3389/fnins.2020.00345.

Shadmehr, R., & Mussa-Ivaldi, F. A. (2012). *Biological learning and control: How the brain builds representations, predicts events, and makes decisions*. MIT Press.

Shimada, S., Qi, Y., & Hiraki, K. (2010). Detection of visual feedback delay in active and passive self-body movements. *Experimental Brain Research*, *201*(2), 359–364. Available from https://doi.org/10.1007/s00221-009-2028-6.

Smail, L. C., Neal, C., Wilkins, C., & Packham, T. L. (2020). Comfort and function remain key factors in upper limb prosthetic abandonment: Findings of a scoping review. *Disability and Rehabilitation: Assistive Technology*, 1–10. Available from https://doi.org/10.1080/17483107. 2020.1738567.

Stephens-Fripp, B., Jean Walker, M., Goddard, E., & Alici, G. (2020). A survey on what Australians with upper limb difference want in a prosthesis: Justification for using soft robotics and additive manufacturing for customized prosthetic hands. *Disability and Rehabilitation: Assistive Technology*, *15*(3), 342–349. Available from https://doi.org/10.1080/17483107.2019.1580777.

Tan, D. W., Schiefer, M. A., Keith, M. W., Anderson, J. R., Tyler, J., & Tyler, D. J. (2014). A neural interface provides long-term stable natural touch perception. *Science Translational Medicine*, *6*(257). Available from https://doi.org/10.1126/scitranslmed.3008669.

Todorov, E., & Jordan, M. I. (2002). Optimal feedback control as a theory of motor coordination. *Nature Neuroscience*, *5*(11), 1226–1235. Available from https://doi.org/10.1038/nn963.

Valle, G., Petrini, F. M., Strauss, I., Iberite, F., D'Anna, E., Granata, G., Controzzi, M., Cipriani, C., Stieglitz, T., Rossini, P. M., Mazzoni, A., Raspopovic, S., & Micera, S. (2018). Comparison of linear frequency and amplitude modulation for intraneural sensory feedback in bidirectional hand prostheses. *Scientific Reports*, *8*(1), 1–13. Available from https://doi.org/10.1038/s41598-018-34910-w.

Van Doren, C. L. (1997). Contours of equal perceived amplitude and equal perceived frequency for electrocutaneous stimuli. *Perception and Psychophysics*, *59*(4), 613–622. Available from https://doi.org/10.3758/BF03211869.

Vincent Systems GmbH. (2020). *Vincent evolution 2: The touch sensing hand prosthesis of the next generation*. https://vincentsystems.de/en/prosthetics/vincent-evolution-2/

Wesselink, W. A., Holsheimer, J., & Boom, H. B. K. (1999). A model of the electrical behaviour of myelinated sensory nerve fibres based on human data. *Medical and Biological Engineering and Computing*, *37*(2), 228–235. Available from https://doi.org/10.1007/BF02513291.

Wolpert, D. M., Ghahramani, Z., & Jordan, M. (1995). An internal model for sensorimotor integration. *Science*, *269*(5232), 1880–1882.

Part II

Non-invasive methods for somatosensory feedback and modulation

Chapter 5

Supplementary feedback for upper-limb prostheses using noninvasive stimulation: methods, encoding, estimation-prediction processes, and assessment

Jakob Dideriksen and Strahinja Dosen
Department of Health Science and Technology, Aalborg University, Aalborg, Denmark

ABSTRACT
In this chapter the authors provide an overview of the concept of noninvasive supplementary feedback for closed-loop control of upper-limb prostheses. Although such feedback has been investigated for decades with encouraging results, its implementation in commercial devices is limited and its clinical utility is yet to be proven. In this context, the authors present and discuss novel approaches for the design and evaluation of noninvasive supplementary feedback, that they believe have the potential to improve the efficiency of feedback systems. The application of these approaches may facilitate the implementation of supplementary feedback in commercial prosthetic devices.

Keywords: Supplementary feedback; prosthesis control; closed-loop control; motor control; noninvasive stimulation; sensory substitution; electrotactile stimulation; vibrotactile stimulation

5.1 Motivation

Consider for a moment the role of your hands in the activities that you carry out throughout the course of a typical day and your ability to perform those activities with the absence of hands. Most will quickly come to the realization that hands are essential tools in our everyday lives. Specifically, hands are used to carry out motor tasks by allowing dexterous manipulation and

Somatosensory Feedback for Neuroprosthetics. DOI: https://doi.org/10.1016/B978-0-12-822828-9.00019-8

exploration of objects, but they also play an important role in social communication by conveying emotions via gestures and affective touch (Löken et al., 2009). Consequently, the restoration of these functions for those who have lost a hand due to disease or trauma has gained massive attention in rehabilitation research. Naturally, the primary emphasis of this line of research has been the development of bionic prosthetic hands (Piazza et al., 2019) and the interface by which humans can control their actions (Yang et al., 2019).

Further reflection upon the instrumental role of hands in everyday life, however, may lead to the realization of what may seem like a paradoxical mismatch between the complexity of highly diverse manual tasks and the low cognitive effort required to carry them out. Many seemingly simple tasks that can be carried out without paying much attention to the hands actually require a complex and accurate interplay between the movement and forces generated independently by multiple digits (MacKenzie & Iberall, 2010). For example, grasping a keychain in your pocket, selecting the correct key, positioning it appropriately between your fingers, and inserting it in a keyhole can usually be done successfully while thinking about other things even in conditions where visual information is impaired (e.g., a dimly lighted hallway). Such considerations indicate that a mechanically dexterous prosthetic hand and a robust control interface is likely not enough to enable prosthesis movements that approximate those of its natural counterpart. This also requires a restoration of the ability of prosthesis users to appropriately select and adjust the commands that are sent to the hand to complete a given task. This ability relies primarily on the existence of high-fidelity somatosensory feedback (Flanagan et al., 2006).

Natural hands are equipped with hundreds of thousands of sensors that convey information about many sensory modalities including touch and joint kinematics (Gesslbauer et al., 2017). In the central nervous system, this vast amount of information is fused with other relevant sensory information (e.g., vision) and the output of so-called internal models representing the experience-based expectations to the sensory consequences of a certain action in a certain context, to give rise to an estimate of the state of the limb (Diedrichsen et al., 2010). This state estimate is a central component in human motor control. To put it in simple, logical terms, even with perfect controllability, any task could not be accomplished successfully if the instantaneous state of the limb at the beginning of the task is not known. Furthermore, to enable compensations for a less than perfect movement plan (i.e., a set of muscle activation commands that generate movements to fulfill the task), a continuously updated state estimate is often desirable during the execution of a movement. For example, when grasping and lifting an object whose weight is not known a priori, it is likely that adjustment in grasp strength and force by which it is lifted are required. Furthermore, when acting in an unpredictable environment, perturbations may occur that require rapid adjustments in movement in order to achieve the task goal. The importance of somatosensory feedback in such motor control processes is well-illustrated by investigations in the control abilities of so-called deafferented individuals. Deafferentation implies an almost complete loss of the sense of touch and proprioception, while the ability to activate muscles is intact. In order to generate accurate movements even in simple tasks, these patients, however, require large amounts of cognitive effort and training (Hermsdörfer et al., 2008). Another remarkable evidence for the importance of somatosensory feedback is provided by a recent anatomical study (Gesslbauer et al., 2017), which reported that the sensory fibers in the peripheral nerves innervating the hand and arm outnumber the motor fibers by a ratio of at least 9:1.

The natural hand has 23 degrees of freedom (DoF) and it is controlled by 10 and 19 muscles located in the hand and forearm, respectively. Consequently, a complete estimate of the state of a natural hand is highly complex. As mentioned above, continuous information from a very high number of sensory receptors is needed for the central nervous system to estimate different aspects of this state. Therefore, if a prosthesis was to have similar complexity to that of a human hand, the restoration of an acceptable state estimate would likely be impossible to achieve. The state of a prosthesis, however, is considerably simpler. Most commercial prostheses have two DoFs (opening/closing and wrist rotation), which implies that the prosthesis can be at any point in a limited number of states. Even the most advanced systems with individually controllable fingers are still far simpler than their biological counterparts (Vujaklija et al., 2016). Furthermore, since prosthetic joints are usually nonbackdrivable, joint angles are not affected by external perturbations or gravity. Therefore the hand will remain in the same state unless activated by the voluntary command signal of the user via a prosthesis motor. Consequently, it is reasonable to assume that in case of a prosthesis, a meaningful state estimate can be generated by a limited number of feedback channels.

In spite of not having explicit somatosensory feedback from the prosthetic hand, an amputee does have a number of relevant sources of sensory information regarding its state (Wilke et al., 2019). First, vision can provide detailed information about the state of a prosthetic hand in many situations. For example, the aperture of the hand can readily be assessed by looking at it. On the other hand, the grasp force that is being applied on a rigid object is harder to identify visually and application of excessive grasp force can only be seen when the object deforms or breaks. However, while visual feedback is informative in many conditions, it implies that the visual attention must be directed toward the hand during the task, which is indeed a behavior that can be seen in prosthesis users (Sobuh et al., 2014). Conversely, able-bodied individuals can perform even relatively complex manual tasks while directing the visual attention elsewhere (due to the availability of somatosensory feedback). Second, auditory noise from the motors of a prosthesis can reveal to some degree changes in the state of the prosthesis. Auditory feedback, however, is generally not preferred by users (Lewis et al., 2012), likely since continuous or repetitive loud noises are not desirable in a social context. Finally, vibrations from the motors can provide similar information as the auditory feedback, but its richness and thus usability for state estimation is limited (Wilke et al., 2019). These considerations indicate that although some sources of feedback are available to prosthesis users, it is likely that supplementary feedback can improve the quality of the state estimate and thus the quality of prosthetic control.

Another relevant aspect that may be promoted by the supplementary somatosensory feedback is prosthesis embodiment. In this context, embodiment of an object refers to the sensation of experiencing it as an integrated part of the body schema as opposed to an external tool (Giummarra et al., 2008). Although a prosthesis that is merely regarded as an external tool may still be very useful in the restoration of the motor functions lost due to an amputation, it is likely that prosthesis embodiment may be a worthwhile aim to pursue. Many prosthesis users report low usage or even abandonment of their prostheses (Biddiss & Chau, 2007). It is likely that the sensation of embodiment may promote a higher usage. In turn, this may lead to more practice and improvements in the control skills of the prosthesis user. Importantly, it has been shown that even simple ways to provide an illusion of sensory feedback from the prosthesis such as mechanical vibration applied synchronously on the skin and on the prosthesis, when only the prosthesis was visible to the user, substantially improved the sense of ownership and embodiment (D'Alonzo & Cipriani, 2012; Shehata et al., 2020).

To summarize, it seems reasonable to assume that some degree of restoration of somatosensory feedback from the prosthesis may enhance the ability of amputees to control it and regard it as an integral part of the body. This reasoning is confirmed by multiple surveys of prosthesis users' needs and wishes for future development. In one of the earliest surveys published in 1996, approximately 900 users of electrically powered prostheses indicated less reliance on visual feedback as the third highest ranked priority for future development (Atkins et al., 1996). As mentioned above, it is reasonable to assume that reliance on visual feedback may be reduced by the supplementary somatosensory feedback. The other highly ranked priorities were associated primarily to hand dexterity (i.e., number of DoFs). In a more recent survey from 2007 on 242 prosthesis users of which 108 used electrically powered prostheses, sensory feedback was ranked as the fourth highest priority for future development after reduction of weight, increase in durability, and reduction in cost (Biddiss et al., 2007). Other studies have confirmed the wish for supplementary sensory feedback among prosthesis users (Lewis et al., 2012; Pylatiuk et al., 2007; Walker et al., 2020). These priorities may indicate that the development in prosthetic hands since 1996 had mainly addressed the need for improved dexterity. Restoration of the ability to "feel" the prosthesis, however, remained an unmet need.

5.2 Restoration of somatosensory feedback

The first research studies presenting prototype systems for the provision of supplementary, noninvasive feedback from prostheses were conducted more than 50 years ago. However, only recently commercial prosthetic hands with simple systems for such feedback are beginning to emerge (Vincent Systems GmbH, Mobius Bionics and Psyonic Inc.). However, these sensate prostheses are yet to prove their clinical utility. Typically, these systems include a vibration motor built into the socket to indicate the magnitude of grasp force. The intensity of the vibration can easily be perceived by the prosthesis user and, in principle, used to adjust the muscle activation levels to control the grasp. This approach to supplementary feedback, however, is far from reflecting the vast diversity of systems proposed in research. These research systems have been extensively reviewed elsewhere (Antfolk et al., 2013; Schofield et al., 2014; Sensinger & Dosen, 2020; Svensson et al., 2017). Instead, here we present three examples of studies that illustrate this diversity, but also highlight common underlying principles for the design of the systems as well as their evaluation.

Jorgovanovic et al. (2014) provided feedback on grasp force magnitude in a grasping task with a virtual prosthesis. Feedback was delivered via electrotactile stimulation on the skin surface, with an intensity (pulse width) that reflected the grasp force applied by the prosthesis using a linear mapping. In this way, a weak sensation (slightly above sensation threshold) indicated a low force, while a strong sensation (slightly below discomfort threshold) indicated the maximum grasp force. In the experiment, the grasping of the virtual prosthesis was controlled with a joystick as opposed to an electromyography (EMG) signal. This interface provides a more accurate control signal, thereby eliminating the effect of control uncertainty on the outcome. Furthermore, with EMG as a control signal, the electrical stimulation may have contaminated the EMG recording, compromising thereby control accuracy, although studies have presented methods to circumvent this problem (Hartmann et al., 2014). In the experiment, the participant was asked to grasp a series of virtual objects. Each object required a specific range of forces to be successfully grasped without slipping or losing the object. For example, a virtual egg required 5%–20% of maximum force to be successfully grasped, while a virtual hammer required >75% of maximum force. Across a series of repetitions of this set of tasks, it was shown that the electrotactile feedback significantly improved the grasp success rate with respect to no-feedback conditions. Furthermore, it was shown that this advantage was maintained even when the gain of the system changed. Specifically, in this condition a given movement of the joystick implied a grasp force twice as high as before. These functional advantages, however, came with the price of increases in the time it took to complete the tasks.

Clemente et al. (2016) used vibrotactile stimulation on the upper arm of a prosthesis user to signal the timing of discrete events during an object

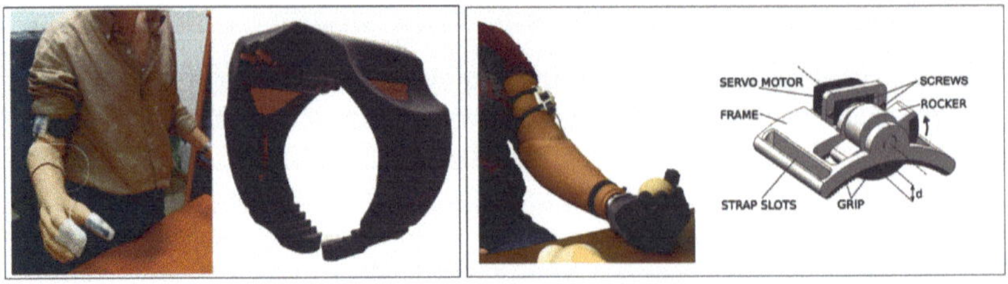

FIGURE 5.1 Two example designs to provide sensory substitution feedback in upper-limb prostheses: (left panel) a brace with embedded vibration motor mounted on the upper arm to deliver vibration bursts indicating important events during grasping, (right panel) a miniature device to produce skin stretch through rotation of a rocker pressed against the skin. *Adapted from (left) Clemente, F., D'Alonzo, M., Controzzi, M., Edin, B. B., & Cipriani, C. (2016). Non-invasive, temporally discrete feedback of object contact and release improves grasp control of closed-loop myoelectric transradial prostheses.* IEEE Transactions on Neural Systems and Rehabilitation Engineering, 24*(12), 1314–1322. https://doi.org/10.1109/TNSRE.2015.2500586; (right) Battaglia, E., Clark, J. P., Bianchi, M., Catalano, M. G., Bicchi, A., & O'malley, M. K. (2019). Skin stretch haptic feedback to convey closure information in anthropomorphic, under-actuated upper limb soft prostheses.* IEEE Transactions on Haptics, 12*(4), 508–520. https://doi.org/10.1109/TOH.2019.2915075.*

grasp-and-lift task. Specifically, a brief burst of vibration was given when the prosthesis made the first contact with or released the object, and when the object was lifted or replaced on the surface (Fig. 5.1, left panel). In this way, the feedback indicated when one phase of the task was completed and thus when a new set of movement commands could be initiated. Although the feedback strategy was biologically inspired, the feedback modality (vibration) implied that the evoked sensation is not equivalent to its natural counterpart (e.g., changes in contact force). This is commonly referred to as sensory substitution since the user must substitute the tingling sense of vibration with mechanical events in order to exploit the feedback. In the experiment, the prosthesis users were asked to lift and replace a so-called virtual egg. If too much pressure was applied on this virtual egg, it broke (virtually) and the trial was counted as a failed repetition. To minimize the generation of excessive force, the participants could exploit the feedback to determine the moment of object contact, which in some cases is not easily accessible by vision (e.g., due to occlusion). This can be then used as a cue to decrease the muscle activation level or relax the muscle fully, thereby slowing or stopping the increase in grasp force. The task was repeated in multiple sessions across 5 weeks. In each session the participant repeated the test with and without vibration feedback. Although the participants had different success rates in the task, the fewest failed trials tended to occur when vibrotactile feedback was activated; both at the beginning and at the end of the period. Overall, the participants performed significantly better with the vibrotactile feedback, suggesting that this type of feedback can improve prosthesis control in realistic conditions.

Battaglia et al. (2019) presented a system that aimed to restore proprioception. Proprioception refers to the sense of the position and movement of limbs. The proprioception therefore enables able-bodied individuals to carry out manual tasks without visual attention. The neural basis for this sense is mainly sensory feedback from muscle spindle receptors and cutaneous receptors detecting muscle and skin stretch, respectively (Proske & Gandevia, 2009). Drawing inspiration from these physiological mechanisms, the authors developed a system that stretched the skin of the upper arm when the fingers of the prosthesis closed to grasp an object (Fig. 5.1, right panel). Therefore, contrary to the sensory substitution system presented by (Clemente et al., 2016), this system can be characterized as a modality-matched solution. Namely, the finger movements that would be detected by the nervous system due to skin stretch in natural conditions are also encoded as skin stretch when the prosthetic hand opens and closes, albeit this sensation is elicited at a different location (homologous but nonsomatotopic feedback). It is commonly assumed that this strategy makes it less cognitively demanding for the user to interpret and exploit the feedback since the nervous system does not have to "translate" the stimulation sensation (e.g., vibration) into the physical variable that would naturally characterize the prosthesis action. The system was evaluated by two tests that were performed on able-bodied participants. In both tests, the closing of the prosthetic hand was controlled by EMG signals from the forearm and the hand was positioned on a table near the participant. In the first test, the participants had to determine if the prosthesis closed around a 6-cm sphere placed in the hand by the experimenter or if the hand was empty, without being able to see or hear it. While performing this task, an additional cognitive task was administered. The skin stretch feedback was on in some trials and off in others. The outcome was that participants' ability to recognize if the hand grasped an object or not was significantly better with feedback. This encouraging result, however, was not matched by the second test where the participant had to actively grasp the sphere and estimate its size with the prosthesis attached to the lower arm. Here, the visual feedback was blurred so the location of the object but not its size could be seen. In this case, the presence of feedback had no significant impact on performance. The authors explained this discrepancy by the fact that a prosthesis was underactuated and with compliant fingers. Therefore the closing of the hand around a sphere could be achieved in many ways, depending on its relative position with regards to the hand. Consequently, grasping the same sphere could generate different feedback signals across repetitions, making it difficult to reliably recognize small differences in sphere size.

These three representative examples demonstrate that noninvasive methods for supplementary feedback have the potential to provide meaningful information to the user and thereby improve prosthetic control. Furthermore, these studies illustrate a large diversity across research systems for supplementary feedback. For example, different stimulation technologies are used. The most common methods are electrotactile and vibrotactile stimulation; however, even within each of these methods, the feedback can be encoded in different ways, for instance, in the amplitude, frequency, or activation of different stimulation units (spatial encoding). On the other hand, there are multiple shared characteristics of most previous studies. For example, most studies involve a single sensory modality that typically directly reflects the mechanical actions of the prosthesis (e.g., grasp force or joint angles). Furthermore, the way that system performance is evaluated is largely similar across studies. A functional task is designed to simulate a realistic manual task and other natural sensory sources of information (e.g., vision) are usually to some degree impaired, to force participants to rely on the supplementary feedback.

In the following, we introduce a series of novel concepts arising from ongoing work in our laboratory that challenge the typical choices for feedback variables and for the design of the protocols for system evaluation. Specifically, we present alternative approaches with regards to which signals are being fed back to the user, how those signals are encoded into stimulation profiles, as well as how the evaluation procedure for feedback assessment can be designed.

5.3 Encoding feedback variables using multichannel electrotactile stimulation

To transmit feedback variables from the prosthesis to the user, they need to be encoded into stimulation profiles. The feedback information can be transmitted by modulating stimulation parameters, such as intensity and frequency, and/or location of stimulation delivery (active channel, spatial coding). There are several factors to consider when designing the encoding scheme. The sensory substitution approaches are noninvasive, but they also produce nonhomologous and nonsomatotopic feedback. Hence, the user needs to learn to interpret feedback information from elicited tactile sensations (e.g., vibration intensity/frequency "means" magnitude of grasping force). To facilitate this process, the encoding scheme should be easy to perceive and intuitive to understand. The latter is also critically important when considering that a prosthesis user estimates the prosthesis state by fusing information from multiple sources, including those that are intrinsically available in the prosthesis (audition, vision). If the supplementary feedback provides information with higher uncertainty compared to those intrinsic sources, the impact of the artificial feedback on prosthesis performance will be at best limited. Finally, modern prostheses normally include several degrees of freedom and the feedback should ideally transmit the full state of the device (e.g., hand aperture, wrist orientation, and grasping force). For instance, several studies (Schiefer et al., 2018; Schofield et al., 2020) have demonstrated that combined information on hand aperture and grasping force enables subjects to recognize the size and stiffness of an object during grasping without looking at the prosthesis. Considering these requirements, it is easy to conclude that a simple feedback interface based on a single stimulation channel (vibration motor and/or electrode) may not provide sufficient functionality. There are efforts to increase the information bandwidth of a single channel interface by combining modalities (hybrid approach) (D'Alonzo et al., 2014), using simultaneous modulation of parameters (Mayer et al., 2020), or by superimposing feedback on discrete events and information describing grasp force magnitude or movement direction in two channels (Karakuş & Güçlü, 2020). To provide feedback that can transmit a larger quantity of information noninvasively and in a clear manner, our group relies on multichannel electrotactile technology.

In this approach, tactile sensations are produced by delivering low-intensity electrical pulses to the skin of the residual limb. The pulses activate skin afferents, and hence the stimulation is nonspecific, in the sense that it cannot selectively target specific types of mechanoreceptive psychophysical channels, contrary to mechanical stimulation (e.g., using high-frequency vibration motors to excite the Pacinian channel predominantly). In addition, the stimulation needs to be calibrated to a specific individual since inappropriate stimulation parameters can elicit uncomfortable sensations. Nevertheless, the electrotactile interface does not have any moving parts while the electrodes can be made of thin and flexible material, providing a compact design and fast response. This allows fitting many channels (conductive pads) into an arbitrary arrangement within a compliant sheet customized for different anatomies. Such an electrode matrix (array) coupled to a miniature multichannel stimulator enables high flexibility in encoding the feedback information (Fig. 5.2). The feedback variables can be transmitted by simultaneously changing intensity, frequency, and/or active pad leading to dynamic patterns that can be used to convey information on multiple degrees of freedom of a prosthesis (Štrbac et al., 2016). Alternatively, such a

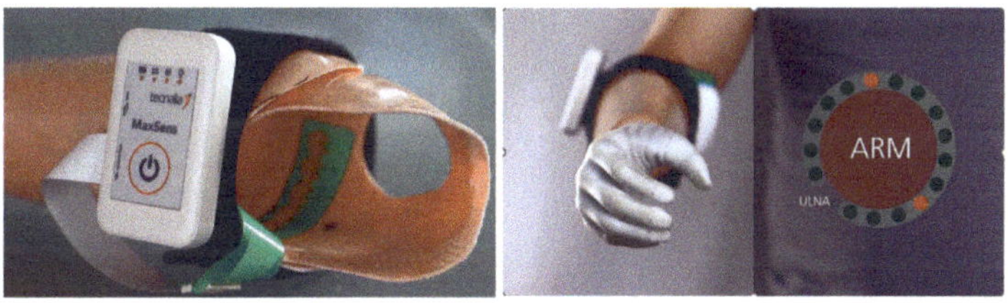

FIGURE 5.2 A compact multichannel electrotactile stimulation system MaxSens (Tecnalia, Spain) used in Garenfeld et al. (2020). A 16-pad array electrode is embedded in the prosthesis socket (left panel) while the stimulator is attached on the outer side. The pads can be independently activated with adjustable frequency and intensity to transmit rich electrotactile feedback on the prosthesis state. *Courtesy Tecnalia Innovation and Research.*

mixed encoding approach can be used to provide feedback on a single variable but with increased resolution. Mixed spatial and frequency coding has been applied (Dosen et al., 2017) to communicate 15 levels of grasping force of a Michelangelo prosthesis (Otto Bock, Germany) using an array electrode with 16 pads by activating five pad groups at three different frequencies. As demonstrated in the study, it was much easier for the subjects to identify two decoupled parameters, first recognizing location (pad group) and then frequency (low, middle or high), than to discriminate between 15 individual pads when using pure spatial coding to transmit the same information. As shown in Seminara et al. (2019), dual-parameter modulation with simultaneous change of intensity and frequency improved the spatial discrimination of individually activated pads arranged into a 4×6 electrode matrix. Using electrode arrays was shown to be beneficial for selecting best pad combination when applying transcutaneous nerve stimulation to provide somatotopic feedback noninvasively (Huang et al., 2019).

The encoding flexibility of multichannel electrotactile stimulation is illustrated in Fig. 5.3 for an array electrode. The whole electrode can be mapped to a single feedback variable using spatial coding or divided into sectors allocated to different variables. In addition, parameter modulation can be superposed to spatial coding to increase the number of levels (resolution). In this manner, an array electrode placed around the circumference of the forearm can be used to communicate (simultaneously) a complete state of a multiple degrees of freedom prosthesis. Recently, two coding schemes, one based on "sectorized" spatial coding and the other using "sectorized" amplitude coding to convey two degrees of freedom (hand aperture and wrist rotation) were compared using an online myoelectric target-reaching task (Garenfeld et al., 2020). The results demonstrated that the subjects achieved similarly high success rates when using both approaches. This encouraging result implies that substantially different encoding methods can lead to similar performance. Therefore, they can be selected according to subject preferences;

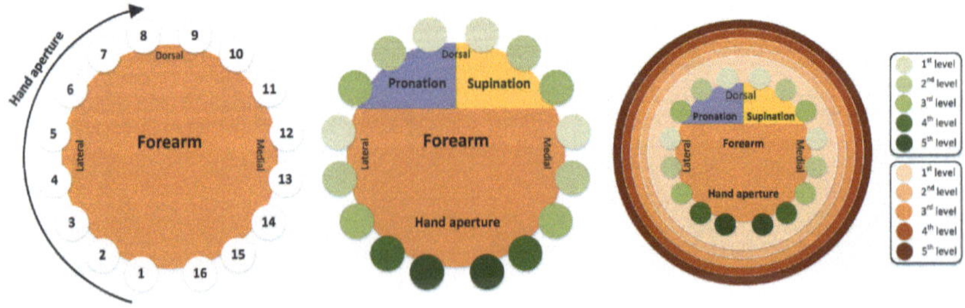

FIGURE 5.3 Different coding approaches using an array electrode with 16 conductive pads: (left) spatial coding to convey levels (pads) of hand aperture, (middle) spatial coding to communicate wrist rotation and hand aperture simultaneously, and (right) mixed coding where each pad is activated at several frequencies to increase the resolution (number of levels).

for instance, if a subject has low tolerance to amplitude modulation, he/she could use the spatial feedback. If the pads of a multichannel electrode are arranged in a matrix rather than array, such an interface can be used to provide tactile information captured by an electronic skin sensor covering a prosthesis (Franceschi et al., 2017; Yang et al., 2019), mimicking thereby the spatially distributed nature of biological skin feedback.

There are efforts to achieve similar capabilities using vibrotactile technology. As shown in Markovic et al. (2018), mixed coding can be also applied in this case but with a reduced number of channels due to spatial constraints. In that study, four pairs of vibrotactors were used to convey the grasping force, contact, and active function of a prosthesis. An array of vibrators positioned longitudinally or circumferentially have been used to communicate hand aperture (Witteveen et al., 2012) and grasping force (Saunders & Vijayakumar, 2011). In most commonly used vibration motors, the stimulation parameters are coupled through mechanical construction (Azadi & Jones, 2014); however, there are methods that allow generating vibrations with independently adjustable intensity and frequency (Dosen et al., 2016). In addition, a recent study presented a matrix of miniature vibrators embedded into a flexible sheet (Yu et al., 2019), but this approach still needs to be proven in functional applications.

5.4 Feeding back the command signal as opposed to its consequences

One common feature of most published systems for supplementary feedback is that the signal is determined by the actions of the prosthesis. The most common feedback variable is the grasp force magnitude, as used, for example, in the abovementioned study by Jorgovanovic et al. (2014). This seems like an obvious choice; however, when considering the mechanical properties of most prostheses, this choice also implies some limitations. Most commercial prostheses are controlled by proportionally mapping a myoelectric signal into prosthesis velocity. Therefore, when a prosthesis user wishes to grasp an object, a certain EMG level commands the closing speed of the hand and the EMG level generated upon contact with the object will determine the force that will be produced. As most prostheses are nonbackdrivable, this force will be maintained in the absence of opposing movement commands (e.g., a command to open the hand by activating another muscle). In this way, a further increase in the EMG level can increase the grasp force, but the grasp force cannot be easily decreased, as the opening command will often lead to the object slipping from the grasp, especially in the case of rigid objects. This behavior limits the degree to which the user can adjust the grasp force, even when receiving grasp force feedback. For example, once the user receives the feedback (i.e., when the prosthetic hand is fully closed around the object), the feedback signal may indicate that excessive force is being produced. However, when the user realizes this, it is too late to compensate and adjust the force. Since the EMG is a noisy signal, the risk of generating such undesired force levels is substantial (Dosen, Markovic, Wille, et al., 2015; Dosen, Markovic, Somer, Graimann, & Farina, 2015).

As an alternative feedback strategy, it was recently proposed to provide the prosthesis command signal as the feedback, as opposed to feeding back the mechanical prosthesis actions (aperture and grasping force) resulting from that command (Dosen, Markovic, Somer et al., 2015). In this strategy the processed (smoothed) EMG signal serves two independent purposes: it drives the motors of the prosthesis but it is also encoded in the feedback signal sent to the user. Although this may seem like an unintuitive strategy, it may hold multiple functional advantages. Specifically, the feedback can signal different types of information to the user depending on the context. If the prosthesis is open, the feedback will indicate closing velocity. As in the abovementioned study by Battaglia et al. (2019), this restores a sense of proprioception, as information about muscle velocity is an important element in natural proprioception (Proske & Gandevia, 2009). While the prosthesis is closing, however, EMG feedback also predicts the force that will be produced once the object is grasped. This can enable participants to predictively adjust the EMG to an appropriate level before the hand is closed to minimize the risk of unexpectedly generating excessive force upon contact. Finally, once an object is grasped, feedback will only be provided as long as the user generates an EMG signal, which may also hold advantages in itself. Consider, for example, a prosthesis user who lifts the lid of a cooking pot with the prosthetic hand while stirring and adding ingredients with the natural hand. Once the prosthetic hand has successfully grasped and lifted the lid, the user in principle needs no further information about the force that is being produced. The absence of stimulation in such steady-state periods not only minimizes the intrusiveness of the feedback (active only when needed), but it can also decrease potential accommodation to the sensation (i.e., a gradual loss of sensitivity), which would likely occur after long-term exposure to the stimulation with constant intensity. Fig. 5.4 illustrates the concept of EMG feedback in relation to a traditional force feedback scheme during a simple grasp force matching task with a prosthesis.

Schweisfurth et al. (2016) compared EMG feedback with traditional force feedback during so-called routine grasping in able-bodied and amputee participants. Routine grasping refers to closing the prosthetic hand using a smooth, continuous contraction without adjusting the force after hand closure. This approach is meant to represent the way by which one would typically grasp objects with a natural hand. In practice, it was ensured that the participants used a routine grasping strategy by stalling the prosthesis force shortly after contact, preventing thereby subsequent corrections in grasp force. A total of 150 repetitions with different target forces were performed with both types of feedback. Feedback was provided by stimulating one of

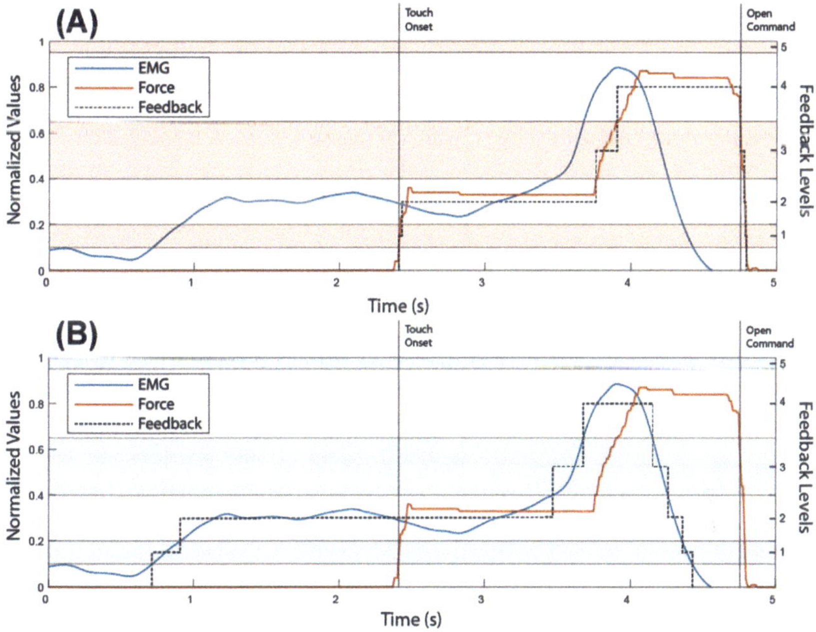

FIGURE 5.4 The behavior of force feedback compared to EMG feedback generated by the same muscle contraction and EMG signal. Feedback was given as a discrete signal with five levels indicated with alternating white and red/blue-shaded areas. Panel (A) displays the behavior of a system implementing force feedback. The feedback is being transmitted to the user only while the prosthesis is in contact with the object. Panel (B) displays the corresponding EMG feedback. The feedback is transmitted while there is a muscle contraction and it is thus independent of grasp force and available even during prosthesis closing. Note the correspondence between the EMG signal before and after contact and the generated force after contact. A simple approach to transmit such feedback to the subject would be to use spatial coding with five stimulation channels (each channel indicates a specific signal interval, i.e., red/blue shaded areas).

four different electrodes positioned around the arm with one of two different stimulation frequencies (mixed coding). In this way, eight easily discriminable signals (electrode × frequency combination) could be generated, where each represented a different level of EMG. The results indicated that the error with respect to the target force was significantly lower with EMG feedback across most target force levels. This demonstrates that EMG feedback improves the predictive control of prosthesis grasp force. In trials with force feedback, however, the participants also exhibited a good ability to generate the target force, although there was no time to exploit the feedback signal within each trial, as the feedback was available only after contact. Nevertheless, the force feedback could be used to train the internal models (the predicted consequences—i.e., grasp force magnitude—associated with a given command signal) as an error in one trial can be used to adjust the magnitude of the command signal in the next trial. In this way, the force feedback may represent a more accurate version of the type of feedback that could also be obtained by, for example, visual feedback of potential deformation of an object and/or prosthesis silicone fingers after the object has been grasped.

Furthermore, humans have a natural sense of muscle activation level. This sense of activation of the muscles from which the EMG signal was recorded could be exploited for control. These factors likely explain the fact that routine grasping tasks can be carried out with some precision by participants with force feedback, or even without supplementary feedback (Sensinger & Dosen, 2020). However, as shown in the study, providing EMG feedback improves the accuracy of such control. Using auditory feedback, Shehata et al. (2018) confirmed that feedback on the processed EMG levels improved the quality of the predictive control. Although it can be argued that auditory feedback is a less practical solution than tactile feedback, it is possible that the internal models of the participants can be acquired and maintained through frequent training sessions, and then exploited during everyday life tasks without feedback. The robustness of the models during prolonged periods without feedback, however, is not known. Finally, Markovic et al. (2020) incorporated the concept of EMG feedback in a scheme for prosthetic control in three degrees of freedom. It was shown that the proposed scheme substantially minimized training time while providing similar performance as the state-of-the-art control schemes.

One important and largely unexplored aspect of EMG feedback is how the raw EMG signal should be processed to facilitate interpretation and use of the feedback. The processing usually consists of two steps, which have important implications for the characteristics of the signal that is being fed back to the user, as illustrated in Fig. 5.5. Here, the feedback was provided through four vibration motors, that each were activated when the EMG signal was within a specific range. The raw EMG signal is first rectified and smoothed using a low-pass filter. A low cut-off frequency will imply a highly stable signal, which may make it easier for the user to accurately perceive and maintain the magnitude of the signal using online feedback. However, during dynamic conditions where the user wishes to rapidly change the prosthesis command signals, a low cut-off frequency implies a substantial delay, which may impair control. In this way, the choice of cut-off frequency likely represents a tradeoff between these two signal properties. The second processing step involves the mapping of the EMG signal [typically expressed as the percentage of the EMG level obtained during maximum effort: maximum voluntary contraction (MVC)] into the feedback/command signal. Usually, this mapping will consist of a linear relation, but the EMG level required to generate maximum command/feedback signal may vary. Again, this choice has several potential functional implications. If, on one hand, a high EMG amplitude is required to generate maximum signal, the subject may quickly experience muscle fatigue, which may be unpleasant

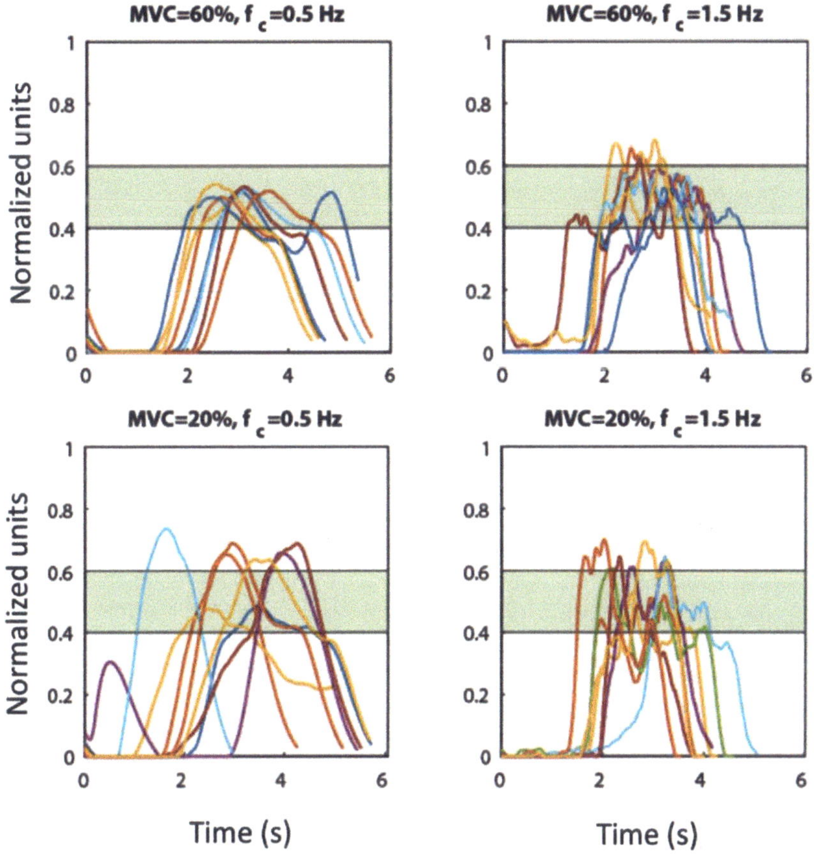

FIGURE 5.5 EMG envelopes generated with different settings for (i) the cut-off frequency (f_c) of the low-pass filter used to smooth the rectified EMG signal and (ii) the percentage of the maximum voluntary contraction (MVC) value to which the resulting command/feedback signal is normalized. In these examples, the subjects were asked to generate an EMG level within the range indicated in green. The online EMG feedback was provided to the subjects using vibration motors. The x-axes indicate time in seconds. Note how the subjects' ability to reach and maintain the myoelectric signal using online feedback depends on the processing parameters.

and can imply that the maximum signal amplitude can no longer be produced. Furthermore, the EMG signal is more variable at high contraction levels due to multiplicative signal-dependent noise (Harris & Wolpert, 1998). On the other hand, if a relatively low EMG amplitude is set to generate the maximum signal, the risk of fatigue is minimal, but the resolution (i.e., the number of EMG levels that can be reliably produced) and thus controllability would be lower. It should be noted that these questions are also largely unresolved for open-loop EMG control (i.e., prosthetic control without supplementary feedback) and that these parameters are often heuristically selected.

5.5 Feedback can support predictive and corrective strategies

As described previously, the control of manual motor tasks in able-bodied individuals may rely on the information received through sensory feedback, but also on predictions from previous experiences, which is referred to as internal models (Davidson & Wolpert, 2005; Shadmehr & Krakauer, 2008). Relying exclusively on internal model predictions can be referred to as feedforward control, since the full set of motor commands to execute the task is preplanned and executed outside the influence of feedback. Although most manual tasks are carried out with a combined feedback and feedforward strategy, many simple, brief tasks, such as grasping and lifting an object, rely heavily on feedforward control triggered by discrete sensory events (Flanagan et al., 2006). This is evident, for example, when lifting an empty milk carton assumed to be full. Here, the internal model for the task of lifting an object with the weight of a full milk carton in a well-controlled way determines the motor commands that are sent to the relevant muscles. As a result, however, the empty milk carton is lifted with unexpectedly high speed to a much higher location than planned before compensatory movements can be executed based on, for example, proprioceptive and visual feedback. If, on the other hand, a strategy relying exclusively on feedback was used in this task, the person would slowly increase the force, evaluating for each step if the lifting force is appropriate or needs adjustments which would likely take several seconds. This example illustrates that although the feedforward control implies a certain risk of failure, if an inappropriate internal model is used, it is the most efficient strategy in many, regular everyday life tasks.

The importance and interplay of such feedforward and feedback processes in prosthesis control has been demonstrated in the literature by showing that amputees rely on internal models (Lum et al., 2014) and that simple tasks can be performed by operating a prosthesis in a purely feedforward fashion (Saunders & Vijayakumar, 2011). When asking naïve subjects to perform a simple functional task such as grasping and lifting an object using a prosthesis with supplementary force feedback, however, it is reasonable to expect that the subjects will not intuitively adopt a feedforward control strategy, as would be expected if carrying out the same task with a natural hand. If receiving tactile feedback that reflects grasp force magnitude is an unfamiliar concept, robust internal models (i.e., the ability to predict the sensory consequences of a given motor command) are not yet available. Unless specifically instructed to do otherwise, it is more likely that the subjects will adopt a control strategy relying primarily on feedback, carefully evaluating the feedback to extract the information it carries for every small adjustment in the motor command. This is illustrated in the examples shown in Fig. 5.6. Here, two able-bodied subjects controlled the grasping of a prosthesis using EMG with vibrotactile feedback indicating

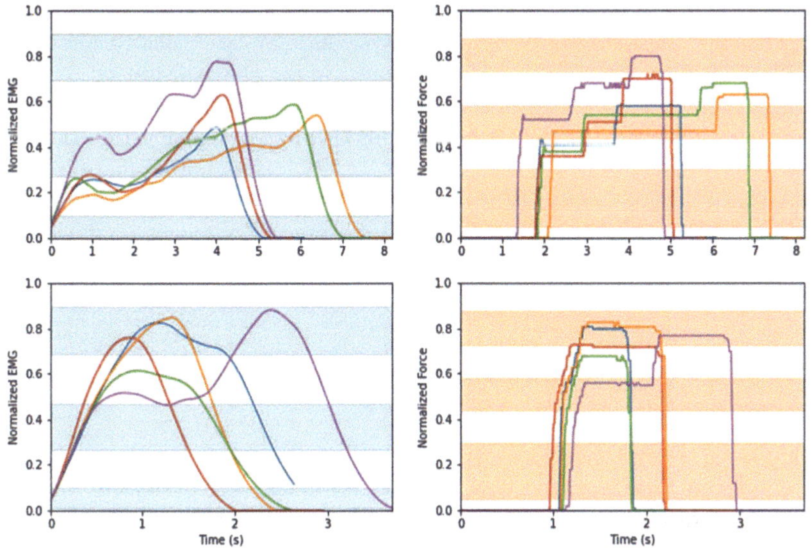

FIGURE 5.6 EMG and prosthesis grasp force across five repetitions in two subjects (top and bottom rows, respectively). Both subjects were asked to reach the target EMG signal/force magnitude (third interval from the top; white background). In contrast to the first subject (top row), the second subject (bottom) had a time constrain of 1 s to reach the force after hand closure (the first time instance where force was produced). The subjects received six levels of force feedback delivered using four vibration motors.

grasp force. Both subjects were told to generate a specific force, but while the first subject was asked to produce this force accurately, the second was told to do it rapidly. For the second subject, these instructions were enforced by automatically opening the prosthesis shortly after closing of the hand. From the representative EMG and force traces, it is clear that this simple difference in the instructions implied that fundamentally different control strategies were adopted. Without having an explicit constraint on the task duration, the first subject generated an initial grasp force well below the target, after which slow, small steps toward the target were performed. This implied that this relatively simple task took 5–7 seconds to complete. On the other hand, the second subject that performed the task with substantial time constraints completed the task in 2–3 seconds but with a lower accuracy. It is likely, however, that prolonged training may form more robust internal models, that would allow the second subject to improve the accuracy without compromising time (Dosen, Markovic, Wille et al., 2015; Dosen, Markovic, Somer et al., 2015). The feedback in this case, although less relevant within the trial, can directly facilitate the process of trial-by-trial adaptation of internal models, as explained in the previous section and demonstrated in Štrbac et al. (2017).

The potential implications of the example shown in Fig. 5.6 are that without proper task instructions, feedback systems may be evaluated in context with less functional relevance. For example, in the abovementioned study by (Jorgovanovic et al., 2014), subjects successfully exploited the feedback to improve performance, but used long durations to complete each grasping task. Although it is an encouraging finding, it clearly showed that the strategy used by the subjects was very different from the one that would be applied under natural conditions (routine grasping). Consequently, this evaluation is limited in the sense that it does not show whether the proposed feedback system is able to support the "normal" control strategy.

5.6 Evaluating the role of feedback in the state estimation process

One other common feature of most published systems for supplementary feedback is related to the way system performance is evaluated. Usually, once systems are implemented, the participant is asked to perform a functional task. Across different studies, the task may be very simple (generating a desired grasp force level) or it may be more representative of actual daily-life functional tasks (e.g., box-and-blocks test). The selected task is typically carried out under different conditions, for instance, with and without feedback activated, or with the participant deprived of other sources of feedback such as visual or auditory inputs. There are promising developments of the approaches to assess sensate prostheses, and recently several interesting tests have been proposed, such as, the virtual egg test (Clemente et al., 2016), block-foraging stiffness discrimination task, and psychophysical Fitts' law grasp force task (GRIP test) (Schofield et al., 2020). These developments compensate for the fact that standard clinical tests such as Box and Blocks and/or SHAP are not suited for the evaluation of closed-loop control, as they do not require careful control of grasping forces. However, there are only a few standardized tasks that are used across multiple studies and the same feedback system may have different effects across different tasks (Markovic et al., 2018). Furthermore, it can be questioned if depriving participants of, for example, visual feedback, provides a relevant representation of the usefulness of the supplementary feedback, since when executing many natural manual tasks, some, but not full, visual attention is anyway on the hands.

As an alternative to evaluating the performance in functional tasks, some recent research has focused on quantifying the role of the supplementary feedback in the state estimation process of the nervous system. As described above, the nervous system fuses information from incoming sensory feedback and internal models to estimate the current state of the body when executing a task. This state estimate is used to plan the movement commands required to successfully complete the task. The sensory fusion can be described as a weighted average of the information carried by each relevant sensory modality, where the weights assigned to each modality reflect their reliability. This principle was demonstrated in the seminal study by Ernst and Banks (2002). In this study, the participants were asked to determine which of two objects were largest across a series of trials. To answer this question haptic as well as visual feedback was available to the participants. For the haptic feedback the participants grasped the objects between two fingers, whereas visual information was provided virtually. Unknown to the participants, the visual feedback was manipulated so that one of the two objects appeared as having a slightly different size. The incongruence between the two sources of information and the fact that the subject was forced to choose which of the objects was larger enabled the authors to deduce the degree to which each of the two sensory modalities affected the state estimate. For example, if object #1 was smaller than object #2 (as perceived by the haptic feedback) but appeared larger (as indicated by visual feedback), the subjective choice of which object was larger indicated which of the two sensory sources of information was considered most reliable. Specifically, it was found that subjects tended to estimate the state (object size) with weights based primarily on vision (relative weight: approximately 80%), while haptic feedback had a smaller impact on the decision (relative weight: approximately 20%). However, as the quality of the visual feedback was diminished by adding noise, the weights were gradually shifted and, at a certain noise level, haptic feedback became the dominant source of information in the state estimate. In this way, the sensory integration is referred to as "optimal" since the weighting of different sensory inputs depends on their reliability according to maximum likelihood estimation (Shadmehr & Mussa-Ivaldi, 2012).

This principle of sensory fusion has been demonstrated for other sensory modalities [e.g., vision vs. proprioception (Van Beers et al., 2002)], but also recently for supplementary feedback in a prosthesis user (Risso et al., 2019). In this study, Risso et al. (2019) implanted an electrode in nerve bundles in the forearm, which previously innervated the natural sensory receptors in an amputated hand. By appropriately stimulating the nerve, natural sensations of touch in the phantom hand were evoked. Using these evoked sensations to restore haptic information, the study showed that the relative weights assigned to the haptic feedback evoked by stimulation and the visual feedback were predicted by their reliability. Here, the reliability was estimated as the just-noticeable difference in intensity for each modality.

Recent work from our laboratory aimed to exploit this methodology to quantify the efficiency of supplementary feedback in grasping with natural hands. The underlying assumption was that the higher the weight assigned by the nervous system in the state estimation process, the more efficient the supplementary feedback. A method based on this principle may be used as an objective figure of merit, compared to previous evaluation methods. As discussed above, most studies investigating such sensory fusion used visual feedback as the reference modality, relative to which the weight of another modality (e.g., haptic, proprioceptive) was evaluated. Instead, our approach aimed to quantify the weight of the sense of exerted force measured through the natural channels (exteroception) to the weight of sense of force as perceived through supplementary feedback. This was done by asking the subject to generate a target force (range: 15%–30% MVC), relax, and repeat the same force. The subject was told that the primary outcome measure was the precision by which the force was reproduced. Throughout the task, the subject received electrotactile feedback with a frequency proportional to the generated force. However, in some repetitions of the second part of the task (reproducing the target force), the mapping between force and stimulation frequency was temporarily altered to generate a mismatch between the information about the magnitude of the instantaneous force indicated by the two modalities. Importantly, this mismatch was subtle and was not noticed by the subject. In this way, as in the study by Ernst and Banks (2002), the subject received seemingly similar, but slightly different, information from two feedback sources, prompting the nervous system to subconsciously estimate the true force as a weighted average between the two. To infer these weights, the difference between the target force and the force generated with biased force–frequency mapping was analyzed. A perfect reproduction of the target force would indicate zero weight of the electrotactile feedback, while a systematic error would indicate an influence of electrotactile feedback on the state estimate.

Fig. 5.7 illustrates the outcome of the experiment. In trials where no mismatch in the force–frequency mapping was imposed (baseline condition), the force was accurately reproduced. However, in the trials where this mapping was biased, the generated forces were systematically shifted, even though the subject did not notice this bias and assumed that the mapping was the same. This suggests that the subject subconsciously compromised the degree to which the natural sensation of force and the sensation of electrotactile stimulation frequency were reproduced to determine the appropriate force commands based on a weighted average. Furthermore, the magnitude of the error in matching task force in the biased condition indicated that relative weight assigned to the supplementary feedback was >80%. The latter is an encouraging outcome since it indicates that sensory substitution feedback can be naturally (subconsciously) integrated into the state estimation process, even in the presence of other feedback sources (natural exteroception).

5.7 Concluding remarks

A variety of noninvasive feedback systems have been proposed, so far with promising results. Most of the concepts for delivering such feedback (e.g., electrotactile/vibrotactile) are not new, as similar developments can be found in the literature from several decades ago (Childress, 1980; Meek et al.,

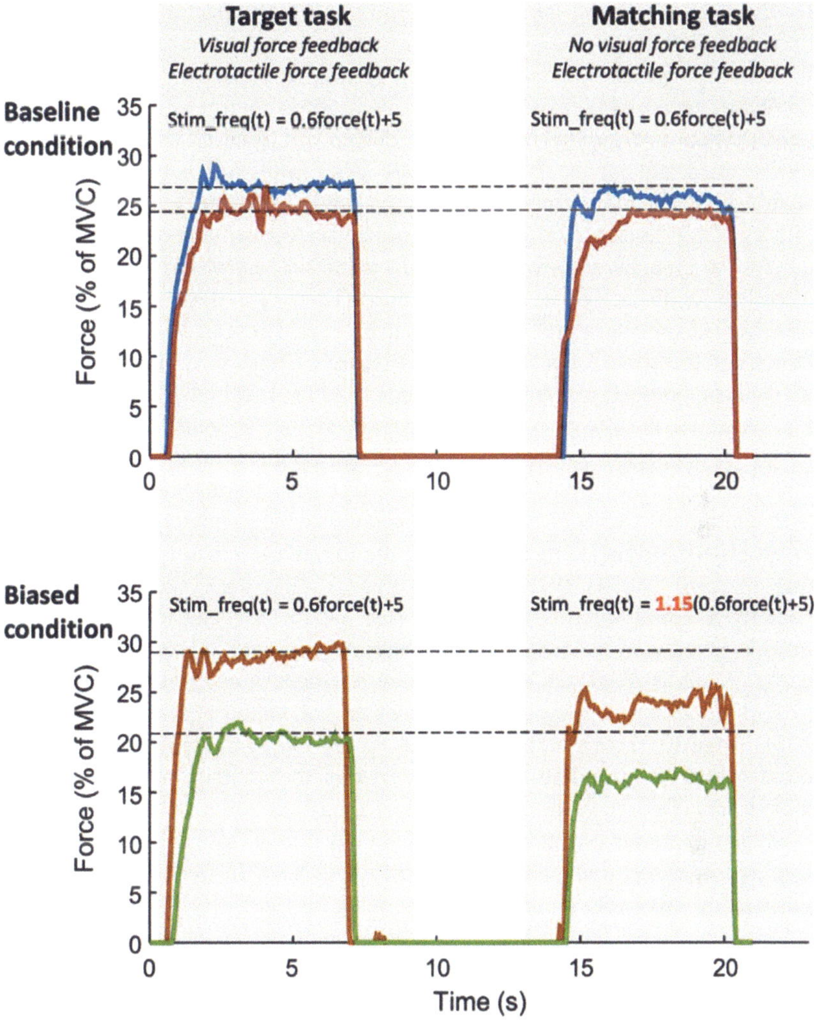

FIGURE 5.7 The subject was asked to generate a target force (randomly selected in the range 15%–30% MVC) and hold it for approximately 5 s (target task) with visual and electrotactile force feedback, as well as natural force feedback. After a break, the subject was asked to reproduce this target force (matching task). In the baseline condition (top panel) the subject could do this with high accuracy (here, two repetitions are shown). In the biased condition, the mapping of force to stimulation frequency (stim_freq) was slightly altered. The subject did not notice this change, but the forces produced in the matching task were systematically suppressed, indicating that the generated force was selected as a compromise between matching the natural sense of force and the sensation elicited by the supplementary feedback.

1989; Shannon, 1979). They are also sufficiently simple from a technical point of view to be implemented in the socket of a prosthesis. Nevertheless, only a few commercial systems, presented recently, integrate some form of feedback interface and their clinical utility and the impact is yet to be demonstrated. This may suggest that a critical revision of the approaches used for developing and evaluating feedback systems is needed.

We advocate that insights from human motor control and computational neuroscience can greatly facilitate the design of closed-loop prostheses. The feedforward and feedback pathways are practically indivisible as they constantly interact, and their properties (e.g., reliability) together with the capabilities of the human controller (e.g., predictive control and state estimation) need to be elucidated in the context of prosthesis control in order to develop effective feedback for a prosthesis user. There are developments in this direction in the literature, as evidenced by recent research addressing basic human motor control questions in this context (Johnson et al., 2014, 2017), as well as proposing feedback methods directly inspired by such insights. For instance, the discrete feedback presented in Clemente et al. (2016) is based on the sensory-driven control of grasping (Johansson & Flanagan, 2009). Recent results have demonstrated that such feedback can even dominate continuous feedback when the two are provided simultaneously (Engels et al., 2019). Similarly, by recognizing the role of anticipatory control in routine grasping, it has been proposed that feeding back prosthesis closing velocity (Ninu et al., 2014) and EMG (see Section 5.4) can improve the control of grasping force and outperform the conventional approach (force feedback). Auditory feedback on joint speed has been investigated as a method to reinforce state estimation in the presence of vision (Earley et al., 2018, 2021). Finally, some authors proposed biologically inspired encoding methods for invasive stimulation and showed that they outperform conventional approaches (George et al., 2019; Valle et al., 2018).

When evaluating system efficiencies, it is critical to enforce a user strategy that is compatible with normal prosthesis use. In Section 5.5, we discussed how the same feedback interface can be used very differently by the subject depending on the task instructions, namely, the feedback can facilitate slow online modulation or fast feedforward control with adaptation across trials. From this perspective, feedforward control in routine grasping is likely a more relevant approach as it mimics how the prosthesis should be ideally used in daily life (i.e., to grasp objects quickly with minimal cognitive load). In addition, we proposed that feedback assessment might depend not only on performance in a functional task, but also on quantifying the degree to which the central nervous system uses the feedback in a natural way in the state estimation process. Closely related to the development of novel assessment methods is the need to quantify the effects of training on perceiving and interpreting the feedback. Both short-term (several sessions/days) (Marini et al., 2014; Stepp et al., 2012; Štrbac et al., 2017) as well as long-term assessments possibly with home use (Cuberovic et al., 2019; Graczyk et al., 2018; Schofield et al., 2020) have been reported, and such developments are critically important for estimating the clinical impact of sensate prostheses. The latter in turn requires the development of appropriate technology, and some examples are demonstrated in recent literature (Brinton et al., 2020; Wu et al., 2021).

Acknowledgments

The authors thank Shima Gholinezhad, Jack Tchimino, Pranav Mamidanna, and Martin Garenfeld for their contributions to this chapter. The work presented in this chapter has been supported by the projects 8022-00243A (ROBIN) and 8022-00226B funded by the Independent Research Fund Denmark.

References

Antfolk, C., D'Alonzo, M., Rosén, B., Lundborg, G., Sebelius, F., & Cipriani, C. (2013). Sensory feedback in upper limb prosthetics. *Expert Review of Medical Devices, 10*(1), 45–54. Retrieved from http://www.ncbi.nlm.nih.gov/pubmed/23278223.

Atkins, D. J., Heard, D. C. Y., & Donovan, W. H. (1996). Epidemiologic overview of individuals with upper-limb loss and their reported research priorities. *Journal of Prosthetics and Orthotics*. Available from https://doi.org/10.1097/00008526-199600810-00003.

Azadi, M., & Jones, L. A. (2014). Vibrotactile actuators: Effect of load and body site on performance. In *IEEE haptics symposium, HAPTICS* (pp. 351–356). https://doi.org/10.1109/HAPTICS.2014.6775480

Battaglia, E., Clark, J. P., Bianchi, M., Catalano, M. G., Bicchi, A., & O'malley, M. K. (2019). Skin stretch haptic feedback to convey closure information in anthropomorphic, under-actuated upper limb soft prostheses. *IEEE Transactions on Haptics, 12*(4), 508–520. Available from https://doi.org/10.1109/TOH.2019.2915075.

Biddiss, E., & Chau, T. (2007). Upper-limb prosthetics: Critical factors in device abandonment. *American Journal of Physical Medicine & Rehabilitation/Association of Academic Physiatrists, 86*(12), 977–987.

Biddiss, E., Beaton, D., & Chau, T. (2007). Consumer design priorities for upper limb prosthetics. *Disability and Rehabilitation: Assistive Technology, 2*(6), 346–357. Available from https://doi.org/10.1080/17483100701714733.

Brinton, M. R., Barcikowski, E., Davis, T., Paskett, M., George, J. A., & Clark, G. A. (2020). Portable take-home system enables proportional control and high-resolution data logging with a multi-degree-of-freedom bionic arm. *Frontiers in Robotics and AI, 7*(559034). Available from https://doi.org/10.3389/frobt.2020.559034.

Childress, D. S. (1980). Closed-loop control in prosthetic systems: Historical perspective. *Annals of Biomedical Engineering, 8*(4–6), 293–303.

Clemente, F., D'Alonzo, M., Controzzi, M., Edin, B. B., & Cipriani, C. (2016). Non-invasive, temporally discrete feedback of object contact and release improves grasp control of closed-loop myoelectric transradial prostheses. *IEEE Transactions on Neural Systems and Rehabilitation Engineering, 24*(12), 1314–1322. Available from https://doi.org/10.1109/TNSRE.2015.2500586.

Cuberovic, I., Gill, A., Resnik, L. J., Tyler, D. J., & Graczyk, E. L. (2019). Learning of artificial sensation through long-term home use of a sensory-enabled prosthesis. *Frontiers in Neuroscience, 13*(853). Available from https://doi.org/10.3389/fnins.2019.00853.

D'Alonzo, M., & Cipriani, C. (2012). Vibrotactile sensory substitution elicits feeling of ownership of an alien hand. *PLoS One*, *7*(11), e50756. Available from https://doi.org/10.1371/journal.pone.0050756.

D'Alonzo, M., Dosen, S., Cipriani, C., & Farina, D. (2014). HyVE: Hybrid vibro-electrotactile stimulation for sensory feedback and substitution in rehabilitation. *IEEE Transactions on Neural Systems and Rehabilitation Engineering*, *22*(2), 290–301.

Davidson, P. R., & Wolpert, D. M. (2005). Widespread access to predictive models in the motor system: A short review. *Journal of Neural Engineering*, *2*(3), 313–319. Available from https://doi.org/10.1088/1741-2560/2/3/S11.

Diedrichsen, J., Shadmehr, R., & Ivry, R. B. (2010). The coordination of movement: Optimal feedback control and beyond. *Trends in Cognitive Sciences*.

Dosen, S., Markovic, M., Somer, K., Graimann, B., & Farina, D. (2015). EMG Biofeedback for online predictive control of grasping force in a myoelectric prosthesis. *Journal of Neuroengineering and Rehabilitation*, *12*(55). Available from https://doi.org/10.1186/s12984-015-0047-z.

Dosen, S., Markovic, M., Wille, N., Henkel, M., Koppe, M., Ninu, A., Cornelius, F., & Farina, D. (2015). Building an internal model of a myoelectric prosthesis via closed-loop control for consistent and routine grasping. *Experimental Brain Research*, *233*(6), 1855–1865.

Dosen, S., Markovic, M., Strbac, M., Belic, M., Kojic, V., Bijelic, G., Thierry, K., & Farina, D. (2017). Multichannel electrotactile feedback with spatial and mixed coding for closed-loop control of grasping force in hand prostheses. *IEEE Transactions on Neural Systems and Rehabilitation Engineering*, *25*(3), 183–195. Available from https://doi.org/10.1109/TNSRE.2016.2550864.

Dosen, S., Ninu, A., Yakimovich, T., Dietl, H., & Farina, D. (2016). A novel method to generate amplitude-frequency modulated vibrotactile stimulation. *IEEE Transactions on Haptics*, *9*(1), 3–12. Available from https://doi.org/10.1109/TOH.2015.2497229.

Earley, E. J., Johnson, R. E., Hargrove, L. J., & Sensinger, J. W. (2018). Joint speed discrimination and augmentation for prosthesis feedback. *Scientific Reports*, *8*(1), 17752. Available from https://doi.org/10.1038/s41598-018-36126-4.

Earley, E. J., Johnson, R. E., Sensinger, J. W., & Hargrove, L. J. (2021). Joint speed feedback improves myoelectric prosthesis adaptation after perturbed reaches in non amputees. *Scientific Reports*, *11*(1), 5158.

Engels, L. F., Shehata, A. W., Scheme, E. J., Sensinger, J. W., & Cipriani, C. (2019). When less is more-discrete tactile feedback dominates continuous audio biofeedback in the integrated percept while controlling a myoelectric prosthetic hand. *Frontiers in Neuroscience*, *13*(578). Available from https://doi.org/10.3389/fnins.2019.00578.

Ernst, M. O., & Banks, M. S. (2002). Humans integrate visual and haptic information in a statistically optimal fashion. *Nature*, *415*(6870), 429–433. Available from https://doi.org/10.1038/415429a.

Flanagan, J. R., Bowman, M. C., & Johansson, R. S. (2006). Control strategies in object manipulation tasks. *Current Opinion in Neurobiology*, *16*, 1–10. Available from https://doi.org/10.1016/j.conb.2006.10.005.

Franceschi, M., Seminara, L., Dosen, S., Strbac, M., Valle, M., & Farina, D. (2017). A system for electrotactile feedback using electronic skin and flexible matrix electrodes: Experimental evaluation. *IEEE Transactions on Haptics*, *10*(2), 162–172. Available from https://doi.org/10.1109/TOH.2016.2618377.

Garenfeld, M. A., Mortensen, C. K., Strbac, M., Dideriksen, J. L., & Dosen, S. (2020). Amplitude vs spatially modulated electrotactile feedback for myoelectric control of two degrees of freedom. *Journal of Neural Engineering*, *17*(4). Available from https://doi.org/10.1088/1741-2552/aba4fd.

George, J. A., Kluger, D. T., Davis, T. S., Wendelken, S. M., Okorokova, E. V., He, Q., Duncan, C. C., Hutchinson, D. T., Thumser, Z. C., Beckler, D. T., Marasco, P. D., & Bensmaia, S. J. (2019). Biomimetic sensory feedback through peripheral nerve stimulation improves dexterous use of a bionic hand. *Science Robotics*, *4*(32). Available from https://doi.org/10.1126/scirobotics.aax2352.

Gesslbauer, B., Hruby, L. A., Roche, A. D., Farina, D., Blumer, R., & Aszmann, O. C. (2017). Axonal components of nerves innervating the human arm. *Annals of Neurology*, *82*(3), 396–408. Available from https://doi.org/10.1002/ana.25018.

Giummarra, M. J., Gibson, S. J., Georgiou-Karistianis, N., & Bradshaw, J. L. (2008). Mechanisms underlying embodiment, disembodiment and loss of embodiment. *Neuroscience and Biobehavioral Reviews*, *32*(1), 143–160. Available from https://doi.org/10.1016/j.neubiorev.2007.07.001.

Graczyk, E. L., Resnik, L., Schiefer, M. A., Schmitt, M. S., & Tyler, D. J. (2018). Home use of a neural-connected sensory prosthesis provides the functional and psychosocial experience of having a hand again. *Scientific Reports*, *8*(1), 9866. Available from https://doi.org/10.1038/s41598-018-26952-x.

Harris, C. M., & Wolpert, D. M. (1998). Signal-dependent noise determines motor planning. *Nature*, *394*(6695), 780–784. Available from https://doi.org/10.1038/29528.

Hartmann, C., Dosen, S., Amsuess, S., & Farina, D. (2014). Closed-loop control of myoelectric prostheses with electrotactile feedback: Influence of stimulation artifact and blanking. *IEEE Transactions on Neural Systems and Rehabilitation Engineering, PP* (99).

Hermsdörfer, J., Elias, Z., Cole, J. D., Quaney, B. M., & Nowak, D. A. (2008). Preserved and impaired aspects of feed-forward grip force control after chronic somatosensory deafferentation. *Neurorehabilitation and Neural Repair*, *22*(4), 374–384. Available from https://doi.org/10.1177/1545968307311103.

Huang, H., Vargas, L., Shin, H., Huang, H., Zhu, Y., & Hu, X. (2019). Object stiffness recognition using haptic feedback delivered through transcutaneous proximal nerve stimulation. *Journal of Neural Engineering*, *17*(1). Available from https://doi.org/10.1088/1741-2552/ab4d99.

Johansson, R. S., & Flanagan, J. R. (2009). Sensorimotor control of manipulation. *Encyclopedia of neuroscience*. Available from https://doi.org/10.1016/B978-008045046-9.01920-3.

Johnson, R. E., Kording, K. P., Hargrove, L. J., & Sensinger, J. W. (2014). Does EMG control lead to distinct motor adaptation? *Frontiers in Neuroscience*, *8*(302). Available from https://doi.org/10.3389/fnins.2014.00302.

Johnson, R. E., Kording, K. P., Hargrove, L. J., & Sensinger, J. W. (2017). Adaptation to random and systematic errors: Comparison of amputee and non-amputee control interfaces with varying levels of process noise. *PLoS One*, *12*(3), e0170473. Available from https://doi.org/10.1371/journal.pone.0170473.

Jorgovanovic, N., Dosen, S., Djozic, D. J., Krajoski, G., & Farina, D. (2014). Virtual grasping: Closed-loop force control using electrotactile feedback. *Computational and Mathematical Methods in Medicine*, 2014.

Karakuş, İ., & Güçlü, B. (2020). Psychophysical principles of discrete event-driven vibrotactile feedback for prostheses. *Somatosensory & Motor Research*. Available from https://doi.org/10.1080/08990220.2020.1769055.

Lewis, S., Russold, M.F., Dietl, H., & Kaniusas, E. (2012). User demands for sensory feedback in upper extremity prostheses. In *MeMeA 2012–2012 IEEE symposium on medical measurements and applications, proceedings*. https://doi.org/10.1109/MeMeA.2012.6226669

Löken, L. S., Wessberg, J., Morrison, I., McGlone, F., & Olausson, H. (2009). Coding of pleasant touch by unmyelinated afferents in humans. *Nature Neuroscience, 12*(547). Available from https://doi.org/10.1038/nn.2312.

Lum, P. S., Black, I., Holley, R. J., Barth, J., & Dromerick, A. W. (2014). Internal models of upper limb prosthesis users when grasping and lifting a fragile object with their prosthetic limb. *Experimental Brain Research, 232*(12), 3785–3795.

MacKenzie, C., & Iberall, T. (2010). *The grasping hand.* Amsterdam: Elsevier B.V..

Marini, F., Tagliabue, C. F., Sposito, A. V., Hernandez-Arieta, A., Brugger, P., Estévez, N., & Maravita, A. (2014). Crossmodal representation of a functional robotic hand arises after extensive training in healthy participants. *Neuropsychologia, 53*, 178–186. Available from https://doi.org/10.1016/j.neuropsychologia.2013.11.017.

Markovic, M., Schweisfurth, M. A., Engels, L. F., Bentz, T., Wüstefeld, D., Farina, D., & Dosen, S. (2018). The clinical relevance of advanced artificial feedback in the control of a multi-functional myoelectric prosthesis. *Journal of Neuroengineering and Rehabilitation, 15* (1). Available from https://doi.org/10.1186/s12984-018-0371-1.

Markovic, M., Varel, M., Schweisfurth, M. A., Schilling, A. F., & Dosen, S. (2020). Closed-loop multi-amplitude control for robust and dexterous performance of myoelectric prosthesis. *IEEE Transactions on Neural Systems and Rehabilitation Engineering, 28*(2), 498–507. Available from https://doi.org/10.1109/TNSRE.2019.2959714.

Mayer, R. M., Garcia-Rosas, R., Mohammadi, A., Tan, Y., Alici, G., Choong, P., & Oetomo, D. (2020). Tactile feedback in closed-loop control of myoelectric hand grasping: Conveying information of multiple sensors simultaneously via a single feedback channel. *Frontiers in Neuroscience, 14*(348). Available from https://doi.org/10.3389/fnins.2020.00348.

Meek, S. G., Jacobsen, S. C., & Goulding, P. P. (1989). Extended physiologic taction: Design and evaluation of a proportional force feedback system. *Journal of Rehabilitation Research and Development, 26*(3), 53–62.

Ninu, A., Dosen, S., Muceli, S., Rattay, F., Dietl, H., & Farina, D. (2014). Closed loop control of grasping with a myoelectric hand prosthesis: Which are the relevant feedback variables for force control? *IEEE Transactions on Neural Systems and Rehabilitation Engineering, 22* (5), 1041–1052.

Piazza, C., Grioli, G., Catalano, M. G., & Bicchi, A. (2019). A century of robotic hands. *Annual Review of Control, Robotics, and Autonomous Systems, 2*, 1–32. Available from https://doi.org/10.1146/annurev-control-060117-105003.

Proske, U., & Gandevia, S. C. (2009). The kinaesthetic senses. *The Journal of Physiology, 587* (Pt 17), 4139–4146.

Pylatiuk, C., Schulz, S., & Döderlein, L. (2007). Results of an internet survey of myoelectric prosthetic hand users. *Prosthetics and Orthotics International, 31*(4), 362–370. Available from https://doi.org/10.1080/03093640601061265.

Risso, G., Valle, G., Iberite, F., Strauss, I., Stieglitz, T., Controzzi, M., Clemente, F., Granata, G., & Baud-Bovy, G. (2019). Optimal integration of intraneural somatosensory feedback with visual information: A single-case study. *Scientific Reports, 9*(1), 7916. Available from https://doi.org/10.1038/s41598-019-43815-1.

Saunders, Ian, & Vijayakumar, S. (2011). The role of feed-forward and feedback processes for closed-loop prosthesis control. *Journal of Neuroengineering and Rehabilitation, 8*(60). Available from https://doi.org/10.1186/1743-0003-8-60.

Schiefer, M. A., Graczyk, E. L., Sidik, S. M., Tan, D. W., & Tyler, D. J. (2018). Artificial tactile and proprioceptive feedback improves performance and confidence on object identification tasks. *PLoS One, 13*(12). Available from https://doi.org/10.1371/journal.pone.0207659.

Schofield, J. S., Evans, K. R., Carey, J. P., & Hebert, J. S. (2014). Applications of sensory feedback in motorized upper extremity prosthesis: A review. *Expert Review of Medical Devices*, 1–13.

Schofield, J. S., Shell, C. E., Beckler, D. T., Thumser, Z. C., & Marasco, P. D. (2020). Long-term home-use of sensory-motor-integrated bidirectional bionic prosthetic arms promotes functional, perceptual, and cognitive changes. *Frontiers in Neuroscience*, *14*(120). Available from https://doi.org/10.3389/fnins.2020.00120.

Schweisfurth, M. A., Markovic, M., Dosen, S., Teich, F., Graimann, B., & Farina, D. (2016). Electrotactile EMG feedback improves the control of prosthesis grasping force. *Journal of Neural Engineering*, *13*(5). Available from https://doi.org/10.1088/1741-2560/13/5/056010.

Seminara, L., Fares, H., Franceschi, M., Valle, M., Strbac, M., Farina, D., & Dosen, S. (2019). Dual-parameter modulation improves stimulus localization in multichannel electrotactile stimulation. *IEEE Transactions on Haptics*, *13*(2), 393–403. Available from https://doi.org/10.1109/TOH.2019.2950625.

Sensinger, J. W., & Dosen, S. (2020). A review of sensory feedback in upper-limb prostheses from the perspective of human motor control. *Frontiers in Neuroscience*, *14*(345). Available from https://doi.org/10.3389/fnins.2020.00345.

Shadmehr, R., & Krakauer, J. W. (2008). A computational neuroanatomy for motor control. *Experimental Brain Research*, *185*(3), 359–381. Available from https://doi.org/10.1007/s00221-008-1280-5.

Shadmehr, R., & Mussa-Ivaldi, S. (2012). *Biological learning and control: How the brain builds representations, predicts events, and makes decisions. Biological learning and control.* Cambridge, MA: The MIT Press, https://doi.org/10.7551/mitpress/9780262016964.001.0001.

Shannon, G. F. (1979). Sensory feedback for artificial limbs. *Medical Progress Through Technology*, *6*(2), 73–79.

Shehata, A. W., Rehani, M., Jassat, Z. E., & Hebert, J. S. (2020). Mechanotactile sensory feedback improves embodiment of a prosthetic hand during active use. *Frontiers in Neuroscience*, *14*(263). Available from https://doi.org/10.3389/fnins.2020.00263.

Shehata, A. W., Scheme, E. J., & Sensinger, J. W. (2018). Audible feedback improves internal model strength and performance of myoelectric prosthesis control. *Scientific Reports*, *8*(1), 8541. Available from https://doi.org/10.1038/s41598-018-26810-w.

Sobuh, M. M. D., Kenney, L. P. J., Galpin, A. J., Thies, S. B., McLaughlin, J., Kulkarni, J., & Kyberd, P. (2014). Visuomotor behaviours when using a myoelectric prosthesis. *Journal of Neuroengineering and Rehabilitation*, *11*(72). Available from https://doi.org/10.1186/1743-0003-11-72.

Stepp, C. E., An, Q., & Matsuoka, Y. (2012). Repeated training with augmentative vibrotactile feedback increases object manipulation performance. *PLoS One*, *7*(2), e32743. Available from https://doi.org/10.1371/journal.pone.0032743.

Štrbac, M., Belić, M., Isaković, M., Kojić, V., Bijelić, G., Popović, I., Radotic, M., Dosen, S., Markovic, M., Farino, D., & Keller, T. (2016). Integrated and flexible multichannel interface for electrotactile stimulation. *Journal of Neural Engineering*, *13*(4). Available from https://doi.org/10.1088/1741-2560/13/4/046014.

Štrbac, M., Isaković, M., Belić, M., Popović, I., Simanić, I., Farina, D., Thierry, K., & Došen, S. (2017). Short-and long-term learning of feedforward control of a myoelectric prosthesis with sensory feedback by amputees. *IEEE Transactions on Neural Systems and Rehabilitation Engineering*, *25*(11), 2133–2145. Available from https://doi.org/10.1109/TNSRE.2017.2712287.

Svensson, P., Wijk, U., Björkman, A., & Antfolk, C. (2017). A review of invasive and non-invasive sensory feedback in upper limb prostheses. *Expert Review of Medical Devices*. Available from https://doi.org/10.1080/17434440.2017.1332989.

Valle, G., Mazzoni, A., Iberite, F., D'Anna, E., Strauss, I., Granata, G., Controzzi, M., Clemente, F., Rognini, G., Cipriani, C., Stieglitz, T., Petrini, F. M., Rossini, P. M., & Micera, S. (2018). Biomimetic intraneural sensory feedback enhances sensation naturalness, tactile sensitivity, and manual dexterity in a bidirectional prosthesis. *Neuron, 100*(1), 37–45. Available from https://doi.org/10.1016/j.neuron.2018.08.033.

Van Beers, R. J., Wolpert, D. M., & Haggard, P. (2002). When feeling is more important than seeing in sensorimotor adaptation. *Current Biology, 12*(10), 834–837.

Vujaklija, I., Farina, D., & Aszmann, O. C. (2016). New developments in prosthetic arm systems. *Orthopedic Research and Reviews, 8*, 31–39. Available from https://doi.org/10.2147/ORR.S71468.

Walker, M. J., Goddard, E., Stephens-Fripp, B., & Alici, G. (2020). Towards including end-users in the design of prosthetic hands: Ethical analysis of a survey of Australians with upper-limb difference. *Science and Engineering Ethics, 26*(2), 981–1007. Available from https://doi.org/10.1007/s11948-019-00168-2.

Wilke, M. A., Niethammer, C., Meyer, B., Farina, D., & Dosen, S. (2019). Psychometric characterization of incidental feedback sources during grasping with a hand prosthesis. *Journal of Neuroengineering and Rehabilitation, 16*(155). Available from https://doi.org/10.1186/s12984-019-0622-9.

Witteveen, H. J. B., Droog, E. A., Rietman, J. S., & Veltink, P. H. (2012). Vibro- and electrotactile user feedback on hand opening for myoelectric forearm prostheses. *IEEE Transactions on Biomedical Engineering, 59*(8), 2219–2226. Available from https://doi.org/10.1109/TBME.2012.2200678.

Wu, H., Dyson, M., & Nazarpour, K. (2021). Arduino-based myoelectric control: Towards longitudinal study of prosthesis use. *Sensors (Switzerland), 21*(3), 763. Available from https://doi.org/10.3390/s21030763.

Yang, D., Gu, Y., Thakor, N. V., & Liu, H. (2019). Improving the functionality, robustness, and adaptability of myoelectric control for dexterous motion restoration. *Experimental Brain Research*. Available from https://doi.org/10.1007/s00221-018-5441-x.

Yang, J. C., Mun, J., Kwon, S. Y., Park, S., Bao, Z., & Park, S. (2019). Electronic skin: Recent progress and future prospects for skin-attachable devices for health monitoring, robotics, and prosthetics. *Advanced Materials, 3*(48), e1904765. Available from https://doi.org/10.1002/adma.201904765.

Yu, X., Xie, Z., Yu, Y., Lee, J., Vazquez-Guardado, A., Luan, H., Ruban, J., Ning, X., Akhtar, A., Li, D., Ji, B., Liu, Y., Sun, R., Cao, J., Huo, Q., Zhong, Y., Lee, C., Kim, S., Gutruf, P., ... Rogers, J. A. (2019). Skin-integrated wireless haptic interfaces for virtual and augmented reality. *Nature, 575*(7783), 473–479. Available from https://doi.org/10.1038/s41586-019-1687-0.

Chapter 6

Noninvasive augmented sensory feedback in poststroke hand rehabilitation approaches

Leonardo Cappello[1,2], Rebecca Baldi[1,2], Leonard Frederik Engels[1,2] and Christian Cipriani[1,2]
[1]*The BioRobotics Institute, Scuola Superiore Sant'Anna, Pisa, Italy,* [2]*Department of Excellence in Robotics and AI, Scuola Superiore Sant'Anna, Pisa, Italy*

ABSTRACT
Stroke is one of the most common and devastating neurological disorders, often leading to major motor impairments. Not infrequently, these come along with sensory deficits that further impact neuromotor control and might hinder rehabilitation outcomes. Intact processing of sensory information is in fact crucial for planning and executing both coarse and fine movements and to regain natural motor function. Although the importance of motor therapy in both the acute and chronic stages is well-established, the sensory aspect of poststroke therapeutic interventions is often overlooked. We perused the literature for studies that investigated noninvasive techniques to artificially reinstate this essential component of motor control and to enhance it with rehabilitative purposes. We summarize important approaches and related findings and conclude that knowledge on the effectiveness of somatosensory stimulation in long-term rehabilitation therapies remains incomplete.

Keywords: Stroke rehabilitation; internal model; noninvasive somatosensory feedback; vibrotactile stimulation; hand control

6.1 Introduction: sensory information in hand motor performance

A stroke is like a heart attack for the brain: a disruption of the blood supply leads to tissue damage and disruption of normal brain function. Approximately 80 million people worldwide have experienced stroke, and over 13 million new strokes occur every year (Lindsay et al., 2019). Stroke is one of the major causes of disability worldwide (Katan & Luft, 2018) and can have long-lasting consequences for those affected. Impairments in parts

Somatosensory Feedback for Neuroprosthetics. DOI: https://doi.org/10.1016/B978-0-12-822828-9.00006-X

of the body contralateral to the affected side of the brain can lead to loss of hand motor function, which poses a dramatic limitation in performing everyday life activities and reduces the patient's independence. Abnormal voluntary motor function, uncontrolled muscle contraction, poor movement coordination, and weakness are common consequences of this neurological impairment (Byl et al., 2003). As a consequence of brain damage, tactile sensitivity is often compromised, negatively affecting skilled movements and the ability to manipulate objects (Blennerhassett et al., 2006).

Depending on the location and extent of the damage that the stroke caused in the brain, the consequences vary greatly. The common obstruction of the medial cerebral artery, for example, can lead to muscle weakness, hemianesthesia, hemianopsia, hemineglect, apraxia, and disturbed spatial perception, to name just a few. Damage to the pons, cerebellum, or midbrain can lead to hemiataxia, (partial) loss of pain and temperature sensation in the face and body, hemiparesis, hemiplegia, and tetraplegia. At a lower level of processing, damage leads to primary sensory deficits, such as anopsia or tactile impairments, whereas higher level damage, at a later stage of processing, leads to more cognitive defects, such as neglect or agnosia (Tinga et al., 2016).

6.1.1 Upper limb impairment

In the majority of stroke survivors, an upper limb is affected (van Dijk et al., 2005), with more than 80% experiencing acute hemiparesis of the contralesional limb, and more than 40% experiencing it chronically (Doyle et al., 2010). Sensory deficits, such as impairments of touch, mechanical, temperature, and pain sensations, two-point discrimination, and proprioception, are also common in the contralesional arm (Bowen et al., 2011; Connell et al., 2008; Doyle et al., 2010; Raghavan & Shah, 2015). However, in some, grasping with the ipsilesional upper limb is affected also (Quaney et al., 2005; Quaney et al., 2010), in part because some corticospinal fibers do not decussate in the medulla, meaning that both cerebral hemispheres are needed for optimal function of either limb (Connell et al., 2008).

Rehabilitation after stroke mostly focuses on motor rehabilitation, but it is well-known that primary and secondary motor areas in the brain also respond to sensory stimulation (Bolognini et al., 2016). Indeed, impairments of grasp force control, prehension patterns, and fine motor manipulation have been associated with sensory impairments (Bolognini et al., 2016; Dobkin, 2004; Doyle et al., 2010). Chronic loss of tactile sensitivity and proprioception decreases awareness of limb, shoulder, and trunk position and movement, which is essential for spontaneous use of the whole limb (Bolognini et al., 2016; Doyle et al., 2010; Wang et al., 2017). Conversely, lesions to the primary motor cortex have also been reported to lead to tactile agnosia—which is typically associated with primary sensory cortex lesions—and severe motor lesions lead to changes in connectivity of all sensory areas (Bolognini et al., 2016). Motor and sensory properties are tightly intertwined and need to be integrated for optimal motor behavior (Bolognini et al., 2016; Dobkin, 2004; Tinga et al., 2016).

6.1.2 Sensorimotor control of the upper limb

Any movement necessarily starts *open-loop*, that is, without sensory feedback, since afferent information from the eyes, tactile receptors, and proprioception are always processed with some delay (Johansson & Flanagan, 2007; Wolpert et al., 2011). This is made possible by *internal models* of movements (Fig. 6.1): the *inverse model* produces motor commands by estimating how a limb needs to be moved to grasp an object, for example, based on its position, length, and weight (Kawato, 1999; Shadmehr & Mussa Ivaldi, 1994; Wolpert et al., 2011). An *efference copy* of these motor commands is used by the *forward model* to predict the sensory consequences of the movement. For example, efficient grasping of an object is separated by the central nervous system (CNS) at *sensorimotor control points* into a series of simpler subtasks (Johansson & Flanagan, 2007). Discrepancies between predicted and actual sensory information, so-called sensory prediction errors, become evident at these control points and allow the intact CNS to correct movements—rendering them *closed-loop*—and to improve the internal models (Johansson & Flanagan, 2007; Wolpert et al., 2011).

As the part of the CNS that is mainly responsible for smooth movements, these internal models, too, are likely based in the cerebellum—so damage to it would directly affect the ability to use and maintain internal models (Imamizu et al., 2000; Kawato et al., 2003; Person, 2019; Popa & Ebner, 2019). However, as would be expected, there is a strong interplay directly related to the formation and refinement of internal models also with other regions of the CNS, such as the primary motor and sensory cortices (Jenmalm et al., 2006; Loh et al., 2010; Person, 2019). Lesions in parietal or premotor areas, for example, may also negatively affect the ability to integrate sensory feedback during movements, likely affecting efficient movement planning and execution (Bolognini et al., 2016).

6.1.3 Sensory input for optimal movement

For efficient grasping, sensory feedback depends crucially on mechanotactile receptors in the skin of the hand, particularly the discrete sensory events upon contacting, lifting, replacing, and releasing an object (Johansson & Cole, 1992; Johansson & Flanagan, 2007; Johnson et al., 2001), but surface friction recognition, too, is important for accurate force scaling (Blennerhassett et al., 2007). Visual information and proprioception are especially important during the reaching phase toward an object to accurately position the arm and hand (Raghavan & Shah, 2015). Chronic impairment of any of these sensory properties after stroke leads to motor impairment, in part due to inaccurate internal models that cannot

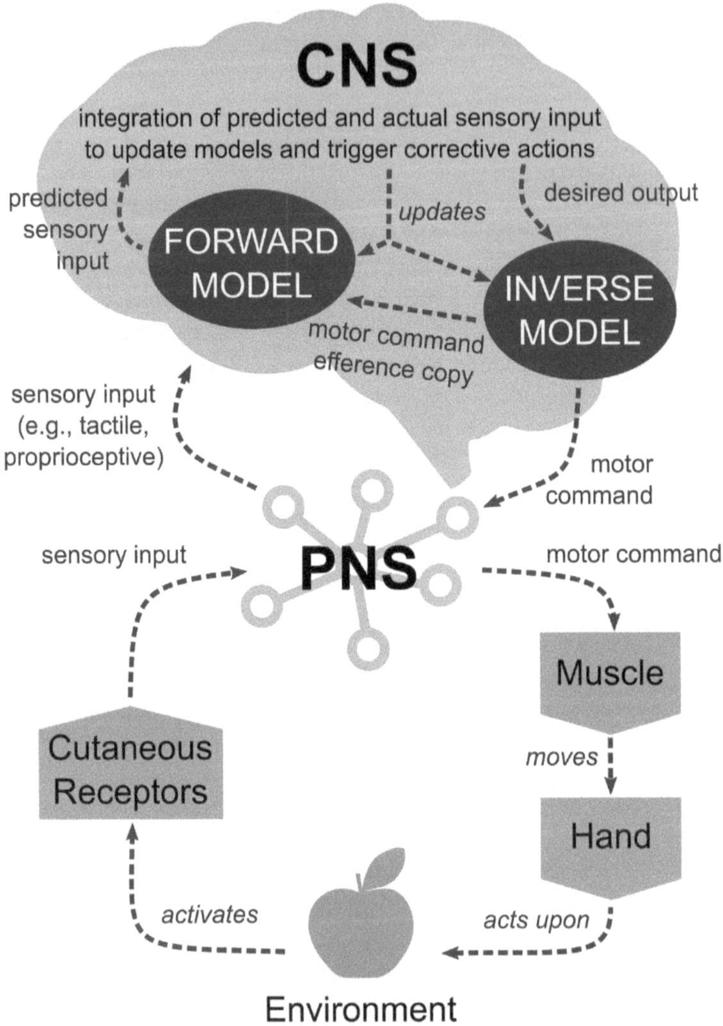

FIGURE 6.1 Schematic of sensorimotor control. To grasp an object, the desired output is translated into motor commands by the inverse model in the CNS. These motor commands are sent to the PNS to activate the muscles that move the hand to the desired position. Activation of sensory afferents, for example, through contact with an object, are sent back to the CNS where these sensory events are compared to the predicted sensory events that the forward model created based on the efference copy of the motor commands. In the case of a mismatch between prediction and actual event, the internal models are updated for future repetitions of the same task, and, if necessary, corrective motor actions are initiated. *CNS*, Central nervous system; *PNS*, peripheral nervous system.

be updated properly (Bolognini et al., 2016; Raghavan & Shah, 2015). The inability to detect mismatches of predicted and actual sensory information can cause anosognosia, leading to significantly poorer recovery (Bolognini et al., 2016).

Vision and even auditory stimulation also activate mirror neurons, which play an important role in learning and mimicking movement (Bolognini et al., 2016). This may serve as a means to stimulate the motor system after stroke, even if other pathways are damaged (Bolognini et al., 2016).

Patients who have impaired sensorimotor or visual capabilities are more dependent on caregivers than those with motor impairments alone (Dobkin, 2004), and they have difficulties exploring and relating to their environment, making them less independent and affecting long-term participation (Doyle et al., 2010). In addition, impaired temperature or pain sensation, balance, or proprioception affect the safety of the stroke survivor even if there is adequate motor recovery (Doyle et al., 2010).

6.1.4 Augmented feedback to stimulate neural plasticity

Therapy intended to recover lost sensation may aid in compensating for and relearning lost motor abilities, especially for the upper limb (Bolognini et al., 2016). Compensation may lead to positive short-term effects; long-term improvements are possible after restoration of sensory function due to neural plasticity, following the strengthening of existing connections and the recruitment of new ones (Tinga et al., 2016). Plasticity in intact areas of the brain is related to the damage to intracortical connections. It is important to enhance ipsilesional recovery during therapy, as increased activity in the contralesional hemisphere relates to poorer recovery (Johansson & Birznieks, 2004).

Based on this knowledge, it can easily be argued that restoring sensory functions would directly and indirectly aid in recovering motor functions (Bolognini et al., 2016). However, although it is generally agreed upon that feedback is beneficial for motor (re-)learning, the way to most effectively provide it is still debated (Sigrist et al., 2012). *Task-inherent* or *intrinsic* feedback, that is, information obtained as the natural result of a process or an action, may not be used by patients with cognitive or more severe perceptual impairments (van Dijk et al., 2005). In those cases, *augmented* or *task-external* feedback—such as that delivered from a therapist, a timer, or some other device—could provide the sensory input that is missing to improve the performance of a given task.

In the following, we will provide an overview of current rehabilitation protocols, followed by an in-depth review of the various approaches to providing augmented sensory feedback for upper limb rehabilitation after stroke, partitioned by sensory channels. A number of studies have found positive effects of augmented sensory feedback, but it is as yet too early to pass a final verdict on its usefulness, or even on how best to implement this form of feedback. Thus this chapter will end with a section providing an outlook on and suggestions for future research and developments.

6.2 Current rehabilitation techniques

Several rehabilitation techniques have been developed to promote the recovery of impaired upper extremity function after stroke, with the ultimate goal of restoring independence and quality of life of the affected. In general, rehabilitation exercises aim at reactivating the neural pathways involved in the completion of a task and provoking neural plasticity and motor learning (Dobkin, 2008). The latter is fundamental for a rehabilitative process: indeed, it is necessary that the functional gains of the training sessions persist over time and that their effects can be generalized to untrained motor actions (Krakauer, 2006).

6.2.1 Approach to rehabilitation

General guidelines on the criteria that rehabilitation protocols should meet in order to be maximally effective for motor learning are described in the literature (Hochstenbach-Waelen & Seelen, 2012): they should be patient-tailored and goal-tailored, taking into consideration the level of individual cognitive impairments. Other important characteristics are that they should resemble real-life contexts as much as possible and be motivating for the patient. Motivation is increased by designing exercises relevant to the activities of daily living (ADLs) (Dobkin, 2004). An appropriate level of involvement and challenge, too, is fundamental to promote motivation during training. The difficulty and intensity of the exercises should increase over time, and tasks should vary. Further, it is important to introduce rest periods and variability among exercises even during massed practice, that is, when a single task is performed repeatedly. This could result in a performance decrease during the acquisition session, but it ultimately leads to improvements in the long-term, which can be seen in retention tests. Variability among training exercises will also facilitate generalization of motor learning to untrained movement (Krakauer, 2006).

In the first stage of rehabilitation, if the patient shows severe motor impairments, it is common practice to passively stimulate the sensorimotor system through massages, passive joint or soft tissue mobilization techniques, or assisted movements (Hunter et al., 2011). In the second stage, sensory training is often addressed more explicitly, for example, with sensory recognition exercises, which train to reconnect stimulation of mechanoreceptors and proprioceptors with conscious percepts (Schabrun & Hillier, 2009). The unmasking of already-present afferent sensory pathways can promote neuroplasticity and increase hand sensitivity (Peurala et al., 2002). Nonetheless, the importance of sensory stimulation is often underestimated in rehabilitation procedures, which usually focus mainly on motor function recovery (Bolognini et al., 2016). In the following, we describe a selection of the most common rehabilitation procedures used today (Fig. 6.2).

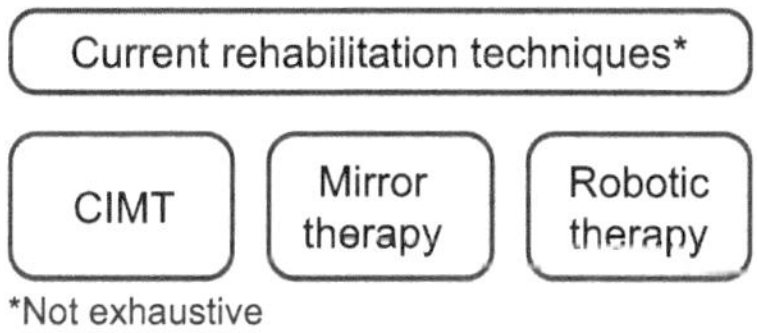

FIGURE 6.2 A selection of the most common rehabilitation techniques.

6.2.2 Constraint-induced movement therapy

Constraint-induced movement therapy (CIMT) consists in the restraint of the unaffected arm while training the impaired limb. The restriction can be applied for the greater part of the waking hours (usually ca. 90%), while the patient is performing daily life activities, and can also be included during massed practice sessions, ideally for at least 6 hours a day for 2 weeks (Dobkin, 2004, 2008; Krakauer, 2006). After an injury, patients tend to neglect the use of the affected extremity because of the resulting pain and fatigue. Using the unaffected arm for most of the daily tasks results in decreased effort and improved timing for achieving such tasks. Hence, the concept of CIMT is to force the use of the impaired limb to perform standard tasks, such as reaching, grasping, and pinching. However, this practice has strict requirements: patients involved in CIMT protocols must have at least 10 degrees of active wrist, thumb, and any other two fingers extension on the affected hand. These requirements are necessary for the therapy to be beneficial, but unfortunately around 90% of stroke patients do not satisfy these eligibility criteria (Dobkin, 2004).

6.2.3 Mirror therapy

CIMT stands in theoretical contrast to mirror therapy, where synchronized movements are performed with both arms to increase coordination. Bimanual training techniques have been proven effective in promoting motor learning by generating a mirror-induced movement in the paretic limb that reduces the inhibition of the damaged cerebral hemisphere (Gandolfi, 2018). Although mirror therapy has demonstrated fewer improvements compared to the unilateral practice (Lum et al., 2006), it is still widely used in combination with functional repetitive training, robotic therapy, and other rehabilitation techniques.

6.2.4 Robot-assisted therapy

Robot-assisted therapies have gained attention in recent decades as they allow to perform highly repetitive, intensive, and adaptive physical training. Robotic devices can be either exoskeletons, which match the kinematics of the impaired limb's joints, or end-effector robots acting only on the distal part of the arm. Robotic trainers are sometimes used in *virtual reality* (VR) settings, combined with visual, auditory, and haptic feedback to promote involvement in the task. These robotic technologies can operate in active assist mode, guiding the hemiparetic limb following an assist-as-needed principle (Duret et al., 2019; Mehrholz et al., 2017). In this modality, many repetitions of the same action are performed, while robot assistance is reduced over the course of the training to stimulate an increase in the patient's active participation. In contrast, passive-mode robotic training enhances poststroke recovery by providing only intrinsic feedback (Oujamaaa et al., 2009; Takahashi et al., 2008). Finally, both the impaired and the sound limb can be involved in the therapy, for example, in bimanual mode robotic training, which exploits the mirror therapy concept.

A great advantage of robot-assisted therapies is the potential they offer to quantitatively monitor the recovery through accurate kinematic recordings. In fact, they allow therapists to assess the smoothness and correctness of the movements at the end of each training session. Moreover, task variability can be easily introduced in several ways when using a robot, for example, by introducing a force-field in the exercise to challenge the patients and to induce them to adapt their internal models to varying environments (Krakauer, 2006). A further advantage of robotic therapy is the possible increase in motivation due to the gamification of the exercises, which can be designed to challenge the patients with short-term goals. This can ultimately lead them to voluntarily increase the dose of the treatment.

The main issues with the use of robotic rehabilitative techniques are their high cost and poor portability to home settings. The use of these therapies requires technical assistance that only a medical center can offer. Moreover, no significant advantage is seen unless a high number of repetitions in the same session are performed (Carey et al., 2007). The risk with such intensive training sessions is that patients and therapists may feel exhausted and unable to participate effectively in the rest of their daily activities, which may compromise the beneficial effects of the overall therapeutic treatment. On the other hand, this increased training intensity and motivation may be the key for improving the outcomes of the rehabilitation therapy and reducing the therapist's workload (Mehrholz et al., 2017). Importantly, robotic training might also allow one therapist to treat several patients at the same time. The therapist can provide assistance and explanation as needed, while the repetitive training itself is performed by the patients independently. This could lead to an overall increase in the hours each patient can spend training under the supervision of a therapist and thus to an increase in quality rehabilitation.

6.3 Augmented sensory feedback

Beyond practice, feedback is generally considered to be a fundamental factor in motor (re-)learning. The term feedback indicates the influx of sensory information related to the performed movement or task. We already differentiated two types of feedback according to the source that generates it: *inherent feedback* is information that is naturally perceived during a process or action, for example, coming from proprioception, vision, hearing, and mechanoreception at the hand. The second is artificially added or enhanced via an external source; it is therefore named *augmented feedback* (often also referred to as supplementary, explicit, extrinsic, or artificial feedback). It can be produced either by a human or an artificial device (van Dijk et al., 2005). Augmented sensory feedback can be delivered using different strategies, which, according to Molier et al. (2010), can be described by their *aspect* and *modality*.

6.3.1 Aspects of feedback

We define aspect as the way the sensory feedback is delivered, that is, its *nature* and *timing*. The nature of the feedback concerns its content, which can either be about the action, commonly referred to as *knowledge of performance*, or about the outcome of the action, referred to as *knowledge of results* (Molier et al., 2010; van Dijk et al., 2005). Timing is the second fundamental aspect of feedback, which defines the moment in which feedback is provided and its duration. The stimulus could be delivered during the execution of the motor task (concurrent feedback) or after its conclusion (terminal feedback) (Sigrist et al., 2012). As for the duration, the stimulus can be either delivered as instantaneous information (discrete feedback) or in a prolonged fashion (continuous feedback). *Biofeedback* is a particular type of feedback that delivers information about the user's own actions (e.g., finger movement, muscle activation, and heart rate); it is concurrent with the task and continuous (*real time*) (Giggins et al., 2013).

6.3.2 Feedback modalities

Modalities are subtypes of sensory experience that the augmented stimulus elicits, and they depend on the types of activated sensory receptors. Augmented feedback can be delivered through the following modalities: (1) auditory, which comprises all the stimuli that can be captured by the hearing system, such as the sound of a metronome or verbal signals; (2) visual, which includes all the visible cues, such as VR displays, signaling lights, and gestures; and (3) haptic, which refers to the sense of touch and can further be split into skin pressure, vibration, skin stretch, etc. In addition, different feedback modalities can be combined to provide multimodal feedback. If the modality of the augmented feedback is identical to the natural one, the feedback is *modality-matched* (e.g., when grasping force is provided as pressure feedback) (Antfolk et al., 2013). If, instead, the sensory information is conveyed through a different modality (e.g., an increasing contact force represented as a sound with increasing volume), the feedback is *modality-mismatched.* The means by which feedback is provided is often called *display*, but it is not confined to the visual domain. In fact, a computer screen can be a display as much as a vibration motor or a speaker (Sigrist et al., 2012).

6.3.3 Strategies for error feedback

Augmented sensory feedback can be tuned to target motor learning and rehabilitation. Especially in the field of robotic rehabilitation, two particular, opposite paradigms of feedback modulation have been extensively studied, namely, *error reduction* (ER)—also referred to as haptic guidance—and *error augmentation* (EA)—also known as motor amplification. To restore motor function after stroke, the ER paradigm advocates for a reduction of the performance errors of participants while performing a motor task, normally by means of a robotic tool that guides the participant's limbs in accomplishing the task (Mehrholz et al., 2017). Since the motor system has been demonstrated to behave as a "greedy" optimizer that rapidly incorporates the robot's assistive forces to reduce the degree of voluntary control and muscle activation (Emken et al., 2007), it was suggested that the assistance should be provided "as needed," that is, with the minimal necessary amount of external assistance (Marchal-Crespo & Reinkensmeyer, 2009; Sanguineti et al., 2013). Conversely, EA paradigms tend to amplify the movement error by means of disturbing force fields, respecting the knowledge that motor learning is, in fact, driven by error (Emken et al., 2007). However, it was demonstrated that skilled participants benefit more from learning with EA than less skilled participants (Milot et al., 2010; Sigrist et al., 2012), which ultimately translates to the recommendation of designing the feedback strategy based on the specific skill level of each patient.

6.3.4 Developing a reliance on extrinsic feedback

When providing augmented feedback, there may be a risk of training the patient to rely on it more than intended. The guidance hypothesis stipulates that the salience of concurrent extrinsic feedback may lead the patient to depend on it to the point of ignoring any natural intrinsic feedback. Improvements that are shown during the performance of the task with extrinsic feedback are then lost in retention tests (Sigrist et al., 2012). In other experiments, however, concurrent feedback seemed beneficial during the performance of more complex tasks (Bolognini et al., 2016).

Several studies on this issue have shown that intermittent trials without feedback allowed the participants to integrate external and internal feedback rather than coming to depend exclusively on the external feedback. In those studies, the participants could replicate the learned movements in retention tests without external feedback, probably because the augmented feedback had allowed them to refine the internal model of the task (Bark et al., 2015; Crowell & Davis, 2011; Shea et al., 2011; Winstein, 1991).

6.3.5 The sensory side of rehabilitation is an open question

Many authors argue that, as yet, there is insufficient literature evidence to support (or refute) the effectiveness of interventions specifically focused on the sensory aspect of stroke rehabilitation (Bolognini et al., 2016; Bowen et al., 2011; Doyle et al., 2010; Liu et al., 2018; van Dijk et al., 2005). Nonetheless, several recent studies encourage further investigation of this topic, as supported by emerging technologies and rehabilitation techniques (Chen et al., 2018; Laffont et al., 2014). Indeed, it has recently been suggested that combined sensorimotor training outperforms conventional motor-oriented therapies (Chen et al., 2018). In the following, several approaches that use different sensory channels to provide specific sensory training are described (Fig. 6.3). Notably, according to our current knowledge of sensory cortical processing, the different sensory channels strongly interact with each other in the CNS and they cannot be treated as separate moduli (Bach-Y-Rita, 2003; Proulx et al., 2014; Shimojo & Shams, 2001). Therefore, it is theoretically impossible to aim at targeting *only* one channel but rather *mainly* one channel.

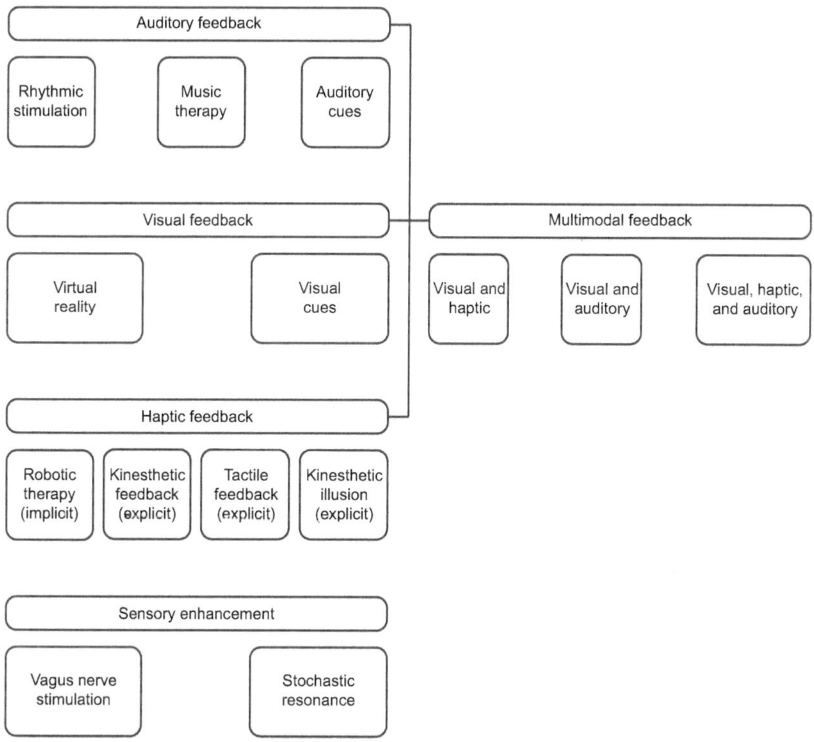

FIGURE 6.3 Augmented sensory feedback approaches for rehabilitation, grouped by main sensory channel (auditory, visual, and haptic, and their multimodal combination). Below, the main sensory enhancement techniques for rehabilitation.

6.3.6 Auditory feedback

6.3.6.1 Relevance of auditory information in motor learning

Verbal encouragements are the most common form of auditory stimulation delivered by physical therapists in clinical practice (Molier et al., 2010). Due to the variable nature of this form of feedback, it has been difficult to evaluate its effectiveness in sensorimotor rehabilitation. However, it is known that auditory stimuli play an important role in motor learning (Sigrist et al., 2012) and induce brain plasticity. This is further exemplified by the fact that trained musicians show anatomical differences in their motor and premotor cortex and cerebellum compared to non-musicians (Bolognini et al., 2016). Playing a musical instrument is, in fact, a complex task that requires fine motor coordination and the association of different sensory stimuli with motor actions. Based on this potential interaction between motor coordination and music, it has been suggested that music can be leveraged to provoke neuroplasticity to facilitate rehabilitation of stroke patients.

6.3.6.2 Types of augmented auditory feedback

A common way of providing auditory feedback in rehabilitation therapies is through rhythmic stimulation, which can improve motor function. It has proven particularly beneficial for gait training (Bolognini et al., 2016; Chen et al., 2018), possibly due to the rhythmic nature of gait. However, music-based interventions produced beneficial effects also in interventions targeting the upper limb (Bolognini et al., 2016; Chen et al., 2018; Parker et al., 2011), suggesting that music (and rhythm in particular) can guide motor timing and improve motor control. Beyond music therapy and rhythmic stimulation, three possible ways of displaying auditory feedback have been identified by Sigrist and colleagues (Sigrist et al., 2012): (1) auditory alarms, (2) sonification of movement variables, and (3) sonification of movement error. While the first type of auditory feedback implies the delivery of auditory stimuli to signal that some variable exceeded a threshold without any form of modulation, the other two involve the use of data values to change the parameters of the auditory stimuli. These data can be kinematic and dynamic variables of the body (e.g., limb velocity, grasp force, and body position), or deviations of these variables from a desired value, for the second and third types, respectively.

6.3.6.3 Auditory feedback devices

It is worth noting that auditory feedback is very often displayed in combination with other sensory modalities, which hinders the possibility of drawing general conclusions regarding the impact of stimulating this sensory modality in the improvement of sensorimotor performance (Molier et al., 2010; Parker et al., 2011; Tinga et al., 2016; Wang et al., 2017). In the remainder of this section, we report a number of studies that attempted to provide evidence for the efficacy of auditory feedback for neurorehabilitation.

6.3.6.3.1 Improvements in motor performance

In a study by Maulucci and Eckhouse, concurrent auditory feedback was displayed to a group of stroke patients to guide their reaching trajectory of the upper limb, while the rest of the patients received physical training alone. Modifications of the trajectory were seen both in the group undergoing practice alone and the one with auditory feedback; however, the auditory feedback group obtained better path performance (Maulucci & Eckhouse, 2001). Another approach that has been tested consists in using musical instruments to practice music-supported motor rehabilitation, in which the patients learn to play simple songs on a piano or rhythms on a drum. The effect on motor performance of auditory feedback delivered while playing an instrument was attributed to error-based learning (Ripollés et al., 2016; Zhang et al., 2016). Van Vugt et al. (2016) aimed to test the error-based learning hypothesis by providing *music-supported therapy* with and without delayed auditory feedback. Surprisingly, the group receiving delayed feedback outperformed the one receiving real-time concurrent feedback in the well-known 9-Hole Pegboard test. The authors argued that auditory–motor integration may be impaired in stroke patients so that concurrent audio feedback led to overcorrection of nonexistent errors.

6.3.6.3.2 Improvements in sensory awareness

Another application of auditory feedback to improve upper limb motor performance was presented by Thielman (2010). Auditory feedback on trunk position aimed to reduce trunk compensation movements during reaching in poststroke patients. The feedback was faded over time in accordance with the guidance hypothesis. Both the feedback group and a control group of participants restrained to a chair improved their performance, but the feedback group improved significantly more than the restrained group. This result suggests that extrinsic feedback can increase sensory awareness during the performance of a motor task, which could ultimately favor motor learning.

6.3.6.4 Conclusions on auditory sensory feedback

The scarcity of evidence on the effectiveness of auditory feedback in sensorimotor rehabilitation and the lack of a standardized procedure prevent us from drawing a satisfactory conclusion on this method. However, promising results have been obtained in improving both motor control and sensory awareness, which confirm the high potential of auditory feedback and encourage further studies in this direction. Notably, auditory feedback is often multimodally combined with other types of augmented feedback (Kiper et al., 2018; Lehrer et al., 2011; Masiero et al., 2006), which are described below.

6.3.7 Visual feedback

6.3.7.1 Relevance of visual information in motor learning

Vision is considered the most important sense for interaction with the environment in activities of daily living (Enoch et al., 2019). Its importance for motor learning and the performance of complex motor tasks is widely acknowledged (Sigrist et al., 2012). Visual information can be leveraged to provide augmented feedback related to motor performance or movement-related parameters with rehabilitative purposes. In motor learning, the effectiveness of visual feedback largely depends on task complexity and skill level (Sigrist et al., 2012). Optimal visual feedback should therefore be tailored to the specific user's needs to avoid cognitive overload. It is suggested that adding visual stimuli to rehabilitation therapy promotes the learning process (Molier et al., 2010) and increases motivation during the training (Oujamaaa et al., 2009). Beyond mirror therapy, which has shown beneficial effects in stroke rehabilitation and is regarded by some authors as a form of augmented feedback (Bolognini et al., 2016; Doyle et al., 2010; Oujamaaa et al., 2009; Parker et al., 2011), visual feedback is effectively delivered in the form of VR (Fu et al., 2015; Laver et al., 2017; Oujamaaa et al., 2009; Parker et al., 2011; Wang et al., 2017). This type of rehabilitation involves the use of simulated environments that are displayed through different devices, such as VR headsets, projectors, or monitors, and that aim at interactively engaging the patient (Laver et al., 2017). The virtual environments are characterized by different levels of *presence* and *immersion* (Bolognini et al., 2016; Laver et al., 2017). While the former is a subjective measure of how much the person experiences actively being in the virtual environment, the latter objectively factors how much the virtual system is intended to capture the user's attention (Laver et al., 2017). A variety of devices can be used to interact with virtual environments, ranging from the computer mouse to state-of-the-art tracking systems used to capture the kinematics of the limbs, which are then represented in the virtual workspace, while yet other sensors can be employed to capture the dynamics of the actions, such as forces and torques.

6.3.7.2 Benefits of virtual reality rehabilitation

Several aspects support the effectiveness of VR-based rehabilitation. First, the recent advances in VR and serious gaming technologies, especially from the entertainment industry, provide great potential for increasing users' participation in rehabilitation exercises, thus encouraging more repetitions of goal-oriented tasks (Fu et al., 2015; Laver et al., 2017; Wang et al., 2017), which is paramount in neurological rehabilitation (Langhorne et al., 2011). Increasing visual quality and immersion of virtual environments results in larger recruitment of visuomotor networks, which may benefit motor performance and learning (Fu et al., 2015). Furthermore, VR-based training produced evidence of provoking neuroplasticity, increasing ipsilesional representation, and stronger activation of the primary motor cortex (Laver et al., 2017). The effectiveness of VR practice on motor learning depends on task complexity: complex tasks lead to marked positive effects, while concurrent visual feedback during simple virtual motor tasks proved rather unfavorable, possibly explained by the guidance hypothesis—that is, permanent feedback during learning leads to a dependency on the feedback (Sigrist et al., 2012).

6.3.7.3 *General features of a virtual reality setup*

In the following subsections, clinical studies reporting the use of augmented visual feedback in the form of VR training are reported (Fig. 6.4).

6.3.7.3.1 Movement representation

In some studies, the movement of the whole body is captured by motion-tracking systems (Sung et al., 2005). This information is used to reproduce the movements in space, usually projected in real time on a large screen. The

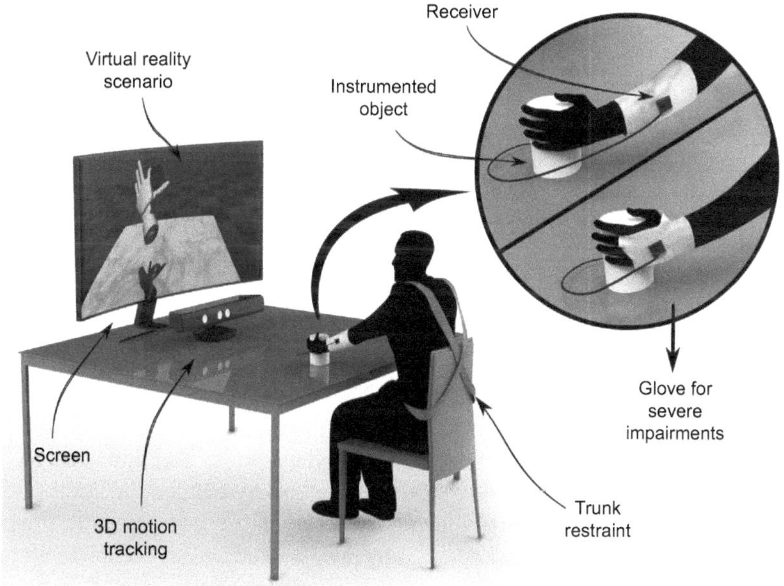

FIGURE 6.4 Stereotypical example of a VR setup for hand rehabilitation purposes. It features a motion-tracking system (e.g., an RGBD sensor) and a screen or a projector to display the VR scenario. The motion of instrumented objects can be coupled with virtual objects by means of a receiver placed on the patient's limb. These objects can be attached to a glove worn by the patients in the case of severe motor impairments that would otherwise prevent them from grasping the object. Methods to restrain the patients' trunk can be included to avoid compensatory movements. *VR*, Virtual reality.

representation of the biomechanical system of the end effector can resemble human aspects, assuming the form of an avatar (Sung et al., 2005), or appear through more abstract shapes. Examples of such abstract representations are arrows depicting the direction of motion in a 3D space (Masiero et al., 2006) or bars varying in color and height according to the arm's movement (Hayward et al., 2013). Even in case of a more realistic and complete representation of the environment, feedback-related aspects are usually enhanced for a clear understanding from the user and to avoid distractions. Another important characteristic of VR systems is the involvement of the participant in the execution of the task: motivation increases when the task appears to be either useful in a real environment, or when it is amusing, in a game-like setting.

6.3.7.3.2 Interaction with objects during task performance/training

In VR rehabilitation, stroke patients are usually asked to perform both simple and complex tasks. Simple tasks can be simply reaching for and grasping an object. More complex tasks, such as pouring water from a bottle into a glass, involve the activation of many muscles and can be decomposed into several submovements. In both cases, it is common to use instrumented objects, synchronized with virtual objects in the virtual environment (Kiper et al., 2014). The virtual objects should resemble the real ones in shape as much as possible and follow their displacements in real time, thanks to electromagnetic, piezoelectric, capacitive, or resistive tracking sensors placed on the instrumented objects. In case the patient presents severe grasping deficits, gloves are often used, and real objects can be forcedly attached to these so as to make grasping possible (Kiper et al., 2014, 2018; Lehrer et al., 2011; Piron et al., 2010). The same motor tasks can be performed with increased complexity by adding extra virtual elements that do not correspond to real objects (Kiper et al., 2014; Piron et al., 2010).

6.3.7.3.3 Kinematic features recording

The kinematic features extracted by the VR experimental setups can be beneficial both for clinicians to evaluate the motor performance and for patients to increase their awareness of the sensory events while executing the movements. These features consist of several kinematic variables recorded by the tracking systems, for example, mean velocity of the end effector, time to complete the task, or number of submovements executed to complete a complex task (peaks) (Broeren et al., 2007; Kiper et al., 2014, 2018; Piron et al., 2010). Often, more complex features are also represented, such as the torque or force applied by the patient's upper limb, the joint range of motion, trunk compensation movements, or movement trajectory (Cordo et al., 2008; Lehrer et al., 2011; Masiero et al., 2006; Sung et al., 2005). Additional information on the error and error rate can be provided as knowledge of results feedback (Sung et al., 2005).

6.3.7.4 Studies in virtual reality for rehabilitation purposes

Studies using VR systems that provided solely visual feedback compared the effects of *reinforced feedback in virtual environment* (RFVE) to standard rehabilitative protocols. Significantly larger improvements in the motor function assessment measures were seen in the RFVE groups compared to the control groups receiving the same amount of personalized or standard traditional therapy (Kiper et al., 2014; Piron et al., 2010). The VR scenarios in these two studies consisted of a motion-tracking system, a large screen, and instrumented objects. These objects were coupled with an electromagnetic sensor on the dorsal part of the hand, or on a glove worn by the patient in the case of severe motor impairments. These studies also observed improvements in the Fugl–Meyer upper extremity scale for the RFVE group. Moreover, Kiper et al. (2014) reported an increase in Functional Independence Measure with respect to the control group. These works demonstrated that motor functions improved after VR-based rehabilitation exercises in addition to regular therapy.

6.3.7.5 Other visual feedback delivery methods

In addition to VR, augmented visual feedback can be added to traditional rehabilitation to provide information on the level of the force applied during grasping, as studied by Quaney et al. (2010). In their study, augmented visuomotor training was administered to patients performing a lifting task by continuously providing information on the target and the current magnitude of the grip force (GF) during object manipulation. After visuomotor training, the patients were able to pick and lift with higher GF accuracy, and the improvements remained stable after the removal of feedback. In fact, the authors hypothesized that stroke patients appropriately used predictive GF scaling by recalling their prior knowledge on motor commands. Another example of providing augmented visual feedback is the display of information on the direction and intensity of normal and tangential forces applied by stroke participants during a manipulation task where the participants were asked to reach and maintain target levels of normal and shear forces (Seo et al., 2011). The training resulted in increased scores in Box and Block test and Action Research Arm test between pre- and posttest measurements. The tangential to normal force ratio decreased after training, indicating a higher level of hand motor performance. Notably, the improvements persisted even after the visual feedback was removed.

6.3.7.6 Conclusions on visual feedback

Visual feedback has been thoroughly investigated for neuromotor rehabilitation, and the level of evidence of its effectiveness is significant. Both VR and visual cues have been demonstrated to provide neurological patients with valuable information that they can leverage to improve their motor performance. VR-based approaches in particular are promising, leading to increases in performance and engagement, also thanks to the continuous advances in this field promoted by the entertainment industry. Similar to auditory feedback, vision-based approaches are often coupled with other types of augmented information in a multimodal fashion, further extending the potential of this method.

6.3.8 Haptic feedback

6.3.8.1 Relevance of haptic information in motor learning

Haptic feedback—information relayed by the sense of touch and proprioception—enables us to interact bidirectionally with the environment, that is, to act on it and perceive these actions (Sigrist et al., 2012). The study of haptic feedback has attracted many researchers worldwide, and the impact of haptics on motor learning and rehabilitation has been widely investigated. These studies have contributed to our understanding of how we use touch to explore the environment and how it impacts our motor function (Demain et al., 2012). It was discovered that motor performance can be improved through haptic stimulation, which increases corticospinal excitability and expands the cortical representation of the body part that is stimulated (Bolognini et al., 2016). Moreover, substantial processing of somatosensory afferent stimuli occurs in the motor cortex, demonstrating the great interconnection between motion and sensation (Chen et al., 2018; Sigrist et al., 2012). Notably, haptic feedback does not require visual attention during exercise (Wang et al., 2017), and indeed, it is used to reduce the workload of visual and auditory systems (Sigrist et al., 2012). Haptic feedback can be broadly divided into two different categories: kinesthetic and tactile feedback (Demain et al., 2012). The former is used by the CNS to infer the position and motion of the body parts (referred to as kinesthetic awareness or proprioception), and the latter consists of information from the external environment. Muscle spindles, Golgi tendon organs, and mechanoreceptors in the joint capsules, ligaments, and skin surrounding the joints collect kinesthetic stimuli, while thermoreceptors, nociceptors, and different types of low-threshold mechanoreceptors provide tactile information from the glabrous and hairy skin, although the latter has much lower spatial acuity (Zimmerman, Bai, & Ginty, 2014). This information is transmitted by various Aβ, Aδ, and C-afferents with a variety of end receptors (e.g., hair follicle receptors). Those receptors (innervated by Aβ fibers) that are most important for tactile perception in the hand are divided into fast-adapting (FA) and slowly-adapting (SA) types, in particular: Meissner corpuscles (FA I), Pacinian corpuscles (FA II), Merkel cell−neurite complexes (SA I), and Ruffini endings (SA II).

6.3.8.2 Movement-based (implicit) and sensory-based (explicit) haptic feedback

In haptic research, it is possible to differentiate between devices that primarily target the sensory function with explicit feedback—which therefore provide only references for the completion of a specified movement—and devices that instead guide the user through the movement and do not provide haptic feedback as a primary goal but nevertheless provide some implicit sensory feedback by promoting motor activity (Demain et al., 2012).

6.3.8.2.1 Implicit haptic feedback

Robotic therapists—such as exoskeletons or wall-mounted systems—guide the limb motion, providing assistive or resistive actions. Notably, they do not only promote rehabilitation due to the implicit improvements in muscle strength or joint range of motion but purposely facilitate motor learning through sensorimotor integration. Despite the increasing adoption of these devices in clinical practice and their proven efficacy in upper limb rehabilitation (Mehrholz et al., 2017), their intervention cannot be considered as primarily sensory and will thus not be extensively treated in this chapter.

6.3.8.2.2 Explicit haptic feedback: kinesthetic and tactile

On the other hand, devices that primarily target sensory function and do not impose motion can either be based on kinesthetic feedback or tactile feedback. Kinesthetic feedback can be delivered through an electromechanical or pneumatic apparatus, such as a force feedback joystick or mouse. Other devices that primarily target the sensory function consist of vibrotactile displays and pressure displays, namely, *tactors*. Vibrotactile systems represent the most common way of providing haptic cues, as testified by their pervasive adoption in technological devices, such as phones or wearable gadgets. These devices have been widely used to support motor learning (Sigrist et al., 2012) and rehabilitation (Wang et al., 2017). Notably, several parameters can be tuned in a vibrotactile display, similar to auditory displays, such as vibration frequency, intensity, duration (Sigrist et al., 2012), and application site (Kaczmarek et al., 1991). However, it is not yet possible to draw conclusions on the ideal tuning of these parameters.

Tactile afferent information has been proven to be central in object manipulation and is used by the CNS to mark the task phases and link them to subgoals (Johansson & Flanagan, 2009). The FA nerve endings in the human skin are very sensitive to vibrations, with a peak in sensitivity in the range of 200–300 Hz for Pacinian corpuscles and 40–50 Hz for Meissner endings (Johansson & Flanagan, 2009). The delivery of discrete vibrotactile cues has been demonstrated to be successful in restoring important tactile information in the field of prosthetics, where vibrotactile feedback was delivered to a sensitive part of the patient's body—the arm—when tactile events at the fingertips occurred, and it was quickly and seamlessly integrated into the user's motor control (Cipriani et al., 2014; Clemente et al., 2015).

6.3.8.2.3 Feedback for kinesthetic illusion

It has been demonstrated that vibrations in the range of 70–110 Hz delivered to musculotendinous regions can induce an illusory limb movement, namely, the kinesthetic illusion (Ferrari et al., 2021; Goodwin, Mccloskey, & Matthews, 1972; Marasco, Bourbeau, Shell, Granja-Vazquez, & Ina, 2017; Schofield, Dawson, Carey, & Hebert, 2015). The rehabilitative potential of the kinesthetic illusion is suggested by the fact that it can elicit proprioceptive sensations in patients regardless of their motor abilities and without requiring actual limb motion. Its efficacy in clinical practice has not been thoroughly assessed (Ferrari et al., 2021; Schofield, Dawson, Carey, & Hebert, 2015), but a recent effort to define a standardized approach may change this in the future (Beaulieu et al., 2020).

6.3.8.3 Devices for haptics

A variety of technologies can be employed to devise tactors, ranging from motors that constrict bands worn around the limbs and skin indentation or stretching mechanisms, to more exotic solutions employing ultrasound, magnetorheological fluids, and shape-memory alloys (Fig. 6.5). However, their efficacy in rehabilitation practice is limited by the reduced sensitivity of individuals affected by somatosensory disorders (Demain et al., 2012). In the remaining part of this section, we describe clinical studies that focused on delivering haptic feedback as a primary intervention for rehabilitation.

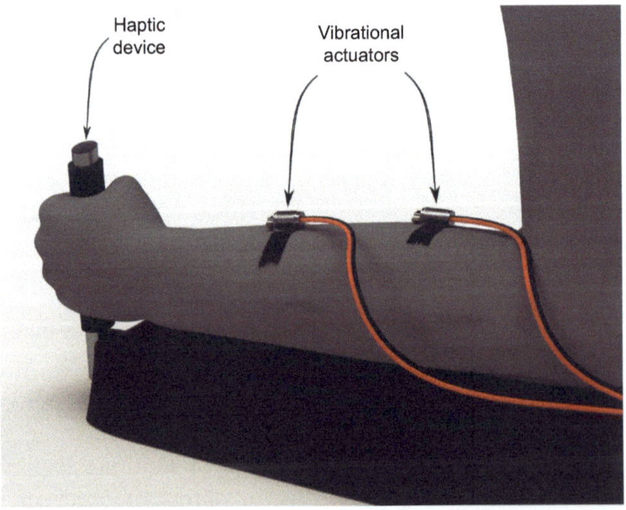

FIGURE 6.5 Example of a setup for delivering augmented haptic feedback. Tactile information can be delivered through vibrational actuators or tactors placed on the users' skin, while kinesthetic information can be delivered through forces or torques applied by a haptic device.

6.3.8.3.1 Types of augmented haptic stimulation

Tactile stimuli can be delivered through many types of motors and devices on the skin, even though the most common technique is to use vibrations generated by eccentric mass, voice coil, piezoelectric or servo motors (Bark et al., 2008; Elangovan et al., 2019; Lakshminarayanan et al., 2017; Liu et al., 2009). Other methods consist of applying mechanical pressure or stretch to the patient's skin (Bark et al., 2008; Norman et al., 2014), or exerting a resistive force through the handle of a haptic device proportional to the information that is to be augmented (Abdollahi et al., 2018; Broeren et al., 2007). The information provided through extrinsic haptic feedback can be of the same type (i.e., tactile) (Cameirão et al., 2012; Liu et al., 2009), or modality-mismatched (Elangovan et al., 2019).

6.3.8.3.2 Vibrotactile sensory substitution

When applying a vibrational stimulus to substitute the fingertips' tactile sensation, particular attention should be paid to the location and the duration of the applied vibration: if the actuator is placed directly on the fingertips, it can interfere with object manipulation, and the user feels like touching a vibrating object (Merrett et al., 2011). Moreover, exposing users to prolonged stimulation could cause adaptation of the receptors to the stimulus, or even tissue damage (Demain et al., 2012). Adaptation to a stimulus can be prevented if the stimulation is interrupted intermittently (Markovic et al., 2018; Shannon, 1976). It is interesting to note, though, that concurrent feedback could also be detrimental for motor learning. In a study where participants had to learn six different movements, concurrent vibrotactile feedback distracted the participants and led to poorer outcomes (Bark et al., 2015).

6.3.8.3.3 Proprioceptive feedback

It has also been shown that vibrotactile feedback can be employed to effectively provide proprioceptive information. In a study by Elangovan et al. (2019), the patients controlled a wrist robot to complete a virtual task that was simultaneously displayed on a screen and haptically rendered by the robot. Beyond the visual and haptic feedback, they received vibrational cues on their forearm regarding the position of the virtual target. The robotic rehabilitation session resulted in a significant reduction of the *just noticeable difference* threshold of the wrist joint angles, demonstrating that persons with impaired proprioception can benefit from augmented haptic feedback.

6.3.8.3.4 Dynamic and performance feedback

Information on dynamic interactions can be conveyed by means of haptic displays with rehabilitation purposes. In a study by Cameirão et al. (2012), the force that patients applied to virtual objects in a rehabilitation task was fed back to them through the handle of a robotic device, which led to better clinical outcomes with respect to other procedures without haptic stimulation. In a study by Broeren et al. (2007), the reaction force generated by the interaction with objects in virtual environments was rendered to chronic stroke patients through a haptic stylus, resulting in improved clinical outcomes after the treatment. Haptic feedback can also provide knowledge of performance of the ongoing motor action: in a recent study, the force applied by a haptic device was proportional to the trajectory error in target-reaching tasks, which led to improvements in clinical scales similar to conventional therapy (Abdollahi et al., 2018).

6.3.8.4 Conclusions on haptic feedback

The tactile sense is a fundamental channel for neuromotor rehabilitation. We reported several studies that leveraged different types of augmented haptic stimulation to promote recovery after stroke. The haptic modality is often combined with other types of feedback, especially with vision in VR-based scenarios, which makes it particularly difficult to draw conclusions on the effectiveness of this sensory channel. However, we can argue that adding haptic stimulation to the sensorimotor rehabilitative training generally has beneficial effects.

6.3.9 Multimodal feedback

6.3.9.1 Multisensory integration in the human brain

The CNS can integrate information coming from different sensory modalities with prior knowledge on the task, minimizing the uncertainty of sensory signals (Bays & Wolpert, 2007; Franklin & Wolpert, 2011). These signals are often inaccurate and noisy, especially in poststroke convalescence, when afferent pathways are corrupted. Bays and Wolpert (2007) suggested that the brain normally integrates the information coming from distinct sensory modalities through the use of weights, modulated according to the accuracy of the sensory signals, so-called Bayesian integration. The lower the variance of the signal coming from a specific sensory modality, the higher the weight used by the CNS when integrating this information (Fig. 6.6). The same concept applies to the integration of prior knowledge the participant has on the task, which is weighted more if the feedback is degraded. Hence, in humans, multisensory integration facilitates the processing of sensory information, ultimately improving motor functions.

6.3.9.2 Studies on multimodal feedback

Many examples of multimodal augmented feedback in rehabilitation tasks can be found in the literature. In fact, visual feedback and VR systems are very often combined with other types of feedback to increase realism and immersion.

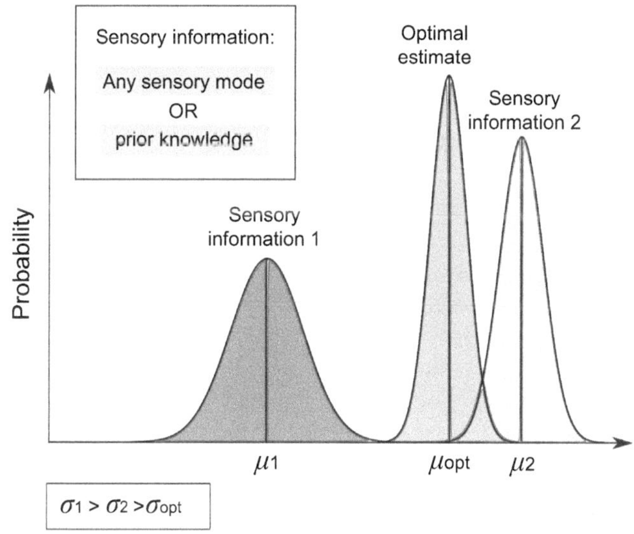

FIGURE 6.6 Integration of the information coming from multiple sensory modalities (i.e., intrinsic feedback) and prior knowledge. Greater importance is given to the sensory information with lower variance resulting in an optimal estimate with a variance lower than both of the two original distributions processed and integrated.

6.3.9.2.1 Visual and haptic feedback

Several studies have reported the simultaneous use of heterogeneous sources of visual and haptic stimulations. For example, the combination of haptic devices and VR for rehabilitation proved effective in a study by Broeren et al. (2007), leading to increased mean velocity at which the task was performed and shorter execution time, as well as a decrease of superfluous movement. Furthermore, Cameirão et al. (2012) compared three setups in the same rehabilitation procedure and showed that merging VR and haptic feedback is more effective than either VR alone or VR combined with a bimanual exoskeleton. Coupling the haptic device with a VR scenario led to significantly better outcomes on Fugl–Meyer upper limb assessment and Box and Block test scores.

The field of rehabilitation robotics represents an exemplary case of beneficial integration of haptic and visual feedback: robotic trainers, which usually feature visual displays in addition to the actual robot, outperform other therapeutic tools lacking sensory feedback (see above for a more detailed discussion on this topic). Furthermore, rehabilitative robotic platforms with dedicated sensory feedback, such as *Sensory-Enhanced Robot-aided Motor Training* proved more effective than robotic training alone in increasing the motor status score for shoulder/elbow and wrist/hand (Liu et al., 2009). In this study, vibratory stimuli were delivered on the participants' hands according to the location and direction of the pushing/pulling force applied to the handle of a robot during hand movements. The participants in this condition outperformed the group that performed the reaching task with the robotic assistance alone.

A similar procedure, performed in a bimanual modality, was used to compare robotic rehabilitation to the combination of robot-aided therapy and EA techniques immersed in a VR environment (Abdollahi et al., 2018). In this experiment, the patients used an upper limb robotic tool for bimanual practice, where the location of the nonaffected wrist was recorded and displayed together with the affected wrist position as visual feedback on a screen. The robot delivered haptic feedback based on EA, which significantly increased the upper extremity Fugl–Meyer assessment score with respect to the control group, although it did not significantly change the Motor Activity Log and Wolf Motor Function test scores.

Cordo et al. (2008) used a robotic device for providing hand-assisted movement therapy combined with tendon vibrators to five patients. Training with this setup led to improvements in ADLs, hand strength and function, observed both in posttest measurements and in a follow-up after 6 months, demonstrating benefits of assisted movement therapy combined with sensory augmentation in motor recovery with long-term effects. In another study, visual information regarding the passive force exerted by a grounded end-effector robot on the participant's limb was provided during the performance of repetitive movements (Masiero et al., 2006). This rehabilitation therapy led to statistically significant improvements in Fugl–Meyer and Functional Independence Measurements scores in the experimental group with respect to a control group receiving the same treatment on the unaffected limb. The improvements were retained in a follow-up after 3 months.

6.3.9.2.2 Visual and auditory feedback

RFVE proved more effective than conventional one-to-one therapy when combined with auditory feedback providing information on the correctness of movement performance (Kiper et al., 2018). In this large study, Kiper and colleagues compared functional and kinematics outcome measures before and after executing simple and complex tasks both in a VR environment and in a traditional one-to-one hospital treatment setting. The VR setup was similar to their previous study described above (Kiper et al., 2014), although it featured additional feedback modalities. A score, provided through acoustic signals and a digital voice, informed the participants on their spatial error while they performed the tasks. Moreover, additional visual feedback was provided in the form of a virtual teacher on the screen. This multisensory environment led to significant improvements in the functional independence measurements and Fugl–Meyer upper extremity scores as well as in kinematics performance parameters with respect to traditional therapy.

6.3.9.2.3 Combination of visual, haptic, and auditory feedback

An illustrative example of multisensory stimulation is presented by *adaptive mixed reality rehabilitation* (AMRR) (Lehrer et al., 2011). The study featured a VR environment where abstract feedback was delivered through visual, auditory, and haptic modalities to promote independent movements during reaching and grasping tasks. Each sensory modality was used to provide information on a specific set of variables: space-related measures were represented by visual feedback, time-related variables by auditory signals, and grasp-related parameters by haptic feedback. A combination of auditory and visual feedback was also shown to provide proprioceptive information, such as joint positions and compensatory movements of the trunk. Specifically, these feedback stimuli, adaptable according to the needs of each participant, were selectively activated to indicate an inefficient movement and provided information on the direction and magnitude of movement errors. The results of this pilot study on three stroke participants stressed the potential benefits of multisensory stimulation in poststroke rehabilitation: Wolf Motor Function test score increased in all patients following the AMRR treatment. Possibly, information on the performance fed back through an integrated multimodal stimulation could have had a central role. The authors underlined the importance of avoiding dependence on the feedback, delivering it only when the patient deviated by a certain margin from the optimal movement trajectory, instead of giving continuous feedback.

6.3.9.3 Conclusions on multimodal feedback

Despite the large number of studies reporting beneficial effects of multisensory stimulation, many authors agree that valid conclusions on their efficacy cannot be drawn. This uncertainty can be attributed to the extremely wide variety of combinations of feedback modalities that have been tested so far (Bolognini et al., 2016; Doyle et al., 2010; Molier et al., 2010; Sigrist et al., 2012; Tinga et al., 2016). For the same reason, we cannot conclude whether multisensory rehabilitation strategies can outperform interventions that target a single sensory modality and affirm that further investigations are needed.

6.3.10 Sensory information enhancement

Along with the delivery of additional sensory feedback stimuli, another way of augmenting intrinsic sensory information in poststroke patients is to strengthen the action of the intact afferent pathways. To date, we have identified two noninvasive methods that aim at increasing sensitivity by enhancing intrinsic feedback mechanisms and brain plasticity: vagus nerve stimulation (VNS) and stochastic resonance.

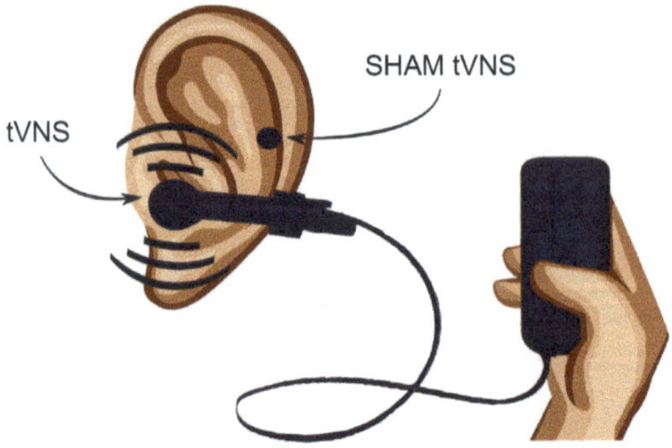

FIGURE 6.7 An illustrative example of transcutaneous vagus nerve stimulation. The tragus is electrically stimulated through a pin electrode, while sham stimulation can be done by moving the pin to the auricle, the outer part of the ear.

6.3.10.1 Vagus nerve stimulation

Noninvasive *transcutaneous VNS* (tVNS) consists of electrically stimulating the inner side of the tragus in the ear, as reported in a representative study (Capone et al., 2017), where the stimulation—consisting of pulse trains—was delivered through two surface electrodes (Fig. 6.7). After this stimulation period, the patients participated in a robotic therapy session. The results obtained after 10 days of tVNS and robotic therapy showed a significant improvement in Fugl–Meyer score. The authors hypothesized that tVNS stimulation induced neuroplasticity and therefore promoted recovery of motor functions, as supported by previous studies that showed an increase in brain-derived neurotrophic factor and neurotransmitters associated with neuroplasticity and recovery after VNS therapy for brain lesions (Follesa et al., 2007).

6.3.10.2 Stochastic resonance

Another approach to increase sensitivity consists in applying subthreshold vibrational noise. Low-intensity noise was found to enhance the afferent tactile signals due to a physical process called *stochastic resonance* (Enders et al., 2013; Liu et al., 2002; Seo et al., 2014; Seo et al., 2019). According to this process, when noise is infused into a nonlinear system characterized by an activation threshold (akin to many biological and physical systems), it can improve the detection of weak input signals. It is common to consider noise as an adverse factor in signal processing because it corrupts the signal, depriving it of its informative content. Conversely, according to the aforementioned process, random energetic fluctuations injected in a system can improve the quality of its output (Hänggi, 2002). However, this happens only for optimal amplitudes of the injected noise: the output performance is in fact an inverted U-shape function of the noise amplitude; it increases

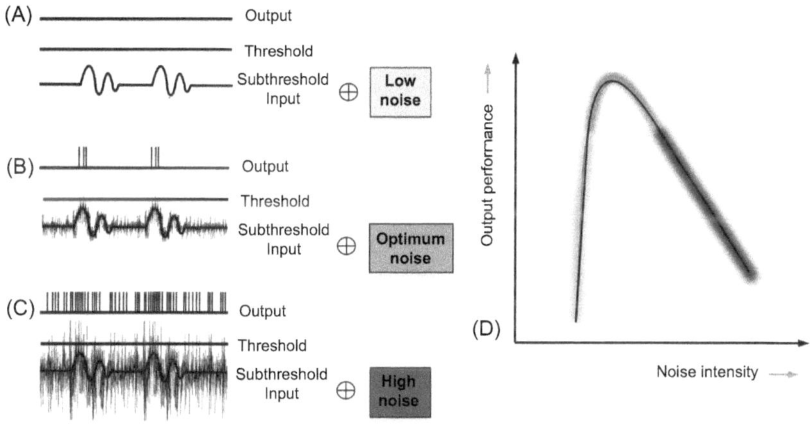

FIGURE 6.8 The mechanism of stochastic resonance. (A) If the amplitude of noise is too low, the output is still below threshold. (B) When the noise intensity is optimal, the signal-to-noise ratio (SNR) increases the output activated by the peaks of the input signal (resonance peak on the graph). (C) When the noise amplitude is too large, the quality of the output degrades. (D) The intensity of the noise is plotted with respect to the output performance.

toward a maximum corresponding to the resonance peak—the optimum level of noise—and decreases for larger noise intensities. When optimum noise is injected, the system can detect subthreshold inputs. It therefore increases the signal-to-noise ratio and lowers the detection threshold (Fig. 6.8).

An interesting feature of stochastic resonance is the possibility of applying any type of noise (random, periodic, aperiodic, etc.), regardless of the input signal (Liu et al., 2002; McDonnell & Abbott, 2009).

6.3.10.2.1 Optimal noise may benefit rehabilitation

Most of the neural processes taking place in the human body, such as sensation, require the stimulus to surpass a threshold. It is suggested that subthreshold noise promotes tactile sensitivity by increasing cortical sensorimotor activity and neural synchronization between cortical regions (Seo et al., 2019). Consequently, the use of low-intensity noise started to gain attention in the biomedical field, particularly for rehabilitation techniques, given the crucial importance of sensory information in motor learning. Subthreshold noise produced comparable results in fingertip sensitivity enhancement, regardless of where on the hand or wrist it was applied (Enders et al., 2013; Lakshminarayanan et al., 2015). The benefits obtained by exploiting stochastic resonance do not depend on the distance between the application area of the subthreshold stimulus and the target site where sensitivity is to be enhanced, presumably as long as the application area and target site pertain to the same body area and they present interneuronal connections at a spinal or supraspinal level (Lakshminarayanan et al., 2015).

In practice, subthreshold noise is commonly delivered via voice coil actuators, which allow to decouple frequency and amplitude of the stimulus signal and therefore to modulate the amplitude of the vibration according to each patient's sensory threshold.

6.3.10.2.2 Studies on stochastic resonance for rehabilitation

Some studies have explored the use of stochastic resonance to enhance tactile sensitivity and improve the motor functions of stroke patients. Liu et al. reported a significant decrease in the vibrational threshold following the application of mechanical noise randomly superimposed on a vibrational stimulus (Liu et al., 2002). Another study reported improvements in motor function and responsiveness to sensory stimulation, measured through *somatosensory evoked potentials*, after brief applications (20 minutes) of mechanical noise (Peurala et al., 2002). Enders and collaborators reported that light touch sensation of the fingertips increased after the application of subthreshold white noise in four locations on the hand and wrist (Enders et al., 2013). Monofilament test scores significantly increased, that is, the sensitivity threshold decreased, while the two-point discrimination remained unaltered, meaning that the spatial resolution was not affected. A year later, the same group found, however, that the monofilament test score was unaltered by the application of subthreshold noise, but it did lead to improvements in hand dexterity, as well as increased movement coordination and GF during manipulation (Seo et al., 2014). Seo and colleagues recently prototyped a wearable device and tested it on stroke survivors (Seo et al., 2019). The treatment group significantly improved their hand motor function score in Box and Block test and Wolf Motor Function test compared to baseline and retained these improvements in the 19-day follow-up. Interestingly, the control group did not present significant improvements compared to baseline, probably because of the limited therapy dose provided during the trial.

6.3.10.2.3 Possible implications in feedback evaluations

Curiously, some authors have employed subthreshold stimuli as a sham baseline feedback to test the efficacy of other experimental techniques. The effects of stochastic resonance stimuli certainly reflected in the baseline group and possibly affected the outcomes of the experimental procedures. For example, Johansson et al. tested the effects of acupuncture and transcutaneous electrical stimulation compared to a control group, to whom they delivered a low-intensity electrical stimulation at 80 Hz through surface electrodes. No significant difference was found in ADL measurements between groups, which all improved their motor performance after the test conditions and kept improving in the 3-month and 12-month follow-ups (Johansson et al., 2001). This result, instead of pointing to the ineffective outcomes of the proposed interventions, may support that subthreshold stimulation can produce long-term beneficial effects. However, this hypothesis cannot be validated due to the lack of conclusive evidence, and further studies on stroke patients with follow-ups are needed.

6.3.10.3 Conclusion on sensory enhancement

According to the guidance hypothesis, augmented feedback can in some cases be distracting and hinder motor learning. This may happen because of the possible dependency on extrinsic feedback, which leads to intact intrinsic feedback being ignored. This effect might be prevented by amplifying the intrinsic feedback stimuli by using stochastic resonance or VNS.

6.4 Future directions for augmented feedback

Many devices have been designed with the aim of delivering augmented feedback over the years. They were not included in the previous sections as they were not validated in rehabilitative scenarios and their effectiveness was tested only on healthy participants. However, they may offer inspiration for or hint at future developments relevant for stroke rehabilitation. Some representative examples are reported hereafter. Since an exhaustive review cannot be accomplished here, the authors wish to apologize to those researchers engaged in important work that was not mentioned.

Vibrotactile feedback could be beneficial to improve the effect of robotic trainers, as suggested by Cuppone et al. (2016). In their work, vibrations were delivered to the arm of participants to provide information on the deviation from the ideal path during the performance of discrete, goal-directed reaching movements while using a wrist robot. Interestingly, this preliminary study found no difference if the stimulation was applied to the same or the opposite arm with respect to the one receiving proprioceptive training. In fact, regardless of whether the stimuli were applied on the treated arm or the resting one, the participants who received additional vibrotactile feedback on movement error significantly reduced their need for guidance from the haptic device, in contrast with the group receiving robot-assisted therapy only.

Merrett et al. (2011) designed and tested three different haptic devices to be worn on the dorsum or ventrum of the thumb and fingers. The tactile devices aimed to augment the sense of touch through vibrations originating from a miniature cylindrical motor attached to the fingertips. This configuration, however, did not give the perception of holding something in the hand; instead participants stated that it felt like touching a vibrating object. This result could suggest that providing tactile feedback directly at the fingertips is unsuitable.

In a study by Norman et al. (2014) with healthy participants, haptic skin stretch feedback proved effective at improving the motor performance in planar movements. They applied haptic feedback through a tactor placed on the fingertips, which delivered directional cues by stretching the skin in up to eight directions. Thanks to the information provided, the users were able to match their movements with the stimulus direction.

An interesting approach for delivering augmented vibrotactile feedback is to deliver short-lasting stimuli corresponding to relevant mechanical events that delimit the manipulation phases, akin to how humans naturally organize motor tasks according to the *Discrete Event-driven Sensory feedback Control* (DESC) model (Clemente et al., 2015; Johansson & Edin, 1993). Our group previously showed that healthy participants can temporally integrate DESC vibrotactile feedback in their sensorimotor control during manipulation tasks using an artificial hand (Aboseria et al., 2018; Barone et al., 2017; Cipriani et al., 2014; Clemente et al., 2015). DESC feedback could be integrated in upper limb rehabilitation procedures to guide the patients through the execution of manipulative tasks.

Novel developments in tele-health technology could potentially translate supervised upper limb stroke rehabilitation therapies to the home setting with low-cost devices. An example is the *Mechanical Muscle Activity with Real-time Kinematics* (M-MARK), a wearable device comprising two elasticized garments equipped with sensors worn by stroke patients while practicing ADLs, such as reaching, grasping, and manipulating objects (Burridge et al., 2017). The data recorded by the sensors are used to provide feedback about movements and muscle activity on a screen in the form of an avatar. The M-MARK system also provides auditory feedback on the correctness of movement strategies and muscle activation timing. An advantageous characteristic of this device is the possibility to be used by the therapists as a system to diagnose movement deficits. Current studies are being conducted to identify the best choice of sensors to be integrated in this device, such as a combination of inertial measurement units and mechanomyographic sensors to measure the quality and quantity of movement. Clinical studies on stroke patients are still needed to evaluate the effectiveness of this promising wearable device.

Several other techniques and devices have been devised with the aim of delivering augmented feedback for rehabilitation; however, only preliminary studies are available, and their effectiveness still needs to be thoroughly assessed. Despite their preliminary status, they can certainly be of inspiration for future research in poststroke rehabilitation and demonstrate the increasing interest in the field of sensory-enhanced rehabilitation.

References

Abdollahi, F., Corrigan, M., Lazzaro, E. D. C., Kenyon, R. V., & Patton, J. L. (2018). Error-augmented bimanual therapy for stroke survivors. *NeuroRehabilitation*. Available from https://doi.org/10.3233/NRE-182413.

Aboseria, M., Clemente, F., Engels, L. F., & Cipriani, C. (2018). Discrete vibro-tactile feedback prevents object slippage in hand prostheses more intuitively than other modalities. *IEEE Transactions on Neural Systems and Rehabilitation Engineering*. Available from https://doi.org/10.1109/TNSRE.2018.2851617.

Antfolk, C., D'alonzo, M., Rosén, B., Lundborg, G., Sebelius, F., & Cipriani, C. (2013). Sensory feedback in upper limb prosthetics. *Expert Review of Medical Devices*. Available from https://doi.org/10.1586/erd.12.68.

Bach-Y-Rita, P. (2003). Theoretical basis for brain plasticity after a TBI. *Brain Injury*. Available from https://doi.org/10.1080/0269905031000107133.

Bark, K., Hyman, E., Tan, F., Cha, E., Jax, S. A., Buxbaum, L. J., & Kuchenbecker, K. J. (2015). Effects of vibrotactile feedback on human learning of arm motions. *IEEE Transactions on Neural Systems and Rehabilitation Engineering*, *23*(1), 51–63.

Bark, K., Wheeler, J.W., Premakumar, S., & Cutkosky, M.R. (2008). Comparison of skin stretch and vibrotactile stimulation for feedback of proprioceptive information. In: *Symposium on haptics interfaces for virtual environment and teleoperator systems 2008 – Proceedings, haptics*. https://doi.org/10.1109/HAPTICS.2008.4479916

Barone, D., D'Alonzo, M., Controzzi, M., Clemente, F., & Cipriani, C. (2017). A cosmetic prosthetic digit with bioinspired embedded touch feedback. *IEEE International Conference on Rehabilitation Robotics*. Available from https://doi.org/10.1109/ICORR.2017.8009402.

Bays, P. M., & Wolpert, D. M. (2007). Computational principles of sensorimotor control that minimize uncertainty and variability. *Journal of Physiology*. Available from https://doi.org/10.1113/jphysiol.2006.120121.

Beaulieu, L.-D., Schneider, C., Massé-Alarie, H., & Ribot-Ciscar, E. (2020). A new method to elicit and measure movement illusions in stroke by means of muscle tendon vibration: The Standardized Kinesthetic Illusion Procedure (SKIP). *Somatosensory & Motor Research*. Available from https://doi.org/10.1080/08990220.2020.1713739.

Blennerhassett, J. M., Carey, L. M., & Matyas, T. A. (2006). Grip force regulation during pinch grip lifts under somatosensory guidance: Comparison between people with stroke and healthy controls. *Archives of Physical Medicine and Rehabilitation*, *87*, 418–429.

Blennerhassett, J. M., Matyas, T. A., & Carey, L. M. (2007). Impaired discrimination of surface friction contributes to pinch grip deficit after stroke. *Neurorehabilitation and Neural Repair*. Available from https://doi.org/10.1177/1545968306295560.

Bolognini, N., Russo, C., & Edwards, D. J. (2016). The sensory side of post-stroke motor rehabilitation. *Restorative Neurology and Neuroscience*. Available from https://doi.org/10.3233/RNN-150606.

Bowen, A., Knapp, P., Gillespie, D., Nicolson, D. J., & Vail, A. (2011). Non-pharmacological interventions for perceptual disorders following stroke and other adult-acquired, non-progressive brain injury. *Cochrane Database of Systematic Reviews*. Available from https://doi.org/10.1002/14651858.cd007039.pub2.

Broeren, J., Rydmark, M., Björkdahl, A., & Sunnerhagen, K. S. (2007). Assessment and training in a 3-dimensional virtual environment with haptics: A report on 5 cases of motor rehabilitation in the chronic stage after stroke. *Neurorehabilitation and Neural Repair*. Available from https://doi.org/10.1177/1545968306290774.

Burridge, J. H., Lee, A. C. W., Turk, R., Stokes, M., Whitall, J., Vaidyanathan, R., Clatworthy, P., Hughes, A. M., Meagher, C., Franco, E., & Yardley, L. (2017). Telehealth, wearable sensors, and the internet: Will they improve stroke outcomes through increased intensity of therapy, motivation, and adherence to rehabilitation programs? *Journal of Neurologic Physical Therapy*. Available from https://doi.org/10.1097/NPT.0000000000000183.

Byl, N., Roderick, J., Mohamed, O., Hanny, M., Kotler, J., Smith, A., Tang, M., & Abrams, G. (2003). Effectiveness of sensory and motor rehabilitation of the upper limb following the principles of neuroplasticity: Patients stable poststroke. *The American Society of Neurorehabilitation*, *17*, 176–191.

Cameirão, M. S., Badia, S. B. I., Duarte, E., Frisoli, A., & Verschure, P. F. M. J. (2012). The combined impact of virtual reality neurorehabilitation and its interfaces on upper extremity functional recovery in patients with chronic stroke. *Stroke*. Available from https://doi.org/10.1161/STROKEAHA.112.653196.

Capone, F., Miccinilli, S., Pellegrino, G., Zollo, L., & Simonetti, D. (2017). Transcutaneous vagus nerve stimulation combined with robotic rehabilitation improves upper limb function after stroke. *Hindawi Neural Plasticity*, *2017*, 1–6.

Carey, J. R., Durfee, W. K., Bhatt, E., Nagpal, A., Weinstein, S. A., Anderson, K. M., & Lewis, S. M. (2007). Comparison of finger tracking versus simple movement training via telerehabilitation to alter hand function and cortical reorganization after stroke. *Neurorehabilitation and Neural Repair*, *21*, 216–232.

Chen, X., Liu, F., Yan, Z., Cheng, S., Liu, X., Li, H., & Li, Z. (2018). Therapeutic effects of sensory input training on motor function rehabilitation after stroke. *Medicine (United States)*. Available from https://doi.org/10.1097/MD.0000000000013387.

Cipriani, C., Segil, J. L., Clemente, F., Richard, R. F., & Edin, B. (2014). Humans can integrate feedback of discrete events in their sensorimotor control of a robotic hand. *Experimental Brain Research*. Available from https://doi.org/10.1007/s00221-014-4024-8.

Clemente, F., D'Alonzo, M., Controzzi, M., Edin, B., & Cipriani, C. (2015). Non-invasive, temporally discrete feedback of object contact and release improves grasp control of closed-loop myoelectric transradial prostheses. *IEEE Transactions on Neural Systems and Rehabilitation Engineering*, *24*(12), 1314–1322.

Connell, L. A., Lincoln, N. B., & Radford, K. A. (2008). Somatosensory impairment after stroke: Frequency of different deficits and their recovery. *Clinical Rehabilitation*. Available from https://doi.org/10.1177/0269215508090674.

Cordo, P., Lutsep, H., Cordo, L., Wright, W., Cacciatore, T., & Skoss, R. (2008). Assisted movement with enhanced sensation (AMES): Coupling motor and sensory to remediate motor deficits in chronic stroke patients. *Neurorehabilitation and Neural Repair*, *23*(1), 67–77.

Crowell, H. P., & Davis, I. S. (2011). Gait retraining to reduce lower extremity loading in runners. *Clinical Biomechanics*. Available from https://doi.org/10.1016/j.clinbiomech.2010.09.003.

Cuppone, A. V., Squeri, V., Semprini, M., Masia, L., & Konczak, J. (2016). Robot-assisted proprioceptive training with added vibro-tactile feedback enhances somatosensory and motor performance. *PLoS One*, *11*, 1–18.

Demain, S., Metcalf, C. D., Merrett, G. V., Zheng, D., & Cunningham, S. (2012). A narrative review on haptic devices: Relating the physiology and psychophysical properties of the hand to devices for rehabilitation in central nervous system disorders. *Disability and Rehabilitation: Assistive Technology*, *8*(3), 181–189.

Dobkin, B. H. (2004). Neurobiology of rehabilitation. *Annals of the New York Academy of Sciences*. Available from https://doi.org/10.1196/annals.1315.024.

Dobkin, B. H. (2008). Training and exercise to drive poststroke recovery. *Nature Clinical Practice Neurology*. Available from https://doi.org/10.1038/ncpneuro0709.

Doyle, S., Bennett, S., Fasoli, S. E., & McKenna, K. T. (2010). Interventions for sensory impairment in the upper limb after stroke. *Cochrane Database of Systematic Reviews*. Available from https://doi.org/10.1002/14651858.cd006331.pub2.

Duret, C., Grosmaire, A., & Krebs, H. I. (2019). Robot-assisted therapy in upper extremity hemiparesis: Overview of an evidence-based approach. *Frontiers in Neurology*, *10*(412), 1–8.

Elangovan, N., Yeh, I. L., Holst-Wolf, J., & Konczak, J. (2019). A robot-assisted sensorimotor training program can improve proprioception and motor function in stroke survivors. *IEEE*

International Conference on Rehabilitation Robotics. Available from https://doi.org/10.1109/ICORR.2019.8779409.

Emken, J. L., Benitez, R., Sideris, A., Bobrow, J. E., & Reinkensmeyer, D. J. (2007). Motor adaptation as a greedy optimization of error and effort. *Journal of Neurophysiology*. Available from https://doi.org/10.1152/jn.01095.2006.

Enders, L. R., Hur, P., Johnson, M. J., & Seo, N. J. (2013). Remote vibrotactile noise improves light touch sensation in stroke survivors' fingertips via stochastic resonance. *Journal of NeuroEngineering and Rehabilitation*. Available from https://doi.org/10.1186/1743-0003-10-105.

Enoch, J., McDonald, L., Jones, L., Jones, P. R., & Crabb, D. P. (2019). Evaluating whether sight is the most valued sense. *JAMA Ophthalmology*. Available from https://doi.org/10.1001/jamaophthalmol.2019.3537.

Ferrari, F., Shell, C. E., Thumser, Z. C., Clemente, F., Plow, E. B., Cipriani, C., & Marasco, P. D. (2021). Proprioceptive augmentation with illusory kinaesthetic sensation in stroke patients improves movement quality in an active upper limb reach-and-point task. *Frontiers in Neurorobotics*, *15*, 610673. Available from https://doi.org/10.3389/fnbot.2021.610673.

Follesa, P., Biggio, F., Gorini, G., Caria, S., Talani, G., Dazzi, L., Puligheddu, M., Marrosu, F., & Biggio, G. (2007). Vagus nerve stimulation increases norepinephrine concentration and the gene expression of BDNF and bFGF in the rat brain. *Brain Research*. Available from https://doi.org/10.1016/j.brainres.2007.08.045.

Franklin, D. W., & Wolpert, D. M. (2011). Computational mechanisms of sensorimotor control. *Neuron*, *72*, 425–442.

Fu, M. J., Knutson, J. S., & Chae, J. (2015). Stroke rehabilitation using virtual environments. *Physical Medicine and Rehabilitation Clinics of North America*. Available from https://doi.org/10.1016/j.pmr.2015.06.001.

Gandolfi, M. (2018). Quantification of upper limb motor recovery and EEG power changes after robot-assisted bilateral arm training in chronic stroke patients: A prospective pilot study. *Hindawi Neural Plasticity*, *2018*, 1–15.

Giggins, O. M., Persson, U. M. C., & Caulfield, B. (2013). Biofeedback in rehabilitation. *Journal of NeuroEngineering and Rehabilitation*. Available from https://doi.org/10.1186/1743-0003-10-60.

Goodwin, G. M., Mccloskey, D. I., & Matthews, P. B. C. (1972). The contribution of muscle afferents to keslesthesia shown by vibration induced illusions of movement and by the effects of paralysing joint afferents. *Brain*. Available from https://doi.org/10.1093/brain/95.4.705.

Hänggi, P. (2002). Stochastic resonance in biology: How noise can enhance detection of weak signals and help improve biological information processing. *ChemPhysChem*, *3*, 285–290.

Hayward, K. S., Barker, R. N., Brauer, S. G., Lloyd, D., Horsley, S. A., & Carson, R. G. (2013). SMART arm with outcome-triggered electrical stimulation: A pilot randomized clinical trial. *Topics in Stroke Rehabilitation*. Available from https://doi.org/10.1310/tsr2004-289.

Hochstenbach-Waelen, A., & Seelen, H. A. M. (2012). Embracing change: Practical and theoretical considerations for successful implementation of technology assisting upper limb training in stroke. *Journal of NeuroEngineering and Rehabilitation*. Available from https://doi.org/10.1186/1743-0003-9-52.

Hunter, S. M., Hammett, L., Ball, S., Smith, N., Anderson, C., Clark, A., Tallis, R., Rudd, A., & Pomeroy, V. M. (2011). Dose-response study of mobilisation and tactile stimulation therapy for the upper extremity early after stroke: A phase I trial. *Neurorehabilitation and Neural Repair*. Available from https://doi.org/10.1177/1545968310390223.

Imamizu, H., Miyauchi, S., Tamada, T., Sasaki, Y., Takino, R., Pütz, B., Yoshioka, T., & Kawato, M. (2000). Human cerebellar activity reflecting an acquired internal model of a new tool. *Nature*. Available from https://doi.org/10.1038/35003194.

Jenmalm, P., Schmitz, C., Forssberg, H., & Ehrsson, H. H. (2006). Lighter or heavier than predicted: Neural correlates of corrective mechanisms during erroneously programmed lifts. *Journal of Neuroscience*. Available from https://doi.org/10.1523/JNEUROSCI.5045-05.2006.

Johansson, B. B., Haker, E., Von Arbin, M., Britton, M., Långström, G., Terént, A., Ursing, D., Asplund, K., Berggren, A. L., Degerman, E., Flemström, B., Fondelius, A., Gissen, K., Lindqvist, L., Ljunglöf, A. C., Nylund, R., Sandström, C., Stålhandske, M., Terner, A., ... Törne, L. (2001). Acupuncture and transcutaneous nerve stimulation in stroke rehabilitation: A randomized, controlled trial. *Stroke*. Available from https://doi.org/10.1161/01.STR.32.3.707.

Johansson, R., & Edin, B. (1993). Predictive feed-forward sensory control during grasping and manipulation in man. *Biomedical Research*, *14*(4), 95–106.

Johansson, R. S., & Birznieks, I. (2004). First spikes in ensembles of human tactile afferents code complex spatial fingertip events. *Nature Neuroscience*. Available from https://doi.org/10.1038/nn1177.

Johansson, R. S., & Cole, K. J. (1992). Sensory-motor coordination during grasping and manipulative actions. *Current Opinion in Neurobiology*. Available from https://doi.org/10.1016/0959-4388(92)90139-C.

Johansson, R. S., & Flanagan, J. R. (2007). *Tactile sensory control of object manipulation in humans. The Senses: A Comprehensive Reference*. Elsevier Academic Press. Available from https://doi.org/10.1016/B978-012370880-9.00346-7.

Johansson, R. S., & Flanagan, J. R. (2009). Coding and use of tactile signals from the fingertips in object manipulation tasks. *Nature Reviews Neuroscience*, *10*, 345–359.

Johnson, S. H., Corballis, P. M., & Gazzaniga, M. S. (2001). Within grasp but out of reach: Evidence for a double dissociation between imagined hand and arm movements in the left cerebral hemisphere. *Neuropsychologia*. Available from https://doi.org/10.1016/S0028-3932(00)00096-8.

Kaczmarek, K. A., Webster, J. G., Bach-y-Rita, P., & Tompkins, W. J. (1991). Electrotactile and vibrotactile displays for sensory substitution systems. *IEEE Transactions on Biomedical Engineering*. Available from https://doi.org/10.1109/10.68204.

Katan, M., & Luft, A. (2018). Global burden of stroke. *Seminars in Neurology*, *38*, 208–211.

Kawato, M. (1999). Internal models for motor control and trajectory planning. *Current Opinion in Neurobiology*. Available from https://doi.org/10.1016/S0959-4388(99)00028-8.

Kawato, M., Kuroda, T., Imamizu, H., Nakano, E., Miyauchi, S., & Yoshioka, T. (2003). Internal forward models in the cerebellum: fMRI study on grip force and load force coupling. *Progress in Brain Research*. Available from https://doi.org/10.1016/S0079-6123(03)42013-X.

Kiper, P., Agostini, M., Luque-Moreno, C., Tonin, P., & Turolla, A. (2014). Reinforced feedback in virtual environment for rehabilitation of upper extremity dysfunction after stroke: Preliminary data from a randomized controlled trial. *BioMed Research International*. Available from https://doi.org/10.1155/2014/752128.

Kiper, P., Szczudlik, A., Agostini, M., Opara, J., Nowobilski, R., Ventura, L., Tonin, P., & Turolla, A. (2018). Virtual reality for upper limb rehabilitation in subacute and chronic stroke: A randomized controlled trial. *Archives of Physical Medicine and Rehabilitation*. Available from https://doi.org/10.1016/j.apmr.2018.01.023.

Krakauer, J. W. (2006). Motor learning: Its relevance to stroke recovery and neurorehabilitation. *Current Opinion in Neurology*. Available from https://doi.org/10.1097/01.wco.0000200544.29915.cc.

Laffont, I., Bakhti, K., Coroian, F., van Dokkum, L., Mottet, D., Schweighofer, N., & Froger, J. (2014). Innovative technologies applied to sensorimotor rehabilitation after stroke. *Annals of Physical and Rehabilitation Medicine*. Available from https://doi.org/10.1016/j.rehab.2014.08.007.

Lakshminarayanan, K., Lauer, A. W., Ramakrishnan, V., Webster, J. G., & Seo, N. J. (2015). Application of vibration to wrist and hand skin affects fingertip tactile sensation. *Physiological Reports*. Available from https://doi.org/10.14814/phy2.12465.

Lakshminarayanan, K., Wang, F., Webster, J. G., & Seo, N. J. (2017). Feasibility and usability of a wearable orthotic for stroke survivors with hand impairment. *Disability and Rehabilitation: Assistive Technology*. Available from https://doi.org/10.3109/17483107.2015.1111945.

Langhorne, P., Bernhardt, J., & Kwakkel, G. (2011). Stroke rehabilitation. *The Lancet*. Available from https://doi.org/10.1016/S0140-6736(11)60325-5.

Laver, K. E., Lange, B., George, S., Deutsch, J. E., Saposnik, G., & Crotty, M. (2017). Virtual reality for stroke rehabilitation. *Cochrane Database of Systematic Reviews*. Available from https://doi.org/10.1002/14651858.CD008349.pub4.

Lehrer, N., Chen, Y., Duff, M., Wolf, S. L., & Rikakis, T. (2011). Exploring the bases for a mixed reality stroke rehabilitation system, Part II: Design of interactive feedback for upper limb rehabilitation. *Journal of NeuroEngineering and Rehabilitation*. Available from https://doi.org/10.1186/1743-0003-8-54.

Lindsay, M. P., Norrving, B., Sacco, R. L., Brainin, M., Hacke, W., Martins, S., Pandian, J., & Feigin, V. (2019). World Stroke Organization (WSO): Global stroke fact sheet 2019. *International Journal of Stroke*. Available from https://doi.org/10.1177/1747493019881353.

Liu, L. Y., Li, Y., & Lamontagne, A. (2018). The effects of error-augmentation versus error-reduction paradigms in robotic therapy to enhance upper extremity performance and recovery post-stroke: A systematic review. *Journal of NeuroEngineering and Rehabilitation*. Available from https://doi.org/10.1186/s12984-018-0408-5.

Liu, W., Lipsitz, L., Montero-Odasso, M., Bean, J., Kerrigan, D., & Collins, J. (2002). Noise-enhanced vibrotactile sensitivity in older adults, patients with stroke, and patients with diabetic neuropathy. *Archives of Physical Medicine and Rehabilitation*, *83*, 171–176.

Liu, W., Mukherjee, M., Tsaur, Y., Kim, S.H., Liu, H., Natarajan, P., & Agah, A. (2009). Development and feasibility study of a sensory-enhanced robot-aided motor training in stroke rehabilitation. In: *Proceedings of the 31st annual international conference of the IEEE engineering in medicine and biology society: Engineering the future of biomedicine, EMBC 2009*. https://doi.org/10.1109/IEMBS.2009.5334526

Loh, M. N., Kirsch, L., Rothwell, J. C., Lemon, R. N., & Davare, M. (2010). Information about the weight of grasped objects from vision and internal models interacts within the primary motor cortex. *Journal of Neuroscience*. Available from https://doi.org/10.1523/JNEUROSCI.6207-09.2010.

Lum, P. S., Burgar, C. G., der Loos, M., Shor, P. C., Majmundar, M., & Yap, R. (2006). MIME robotic device for upper-limb neurorehabilitation in subacute stroke subjects: A follow-up study. *Journal of Rehabilitation Research and Development*, *43*, 631–642.

Marasco, P. D., Bourbeau, D. J., Shell, C. E., Granja-Vazquez, R., & Ina, J. G. (2017). The neural response properties and cortical organization of a rapidly adapting muscle sensory group response that overlaps with the frequencies that elicit the kinesthetic illusion. *PLoS One*. Available from https://doi.org/10.1371/journal.pone.0188559.

Marchal-Crespo, L., & Reinkensmeyer, D. J. (2009). Review of control strategies for robotic movement training after neurologic injury. *Journal of NeuroEngineering and Rehabilitation*. Available from https://doi.org/10.1186/1743-0003-6-20.

Markovic, M., Schweisfurth, M. A., Engels, L. F., Bentz, T., Wüstefeld, D., Farina, D., & Dosen, S. (2018). The clinical relevance of advanced artificial feedback in the control of a multi-functional myoelectric prosthesis. *Journal of NeuroEngineering and Rehabilitation*. Available from https://doi.org/10.1186/s12984-018-0371-1.

Masiero, S., Celia, A., Armani, M., & Rosati, G. (2006). A novel robot device in rehabilitation of post-stroke hemiplegic upper limbs. *Aging Clinical and Experimental Research*. Available from https://doi.org/10.1007/BF03324854.

Maulucci, R. A., & Eckhouse, R. H. (2001). Retraining reaching in chronic stroke with real-time auditory feedback. *NeuroRehabilitation*. Available from https://doi.org/10.3233/nre-2001-16306.

McDonnell, M. D., & Abbott, D. (2009). What is stochastic resonance? Definitions, misconceptions, debates, and its relevance to biology. *PLoS Computational Biology*, *5*(5), 1–9.

Mehrholz, J., Thomas, S., Werner, C., Kugler, J., Pohl, M., & Elsner, B. (2017). Electromechanical-assisted training for walking after stroke. *Cochrane Database of Systematic Reviews*. Available from https://doi.org/10.1002/14651858.CD006185.pub4.

Merrett, G.V., Metcalf, C.D., Zheng, D., Cunningham, S., Barrow, S., & Demain, S.H. (2011). Design and qualitative evaluation of tactile devices for stroke rehabilitation. In: *IET seminar on assisted living* (pp. 1–6).

Milot, M. H., Marchal-Crespo, L., Green, C. S., Cramer, S. C., & Reinkensmeyer, D. J. (2010). Comparison of error-amplification and haptic-guidance training techniques for learning of a timing-based motor task by healthy individuals. *Experimental Brain Research*. Available from https://doi.org/10.1007/s00221-009-2014-z.

Molier, B. I., Van Asseldonk, E. H. F., Hermens, H. J., & Jannink, M. J. A. (2010). Nature, timing, frequency and type of augmented feedback; Does it influence motor relearning of the hemiparetic arm after stroke? A systematic review. *Disability and Rehabilitation*. Available from https://doi.org/10.3109/09638281003734359.

Norman, S. L., Doxon, A. J., Gleeson, B. T., & Provancher, W. R. (2014). Planar hand motion guidance using fingertip skin-stretch feedback. *IEEE Transactions on Haptics*. Available from https://doi.org/10.1109/TOH.2013.2296306.

Oujamaaa, L., Relavea, I., Frogera, J., Mottetb, D., & Pelissiera, J. Y. (2009). Rehabilitation of arm function after stroke. Literature review. *Journal of Neuroscience Nursing*, *29*, 179–186.

Parker, J., Mountain, G., & Hammerton, J. (2011). A review of the evidence underpinning the use of visual and auditory feedback for computer technology in post-stroke upper-limb rehabilitation. *Disability and Rehabilitation: Assistive Technology*. Available from https://doi.org/10.3109/17483107.2011.556209.

Person, A. L. (2019). *Corollary discharge signals in the cerebellum. Biological psychiatry: Cognitive neuroscience and neuroimaging*. Elsevier. Available from https://doi.org/10.1016/j.bpsc.2019.04.010.

Peurala, S. H., Pitkänen, K., Sivenius, J., & Tarkka, I. M. (2002). Cutaneous electrical stimulation may enhance sensorimotor recovery in chronic stroke. *Clinical Rehabilitation*. Available from https://doi.org/10.1191/0269215502cr543oa.

Piron, L., Turolla, A., Agostini, M., Zucconi, C. S., Ventura, L., Tonin, P., & Dam, M. (2010). Motor learning principles for rehabilitation: A pilot randomized controlled study in post-stroke patients. *Neurorehabilitation and Neural Repair*. Available from https://doi.org/10.1177/1545968310362672.

Popa, L. S., & Ebner, T. J. (2019). Cerebellum, predictions and errors. *Frontiers in Cellular Neuroscience*. Available from https://doi.org/10.3389/fncel.2018.00524.

Proulx, M. J., Brown, D. J., Pasqualotto, A., & Meijer, P. (2014). Multisensory perceptual learning and sensory substitution. *Neuroscience and Biobehavioral Reviews*. Available from https://doi.org/10.1016/j.neubiorev.2012.11.017.

Quaney, B. M., Jianghua, H., Timberlake, G., Dodd, K., & Carr, C. (2010). Visuomotor training improves stroke-related ipsilesional upper extremity impairments. *Neurorehabilitation and Neural Repair*. Available from https://doi.org/10.1177/1545968309341646.

Quaney, B. M., Perera, S., Maletsky, R., Luchies, C. W., & Nudo, R. J. (2005). Impaired grip force modulation in the ipsilesional hand after unilateral middle cerebral artery stroke. *Neurorehabilitation and Neural Repair*. Available from https://doi.org/10.1177/1545968305282269.

Raghavan, A., & Shah, Z. A. (2015). Withania somnifera: A pre-clinical study on neuroregenerative therapy for stroke. *Neural Regeneration Research*. Available from https://doi.org/10.4103/1673-5374.152362.

Ripollés, P., Rojo, N., Grau-Sánchez, J., Amengual, J. L., Càmara, E., Marco-Pallarés, J., Juncadella, M., Vaquero, L., Rubio, F., Duarte, E., Garrido, C., Altenmüller, E., Münte, T. F., & Rodríguez-Fornells, A. (2016). Music supported therapy promotes motor plasticity in individuals with chronic stroke. *Brain Imaging and Behavior*. Available from https://doi.org/10.1007/s11682-015-9498-x.

Sanguineti, V., Casadio, M., Masia, L., Squeri, V., & Morasso, P. G. (2013). *Robot-assisted neurorehabilitation. Introduction to neural engineering for motor rehabilitation*. Wiley. Available from https://doi.org/10.1002/9781118628522.ch26.

Schabrun, S. M., & Hillier, S. (2009). Evidence for the retraining of sensation after stroke: A systematic review. *Clinical Rehabilitation*. Available from https://doi.org/10.1177/0269215508098897.

Schofield, J. S., Dawson, M. R., Carey, J. P., & Hebert, J. S. (2015). Characterizing the effects of amplitude, frequency and limb position on vibration induced movement illusions: Implications in sensory-motor rehabilitation. *Technology and Health Care*. Available from https://doi.org/10.3233/THC-140879.

Seo, N. J., Fischer, H. W., Bogey, R. A., Rymer, W. Z., & Kamper, D. G. (2011). Use of visual force feedback to improve digit force direction during pinch grip in persons with stroke: A pilot study. *Archives of Physical Medicine and Rehabilitation*. Available from https://doi.org/10.1016/j.apmr.2010.08.016.

Seo, N. J., Kosmopoulos, M. L., Enders, L. R., & Hur, P. (2014). Effect of remote sensory noise on hand function post stroke. *Frontiers in Human Neuroscience*. Available from https://doi.org/10.3389/fnhum.2014.00934.

Seo, N. J., Lakshminarayanan, K., Lauer, A. W., Ramakrishnan, V., Schmit, B. D., Hanlon, C. A., George, M. S., Bonilha, L., Downey, R. J., DeVries, W., & Nagy, T. (2019). Use of imperceptible wrist vibration to modulate sensorimotor cortical activity. *Experimental Brain Research*. Available from https://doi.org/10.1007/s00221-018-05465-z.

Shadmehr, R., & Mussa-Ivaldi, F. A. (1994). Adaptive representation of dynamics during learning of a motor task. *Journal of Neuroscience*. Available from https://doi.org/10.1523/jneurosci.14-05-03208.1994.

Shannon, G. F. (1976). A comparison of alternative means of providing sensory feedback on upper limb prostheses. *Medical & Biological Engineering*. Available from https://doi.org/10.1007/BF02478123.

Shea, C. H., Kovacs, A. J., & Panzer, S. (2011). The coding and inter-manual transfer of movement sequences. *Frontiers in Psychology*. Available from https://doi.org/10.3389/fpsyg.2011.00052.

Shimojo, S., & Shams, L. (2001). Sensory modalities are not separate modalities: Plasticity and interactions. *Current Opinion in Neurobiology*. Available from https://doi.org/10.1016/S0959-4388(00)00241-5.

Sigrist, R., Rauter, G., Riener, R., & Wolf, P. (2012). Augmented visual, auditory, haptic, and multimodal feedback in motor learning: A review. *Psychonomic Bulletin and Review*. Available from https://doi.org/10.3758/s13423-012-0333-8.

Sung, H. J., You, S. H., Hallett, M., Yun, W. C., Park, C. M., Cho, S. H., Lee, H. Y., & Kim, T. H. (2005). Cortical reorganization and associated functional motor recovery after virtual reality in patients with chronic stroke: An experimenter-blind preliminary study. *Archives of Physical Medicine and Rehabilitation*. Available from https://doi.org/10.1016/j.apmr.2005.04.015.

Takahashi, C. D., Der-Yeghiaian, L., Le, V., Motiwala, R. R., & Cramer, S. C. (2008). Robot-based hand motor therapy after stroke. *Brain, 131*, 425–437.

Thielman, G. (2010). Rehabilitation of reaching poststroke: A randomized pilot investigation of tactile versus auditory feedback for trunk control. *Journal of Neurologic Physical Therapy*. Available from https://doi.org/10.1097/NPT.0b013e3181efa1e8.

Tinga, A. M., Visser-Meily, J. M. A., van der Smagt, M. J., Van der Stigchel, S., van Ee, R., & Nijboer, T. C. W. (2016). Multisensory stimulation to improve low- and higher-level sensory deficits after stroke: A systematic review. *Neuropsychology Review*. Available from https://doi.org/10.1007/s11065-015-9301-1.

van Dijk, H., Jannink, M. J. A., & Hermens, H. J. (2005). Effect of augmented feedback on motor function of the affected upper extremity in rehabilitation patients: A systematic review of randomized controlled trials. *Journal of Rehabilitation Medicine*. Available from https://doi.org/10.1080/16501970510030165.

Van Vugt, F. T., Kafczyk, T., Kuhn, W., Rollnik, J. D., Tillmann, B., & Altenmüller, E. (2016). The role of auditory feedback in music-supported stroke rehabilitation: A single-blinded randomised controlled intervention. *Restorative Neurology and Neuroscience*, *34*(2), 297–311.

Wang, Q., Markopoulos, P., Yu, B., Chen, W., & Timmermans, A. (2017). Interactive wearable systems for upper body rehabilitation: A systematic review. *Journal of NeuroEngineering and Rehabilitation*. Available from https://doi.org/10.1186/s12984-017-0229-y.

Winstein, C. J. (1991). Knowledge of results and motor learning – Implications for physical therapy. *Physical Therapy*. Available from https://doi.org/10.1093/ptj/71.2.140.

Wolpert, D. M., Diedrichsen, J., & Flanagan, J. R. (2011). Principles of sensorimotor learning. *Nature Reviews Neuroscience*. Available from https://doi.org/10.1038/nrn3112.

Zhang, Y., Cai, J., Zhang, Y., Ren, T., Zhao, M., & Zhao, Q. (2016). Improvement in stroke-induced motor dysfunction by music-supported therapy: A systematic review and meta-analysis. *Scientific Reports*. Available from https://doi.org/10.1038/srep38521.

Zimmerman, A., Bai, L., & Ginty, D. D. (2014). The gentle touch receptors of mammalian skin. *Science*. Available from https://doi.org/10.1126/science.1254229.

Chapter 7

Targeted reinnervation for somatosensory feedback

Jacqueline S. Hebert[1] and Paul D. Marasco[2]
[1]*Division of Physical Medicine & Rehabilitation, Department of Medicine, Faculty of Medicine & Dentistry, University of Alberta 5-005 Katz Group Centre Edmonton, Alberta, AB, Canada,* [2]*Department of Biomedical Engineering, Lerner Research Institute, Cleveland Clinic, Cleveland, OH, United States*

ABSTRACT
Targeted reinnervation was introduced as a surgical procedure to improve myoelectric control signals for proximal upper limb amputation. From the first case studies in patients with shoulder disarticulation and transhumeral amputation, it was noted that in addition to increased motor control signals, there was restoration of sensations related to the missing arm and hand. The neural remapping of hand sensation has been explored as an avenue for tactile feedback related to the cutaneous reinnervation, and kinesthetic feedback related to the reinnervation of the muscle sensory fibers. Advances in robotic haptic devices and prosthesis sensorization have led to fully integrated wearable prosthesis systems that noninvasively provide real-time somatotopically and modality-matched feedback. These bidirectional neural human–machine interface systems provide insight into the underlying neural mechanisms of sensory processing and integration. Planning for targeted reinnervation should include a thorough assessment of potential motor sites, nerve management, and sensory restoration options.

Keywords: Amputation; reinnervation; targeted reinnervation; sensory restoration; sensory–motor integration

7.1 Introduction

Targeted reinnervation (TR) was introduced as a surgical procedure to improve myoelectric control signals for individuals with proximal upper limb amputation in 2003 (Kuiken et al., 2004). Early work with hyper-reinnervation models demonstrated the robust ability of amputated nerves to reinnervate new muscle sites and provide physiologically useful motor units for myoelectric control (Kuiken et al., 1995). The translation of this

Somatosensory Feedback for Neuroprosthetics. DOI: https://doi.org/10.1016/B978-0-12-822828-9.00003-4

technique to the human model was demonstrated with initial case reports for shoulder level disarticulation (Kuiken et al., 2004, 2007b) and transhumeral amputation (Dumanian et al., 2009; O'Shaughnessy et al., 2008). TR surgery is now widely employed after upper extremity amputation to increase neural control of complex myoelectric devices, and at other levels of amputation to reduce neuropathic pain both as a revision procedure (Dumanian et al., 2019; Kuiken et al., 2017) and at time of primary amputation (Cheesborough et al., 2014; Valerio et al., 2019). From the first case studies in patients with shoulder disarticulation, it was noted that, in addition to increased motor control signals, there was restoration of sensations related to the missing arm and hand (Kuiken et al., 2007a, 2007b). This neural remapping of hand sensation has been explored as an avenue for both tactile feedback related to the cutaneous reinnervation (Hebert et al., 2014a; Marasco et al., 2011; Schofield et al., 2020), and kinesthetic feedback related to the reinnervation of the muscle sensory fibers (Marasco et al., 2018).

7.2 Targeted reinnervation surgery and mechanisms of somatosensory restoration

A review of the surgical procedure and neuroplasticity of the brain is instructive to understanding the mechanisms of sensory restoration resulting from TR. At its core, reinnervation surgery involves redirecting a peripheral nerve that has been severed by the limb amputation to new biological muscle targets in the remaining proximal limb. The mixed nerve trunk severed by the amputation is coapted to a newly transected intact motor nerve branch to a muscle belly made redundant by the loss of the next functional joint; it is essential that sacrificing the chosen muscle target will not result in loss of remaining functional joint movement distal to the site. The target muscle is denervated from its original innervation, and the neurotrophic factors guiding the reinnervation of axons ensure that the newly coapted motor branches efficiently and competitively reinnervate the acutely denervated muscle (Tsao et al., 2012). The selected large donor mixed nerve that is transferred onto the smaller target motor branch ensures hyper-reinnervation of the muscle due to the high ratio of donor fibers to the smaller motor branch, with a shift in muscle fiber type (Bergmeister et al., 2019) and a corresponding high rate of muscle reinnervation success (Kuiken et al., 1995). These nerve transfers result in new pathways for triggering muscle contractions that are linked to the original motor intent, convenient for intuitive myoelectric prosthetic control (Hijjawi et al., 2006; Kuiken et al., 2009, 2016; Miller et al., 2008). When the person initiates the intent to activate the hand muscles, which have been amputated, those nerve signals travel from the primary motor cortex and descending motor axons to a new section of remaining muscle now transformed as a new end organ that serves as a biological amplifier for the neural control signals.

TABLE 7.1 Recommended targeted reinnervation nerve transfers for myoelectric control for upper limb amputation (Kuiken et al., 2017).

Amputation level	Transferred nerve	Muscle target
Shoulder disarticulation	Musculocutaneous n.	Clavicular head of the pectoralis
	Median n.	Sternal head of the pectoralis
	Radial n.	Thoracodorsal nerve (to latissimus)
	Ulnar n.	Lateral pectoralis minor or long thoracic nerve (to serratus anterior)
Transhumeral	Median n.	Biceps, short head
	Distal radial n.	Triceps, lateral head
	Ulnar n.	Brachialis
Transradial	Median n.	Flexor digitorum superficialis
	Ulnar n.	Flexor carpi ulnaris

Surgical TR techniques for motor reinnervation at various levels of amputation have been summarized by Kuiken et al. (2017). In the case of shoulder disarticulation, up to four nerve transfers have been described (Table 7.1), typically to provide signals for hand open/close, and elbow flexion/extension, although additional signals (such as for wrist control) can be extracted with more advanced signal processing. At the transhumeral level of amputation, in addition to the nerve transfers for the major hand nerves, care must be taken to preserve the original innervations of the long head of the biceps (musculocutanous nerve) and the long head of the triceps (proximal branches of the radial nerve) so that natural elbow flexion/extension muscle control signals are preserved. If there are insufficient muscle targets due to extensive injury to the arm or chest, pedicled serratus anterior flaps have been used as alternative muscle targets through free flap transfers (Lu et al., 2019). More creative nerve–muscle transfers have also been described in patients with brachial plexus injury by performing amputation in combination with the use of donor grafts from free-functioning muscle transfers (Aszmann et al., 2015, 2016).

The large transferred nerve trunks contain both motor and sensory fibers, and indeed a substantial portion of fibers in the major nerves of the arm carry sensory information. During the surgical procedure, denervation of sensory organs occurs in several ways. The skin is commonly thinned of subcutaneous and adipose tissue over the anticipated motor sites in order to improve the myoelectric signal detection (Kuiken et al., 2003, 2004). This procedure disrupts local cutaneous afferent connections to the skin, which renders them receptive for reinnervation. Sensory branches to the skin can be severed during the surgical approach or deliberately transected, and connected to the larger mixed nerves (which have substantial sensory afferents) by end-to-side or end-to-end coaptation. Separate fascicular bundles of the larger donor nerve can also be dissected out and rerouted to a specific cutaneous branch area (Hebert et al., 2014b). These approaches result in a spontaneous or deliberate reconnection of sensory pathways from the donor mixed nerve to cutaneous territories (see Fig. 7.1).

Two distinct sensory reinnervation phenomena have been realized after targeted motor and sensory reinnervation procedures, and exploited to

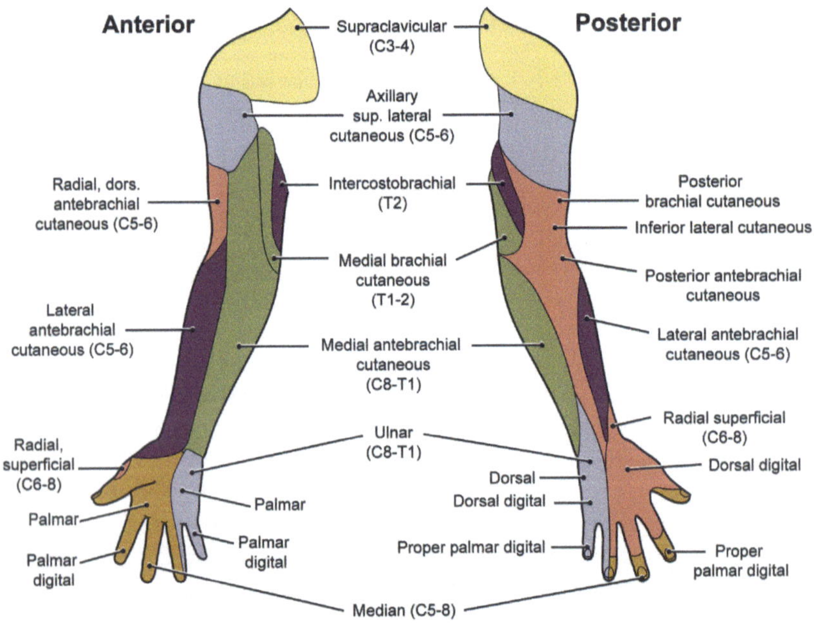

FIGURE 7.1 Peripheral nerve cutaneous territories of the upper limb and potential cutaneous branches for sensory reinnervation. For shoulder disarticulation, sensory reinnervation has been performed for the supraclavicular cutaneous nerve (ulnar donor nerve) and intercostobrachial cutaneous nerve (median donor nerve). For transhumeral amputation, end-to-end coaptation has been performed for the intercostobrachial cutaneous (T2) nerve (median donor) and the cutaneous branch of the axillary nerve (ulnar donor). Note the intercostobrachial cutaneous nerve originates from a branch of T2 and can have a variable course to the axillary area and upper medial arm, and may be indistinguishable from the proximal medial brachial cutaneous branch (T1−T2). The cutaneous branches to the skin are dissected and isolated during the surgery to identify the specific dermatomal location. *Courtesy of Wikimedia Commons, file (Gray's Anatomy 812 and 814.PNG; https://www.ncbi.nlm.nih.gov/books/NBK545249/). Creative Commons Attribution 4.0 International License (http://creativecommons.org/licenses/by/4.0/).*

provide somatosensory feedback within a prosthetic system—cutaneous tactile (touch transients and pressure) and kinesthetic (movement) sensation.

7.3 Cutaneous reinnervation: tactile sensation

Cutaneous reinnervation of new skin sites resulting from the reinnervation procedure has resulted in extensive tactile mapping of hand sensations over the reinnervated muscle transfer sites for shoulder disarticulation (Kuiken et al., 2004, 2007a; Marasco et al., 2009) and transhumeral amputation (Hebert et al., 2014a, 2016; Schofield et al., 2020) reinnervation procedures. Regardless of the specifics of the surgical approach, so long as there is resulting cutaneous denervation and nearby sensory fascicles, a representation of the hand map can be restored on the overlying skin (Hebert et al., 2014b). The hand map may show mixed findings from different nerve distributions (Kuiken et al., 2004) or discrete separated hand maps corresponding to expected peripheral nerve distributions (Hebert et al., 2014a). The hand map may be more precisely controlled when a specific cutaneous nerve is deliberately transected and coapted with a donor nerve (Hebert et al., 2014b).

7.3.1 Neurophysiology of cutaneous targeted sensory reinnervation

TR rebuilds a multifaceted sensory complex in the target skin. In the earliest surgical case, sensory nerves from the large transferred median, ulnar, radial, and musculocutaneous mixed nerves grew through the reinnervated muscle to establish connections in the skin (Kuiken et al., 2007a). Upon the realization of the potential for utilizing this compelling sensory phenomenon for prosthetic sensory feedback, the next patient underwent purposeful sensory nerve transfers in conjunction with motor nerve transfers (Kuiken et al., 2007a, 2007b). The reinnervated sensory terminals in the chest skin in these individuals appeared to show a return of modality-specific functionality. For example, individuals with targeted sensory reinnervation regained temperature and pain sensation at levels of near-normal perception thresholds (Kuiken et al., 2007a). Similarly, mechanosensation also returned to the reinnervated skin, which was shown to have vibration sensitivity at near-normal levels with psychophysical evidence of intact receptor modalities including Merkel cells and Pacinian corpuscles (Schultz et al., 2009).

Experiments with grating orientation thresholds applied to the target skin with grooved spatial discrimination domes revealed a number of important facets of post-reinnervation mechanosensory functionality. During these experiments, domes with parallel grooves of different widths are placed against the skin in one of two orientations that are perpendicular to each other. After each application the participant tells the experimenter which orientation they think the domes have been placed in. The distance between the grooves on the domes that can be oriented reliably (at significantly better than chance) is an indicator of the distance between mechanoreceptors. It has been shown that individuals with TR can resolve grating orientation on their reinnervated skin surface at the same level as their intact contralateral skin surface (Marasco et al., 2009). This result is interesting for two reasons. First, because it provides evidence that the hyper-reinnervation approach provides restoration of innervation density that is similar to the native contralateral skin (Phillips & Johnson, 1981). Second, the ability to discern grating orientation depends on intact communication between the somatosensory and visual cortices (Zangaladze et al., 1999). What is particularly compelling about this second result is that the original somatotopy of the fingers and hands that was established during development in these individuals is completely disrupted after their sensory nerves are surgically redirected to the new target skin sites. Although the reinnervated skin reflects the placement of the surgical nerve transfers (Kuiken et al., 2007a), the former placements of sensory areas within each nerve distribution now reside in a patchwork of different fingers, finger pads, and palmar and dorsal areas from both sides of the hand. This new random somatotopy would be expected to disrupt the ability to discern grating orientation for an individual with TR; however, that was shown not to be the case. Instead, the reinnervated chest skin performed at near-normal levels, with one participant showing reinnervated chest skin grating orientation thresholds that were equal to her normal contralateral side and also equal to the same skin area on a cohort of able-bodied individuals (Marasco et al., 2009). The reestablishment of functional sensory capacity occurred without training and suggests that mechanisms of neural plasticity accounted for the scrambled sensory nerves adjusted accordingly over time to rebuild a new representational somatotopy that contributed to effective sensory visualization of the grating orientations.

Brain processing appears to play a key role in the return of functional mechanoreceptive capacity to the reinnervated skin of individuals with TR. More cortical surface area in the touch-processing regions of the brain is devoted to skin that is involved in behaviorally relevant tactile exploration with the hands and fingers. Conversely, less cortical surface is devoted to body surfaces that are not as involved in touch, such as the arms, back, and chest (Sur et al., 1980). TR provides insight into sensory brain processing because the amputated nerves communicate with the behaviorally important hand areas of the brain; yet, their reconnected terminal receptors reside in skin areas that were formerly far less relevant to touch.

The effect of connecting a skin surface with fewer mechanoreceptors to a large behaviorally relevant sensory brain area can be visualized with a point localization test. In this test an experimenter uses a handheld probe to touch the skin of a participant twice in succession. The first touch is the reference touch and the second touch either occurs at the same place as the reference touch or at a random position on a grid at varying distances away from the reference touch (Weinstein, 1968). The participant reports whether the two touches were the same or different and the distances. Where they can reliably (at significantly better than chance) determine that they were touched in two different places is their point localization threshold. With this simple test it was revealed that TR appears to increase the functional mechanosensitivity of the target skin. The participants with TR had point localization thresholds that were far better than those for their contralateral native skin and equivalent to the palm of the hand for able-bodied individuals. Even though the receptor density of the target skin was relatively low, as revealed by the grating orientation threshold experiments described above, it had higher than normal sensory function. The increased sensory capacity of the reinnervated skin was likely a result of the increased processing power conferred to the receptors by being connected to a larger cortical area in the brain (Dinse, et al., 1997; Godde et al., 1996). This result was also reflected in animal studies of TR. Electrophysiological brain mapping revealed that large behaviorally relevant sensory processing forepaw areas in rats were extensively reactivated by relatively few sensory nerve terminals that were surgically transferred to the proximal shoulder skin (Marasco & Kuiken, 2010).

From grating orientation, to point localization, to vibration sensitivity, multiple lines of evidence point to the ability of TR to restore mechanosensory function. The ability to detect gradations in force at near-normal levels (Sensinger et al., 2009) further opened up the intriguing possibility of matching pressure and force sensation to that of a prosthetic hand, to provide somatotopically matched tactile feedback.

7.3.2 Functional use of cutaneous sensory reinnervated sites

The ability to use reinnervated cutaneous sites for functional matching to a prosthesis was investigated with mechanical haptic devices that provided tactile transients and pressure to the skin, taking advantage of the restored force gradation ability. An early study of participants with TR on the chest or upper arm demonstrated the normal ability to discriminate gradations in force over a range of 1–4 N (Sensinger et al., 2009) when touch and pressure feedback was provided to the reinnervated skin. In subsequent virtual experiments, somatotopically matched haptic feedback was provided along with active surface electromyography (sEMG) control of a virtual hand to enhance grip force control (Kim & Colgate, 2012). The authors utilized a miniature haptic device capable of delivering touch, pressure, shear, and temperature sensation, and placed it on the areas of the skin that corresponded to mapped sensations projected to the missing hand. The participants used the reinnervated muscle signals to operate a virtual hand via sEMG placed over the reinnervated muscles. Haptic feedback was linked to contact of the virtual hand with the virtual object, and task performance was measured by grip force and task completion time. The participants were provided with three conditions of feedback: pressure (normal to the skin), shear (tangential to the skin), or both feedback types simultaneously. They found that provision of either type of feedback alone was superior to delivering two feedback modalities simultaneously, and single modal feedback enhanced the ability to control grip force compared to no feedback. These early papers discussed the limitations of participants to simultaneously activate the muscles and feel the feedback at the same time, with movement of the muscle likely causing inadvertent changes in tactor contact against the skin. Subsequent experiments and technology improved on these limitations in two main ways.

One solution was to separate the sites of muscle control signals and tactile feedback to spatially distinct locations to avoid interference of muscle contraction with tactor contact. Functional experiments involving a physical robotic testing platform used active reinnervated muscle control with simultaneous somatotopic and modality-matched feedback (Hebert et al., 2014a). The participant had received targeted motor and sensory reinnervation with intraoperative fascicular dissection which allowed the cutaneous reinnervation territory to be selectively mapped to a cutaneous location separate from the muscle sites. Specifically, a fascicle of the ulnar nerve was rerouted to the cutaneous branch of the axillary nerve, and a fascicle of the median nerve was coapted to the cutaneous intercostal brachial cutaneous nerve. The main nerve bundles were preserved for standard muscle reinnervation. The resulting cutaneous reinnervation preserved isolated representations of the median innervated hand and digits to the chosen medial arm cutaneous territory, and ulnar hand representation to the lateral outer arm. Two tactile actuating devices were mounted over the areas corresponding to the index finger (medial arm) and small finger (lateral arm). Force sensors on the robotic gripper drove proportionally mapped pressure feedback supplied by the tactile actuator device to the corresponding skin sites. There was no overlap or interference with the four sEMG sites located anteriorly (elbow flexion and hand close sites) and posteriorly (hand open and elbow extend sites). The participant was able to actively operate two degrees of freedom of a physical robotic system with simultaneous haptic touch and pressure feedback. Provision of this feedback allowed accurate object stiffness discrimination of three levels, and object size identification via single versus dual tactor activation, relying on the distinct somatotopy of the two discretely innervated areas (median index digit and ulnar small digit; Hebert et al., 2014a).

A second method to improve the delivery of tactile feedback was to develop a force-following capability of the haptic device to adjust to changes in skin contact occurring with muscle contractions. Schofield et al. (2020) integrated multiple mechanotactile haptic feedback devices to provide touch and pressure to six reinnervated hand sites on the residuum, matched to prosthetic fingertip and palm sensors. Strain gauges and force-sensitive resistors

(FSRs) in each of the fingertips were used to duplicate and transfer relevant features of touch (rapid contact transients and variable proportional pressures) to small four-bar linkage touch tactor devices that pressed into the skin sites, maintaining constant contact through the closed-loop force-following mechanism. In a long-term take-home trial with the touch-integrated robotic prostheses, providing touch sensation had an immediate impact on how the users operated their prostheses. Block foraging and psychophysical tests specifically designed to assess the impact of touch feedback were applied. At the baseline evaluation, with touch sensation the participants took more time, but were more accurate in a block-handling discrimination task, and demonstrated more accurate attainment of moderate grip forces (Schofield et al., 2020). There were smaller changes reflective of learning and adaptation observed in follow-up after 1 year of use, demonstrating a dynamic relationship between users and sensate neural–machine interface prostheses with long-term use.

7.3.3 The importance of matched feedback: embodiment

The importance of touch feedback within a prosthetic system is not limited to functional improvements in control of the prosthesis. An important negative aspect of upper limb amputation is the sense that a prosthetic limb is a disconnected tool and not a part of the body as the natural arm once was. Our brains continuously distinguish our own body from the bodies of others and the objects around us. This basic sense of body ownership is mediated by touch and vision (visual–tactile integration). For example, when you see someone touch a hand and you feel that same touch, your brain knows it was your hand that was touched. When you see a touch to a hand but do not feel it, your brain knows that is not your hand. This concept of body ownership was first definitively explored with a cognitive manipulation called the rubber hand illusion in which healthy individuals can be made to experience a fake hand as if it was their own hand (Armel & Ramachandran, 2003; Botvinick & Cohen, 1998). To make this happen, the participant's real hand is hidden from their view behind a screen. A rubber hand is posed in front of the screen to visually replace their hidden hand. The experimenter then strokes both the rubber (seen) and real (hidden) hands simultaneously. After a brief time, the participants report feeling like the touch is coming from the rubber hand, which feels like a part of their body (embodied). This perceptual shift is involuntary and occurs even though the participant knows the rubber hand is fake. Sensory neural–machine interfaces established through TR provide a new fascinating avenue to further explore the constructs of body ownership through this illusion.

A typical prosthetic limb does not provide the user with tactile sensation from the hand, which breaks the normal links between vision and touch that the brain uses to feel whole. Targeted sensory reinnervation can restore natural sensations of the hand, that can be connected to prosthetic limbs through neuro–robotic interfaces so that the person can feel the touch of the device as if it is their own hand. Touch sensors on the fingertips of the prosthetic hand are connected to small robotic touch devices (tactors) that are positioned over the target skin where the individuals with TR feel their different fingers (Schofield et al., 2020). When touching objects with the prosthetic hand it feels like the touch is occurring in their missing hand. Providing touch for bionic limbs with TR provides the appropriate visual and tactile cues that the brain needs to establish ownership. Now when these individuals with amputation feel sensation from their prosthetic limbs, they are able to incorporate the device into their self-image. An earlier study by Marasco et al. provides evidence that with even a simple touch interface an individual with TR can be made to feel like a prosthetic limb was part of their body (Marasco et al., 2011). The acceptance of the mechanical device into their self-image was shown with a number of limb ownership markers including agreeing more strongly with statements of ownership of the prosthesis, elevated residual limb temperature, and embodiment-specific alterations in brain processing of tactile events (Marasco et al., 2011).

Important insights into the phenomenon of prosthetic limb ownership were discovered in a take-home trial with two participants who used sensory-integrated prostheses for 2 years with physiologically matched tactile feedback (Schofield et al., 2020). At baseline assessment after being provided with touch sensation within the actively controlled prosthetic system, the participants showed embodiment responses to the "touch-on" condition, as well as during situations where the timings of the touches were lagged by 500 ms and when their fingers were scrambled (by connecting the finger sensors on their prosthetic hands to the wrong finger sites on their reinnervated skin). This interesting finding provided evidence that the mechanisms of visual–tactile integration were over-inclusive of simple correlations between touch and vision that lacked appropriate spatial and temporal context; the brain appeared "eager" to incorporate any feedback at all into a feeling of ownership of the prosthesis. However, after 2 years of use of modality and somatotopically matched touch feedback in the take-home prosthesis, the embodiment feelings were present only for the fully contextually appropriate touch-on condition. This suggests that long-term exposure to matched touch feedback helps recovery from maladaptive neural plasticity effects and can fine tune and focus embodiment of a sensorized prosthesis (Schofield et al., 2020).

Evidence of altered cortical adaptation induced by the new peripheral connections has been confirmed in other studies. Upper limb amputation has been shown to result in functional reorganization and altered cortical mapping of both the motor and sensory cortices (Schwenkreis et al., 2001). This maladaptive cortical plasticity is thought to be associated with resulting phantom limb pain (Flor et al., 1995, 2006), although cortical plasticity can be altered and influenced by adaptive patterns of limb usage after amputation (Foell et al., 2014; Makin et al., 2013). Studies using high-density electroencephalography after TR have shown a normalization of cortical remapping of the motor representation (Chen et al., 2013). Neural remapping and brain reorganization occur not only in the motor pathways but also the somatosensory system. Evidence for a restored cortical pathway in TR participants with sensory transfer maps was demonstrated in the reorganization of the sensory hand map in the S1 hand area (Yao et al., 2015), supporting the existence of long-term sensory cortical plasticity. Functional MRI after TR has further shown normalization of primary motor cortex activation patterns similar to healthy controls when reinnervated participants attempt to activate a prosthetic hand, as well as similar patterns of activation of the primary sensory cortex in response to touch on the reinnervated skin (Serino et al., 2017). Normal local functional connectivity between the M1−S1 upper limb regions was found uniquely for participants with TR, and not in individuals with amputation and no reinnervation procedure. This adaptive neuroplasticity emphasizes the importance of restored multisensory integration not only for closed-loop control but also for somatosensory experiences.

7.3.4 Variability in cutaneous reinnervation

Long-term changes in hand maps after TR have been reported to show significant intersubject variability (Hebert et al., 2016), which is not surprising given the nature of the expected random competitive reinnervation. The impact of this variability across subjects however, is not likely of clinical importance, as for functional use, each patient would require an individualization of their cutaneous reinnervated site mapping in the same manner that individualization of myoelectric sites is required. There is some suggestion that the hand maps may attenuate over time, which may reflect a lack of functional use in participants studied up to 5 years after reinnervation (Hebert et al., 2016). However, there is alternate evidence that with regular stimulation of reinnervated skin matched to a functional prosthesis, the representations of a participant's missing hand may expand and strengthen over time (Schofield et al., 2020). In this long-term study, cortical adaptation to the extended exposure to touch feedback within a prosthetic system was corroborated by findings of more comparable symmetric processing of sensory information from the sound limb side and the prosthetic side when accessing the reinnervated skin sites.

Spontaneous tactile reinnervation with phantom hand mapping has also been noted by others in the absence of specific reinnervation techniques, however, it is notable that this has only been reported at the transradial level of amputation (Ehrsson et al., 2008; Wijk, Björkman, Antfolk, Björkman-Burtscher, & Rosén, 2016) but not at more proximal levels. In the presence of incidental reinnervation, the occurrence of cutaneous reinnervation directly over muscle sites may limit use, as previously noted. Therefore a planned sensory reinnervation procedure may provide more control over the otherwise random occurrence of mapping sensation that is projected to the missing hand.

7.3.5 State of technology for providing haptic feedback

Advances in robotic haptic devices and prosthesis sensorization have provided systems that are feasible for providing touch and pressure feedback on cutaneous skin sites for reliable long-term use within a prosthetic system (Kim & Colgate, 2012; Kim et al., 2010; Schofield et al., 2020). Several other systems have been developed for non-invasive sensory feedback explorations in intact arms and those with amputated limbs (Schofield et al., 2014; Wijk, Carlsson, Antfolk, Björkman, & Rosén, 2020), however here we focus on those demonstrated after TR with the capability for modality and somatotopic matching.

Kim and Colgate (2012) developed a multifunctional tactor that was capable of providing touch, pressure, vibration, shear force, and temperature sensation to the skin. When utilized in lab experiments with reinnervation participants for virtual experiments, the additional feedback modalities did not appear to be superior to a single modality (Kim & Colgate, 2012). A simpler two degree of freedom robotic, four-bar haptic pushing device (touch tactor) (Global, Fredericksburg, VA, United States) was also developed (Kim et al., 2010) and used for touch feedback in take-home socket trials (Schofield et al., 2020). The device is small enough that six tactors were able to be mounted within a shoulder disarticulation and a transhumeral socket to represent six discrete sites of feedback related to hand sensation, and did not interfere with the sEMG sites for motor control (Schofield et al., 2020). The tactors were placed on the reinnervated skin at matching sites for touch, and mapped to corresponding sensorized sites on the prosthetic hand. Other simpler and affordable mechanotactile devices using servo motors and cabling have been developed and tested in a lab setting (Hebert et al., 2014a), but require some refinement for take-home use to ensure robustness in a daily life setting (Schoepp et al., 2018).

Various methods of sensorizing prosthetic hands have been utilized in these experiments, mainly by retrofitting existing commercial devices with sensors. Methods include the use of strain gauges on the modified digits of a SensorHand Speed (Ottobock, Duderstadt, Germany), and FSRs on the cosmetic hand shell (Schofield et al., 2020), or capacitance sensors on the fingertips of the prosthetic hand (Schoepp et al., 2018). Full digit replacement of a commercial hand is available with the use of barometric sensors (Segil et al., 2019), and more manufacturers are releasing prosthetic hands with built-in sensor capacity. As advancements continue, there is anticipated to be an expanding arsenal of instrumented prosthetic hands and haptic feedback devices to facilitate providing various feedback methods.

7.4 Muscle sensory reinnervation: kinesthesia

Kinesthesia refers to the sensation of joint movement. This movement sense is crucial to coordinated arm and hand function and is an entirely distinct sensory modality from touch (contact, vibration, and force sensation). Despite the benefits of touch feedback when interacting with objects, without knowing where and how the hand is moving in space, motor control is left largely as an open-loop system that requires visual confirmation for positioning and confirmation of action. The lack of relevant movement feedback is a clear limitation of current prosthetic interventions and is one of the reasons that cable-driven body-powered prosthetic limbs continue to see widespread use because the user can intuitively feel the movement of their device through the excursion of the cables (Bongers et al., 2012; Lee, 1987).

TR has provided a new pathway for recent advances in proprioception-mediated closed-loop control of prosthetic upper limbs (Marasco et al., 2018). In addition to reinnervation of motor control targets in the muscles and sensory targets in the skin, TR also reassigns the sensory neural structure of the target muscles themselves. The proprioceptive muscle afferents for movement sensation can be accessed by using a vibration-induced perceptual illusion of movement called the kinesthetic illusion (Roll & Vedel, 1982). In this illusion, vibrating limb tendons at 70–115 Hz generate a joint-specific perception of movement, even though the joint crossed by the tendon is not physically changing position. Both single-joint and complex three-dimensional arm movements have been simulated in able-bodied individuals by inducing multijoint kinesthetic illusions (Thyrion & Roll, 2010). Parameters to optimize the illusion by altering the frequency and amplitude of vibration have also been explored (Schofield et al., 2015).

In participants with TR, it was found that the use of the kinesthetic illusion did not provide simple single-joint illusions as it does in able-bodied individuals. Instead the vibration-induced percepts were composed of highly complex and synergistic kinesthetic percepts of functional grip patterns. The kinesthetic illusory percepts were generated in the deep muscles without cutaneous tactile correspondence in the overlying skin. The illusionary movements had high functional relevance and when paired in a closed loop with active movement control, the kinesthetic feedback improved grip aperture control to levels rivaling able-bodied performance (Marasco et al., 2018). Furthermore, the kinesthetic feedback provided a cognitive sense of agency (a sense of authorship of one's movements) over the movements of the user's prosthesis, which is related to but separate from limb ownership (the sense of ownership of one's limbs) (Schofield et al., 2019). Interestingly, kinesthesia alone was not enough to provide a cognitive sense of prosthetic ownership, which suggests that pairing it to touch sensation will likely fully potentiate a broad sense of limb embodiment (the sense that one's limbs belong to them and are under their control). Importantly, the use of kinesthetic perception for closed-loop control in upper limb prosthetics was shown to effectively operate within the fitting constraints of a standard-of-care prosthetic limb.

7.5 Neuropathic pain

An unanticipated, but greatly beneficial, effect of TR has been its role in the treatment of neuropathic pain. Transected nerves typically form neuromas with or without associated scar tissue, that can lead to neuropathic pain and dysfunction in the residual limb. Initial reports of a retrospective cohort of patients that had undergone reinnervation surgery showed a reduction in post-amputation pain in the majority of patients with no new neuroma pain reported (Souza et al., 2014). A prospective cohort study of 51 patients found that pre-emptive TR surgery at the time of major limb amputation reduced phantom limb pain and symptomatic neuroma-related residual limb pain compared to the general incidence in a population of persons with amputation (Valerio et al., 2019). A subsequent randomized controlled trial comparing TR to standard neuroma excision showed that reinnervation significantly improved phantom limb pain and trended toward improved residual limb pain at 1 year follow-up, compared with conventional neurectomy (Dumanian et al., 2019). These finding have expanded the use of TR not only for motor control and somatosensory feedback, but for all levels of amputation as treatment or prevention of chronic amputation-related neuropathic pain.

7.6 Conclusion

With the expansion of TR techniques to all levels of amputation for nerve management, the original surgical principles and techniques require re-emphasis, in order to realize the full potential of sensory–motor integration for these additional levels of limb amputation. The ability to restore tactile cutaneous information within wearable standard-of-care sensorized prostheses for transhumeral and shoulder disarticulation levels of amputation has been demonstrated. Evaluation of patients after TR should routinely explore tactile thresholds along with new muscle control sites, to identify opportunities to provide matched sensory feedback early on in the rehabilitation process. Early application of closed-loop feedback could alter the natural progression of cortical reorganization and encourage feelings of ownership and embodiment of the prosthesis that could reduce rejection of the prosthesis and may address neuropathic phantom limb pain. Provision of kinesthetic feedback adds a layer of complexity with the need for greater miniaturization of vibration tactors capable of eliciting an illusion, however it could be further explored in the training phase to enhance motor control learning by closing the feedback loop with real-time feedback when learning muscle control signals. The assessment and follow-up of patients undergoing TR surgery should encompass a thorough plan to address neuroma management, muscle reinnervation for control signals, and sensory reinnervation for potential somatotopically matched closed-loop feedback.

References

Armel, K. C., & Ramachandran, V. S. (2003). Projecting sensations to external objects: Evidence from skin conductance response. *Proceedings Biological Science*, *270*(1523), 1499–1506. Available from https://doi.org/10.1098/rspb.2003.2364.

Aszmann, O. C., Roche, A. D., Salminger, S., Paternostro-Sluga, T., Herceg, M., Sturma, A., & Farina, D. (2015). Bionic reconstruction to restore hand function after brachial plexus injury: A case series of three patients. *Lancet*, *385*(9983), 2130–2132. Available from https://doi.org/10.1016/s0140-6736(14)61776-1.

Aszmann, O. C., Vujaklija, I., Roche, A. D., Salminger, S., Herceg, M., & Farina, D. (2016). Elective amputation and bionic substitution restore functional hand use after critical soft tissue injuries. *Scientific Reports*, *6*, 34960. Available from https://doi.org/10.1038/srep34960.

Bergmeister, K. D., Aman, M., Muceli, S., Vujaklija, I., Manzano-Szalai, K., Unger, E., et al. (2019). Peripheral nerve transfers change target muscle structure and function. *Science Advances*, *5*(1), eaau2956. Available from https://doi.org/10.1126/sciadv.aau2956.

Bongers, R. M., Kyberd, P. J., Bouwsema, H., Kenney, L. P. J., & Plettenburg, D. H. (2012). Bernstein's levels of construction of movements applied to upper limb prosthetics. *Journal of Prosthetics and Orthotics*, *24*, 67–76. Available from https://doi.org/10.1097/JPO.0b013e3182532419.

Botvinick, M., & Cohen, J. (1998). Rubber hands feel touch that eyes see. *Nature*, *391*, 756. Available from https://doi.org/10.1038/35784.

Cheesborough, J. E., Souza, J. M., Dumanian, G. A., & Bueno, R. A., Jr. (2014). Targeted muscle reinnervation in the initial management of traumatic upper extremity amputation injury. *Hand (N Y)*, *9*(2), 253–257. Available from https://doi.org/10.1007/s11552-014-9602-5.

Chen, A., Yao, J., Kuiken, T., & Dewald, J. P. (2013). Cortical motor activity and reorganization following upper-limb amputation and subsequent targeted reinnervation. *NeuroImage: Clinical*, *3*, 498–506. Available from https://doi.org/10.1016/j.nicl.2013.10.001.

Dinse, H. R., Godde, B., Hilger, T., Haupt, S. S., Spengler, F., & Zepka, R. (1997). Short-term functional plasticity of cortical and thalamic sensory representations and its implication for information processing. *NeuroImage*, *73*, 159–178. Available from https://doi.org/10.1097/prs.0b013e3181b038c9.

Dumanian, G. A., Ko, J. H., O'Shaughnessy, K. D., Kim, P. S., Wilson, C. J., & Kuiken, T. A. (2009). Targeted reinnervation for transhumeral amputees: Current surgical technique and update on results. *Plastic and Reconstructive Surgery*, *124*(3), 863–869. Available from https://doi.org/10.1097/prs.0b013e3181b038c9.

Dumanian, G. A., Potter, B. K., Mioton, L. M., Ko, J. H., Cheesborough, J. E., Souza, J. M., & Jordan, S. W. (2019). Targeted muscle reinnervation treats neuroma and phantom pain in major limb amputees: A randomized clinical trial. *Annals of Surgery*, *270*(2), 238–246. Available from https://doi.org/10.1097/sla.0000000000003088.

Ehrsson, H. H., Rosén, B., Stockselius, A., Ragnö, C., Köhler, P., & Lundborg, G. (2008). Upper limb amputees can be induced to experience a rubber hand as their own. *Brain*, *131*(12), 3443–3452. Available from https://doi.org/10.1093/brain/awn297.

Flor, H., Elbert, T., Knecht, S., Wienbruch, C., Pantev, C., & Taub, E. (1995). Phantom-limb pain as a perceptual correlate of cortical reorganization following arm amputation. *Nature, 375*(6531), 482–484. Available from https://doi.org/10.1038/375482a0.

Flor, H., Nikolajsen, L., & Staehelin Jensen, T. (2006). Phantom limb pain: A case of maladaptive CNS plasticity? *Nature Reviews Neuroscience, 7*((11), 873–881. Available from https://doi.org/10.1038/nrn1991.

Foell, J., Bekrater-Bodmann, R., Diers, M., & Flor, H. (2014). Mirror therapy for phantom limb pain: Brain changes and the role of body representation. *European Journal of Pain, 18*(5), 729–739. Available from https://doi.org/10.1002/j.1532-2149.2013.00433.x.

Godde, B., Spengler, F., & Dinse, H. R. (1996). Associative pairing of tactile stimulation induces somatosensory cortical reorganization in rats and humans. *NeuroReport, 8*(1), 281–285. Available from https://doi.org/10.1097/00001756-199612200-00056.

Hebert, J. S., Chan, K. M., & Dawson, M. R. (2016). Cutaneous sensory outcomes from three transhumeral targeted reinnervation cases. *Prosthetics Orthotics International, 40*(3), 303–310. Available from https://doi.org/10.1177/0309364616633919.

Hebert, J. S., Chan, K. M., Elzinga, K., Morhart, M. J., & Olson, J. L. (2014b). Updates in targeted sensory reinnervation for upper limb amputation. *Current Surgery Reports, 2*(3), 45–53. Available from https://doi.org/10.1007/s40137-013-0045-7.

Hebert, J. S., Olson, J. L., Morhart, M. J., Dawson, M. R., Marasco, P. D., & Chan, K. M. (2014a). Novel targeted sensory reinnervation technique to restore functional hand sensation after transhumeral amputation. *IEEE Transactions on Neural Systems and Rehabilitation Engineering, 22*(4), 765–773. Available from https://doi.org/10.1109/tnsre.2013.2294907.

Hijjawi, J. B., Kuiken, T. A., Lipschutz, R. D., Miller, L. A., Stubblefield, K. A., & Dumanian, G. A. (2006). Improved myoelectric prosthesis control accomplished using multiple nerve transfers. *Plastic and Reconstructive Surgery, 118*(7), 1573–1578. Available from https://doi.org/10.1097/01.prs.0000242487.62487.fb.

Kim, K., & Colgate, J. E. (2012). Haptic feedback enhances grip force control of sEMG-controlled prosthetic hands in targeted reinnervation amputees. *IEEE Transactions on Neural Systems and Rehabilitation Engineering, 20*(6), 798–805. Available from https://doi.org/10.1109/TNSRE.2012.2206080.

Kim, K., Colgate, J. E., Santos-Munne, J. J., Makhlin, A., & Peshkin, M. A. (2010). On the design of miniature haptic devices for upper extremity prosthetics. *IEEE/ASME Transactions on Mechatronics, 15*(1), 27–39. Available from https://doi.org/10.1109/TMECH.2009.2013944.

Kuiken, T. A., Barlow, A. K., Hargrove, L. J., & Dumanian, G. A. (2017). Targeted muscle reinnervation for the upper and lower extremity. *Techniques in Orthopaedics, 32*(2), 109–116. Available from https://doi.org/10.1097/bto.0000000000000194.

Kuiken, T. A., Childress, D. S., & Rymer, W. Z. (1995). The hyper-reinnervation of rat skeletal-muscle. *Brain Research, 676*(1), 113–123. Available from https://doi.org/10.1016/0006-8993(95)00102-v.

Kuiken, T. A., Dumanian, G. A., Lipschutz, R. D., Miller, L. A., & Stubblefield, K. A. (2004). The use of targeted muscle reinnervation for improved myoelectric prosthesis control in a bilateral shoulder disarticulation amputee. *Prosthetics and Orthotics International, 28*(3), 245–253. Available from https://doi.org/10.3109/03093640409167756.

Kuiken, T. A., Li, G., Lock, B. A., Lipschutz, R. D., Miller, L. A., Stubblefield, K. A., & Englehart, K. B. (2009). Targeted muscle reinnervation for real-time myoelectric control of multifunction artificial arms. *JAMA, 301*(6), 619–628. Available from https://doi.org/10.1001/jama.2009.116.

Kuiken, T. A., Lowery, M. M., & Stoykonv, S. (2003). The effect of subcutaneous fat on myoelectric signal amplitude and cross-talk. *Prosthetics and Orthotics International*, *2*(1), 48–54. Available from https://doi.org/10.3109/03093640309167976.

Kuiken, T. A., Marasco, P. D., Lock, B. A., Harden, R. N., & Dewald, J. P. A. (2007a). Redirection of cutaneous sensation from the hand to the chest skin of human amputees with targeted reinnervation. *Proceedings of the National Academy of Sciences of the United States of America*, *104*(50), 20061–20066. Available from https://doi.org/10.1073/pnas.0706525104.

Kuiken, T. A., Miller, L. A., Lipschutz, R. D., Lock, B. A., Stubblefield, K., Marasco, P. D., & Dumanian, G. A. (2007b). Targeted reinnervation for enhanced prosthetic arm function in a woman with a proximal amputation: A case study. *Lancet*, *369*(9559), 371–380. Available from https://doi.org/10.1016/s0140-6736(07)60193-7.

Kuiken, T. A., Miller, L. A., Turner, K., & Hargrove, L. J. (2016). A comparison of pattern recognition control and direct control of a multiple degree-of-freedom transradial prosthesis. *IEEE Journal of Translational Engineering in Health and Medicine*, *4*. Available from https://doi.org/10.1109/jtehm.2016.2616123, Article 2100508.

Lee, R. E. (1987). Reassessing myoelectric control - is it time to look at alternatives. *Canadian Medical Association Journal*, *136*(5), 467–469. Available from http://www.ncbi.nlm.nih.gov/pmc/articles/pmc1491640/.

Lu, D., Myers, H., & Bruscino-Raiola, F. (2019). Pedicled serratus anterior flap as an alternative muscle target for targeted muscle reinnervation in transhumeral amputees. *Journal of Hand Surgery-American*, *44*(11). Available from https://doi.org/10.1016/j.jhsa.2019.07.008, 997.E1.

Makin, T. R., Cramer, A. O., Scholz, J., Hahamy, A., Slater, D. H., & Johansen-Berg, H. (2013). Deprivation-related and use-dependent plasticity go hand in hand. *eLife*, *2*, e01273. Available from https://doi.org/10.7554/elife.01273.

Marasco, P. D., Hebert, J. S., Sensinger, J. W., Shell, C. E., Schofield, J. S., Thumser, Z. C., & Orzell, B. M. (2018). Illusory movement perception improves motor control for prosthetic hands. *Science Translational Medicine*, *10*(432), eaao6990. Available from https://doi.org/10.1126/scitranslmed.aao6990.

Marasco, P. D., Kim, K., Colgate, J. E., Peshkin, M. A., & Kuiken, T. A. (2011). Robotic touch shifts perception of embodiment to a prosthesis in targeted reinnervation amputees. *Brain*, *134*(3), 747–758. Available from https://doi.org/10.1093/brain/awq361.

Marasco, P. D., & Kuiken, T. A. (2010). Amputation with median nerve redirection (targeted reinnervation) reactivates forepaw barrel subfield in rats. *Journal of Neuroscience*, *30*(47), 16008–16014. Available from https://doi.org/10.1523/jneurosci.4044-10.2010.

Marasco, P. D., Schultz, A. E., & Kuiken, T. A. (2009). Sensory capacity of reinnervated skin after redirection of amputated upper limb nerves to the chest. *Brain*, *132*(6), 1441–1448. Available from https://doi.org/10.1093/brain/awp082.

Miller, L. A., Stubblefield, K. A., Lipschutz, R. D., Lock, B. A., & Kuiken, T. A. (2008). Improved myoelectric prosthesis control using targeted reinnervation surgery: A case series. *IEEE Transactions on Neural Systems and Rehabilitation Engineering*, *16*(1), 46–50. Available from https://doi.org/10.1109/tnsre.2007.911817.

O'Shaughnessy, K. D., Dumanian, G. A., Lipschutz, R. D., Miller, L. A., Stubblefield, K., & Kuiken, T. A. (2008). Targeted reinnervation to improve prosthesis control in transhumeral amputees. A report of three cases. *Journal of Bone and Joint Surgery-American Volume*, *90*(2), 393–400. Available from https://doi.org/10.2106/jbjs.g.00268.

Phillips, J. R., & Johnson, K. O. (1981). Tactile spatial resolution. II. Neural representation of Bars, edges, and gratings in monkey primary afferents. *Journal of Neurophysiology*, *46*(6), 1192–1203. Available from https://doi.org/10.1152/jn.1981.46.6.1192.

Roll, J. P., & Vedel, J. P. (1982). Kinaesthetic role of muscle afferents in man, studied by tendon vibration and microneurography. *Experimental Brain Research*, *47*(2), 177–190. Available from https://doi.org/10.1007/bf00239377.

Schoepp, K. R., Dawson, M. R., Schofield, J. S., Carey, J. P., & Hebert, J. S. (2018). Design and integration of an inexpensive wearable mechanotactile feedback system for myoelectric prostheses. *IEEE Journal of Translational Engineering in Health and Medicine*, *6*, 2100711. Available from https://doi.org/10.1109/jtehm.2018.2866105.

Schofield, J. S., Dawson, M. R., Carey, J. P., & Hebert, J. S. (2015). Characterizing the effects of amplitude, frequency and limb position on vibration induced movement illusions: Implications in sensory-motor rehabilitation. *Technology and Health Care*, *23*(2), 129–141. Available from https://doi.org/10.3233/thc-140879.

Schofield, J. S., Evans, K. R., Carey, J. P., & Hebert, J. S. (2014). Applications of sensory feedback in motorized upper extremity prosthesis: A review. *Expert Review of Medical Devices*, *11*(5), 499–511. Available from https://doi.org/10.1586/17434440.2014.929496.

Schofield, J. S., Shell, C. E., Beckler, D. T., Thumser, Z. C., & Marasco, P. D. (2020). Long-term home-use of sensory-motor-integrated bidirectional bionic prosthetic arms promotes functional, perceptual, and cognitive changes. *Frontiers in Neuroscience*, *14*, 120. Available from https://doi.org/10.3389/fnins.2020.00120.

Schofield, J. S., Shell, C. E., Thumser, Z. C., Beckler, D. T., Nataraj, R., & Marasco, P. D. (2019). Characterization of the sense of agency over the actions of neural-machine interface-operated prostheses. *Journal of Visualized Experiments*, *143*, e58702. Available from https://doi.org/10.3791/58702.

Schultz, A. E., Marasco, P. D., & Kuiken, T. A. (2009). Vibrotactile detection thresholds for chest skin of amputees following targeted reinnervation surgery. *Brain Research*, *125*, 121–129. Available from https://doi.org/10.1016/j.brainres.2008.11.039.

Schwenkreis, P., Witscher, K., Janssen, F., Pleger, B., Dertwinkel, R., Zenz, M., & Tegenthoff, M. (2001). Assessment of reorganization in the sensorimotor cortex after upper limb amputation. *Clinical Neurophysiology*, *112*(4), 627–635. Available from https://doi.org/10.1016/s1388-2457(01)00486-2.

Segil, J., Patel, R., Klingner, J., Weir, R. F. F., & Correll, N. (2019). Multi-modal prosthetic fingertip sensor with proximity, contact, and force localization capabilities. *Advances in Mechanical Engineering*, *11*(4). Available from https://doi.org/10.1177/1687814019844643.

Sensinger, J. W., Schultz, A. E., & Kuiken, T. A. (2009). Examination of force discrimination in human upper limb amputees with reinnervated limb sensation following peripheral nerve transfer. *IEEE Transactions on Neural Systems and Rehabilitation Engineering*, *17*(5), 438–444. Available from https://doi.org/10.1109/tnsre.2009.2032640.

Serino, A., Akselrod, M., Salomon, R., Martuzzi, R., Blefari, M. L., Canzoneri, E., & Blanke, O. (2017). Upper limb cortical maps in amputees with targeted muscle and sensory reinnervation. *Brain*, *140*(11), 2993–3011. Available from https://doi.org/10.1093/brain/awx242.

Souza, J. M., Cheesborough, J. E., Ko, J. H., Cho, M. S., Kuiken, T. A., & Dumanian, G. A. (2014). Targeted muscle reinnervation: A novel approach to postamputation neuroma pain. *Clinical Orthopaedics and Related Research*, *472*(10), 2984–2990. Available from https://doi.org/10.1007/s11999-014-3528-7.

Sur, M., Merzenich, M. M., & Kaas, J. H. (1980). Magnification, receptive-field area, and hypercolumn size in area-3b and area-1 of somatosensory cortex in owl monkeys. *Journal of Neurophysiology*, *44*(2), 295–311. Available from https://doi.org/10.1152/jn.1980.44.2.295.

Thyrion, C., & Roll, J. P. (2010). Predicting any arm movement feedback to induce three-dimensional illusory movements in humans. *Journal of Neurophysiology*, *104*(2), 949–959. Available from https://doi.org/10.1152/jn.00025.2010.

Tsao, B., Boulis, N., Bethoux, F., & Murray, B. (2012). Trauma of the nervous system, peripheral nerve trauma. In R. B. Daroff, G. M. Fenichel, J. Jankovic, & J. Mazziotta (Eds.), *Bradley's neurology in clinical practice* (pp. 984–1001). Philadelphia: Elsevier.

Valerio, I. L., Dumanian, G. A., Jordan, S. W., Mioton, L. M., Bowen, J. B., West, J. M., ... Potter, B. K. (2019). Preemptive treatment of phantom and residual limb pain with targeted muscle reinnervation at the time of major limb amputation. *Journal of the American College of Surgeons*, *228*(3), 217–226. Available from https://doi.org/10.1016/j.jamcollsurg.2018.12.015.

Weinstein, S. (1968). Intensive and extensive aspects of tactile sensitivity as a function of body part, sex and laterality. In D. R. Kenshalo (Ed.), *The skin senses* (pp. 195–222). Springfield, IL: Thomas.

Wijk, U., Björkman, A., Antfolk, C., Björkman-Burtscher, I., Rosén, B. Superior Tactile Discrimination in the Phantom Hand Map in Forearm Amputees. (2016). Hand, 11(1_suppl), 131S-131S. Available from https://doi.org/10.1177/1558944716660555ja.

Wijk, U., Carlsson, I. K., Antfolk, C., Björkman, A., & Rosén, B. (2020). Sensory Feedback in Hand Prostheses: A Prospective Study of Everyday Use. *Frontiers in Neuroscience*, *14*. Available from https://doi.org/10.3389/fnins.2020.00663.

Yao, J., Chen, A., Kuiken, T., Carmona, C., & Dewald, J. (2015). Sensory cortical re-mapping following upper-limb amputation and subsequent targeted reinnervation: A case report. *NeuroImage: Clinical*, *8*, 329–336. Available from https://doi.org/10.1016/j.nicl.2015.01.010.

Zangaladze, A., Epstein, C. M., Grafton, S. T., & Sathian, K. (1999). Involvement of visual cortex in tactile discrimination of orientation. *Nature*, *401*(6753), 587–590. Available from https://doi.org/10.1038/44139.

Chapter 8

Transcranial electrical stimulation for neuromodulation of somatosensory processing

Sacit Karamursel and Ezgi Tuna Erdogan
Department of Physiology, School of Medicine, Koç Universitesi, Istanbul, Turkey

ABSTRACT
Transcranial electrical stimulation is a promising neuromodulation technique that has gained interest in the last 20 years after Nitsche and Paulus showed the polarity-specific cortical modulation effect of the weak direct current stimulation over the scalp in 2000. In the following years, new electrical stimulation techniques were introduced, such as alternative current stimulation, random noise stimulation, and pulse current stimulation. According to the literature, transcranial electrical stimulation may be a potential tool to modulate inhibitory circuits to improve the sensory discrimination function and reduce the perception threshold. Furthermore, studies investigating the modulation of multisensory integration support that modulation of sensory networks may increase the perception of body ownership and sensory perception in patients with the prosthesis. Based on the current findings, the electrical modulation of somatosensory processing should be further investigated regarding the type of stimulation, target cortical areas, and optimal parameters of the application.

Keywords: Transcranial direct current stimulation; transcranial alternating current stimulation; transcranial random noise stimulation; transcranial pulse current stimulation; cortical neuromodulation; sensory processing

8.1 Introduction

Transcranial electrical stimulation (TES) is a noninvasive neuromodulation method that can inhibit or facilitate the activity in targeted cortical neurons. There is a growing body of literature on improving motor and cognitive functions induced by various types of TES. However, there is a limited number of well-controlled studies for sensory modulation by TES, and the results of those

Somatosensory Feedback for Neuroprosthetics. DOI: https://doi.org/10.1016/B978-0-12-822828-9.00004-6

studies are conflicting and inconsistent compared with those from the motor and cognitive function modulation. Today, research in this field continues to grow with the aim of understanding the neurophysiology of perception and finding the optimal modulation method for the rehabilitation of patients.

8.2 Chapter objectives

In this chapter, we review the types of TES and their use in the modulation of sensory processing. We also discuss the results of experimental studies in humans and the future of TES in neuroprosthetics.

8.3 Methods of transcranial electrical stimulation and mechanism of action

The use of electricity in brain stimulation has a very long history, with ups and downs starting from the use of electric fish for medical cures in Egyptian and Roman periods (Finger & Piccolino, 2011). The evolution of electrical brain stimulation proceeded in two different pathways. One branch was invasive direct cortical stimulation studies, and the other was noninvasive transcranial applications. The latter took a huge step with the studies of Volta, Galvani, and Aldini in the 1700s and is now in the form of modern transcranial brain stimulation techniques. The first study on the direct cortical electrical stimulation of the exposed human brain was carried out by a surgeon in 1874 (Harris & Almerigi, 2009). On the other hand, the application of electricity to the outer surface of the scalp failed to induce any body responses because of the current flow on the skin due to the high resistance of the skull. Further studies showed that the intensity of current in the transcranial method needs to be higher than direct cortical stimulation to reach the cortical neurons. In 1980, Merton and Morton applied high-voltage stimulation on the scalp surface and succeeded in eliciting a contralateral body response (Merton & Morton, 1980). Today, neurosurgeons use both intraoperative direct cortical stimulation and transcranial high-voltage stimulation to map eloquent areas and monitor other cortical functions in their routine practice. Therefore the term "TES" indicates both intraoperative high-voltage use in neurosurgery and low-current use in cortical neuromodulation techniques which originated from the very early studies of the 1700s. The aim of use in neurosurgery is to induce action potentials (i.e., direct neural activation) while low-intensity TES is used to induce an increase or decrease in the excitability of neurons under the electrodes without triggering any action potentials directly. Another goal of TES from a clinical perspective is to induce neuroplasticity with repeated sessions for therapeutic use. In this chapter, we focus on low-intensity TES in neuromodulation.

Animal studies in the 1960s showed both increasing and decreasing activity in cortical neurons induced by direct current polarization (Bindman et al., 1962).

However, currently transcranial electrical neuromodulation in humans has come to the forefront after the study of Nitsche and Paulus in 2000 (Nitsche & Paulus, 2000). They applied a weak direct current from the scalp between two sponge electrodes placed over the human motor cortex. The excitability of cortex was measured by transcranial *magnetic* stimulation during and after each electrical stimulation. They found an increase in cortical excitability during and after anodal stimulation, and a decrease after cathodal stimulation. Furthermore, the changes in the excitability lasted for several minutes after the stimulation, and they showed a correlation between the intensity of current and the change in cortical excitability. These results were the first evidence of the polarity-specific effect and the potential of a long-lasting effect in human electrical cortical neuromodulation.

There are two distinct effects of transcranial direct current stimulation (tDCS) regarding the duration of the desired change; a short-term effect and a long-term effect. The short-term effect indicates acute changes measured immediately after stimulation, and lasts for hours. Even though the exact mechanism of tDCS is not clear, it is thought that the basis of the short-term effect is related to the membrane polarization induced during stimulation. The current flow between anode and cathode electrodes leads to depolarization in the resting membrane potential of cortical neurons under the anode. Contrarily, it leads to hyperpolarization in the resting membrane potential of cortical neurons under the cathode (Fig. 8.1). The level of depolarization stays below the threshold for action potential generation in neurons. That subthreshold polarization leads to changes in ion channels on the membrane,

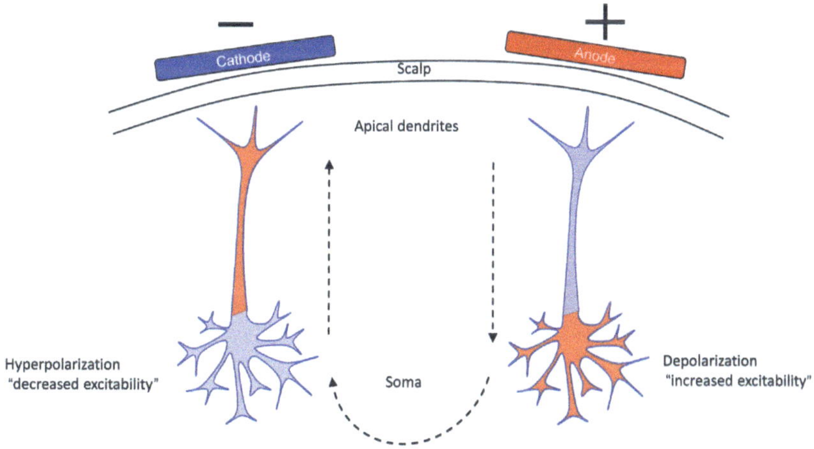

FIGURE 8.1 The somatic subthreshold depolarization of cortical neurons below the anode electrode and somatic hyperpolarization of cortical neurons below the cathode electrode. Depolarization indicates increased excitability, while hyperpolarization indicates decreased excitability.

firing rate, and modulates the spontaneous activity of relevant cortical neurons and synaptic transmission accordingly (Lefaucheur & Wendling, 2019).

The long-term effect means prolonged change, which lasts for days/months and carries a significant potential for therapeutic use. The possible mechanisms underlying long-term effects include strengthening of the synaptic transmission, modulation of receptor activity, changes in neurotransmitter release, plasticity, and structural changes in neurons and non-neuronal tissue. It has been clearly shown that there is an involvement of glutamatergic *N*-methyl-D-aspartate receptors for the long-lasting effect (Liebetanz et al., 2002). To date, there have been various theories about the mechanisms of tDCS; however, all those theories are secondary to the polarization of neurons and non-neuronal tissue basically (Jackson et al., 2016; Lefaucheur & Wendling, 2019).

Since there are multiple technical and practical parameters associated with TES, a particular nomenclature has eventually been established in time. First, tDCS will be discussed in detail; and then, other stimulation techniques will be explained based on that.

8.3.1 Transcranial direct current stimulation

Today, tDCS is a more common and better established electrical stimulation method compared to other TES techniques for neuromodulation. It is applied from the scalp surface by two saline-soaked sponge electrodes: one anode and one cathode electrode. The currently used sizes of electrodes are 5×7 cm (35 cm^2). The term "electrode montage" indicates the exact location of electrodes over the head, which depends on the site of desired cortical stimulation. If both electrodes are placed over the head, it is called a bicephalic montage. If one of the electrodes (which is also called the "return" electrode) is placed out of the head, such as arm, shoulder, etc., it is called an extracephalic montage (Fig. 8.2). The electrodes are directly connected to a battery-driven stimulator, which is developed specifically for tDCS (Fig. 8.3). The current is a direct current without any polarity change during the stimulation period. Even though studies show the safety of higher currents in human applications, the upper limit for current intensity in common use remains 2 mA, and session durations these change between 20 and 30 minutes. The number of sessions depends on the aim of the research. If a prolonged therapeutic effect is desired, then the repetition of sessions is required to be higher, such as five times a week or more.

Antal et al. published safety recommendations for tDCS in 2017. In that paper, it is indicated that the safety limit for low-intensity "conventional" TES is <4 mA, up to 60 minutes duration per day (Antal et al., 2017). The polarity-dependent effect of tDCS for the application at particular cortical sites is shown repeatedly in the literature, and a group of European experts published evidence-based guidelines for the therapeutic use of tDCS in 2017 (Lefaucheur et al., 2017).

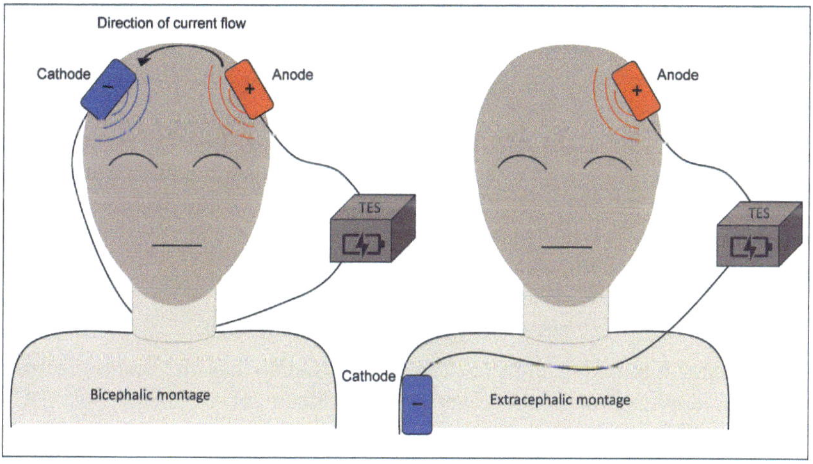

FIGURE 8.2 The bicephalic and extracephalic montages. Red indicates the anode electrode, and blue indicates the cathode electrode. The current flow is shown between the electrodes. *TES*, transcranial electrical stimulation.

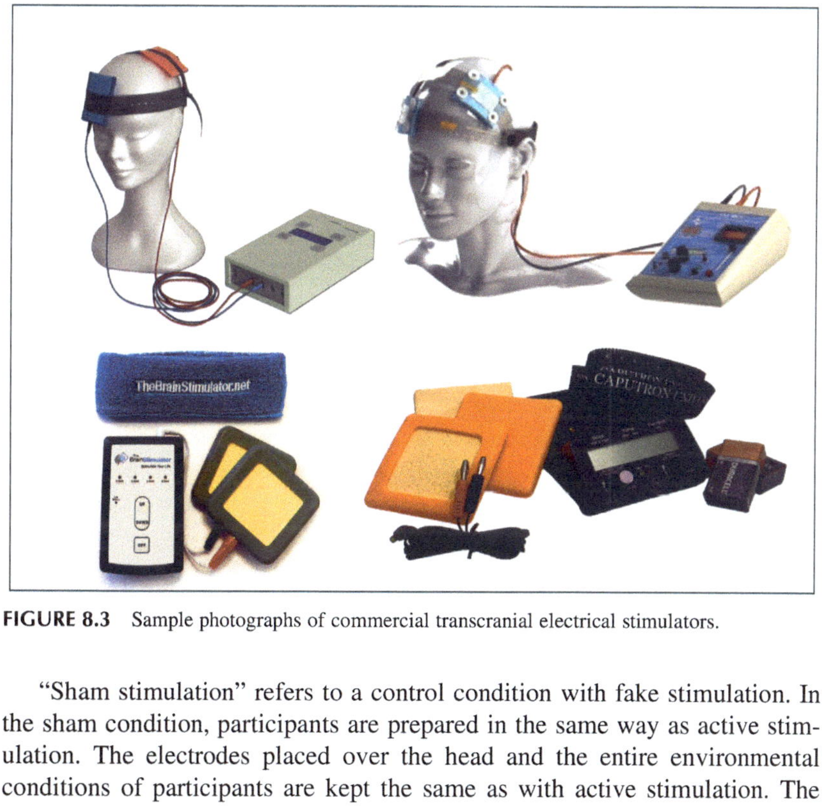

FIGURE 8.3 Sample photographs of commercial transcranial electrical stimulators.

"Sham stimulation" refers to a control condition with fake stimulation. In the sham condition, participants are prepared in the same way as active stimulation. The electrodes placed over the head and the entire environmental conditions of participants are kept the same as with active stimulation. The electric current starts to increase step by step to the maximum intensity of application (usually 2 mA), and, depending on the sham type, it stays at that level for a few seconds, and then it ramps down to 0 mA. These fade-in and fade-out periods last 30 seconds in common use for both active and sham stimulations (Fig. 8.4). The aim of the sham condition is to induce the same itching-tingling feeling generated by the current flow on the skin. Hence, participants cannot distinguish the active stimulation from sham stimulation, which prevents a possible placebo effect.

8.3.2 Transcranial alternating current stimulation

Transcranial alternating current stimulation (tACS) has the same electrode montage, application practice, and general electrical parameters as tDCS, except for the waveform of the current (Antal et al., 2008b). In tACS, the current is a sinusoidal wave at a particular frequency, which oscillates between positive and negative polarities (Fig. 8.4). In other words, the anode and cathode electrodes change their polarity in the mid-time of every sine wave. Since the direction of the current flow is not stable and alternates at each sine wave, it does not induce a steady depolarization on the membranes of the cortical neurons. Instead, an alternating current influences the oscillatory activity of cortical areas under and between the electrodes (Woods et al.,

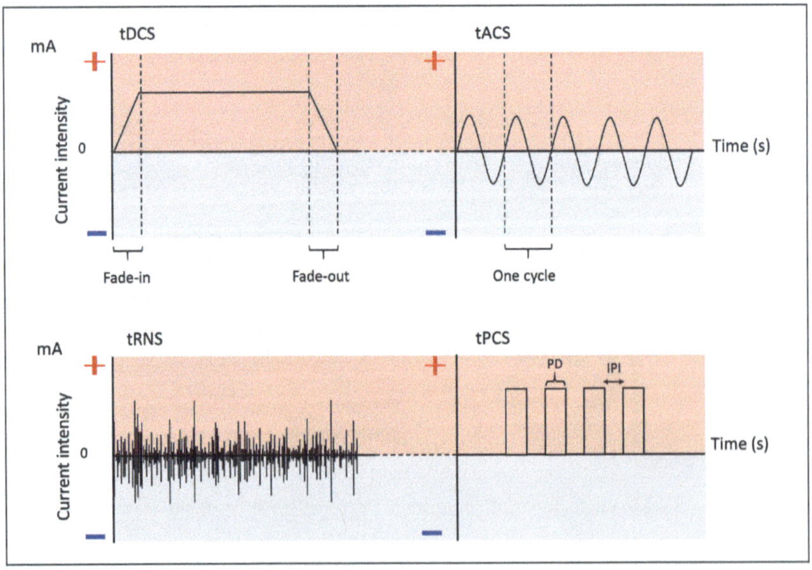

FIGURE 8.4 The different types of current waveforms in tDCS, tACS, tRNS, and tPCS. *IPI*, interpulse interval; *PD*, pulse duration; *tACS*, transcranial alternating current stimulation; *tDCS*, transcranial direct current stimulation; *tPCS*, transcranial pulsed current stimulation; *tRNS*, transcranial random noise stimulation.

2016). tACS in low frequencies (EEG range) entrains the networks and oscillatory brain activity. However, higher frequency applications do not alter the oscillatory activity; instead, they interfere with the biochemical mechanisms of the synapse and induce short-term synaptic plasticity (Antal & Paulus, 2013). Thus, there are variable results (inhibition–excitation) in the literature depending on the current intensity and the frequency of applied tACS.

8.3.3 Transcranial random noise stimulation

This relatively novel stimulation method was developed and introduced by Terney et al. (2008b). It is a type of alternating current stimulation with stimulation frequencies changing typically between 0.1 and 640 Hz randomly throughout the stimulation session (Fig. 8.4). High frequencies (100–640 Hz) are more efficient in generating an excitability increase on cortical neurons. In a recent study, three different frequency ranges (100–400 Hz; 400–700 Hz; 100–700 Hz) were compared, and only the 100–700 Hz band range was able to significantly modulate cortical excitability after stimulation, which supports that the efficacy of transcranial random noise stimulation (tRNS) also needs a wide range of frequencies (Moret et al., 2019). tRNS can lead to an increase in cortical excitability at an intensity of 1 mA; however, it induces inhibition when the intensity is set to 0.4 mA (Moliadze et al., 2012). The mechanism of tRNS is not understood precisely. It can be explained by two hypotheses: stochastic resonance and temporal summation of neural activity (Fertonani et al., 2011). Stochastic resonance is based on adding subthreshold noise to sensitize the neurons; thus, weak signals can reach the threshold, and a subthreshold oscillatory activity can be amplified (Wiesenfeld & Moss, 1995).

Furthermore, in a recent study, the mechanism of tRNS was found to be independent of NMDA receptors and can be suppressed by Na^+-channel blockers, which supports its membrane modulation effect (Chaieb et al., 2015).

Based on these studies, we can conclude that the short-term effect of tDCS is related to transient changes in neuron membrane resting potential, while the long-term effect includes possible mechanisms such as structural changes in neuronal and non-neuronal cells which modulate synaptic transmission, receptor activity, and neurotransmitter activity in the course of time. tACS has no polarity-specific effect, instead it drives the cortical oscillatory activity and strengthens the networks. Finally, tRNS has a possible direct effect on neuronal excitability depending on the applied current intensity. Therefore the mechanisms of action underlying tRNS, tACS, and tDCS are probably different from each other.

8.3.4 Transcranial pulsed current stimulation

Transcranial pulsed current stimulation (tPCS) is a novel technique to modulate cortical excitability. It is the application of direct current pulses (square waves) at a certain frequency with a constant interval between pulses

(Jaberzadeh et al., 2014). Two important parameters are added in this technique: pulse duration (PD) and interpulse interval (IPI). The pulses are unidirectional but separated with IPI (Fig. 8.4). Therefore it is hypothesized that besides the continuous modulation of polarity-specific direct current, this technique also alters the voltage-gated membrane proteins by the on–off feature of the pulses. There are two subtype stimulation modes according to IPI: short IPI tPCS and long IPI tPCS. It was shown that short IPI anodal tPCS of M1 induced increased corticospinal excitability compared to long IPI anodal tPCS. In a more recent study, the effect of PD on corticospinal excitability was investigated and larger changes were induced with PCS with the longest PD compared to anodal tDCS (Jaberzadeh et al., 2015). Despite the fact that the underlying physiological mechanism is unknown, and additional experiments are required; tPCS is a valuable, promising novel technique for the development of new therapeutic approaches for neurorehabilitation.

8.4 Experiment results and discussion

8.4.1 Neuromodulation of somatosensory processing by transcranial electrical stimulation

This part of the chapter presents a review of TES literature about somatosensory function including pain. In the first section, we start to examine the studies investigating tactile perception alone and continue with the studies about haptic perception which involves a combination of tactile perception and proprioception. In the next section, studies solely focusing on proprioception are examined. The last section is a short review of the literature regarding pain perception studies.

8.4.1.1 Modulation of tactile senses and haptic perception

Several studies indicate that modulation of sensory processing can help in the rehabilitation of neurological disorders. One of the measurement methods of sensory processing is recording the somatosensory evoked potential (SEP). This evoked potential is recorded from the scalp over the somatosensory cortex, and usually is caused by peripheral electrical stimulation of the median or tibial nerve. Basically, the amplitude of cortical SEP responses indicates the excitability of cortex under the electrodes. In addition to evaluating the excitability of the somatosensory cortex (S1), it also shows the integrity of the whole ascending posterior column–medial lemniscal pathway. There are also behavioral measurements for assessing the sensory function. The most common ones used in sensory studies are the measurement of the cold/heat detection threshold, mechanical detection threshold, electrical perception threshold, pain threshold, and tactile discrimination tasks (vibration, two-point, spatial acuity, grating orientation).

The primary somatosensory cortex (S1) is the major target for TES studies with the aim of sensory modulation. However, there are only a few studies that have investigated the changes in SEP responses during and after stimulation of the primary sensory cortex. Dieckhofer et al. found significantly decreased SEP amplitudes after cathodal tDCS applied over S1, while anodal tDCS did not induce a significant change (they reported a nonsignificant increasing trend) (Dieckhöfer et al., 2006). The results of Dieckhofer et al. were in line with the first behavioral study published by Rogalewski et al. (2004). They found a disruption of tactile discrimination of vibration during and after cathodal stimulation of S1 without any effect of anodal stimulation (Rogalewski et al., 2004). In another study, cathodal tDCS of S1 was found to induce a reduction in laser evoked potentials and pain perception (Antal et al., 2008a). Finally, an SEP study showed the excitatory effect of anodal tDCS over S1 and the inhibitory effect of cathodal tDCS, compared to sham stimulation (Rehmann et al., 2016). The results of behavioral measurements are more inconsistent in the literature regarding TES. Ragert et al. (2008) showed the positive effect of anodal stimulation of S1 on the grating orientation task (GOT) (tactile orientation) which was not found to be significant, however, in previous studies. The major differences from previous studies were tDCS parameters with longer stimulation duration and higher current intensity. These results support the importance of finding optimal parameters to achieve the highest efficacy of tDCS.

From the point of view of changes in thermal, mechanical, and discrimination thresholds induced by tDCS of S1, we found relatively more and conflicting studies. In 2011, Grundmann et al. used quantitative sensory testing to evaluate the thermal and mechanical detection thresholds (and also pain thresholds) after anodal/cathodal and sham tDCS of S1. They found only a significant increase in cold detection threshold compared to sham, while no significant change was found in other sensory measurements (Grundmann et al., 2011). The modulation of vibrotactile detection and discrimination thresholds induced by tDCS of S1 was investigated in another study (Labbé et al., 2016). Significant reductions in both thresholds were found with anodal stimulation compared to sham and cathodal stimulations. A cathodal high-definition tDCS (HD-tDCS) study found a differential involvement of S1 in the sensory processing of vibrotactile stimuli and nociception (Lenoir et al., 2017). In HD-tDCS, one active electrode was placed over the target cortical area and it was surrounded by four (or more) return electrodes. In a 4×1 ring HD-tDCS electrode montage, it was shown that current flow stays much more focal on the cortex than a conventional tDCS electrode montage (two 5×7 cm rectangular electrodes) (Datta et al., 2009; Villamar et al., 2013). Lenoir et al. showed that cathodal HD-tDCS over S1 reduced the short-latency SEP responses to the electrical stimuli delivered to the contralateral hand and the long latency event-related cortical responses (N120) to the contralateral vibrotactile stimulation. On the contrary, HD-tDCS of S1 altered the nociceptive laser-induced evoked responses (N240) symmetrically to both stimuli delivered to the ipsilateral and contralateral hand, which supports the absence of direct side-specific involvement of S1 in nociception. As a result, the researchers concluded that S1 is a fundamental area for processing non-nociceptive tactile stimuli coming from the contralateral side of the body, while it is not a primary projection area for nociception which projects to other cortical areas such as insula, secondary somatosensory cortex, and cingulate cortex.

The conventional tDCS applications were pure anodal or cathodal, with an extracortical return electrode (mostly over the contralateral supraorbital area). However, Fujimoto et al. (2014) used a different stimulation montage for S1, called the dual-hemisphere tDCS. They placed the anode electrode over the left S1, while the cathode electrode was over the right S1 cortex. By this montage, they elicited a dual effect of tDCS-excitatory and inhibitory, simultaneously, on bilateral S1 cortex. They obtained a more significant improvement in the GOT in the right index finger (contralateral to the anodal stimulation) (Fujimoto et al., 2014). In a more recent study by the same group, dual-hemisphere tDCS of parietal operculum improved the discrimination threshold of a GOT compared to sham and unihemisphere anodal stimulation (Fujimoto et al., 2017). The result of this study supports the inhibition of ipsilateral parietal operculum as enhancing tactile discrimination, which indicates dual hemisphere tDCS may be an important contribution to the rehabilitation of sensory function in patients.

Previous research showed that the primary somatosensory cortex is involved with tactile spatial discrimination and inhibitory circuits in the primary somatosensory cortex can alter the tactile spatial discrimination performance (Zhang et al., 2005; Li Hegner et al., 2015). For instance, a decrease in the inhibitory activity of the somatosensory cortex resulted in the improvement in tactile spatial discrimination performance (Saito et al., 2018). SEP−paired-pulse depression (SEP-PPD) is a method used to show the inhibitory circuit activity in the somatosensory cortex. It is a classical SEP recording over the somatosensory cortex against a paired electrical stimulation of the median nerve separated with a very short interval. The reduction in the SEP response to the second stimulus reflects the activity of inhibitory circuits in the somatosensory cortex. In a previous study mentioned above, SEP-PPD was used and the results showed the excitatory effect of the anodal and inhibitory effect of the cathodal tDCS on the primary somatosensory cortex (Rehmann et al., 2016). A more recent study tested the efficacy of other types of TES in healthy subjects; tACS, tRNS, and tPCS were applied over the primary somatosensory cortex in a tactile orientation discrimination task which was never investigated previously (Saito et al., 2019). Moreover, SEP-PPD was measured to evaluate the inhibitory circuit activity. Different TES protocols applied over S1 resulted in complicated and distinct results which prevented firm conclusions. Anodal tDCS showed no effect on tactile discrimination performance in contrast with previous findings, while there was a decrease in SEP-PPD. Anodal tPCS induced an improvement in tactile discrimination tasks and an increased second SEP N20 wave (reduced SEP-PPD). tRNS induced improved perceptual discrimination, together with an increase in the first SEP N20 amplitude. The authors explained the result of anodal tDCS with a possible insufficient current density compared to positive results of previous studies in the literature, while anodal tDCS was enough to decrease the SEP-PPD. Based on these findings it was suggested that the most optimal and precise application parameters of tRNS and tPCS need to be investigated for an improvement in sensory function.

Despite several studies which investigated the short-term effects of S1 tDCS on somatosensory thresholds, very few studies focused on repeated S1 tDCS effects. Interestingly, only one study, which included multiple sclerosis patients with sensory deficits, found an improvement in tactile discrimination thresholds after repeated anodal S1 tDCS, and also within their follow-up period (4 weeks) (Mori et al., 2013).

Repeated application of anodal tDCS over primary somatosensory cortex was investigated in healthy subjects, with an expectation of a more profound improvement in GOT either during the tDCS or after a 4-week follow-up period (Hilgenstock et al., 2016). These authors examined the online (during the tDCS) learning and offline (between sessions) learning by comparing the GOT thresholds. Online learning was defined as the difference between the baseline GOT performance before tDCS and GOT performance measured 30 minutes after tDCS for each trial day. Offline learning was defined as the difference between baseline GOT performance prior to tDCS application and the last GOT performance in the day preceding the trial. The results showed that 5 days repeated anodal tDCS enhanced offline tactile learning measured between tDCS sessions and the following 4-week period. Significant online tactile learning was also found in the same group; however, it reached a limit very early and did not change more. In conclusion, these results supported the long-term boosting effect of repeated anodal S1 tDCS in tactile skill learning and in decreasing the discrimination threshold.

Based on the tDCS literature, we can conclude that there is still no clear consensus on how tDCS of S1 cortex or which type of electrical stimulation modulates the sensory processing optimally.

Over time, a considerable body of literature has developed on the effect of motor cortex stimulation for somatosensory modulation. The first electrophysiological study of somatosensory modulation by tDCS was a motor cortex stimulation. The anodal stimulation applied over the left primary motor cortex (M1) induced an increase in SEP amplitudes lasting for 60 minutes after stimulation, while cathodal stimulation had no effect (Matsunaga et al., 2004). Nonetheless, the majority of M1 studies with the aim of sensory modulation examined the behavioral results. After the findings of Matsunaga et al. about SEP and the positive results of previous studies showing the effect of S1 tDCS on tactile discrimination and orientation, further studies continued to investigate the M1 tDCS. Several studies demonstrated that the M1 anodal tDCS significantly reduced the pain levels in chronic pain patients (Fregni et al., 2006; Antal et al., 2010). Regarding the modulation of pain in healthy volunteers, there is an inconsistency and heterogeneity of methods to examine the pain thresholds. In those studies, heat, cold, pressure, electrical current, and laser were the most used stimulation techniques to induce pain. However, this diversity in pain induction also causes difficulty in comparing the results and making clear neurophysiological explanations. The studies which investigated laser-induced pain showed a reduction in pain levels by cathodal tDCS of M1 (Terney et al., 2008a; Csifcsak et al., 2009). The mechanical pain threshold also showed, somewhat inconsistent, changes after either anodal or cathodal tDCS of M1 (Bachmann et al., 2010; Jürgens et al., 2012). Boggio et al. used peripheral electrical stimulation of the index finger for determining the perception and pain thresholds in healthy subjects and reported an increase in both thresholds after anodal M1 tDCS (Boggio et al., 2008). We replicated this study with the same stimulation montages; however, we could not show a significant difference in pain and perception thresholds between M1, dorsolateral prefrontal cortex (DLPFC), and occipital cortex stimulations in healthy volunteers (unpublished data). Even though they are not common targets, somatosensory cortex and DLPFC were also investigated to modulate pain perception. Some researchers proposed the possible effect of DLPFC on the emotional process of pain perception and modulation of DLPFC, might cause a reduction in pain thresholds. In addition, some of the neuromodulation studies of S1 showed significant changes in tactile perception thresholds. However, it is not clear yet whether modulation of S1 has an effect on pain perception or not; and the results of pain studies are not consistent. There are other studies of S1 tDCS conducted by the same group, which showed reduced sensitivity to painful stimuli after cathodal, but not anodal, tDCS (Antal et al., 2008a; Grundmann et al., 2011). Mylius et al. (2012) reviewed the experimental pain studies in healthy volunteers and concluded that the effects of tDCS depend on the fiber type ($A\delta$ or C fiber) carrying the signal provoked by experimental pain methods. Furthermore, anodal tDCS of M1 was found to be more effective in patients suffering from pain, while cathodal tDCS of M1 seems to be more effective in experimental pain. This is due to differences in pain induction, analgesic mechanisms of experimental acute pain, and pain network between healthy individuals and chronic pain patients.

On the other hand, there is a lack of evidence about the effect of M1 tDCS on somatic senses apart from pain. In 2010, Bachmann et al. found the cathodal tDCS of M1 increased the thermal and mechanical detection thresholds on the contralateral body side (Bachmann et al., 2010). There are two meta-analysis studies of the same group that evaluated the effectiveness of anodal and cathodal tDCS separately on sensory perception and pain (Vaseghi et al., 2014; Vaseghi et al., 2015). All citations included in the meta-analysis are the studies mentioned above. The results of the meta-analysis provide evidence for the effectiveness of cathodal tDCS of both S1 and M1 in increasing the sensory thresholds significantly in healthy individuals immediately after stimulation, with a stronger effect of M1 stimulation (Vaseghi et al., 2015). In the meta-analysis of anodal tDCS, they indicated that M1 anodal tDCS increased the sensory thresholds significantly compared to sham stimulation (Vaseghi et al., 2014). The results of two meta-analyses were in line with each other in terms of the short-term effects of tDCS. Long-term effects were not evaluated since measurement time-points of studies were mismatched. Furthermore, due to the limited number of studies and small sample sizes, it was difficult to reach firm conclusions about the efficacy of M1 tDCS in sensory modulation and more controlled studies are required.

8.4.1.2 Modulation of proprioception

Proprioceptive senses include the position and movement information regarding our body and limbs. The receptors of proprioception are located widely in muscles, joints, and skin. The studies of transcranial modulation of proprioception mostly investigated body functions which require the integration of different sensory modalities. Therefore, on this basis, it is not possible to investigate and assess the modulation of proprioception independent from other modalities. The sense of limb position, movement, force, and heaviness require different sensory signals arising from different proprioceptors. These signals are also combined to maintain the dynamic and static postural control and balance of the body and movements. For example, postural control requires the integration of three main sensory modalities of information: vestibular, somatosensory (proprioception), and visual. The majority of tDCS studies related to proprioception examine postural control, gait, and balance. They focus on facilitating multisensory feedback and integration. Proprioception deteriorates and the adaptation and reintegration of proprioceptive information decline with age, which is an important risk factor for increased falls in the elderly (Doumas & Krampe, 2010). The use of tDCS with the aim of enhancement of postural control and balance has gradually increased in recent years. Some of the research focused on the cerebellum as a cortical target to improve postural control, since it is the main relay of three types of sensory information (vestibular, proprioception, and visual). The decline in postural control and balance in older individuals is suggested to be correlated with diminished cerebellar white matter and cerebellar vermis volume (Cavallari et al., 2013; Rosano et al., 2007). Pijnenburg et al. found that decreased integrity of the superior cerebellar peduncle was associated with a decline in weighting the proprioceptive signals in patients with low-back pain, which is thought to be related to impaired postural control (Pijnenburg et al., 2014). This result indicates the important role of the cerebellum in proprioceptive integration. In healthy volunteers, cathodal stimulation of cerebellum induced an improvement in balance performance in the study of Inukai et al., while an impairment was found in the studies of Foerster et al. (2017) and Inukai et al. (2016). Anodal stimulation of cerebellum resulted in increased postural control, postural steadiness during Achilles tendon vibration, and improvement in motor learning (Jayaram et al., 2012; Ehsani et al., 2016; Poortvliet et al., 2018), while a recent study showed no positive effect for a dynamic balance task in young healthy subjects (Steiner et al., 2016). We conducted a study of cerebellar tDCS in our research center and compared the dynamic and static balance scores between anodal and sham conditions in young healthy subjects. Even though there was an increase in the scores after tDCS, the difference was not significant between sham and anodal groups. Having said that, however, the healthy young adults may have their maximum level of balance or motor learning function which prevents further improvement by TES. This is an example of a ceiling effect. In older subjects, anodal stimulation of the cerebellum resulted in an improvement of balance and postural control in the study of Ehsani et al. (2017). On the other hand, another work published in the same year included both young and older adults in the study. They arranged the level of task difficulty according to age to exclude the baseline differences in postural control assessment. There were only minimal effects that were dependent on the measures assessed, age, and visual conditions (eyes opened/closed) (Craig & Doumas, 2017). A recent study investigated the anodal stimulation of cerebellum in stroke patients and healthy age-matched elderly (Zandvliet et al., 2018). The contralesional anodal tDCS of cerebellum resulted in an improvement in one of the three standing balance conditions, while no change was observed in other conditions and in the healthy group. Two previous studies testing the effect of anodal cerebellar stimulation on balance in ataxic patients also showed conflicting results (Grimaldi & Manto, 2013; Grecco et al., 2017). Those recent findings lead one to question the efficacy of cerebellar modulation on postural control and balance. As is mentioned above, the literature on cerebellar TES has divergent results and is nonhomogeneous in terms of subject groups (healthy/patients), age (old/young), TES parameters (anodal/cathodal, duration), assessment methods (dynamic/static balance, postural sway) and time point of assessment (during/after). Therefore larger sampled, homogeneous studies controlling the task difficulty to remove age effects are required.

Indeed, cerebellar TES may affect multiple aspects of postural control and may be a promising tool to facilitate the connectivity of cerebellum and the integration of incoming sensory information. This would improve the cerebellar functions including proprioception.

8.4.1.3 Sensory modulation in stroke patients

tDCS is a common noninvasive neuromodulation method for clinical research targeting the improvement of motor functions in stroke patients. Additionally, researchers also continue to investigate the effect of tDCS on the other symptoms of stroke such as aphasia, pain, neglect, and cognitive functions. There is growing interest in the potential of tDCS-induced motor recovery. Even though previous studies mostly reported changes in upper limb functions induced by tDCS, there has been a trend in recent studies to focus on lower limb functions after stroke. Above all, M1 tDCS seems to be complementary to classical rehabilitation programs and has high potential in combined therapies (Kaski et al., 2014). The application of M1 tDCS in stroke patients is based on the hemispheric imbalance model (Liepert et al., 2000; Shimizu et al., 2002). Increased excitability in the unaffected hemisphere and accordingly increased inhibition from the unaffected hemisphere on the ipsilesional hemisphere have been reported. Therefore enhancing the excitability of the ipsilesional hemisphere or inhibiting the contralesional hemisphere are suggested to improve the interhemispheric imbalance and motor functions. Nevertheless, strong evidence is still not available for the clinical recommendation of tDCS due to the heterogeneity of clinical studies in terms of parameters, electrode montages, and patient groups (acute, subacute, chronic) (Lefaucheur et al., 2017). Although there have been many studies, the research into sensory function in stroke patients remains limited in scope. A randomized, sham-controlled double-blind study was conducted in subacute stroke patients with somatosensory deficits (Koo et al., 2018). Anodal tDCS was applied over ipsilesional S1 for the purpose of improving somatosensation. A somatosensation assessment including light touch, pressure, pinprick, proprioception, and cortical sensation, together with stereognosis were evaluated before and after 10-day sessions of tDCS. There was no statistically significant difference in comparisons of before–after assessments between sham and anodal groups. However, in the anodal stimulation group, comparing the baselines and post-stimulation assessments showed that the tactile sense, pinprick, and kinesthesia scores were significantly improved in the affected side. Kinesthesia was also somewhat improved in the sham group, but the anodal group showed higher significance. Furthermore, the anodal stimulation group showed a higher significance of improvement in stereognosis in the affected side compared to the sham group. Interestingly, five patients who did not perceive any sensation on the affected side showed a recovery after the anodal group, but only one patient recovered in the sham group. There was additionally a tendency for a decrease in sensory thresholds in the unaffected side.

It is known that bilateral tactile sensory deficits can occur after unilateral stroke and the mechanism is unclear. There are various explanations, however, one of them stands out which considers a possible sensory pathway/fibers projecting to the ipsilateral hemisphere. The study of Koo et al. also supports this explanation. In a very recent well-controlled study, Bornheim et al. applied anodal tDCS over the ipsilesional motor cortex in the acute period of 50 stroke patients (Bornheim et al., 2020). tDCS was applied five times a week for 4 weeks prior to daily routine rehabilitation with a 1-year follow-up period. This was the first study to look at the repetitive sessions of tDCS in acute stroke patients and to evaluate the first-year results. They found an improvement both in motor and somatosensory functions (Semmes Weinstein Monofilament Test, the somatosensory section of the Fugl Meyer Test) in the anodal group compared to the sham group; and this improvement started at the fourth week and lasted to the end of the first year. The effect of single-session dual-hemisphere (i.e., S1 and S2) tDCS on GOT in chronic stroke patients was investigated in a double-blind sham-controlled study (Fujimoto et al., 2016). An anodal electrode was placed over the lesioned hemisphere, while the cathodal electrode was placed over the opposite side based on the above-mentioned interhemispheric imbalance model. GOT discriminative thresholds at the affected index finger measured 10 minutes after tDCS were lower after active stimulation compared to the sham condition. This study showed the immediate enhancement of tactile discrimination by dual tDCS of S1 and S2 cortex in chronic stroke patients with sensory deficits.

8.4.2 *Modulating multisensory integration*

Illusions are misinterpretations of the brain because of inputs conflicting with the brain templates (created by experience) (Knill & Richards, 1996). One widely used way to investigate multisensory integration is the rubber hand illusion (RHI). In this illusion, one hand of the subject is hidden, and is replaced by a same-sized artificial rubber hand. During the illusion, two brushes strike both hands simultaneously and subjects feel the artificial hand as their own hand (Botvinick & Cohen, 1998). In this way, a conflict is created between visual, somatosensory, and proprioceptive senses. Integration of these senses creates body ownership feeling of a fake hand. In the study of Collins et al., primary somatosensory cortex was stimulated by invasive subdural electrodes in upper limb amputees (Collins et al., 2017). They found an induced ownership feeling of an artificial hand prosthetic in RHI. The studies into the modulation of multisensory integration by tDCS have been gradually increasing in recent years.

Although there is no direct evidence, it is hypothesized that intersensory conflict in RHI is resolved by suppression of the somatosensory cortex in favor of visual sensory information (Limanowski & Blankenburg, 2015; Zeller et al., 2015). tDCS over contralateral S1 was used to elicit inhibition of the somatosensory cortex for the purpose of enhancement in RHI (Hornburger et al., 2019).

tDCS was applied for 20 minutes with 1 mA. During tDCS application, RHI was retested at six different distances of the artificial hand from the real hand. In this way, they prevented the ceiling effect at short distances (illusion effect expected to be higher). Proprioceptive drift (PD) is defined as the distance between the perceived hand position before and after brush strokes. Since the researchers used different distances for the artificial hand, they calculated relative proprioceptive drift (RD) for each interval (PD divided by the distance between the artificial hand and real hand). They could not show any difference in the illusion score between cathodal and sham tDCS; however, the score was higher in cathodal tDCS as compared to anodal tDCS. There was no correlation between illusion scores and RD. Therefore, in conclusion, the sensory integration of PD and illusion score, or in other words the ownership feeling, were due to different mechanisms in the somatosensory network.

RHI requires the sensory stimuli to be integrated with the body representation and the subjective feeling of body ownership (Tsakiris, 2010). fMRI studies showed the involvement of posterior parietal cortex (PPC), ventral premotor cortex (PMv), inferior parietal lobe, and right temporoparietal junction (rTPJ) in the process of multisensory integration during the RHI (Ehrsson et al., 2004; Tsakiris et al., 2008; Petkova et al., 2011; Bekrater-Bodmann et al., 2012; Limanowski & Blankenburg, 2015). Based on this information, Lira et al. (2018) conducted a tDCS study targeting both the right PPC and right PMv. They revealed that anodal tDCS of right PPC shortens the onset time of RHI perception and increased subjective body ownership feeling compared to cathodal and sham. There was no difference induced by anodal tDCS of right PMv. The researchers suggested that decreased onset time of RHI might be related to the facilitated speed of process in visual–tactile sensory integration. Accordingly, enhanced body ownership might be due to shortened onset time, leading to a longer duration of ownership feeling. Convento et al. investigated the modulation of the right premotor cortex (rPMc) and rTPJ by anodal tDCS (Convento et al., 2018). Proprioceptive drift was increased in tDCS of both areas, while subjective ownership was affected by only rTPJ tDCS. The difference in proprioceptive drift due to congruent and incongruent brush strokes was also analyzed and showed an increased difference in rTPJ tDCS, while there was a decrease in rPMc tDCS. It was concluded that rTPJ modulates RHI, while rPMc was related to the recalibration of hand coordinates.

Those RHI studies are promising for understanding the mechanism of body ownership. Possible modulation techniques may be useful in the rehabilitation of patients for their acceptance of prosthetics and their integration into the body model within the brain. In terms of tDCS, especially in multisensory integration studies, it seems that smaller electrodes are needed for the application of currents that are more focal on the cortex and for eliminating the costimulation of other brain regions outside the target area.

8.5 Future opportunities

Taken together with the studies considered in this chapter, the current overview of tDCS research supports the potential clinical use of tDCS in somatosensory rehabilitation. The long-term effects of TES have crucial potential for facilitating the plasticity of the intact cortex to take over the functions of damaged cortical areas for sensory processing. To use tDCS in combination with the different types of neuroprosthetic devices as feedback information may also have potential benefits for the adaptation, learning, and changing the coding of sensory data by neural plasticity. Conducting more studies into the effect and methodology of tDCS will also provide some valuable clues about neural network dynamics. Modulation of networks, probably applied to relay nuclei, may help the integration of multisensory information at the cortical level. Such techniques may shorten the rehabilitation time course for many patients. Those patients with already implanted sensory stimulators would probably be the first experimental pioneers of augmented reality sensations.

8.6 Conclusions

Several studies have shown that TES can modulate somatosensory processing in humans. However, more controlled, that is, sham-controlled, double-blind, and randomized, studies are required for determining the optimal stimulation parameters, electrode montages, and cortical targets to achieve the highest efficacy of TES for somatosensory modulation. Ultimately, this will offer an insight into the potential of TES for different purposes in neuroprosthetics, including patient adaptation, training, and facilitating the sensory feedback and improving its function. These studies may provide invaluable data in order to replace/bypass certain parts of sensory systems with adequately functioning sensing and stimulating prostheses.

References

Antal, A., Brepohl, N., Poreisz, C., Boros, K., Csifcsak, G., & Paulus W. (2008a) Transcranial direct current stimulation over somatosensory cortex decreases experimentally induced acute pain perception. *Clinical Journal of Pain, 24*(1), 56-63. Available from https://doi.org/10.1097/AJP.0b013e318157233b.

Antal, A., Boros, K., Poreisz, C., Chaieb, L., Terney, D., & Paulus, W. (2008b). Comparatively weak after-effects of transcranial alternating current stimulation (tACS) on cortical excitability in humans. *Brain Stimulation, 1*(2), 97–105. Available from https://doi.org/10.1016/j.brs.2007.10.001.

Antal, A., Terney, D., Kühnl, S., & Paulus, W. (2010). Anodal transcranial direct current stimulation of the motor cortex ameliorates chronic pain and reduces short intracortical inhibition. *Journal of Pain and Symptom Management, 39*(5), 890–903. Available from https://doi.org/10.1016/j.jpainsymman.2009.09.023.

Antal, A., & Paulus, W. (2013). Transcranial alternating current stimulation (tACS). *Frontiers in Human Neuroscience*, *7*, 317. Available from https://doi.org/10.3389/fnhum.2013.00317.

Antal, A., Alekseichuk, I., Bikson, M., Brockmöller, J., Brunoni, A. R., Chen, R., Cohen, L. G., Dowthwaite, G., Ellrich, J., Flöel, A., Fregni, F., George, M. S., Hamilton, R., Haueisen, J., Herrmann, C. S., Hummel, F. C., Lefaucheur, J. P., Liebetanz, D., Loo, C. K., McCaig, C. D., … Paulus, W. (2017). Low intensity transcranial electric stimulation: Safety, ethical, legal regulatory and application guidelines. *Clinical Neurophysiology: Official Journal of the International Federation of Clinical Neurophysiology*, *128*(9), 1774–1809. Available from https://doi.org/10.1016/j.clinph.2017.06.001.

Bachmann, C. G., Muschinsky, S., Nitsche, M. A., Rolke, R., Magerl, W., Treede, R. D., Paulus, W., & Happe, S. (2010). Transcranial direct current stimulation of the motor cortex induces distinct changes in thermal and mechanical sensory percepts. *Clinical Neurophysiology: Official Journal of the International Federation of Clinical Neurophysiology*, *121*(12), 2083–2089. Available from https://doi.org/10.1016/j.clinph.2010.05.005.

Bekrater-Bodmann, R., Foell, J., Diers, M., & Flor, H. (2012). The perceptual and neuronal stability of the rubber hand illusion across contexts and over time. *Brain Research*, *1452*, 130–139. Available from https://doi.org/10.1016/j.brainres.2012.03.001.

Bindman, L. J., Lippold, O. C., & Redfearn, J. W. (1962). Long-lasting changes in the level of the electrical activity of the cerebral cortex produced by polarizing currents. *Nature*, *196*(6), 584–585. Available from https://doi.org/10.1038/196584a0.

Boggio, P. S., Zaghi, S., Lopes, M., & Fregni, F. (2008). Modulatory effects of anodal transcranial direct current stimulation on perception and pain thresholds in healthy volunteers. *European Journal of Neurology*, *15*(10), 1124–1130. Available from https://doi.org/10.1111/j.1468-1331.2008.02270.x.

Bornheim, S., Croisier, J. L., Maquet, P., & Kaux, J. F. (2020). Transcranial direct current stimulation associated with physical-therapy in acute stroke patients - A randomized, triple blind, sham-controlled study. *Brain Stimulation*, *13*(2), 329–336. Available from https://doi.org/10.1016/j.brs.2019.10.019.

Botvinick, M., & Cohen, J. (1998). Rubber hands 'feel' touch that eyes see. *Nature*, *391*(6669), 756. Available from https://doi.org/10.1038/35784, PMID 9486643.

Cavallari, M., Moscufo, N., Skudlarski, P., Meier, D., Panzer, V. P., Pearlson, G. D., White, W. B., Wolfson, L., & Guttmann, C. R. (2013). Mobility impairment is associated with reduced microstructural integrity of the inferior and superior cerebellar peduncles in elderly with no clinical signs of cerebellar dysfunction. *NeuroImage: Clinical*, *2*, 332–340. Available from https://doi.org/10.1016/j.nicl.2013.02.003.

Chaieb, L., Antal, A., & Paulus, W. (2015). Transcranial random noise stimulation-induced plasticity is NMDA-receptor independent but sodium-channel blocker and benzodiazepines sensitive. *Frontiers in Neuroscience*, *9*, 125. Available from https://doi.org/10.3389/fnins.2015.00125.

Collins, K. L., Guterstam, A., Cronin, J., Olson, J. D., Ehrsson, H. H., & Ojemann, J. G. (2017). Ownership of an artificial limb induced by electrical brain stimulation. *Proceedings of the National Academy of Sciences of the United States of America*, *114*(1), 166–171. Available from https://doi.org/10.1073/pnas.1616305114.

Convento, S., Romano, D., Maravita, A., & Bolognini, N. (2018). Roles of the right temporo-parietal and premotor cortices in self-location and body ownership. *The European Journal of Neuroscience*, *47*(11), 1289–1302. Available from https://doi.org/10.1111/ejn.13937.

Craig, C. E., & Doumas, M. (2017). Anodal transcranial direct current stimulation shows minimal, measure-specific effects on dynamic postural control in young and older adults: A double blind, sham-controlled study. *PLoS One*, *12*(1), e0170331. Available from https://doi.org/10.1371/journal.pone.0170331.

Csifcsak, G., Antal, A., Hillers, F., Levold, M., Bachmann, C. G., Happe, S., Nitsche, M. A., Ellrich, J., & Paulus, W. (2009). Modulatory effects of transcranial direct current stimulation on laser-evoked potentials. *Pain Medicine*, *10*(1), 122–132. Available from https://doi.org/10.1111/j.1526-4637.2008.00508.x.

Datta, A., Bansal, V., Diaz, J., Patel, J., Reato, D., & Bikson, M. (2009). Gyri-precise head model of transcranial direct current stimulation: Improved spatial focality using a ring electrode versus conventional rectangular pad. *Brain Stimulation*, *2*(4), 201–207. Available from https://doi.org/10.1016/j.brs.2009.03.005, e1.

Dieckhöfer, A., Waberski, T. D., Nitsche, M., Paulus, W., Buchner, H., & Gobbelé, R. (2006). Transcranial direct current stimulation applied over the somatosensory cortex - differential effect on low and high frequency SEPs. *Clinical Neurophysiology: Official Journal of the International Federation of Clinical Neurophysiology*, *117*(10), 2221–2227. Available from https://doi.org/10.1016/j.clinph.2006.07.136.

Doumas, M., & Krampe, R. T. (2010). Adaptation and reintegration of proprioceptive information in young and older adults' postural control. *Journal of Neurophysiology*, *104*(4), 1969–1977. Available from https://doi.org/10.1152/jn.00345.2010.

Ehrsson, H. H., Spence, C., & Passingham, R. E. (2004). That's my hand! Activity in premotor cortex reflects feeling of ownership of a limb. *Science*, *305*(5685), 875–877. Available from https://doi.org/10.1126/science.1097011.

Ehsani, F., Bakhtiary, A. H., Jaberzadeh, S., Talimkhani, A., & Hajihasani, A. (2016). Differential effects of primary motor cortex and cerebellar transcranial direct current stimulation on motor learning in healthy individuals: A randomized double-blind sham-controlled study. *Neuroscience Research*, *112*, 10–19. Available from https://doi.org/10.1016/j.neures.2016.06.003.

Ehsani, F., Samaei, A., Zoghi, M., Hedayati, R., & Jaberzadeh, S. (2017). The effects of cerebellar transcranial direct current stimulation on static and dynamic postural stability in older individuals: A randomized double-blind sham-controlled study. *The European Journal of Neuroscience*, *46*(12), 2875–2884. Available from https://doi.org/10.1111/ejn.13731.

Fertonani, A., Pirulli, C., & Miniussi, C. (2011). Random noise stimulation improves neuroplasticity in perceptual learning. *The Journal of Neuroscience: The Official Journal of the Society for Neuroscience*, *31*(43), 15416–15423. Available from https://doi.org/10.1523/JNEUROSCI.2002-11.2011.

Finger, S., & Piccolino, M. (2011). *The shocking history of electric fishes: From ancient epochs to the birth of modern neurophysiology*. New York, NY: Oxford University Press.

Foerster, Á., Melo, L., Mello, M., Castro, R., Shirahige, L., Rocha, S., & Monte-Silva, K. (2017). Cerebellar transcranial direct current stimulation (ctDCS) impairs balance control in healthy individuals. *Cerebellum*, *16*(4), 872–875. Available from https://doi.org/10.1007/s12311-017-0863-8.

Fregni, F., Boggio, P. S., Lima, M. C., Ferreira, M. J., Wagner, T., Rigonatti, S. P., Castro, A. W., Souza, D. R., Riberto, M., Freedman, S. D., Nitsche, M. A., & Pascual-Leone, A. (2006). A sham-controlled, phase II trial of transcranial direct current stimulation for the treatment of central pain in traumatic spinal cord injury. *Pain*, *122*(1-2), 197–209. Available from https://doi.org/10.1016/j.pain.2006.02.023.

Fujimoto, S., Yamaguchi, T., Otaka, Y., Kondo, K., & Tanaka, S. (2014). Dual-hemisphere transcranial direct current stimulation improves performance in a tactile spatial discrimination task. *Clinical Neurophysiology: Official Journal of the International Federation of Clinical Neurophysiology*, *125*(8), 1669–1674. Available from https://doi.org/10.1016/j.clinph.2013.12.100.

Fujimoto, S., Kon, N., Otaka, Y., Yamaguchi, T., Nakayama, T., Kondo, K., Ragert, P., & Tanaka, S. (2016). Transcranial direct current stimulation over the primary and secondary somatosensory cortices transiently improves tactile spatial discrimination in stroke patients. *Frontiers in Neuroscience*, *10*, 128. Available from https://doi.org/10.3389/fnins.2016.00128.

Fujimoto, S., Tanaka, S., Laakso, I., Yamaguchi, T., Kon, N., Nakayama, T., Kondo, K., & Kitada, R. (2017). The effect of dual-hemisphere transcranial direct current stimulation over the parietal operculum on tactile orientation discrimination. *Frontiers in Behavioral Neuroscience*, *11*, 173. Available from https://doi.org/10.3389/fnbeh.2017.00173.

Grecco, L. A., Oliveira, C. S., Duarte, N. A., Lima, V. L., Zanon, N., & Fregni, F. (2017). Cerebellar transcranial direct current stimulation in children with ataxic cerebral palsy: A sham-controlled, crossover, pilot study. *Developmental Neurorehabilitation*, *20*(3), 142–148. Available from https://doi.org/10.3109/17518423.2016.1139639.

Grimaldi, G., & Manto, M. (2013). Anodal transcranial direct current stimulation (tDCS) decreases the amplitudes of long-latency stretch reflexes in cerebellar ataxia. *Annals of Biomedical Engineering*, *41*(11), 2437–2447. Available from https://doi.org/10.1007/s10439-013-0846-y.

Grundmann, L., Rolke, R., Nitsche, M. A., Pavlakovic, G., Happe, S., Treede, R. D., Paulus, W., & Bachmann, C. G. (2011). Effects of transcranial direct current stimulation of the primary sensory cortex on somatosensory perception. *Brain Stimulation*, *4*(4), 253–260. Available from https://doi.org/10.1016/j.brs.2010.12.002.

Harris, L. J., & Almerigi, J. B. (2009). Probing the human brain with stimulating electrodes: The story of Roberts Bartholow's (1874) experiment on Mary Rafferty. *Brain and Cognition*, *70* (1), 92–115. Available from https://doi.org/10.1016/j.bandc.2009.01.008.

Hilgenstock, R., Weiss, T., Huonker, R., & Witte, O. W. (2016). Behavioural and neurofunctional impact of transcranial direct current stimulation on somatosensory learning. *Human Brain Mapping*, *37*(4), 1277–1295. Available from https://doi.org/10.1002/hbm.23101.

Hornburger, H., Nguemeni, C., Odorfer, T., & Zeller, D. (2019). Modulation of the rubber hand illusion by transcranial direct current stimulation over the contralateral somatosensory cortex. *Neuropsychologia*, *131*, 353–359. Available from https://doi.org/10.1016/j.neuropsychologia.2019.05.008.

Inukai, Y., Saito, K., Sasaki, R., Kotan, S., Nakagawa, M., & Onishi, H. (2016). Influence of transcranial direct current stimulation to the cerebellum on standing posture control. *Frontiers in Human Neuroscience*, *10*, 325. Available from https://doi.org/10.3389/fnhum.2016.00325.

Jaberzadeh, S., Bastani, A., & Zoghi, M. (2014). Anodal transcranial pulsed current stimulation: A novel technique to enhance corticospinal excitability. *Clinical Neurophysiology: Official Journal of the International Federation of Clinical Neurophysiology*, *125*(2), 344–351. Available from https://doi.org/10.1016/j.clinph.2013.08.025.

Jaberzadeh, S., Bastani, A., Zoghi, M., Morgan, P., & Fitzgerald, P. B. (2015). Anodal transcranial pulsed current stimulation: The effects of pulse duration on corticospinal excitability. *PLoS One*, *10*(7), e0131779. Available from https://doi.org/10.1371/journal.pone.0131779.

Jackson, M. P., Rahman, A., Lafon, B., Kronberg, G., Ling, D., Parra, L. C., & Bikson, M. (2016). Animal models of transcranial direct current stimulation: Methods and mechanisms. *Clinical Neurophysiology: Official Journal of the International Federation of Clinical Neurophysiology*, *127*(11), 3425–3454. Available from https://doi.org/10.1016/j.clinph.2016.08.016.

Jayaram, G., Tang, B., Pallegadda, R., Vasudevan, E. V., Celnik, P., & Bastian, A. (2012). Modulating locomotor adaptation with cerebellar stimulation. *Journal of Neurophysiology*, *107*(11), 2950–2957. Available from https://doi.org/10.1152/jn.00645.2011.

Jürgens, T. P., Schulte, A., Klein, T., & May, A. (2012). Transcranial direct current stimulation does neither modulate results of a quantitative sensory testing protocol nor ratings of suprathreshold heat stimuli in healthy volunteers. *European Journal of Pain*, *16*(9), 1251–1263. Available from https://doi.org/10.1002/j.1532-2149.2012.00135.x.

Kaski, D., Dominguez, R. O., Allum, J. H., Islam, A. F., & Bronstein, A. M. (2014). Combining physical training with transcranial direct current stimulation to improve gait in Parkinson's disease: A pilot randomized controlled study. *Clinical Rehabilitation*, *28*(11), 1115–1124. Available from https://doi.org/10.1177/0269215514534277.

Knill, D. C., & Richards, W. (1996). *Perception as Bayesian inference*. Cambridge: Cambridge University Press, 978-0-521-06499-6.

Koo, W. R., Jang, B. H., & Kim, C. R. (2018). Effects of anodal transcranial direct current stimulation on somatosensory recovery after stroke: A randomized controlled trial. *American Journal of Physical Medicine & Rehabilitation*, *97*(7), 507–513. Available from https://doi.org/10.1097/PHM.0000000000000910.

Labbé, S., Meftah, E., & Chapman, C. E. (2016). Effects of transcranial direct current stimulation of primary somatosensory cortex on vibrotactile detection and discrimination. *Journal of Neurophysiology*, *115*(4), 1978–1987. Available from https://doi.org/10.1152/jn.00506.2015.

Lefaucheur, J. P., Antal, A., Ayache, S. S., Benninger, D. H., Brunelin, J., Cogiamanian, F., Cotelli, M., De Ridder, D., Ferrucci, R., Langguth, B., Marangolo, P., Mylius, V., Nitsche, M. A., Padberg, F., Palm, U., Poulet, E., Priori, A., Rossi, S., Schecklmann, M., Vanneste, S., … Paulus, W. (2017). Evidence-based guidelines on the therapeutic use of transcranial direct current stimulation (tDCS). *Clinical Neurophysiology: Official Journal of the International Federation of Clinical Neurophysiology*, *128*(1), 56–92. Available from https://doi.org/10.1016/j.clinph.2016.10.087.

Lefaucheur, J. P., & Wendling, F. (2019). Mechanisms of action of tDCS: A brief and practical overview. *Neurophysiologie Clinique = Clinical Neurophysiology*, *49*(4), 269–275. Available from https://doi.org/10.1016/j.neucli.2019.07.013.

Lenoir, C., Huang, G., Vandermeeren, Y., Hatem, S. M., & Mouraux, A. (2017). Human primary somatosensory cortex is differentially involved in vibrotaction and nociception. *Journal of Neurophysiology*, *118*(1), 317–330. Available from https://doi.org/10.1152/jn.00615.2016.

Li Hegner, Y., Lindner, A., & Braun, C. (2015). Cortical correlates of perceptual decision making during tactile spatial pattern discrimination. *Human Brain Mapping*, *36*(9), 3339–3350. Available from https://doi.org/10.1002/hbm.22844.

Liebetanz, D., Nitsche, M. A., Tergau, F., & Paulus, W. (2002). Pharmacological approach to the mechanisms of transcranial DC-stimulation-induced after-effects of human motor cortex excitability. *Brain: A Journal of Neurology*, *125*(Pt 10), 2238–2247. Available from https://doi.org/10.1093/brain/awf238.

Liepert, J., Hamzei, F., & Weiller, C. (2000). Motor cortex disinhibition of the unaffected hemisphere after acute stroke. *Muscle & Nerve*, *23*(11), 1761–1763. Available from https://doi.org/10.1002/1097-4598(200011)23:11 < 1761::aid-mus14 > 3.0.co;2-m.

Limanowski, J., & Blankenburg, F. (2015). Network activity underlying the illusory self-attribution of a dummy arm. *Human Brain Mapping*, *36*(6), 2284–2304. Available from https://doi.org/10.1002/hbm.22770.

Lira, M., Pantaleão, F. N., de Souza Ramos, C. G., & Boggio, P. S. (2018). Anodal transcranial direct current stimulation over the posterior parietal cortex reduces the onset time to the rubber hand illusion and increases the body ownership. *Experimental Brain Research*, *236*(11), 2935–2943. Available from https://doi.org/10.1007/s00221-018-5353-9.

Matsunaga, K., Nitsche, M. A., Tsuji, S., & Rothwell, J. C. (2004). Effect of transcranial DC sensorimotor cortex stimulation on somatosensory evoked potentials in humans. *Clinical Neurophysiology: Official Journal of the International Federation of Clinical Neurophysiology*, *115*(2), 456–460. Available from https://doi.org/10.1016/s1388-2457(03)00362-6.

Merton, P. A., & Morton, H. B. (1980). Stimulation of the cerebral cortex in the intact human subject. *Nature*, *285*(5762), 227. Available from https://doi.org/10.1038/285227a0.

Moliadze, V., Atalay, D., Antal, A., & Paulus, W. (2012). Close to threshold transcranial electrical stimulation preferentially activates inhibitory networks before switching to excitation with higher intensities. *Brain Stimulation*, *5*(4), 505–511. Available from https://doi.org/10.1016/j.brs.2011.11.004.

Moret, B., Donato, R., Nucci, M., Cona, G., & Campana, G. (2019). Transcranial random noise stimulation (tRNS): A wide range of frequencies is needed for increasing cortical excitability. *Scientific Reports*, *9*(1), 15150. Available from https://doi.org/10.1038/s41598-019-51553-7.

Mori, F., Nicoletti, C. G., Kusayanagi, H., Foti, C., Restivo, D. A., Marciani, M. G., & Centonze, D. (2013). Transcranial direct current stimulation ameliorates tactile sensory deficit in multiple sclerosis. *Brain Stimulation*, *6*(4), 654–659. Available from https://doi.org/10.1016/j.brs.2012.10.003.

Mylius, V., Borckardt, J. J., & Lefaucheur, J. P. (2012). Noninvasive cortical modulation of experimental pain. *Pain*, *153*(7), 1350–1363. Available from https://doi.org/10.1016/j.pain.2012.04.009.

Nitsche, M. A., & Paulus, W. (2000). Excitability changes induced in the human motor cortex by weak transcranial direct current stimulation. *The Journal of Physiology*, *527*(Pt 3), 633–639. Available from https://doi.org/10.1111/j.1469-7793.2000.t01-1-00633.x, Pt 3.

Petkova, V. I., Björnsdotter, M., Gentile, G., Jonsson, T., Li, T. Q., & Ehrsson, H. H. (2011). From part- to whole-body ownership in the multisensory brain. *Current Biology*, *21*(13), 1118–1122. Available from https://doi.org/10.1016/j.cub.2011.05.022.

Pijnenburg, M., Caeyenberghs, K., Janssens, L., Goossens, N., Swinnen, S. P., Sunaert, S., & Brumagne, S. (2014). Microstructural integrity of the superior cerebellar peduncle is associated with an impaired proprioceptive weighting capacity in individuals with non-specific low back pain. *PLoS One*, *9*(6), e100666. Available from https://doi.org/10.1371/journal.pone.0100666.

Poortvliet, P., Hsieh, B., Cresswell, A., Au, J., & Meinzer, M. (2018). Cerebellar transcranial direct current stimulation improves adaptive postural control. *Clinical Neurophysiology: Official Journal of the International Federation of Clinical Neurophysiology*, *129*(1), 33–41. Available from https://doi.org/10.1016/j.clinph.2017.09.118.

Ragert, P., Vandermeeren, Y., Camus, M., & Cohen, L. G. (2008). Improvement of spatial tactile acuity by transcranial direct current stimulation. *Clinical Neurophysiology: Official Journal of the International Federation of Clinical Neurophysiology*, *119*(4), 805–811. Available from https://doi.org/10.1016/j.clinph.2007.12.001.

Rehmann, R., Sczesny-Kaiser, M., Lenz, M., Gucia, T., Schliesing, A., Schwenkreis, P., Tegenthoff, M., & Höffken, O. (2016). Polarity-specific cortical effects of transcranial direct current stimulation in primary somatosensory cortex of healthy humans. *Frontiers in Human Neuroscience*, *10*, 208. Available from https://doi.org/10.3389/fnhum.2016.00208.

Rogalewski, A., Breitenstein, C., Nitsche, M. A., Paulus, W., & Knecht, S. (2004). Transcranial direct current stimulation disrupts tactile perception. *The European Journal of Neuroscience*, *20*(1), 313–316. Available from https://doi.org/10.1111/j.0953-816X.2004.03450.x.

Rosano, C., Aizenstein, H. J., Studenski, S., & Newman, A. B. (2007). A regions-of-interest volumetric analysis of mobility limitations in community-dwelling older adults. *The Journals of Gerontology: Series A, Biological Sciences and Medical Sciences*, *62*(9), 1048–1055. Available from https://doi.org/10.1093/gerona/62.9.1048.

Saito, K., Otsuru, N., Inukai, Y., Kojima, S., Miyaguchi, S., Tsuiki, S., Sasaki, R., & Onishi, H. (2018). Inhibitory mechanisms in primary somatosensory cortex mediate the effects of peripheral electrical stimulation on tactile spatial discrimination. *Neuroscience*, *384*, 262–274. Available from https://doi.org/10.1016/j.neuroscience.2018.05.032.

Saito, K., Otsuru, N., Inukai, Y., Miyaguchi, S., Yokota, H., Kojima, S., Sasaki, R., & Onishi, H. (2019). Comparison of transcranial electrical stimulation regimens for effects on inhibitory circuit activity in primary somatosensory cortex and tactile spatial discrimination performance. *Behavioural Brain Research*, *375*, 112168. Available from https://doi.org/10.1016/j.bbr.2019.112168.

Shimizu, T., Hosaki, A., Hino, T., Sato, M., Komori, T., Hirai, S., & Rossini, P. M. (2002). Motor cortical disinhibition in the unaffected hemisphere after unilateral cortical stroke. *Brain: A Journal of Neurology*, *125*(Pt 8), 1896–1907. Available from https://doi.org/10.1093/brain/awf183.

Steiner, K. M., Enders, A., Thier, W., Batsikadze, G., Ludolph, N., Ilg, W., & Timmann, D. (2016). Cerebellar tDCS does not improve learning in a complex whole body dynamic balance task in young healthy subjects. *PLoS One*, *11*(9), e0163598. Available from https://doi.org/10.1371/journal.pone.0163598.

Terney, D., Bergmann, I., Poreisz, C., Chaieb, L., Boros, K., Nitsche, M. A., Paulus, W., & Antal, A. (2008a). Pergolide increases the efficacy of cathodal direct current stimulation to reduce the amplitude of laser-evoked potentials in humans. *Journal of Pain and Symptom Management*, *36* (1), 79–91. Available from https://doi.org/10.1016/j.jpainsymman.2007.08.014.

Terney, D., Chaieb, L., Moliadze, V., Antal, A., & Paulus, W. (2008b). Increasing human brain excitability by transcranial high-frequency random noise stimulation. *The Journal of Neuroscience: The Official Journal of the Society for Neuroscience*, *28*(52), 14147–14155. Available from https://doi.org/10.1523/JNEUROSCI.4248-08.2008.

Tsakiris, M. (2010). My body in the brain: A neurocognitive model of body-ownership. *Neuropsychologia*, *48*(3), 703–712. Available from https://doi.org/10.1016/j.neuropsychologia.2009.09.034.

Tsakiris, M., Costantini, M., & Haggard, P. (2008). The role of the right temporo-parietal junction in maintaining a coherent sense of one's body. *Neuropsychologia*, *46*(12), 3014–3018. Available from https://doi.org/10.1016/j.neuropsychologia.2008.06.004.

Vaseghi, B., Zoghi, M., & Jaberzadeh, S. (2014). Does anodal transcranial direct current stimulation modulate sensory perception and pain? A meta-analysis study. *Clinical Neurophysiology: Official Journal of the International Federation of Clinical Neurophysiology*, *125*(9), 1847–1858. Available from https://doi.org/10.1016/j.clinph.2014.01.020.

Vaseghi, B., Zoghi, M., & Jaberzadeh, S. (2015). A meta-analysis of site-specific effects of cathodal transcranial direct current stimulation on sensory perception and pain. *PLoS One*, *10* (5), e0123873. Available from https://doi.org/10.1371/journal.pone.0123873.

Villamar, M. F., Volz, M. S., Bikson, M., Datta, A., Dasilva, A. F., & Fregni, F. (2013). Technique and considerations in the use of 4x1 ring high-definition transcranial direct current stimulation (HD-tDCS). *Journal of Visualized Experiments* (77), e50309. Available from https://doi.org/10.3791/50309.

Wiesenfeld, K., & Moss, F. (1995). Stochastic resonance and the benefits of noise: From ice ages to crayfish and SQUIDs. *Nature*, *373*(6509), 33–36. Available from https://doi.org/10.1038/373033a0.

Woods, A. J., Antal, A., Bikson, M., Boggio, P. S., Brunoni, A. R., Celnik, P., Cohen, L. G., Fregni, F., Herrmann, C. S., Kappenman, E. S., Knotkova, H., Liebetanz, D., Miniussi, C., Miranda, P. C., Paulus, W., Priori, A., Reato, D., Stagg, C., Wenderoth, N., & Nitsche, M. A. (2016). A technical guide to tDCS, and related non-invasive brain stimulation tools. *Clinical Neurophysiology: Official Journal of the International Federation of Clinical Neurophysiology, 127*(2), 1031–1048. Available from https://doi.org/10.1016/j.clinph.2015.11.012.

Zandvliet, S. B., Meskers, C., Kwakkel, G., & van Wegen, E. (2018). Short-term effects of cerebellar tDCS on standing balance performance in patients with chronic stroke and healthy age-matched elderly. *Cerebellum, 17*(5), 575–589. Available from https://doi.org/10.1007/s12311-018-0939-0.

Zeller, D., Litvak, V., Friston, K. J., & Classen, J. (2015). Sensory processing and the rubber hand illusion—An evoked potentials study. *Journal of Cognitive Neuroscience, 27*(3), 573–582. Available from https://doi.org/10.1162/jocn_a_00705.

Zhang, M., Mariola, E., Stilla, R., Stoesz, M., Mao, H., Hu, X., & Sathian, K. (2005). Tactile discrimination of grating orientation: fMRI activation patterns. *Human Brain Mapping, 25*(4), 370–377. Available from https://doi.org/10.1002/hbm.20107.

Part III

Peripheral nerve implants for somatosensory feedback

Chapter 9

Connecting residual nervous system and prosthetic legs for sensorimotor and cognitive rehabilitation

Giacomo Valle, Greta Preatoni and Stanisa Raspopovic

Neuroengineering Laboratory, Department of Health Sciences and Technology, Institute for Robotics and Intelligent Systems, ETH Zürich, Zürich, Switzerland

ABSTRACT

Leg amputees wear commercial prosthetic devices that do not give any sensory information about the interaction of the device with the ground or its movement. Amputees, relying on a very limited and uncomfortable haptic information from the stump–socket interaction, face grave impairments: risk of falls, decreased mobility, perception of the prosthesis as an extraneous body (low embodiment), and increased cognitive burden during walking with consequent psychological distress and device abandonment. Recently, the restoration of sensory feedback was obtained by stimulating the tibial nerve of the amputees through electrodes implanted in the nerve. It has been shown that this allows subjects to restore symmetric walking and confidence in the prosthesis, which enable increased speed of walking over uneven terrains. Touch and proprioception restored through intraneural stimulation diminish the fatigue during ambulation tasks as well as phantom limb pain. This also enhanced prosthesis embodiment and provided cognitive relief to amputees.

Keywords: Sensory feedback; amputees; neural stimulation; neural interfaces; prosthesis; peripheral nervous system; touch; neuroprosthesis; embodiment; cognitive

9.1 Introduction

For centuries, prosthetic devices were used as simple tools attached to the amputees' stump, but in recent decades the field of artificial limbs has become a vibrant field of research (Fig. 9.1).

The key challenge today is to develop prostheses that are not only esthetically more similar to a real limb and functionally efficient, but that are also

Somatosensory Feedback for Neuroprosthetics. DOI: https://doi.org/10.1016/B978-0-12-822828-9.00007-1

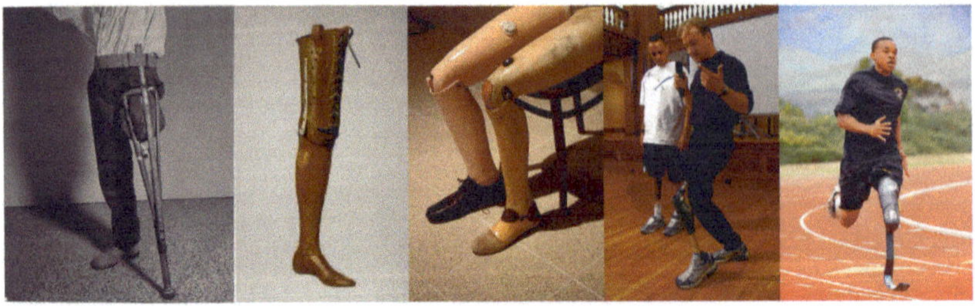

FIGURE 9.1 Evolution of prosthetic devices for lower limb amputees.

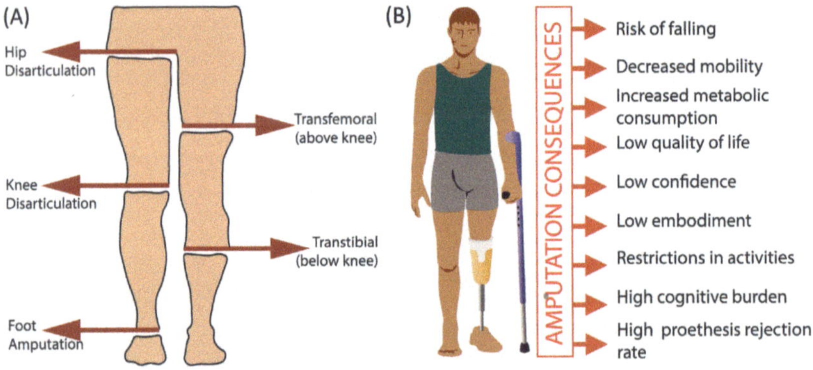

FIGURE 9.2 (A) Most common levels of lower extremity amputations. (B) Consequences of amputation on physical and psychological levels.

perceived as part of users' own bodies (Blanke, 2012; Blanke et al., 2015; Makin et al., 2017). When facing this issue, the main limiting factor is related to something that the majority of us takes for granted: the sense of touch. Although considerable efforts have focused on developing and controlling sophisticated lower limb prostheses (Hargrove et al., 2013, 2015), few trials have been conducted to restore sensory feedback (Raspopovic, 2020). Stepping over an obstacle, walking on uneven terrain, and balancing our weight are all things that might look easy and intuitive. However, for lower limb amputees these simple tasks become unbearable because of the lack of sensory feedback from the foot–ground interaction. This is particularly relevant for transfemoral amputations (Fig. 9.2A), which cause even less mobility and gait symmetry, together with higher energy expenditure compared to transtibial ones (Nolan et al., 2003; Waters et al., 1976).

To get feedback on their movements, leg amputees have to rely only on the very limited haptic information from the stump–socket contact. This results in dangerous consequences impacting on their physical and psychological health (Fig. 9.2B).

First, from a functional perspective, amputees have an increased risk of falling (Miller et al., 2001a) and decreased mobility (Nolan et al., 2003). Besides the obvious potential physical injury that falling may cause, we also have to consider that they have decreased independence, lower confidence, and self-imposed restrictions in activity (Tinetti et al., 1994). Furthermore, the lack of physiological feedback from the remaining extremity to the brain is associated with a higher cognitive burden. The constant need for a visual monitoring of the motor performance forces amputees to direct all their attention to the gait, leaving little or no attentional resources to be dedicated to simultaneous activities (Williams et al., 2006). All this leads to the inevitable consequence of the prosthesis being perceived as an external object, that is the prosthesis is not *embodied* in the subject's body schema (Blanke, 2012; de Vignemont, 2011). Improving the embodiment of prosthetic devices could promote their intuitive control, as well as increasing users' satisfaction and confidence (Makin et al., 2017).

Additionally, it must be considered that limb amputation often causes abnormal sensory experiences (Fig. 9.3), which can be painful (phantom limb pain) or nonpainful (phantom limb syndrome) (Flor, 2002; Flor et al., 2001, 2006; Melzack, 1990).

Clearly, pain, which occurs in 50%–80% of amputees (Ehde et al., 2000), is the most debilitating condition and has major effects on the subjects' quality of life (Van der Schans et al., 2003). The pain symptoms are described as knifelike, striking, pricking, and burning (Desmond & MacLachlan, 2010). On the other hand, phantom limb sensations may include phenomena such as perceiving the phantom limb in an unnatural anatomical position or so-called telescoping (Ramachandran, 1998; Rognini et al., 2019), that is shortening of the phantom limb.

Interestingly, the restoration of sensory feedback through peripheral nerve stimulation has shown great benefits for these problems. For instance, intraneural stimulation has shown great functional advantages, increasing walking speed, preventing falls, and enhancing agility (Petrini, Bumbasirevic, et al., 2019; Petrini, Valle, et al., 2019; Preatoni, Valle,

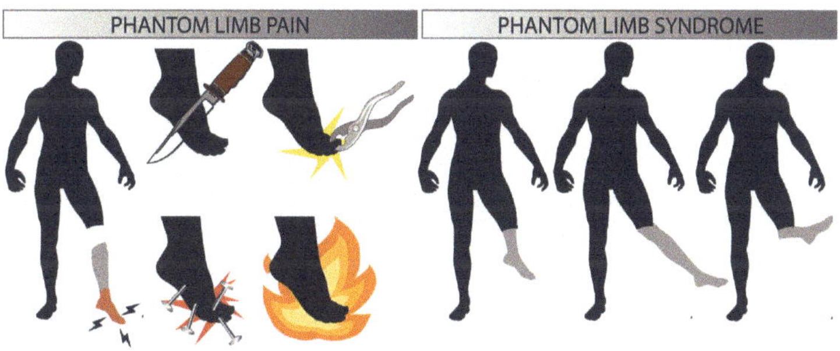

FIGURE 9.3 Phantom limb pain (*left*) and phantom limb syndrome (*right*).

Petrini, & Raspopovic, 2021; Raspopovic, Valle, & Petrini, 2021; Valle et al., 2021). Furthermore, the combination of intraneural stimulation and visual synchronous feedback (Fig. 9.4) has enabled upper limb amputees to reduce distorted phantom limb sensations and increase the embodiment of the prosthetic device (Graczyk et al., 2018; Rognini et al., 2019; Valle et al., 2018; Valle, Petrini et al., 2018). These phenomena are typically investigated by measuring subjective (questionnaires) and objective (proprioceptive drift) outcomes after some time of visuo-tactile stimulation. The proprioceptive drift entails asking the subject to indicate where he perceives his hidden limb in space: the closest he indicates his perception toward the *illusion object* and away from his real limb, the more the *trick* is successful. The hypothesis is that a synchronous stimulation would increase the ownership over the body part/object that is being stimulated (Botvinick & Cohen, 1998). The reasoning behind this can be found in the *unity assumption* (Welch & Warren, 1980), which states that if the subject perceives unisensory cues (in this case visual and tactile) as belonging to the same event, he/she is more prone to optimally integrate the multisensory information. Hence, if the amputee perceives the tactile feedback as congruent with what he experiences through vision, he is more likely to be persuaded that what he sees (the prosthesis) is merged within his body schema because the inputs arriving from the device are integrated.

However, while the majority of studies available in the literature has focused on developing (Windrich et al., 2016) and controlling (Hargrove et al., 2013, 2015) technologically advanced lower limb prostheses, considerably fewer trials have been conducted to restore sensory feedback (Charkhkar et al., 2018; Clippinger et al., 1982; Clites et al., 2018; Crea et al., 2017; Dietrich et al., 2018; Petrini, Bumbasirevic, et al., 2019; Petrini, Valle, Bumbasirevic, et al., 2019; Petrini et al., 2019; Rusaw et al., 2012).

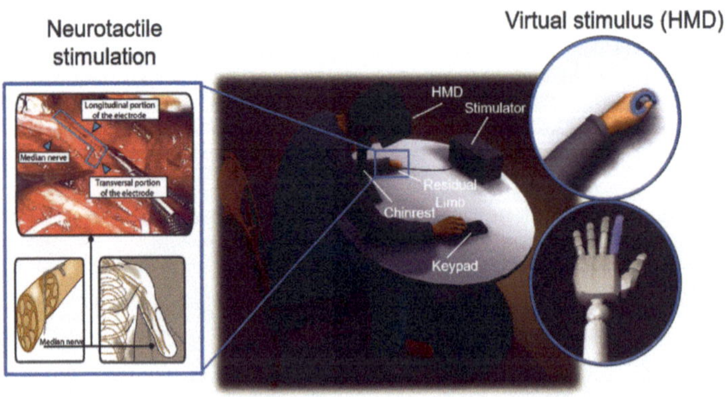

FIGURE 9.4 Multisensory stimulation in upper limb amputees. The subject sees a visual illumination of the prosthetic finger in virtual reality and receives an electrical stimulation (neurotactile stimulation), which can be either synchronous or asynchronous.

Generally, sensory feedback can be restored through invasive or noninvasive approaches (Fig. 9.5).

Surgical techniques (Clites et al., 2018; Petrini et al., 2019; Petrini, Valle, Bumbasirevic, et al., 2019; Petrini et al., 2019) have the benefit of recreating homologous (touch-like sensations) and somatotopic (referred on the phantom) sensations, but they have the drawback of requiring an operation. On the other hand, noninvasive approaches, such as continuous (Rusaw et al., 2012) or time-discrete vibrotactile (Crea et al., 2017) and electrocutaneous stimulation (Dietrich et al., 2018) have only limited benefits. This can be attributed to the fact that they are not selective (they evoke unrefined perceptions) and not homologous, which forces amputees to spend time in training to get used to and to interpret the new feedback strategy.

In addition to all the different approaches existing today for sensory feedback restoration, it is clear that there is a growing interest toward this field because of the many functional and cognitive benefits that it has shown. The remaining sections will discuss these aspects in more detail.

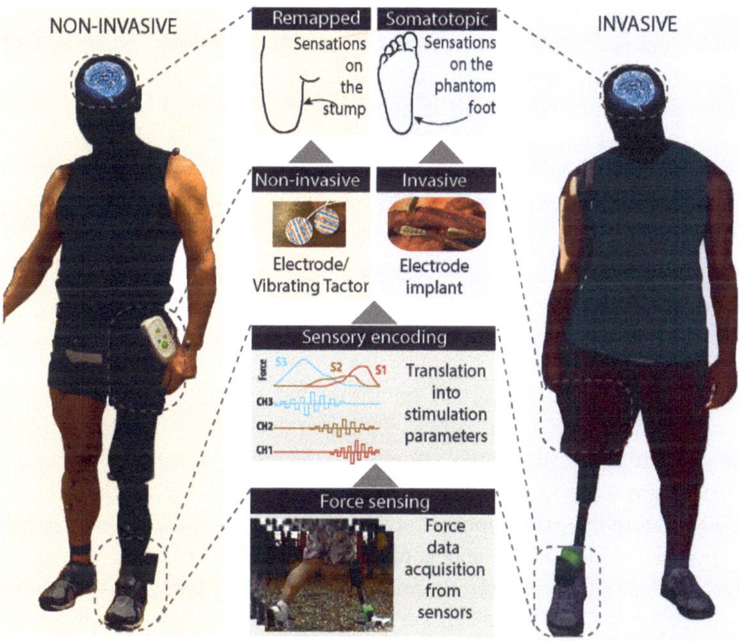

FIGURE 9.5 General schema of approaches to restore sensory feedback in amputees. Left: noninvasive. Right: invasive. The first step is the force sensing, where sensors collect the necessary data from the real world. This information is translated into parameters of stimulation, which can be conveyed in a noninvasive (e.g., surface electrodes, vibrating tactors) or invasive (e.g., TIME electrodes, cuff electrodes) manner. The final perception is a remapped sensation for the first approach and a somatotopic (in the phantom) sensation for the second one.

9.2 Intraneural electrodes

9.2.1 Implantable electrodes

To achieve a selective electrical interface to the peripheral nervous system, various types of microelectrodes have been developed and used over recent years (Navarro et al., 2005; Rutten, 2002). The design usually depends on its application. Often, very small structures are required to meet the high demands on selectivity and to address very small regions of neural tissue, for example, individual fascicles. For this reason, these small electrode structures are usually fabricated by microtechnological processes, for example, the creation of silicon-based structures (Wise et al., 2004). However, the drawback of using silicon-based electrodes is the mismatch between the mechanical properties of the implant and the tissue, leading to increased stress, tissue damage, and encapsulation of the implant by a layer of connective tissue (Polikov et al., 2005).

For this reason, the development of flexible thin-film electrodes based on polymers such as polyimide (Hoffmann et al., 2006) has been promoted. This material can be structured by microtechnological processes and allows a feature size down to 2–4 μm and a thickness of 10–15 μm. For the tracks, noble metals such as gold or platinum are typically used. To improve their electrochemical properties, the individual electrode contacts can be coated with functional materials, such as micro-rough platinum.

The neural interface electrode has long been the limiting technological component for achieving a successful interface to the nervous system. The adequate neural interface should be able to create a selective contact with different fascicles in the nerves in order to restore the efferent and afferent neural pathways in an effective way (e.g., stimulating different afferent nerves to deliver sensory feedback in a cybernetic prosthesis, or to extract kinematic and kinetic information for the control of neuroprostheses; Fig. 9.6).

The current state of the art on implanted interfaces for the peripheral nerve are divided into two types: extraneural (implanted around the nerve trunk) and intraneural, which penetrates the nerve trunk. Extraneural cuff electrodes are reliable, robust, and provide less invasiveness, but suffer from limited selectivity (Raspopovic et al., 2017; Tarler & Mortimer, 2004; Zelechowski et al., 2020) and capability of recording neural signals. With cuff electrodes, it is possible to detect the general activity of the nerve, and they have been used to switch on or off the contraction of muscle groups (Jensen et al., 2001). In order to improve selectivity, intraneural electrodes have been developed and tested for inserting longitudinally (Longitudinal Intra-Fascicular Electrode—LIFE) or transversally (Utah Slanted Electrode Array—USEA; Transversal Intrafascicular Multichannel Electrode—TIME) into the peripheral nerve (Fig. 9.7). This approach seems more promising

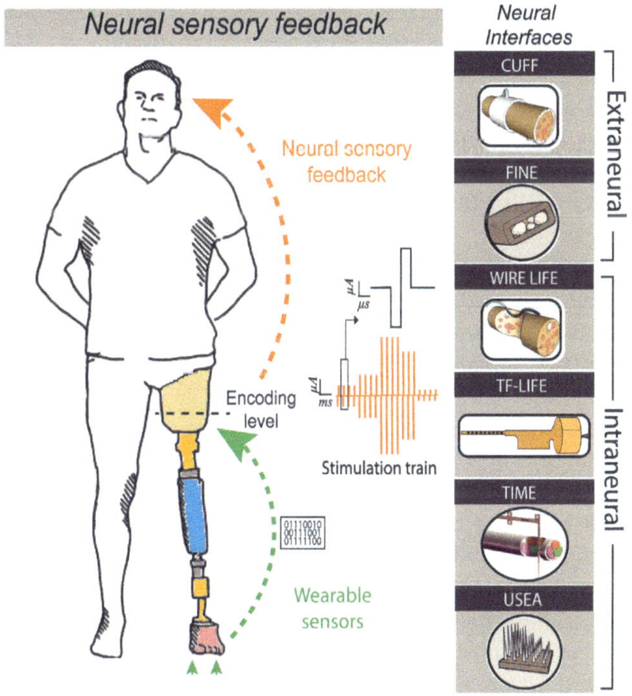

FIGURE 9.6 Neural sensory feedback provided by means of a neural interface (extraneural or intraneural) surgically implanted in the peripheral nervous system.

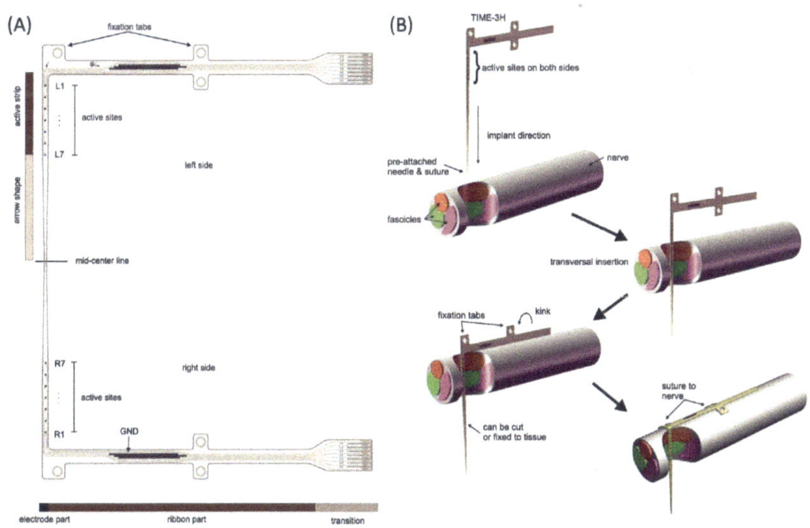

FIGURE 9.7 (A) Schematic representation of the transversal intrafascicular multichannel electrode (TIME); (B) implantation method into a peripheral nerve.

because it combines acceptable invasiveness with good selectivity (Badia et al., 2011). USEAs (Branner & Normann, 2000; Branner et al., 2001) are micromachined multineedle arrays made of silicon structures, originally developed as a neural interface for the brain, but modified for application in the peripheral nerve. USEAs are rigid silicon structures that record from the tips of the needles, transversally inserted in the nerve, which can induce damage in chronic implants. In contrast, LIFEs and TIMEs are flexible polymer structures inserted in the nerve, and thus better suited for the stretch motion of the nerve during limb movement (Badia et al., 2011; Lago et al., 2007; Lawrence et al., 2004).

For selective stimulation of peripheral nerves in the human arm, LIFEs have been applied in recent years (Dhillon & Horch, 2005; Overstreet et al., 2019; Rossini et al., 2010; Zollo et al., 2019). The thin-film LIFE (tf-LIFE) marked the development of a functional multichannel micro-fabricated LIFE structure. This type of microelectrode consists of a polyimide loop with multiple electrode contacts. Thanks to a second loop with an attached tungsten needle, the tf-LIFE can be placed inside a peripheral nerve and be longitudinally drawn through an individual fascicle to achieve a very close and selective contact. The second loop with the needle is removed after the implantation, and the electrode is fixed to the nerve by a suture. The tf-LIFE has already been applied in one human trial with a bidirectional hand prosthesis (Rossini et al., 2010). Some problems were reported regarding the mechanical stability of the electrode, especially at two sites (the point where the second polyimide loop with the needle is attached, and the points where the thin-film structure is connected to the adapter).

Furthermore, the state-of-the-art polyimide-based thin-film electrodes are associated with the major difficulty of approval as an active implantable medical device (required for clinical application), because they employ nonstandard materials to be implanted (polyimide) and are associated with a high risk of failure when the implantation time exceeds years. These electrodes can be coated by a thin-film metal. The latter is prone to delaminating from the polymer carrier and, because of its very thin layer thickness (typically 300 nm), does not allow any sort of electrochemical corrosion without immediate catastrophic failure. However, the latest technological developments in iridium oxide coatings and adhesion promotion layers between metal and polyimide led to promising results with respect to the long-term integrity of devices and electrode stability (Ordonez et al., 2012; Ordonez et al., 2012).

In order to overcome the problems associated with polyimide-based thin-film electrodes, a novel fabrication technology was developed that exclusively uses traditional implant materials such as cobalt-based alloys, platinum, and medical-grade silicone rubber. It requires a very lean (and easy to control) process line consisting of a spin-coater and a laser.

Recently, the use of intraneural solutions showed interesting and effective results in the field of neural prosthetics (Davis et al., 2016; George et al., 2019;

Petrini et al., 2019; Petrini et al., 2019; Petrini et al., 2019; Strauss et al., 2019; Valle et al., 2018; Valle et al., 2018; Wendelken et al., 2017). This type of interface allows very selective stimulation of the nerve fascicles and the sensory fibers eliciting the perception of several sensations referred to the phantom limb (Fig. 9.8). Interestingly, the TIME (Boretius et al., 2010) was adopted for targeting both the arm (median and ulnar) (Clemente et al., 2019; Čvančara et al., 2019; D'Anna et al., 2019; Granata et al., 2018; Mazzoni et al., 2020; Oddo et al., 2016; Petrini, Valle, Strauss, et al., 2019; Raspopovic et al., 2014; Risso et al., 2019; Rognini et al., 2019; Strauss et al., 2019; Valle, et al., 2018; Valle et al., 2018; Valle et al., 2020) and leg (tibial) (Petrini et al., 2019; Petrini et al., 2019) nerves. In this chapter, we focus more on neuroprosthetic legs exploiting the use of TIMEs.

9.2.2 Surgical procedure

To develop a sensorized neuroprothetic leg, the first necessary step is to perform a surgical implantation of the electrodes. In recent studies by Petrini and collaborators (Petrini, Bumbasirevic, et al., 2019; Petrini, Valle, Bumbasirevic, et al., 2019; Petrini, Valle, Strauss, et al., 2019), four TIME electrodes were implanted in the tibial branch of the sciatic nerve (Fig. 9.9), which conveys the majority of somatosensory innervation of the foot and ankle along with sensory–motor innervation of the leg. Implanting the intraneural electrodes was performed in an operating room under general anesthesia. The line of the incision for electrode insertion was over the sulcus between the biceps femoris and semitendinosus muscles, in the middle of the posterior aspect of the thigh, starting 4–5 cm proximally to the end of the amputation stump. The nerve was isolated, by moving the semitendinosus medially and the biceps femoris laterally. The electrode cables were tunneled through the thigh and routed outside the body through four small skin incisions on the level 3–5 cm higher than the pelvis.

9.3 Intraneural electrical stimulation

9.3.1 Characterization of the electrically evoked sensation

After the implantation, each channel from all the electrodes was connected to a stimulator purposely developed to drive the stimulation of TIME electrodes (Andreu et al., 2009). Then the sensation characterization (or mapping) procedure was performed, which allowed exploration of the subjects' sensation related to the stimulation from different electrodes and active sites.

Short trains of current pulse with variable intensity, pulsewidth, and frequency were delivered through every active site. Charge-balanced, biphasic, rectangular stimulation pulses were applied versus the ground electrode in a monopolar configuration. The single pulse varied between 10 and 120 μs

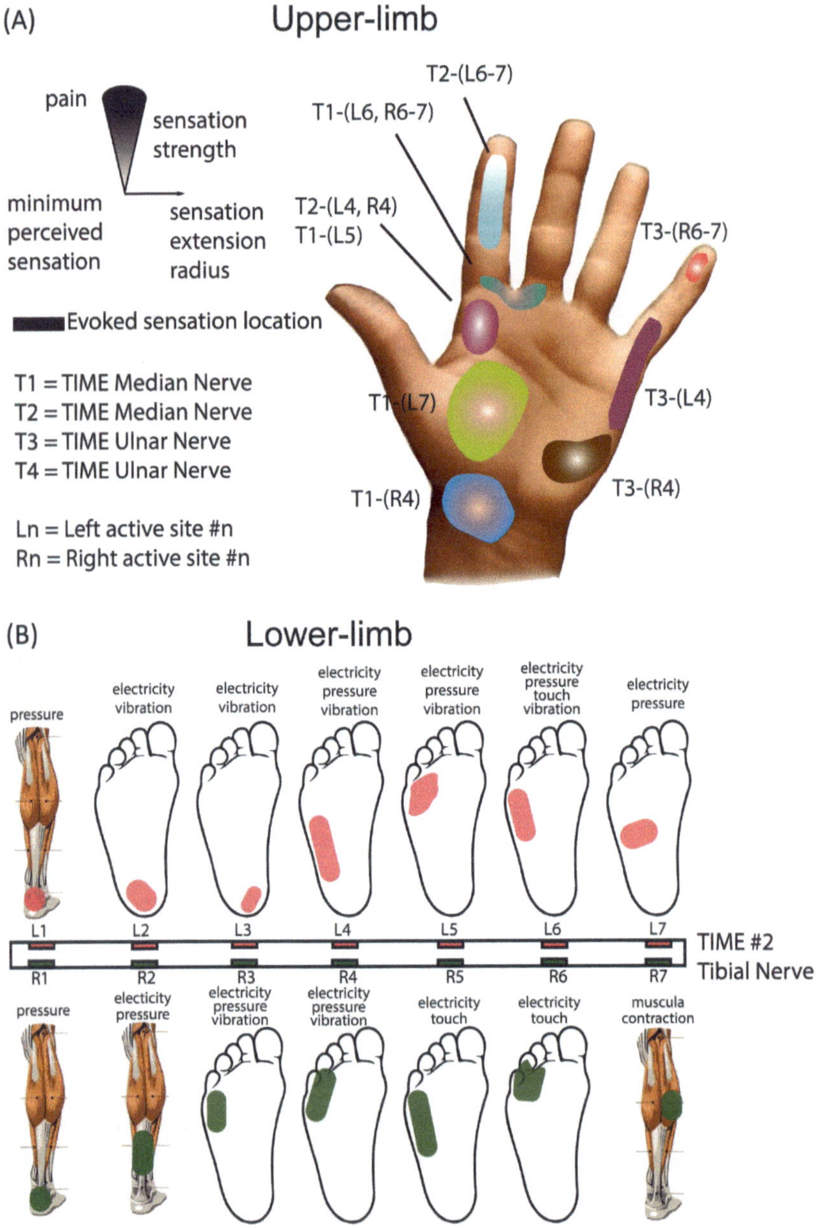

FIGURE 9.8 (A) Map of sensation locations and perceived intensities during stimulation of arm nerves; (B) map of sensation locations and types evoked during stimulation with a TIME implanted in the tibial nerve of an amputee.

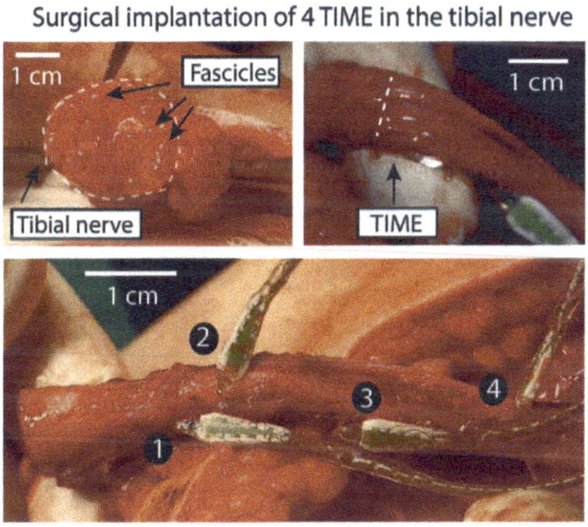

FIGURE 9.9 Surgical implantation of TIME in human tibial nerve for sensory feedback restoration in neuroprosthetics.

(steps of 10 μs), the pulse intensity between 20 and 980 μA (steps of 20 μA), and the pulse frequency between 4 and 500 (steps of 1 Hz). Pulse trains of 1 second duration were delivered. The interval between trains was 2 seconds.

Subjects were asked to report the location, type, and strength of the artificial sensations whenever they perceived them. The minimum threshold to sensation and the saturation values of the electrical charge were defined. The former parameter was considered as the lowest stimulus charge at which the subject reliably feels a sensation and the latter one as the stimulus charge at which the sensation becomes close to uncomfortable or painful. The stimulation parameters and the subjects' reports were controlled and recorded via a custom-made user interface specifically designed for the study (Valle et al., 2020).

A map of the sensations reported referred to the correspondent active sites was obtained and used for the calibration of the sensory feedback restoration system (Fig. 9.10A and B).

9.3.2 Neuroprosthetic leg

The neuroprosthetic leg is implemented with the personalized map of the evoked sensations obtained by the characterization procedure (Fig. 9.10C). The prosthesis is composed by commercial units (prosthetic knee and foot) and a customized socket/liner structure. The prosthetic leg was equipped with an encoder coupled with a Bluetooth unit, which could be used to make communication with devices external to the knee itself. A sensorized insole was placed under the prosthetic foot. The insole is constituted by a substrate of fabric, on which seven pressure sensors were arranged. The sensors have a resolution of 0.05 kg and a maximum measurable weight of 100 kg (Petrini et al., 2019; Petrini et al., 2019). The acquisition and amplification system of the sensorized sole has a sampling frequency

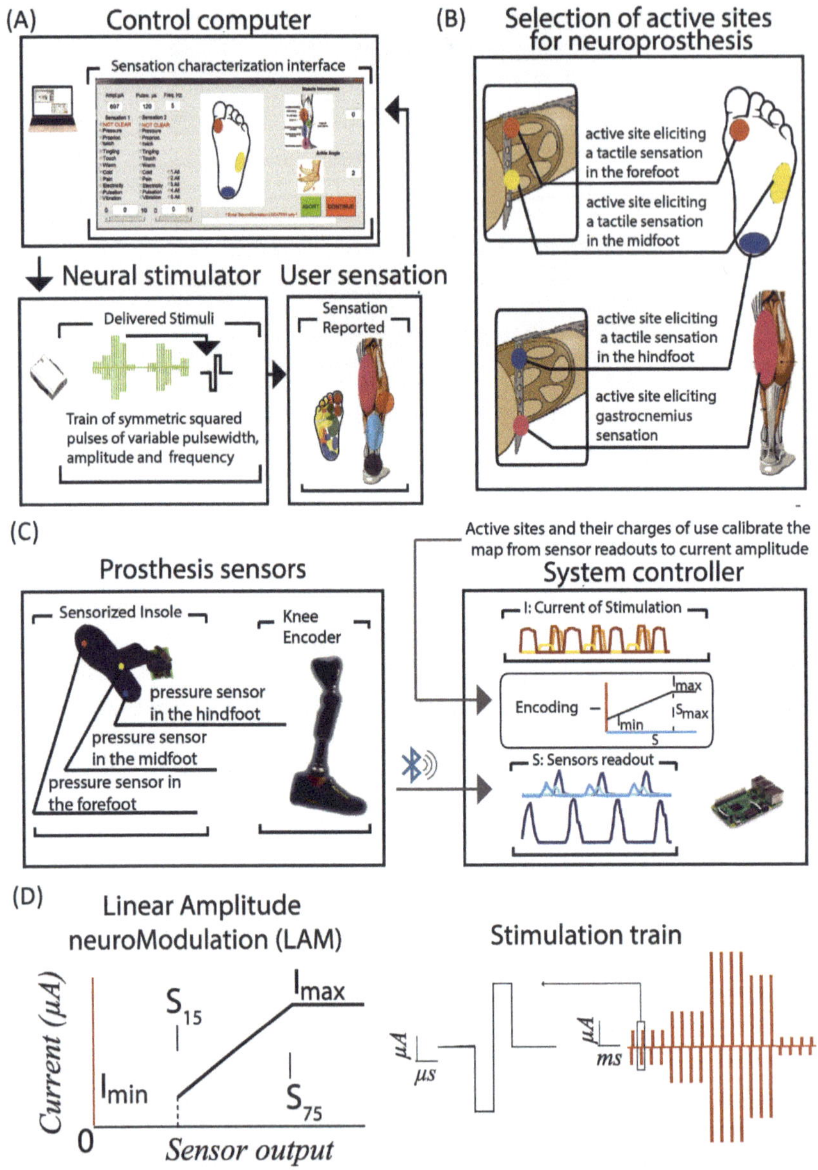

FIGURE 9.10 (A) Characterization is used to identify electrode active sites eliciting artificial sensations referred to the insole and knee locations, (B) and to determine minimum and maximum charges for driving the active sites. (C) The stimulus charge value and active sites are used to calibrate the system. A combination of sensors–active sites are selected to match prosthetic–phantom lower limb locations. (D) Linear amplitude modulation (LAM). The amplitude of the stimulation is linearly modulated according to the sensor values.

of 75 Hz and a Bluetooth module. An external controller (e.g., Raspberry pi 3) was wired to the external stimulator and communicated via Bluetooth with both the sensorized sole and the knee encoder. This portable processor managed the acquisition and recording of the sensor's readout and run the encoding algorithm for transducing the stimulation parameters needed for driving the stimulator. In particular, the results from the characterization procedure were used to couple sensors and active sites, eliciting a sensation in the phantom area corresponding to the position of the sensors themselves.

9.3.3 Sensory encoding strategy

In order to restore sensory feedback to the prosthetic user, the intraneural stimulation was delivered by different active sites (to be spatially matched) and modulated to provide force information (Raspopovic et al., 2014; Valle et al., 2018; Valle, Petrini et al., 2018). Indeed, three force sensors embedded in the sensorized insole and the knee encoder were used as control inputs for the intraneural stimulation of four active sites. One sensor was related to a sensation in the medial metatarsus, one to lateral metatarsus, one to the heel (tactile sensations), and there was one sensor related to muscle contraction of the calf (proprioceptive sensation) for each prosthetic user.

Then, the amplitude of biphasic, symmetric, cathodic-first, and rectangular charge-balanced pulses was modulated (Petrini, Valle, Bumbasirevic, et al., 2019; Petrini et al., 2019; Valle et al., 2018; Valle, Petrini et al., 2018) (Fig. 9.10D) according to the following linear relationship:

$$I = (I_{\max} - I_{\min})\frac{(s - s_{15})}{(s_{75} - s_{15})} + I_{\min} \quad \text{when } s_{15} \leq s \leq s_{75};$$

$$I = 0 \quad \text{when } s < s_{15};$$

$$I = I_{\max} \quad \text{when } s < s_{75};$$

where I is the current amplitude, s is the sensor readout, s_{15} and s_{75} represent the minimum and maximum load applied during walking by the subject in the case of the sensorized sole, and 10 and 55 degrees for the encoder. $I_{\min}$ and $I_{\max}$ are the stimulation current amplitudes that elicited, respectively, the minimum and maximum (i.e., below pain threshold) sensations as reported by the subject according to the sensation characterization procedure. The frequency (50 Hz) and pulsewidth (dependent on the active site) of the stimulation were fixed.

9.3.4 Sensorimotor integration

The restoration of sensory feedback through direct stimulation of somatosensory nerves showed several benefits in transfemoral amputees. In particular, the artificial sensory information provided in real-time in a sensorized prosthetic leg (Fig. 9.11), and also provided an improvement in functional performance of the prosthetic operation.

When the sensory feedback was provided to the prosthetic user, the walking was faster and more efficient. The higher prosthetic confidence allowed for a better functional performance and higher mobility in three lower limb amputees with respect to exploiting the same prosthesis without neural

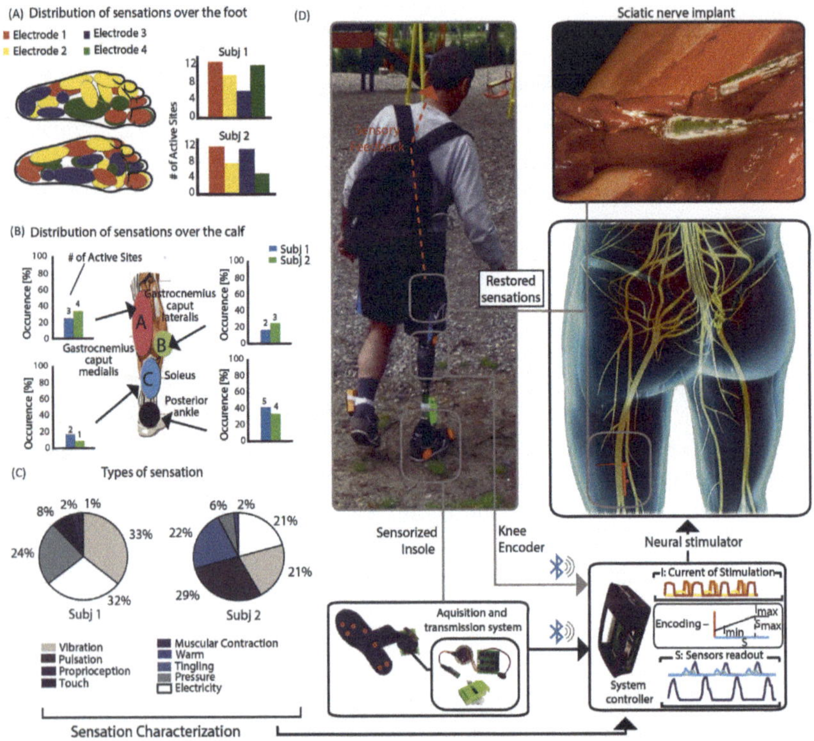

FIGURE 9.11 Sensorized bionic leg. The sensation characterization process is implemented to determine the response of the subject to the stimulation. (A) Distribution of tactile sensations over the foot elicited by the stimulation of the four electrodes (color coded). Bar plots report the number of active electrode active sites evoking a sensation in the foot. (B) Distribution of sensations over the lower leg. Bar plots represent the sensations occurring in the specific area (A, gastrocnemius caput medialis; B, gastrocnemius caput lateralis; C, soleus; d: posterior ankle). The number of active sites eliciting sensations is also reported. (C) Pie graphs represent the percentage of sensation types reported during the trial for each subject. (D) A subject wearing the whole system composed of a sensorized insole placed under the foot, which is fastened to the ankle with the electronics. The subject is using a commercial leg prosthesis where the microprocessor-controlled knee has an integrated knee encoder, an external controller, and an external stimulator. Data from both the insole and the knee encoder are transmitted via Bluetooth communication to the external controller that proportionally converts sensor readouts into currents of stimulation, which are injected into the electrodes implanted into the tibial branch of the sciatic nerve by the neural stimulator.

sensory feedback. Indeed, when the sensory feedback was implemented in the prosthetic leg, the user successfully recognized the localized touched area under the foot, the different flexion angles of the prosthetic leg, and their combination. Thanks to this intimate connection and to the fact that the system worked in real time without any perceivable delay (~50 ms), the subjects walked faster on the stairs (Petrini, Valle, Bumbasirevic, et al., 2019; Petrini et al., 2019) and even outdoors on sand (Petrini et al., 2019). Their agility was also boosted, allowing them to walk on a straight line in a more precise way (Fig. 9.12). This was enabled by the proprioceptive and tactile information provided to the user in real time that enhanced the ownership of the prosthetic device during the motor tasks.

Another interesting benefit shown by the restoration of artificial sensory feedback was the reduced risk of fall (Petrini, Valle, Bumbasirevic, et al., 2019; Petrini, Valle, Strauss, et al., 2019). Indeed, lower limb amputees experience a high risk of dangerous falls (Miller et al., 2001b), but the awareness triggered by the artificial feedback evoked when stepping on an obstacle prompts the user to stabilize walking (for example, by abruptly transferring the weight with a consequent stop of the prosthetic knee or making an extra stabilizing step). In other words, the subjects manage to use this information to drastically reduce their falls. The improvement in balance and

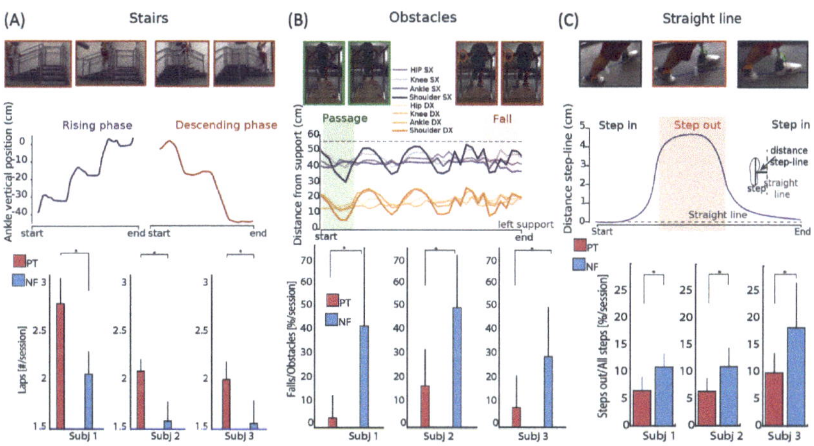

FIGURE 9.12 Sensory feedback improves walking performance of amputees. (A) Bar plots reporting the mean number of laps per minute with proprioception + touch (PT) condition, and without stimulation (NF) during stairs tests for S1, 2, and 3. The pictures depict an ascent and descent of the stairs. (B) Bar plots reporting the mean number of falls with PT, and NF during obstacles tests for S1, 2, and 3. (C) Bar plots reporting the mean number of steps out of the straight line over all the steps performed with feedback restored and without stimulation during straight line tests for S1, 2, and 3. The pictures represent a step performed out of the straight line, while the ones on the right represent a step accurately performed. Data in the figure are represented as means and std. *$p < .05$. Two-tailed ANOVA test with Tukey–Kramer correction for multiple groups was performed.

postural stability was observed also in transtibial amputees provided with neural sensory feedback (Charkhkar et al., 2020; Christie et al., 2020).

9.3.5 Cognitive integration

Although engineers are successfully implementing the concept of body augmentation from a technical point of view, little attention has been given to how the human brain might integrate such technological innovation (Makin et al., 2017). Indeed, human cognitive capacity has a limited nature (Wolfe & Horowitz, 2004). This has to be taken into account while making progress in the field of sensory feedback restoration to avoid cognitive overloading of the user (Makin et al., 2017) and to allow them to perceive the novel technology as a unified and consistent percept (Chen & Spence, 2017). This becomes even more relevant when considering that amputees have the constant need to visually monitor their prosthesis while performing functional tasks (Atkins et al., 1996). Since their attentional resources are mainly spent in controlling the prosthesis, they have a reduction in the resources available to perform other concomitant tasks (Williams et al., 2006). Hence, the sensory feedback should be able to *free* some of the cognitive space, rather than occupy it.

Recent research has shown that this is possible: the restoration of intraneurally plausible sensations allowed transfemoral amputees to allocate sufficient cognitive resources to differentiate between the target and nontarget tones in a three-tone auditory oddball task while walking (Petrini, Valle, Bumbasirevic, et al., 2019; Petrini et al., 2019). Interestingly, the EEG recordings showed that such resources were higher in the sensory feedback compared to the no-feedback condition (Fig. 9.13A). In particular, the P300 event-related potential component had a higher amplitude for the target tones while walking with the artificial feedback, suggesting that the latter provided a cognitive ease to the subjects.

These results suggest that a properly restored sensory feedback can be intuitively integrated by the subjects' central nervous system. This *neural embodiment* (Makin et al., 2017) has important functional and cognitive implications. For example, in a recent study Preatoni and collaborators showed that thanks to an invasive sensory feedback restoration a transfemoral amputee perceived the prosthesis as lighter, reducing its perceved weight of almost half a kg (Preatoni, Valle, Petrini, & Raspopovic, 2021). In addition, the benefits in motor control and learning previously discussed, and the integration of the prosthesis in the body schema may result in a decreased rejection rate. Indeed, not only the electroencephalographic recordings gave an objective measurement of the cognitive ease provided by the artificial feedback, but the amputees subjectively reported also a higher level of confidence (Fig. 9.13B) in the device (Petrini et al., 2019).

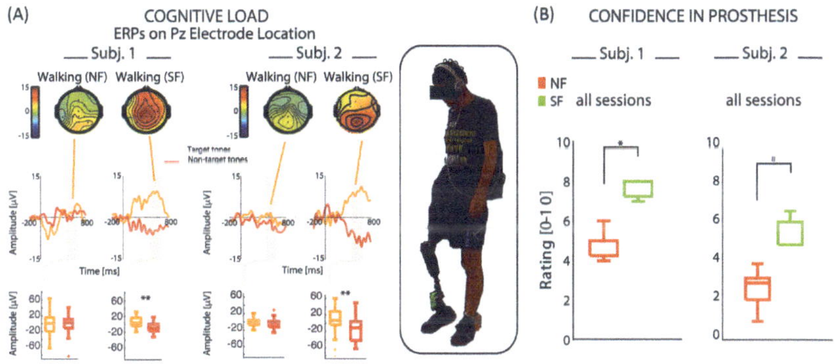

FIGURE 9.13 (A) Cognitive load assessment. The subjects are walking with (SF) and without (NF) sensory feedback and are asked to mentally discriminate target tones (red) from deviant (*orange*) tones. (A) Topographical maps of the voltage distribution in the P300 time window. As expected, the P300 component was most prominent over the parietocentral scalp locations. Significant values are $p < .01$. (B) EEG traces with respect to time. The gray areas represent the time window for the P300 computation, that is between 450 and 600 ms after the acoustic stimulus is presented. (C) Boxplots display the comparison of the amplitudes for the SF and NF conditions in the two subjects. (B) Confidence ratings of two transfemoral amputees while walking with (SF, *green*) and without (NF, *red*) sensory feedback. The scores represent the subjects' rating on a 10 cm visual analog scale (VAS scale) after motor tasks.

This gives promise to represent a solution for the high abandonment rate in prothesis use, which may be connected to a lack of confidence and low comfort (Gailey et al., 2010). Finally, sensory feedback has been shown to be able to enhance the embodiment (Fig. 9.14) toward the prosthesis (Marasco, Kim, Colgate, Peshkin, & Kuiken, 2011; Petrini, Bumbasirevic, et al., 2019; Petrini, Valle, et al., 2019; Preatoni, Valle, Petrini, & Raspopovic, 2021).

The latter refers to the feeling that the artificial device is a part of one's own body schema and has been proposed as a process of great relevance for the design of prostheses and their use for rehabilitation (Holmes & Spence, 2006). Indeed, the loss of afferent inputs following amputation disrupts the subject's body image (Rybarczyk & Behel, 2008). Nevertheless, research has shown that the feeling of ownership and body self-identification is strictly related with cutaneous touch (Armel & Ramachandran, 2003; Botvinick & Cohen, 1998; Ehrsson et al., 2004; Moseley, 2008; Tsakiris & Haggard, 2005). Indeed, there is a very strong relationship between the subject's feeling of ownership over a body part and the processing of tactile and visual inputs in the brain. Studies that focus on this field typically use illusions that are able to explore this dynamic neural representation, known as *body-matrix* (Moseley et al., 2012). Probably, the most famous one is the rubber hand illusion, introduced by Botvinick and Cohen in 1998 (Botvinick & Cohen, 1998). This perceptual illusion of ownership over a fake limb is created by stroking the hidden hand of the subject while synchronously stroking a

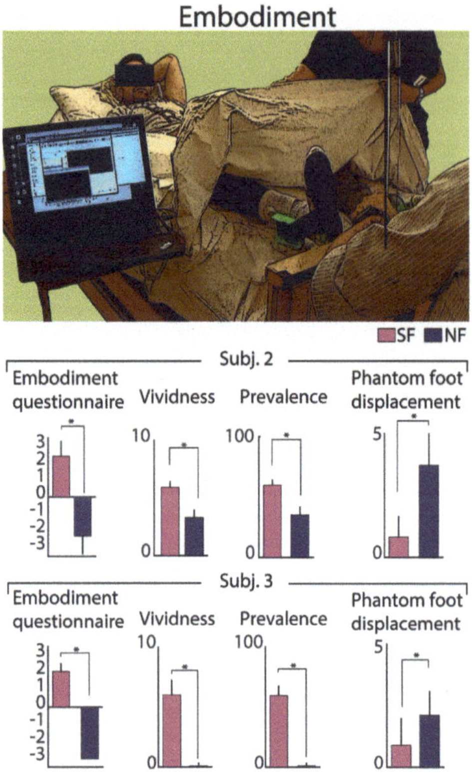

FIGURE 9.14 Top row: Experimental set-up for the objective measurement of embodiment (proprioceptive displacement task). The subject is lying down on a bed and cannot see his feet. The experimenter asks the subject to indicate where the phantom hallux is with the aid of a shaft moving into a rail. Bottom row: Subjects' scores after performing tasks with (SF, *pink*) and without sensory feedback (NF, *blue*). The scores are from the embodiment questionnaire, vividness of the embodiment feeling (how real is the perception), prevalence of the feeling (how long did it last), and of the proprioceptive displacement.

rubber hand placed in his eyesight. The approach of incorporating a rubber hand into the body-image of a subject is particularly interesting, because it opens up the possibility of obtaining a similar effect with a prosthesis able to provide physiologically relevant sensory feedback (Marasco et al., 2011; Schmalzl et al., 2014), which supports in turn the psychological and emotional adjustment to the amputation.

9.3.6 Health benefits

Thanks to the restoration of the connection between the limb and nervous system, some important health benefits were observed in transfemoral amputees. In particular, because of the lack of confidence, amputees produce counterbalancing movements that increase fatigue (Fleury et al., 2013). Thus, users do not manage to maintain symmetry during standing and walking, that is, they tend to shift more weight and to have a prolonged stance phase on the sound limb than on the prosthetic limb. The resulting abnormal kinematics and postural asymmetries can, after long-term use of the prosthesis, lead to musculoskeletal diseases as well as knee and hip osteoarthritis, osteoporosis, and back pain. Moreover, since they exert unnatural compensatory movements with prosthetic and healthy leg and body, they face augmented metabolic cost, then fatigue and occasionally heart failures (Fleury et al., 2013). Thus amputees,

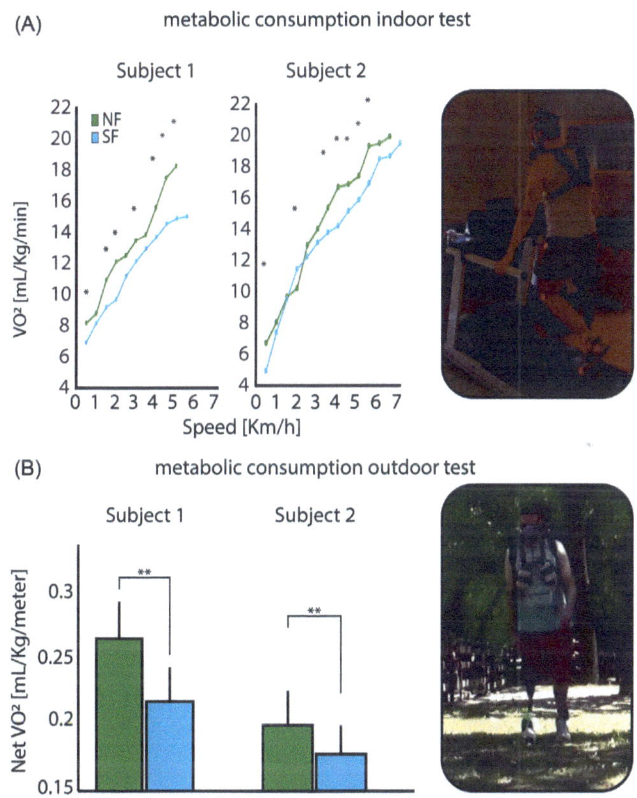

FIGURE 9.15 Metabolic consumption assessment. The cardiorespiratory consumption of the subjects was assessed in a task on a treadmill (indoor) and on a surface with grass (outdoor). (A) Top: oxygen consumption normalized on subject body mass (VO_2) on a treadmill when intraneural feedback is restored (SF, *light blue*) and without stimulation (NF, *green*) is displayed. The data are reported as mean values ± SEM. Top-right: a subject executing the treadmill task. (B) Bottom: net VO_2 in the two feedback conditions when the subjects were walking on the ground. The data are reported as mean values ± SD. Bottom-right: a subject executing the outdoor task. $^{*}p < .05$, $^{**}p < .01$.

especially thigh-level ones (transfemoral), face several challenges: sitting and standing up, performing steps or ramps, running, shuffling, and carrying loads can be a difficult and even dangerous task. The use of neural sensory feedback in leg prostheses showed decreased physical fatigue for participants compared to the no stimulation trials. This is an essential result since the amputees have more than a 100% higher risk of having a heart attack. With continuous use of this system it can be assumed that this risk will be diminished.

Indeed, metabolic consumption tests indoors and outdoors revealed that subjects spend significantly less energy when equipped with the SF with respect to the NF (Fig. 9.15).

Furthermore, participants exhibited reduced phantom limb pain with neural sensory feedback (Petrini, Bumbasirevic, et al., 2019). Indeed, the technology might target both peripheral and central components of pain by using neuromodulation (direct nerve stimulation) and boosting prosthesis cognitive integration (reduction of sensorimotor conflicts and distorted phantom limb representations in the brain). Reduction of pain by 30% or 2 points on the numeric rating scale have been suggested as clinically relevant outcomes (Farrar et al., 2001). The improvements found with direct stimulation of tibial nerve were more than 80% and significant pain suppression was achieved before the electrodes were explanted (Fig. 9.16). The acute reduction of pain (after every session of stimulation) could be explained through the gate control theory (Moayedi & Davis, 2012). Regarding the gradual pain diminishment until the complete vanishing over time, we hypothesized that it was due to the restoration of the physiologically plausible sensory feedback, which triggered beneficial neuroplastic changes in the brain (Flor et al., 2006). This represents the first step toward purposely defined peripheral nerve stimulation therapies for Phantom Limb Pain (PLP) treatment.

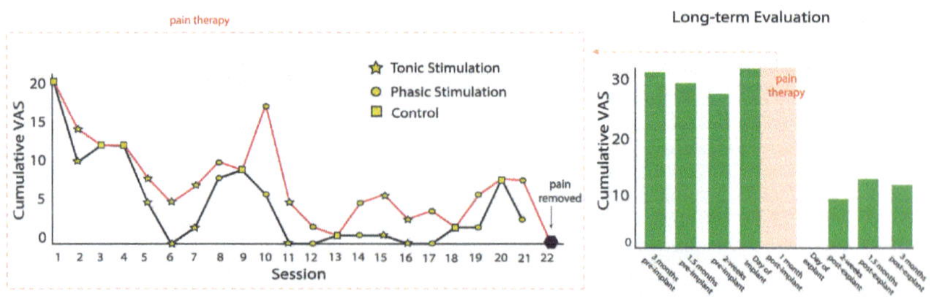

FIGURE 9.16 Pain treatments: VAS measurement. A pain treatment session consisted of 10 minutes of stimulation. Before and after it, the subjects completed the VAS questionnaire. The VAS was also recorded before the implant, before the explant, and 2 weeks, 1.5 months, and 3 months after the explant. The VAS score during the sessions with tonic (TS) and phasic (PS) stimulation treatments is shown for a lower limb amputee. A comparison between the Neuropathic Pain Symptom Inventory (NPSI) scores before and after the different treatments is shows. The NPSI evolution over the weeks is displayed.

9.4 Conclusions

After an amputation, the neural pathways between the remaining periphery and brain are still functional (Makin et al., 2013). Peripheral nerve electrical stimulation of the sensory fibers proximal to limb amputation can reactivate sensations from the missing extremity in the brain. A human–machine system, whereby prosthetic sensor readouts are translated into the language of the nervous system, is able to achieve significant health, cognitive, and functional benefits in leg amputees. The works in this field pave the way for further investigations about how the brain interprets different artificial feedback strategies and for the development of fully implantable sensory-enhanced leg neuroprostheses, which could drastically ameliorate quality of life in people with disability. Bionic legs, integrated with the residual peripheral nervous system of amputees, enable the brain to accept the prosthesis as the continuation of the natural leg, and this is essential for higher confidence of the users, and future widespread availability of these technologies (Valle, 2019).

References

Andreu, D., Guiraud, D., & Souquet, G. (2009). A distributed architecture for activating the peripheral nervous system. *Journal of Neural Engineering*, *6*(2), 026001. Available from https://doi.org/10.1088/1741-2560/6/2/026001.

Armel, K. C., & Ramachandran, V. S. (2003). Projecting sensations to external objects: Evidence from skin conductance response. *Proceedings of the Royal Society B: Biological Sciences*, *270*(1523), 1499–1506. Available from https://doi.org/10.1098/rspb.2003.2364.

Atkins, D., Heard, D., & Donovan, W. (1996). Epidemiologic overview of individuals with upper-limb loss and their reported research priorities. *Journal of Prosthetics and Orthotics*, 2–11.

Badia, J., Boretius, T., Andreu, D., Azevedo-Coste, C., Stieglitz, T., & Navarro, X. (2011). Comparative analysis of transverse intrafascicular multichannel, longitudinal intrafascicular and multipolar cuff electrodes for the selective stimulation of nerve fascicles. *Journal of Neural Engineering*, *8*(3), 036023. Available from https://doi.org/10.1088/1741-2560/8/3/036023.

Badia, J., Boretius, T., Pascual-Font, A., Udina, E., Stieglitz, T., & Navarro, X. (2011). Biocompatibility of chronically implanted transverse intrafascicular multichannel electrode (TIME) in the rat sciatic nerve. *IEEE Transactions on Bio-Medical Engineering*, *58*(8). Available from https://doi.org/10.1109/TBME.2011.2153850.

Blanke, O. (2012). Multisensory brain mechanisms of bodily self-consciousness. *Nature Reviews Neuroscience*, *13*(8), 556–571. Available from https://doi.org/10.1038/nrn3292.

Blanke, O., Slater, M., & Serino, A. (2015). Behavioral, neural, and computational principles of bodily self-consciousness. *Neuron*, *88*(1), 145–166. Available from https://doi.org/10.1016/j.neuron.2015.09.029.

Boretius, T., Badia, J., Pascual-Font, A., Schuettler, M., Navarro, X., Yoshida, K., & Stieglitz, T. (2010). A transverse intrafascicular multichannel electrode (TIME) to interface with the peripheral nerve. *Biosensors and Bioelectronics*, *26*(1), 62–69. Available from https://doi.org/10.1016/j.bios.2010.05.010.

Botvinick, M., & Cohen, J. D. (1998). Rubber hand 'feels' what eyes see. *Nature*, *391*, 756.

Branner, A., & Normann, R. A. (2000). A multielectrode array for intrafascicular recording and stimulation in sciatic nerve of cats. *Brain Research Bulletin*, *51*(4), 293–306. Available from https://doi.org/10.1016/s0361-9230(99)00231-2.

Branner, A., Stein, R. B., & Normann, R. A. (2001). Selective stimulation of cat sciatic nerve using an array of varying-length microelectrodes. *Journal of Neurophysiology*, *85*(4), 1585–1594. Available from https://doi.org/10.1152/jn.2001.85.4.1585.

Charkhkar, H., Christie, B. P., & Triolo, R. J. (2020). Sensory neuroprosthesis improves postural stability during sensory organization test in lower-limb amputees. *Scientific Reports*, *10*(1), 6984. Available from https://doi.org/10.1038/s41598-020-63936-2.

Charkhkar, H., Shell, C. E., Marasco, P. D., Pinault, G. J., Tyler, D. J., & Triolo, R. J. (2018). High-density peripheral nerve cuffs restore natural sensation to individuals with lower-limb amputations. *Journal of Neural Engineering*, *15*(5). Available from https://doi.org/10.1088/1741-2552/aac964.

Chen, Y.-C., & Spence, C. (2017). Assessing the role of the 'unity assumption' on multisensory integration: A review. *Frontiers in Psychology*, *8*. Available from https://doi.org/10.3389/fpsyg.2017.00445.

Christie, B. P., Charkhkar, H., Shell, C. E., Burant, C. J., Tyler, D. J., & Triolo, R. J. (2020). Ambulatory searching task reveals importance of somatosensation for lower-limb amputees. *Scientific Reports*, *10*(1), 10216. Available from https://doi.org/10.1038/s41598-020-67032-3.

Clemente, F., Valle, G., Controzzi, M., Strauss, I., Iberite, F., Stieglitz, T., Granata, G., Rossini, P. M., Petrini, F., Micera, S., & Cipriani, C. (2019). Intraneural sensory feedback restores grip force control and motor coordination while using a prosthetic hand. *Journal of Neural Engineering*, *16*(2), 026034. Available from https://doi.org/10.1088/1741-2552/ab059b.

Clippinger, F. W., Seaber, A. V., McElhaney, J. H., Harrelson, J. M., & Maxwell, G. M. (1982). Afferent sensory feedback for lower extremity prosthesis. *Clinical Orthopaedics and Related Research*, *169*, 202–206.

Clites, T. R., Carty, M. J., Ullauri, J. B., Carney, M. E., Mooney, L. M., Duval, J. F., Srinivasan, S. S., & Herr, H. M. (2018). Proprioception from a neurally controlled lower-extremity prosthesis. *Science Translational Medicine*, *10*(443). Available from https://doi.org/10.1126/scitranslmed.aap8373.

Crea, S., Edin, B. B., Knaepen, K., Meeusen, R., & Vitiello, N. (2017). Time-discrete vibrotactile feedback contributes to improved gait symmetry in patients with lower limb amputations: Case series. *Physical Therapy*, *97*(2), 198–207. Available from https://doi.org/10.2522/ptj.20150441.

Čvančara, P., Valle, G., Müller, M., Guiho, T., Hiairrassary, A., Petrini, F., Raspopovic, S., Strauss, I., Granata, G., Fernandez, E., Rossini, P. M., Barbaro, M., Yoshida, K., Jensen, W., Divoux, J.-L., Guiraud, D., Micera, S., & Stieglitz, T. (2019). On the reliability of chronically implanted thin-film electrodes in human arm nerves for neuroprosthetic applications. *BioRxiv*, 653964. Available from https://doi.org/10.1101/653964.

D'Anna, E., Valle, G., Mazzoni, A., Strauss, I., Iberite, F., Patton, J., Petrini, F. M., Raspopovic, S., Granata, G., Iorio, R. D., Controzzi, M., Cipriani, C., Stieglitz, T., Rossini, P. M., & Micera, S. (2019). A closed-loop hand prosthesis with simultaneous intraneural tactile and position feedback. *Science Robotics*, *4*(27), eaau8892. Available from https://doi.org/10.1126/scirobotics.aau8892.

Davis, T. S., Wark, H. A. C., Hutchinson, D. T., Warren, D. J., O'Neill, K., Scheinblum, T., Clark, G. A., Normann, R. A., & Greger, B. (2016). Restoring motor control and sensory feedback in people with upper extremity amputations using arrays of 96 microelectrodes implanted in the median and ulnar nerves. *Journal of Neural Engineering*, *13*(3), 036001. Available from https://doi.org/10.1088/1741-2560/13/3/036001.

de Vignemont, F. (2011). Embodiment, ownership and disownership. *Consciousness and Cognition*, *20*(1), 82–93. Available from https://doi.org/10.1016/j.concog.2010.09.004.

Desmond, D. M., & MacLachlan, M. (2010). Prevalence and characteristics of phantom limb pain and residual limb pain in the long term after upper limb amputation. *International Journal of Rehabilitation Research*, *33*(3), 279–282. Available from https://doi.org/10.1097/MRR.0b013e328336388d.

Dhillon, G. S., & Horch, K. W. (2005). Direct neural sensory feedback and control of a prosthetic arm. *IEEE Transactions on Neural Systems and Rehabilitation Engineering*, *13*(4), 468–472. Available from https://doi.org/10.1109/TNSRE.2005.856072.

Dietrich, C., Nehrdich, S., Seifert, S., Blume, K. R., Miltner, W. H. R., Hofmann, G. O., & Weiss, T. (2018). Leg prosthesis with somatosensory feedback reduces phantom limb pain and increases functionality. *Frontiers in Neurology*, *9*(Apr). Available from https://doi.org/10.3389/fneur.2018.00270.

Ehde, D. M., Czerniecki, J. M., Smith, D. G., Campbell, K. M., Edwards, W. T., Jensen, M. P., & Robinson, L. R. (2000). Chronic phantom sensations, phantom pain, residual limb pain, and other regional pain after lower limb amputation. *Archives of Physical Medicine and Rehabilitation*, *81*(8), 1039–1044. Available from https://doi.org/10.1053/apmr.2000.7583.

Ehrsson, H. H., Spence, C., & Passingham, R. E. (2004). That's my hand! Activity in premotor cortex reflects feeling of ownership of a limb. *Science*, *305*(5685), 875–877. Available from https://doi.org/10.1126/science.1097011.

Farrar, J. T., Young, J. P., LaMoreaux, L., Werth, J. L., & Poole, R. M. (2001). Clinical importance of changes in chronic pain intensity measured on an 11-point numerical pain rating scale. *Pain*, *94*(2), 149–158. Available from https://doi.org/10.1016/S0304-3959(01)00349-9.

Fleury, A. M., Salih, S. A., & Peel, N. M. (2013). Rehabilitation of the older vascular amputee: A review of the literature. *Geriatrics & Gerontology International*, *13*(2), 264–273. Available from https://doi.org/10.1111/ggi.12016.

Flor, H. (2002). Phantom-limb pain: Characteristics, causes, and treatment. *The Lancet Neurology*, *1*(3), 182–189. Available from https://doi.org/10.1016/S1474-4422(02)00074-1.

Flor, H., Denke, C., Schaefer, M., & Grüsser, S. (2001). Effect of sensory discrimination training on cortical reorganisation and phantom limb pain. *The Lancet*, *357*(9270), 1763–1764. Available from https://doi.org/10.1016/S0140-6736(00)04890-X.

Flor, H., Nikolajsen, L., & Staehelin Jensen, T. (2006). Phantom limb pain: A case of maladaptive CNS plasticity? *Nature Reviews Neuroscience*, *7*(11), 873–881. Available from https://doi.org/10.1038/nrn1991.

Gailey, R., McFarland, L. V., Cooper, R. A., Czerniecki, J., Gambel, J. M., Hubbard, S., Maynard, C., Smith, D. G., Raya, M., & Reiber, G. E. (2010). Unilateral lower-limb loss: Prosthetic device use and functional outcomes in servicemembers from Vietnam war and OIF/OEF conflicts. *The Journal of Rehabilitation Research and Development*, *47*(4), 317. Available from https://doi.org/10.1682/JRRD.2009.04.0039.

George, J. A., Kluger, D. T., Davis, T. S., Wendelken, S. M., Okorokova, E. V., He, Q., Duncan, C. C., Hutchinson, D. T., Thumser, Z. C., Beckler, D. T., Marasco, P. D., Bensmaia, S. J., & Clark, G. A. (2019). Biomimetic sensory feedback through peripheral nerve stimulation improves dexterous use of a bionic hand. *Science Robotics*, *4*(32), eaax2352. Available from https://doi.org/10.1126/scirobotics.aax2352.

Graczyk, E. L., Resnik, L., Schiefer, M. A., Schmitt, M. S., & Tyler, D. J. (2018). Home use of a neural-connected sensory prosthesis provides the functional and psychosocial experience of having a hand again. *Scientific Reports*, *8*(1), 9866. Available from https://doi.org/10.1038/s41598-018-26952-x.

Granata, G., Di Iorio, R., Romanello, R., Iodice, F., Raspopovic, S., Petrini, F., Strauss, I., Valle, G., Stieglitz, T., vančara, P., Andreu, D., Divoux, J.-L., Guiraud, D., Wauters, L., Hiairrassary, A., Jensen, W., Micera, S., & Rossini, P. M. (2018). Phantom somatosensory evoked potentials following selective intraneural electrical stimulation in two amputees. *Clinical Neurophysiology*, *129*(6), 1117–1120. Available from https://doi.org/10.1016/j.clinph.2018.02.138.

Hargrove, L. J., Simon, A. M., Young, A. J., Lipschutz, R. D., Finucane, S. B., Smith, D. G., & Kuiken, T. A. (2013). Robotic leg control with EMG decoding in an amputee with nerve transfers. *New England Journal of Medicine*, *369*(13), 1237–1242. Available from https://doi.org/10.1056/NEJMoa1300126.

Hargrove, L. J., Young, A. J., Simon, A. M., Fey, N. P., Lipschutz, R. D., Finucane, S. B., Halsne, E. G., Ingraham, K. A., & Kuiken, T. A. (2015). Intuitive control of a powered prosthetic leg during ambulation: A randomized clinical trial. *JAMA—Journal of the American Medical Association*, *313*(22), 2244–2252. Available from https://doi.org/10.1001/jama.2015.4527.

Hoffmann, K.P., Koch, K.P., Doerge, T., & Micera, S. (2006). New technologies in manufacturing of different implantable microelectrodes as an interface to the peripheral nervous system. In *The First IEEE/RAS-EMBS International Conference on Biomedical Robotics and Biomechatronics, 2006. BioRob 2006* (pp. 414–419). Available from https://doi.org/10.1109/BIOROB.2006.1639123

Holmes, N., & Spence, C. (2006). *Beyond the body schema: Visual, prosthetic, and technological contributions to bodily perception and awareness.* Oxford University Press.

Jensen, W., Lawrence, S. M., Riso, R. R., & Sinkjaer, T. (2001). Effect of initial joint position on nerve-cuff recordings of muscle afferents in rabbits. *IEEE Transactions on Neural Systems and Rehabilitation Engineering*, *9*(3), 265–273. Available from https://doi.org/10.1109/7333.948454.

Lago, N., Yoshida, K., Koch, K. P., & Navarro, X. (2007). Assessment of biocompatibility of chronically implanted polyimide and platinum intrafascicular electrodes. *IEEE Transactions on Bio-Medical Engineering*, *54*(2), 281–290. Available from https://doi.org/10.1109/TBME.2006.886617.

Lawrence, S. M., Dhillon, G. S., Jensen, W., Yoshida, K., & Horch, K. W. (2004). Acute peripheral nerve recording characteristics of polymer-based longitudinal intrafascicular electrodes. *IEEE Transactions on Neural Systems and Rehabilitation Engineering*, *12*(3), 345–348. Available from https://doi.org/10.1109/TNSRE.2004.831491.

Makin, T. R., De Vignemont, F., & Faisal, A. A. (2017). Neurocognitive barriers to the embodiment of technology. *Nature Biomedical Engineering*, *1*(1). Available from https://doi.org/10.1038/s41551-016-0014.

Makin, T. R., Scholz, J., Filippini, N., Slater, D. H., Tracey, I., & Johansen-Berg, H. (2013). Phantom pain is associated with preserved structure and function in the former hand area. *Nature Communications*, *4*, 1570. Available from https://doi.org/10.1038/ncomms2571.

Marasco, P. D., Kim, K., Colgate, J. E., Peshkin, M. A., & Kuiken, T. A. (2011). Robotic touch shifts perception of embodiment to a prosthesis in targeted reinnervation amputees. *Brain*, *134*(3), 747–758. Available from https://doi.org/10.1093/brain/awq361.

Mazzoni, A., Oddo, C. M., Valle, G., Camboni, D., Strauss, I., Barbaro, M., Barabino, G., Puddu, R., Carboni, C., Bisoni, L., Carpaneto, J., Vecchio, F., Petrini, F. M., Romeni, S., Czimmermann, T., Massari, L., di Iorio, R., Miraglia, F., Granata, G., … Micera, S. (2020).

Morphological neural computation restores discrimination of naturalistic textures in trans-radial amputees. *Scientific Reports*, *10*(1), 1–14. Available from https://doi.org/10.1038/s41598-020-57454-4.

Melzack, R. (1990). Phantom limbs and the concept of a neuromatrix. *Trends in Neurosciences*, *13*(3), 88–92. Available from https://doi.org/10.1016/0166-2236(90)90179-E.

Miller, W. C., Speechley, M., & Deathe, B. (2001a). The prevalence and risk factors of falling and fear of falling among lower extremity amputees. *Archives of Physical Medicine and Rehabilitation*, *82*(8), 1031–1037. Available from https://doi.org/10.1053/apmr.2001.24295.

Miller, W. C., Speechley, M., & Deathe, B. (2001b). The prevalence and risk factors of falling and fear of falling among lower extremity amputees. *Archives of Physical Medicine and Rehabilitation*, *82*(8), 1031–1037. Available from https://doi.org/10.1053/apmr.2001.24295.

Moayedi, M., & Davis, K. D. (2012). Theories of pain: From specificity to gate control. *Journal of Neurophysiology*, *109*(1), 5–12. Available from https://doi.org/10.1152/jn.00457.2012.

Moseley, G. L., Gallace, A., & Spence, C. (2012). Bodily illusions in health and disease: Physiological and clinical perspectives and the concept of a cortical 'body matrix'. *Neuroscience & Biobehavioral Reviews*, *36*(1), 34–46. Available from https://doi.org/10.1016/j.neubiorev.2011.03.013.

Moseley, L. G. (2008). I can't find it! Distorted body image and tactile dysfunction in patients with chronic back pain. *Pain*, *140*(1), 239–243. Available from https://doi.org/10.1016/j.pain.2008.08.001.

Navarro, X., Krueger, T. B., Lago, N., Micera, S., Stieglitz, T., & Dario, P. (2005). A critical review of interfaces with the peripheral nervous system for the control of neuroprostheses and hybrid bionic systems. *Journal of the Peripheral Nervous System: JPNS*, *10*(3), 229–258. Available from https://doi.org/10.1111/j.1085-9489.2005.10303.x.

Nolan, L., Wit, A., Dudziñski, K., Lees, A., Lake, M., & Wychowañski, M. (2003). Adjustments in gait symmetry with walking speed in trans-femoral and trans-tibial amputees. *Gait and Posture*, *17*(2), 142–151. Available from https://doi.org/10.1016/S0966-6362(02)00066-8.

Oddo, C. M., Raspopovic, S., Artoni, F., Mazzoni, A., Spigler, G., Petrini, F., Giambattistelli, F., Vecchio, F., Miraglia, F., Zollo, L., Di Pino, G., Camboni, D., Carrozza, M. C., Guglielmelli, E., Rossini, P. M., Faraguna, U., & Micera, S. (2016). Intraneural stimulation elicits discrimination of textural features by artificial fingertip in intact and amputee humans. *ELife*, *5*, e09148. Available from https://doi.org/10.7554/eLife.09148.

Ordonez, J., Schuettler, M., Boehler, C., Boretius, T., & Stieglitz, T. (2012). Thin films and microelectrode arrays for neuroprosthetics. *MRS Bulletin*, *37*(6), 590–598. Available from https://doi.org/10.1557/mrs.2012.117.

Ordonez, J.S., Boehler, C., Schuettler, M., & Stieglitz, T. (2012). Improved polyimide thin-film electrodes for neural implants. In *Conference Proceedings:...Annual International Conference of the IEEE Engineering in Medicine and Biology Society. IEEE Engineering in Medicine and Biology Society. Annual Conference, 2012* (pp. 5134–5137). Available from https://doi.org/10.1109/EMBC.2012.6347149.

Overstreet, C. K., Cheng, J., & Keefer, E. (2019). Fascicle specific targeting for selective peripheral nerve stimulation. *Journal of Neural Engineering*. Available from https://doi.org/10.1088/1741-2552/ab4370.

Petrini, F. M., Bumbasirevic, M., Valle, G., Ilic, V., Mijović, P., Čvančara, P., Barberi, F., Katic, N., Bortolotti, D., Andreu, D., Lechler, K., Lesic, A., Mazic, S., Mijović, B., Guiraud, D., Stieglitz, T., Alexandersson, A., Micera, S., & Raspopovic, S. (2019). Sensory feedback restoration in leg amputees improves walking speed, metabolic cost and phantom pain. *Nature Medicine*, *25*(9), 1356–1363. Available from https://doi.org/10.1038/s41591-019-0567-3.

Petrini, F. M., Valle, G., Bumbasirevic, M., Barberi, F., Bortolotti, D., Cvancara, P., Hiairrassary, A., Mijovic, P., Sverrisson, A. Ö., Pedrocchi, A., Divoux, J.-L., Popovic, I., Lechler, K., Mijovic, B., Guiraud, D., Stieglitz, T., Alexandersson, A., Micera, S., Lesic, A., & Raspopovic, S. (2019). Enhancing functional abilities and cognitive integration of the lower limb prosthesis. *Science Translational Medicine*, *11*(512), eaav8939. Available from https://doi.org/10.1126/scitranslmed.aav8939.

Petrini, F. M., Valle, G., Strauss, I., Granata, G., Di Iorio, R., D'Anna, E., Čvančara, P., Mueller, M., Carpaneto, J., Clemente, F., Controzzi, M., Bisoni, L., Carboni, C., Barbaro, M., Iodice, F., Andreu, D., Hiairrassary, A., Divoux, J.-L., Cipriani, C., ... Micera, S. (2019). Six-month assessment of a hand prosthesis with intraneural tactile feedback. *Annals of Neurology*, *85*(1), 137–154. Available from https://doi.org/10.1002/ana.25384.

Polikov, V. S., Tresco, P. A., & Reichert, W. M. (2005). Response of brain tissue to chronically implanted neural electrodes. *Journal of Neuroscience Methods*, *148*(1), 1–18. Available from https://doi.org/10.1016/j.jneumeth.2005.08.015.

Preatoni, G., Valle, G., Petrini, F. M., & Raspopovic, S. (2021). Lightening the perceived prosthesis weight with neural embodiment promoted by sensory feedback. *Current Biology*, *31* (5), 1065–1071.e4. Available from https://doi.org/10.1016/j.cub.2020.11.069, 33417885.

Ramachandran, V. (1998). The perception of phantom limbs. The D. O. Hebb lecture. *Brain*, *121*(9), 1603–1630. Available from https://doi.org/10.1093/brain/121.9.1603.

Raspopovic, S. (2020). Advancing limb neural prostheses. *Science*, *370*(6514), 290–291. Available from https://doi.org/10.1126/science.abb1073.

Raspopovic, S., Capogrosso, M., Petrini, F. M., Bonizzato, M., Rigosa, J., Di Pino, G., Carpaneto, J., Controzzi, M., Boretius, T., Fernandez, E., Granata, G., Oddo, C. M., Citi, L., Ciancio, A. L., Cipriani, C., Carrozza, M. C., Jensen, W., Guglielmelli, E., Stieglitz, T., ... Micera, S. (2014). Restoring natural sensory feedback in real-time bidirectional hand prostheses. *Science Translational Medicine*, *6*(222). Available from https://doi.org/10.1126/scitranslmed.3006820, 222ra19-222ra19.

Raspopovic, S., Petrini, F. M., Zelechowski, M., & Valle, G. (2017). Framework for the development of neuroprostheses: From basic understanding by sciatic and median nerves models to bionic legs and hands. *Proceedings of the IEEE*, *105*(1), 34–49. Available from https://doi.org/10.1109/JPROC.2016.2600560.

Raspopovic, S., Valle, G., & Petrini, F. M. (2021). Sensory feedback for limb prostheses in amputees. *Nature Materials*. Available from https://doi.org/10.1038/s41563-021-00966-9, 33859381.

Risso, G., Valle, G., Iberite, F., Strauss, I., Stieglitz, T., Controzzi, M., Clemente, F., Granata, G., Rossini, P. M., Micera, S., & Baud-Bovy, G. (2019). Optimal integration of intraneural somatosensory feedback with visual information: A single-case study. *Scientific Reports*, *9* (1), 7916. Available from https://doi.org/10.1038/s41598-019-43815-1.

Rognini, G., Petrini, F. M., Raspopovic, S., Valle, G., Granata, G., Strauss, I., Solcà, M., Bello-Ruiz, J., Herbelin, B., Mange, R., D'Anna, E., Iorio, R. D., Pino, G. D., Andreu, D., Guiraud, D., Stieglitz, T., Rossini, P. M., Serino, A., Micera, S., & Blanke, O. (2019). Multisensory bionic limb to achieve prosthesis embodiment and reduce distorted phantom limb perceptions. *Journal of Neurology, Neurosurgery & Psychiatry*, *90*(7), 833–836. Available from https://doi.org/10.1136/jnnp-2018-318570.

Rossini, P. M., Micera, S., Benvenuto, A., Carpaneto, J., Cavallo, G., Citi, L., ... Dario, P. (2010). Double nerve intraneural interface implant on a human amputee for robotic hand control. *Clinical Neurophysiology*, *121*(5), 777–783. Available from https://doi.org/10.1016/j.clinph.2010.01.001.

Rusaw, D., Hagberg, K., Nolan, L., & Ramstrand, N. (2012). Can vibratory feedback be used to improve postural stability in persons with transtibial limb loss? *Journal of Rehabilitation Research and Development*, *49*(8), 1239–1253. Available from https://doi.org/10.1682/JRRD.2011.05.0088.

Rutten, W. L. C. (2002). Selective electrical interfaces with the nervous system. *Annual Review of Biomedical Engineering*, *4*(1), 407–452. Available from https://doi.org/10.1146/annurev.bioeng.4.020702.153427.

Rybarczyk, B., & Behel, J. (2008). Limb loss and body image. In P. Gallagher, D. Desmond, & M. MacLachlan (Eds.), *Psychoprosthetics* (pp. 23–31). London: Springer. Available from https://doi.org/10.1007/978-1-84628-980-4_3.

Schmalzl, L., Kalckert, A., Ragnö, C., & Ehrsson, H. H. (2014). Neural correlates of the rubber hand illusion in amputees: A report of two cases. *Neurocase*, *20*(4), 407–420. Available from https://doi.org/10.1080/13554794.2013.791861.

Strauss, I., Valle, G., Artoni, F., D'Anna, E., Granata, G., Iorio, R. D., Guiraud, D., Stieglitz, T., Rossini, P. M., Raspopovic, S., Petrini, F. M., & Micera, S. (2019). Characterization of multi-channel intraneural stimulation in transradial amputees. *Scientific Reports*, *9*(1), 1–11. Available from https://doi.org/10.1038/s41598-019-55591-z.

Tarler, M. D., & Mortimer, J. T. (2004). Selective and independent activation of four motor fascicles using a four contact nerve-cuff electrode. *IEEE Transactions on Neural Systems and Rehabilitation Engineering*, *12*(2), 251–257. Available from https://doi.org/10.1109/TNSRE.2004.828415.

Tinetti, M. E., De Leon, C. F. M., Doucette, J. T., & Baker, D. I. (1994). Fear of falling and fall-related efficacy in relationship to functioning among community-living elders. *Journal of Gerontology*, *49*(3), M140–M147. Available from https://doi.org/10.1093/geronj/49.3.M140.

Tsakiris, M., & Haggard, P. (2005). The rubber hand illusion revisited: Visuotactile integration and self-attribution. *Journal of Experimental Psychology: Human Perception and Performance*, *31*(1), 80–91. Available from https://doi.org/10.1037/0096-1523.31.1.80.

Valle, G. (2019). The connection between the nervous system and machines: Commentary. *Journal of Medical Internet Research*, *21*(11), e16344. Available from https://doi.org/10.2196/16344.

Valle, G., D'Anna, E., Strauss, I., Clemente, F., Granata, G., Di Iorio, R., Controzzi, M., Stieglitz, T., Rossini, P. M., Petrini, F. M., & Micera, S. (2020). Hand control with invasive feedback is not impaired by increased cognitive load. *Frontiers in Bioengineering and Biotechnology*, *8*. Available from https://doi.org/10.3389/fbioe.2020.00287.

Valle, G., Iberite, F., Strauss, I., D'Anna, E., Granata, G., Stieglitz, T., Raspopovic, S., Petrini, F. M., Rossini, P. M., & Micera, S. (2020). A psychometric platform to collect somatosensory sensations for neuroprosthetic use. *BioRxiv*. Available from https://doi.org/10.1101/2020.07.23.218222, 2020, 218–222.

Valle, G., Mazzoni, A., Iberite, F., D'Anna, E., Strauss, I., Granata, G., Controzzi, M., Clemente, F., Rognini, G., Cipriani, C., Stieglitz, T., Petrini, F. M., Rossini, P. M., & Micera, S. (2018). Biomimetic intraneural sensory feedback enhances sensation naturalness, tactile sensitivity, and manual dexterity in a bidirectional prosthesis. *Neuron*. Available from https://doi.org/10.1016/j.neuron.2018.08.033.

Valle, G., Petrini, F. M., Strauss, I., Iberite, F., D'Anna, E., Granata, G., Controzzi, M., Cipriani, C., Stieglitz, T., Rossini, P. M., Mazzoni, A., Raspopovic, S., & Micera, S. (2018).

Comparison of linear frequency and amplitude modulation for intraneural sensory feedback in bidirectional hand prostheses. *Scientific Reports*, *8*(1), 16666. Available from https://doi.org/10.1038/s41598-018-34910-w.

Van der Schans, C. P., Geertzen, J. H. B., Schoppen, T., & Dijkstra, P. U. (2003). Phantom pain and health-related quality of life in lower limb amputees. *Journal of Pain and Symptom Management*, *24*(4), 429–436. Available from https://doi.org/10.1016/S0885-3924(02)00511-0.

Valle, G., Saliji, A., Fogle, E., Cimolato, A., Petrini, F. M., & Raspopovic, S. (2021). Mechanisms of neuro-robotic prosthesis operation in leg amputees. *Science Advances*, *7*(17). Available from https://doi.org/10.1126/sciadv.abd8354, 33883127.

Waters, R. L., Perry, J., Antonelli, D., & Hislop, H. (1976). Energy cost of walking of amputees: The influence of level of amputation. *Journal of Bone and Joint Surgery—Series A*, *58*(1), 42–46. Available from https://doi.org/10.2106/00004623-197658010-00007.

Welch, R. B., & Warren, D. H. (1980). Immediate perceptual response to intersensory discrepancy. *Psychological Bulletin*, *88*(3), 638–667. Available from https://doi.org/10.1037/0033-2909.88.3.638.

Wendelken, S., Page, D. M., Davis, T., Wark, H. A. C., Kluger, D. T., Duncan, C., Warren, D. J., Hutchinson, D. T., & Clark, G. A. (2017). Restoration of motor control and proprioceptive and cutaneous sensation in humans with prior upper-limb amputation via multiple Utah Slanted Electrode Arrays (USEAs) implanted in residual peripheral arm nerves. *Journal of NeuroEngineering and Rehabilitation*, 14. Available from https://doi.org/10.1186/s12984-017-0320-4.

Williams, R. M., Turner, A. P., Orendurff, M., Segal, A. D., Klute, G. K., Pecoraro, J., & Czerniecki, J. (2006). Does having a computerized prosthetic knee influence cognitive performance during amputee walking? *Archives of Physical Medicine and Rehabilitation*, *87*(7), 989–994. Available from https://doi.org/10.1016/j.apmr.2006.03.006.

Windrich, M., Grimmer, M., Christ, O., Rinderknecht, S., & Beckerle, P. (2016). Active lower limb prosthetics: A systematic review of design issues and solutions. *BioMedical Engineering Online*, *15*. Available from https://doi.org/10.1186/s12938-016-0284-9.

Wise, K. D., Anderson, D. J., Hetke, J. F., Kipke, D. R., & Najafi, K. (2004). Wireless implantable microsystems: High-density electronic interfaces to the nervous system. *Proceedings of the IEEE*, *92*(1), 76–97. Available from https://doi.org/10.1109/JPROC.2003.820544.

Wolfe, J. M., & Horowitz, T. S. (2004). What attributes guide the deployment of visual attention and how do they do it? *Nature Reviews Neuroscience*, *5*(6), 495–501. Available from https://doi.org/10.1038/nrn1411.

Zelechowski, M., Valle, G., & Raspopovic, S. (2020). A computational model to design neural interfaces for lower-limb sensory neuroprostheses. *Journal of NeuroEngineering and Rehabilitation*, *17*(1), 24. Available from https://doi.org/10.1186/s12984-020-00657-7.

Zollo, L., Pino, G. D., Ciancio, A. L., Ranieri, F., Cordella, F., Gentile, C., Noce, E., Romeo, R. A., Bellingegni, A. D., Vadalà, G., Miccinilli, S., Mioli, A., Diaz-Balzani, L., Bravi, M., Hoffmann, K.-P., Schneider, A., Denaro, L., Davalli, A., Gruppioni, E., ... Guglielmelli, E. (2019). Restoring tactile sensations via neural interfaces for real-time force-and-slippage closed-loop control of bionic hands. *Science Robotics*, *4*(27), eaau9924. Available from https://doi.org/10.1126/scirobotics.aau9924.

Chapter 10

Biomimetic bidirectional hand neuroprostheses for restoring somatosensory and motor functions

Francesco Iberite[2,3], Vincent Mendez[1], Alberto Mazzoni[2,3,*], Solaiman Shokur[1,*] and Silvestro Micera[1,2,3,*]

[1]*Bertarelli Foundation Chair in Translational Neuroengineering, Center for Neuroprosthetics and Institute of Bioengineering, Ecole Polytechnique Federale de Lausanne, Lausanne, Switzerland,* [2]*The BioRobotics Institute, Scuola Superiore Sant'Anna, Pisa, Italy,* [3]*Department of Excellence in Robotics and AI, Scuola Superiore Sant'Anna, Pisa, Italy*

ABSTRACT

Bidirectional hand neuroprostheses for amputee subjects aim to restore sensory–motor functions via the decoding of motor intentions in efferent channels and the stimulation of afferent nerves with electrical pulses conveying sensory information and closing the control loop. To optimize sensory feedback, neural stimulation should evoke both *informative* and *plausible* sensations. This latter factor affects the level of embodiment, which is crucial to avoid prosthesis rejection, as plausible and naturalistic sensations reduce the patient's discomfort and require less cognitive load. It also affects the steepness of the learning curve as it is more straightforward to associate the stimulation-induced perception with an intact hand. The amount of information conveyed by a given neural stimulation pattern and the similarity of the sensation with the natural ones depend on the choice of the encoding strategy, and there is, at the moment, no consensus on the optimal approach. For example, important elements of sensory feedback can be conveyed simply by modulation of the amplitude of the injected pulses; however, a natural temporal pattern of pulses for the encoded somatosensory feedback leads to more intuitive sensations. The challenges for the motor and bidirectional sensory–motor restoration are similar. Ideally, a seamless and intuitive control of the hand neuroprostheses using simultaneous and proportional decoding of the intended degrees of freedom is the aim.

* Equal contributions as senior authors.

Somatosensory Feedback for Neuroprosthetics. DOI: https://doi.org/10.1016/B978-0-12-822828-9.00011-3

In this chapter, we describe the technical approaches and recent results for both biomimetic encodings of sensory feedback and the decoding of motor functions.

Keywords: Neural stimulation; amputee; biomimetic; somatosensory feedback; bidirectional neuroprosthesis; hand; slow-adapting; fast-adapting; mechanoreceptor

10.1 Introduction

The human hand is an extremely dexterous "tool." To use it, we exploit a sophisticated yet effective and robust control strategy, which allows us to continuously control the fingers, to manipulate objects, and to interact with the external world. Robotic prosthetic hands (RPHs) allow amputee patients to recover part of their lost functions. As we will see in this chapter, the biomimetic approach (i.e., inspired by natural processes) holds the promise for RPHs to reproduce the closed-loop control dynamics of the natural arm. The pursuit of touch restoration through biomimetic encoding aims to design better strategies for tactile representation that are, compared to traditional linear encodings, more informative and easier to interpret for the end-user thanks to their biological inspiration.

The most straightforward neuromorphic strategy encodes dynamic and static components of the force or indentation separately, mimicking the action of slow-adapting (SA) and fast-adapting (FA) mechanoreceptors. More advanced techniques improve this approach by adding, for example, stochastic noise over the signal in order to mimic observed afferent nerve activity or by simulating mechanoreceptor activity with biologically plausible models. In our group, we recently tested a strategy to exploit the model of the whole afferent population and to infer the stimulation parameters (Rongala et al., 2020). Here we focus on biomimetic approaches to peripheral nerve stimulation, aiming to restore touch perception in RPH users. The workflow to develop biomimetic strategies goes from the study of biological systems to the design of a model capable of replicating the features of interest of the observed natural process.

In the motor domain, the decoding generally relies on surface electromyography (sEMG) recording of subjects' forearm muscles for the control, but other techniques using implanted EMG electrodes (Ortiz-Catalan, Mastinu, Sassu, Aszmann, & Brånemark, 2020; Smith & Hargrove, 2013) or intraneural electrodes (Davis et al., 2016; Cracchiolo et al., 2020) are also gaining popularity. In this chapter, we describe the challenges for biomimetic motor decoding and the efforts to increase the dexterity while keeping an intuitive and robust control. As we will see, motor decoding via intraneural recording can, in particular, be of interest to this end.

10.2 Mechanoreceptors and somatosensory pathways

Tactile information is encoded through changes in the membrane potentials of specialized nerve endings placed all over the skin, known as mechanoreceptors

(Johansson & Flanagan, 2009). Volar surfaces of hands and fingertips, in particular, have a high density of those sensors as they are paramount in manipulation and tactile exploration tasks. Mechanoreceptors and their associated nerve fibers are classified according to whether they can respond to the static component of the stimuli (i.e., slow adaptation) or mostly only to the dynamical part through fast adaptation (Johansson et al., 1982). The depth at which the receptor is located in the skin affects its sensitivity in time and also in space, as it determines the area of the skin the receptor collects information from, known as the receptive field (Johansson, 1978). According to this criterion, it is possible to discern type I receptors (located more superficially in the dermis or at the epidermal junction) and type II receptors (located deeper in the dermis). Combining the two classifications, there are four types of mechanoreceptors in the glabrous skin that together deliver tactile information to the brain through their afferent nerve fibers (i.e., FA I, FA II, SA-I, and SA II fibers). This classification is also useful from an engineering perspective, as it helps in understanding how our sensory system decomposes the tactile experience and how to mimic it. SA II fibers, responding to skin stretch (Knibestöl, 1975), also participate in the perception of the position of the limb, together with specialized receptors both in the muscle and tendon (muscle spindles and Golgi tendon organs; Gandevia & McCloskey, 1976).

The spike trains generated by these mechanoreceptive cells, carrying information from the upper limbs, are conveyed by the central branches of the primary sensory neurons in the dorsal root ganglia to the cuneate nucleus located in the medulla (Jörntell et al., 2014). Peripheral nerves have a hierarchical structure as they are divided first into fascicles, then into individual fibers (mostly both somatosensory and motor), which are protected by sheaths of connective tissue. Epineurium surrounds the entire nerve trunk. Perineurium covers the fascicles, and the endoneurium protects and isolates the axons (Afifi, 1991). Peripheral nerves are common targets of electrical stimulation to restore sensory feedback in neuroprosthesis.

10.3 Neural interfaces

All electrical stimulators share the same modality to interface with the nervous system. Electrical current pulses are sent through a conductive surface to the neural tissue. At the level of that interface, charges are redistributed and eventually depolarize the neuron membrane leading to action potentials. Usually, an electrode has more than one of these active stimulation sites. The electric field induced at the conductor–neural interface influences excitable tissue over a volume proportional to the charge delivered, and within a peripheral nerve, potentially triggering many fibers at once. This "indirect" mechanism through which the nerve is stimulated is an essential factor to take into account in the design of a stimulation strategy aiming to induce natural-like activity in the nerve. As we will see here, several electrode types are available for interfacing with the human nervous system, both in the commercial and research fields. Each of the available devices presents a different trade-off between invasiveness, size, and active site count.

Interfacing with the peripheral nervous system to stimulate afferent nerve fibers and to record neural activity requires a stable, reliable, and precise long-term electrical connection with the implanted nerve. The main features of a neural interface are, therefore, selectivity, that is, the capability to target a small subset of fibers, and lower invasiveness within the nerve. Almost all the electrode designs share some common components: an insulated substrate sustaining the structure, one or more active sites, where the stimulation is delivered through conductive areas (or where neural signals are recorded), and cables connecting the electrode with the other devices. As a reference for the most adopted means to deliver sensory feedback to the PNS, we present a selection of works involving bidirectional hand prostheses and their electrode interfaces (Fig. 10.1).

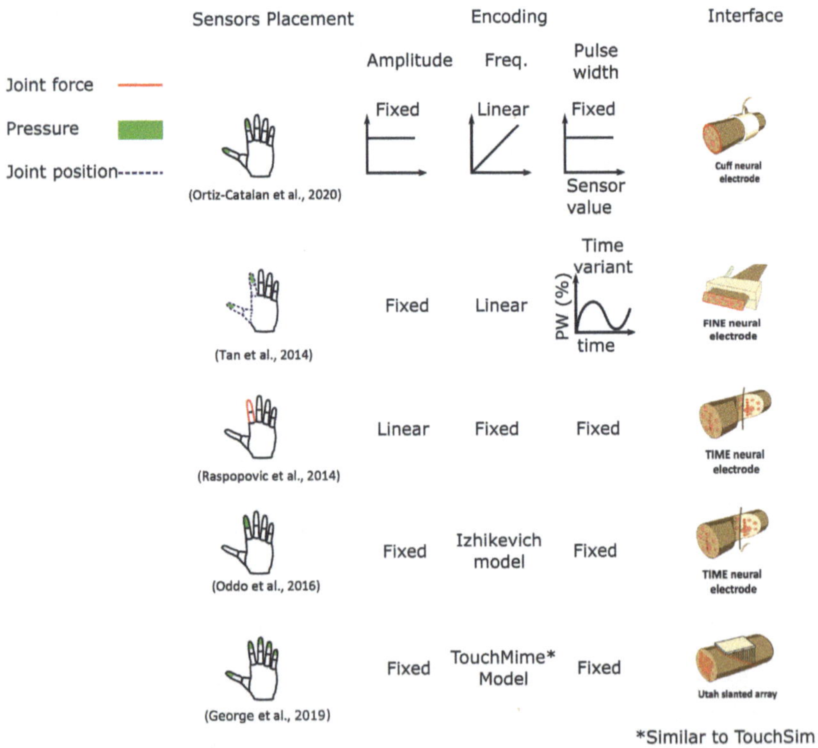

FIGURE 10.1 Summary of the milestones in the development of somatosensory feedback for bidirectional robotic prosthetic hands. The placement configuration of sensors is reported on each RPH and color-coded as follows: green for tactile sensors, red for force sensors, and blue for joint position sensors. The central columns contain information about the modulation of sensory information through the available parameters. At the rightmost column, the electrode interface with the peripheral nerve is shown.

Starting from the least invasive approach, cuff electrodes are implanted around the nerve of interest, wrapping it with an insulating substrate where conductive pads are placed in contact with the epineurium. Multisite cuff electrodes (Tarler & Mortimer, 2004) were proven to be effective in selectively targeting thick myelinated fibers close to the epineurium also in chronic implants (Ortiz-Catalan et al., 2014), while avoiding penetration of the nerve.

Flat interface nerve electrodes (FINEs) (Tyler & Durand, 2002) (Fig. 10.1) have a variation of the cuff design, aiming to improve the selectivity of the latter. In the FINE approach, the nerve is deformed with a stiff substrate to increase the available stimulation surface, thus increasing the selectivity and number of reachable fascicles.

To further improve the selectivity of the interface, longitudinally implanted intrafascicular electrodes (LIFEs) (Lefurge et al., 1991; Malagodi et al., 1989) have an insulated conductive wire inside the nerve to reach fibers within fascicles. While LIFEs were proven to be chronically viable (Lefurge et al., 1991) and functional with upper limb amputees (Dhillon et al., 2004; Navarro et al., 2007), they are limited by the small active site count.

The number of available active sites plays a critical role in the performance of an electrode. It is important to consider that electrodes are implanted with a surgical procedure where it is impossible to assess which fibers are in reach of each active site. Given the difficulty of planning an optimal electrode position a priori, it is then clear that the higher the number of active sites, the higher is the probability to target a suitable subset of neural fibers for the intended application.

Transverse intrafascicular multichannel electrodes (TIMEs) (Boretius et al., 2010; Fig. 10.1) aim to be a neural interface matching the selectivity of LIFE, and at the same time, by increasing the number of available channels (Badia et al., 2011), they can improve viability in neuroprosthesis applications. TIME electrodes are implanted transversally in the nerve and have conductive pads on the surface for delivering electric stimuli or recording neural activity. In our group, we used implanted TIMEs in upper limb amputees to restore sensory feedback (Oddo et al., 2016; Raspopovic et al., 2014; Valle, Mazzoni, et al., 2018).

Instead of implanting a narrow surface transversely, microelectrode arrays (MEAs; Nordhausen et al., 1996; Rutten et al., 1999) have a planar 2D array of inserted needles which have a rigid substrate lying tangentially on the nerve. There is a high number of those micrometer-sized needles in the electrode (Fig. 10.1). In order to reach deeper fascicles, the Utah slanted electrode array (USEA) (Branner et al., 2004) is designed with needles having lengths decreasing along the nerve axis, which maximizes the coverage by the device. MEA was tested as a peripheral interface (Warwick et al., 2003), but the improved reach makes USEA the preferred choice for neuroprostheses (George et al., 2019).

10.4 Neural stimulation

The main aim of neural stimulation in neuroprostheses is to convey information about selected features of tactile sensation, with most studies focusing on conveying applied force or tactile pressure. Note that the combination of forces perceived through different channels can give information about the shape of a grasped object, its stiffness (Raspopovic et al., 2014), or to a finer scale, even surface texture (Mazzoni et al., 2020; Oddo et al., 2016). These sensations can currently be delivered in a biomimetic way, that is, through the peripheral nerves originally conveying tactile information generated by the mechanoreceptors. If multiple stimulation sites are present, the precise location of the elicited sensation can be mapped in the phantom hand of the patient (Valle, Petrini, et al., 2018). It is important to note, however, that due to the spread of the stimulus current, the exact configuration of excited fibers may not be determined. Nevertheless, stimulating tactile afferents can also be used to convey nontactile information, with a procedure called "remapping." One key example is the delivery of proprioceptive information through peripherally stimulated fibers, which mainly consisted of tactile afferents, assessed by mapping in a related study (D'Anna et al., 2019).

Electrical current stimulation is typically delivered through the neural interface as a train of pulses of defined frequency, amplitude, and pulse width. The way these parameters are modulated by the tactile feature defines the particular encoding strategy for somatosensory feedback. From a low-level perspective, it is important to keep in mind that pulses are usually delivered in a biphasic fashion, with a symmetric application of negative and positive current, resetting the charge redistribution induced by a pulse right before the next. Charge balancing is fundamental to avoid tissue or electrode damage, and hence all the stimulation strategies discussed here present biphasic stimulation. We will classify, then, the feedback strategies only according to the way tactile features modulate the three parameters: amplitude, pulse width, and frequency of pulses (or more generally, the temporal structure of the pulse train).

Amplitude and pulse width act together, controlling the charge delivered by the individual pulses at the interface, and thus acting on the size of the corresponding electrical field. These two parameters control the recruitment of the neural fibers, that is, the fraction of the nerves triggered by each pulse. Recruitment in electrical nerve stimulation influences both the sensation magnitude and the area of the evoked perception. From a bio-inspired point of view, the most critical relationship is the latter, as a synchronous activation of multiple fibers at the same time usually causes unnatural and unpleasant sensations. Frequency, on the other side, changes, with some limitations, the firing rate of the recruited fibers. It acts on the sensation magnitude in a more natural way, as the total firing rate in a given population of fibers is strongly related to the magnitude of the perceived sensation. Therefore, for conveying focused intensity information, it is desired to increase the overall activity in a population of fibers, without changing recruitment much. If individual mechanoreceptive fiber axons could be selectively stimulated, a purely biomimetic approach would have fixed amplitude and fixed pulse width (as action potentials do not vary in this respect) and have an event-based temporal structure rather than a time-varying frequency. Since this is not possible, amplitude and pulse width can compensate for interface limitations by mimicking the progressive recruitment of more fibers, and time-varying frequency modulation can mimic the fluctuations of the total firing rate of a given afferent population.

10.5 Closed-loop system

The term "closed-loop" refers to systems that rely on the concept of feedback to adjust their internal parameters for reaching a defined set-point. The feedback is a signal, usually coming from sensors, that informs the system about the effect of its interactions with the external environment. The concept of feedback is on its own borrowed from the natural world, as it originates from the discipline of cybernetics founded by Wiener in 1948 (Wiener, 1948), who was trying to exploit the power of automatic systems imitating self-adjusting processes in nature.

In a closed-loop system, it is then possible to identify three main parts: an actuation section, a sensory section, and a controller adjusting the first with respect to the latter. The main difference with a natural system lies in the controller, as this usually relies in nature on physical or chemical reactions as means of self-adjustment, while in artificial systems, the control for the set-point is usually achieved by a digital processor.

A brief introduction about the inner functioning of closed-loop systems is necessary with the aim to introduce another requirement for a system designed for real-world use: available computational power. For a closed-loop system to work, the update of the controlling signal needs to happen at a faster rate than the changes in the observed variable, or else the performance will degrade. In the case of bidirectional hand prostheses, this principle translates to the need to deliver the feedback to the user without a perceptible delay, posing an upper bound to the computational time available to encode tactile information in stimulus patterns.

10.6 Encoding strategies

10.6.1 Linear modulation

The simplest way to encode sensory information in a stimulation pattern, both mechanically and electrically, is to vary one stimulation parameter according to the quantity of interest, while keeping the other parameters fixed. This approach was proven to be effective for intraneural electrical stimulation by testing all the available parameters (Raspopovic et al., 2014). Such a simple approach for sensory encoding is expected to deliver information about a single feature of interest in the tactile experience, usually the force applied or the position of the hand, conveyed by the magnitude of the evoked sensation.

10.6.2 Amplitude modulation

Amplitude modulation is, currently, one of the most precise strategies to deliver tactile information to the prosthesis user (Valle, Petrini, et al., 2018), which was assessed with psychophysical methods for comparison with healthy subjects. The amplitude of neural stimulation controls the section of the nerve influenced by the electrical field produced by the electrode, in this way modulating the number of fibers activated, known as the recruitment. Recruitment can indeed be used to encode sensation magnitude, but, as it is expected from the underlying physiological phenomena, it has the side effect of also influencing the projective field of the elicited sensation referred to the skin. This can lead to a loss of spatial precision in the sensory encoding. As stated before, pulse width concurs with amplitude in defining the charge delivered at the interface by each pulse, and so they can be considered somewhat equivalent. Neither of these approaches, while simple and precise, evoke natural sensation due to synchronous firing (Valle, Mazzoni, et al., 2018), and may induce paresthesia or unpleasant sensations. In order to limit the risk of such side effects, amplitude linear modulation is always defined in an interval strictly limited by the minimal stimulation amplitude perceived and the maximal stimulation amplitude, not causing any discomfort.

10.6.3 Frequency modulation

Frequency modulation is a straightforward approach for simple biomimetic encoding, as it is strongly related to how information is represented in the spiking rate. Experimental results using pulse frequency modulation to encode applied force confirmed this strategy as valid (Graczyk et al., 2018; Ortiz-Catalan et al., 2014). However, frequency modulation displays shorter adaptation times, namely the time taken by the prosthetic user until he/she stops perceiving a sustained neural stimulation, compared to amplitude modulation (Valle, Petrini, et al., 2018). Minimizing adaptation is important when designing a prosthesis intended for continuous manipulation of objects, for which a slowly changing modulation of force is applied. On the other hand, faster and nonconstant modulation of the frequency is known to deliver reliable tactile information over a manipulation task (Valle, Mazzoni, et al., 2018), suggesting the need for a more biomimetic approach compared to a simple linear frequency modulation. The frequency was also shown to change the quality of the evoked sensation (Graczyk et al., 2020) in a consistent way across multiple stimulation sites, proving its effectiveness as another tool to shape the sensory perception of the prosthesis user, alongside the magnitude. As stated above, both the frequency of pulses and charge per pulse contribute to the sensation magnitude. A step toward disentangling these phenomena was taken by Graczyk and colleagues (Graczyk et al., 2016); they proposed a metric of the activation charge rate which can be related to the sensation magnitude independently from the modulated parameter. This metric is consistent with the models which state that the population spike count is the most likely code for sensation magnitude (Güçlü & Dinçer, 2013; Muniak et al., 2007).

10.6.4 Biomimetic stimulation

Feedback strategies described so far are based on the modulation of a single parameter and rely on deterministic, typically linear relationships between the sensor value and the stimulation parameter. This might lead to synchronous (artificial) neural activations and usually unpleasant sensations. A simple solution to avoid synchronous spiking of the stimulated fibers is to add random noise to the sensor value, and, in this way, a small variation is obtained in the stimulation parameter over time. While this approach has been validated both in models and neural stimulation (Chatterjee & Robert, 2001; Moss, 2004), there is no bidirectional hand prosthesis implementing it at this time.

In the 2014 work of Tan et al. (2014), the problem of synchronous activation was addressed with simultaneous modulation of pulse width and frequency. In a bio-inspired fashion, frequency is modulated proportionally with a pressure sensor on the robotic hand, and pulse width is varied over time following a sinusoidal function to ensure nonconstant recruitment. Experimental results validated this approach, with the time-variant stimulation eliciting more natural sensations, while keeping control of the sensation magnitude and recruitment by the mean delivered charge.

Both the static (constant displacement) and dynamic (varying displacement, such as speed and even acceleration) components of mechanical deformation in the skin are detected through the activation of slow- and fast-adapting receptors. In the work of George et al. (2019), a biomimetic stimulation technique has been devised trying to encode both signals at the same time, mimicking a population activity of tactile fibers. The results showed a significant decrease in the time needed for size and compliance recognition tasks, supporting the intuitiveness of such an encoding.

High-frequency pulses are known to saturate the stimulated nerve fibers, inducing synchronous activation. This phenomenon can be exploited together with an amplitude increase over time to induce an asynchronous spiking activity in the population around the active site (Formento et al., 2020).

10.7 Neuron models

Neurons can be simulated as dynamic systems by mathematical models. A dynamical system is defined by a state vector, containing all the variables of interest, and a set of laws (equations) governing the change of the states over time. Neural dynamical systems, according to their possible trajectories in the state space, can be further analyzed by finding stable and unstable equilibrium points and identifying the onset of oscillations and bifurcations in the parameter space (Izhikevich, 2007). In this work, we only briefly describe a set of systems that has been mostly adopted in model-based tactile restoration.

In a scale measuring the biological plausibility, the neural model devised by Hodgkin and Huxley (HH) (Hodgkin & Huxley, 1952) and similar approaches are placed among those which adhere to experimental observations the most. The HH-type models capture all the dynamics induced by the ion flows through the cell membrane, faithfully representing a variety of different channel proteins on the particular neuron. The great biological plausibility of the HH-type models is counterbalanced by the computational power needed for the simulation of even a single neuron, making it rather unsuitable for large-scale network simulations and real-time applications.

While being still able to reproduce a subset of neuronal behavior, the integrate and fire (IF) (Archibald Vivian Hill, 1936; Lapicque & Lapicque, 1907) model is among the simplest neuron models available. The equations governing this model are:

$$\frac{du}{dt} = \frac{I(t)}{C} \tag{10.1}$$

$$if\ u \geq u^*,\ then\ u \leftarrow u_r \tag{10.2}$$

where u is the membrane potential, I the input current, t the simulation time, and C the value of the membrane capacitance. Whenever the membrane potential u crosses the threshold u^*, there is a spike, defined as a formal event characterized by a spike time $t^{(f)}$, after which the membrane potential is reset to u_r and the integration is suspended for a refractory period Δ^{abs}. In this case, membrane conductances are not considered, and only the membrane potential is represented as the state variable. While very simple, the IF model has been extensively used, in particular for encoding strategies intended for real-time use (Kim et al., 2010).

The IF model is a perfect integrator circuit, as below the threshold value u^* the membrane potential is given by integrating the current due to synaptic inputs over time. An extension of this model is the leaky integrate and fire (LIF) neuron model (Lapicque & Lapicque, 1907), where a resistance is placed in parallel to the capacitor mimicking the return of the membrane potential to its stable equilibrium point when no stimulation is provided. The equations of the LIF model are:

$$RC\frac{du}{dt} = RI(t) - u \tag{10.3}$$

$$if\ u \geq u^*,\ then\ u \leftarrow u_r \tag{10.4}$$

The variables considered are the same as in the IF model besides R that models the leak currents leading the membrane potential to its resting value (here set with value 0 for convenience). RC is the time constant which is sometimes represented as τ_m.

The LIF model has the advantage of simplicity with the added capability of better capturing time-dependent dynamics; it has been implemented by the Bensmaia group for model-based tactile encoding (Saal et al., 2017) and used in software platforms for simulating large-scale network activity, such as Nengo (Bekolay et al., 2014) and Brian (Stimberg et al., 2019).

One of the best trade-offs between simplicity and biological plausibility is given by the Izhikevich model (Izhikevich, 2003). It is suitable for both fast computation and capturing more complex dynamics, besides the regular spiking behavior observed in cortical neurons, as an adaptation to constant stimuli. The equations defining the Izhikevich model consist of an integrating and a resonating component:

$$\frac{du}{dt} = I(t) + 0.04u^2 + 5u + 140 - w \tag{10.5}$$

$$dw/dt = a(bu - w) \tag{10.6}$$

$$if\ u \geq 30\ ,\ then\ u \leftarrow c\ and\ w \leftarrow (w + d) \tag{10.7}$$

The first equation represents the dynamics of the membrane potential as a function of both input current and the recovery variable w that introduces behaviors such as adaptation (when the firing rate is reduced in response to a constant stimulus) and resonance. The dynamics of w is defined in the second equation. As in the previous models, the third equation defines the update of the state variables at the time of each spike. The four parameters a, b, c, and d represent, respectively, the restoration time of the recovery variable, the sensitivity of the recovery variable to the membrane potential, the reset potential, and the after-spike reset of the recovery variable which depends on slow, high-threshold conductances. These parameters can be tuned to fit many types of spiking behavior observed in cortical neurons. However, they do not have direct biological meaning; therefore the model is purely phenomenological. The Izhikevich model, thanks to its viability for real-time systems, has also been implemented in systems restoring the sense of touch peripherally with success (Oddo et al., 2016; Osborn et al., 2018).

10.8 Model-based approaches

Neural bioinspiration, as stated before, usually relies on reproducing the modulation of the variable of interest in a natural neural process. While the linear modulation strategies drastically simplify the input–output function between stimuli and mechanoreceptive fibers, model-based approaches are a class of strategies based on spiking neuron models. These models aim to replicate the dynamics of natural processes more faithfully and precisely, at the cost of increased computational power.

In the seminal papers by the Bensmaia group, they developed a model based on IF neurons, and simulated the input current as a linear combination of the pressure signal and its derivatives (Kim et al., 2009, 2010) to mimic the heterogeneity of mechanoreceptive fibers (Fig. 10.2A). In this model, the three contributions coming from the signal and its first and second derivatives are rectified, filtered, summed, and fed into a LIF neuron model with additive noise. The noise simulates the stochasticity observed in neural recordings. In a complementary approach, Oddo and colleagues developed a neuromorphic simulation in which simple inputs were injected in an adaptive neuronal model (Oddo et al., 2016; Rongala et al., 2017), with a stronger emphasis on the physical embodiment of the sensor in the biomimetic fingertip (Mazzoni et al., 2020). A similar approach has also been used to deliver pain (Osborn et al., 2018). Oddo's model (Fig. 10.2B) relies on embedded hardware implementing a fast simulation of the Izhikevich model; this is in contrast with the majority of biomimetic sensory feedback models, which are usually devised only as offline software simulations.

Recently, the TouchSim (Saal et al., 2017) was proposed as a model capable of simulating the whole afferent fiber activity starting from a tactile event, defined both in space (position and shape on the hand) and time (profile of the indentation). TouchSim extends the model devised by Kim et al. (2010), aiming to simulate the whole afferent activity instead of a single fiber. The tactile experience is simulated from the mechanical deformation of the skin (Fig. 10.2C) to the activity of the mechanoreceptors and with also the delay induced by the afferent nerves. Fig. 10.2C shows a single mechanoreceptive fiber implementation; each fiber is tuned to mimic the recorded activity of SA-I, RA-I, and RA-II fibers.

The work of Valle, Mazzoni et al. (2018) exploited the TouchSim model to determine tactile fibers' response over time both in terms of recruitment and spike rate. To replicate the desired features in the stimulated nerve, the electrical stimulus amplitude was modulated according to recruitment and frequency according to population activity. The sensation evoked with this hybrid strategy was felt to be more natural by the user while delivering reliable and precise force information. In the work of George et al. (2019), the TouchMime model simulated neural responses to the tactile interactions registered by the robotic hand sensors in real time, mimicking the population

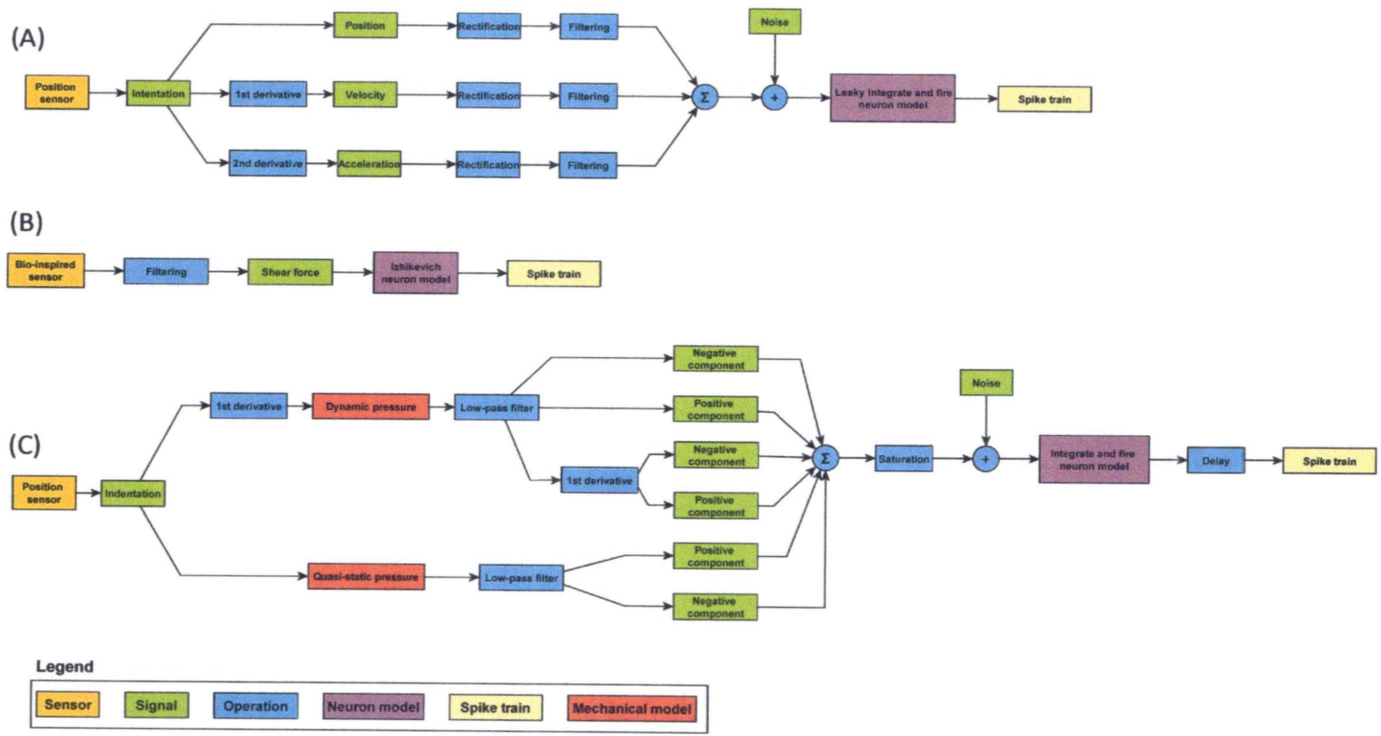

FIGURE 10.2 Model-based approaches for somatosensory feedback. Orange boxes refer to sensors, green boxes to signals, blue boxes to operations, purple to neural models, and red to mechanical models. (A) The model-based approach was proposed in the work of Kim et al. (2010). (B) Neuromorphic apprcach implementing an Izhikevich model encoding information coming from a bio-inspired tactile sensor (Rongala et al., 2017). (C) A model simulating the activity of the whole afferent population of the hand starting from the deformation of the skin (Saal et al., 2017).

response with the stimulation frequency. Performance in object size and compliance discrimination tasks improved significantly, suggesting increased intuitiveness of the encoding. Neuromorphic encoding can be implemented not only as a software solution but as an actual hardware configuration of the whole system, as neural models can be deployed on FPGAs and custom processors (ASICs). Given the advantage of bio-inspired information flow (Lee et al., 2019) and the reduced power need of neuromorphic hardware, this approach is promising for a new generation of prosthetic devices.

10.9 Challenges for bidirectional sensory and motor function restoration

A perfect motor strategy can be described as one that increases subject dexterity to a level comparable to the healthy hand. Given the complexity of the human hand, one can easily understand that this dream is far from our technical reality. Naturally, the proposed solutions should also be robust and reliable. The most widespread clinical solution is based on the activation amplitude of an agonist–antagonist pair of muscles to control the opening/closing of the RPH. The rationale of choice for this strategy is that recording from a large remaining muscle guarantees a good signal-to-noise ratio, and therefore good overall decoding robustness, and comes at a very low learning cost for the patient. In recent years, different strategies have been proposed to increase the number of grasp types available for the user. For example, some companies (e.g., Össur or Ottobock) offer the possibility to select among several kinds of grasp types by iterating (using muscle co-contractions) through a set of predefined configurations. This allows the subject to use several different types of grasp depending on the RPH capability. Such a solution keeps the advantage of a robust technique, but takes time to navigate through the different configurations in order to achieve the desired movement.

Another approach, based on motor learning (Dyson et al., 2020), uses an inverse map to relate motor outputs to arbitrary control variables. The technique consists of a center-out task where each target corresponds to a state (e.g., a certain grasping modality such as pinch or spherical grasp). This approach also uses a simple and robust muscle-decoding technique, and the time to trigger a different grasp does not increase with the size of the grasp catalog. However, it nevertheless necessitates a longer training period for the subject to master.

On the other hand, a machine learning approach based on pattern recognition with a larger number of electrodes (6–12) permits the adaptation of the decoder to the subject and can reduce training time. Indeed, after a short calibration, the algorithm can learn to map a grasp type to the subject's muscle activity. Using this approach, it was possible to obtain good classification rates for different kinds of grasps with performances of 90%–95% accuracy for 4–12 classes and up to 75% on 50 classes (Atzori et al., 2016; Castellini & Van Der Smagt, 2009; Ferguson et al., 2002; Kakoty & Hazarika, 2011; Martelloni et al., 2008). Handcrafted features were extracted to characterize the signal in a discriminative way and classify the type of grasp intended by the user. Thanks to the improvement of decoding robustness and wearable hardware solutions, this solution has become mature enough to reach the market (see COAPT system, http://www.coaptengineering.com; Ottobock Myo plus, https://www.ottobock.com). More advanced paradigms allow even the decoding of single finger movements. Using the classification of flexion or extension (Bhattachargee et al., 2019; Tenore et al., 2009), Bhagwat and Mukherji have shown up to 99.79% classification accuracy on 15 single and multifinger movements (Bhagwat & Mukherji, 2020). Finally, several authors have demonstrated control of the wrist and finger angles simultaneously and proportionally (Jiang et al., 2012; Muceli & Farina, 2012), while other studies also showed the possibility of single finger proportional control (Hioki & Kawasaki, 2012; Zhuang et al., 2019).

To improve decoding performance, high-density EMG (HD-EMG) recording systems were developed, consisting of a grid of closely spaced electrodes. With such a high number of electrodes, recordings are obtained from a wider surface of the subject's forearm, giving access to more detailed information on the underlying muscular activity. Decoding techniques can be borrowed from the field of image processing, as the record signals can be interpreted as spatial images of EMG activity. Several authors took advantage of this image-like representation of the signal to develop deep learning (DL) approaches from raw EMG data for motor intention decoding. Since 2016 (Park & Lee, 2016), we have observed a paradigm shift from feature engineering (handcrafted features) to feature learning using raw EMG data and DL. Several authors have shown superior performance of this DL approach with HD-EMG, as well as with a lower number of electrodes when compared to the standard features used in grasp classification (Atzori et al., 2016; Hu et al., 2018), and regression of arm or wrist motions (Ameri et al., 2019b; Xia et al., 2018). DL can increase decoding performance (Côté-Allard et al., 2019), improve robustness to electrode shift, and reduce the number of required repetitions during training (Ameri et al., 2019a). However, these examples remain at the prototype level, with decoding usually performed offline on a powerful computer. Integration and miniaturization are necessary to utilize these next-generation recording systems and decoding algorithms in commercial RPHs.

An invasive alternative to EMG for movement decoding consists of recording intraneural signals from implants in the peripheral nerves. Several studies have shown grasp classification could be obtained with high accuracy both offline and in real time (Davis et al., 2016; Petrini, Mazzoni, et al., 2019; Rossini et al., 2010; Wendelken et al., 2017). Intraneural recordings are invasive, but are more stable than sEMG over time since donning and doffing of the prosthesis does not change the electrode location as much as sEMG. Cracchiolo and colleagues (Cracchiolo et al., 2020) decoded up to 11 class states using TIME electrodes on an amputee subject, with, among other movements, adduction/abduction of the fingers, which is controlled via hand intrinsic muscles (Backhouse & Catton, 1954). It is important to note that hand intrinsic muscles cannot be targeted in noninvasive approaches. They showed that the active sites chosen on the first day could also be used in the following sessions up to 7 days (80% accuracy compared to 83% by selecting active sites every session). Using slanted Utah arrays, Wendelken et al. (2017) showed an informal 12 DoFs proportional control (5 DoFs formally assessed).

To summarize, motor decoding can be obtained from both sEMG, iEMG, and implanted neural electrodes with different approaches (Mendez, Iberite, Shokur, & Micera, 2021). While the sEMG is a well-studied and mature approach, intraneural decoding is a promising approach, potentially leading to more robust decoding due to higher signal stability, which allows decoding movements performed by muscles that are not present anymore in patients with amputation. As detailed in the previous section, this approach is also particularly interesting for somatosensory feedback restoration (D'Anna et al., 2019; Graczyk et al., 2016; Mazzoni et al., 2020; Petrini, Valle, et al., 2019), with results significantly above noninvasive counterparts. As such, a bidirectional intraneural implant with "read" and "write" capabilities will lead to the next big leap for neuroprosthetic hands that feel and are controlled like a natural limb. As we will see, bidirectional communication with the same implant has some major difficulties in terms of stimulation artifacts, interfering with recording for motor decoding, while delivering sensory feedback. While the technologies for intraneural recording and stimulation both exist, their integration introduces a major technical challenge due to stimulation artifacts. We discuss here the best solutions for this foreseeable issue.

10.9.1 Artifact removal for bidirectional neural systems

When an electrical stimulus is used to evoke afferent action potentials in peripheral nerves for somatosensory feedback, the stimulus causes an artifact that interferes with the measurement of the efferent signals for decoding due to the proximity between stimulating and recording electrodes (Scott et al., 1997).

We discuss several techniques to overcome or reduce stimulation artifacts. The first solution consists of recording and stimulating at a different site, for instance, implanting one or more electrodes both on the median and ulnar nerves. One example of bidirectional control based on intraneural electrodes is shown in Wendelken et al. (2017), where sensory feedback was delivered to the ulnar nerve, and motor intentions were decoded from the median nerve electrode. However, this solution is not completely satisfactory, as it reduces the number of possible channels for sensory feedback and is feasible only when more than one electrode is implanted.

Although there is no example of a bidirectional prosthesis control with only one neural electrode, we can follow the approach of brain–machine interfaces (BMIs), where bidirectional neuromodulation from a single implant has been a subject of interest for several years (Brown et al., 2008; O'Doherty, Shokur, Medina, Lebedev, & Nicolelis, 2019). Design and integration of artifact cancellation techniques are critical to obtaining efficient bidirectional BMIs (Zhou et al., 2018). In general, electrical stimulation will create persistent high-voltage transients during the stimulation period (Butovas & Schwarz, 2003), hiding discriminative biomarkers of the signal (for several milliseconds), and thus limiting efficient motor decoding. The first step consists of artifact prevention to reduce the artifact as much as possible, and to ease the following steps of artifact removal. Ideally, the cathodic and anodic phases of biphasic stimulation should be matched as much as possible to reduce artifacts (Merrill et al., 2005). Chu et al. (2013) decreased the intracortical microstimulation artifact time by 73% with a specially designed adequate stimulation waveform. Another approach to reducing stimulation artifact is the symmetric configuration between stimulation and recording electrodes (Stanslaski et al., 2012). Artifact reduction techniques include specific hardware design (front-end) and algorithmic cancellation methods (back-end). Conventional recording front end circuits (i.e., amplifiers) will saturate and recover slowly from the artifact. With special front ends, saturation can be prevented, lowering signal distortion (Rolston, 2009). Another approach consists of front ends that recover quickly from saturation (Viswam et al., 2016). Johnson et al. (2017) showed a combination of saturation prevention and rapid recovery. Back-end methods consist of the interpolation of data during the stimulation period (Zhou et al., 2018). When performing this approach online, the last value recorded before stimulation is typically held constant (also known as blanking; Hartmann et al., 2015). Other artifact removal techniques consist of subtracting an artifact template (Wichmann & Devergnas, 2011) from the original signal during stimulation or reconstructing the signal after removing the artifactual component obtained with signal decomposition (Al-ani et al., 2011; Lu et al., 2012). However, of all these back-end techniques, only a few were applied online (Culaclii et al., 2016; Limnuson et al., 2014; Mendrela et al., 2016; Wichmann & Devergnas, 2011; Zhou et al., 2019). See Zhou et al. (2018) for a review of artifact removal techniques used in BMIs.

The application of such techniques on peripheral implants is worth considering. Blanking is the simplest approach to reduce artifacts when stimulation time and frequency are relatively low (see Hartmann et al. (2015) for a noninvasive application). However, more complex stimulation strategies for somatosensory feedback (e.g., model-based approaches) require continuous stimulation at high frequency (Formento et al., 2020); therefore a simple blanking approach might deteriorate the signal too much for motor intention decoding. More complex artifact removal procedures, such as the ones described above, might be needed.

10.10 Conclusions

While modern RPH solutions permit patients to recover significant levels of motor and sometimes sensory functions, it is now time to consider solutions that are not only functionally valid, but that also approach the experience of a natural hand. In this chapter, we have presented a comprehensive list of techniques for biomimetic somatosensory feedback and the challenges to integrating them in a bidirectional RPH.

We believe that the pursuit of a biomimetic approach for the design of bidirectional hand prostheses will lead to devices that are more integrated with the information flow in the neural system, and those which substantially improve the experience of the users by providing plausible sensations and easy-to-understand control strategies. A prosthesis that has improved integration with the user is expected to provide increased quality of life to upper limb amputees by narrowing the difference between a healthy hand and the prosthetic device.

References

Afifi, A. K. (1991). The fine structure of the nervous system: Neurons and their supporting cells, 3rd ed. *Neurology*, *41*(9). Available from https://doi.org/10.1212/WNL.41.9.1535-b, 1535-1535-b.

Al-ani, T., Cazettes, F., Palfi, S., & Lefaucheur, J. P. (2011). Automatic removal of high-amplitude stimulus artefact from neuronal signal recorded in the subthalamic nucleus. *Journal of Neuroscience Methods*, *198*(1), 135–146. Available from https://doi.org/10.1016/j.jneumeth.2011.03.022.

Ameri, A., Akhaee, M. A., Scheme, E., & Englehart, K. (2019a). A deep transfer learning approach to reducing the effect of electrode shift in EMG pattern recognition-based control. *IEEE Transactions on Neural Systems and Rehabilitation Engineering*, 1. Available from https://doi.org/10.1109/TNSRE.2019.2962189.

Ameri, A., Akhaee, M. A., Scheme, E., & Englehart, K. (2019b). Regression convolutional neural network for improved simultaneous EMG control. *Journal of Neural Engineering*, *16*(3), 036015. Available from https://doi.org/10.1088/1741-2552/ab0e2e.

Archibald, V. H. (1936). Excitation and accommodation in nerve. *Proceedings of the Royal Society of London. Series B - Biological Sciences*, *119*(814), 305–355. Available from https://doi.org/10.1098/rspb.1936.0012.

Atzori, M., Cognolato, M., & Müller, H. (2016). Deep learning with convolutional neural networks applied to electromyography data: A resource for the classification of movements for prosthetic hands. *Frontiers in Neurorobotics*, *10*, 9. Available from https://doi.org/10.3389/fnbot.2016.00009.

Backhouse, K. M., & Catton, W. T. (1954). An experimental study of the functions of the lumbrical muscles in the human hand. *Journal of Anatomy*, *88*(Pt 2), 133–141.

Badia, J., Boretius, T., Andreu, D., Azevedo-Coste, C., Stieglitz, T., & Navarro, X. (2011). Comparative analysis of transverse intrafascicular multichannel, longitudinal intrafascicular and multipolar cuff electrodes for the selective stimulation of nerve fascicles. *Journal of Neural Engineering*, *8*(3), 036023. Available from https://doi.org/10.1088/1741-2560/8/3/036023.

Bekolay, T., Bergstra, J., Hunsberger, E., DeWolf, T., Stewart, T. C., Rasmussen, D., Choo, X., Voelker, A. R., & Eliasmith, C. (2014). Nengo: A python tool for building large-scale functional brain models. *Frontiers in Neuroinformatics*, *7*. Available from https://doi.org/10.3389/fninf.2013.00048.

Bhagwat, S., & Mukherji, P. (2020). Electromyogram (EMG) based fingers movement recognition using sparse filtering of wavelet packet coefficients. *Sadhana - Academy Proceedings in Engineering Sciences*, *45*(1), 1–11. Available from https://doi.org/10.1007/s12046-019-1231-9.

Bhattachargee, C. K., Sikder, N., Hasan, M. T., & Nahid, A. A. (2019). Finger movement classification based on statistical and frequency features extracted from surface EMG signals. In *5th International Conference on Computer, Communication, Chemical, Materials and Electronic Engineering, IC4ME2 2019*. https://doi.org/10.1109/IC4ME247184.2019.9036671

Boretius, T., Badia, J., Pascual-Font, A., Schuettler, M., Navarro, X., Yoshida, K., & Stieglitz, T. (2010). A transverse intrafascicular multichannel electrode (TIME) to interface with the peripheral nerve. *Biosensors and Bioelectronics*, *26*(1), 62–69. Available from https://doi.org/10.1016/j.bios.2010.05.010.

Branner, A., Stein, R. B., Fernandez, E., Aoyagi, Y., & Normann, R. A. (2004). Long-term stimulation and recording with a penetrating microelectrode array in cat sciatic nerve. *IEEE Transactions on Biomedical Engineering*, *51*(1), 146–157. Available from https://doi.org/10.1109/TBME.2003.820321.

Brown, E. A., Ross, J. D., Blum, R. A., Nam, Y., Wheeler, B. C., & DeWeerth, S. P. (2008). Stimulus-artifact elimination in a multi-electrode system. *IEEE Transactions on Biomedical Circuits and Systems*, *2*(1), 10–21. Available from https://doi.org/10.1109/TBCAS.2008.918285.

Butovas, S., & Schwarz, C. (2003). Spatiotemporal effects of microstimulation in rat neocortex: A parametric study using multielectrode recordings. *Journal of Neurophysiology*, *90*(5), 3024–3039. Available from https://doi.org/10.1152/jn.00245.2003.

Castellini, C., & Van Der Smagt, P. (2009). Surface EMG in advanced hand prosthetics. *Biological Cybernetics*, *100*(1), 35–47. Available from https://doi.org/10.1007/s00422-008-0278-1.

Chatterjee, M., & Robert, M. E. (2001). Noise enhances modulation sensitivity in cochlear implant listeners: Stochastic resonance in a prosthetic sensory system? *Journal of the Association for Research in Otolaryngology*, *2*(2), 159–171. Available from https://doi.org/10.1007/s101620010079.

Chu, P., Muller, R., Koralek, A., Carmena, J. M., Rabaey, J. M., & Gambini, S. (2013). Equalization for intracortical microstimulation artifact reduction. In *Proceedings of the Annual International Conference of the IEEE Engineering in Medicine and Biology Society, EMBS* (pp. 245–248). https://doi.org/10.1109/EMBC.2013.6609483

Côté-Allard, U., Fall, C. L., Drouin, A., Campeau-Lecours, A., Gosselin, C., Glette, K., Laviolette, F., & Gosselin, B. (2019). Deep learning for electromyographic hand gesture signal classification using transfer learning. *IEEE Transactions on Neural Systems and Rehabilitation Engineering*, *27*(4), 760–771. Available from https://doi.org/10.1109/TNSRE.2019.2896269.

Cracchiolo, M., Valle, G., Petrini, F., Strauss, I., Granata, G., Stieglitz, T., Rossini, P. M., Raspopovic, S., Mazzoni, A., & Micera, S. (2020). Decoding of grasping tasks from intraneural recordings in trans-radial amputee. *Journal of Neural Engineering*, *17*(2)026034. Available from https://doi.org/10.1088/1741-2552/ab8277.

Culaclii, S., Kim, B., Lo, Y. K., & Liu, W. (2016). A hybrid hardware and software approach for cancelling stimulus artifacts during same-electrode neural stimulation and recording. In *Proceedings of the Annual International Conference of the IEEE Engineering in Medicine and Biology Society, EMBS, 2016-October* (pp. 6190–6193). https://doi.org/10.1109/EMBC.2016.7592142

D'Anna, E., Valle, G., Mazzoni, A., Strauss, I., Iberite, F., Patton, J., Petrini, F. M., Raspopovic, S., Granata, G., Iorio, R., Di., Controzzi, M., Cipriani, C., Stieglitz, T., Rossini, P. M., & Micera, S. (2019). A closed-loop hand prosthesis with simultaneous intraneural tactile and position feedback. *Science Robotics*, *4*(27). Available from https://doi.org/10.1126/scirobotics.aau8892.

Davis, T. S., Wark, H. A. C., Hutchinson, D. T., Warren, D. J., O'Neill, K., Scheinblum, T., Clark, G. A., Normann, R. A., & Greger, B. (2016). Restoring motor control and sensory feedback in people with upper extremity amputations using arrays of 96 microelectrodes implanted in the median and ulnar nerves. *Journal of Neural Engineering*. Available from https://doi.org/10.1088/1741-2560/13/3/036001.

Dhillon, G. S., Lawrence, S. M., Hutchinson, D. T., & Horch, K. W. (2004). Residual function in peripheral nerve stumps of amputees: Implications for neural control of artificial limbs. *The Journal of Hand Surgery*, *29*(4), 605–615. Available from https://doi.org/10.1016/j.jhsa.2004.02.006.

Dyson, M., Dupan, S., Jones, H., & Nazarpour, K. (2020). Learning, generalization, and scalability of abstract myoelectric control. *IEEE Transactions on Neural Systems and Rehabilitation Engineering*. Available from https://doi.org/10.1109/tnsre.2020.3000310, 1–1.

Ferguson, S., Ferguson, S., & Dunlop, G.R. (2002). Grasp recognition from myoelectric signals. In *Proceedings of the Australasian Conference Robotics and Automation.*

Formento, E., D'Anna, E., Gribi, S., Lacour, S. P., & Micera, S. (2020). A biomimetic electrical stimulation strategy to induce asynchronous stochastic neural activity. *Journal of Neural Engineering*. Available from https://doi.org/10.1088/1741-2552/aba4fc.

Gandevia, S. C., & McCloskey, D. I. (1976). Joint sense, muscle sense, and their combination as position sense, measured at the distal interphalangeal joint of the middle finger. *The Journal of Physiology*, *260*(2), 387–407. Available from https://doi.org/10.1113/jphysiol.1976.sp011521.

George, J. A., Kluger, D. T., Davis, T. S., Wendelken, S. M., Okorokova, E. V., He, Q., Duncan, C. C., Hutchinson, D. T., Thumser, Z. C., Beckler, D. T., Marasco, P. D., Bensmaia, S. J., & Clark, G. A. (2019). Biomimetic sensory feedback through peripheral nerve stimulation improves dexterous use of a bionic hand. *Science Robotics*, *4*(32). Available from https://doi.org/10.1126/scirobotics.aax2352, eaax2352.

Graczyk, E. L., Christie, B. P., He, Q., Tyler, D. J., & Bensmaia, S. J. (2020). Frequency shapes the quality of tactile percepts evoked through electrical stimulation of the nerves. *BioRxiv*.

Graczyk, E. L., Resnik, L., Schiefer, M. A., Schmitt, M. S., & Tyler, D. J. (2018). Home use of a neural-connected sensory prosthesis provides the functional and psychosocial experience of having a hand again. *Scientific Reports*, *8*(1), 9866. Available from https://doi.org/10.1038/s41598-018-26952-x.

Graczyk, E. L., Schiefer, M. A., Saal, H. P., Delhaye, B. P., Bensmaia, S. J., & Tyler, D. J. (2016). The neural basis of perceived intensity in natural and artificial touch. *Science Translational Medicine*, *8*(362). Available from https://doi.org/10.1126/scitranslmed.aaf5187, 362ra142.

Güçlü, B., & Dinçer, Ş. M. (2013). Neural coding in the Non-Pacinian I tactile channel: A psychophysical and simulation study of magnitude estimation. *Somatosensory & Motor Research*, *30*(1), 1–15.

Hartmann, C., Došen, S., Amsuess, S., & Farina, D. (2015). Closed-loop control of myoelectric prostheses with electrotactile feedback: Influence of stimulation artifact and blanking. *IEEE Transactions on Neural Systems and Rehabilitation Engineering*, *23*(5), 807–816. Available from https://doi.org/10.1109/TNSRE.2014.2357175.

Hioki, M., & Kawasaki, H. (2012). Estimation of finger joint angles from sEMG using a neural network including time delay factor and recurrent structure. *ISRN Rehabilitation*, *2012*, 1–13. Available from https://doi.org/10.5402/2012/604314.

Hodgkin, A. L., & Huxley, A. F. (1952). A quantitative description of membrane current and its application to conduction and excitation in nerve. *The Journal of Physiology*, *117*(4), 500–544. Available from https://doi.org/10.1113/jphysiol.1952.sp004764.

Hu, Y., Wong, Y., Wei, W., Du, Y., Kankanhalli, M., & Geng, W. (2018). A novel attention-based hybrid CNN-RNN architecture for sEMG-based gesture recognition. *PLoS ONE*, *13*(10), 1–18. Available from https://doi.org/10.1371/journal.pone.0206049.

Izhikevich, E. M. (2003). Simple model of spiking neurons. *IEEE Transactions on Neural Networks*, *14*(6), 1569–1572. Available from https://doi.org/10.1109/TNN.2003.820440.

Izhikevich, E. M. (2007). *Dynamical systems in neuroscience*. MIT Press.

Jiang, N., Vest-Nielsen, J. L., Muceli, S., & Farina, D. (2012). EMG-based simultaneous and proportional estimation of wrist/hand kinematics in uni-lateral trans-radial amputees. *Journal of NeuroEngineering and Rehabilitation*, *9*(1), 1–11. Available from https://doi.org/10.1186/1743-0003-9-42.

Johansson, R. S. (1978). Tactile sensibility in the human hand: Receptive field characteristics of mechanoreceptive units in the glabrous skin area. *The Journal of Physiology*, *281*(1), 101–125. Available from https://doi.org/10.1113/jphysiol.1978.sp012411.

Johansson, R. S., & Flanagan, J. R. (2009). Coding and use of tactile signals from the fingertips in object manipulation tasks. *Nature Reviews Neuroscience*, *10*(5), 345–359. Available from https://doi.org/10.1038/nrn2621.

Johansson, R. S., Landström, U., & Lundström, R. (1982). Responses of mechanoreceptive afferent units in the glabrous skin of the human hand to sinusoidal skin displacements. *Brain Research*, *244*(1), 17–25. Available from https://doi.org/10.1016/0006-8993(82)90899-X.

Johnson, B. C., Gambini, S., Izyumin, I., Moin, A., Zhou, A., Alexandrov, G., Santacruz, S. R., Rabaey, J. M., Carmena, J. M., & Muller, R. (2017). An implantable 700μW 64-channel neuromodulation IC for simultaneous recording and stimulation with rapid artifact recovery. In *IEEE Symposium on VLSI Circuits, Digest of Technical Papers* (pp. C48–C49). Available from https://doi.org/10.23919/VLSIC.2017.8008543.

Jörntell, H., Bengtsson, F., Geborek, P., Spanne, A., Terekhov, A. V., & Hayward, V. (2014). Segregation of tactile input features in neurons of the cuneate nucleus. *Neuron*, *83*(6), 1444–1452. Available from https://doi.org/10.1016/j.neuron.2014.07.038.

Kakoty, N.M., & Hazarika, S.M. (2011). Recognition of grasp types through principal components of DWT based EMG features. In *IEEE International Conference on Rehabilitation Robotics*. https://doi.org/10.1109/ICORR.2011.5975398

Kim, S. S., Sripati, A. P., & Bensmaia, S. J. (2010). Predicting the timing of spikes evoked by tactile stimulation of the hand. *Journal of Neurophysiology*, *104*(3), 1484–1496. Available from https://doi.org/10.1152/jn.00187.2010.

Kim, S. S., Sripati, A. P., Vogelstein, R. J., Armiger, R. S., Russell, A. F., & Bensmaia, S. J. (2009). Conveying tactile feedback in sensorized hand neuroprostheses using a biofidelic model of mechanotransduction. *IEEE Transactions on Biomedical Circuits and Systems*, *3*(6), 398–404. Available from https://doi.org/10.1109/TBCAS.2009.2032396.

Knibestöl, M. (1975). Stimulus-response functions of slowly adapting mechanoreceptors in the human glabrous skin area. *The Journal of Physiology*, *245*(1), 63–80. Available from https://doi.org/10.1113/jphysiol.1975.sp010835.

Lapicque, L., & Lapicque, M. (1907). Recherches quantitatives sur l'excitation électrique des nerfs traitée comme une polarisation. *Journal of Physiology Pathology*.

Lee, W. W., Tan, Y. J., Yao, H., Li, S., See, H. H., Hon, M., Ng, K. A., Xiong, B., Ho, J. S., & Tee, B. C. K. (2019). A neuro-inspired artificial peripheral nervous system for scalable electronic skins. *Science Robotics*, *4*(32). Available from https://doi.org/10.1126/scirobotics.aax2198, eaax2198.

Lefurge, T., Goodall, E., Horch, K., Stensaas, L., & Schoenberg, A. (1991). Chronically implanted intrafascicular recording electrodes. *Annals of Biomedical Engineering*, *19*(2), 197–207. Available from https://doi.org/10.1007/BF02368469.

Limnuson, K., Lu, H., Chiel, H. J., & Mohseni, P. (2014). Real-time stimulus artifact rejection via template subtraction. *IEEE Transactions on Biomedical Circuits and Systems*, *8*(3), 391–400. Available from https://doi.org/10.1109/TBCAS.2013.2274574.

Lu, Y., Cao, P., Sun, J., Wang, J., Li, L., Ren, Q., Chen, Y., & Chai, X. (2012). Using independent component analysis to remove artifacts in visual cortex responses elicited by electrical stimulation of the optic nerve. *Journal of Neural Engineering*, *9*(2), 026002. Available from https://doi.org/10.1088/1741-2560/9/2/026002.

Malagodi, M. S., Horch, K. W., & Schoenberg, A. A. (1989). An intrafascicular electrode for recording of action potentials in peripheral nerves. *Annals of Biomedical Engineering*, *17*(4), 397–410. Available from https://doi.org/10.1007/BF02368058.

Martelloni, C., Carpaneto, J., & Micera, S. (2008). Classification of upper arm EMG signals during object-specific grasp. In *Proceedings of the 30th Annual International Conference of the IEEE Engineering in Medicine and Biology Society*, EMBS'08—"Personalized Healthcare through Technology" (pp. 5061–5064). Available from https://doi.org/10.1109/iembs.2008.4650351.

Mazzoni, A., Oddo, C. M., Valle, G., Camboni, D., Strauss, I., Barbaro, M., Barabino, G., Puddu, R., Carboni, C., Bisoni, L., Carpaneto, J., Vecchio, F., Petrini, F. M., Romeni, S., Czimmermann, T., Massari, L., Iorio, R., di, Miraglia, F., Granata, G., & Micera, S. (2020). Morphological neural computation restores discrimination of naturalistic textures in trans-radial amputees. *Scientific Reports*. Available from https://doi.org/10.1038/s41598-020-57454-4.

Mendez, V., Iberite, F., Shokur, S., & Micera, S. (2021). Current Solutions and Future Trends for Robotic Prosthetic Hands. *Annual Review of Control, Robotics, and Autonomous Systems*, *4*. Available from https://doi.org/10.1146/annurev-control-071020-104336.

Mendrela, A. E., Cho, J., Fredenburg, J. A., Nagaraj, V., Netoff, T. I., Flynn, M. P., & Yoon, E. (2016). A bidirectional neural interface circuit with active stimulation artifact cancellation and cross-channel common-mode noise suppression. *IEEE Journal of Solid-State Circuits*, *51*(4), 955–965. Available from https://doi.org/10.1109/JSSC.2015.2506651.

Merrill, D. R., Bikson, M., & Jefferys, J. G. R. (2005). Electrical stimulation of excitable tissue: Design of efficacious and safe protocols. *Journal of Neuroscience Methods*, *141*(2), 171–198. Available from https://doi.org/10.1016/j.jneumeth.2004.10.020, Elsevier.

Moss, F. (2004). Stochastic resonance and sensory information processing: A tutorial and review of application. *Clinical Neurophysiology*, *115*(2), 267–281. Available from https://doi.org/10.1016/j.clinph.2003.09.014.

Muceli, S., & Farina, D. (2012). Simultaneous and proportional estimation of hand kinematics from EMG during mirrored movements at multiple degrees-of-freedom. *IEEE Transactions on Neural Systems and Rehabilitation Engineering*, *20*(3), 371–378. Available from https://doi.org/10.1109/TNSRE.2011.2178039.

Muniak, M. A., Ray, S., Hsiao, S. S., Dammann, J. F., & Bensmaia, S. J. (2007). The neural coding of stimulus intensity: Linking the population response of mechanoreceptive afferents with psychophysical behavior. *Journal of Neuroscience*, *27*(43), 11687–11699.

Navarro, X., Lago, N., Vivo, M., Yoshida, K., Koch, K. P., Poppendieck, W., & Micera, S. (2007). Neurobiological evaluation of thin-film longitudinal intrafascicular electrodes as a peripheral nerve interface. In *2007 IEEE 10th International Conference on Rehabilitation Robotics* (pp. 643–649). Available from https://doi.org/10.1109/ICORR.2007.4428492.

Nordhausen, C. T., Maynard, E. M., & Normann, R. A. (1996). Single unit recording capabilities of a 100 microelectrode array. *Brain Research*, *726*(1–2), 129–140.

Oddo, C. M., Raspopovic, S., Artoni, F., Mazzoni, A., Spigler, G., Petrini, F., Giambattistelli, F., Vecchio, F., Miraglia, F., Zollo, L., Di Pino, G., Camboni, D., Carrozza, M. C., Guglielmelli, E., Rossini, P. M., Faraguna, U., & Micera, S. (2016). Intraneural stimulation elicits discrimination of textural features by artificial fingertip in intact and amputee humans. *ELife*, *5*. Available from https://doi.org/10.7554/eLife.09148. (March), e09148.

O'Doherty, J. E., Shokur, S., Medina, L. E., Lebedev, M. A., & Nicolelis, M. A. L. (2019). Creating a neuroprosthesis for active tactile exploration of textures. *Proceedings of the National Academy of Sciences of the United States of America*, *116*(43), 21821–21827. Available from https://doi.org/10.1073/pnas.1908008116, 31591224.

Ortiz-Catalan, M., Håkansson, B., & Brånemark, R. (2014). An osseointegrated human-machine gateway for long-term sensory feedback and motor control of artificial limbs. *Science Translational Medicine*, *6*(257). Available from https://doi.org/10.1126/scitranslmed.3008933, 257re6–257re6.

Ortiz-Catalan, M., Mastinu, E., Sassu, P., Aszmann, O., & Brånemark, R. (2020). Self-Contained Neuromusculoskeletal Arm Prostheses. *New England Journal of Medicine*, *382*(18), 1732–1738. Available from https://doi.org/10.1056/NEJMoa1917537.

Osborn, L. E., Dragomir, A., Betthauser, J. L., Hunt, C. L., Nguyen, H. H., Kaliki, R. R., & Thakor, N. V. (2018). Prosthesis with neuromorphic multilayered e-dermis perceives touch and pain. *Science Robotics*, *3*(19). Available from https://doi.org/10.1126/scirobotics.aat3818, eaat3818.

Park, K., & Lee, S. (2016). Movement intention decoding based on deep learning for multiuser myoelectric interfaces. In *2016 4th International Winter Conference on Brain-Computer Interface (BCI)* (pp. 1–2). Available from https://doi.org/10.1109/IWW-BCI.2016.7457459.

Petrini, F. M., Mazzoni, A., Rigosa, J., Giambattistelli, F., Granata, G., Barra, B., Pampaloni, A., Guglielmelli, E., Zollo, L., Capogrosso, M., Micera, S., & Raspopovic, S. (2019). Microneurography as a tool to develop decoding algorithms for peripheral neuro-controlled hand prostheses. *BioMedical Engineering OnLine*, *18*(1), 44. Available from https://doi.org/10.1186/s12938-019-0659-9.

Petrini, F. M., Valle, G., Strauss, I., Granata, G., Di Iorio, R., D'Anna, E., Čvančara, P., Mueller, M., Carpaneto, J., Clemente, F., Controzzi, M., Bisoni, L., Carboni, C., Barbaro, M., Iodice, F., Andreu, D., Hiairrassary, A., Divoux, J. L., Cipriani, C., … Micera, S. (2019). Six-month assessment of a hand prosthesis with intraneural tactile feedback. *Annals of Neurology*. Available from https://doi.org/10.1002/ana.25384.

Raspopovic, S., Capogrosso, M., Petrini, F. M., Bonizzato, M., Rigosa, J., Pino, G., Di., Carpaneto, J., Controzzi, M., Boretius, T., Fernandez, E., Granata, G., Oddo, C. M., Citi, L., Ciancio, A. L., Cipriani, C., Carrozza, M. C., Jensen, W., Guglielmelli, E., … Micera, S. (2014). Bioengineering: Restoring natural sensory feedback in real-time bidirectional hand prostheses. *Science Translational Medicine*, *6*(222). Available from https://doi.org/10.1126/scitranslmed.3006820, 222ra19–222ra19.

Rolston, J. D. (2009). A low-cost multielectrode system for data acquisition enabling real-time closed-loop processing with rapid recovery from stimulation artifacts. *Frontiers in Neuroengineering*, *2*(Jul), 12. Available from https://doi.org/10.3389/neuro.16.012.2009.

Rongala, U. B., Mazzoni, A., & Oddo, C. M. (2017). Neuromorphic artificial touch for categorization of naturalistic textures. *IEEE Transactions on Neural Networks and Learning Systems*, *28*(4), 819–829. Available from https://doi.org/10.1109/TNNLS.2015.2472477.

Rongala, U. B., Mazzoni, A., Spanne, A., Jörntell, H., & Oddo, C. M. (2020). Cuneate spiking neural network learning to classify naturalistic texture stimuli under varying sensing conditions. *Neural Networks*, *123*, 273–287. Available from https://doi.org/10.1016/j.neunet.2019.11.020.

Rossini, P. M., Micera, S., Benvenuto, A., Carpaneto, J., Cavallo, G., Citi, L., Cipriani, C., Denaro, L., Denaro, V., Di Pino, G., Ferreri, F., Guglielmelli, E., Hoffmann, K. P., Raspopovic, S., Rigosa, J., Rossini, L., Tombini, M., & Dario, P. (2010). Double nerve intraneural interface implant on a human amputee for robotic hand control. *Clinical Neurophysiology*, *121*(5), 777–783. Available from https://doi.org/10.1016/j.clinph.2010.01.001.

Rutten, W. L. C., Smit, J. P. A., Frieswijk, T. A., Bielen, J. A., Brouwer, A. L. H., Buitenweg, J. R., & Heida, C. (1999). Neuro-electronic interfacing with multielectrode arrays. *IEEE Engineering in Medicine and Biology Magazine*, *18*(3), 47–55. Available from https://doi.org/10.1109/51.765188.

Saal, H. P., Delhaye, B. P., Rayhaun, B. C., & Bensmaia, S. J. (2017). Simulating tactile signals from the whole hand with millisecond precision. *Proceedings of the National Academy of Sciences of the United States of America*, *114*(28), E5693–E5702. Available from https://doi.org/10.1073/pnas.1704856114.

Scott, R. N., McLean, L., & Parker, P. A. (1997). Stimulus artefact in somatosensory evoked potential measurement. *Medical and Biological Engineering and Computing*, *35*(3), 211–215. Available from https://doi.org/10.1007/BF02530040.

Smith, L. H., & Hargrove, L. J. (2013). Comparison of surface and intramuscular EMG pattern recognition for simultaneous wrist/hand motion classification. *Annual international conference of the IEEE engineering in medicine and biology society*, *2013*, 4223–4226. Available from https://doi.org/10.1109/EMBC.2013.6610477, 24110664.

Stanslaski, S., Afshar, P., Cong, P., Giftakis, J., Stypulkowski, P., Carlson, D., Linde, D., Ullestad, D., Avestruz, A. T., & Denison, T. (2012). Design and validation of a fully implantable, chronic, closed-loop neuromodulation device with concurrent sensing and stimulation. *IEEE Transactions on Neural Systems and Rehabilitation Engineering*, *20*(4), 410–421. Available from https://doi.org/10.1109/TNSRE.2012.2183617.

Stimberg, M., Brette, R., & Goodman, D. F. (2019). Brian 2, an intuitive and efficient neural simulator. *ELife*, 8. Available from https://doi.org/10.7554/eLife.47314.

Tan, D. W., Schiefer, M. A., Keith, M. W., Anderson, J. R., Tyler, J., & Tyler, D. J. (2014). A neural interface provides long-term stable natural touch perception. *Science Translational Medicine*, *6*(257). Available from https://doi.org/10.1126/scitranslmed.3008669, 257ra138.

Tarler, M. D. M. D. M. D., & Mortimer, J. T. T. (2004). Selective and independent activation of four motor fascicles using a four contact nerve-cuff electrode. *IEEE Transactions on Neural Systems and Rehabilitation Engineering*, *12*(2), 251–257. Available from https://doi.org/10.1109/TNSRE.2004.828415.

Tenore, F. V. G., Ramos, A., Fahmy, A., Acharya, S., Etienne-Cummings, R., & Thakor, N. V. (2009). Decoding of individuated finger movements using surface electromyography. *IEEE*

Transactions on Biomedical Engineering, *56*(5), 1427–1434. Available from https://doi.org/10.1109/TBME.2008.2005485.

Tyler, D. J., & Durand, D. M. (2002). Functionally selective peripheral nerve stimulation with a flat interface nerve electrode. *IEEE Transactions on Neural Systems and Rehabilitation Engineering*, *10*(4), 294–303. Available from https://doi.org/10.1109/TNSRE.2002.806840.

Valle, G., Mazzoni, A., Iberite, F., D'Anna, E., Strauss, I., Granata, G., Controzzi, M., Clemente, F., Rognini, G., Cipriani, C., Stieglitz, T., Petrini, F. M., Rossini, P. M., & Micera, S. (2018). Biomimetic intraneural sensory feedback enhances sensation naturalness, tactile sensitivity, and manual dexterity in a bidirectional prosthesis. *Neuron*. Available from https://doi.org/10.1016/j.neuron.2018.08.033.

Valle, G., Petrini, F. M., Strauss, I., Iberite, F., D'Anna, E., Granata, G., Controzzi, M., Cipriani, C., Stieglitz, T., Rossini, P. M., Mazzoni, A., Raspopovic, S., & Micera, S. (2018). Comparison of linear frequency and amplitude modulation for intraneural sensory feedback in bidirectional hand prostheses. *Scientific Reports*, *8*(1), 16666. Available from https://doi.org/10.1038/s41598-018-34910-w.

Viswam, V., Chen, Y., Shadmani, A., Dragas, J., Bounik, R., Milos, R., Muller, J., & Hierlemann, A. (2016). 2048 action potential recording channels with 2.4 μvrms noise and stimulation artifact suppression. In *Proceedings—2016 IEEE Biomedical Circuits and Systems Conference, BioCAS 2016* (pp. 136–139). Available from https://doi.org/10.1109/BioCAS.2016.7833750.

Warwick, K., Gasson, M., Hutt, B., Goodhew, I., Kyberd, P., Andrews, B., Teddy, P., & Shad, A. (2003). The application of implant technology for cybernetic systems. *Archives of Neurology*, *60* (10), 1369. Available from https://doi.org/10.1001/archneur.60.10.1369.

Wendelken, S., Page, D. M., Davis, T., Wark, H. A. C., Kluger, D. T., Duncan, C., Warren, D. J., Hutchinson, D. T., & Clark, G. A. (2017). Restoration of motor control and proprioceptive and cutaneous sensation in humans with prior upper-limb amputation via multiple Utah Slanted Electrode Arrays (USEAs) implanted in residual peripheral arm nerves. *Journal of NeuroEngineering and Rehabilitation*, *14*(1), 121. Available from https://doi.org/10.1186/s12984-017-0320-4.

Wichmann, T., & Devergnas, A. (2011). A novel device to suppress electrical stimulus artifacts in electrophysiological experiments. *Journal of Neuroscience Methods*, *201*(1), 1–8. Available from https://doi.org/10.1016/j.jneumeth.2011.06.026.

Wiener, N. (1948). *Cybernetics: Or control and communication in the animal and the machine*. Cambridge, MA: MIT Press.

Xia, P., Hu, J., & Peng, Y. (2018). EMG-based estimation of limb movement using deep learning with recurrent convolutional neural networks. *Artificial Organs*, *42*(5), E67–E77. Available from https://doi.org/10.1111/aor.13004.

Zhou, A., Johnson, B. C., & Muller, R. (2018). Toward true closed-loop neuromodulation: Artifact-free recording during stimulation. *Current Opinion in Neurobiology*, *50*, 119–127. Available from https://doi.org/10.1016/j.conb.2018.01.012, Elsevier Ltd.

Zhou, A., Santacruz, S. R., Johnson, B. C., Alexandrov, G., Ali Moin., Burghardt, F. L., Rabaey, J. M., Carmena, J. M., & Muller, R. (2019). A wireless and artefact-free 128-channel neuromodulation device for closed-loop stimulation and recording in non-human primates. *Nature Biomedical Engineering*, *3*(1), 15–26.

Zhuang, K. Z., Sommer, N., Mendez, V., Aryan, S., Formento, E., D'Anna, E., Artoni, F., Petrini, F., Granata, G., Cannaviello, G., Raffoul, W., Billard, A., & Micera, S. (2019). Shared human–robot proportional control of a dexterous myoelectric prosthesis. *Nature Machine Intelligence*, *1*(9), 400–411. Available from https://doi.org/10.1038/s42256-019-0093-5.

Part IV

Cortical implants for somatosensory feedback

Chapter 11

Restoring the sense of touch with electrical stimulation of the nerve and brain

Thierri Callier and Sliman J. Bensmaia
Department of Organismal Biology and Anatomy, University of Chicago, Chicago, IL, United States

ABSTRACT
Though prosthetic limbs controlled with brain–machine interfaces have the potential to confer to patients with tetraplegia the ability to physically interact with the world and regain a level of independence, the dexterity of prosthetic hands is severely limited without sensory feedback. Current brain–machine interfaces can decode motor intent from the brain remarkably well, but movements of brain-controlled anthropomorphic limbs remain slow and clumsy, in part due to this lack of somatosensory feedback. Accordingly, efforts are underway to restore tactile sensation to improve the dexterity of prosthetic limbs. Restoration of touch can also promote embodiment, the integration of the limb into the user's sense of self. Electrical stimulation of somatosensory cortex evokes sensations of touch and can be used to convey information about interactions with objects, but our ability to shape the artificial percept is impeded by an inadequate understanding of the features of stimulation that affect sensation. To date, attempts to restore the sense of touch through intracortical microstimulation (ICMS) used either arbitrary mappings between touched objects and stimulation patterns, which took months to interpret and use, or monotonic mappings between contact force and stimulation amplitude, which led to improved performance. However, even with this feedback, dexterity does not match that of our native hands. Recent results from sensory restoration through nerve stimulation suggest that more natural feedback is associated with better performance in functional tasks. To achieve ever more naturalistic ICMS-based feedback requires that we develop our understanding of how ICMS parameters such as pulse charge, pulse patterning, and stimulation location shape the evoked sensations.

Keywords: Brain–machine interface; somatosensory feedback; prosthetic limb; cerebral cortex; intracortical microstimulation; touch

Somatosensory Feedback for Neuroprosthetics. DOI: https://doi.org/10.1016/B978-0-12-822828-9.00010-1

11.1 Introduction

11.1.1 The importance of touch in manual behavior

State-of-the-art bionic hands are highly sophisticated, approaching the complexity of the human hand, and algorithms to infer intended movements from patterns of activity in residual muscles or neurons have improved (Ajiboye et al., 2017; Collinger et al., 2013; Gilja et al., 2012; Hochberg et al., 2012; Wodlinger et al., 2015). However, prosthetic hands can never confer to their user the dexterity with which natural hands are endowed without restoring somatosensory feedback (Bensmaia, 2015). Indeed, manual interactions with objects rely on a barrage of sensory signals from the hand that convey information about the object itself—its size, shape, texture, etc.—and about the interactions themselves—contact timing, contact force, contact location, etc. Individuals who have lost somatosensation due to a neuropathy struggle to perform activities of daily living because their movements are slow, clumsy, and effortful (Sainburg et al., 1993). Without sensation, not only is the functionality of a bionic hand compromised, but the hand seems disembodied. Indeed, prosthetics users who can volitionally control an arm without being able to feel through it experience the limb as a tool rather than a body part (Page et al., 2018; Valle et al., 2018), which makes it far less appealing. The restoration of touch has the potential to lead to greater embodiment of the prosthetic arm (Flesher et al., 2019; Valle et al., 2018).

11.1.2 Electrical activation of neurons

Two observations in neuroscience that predate even the discovery of the neuron and the action potential (Hodgkin & Huxley, 1952; Ramon y Cajal, 1888) set the stage for current efforts to sensitize bionic hands. The first is that nervous tissue can be activated by delivering electrical currents (Galvani, 1791). The second is that certain regions of the brain are implicated in specific sensory functions (Gennari, 1782; Glickstein & Whitteridge, 1987; Glickstein, 1988; Munk, 1881). These phenomena converged in a pivotal way in the 1930s when, searching for the foci of elliptical seizures, Wilder Penfield and colleagues electrically stimulated the brains of awake patients and discovered that stimulating somatosensory cortex elicits sensations of touch localized to restricted regions of the body (Penfield & Boldrey, 1937). This opened the possibility that artificial sensations might be systematically elicited by electrically activating the sensory apparatus of the nervous system. This phenomenon has been harnessed with great success in the restoration of hearing using cochlear implants and, as discussed herein, is key to the restoration of the sense of touch through bionic hands (Fig. 11.1).

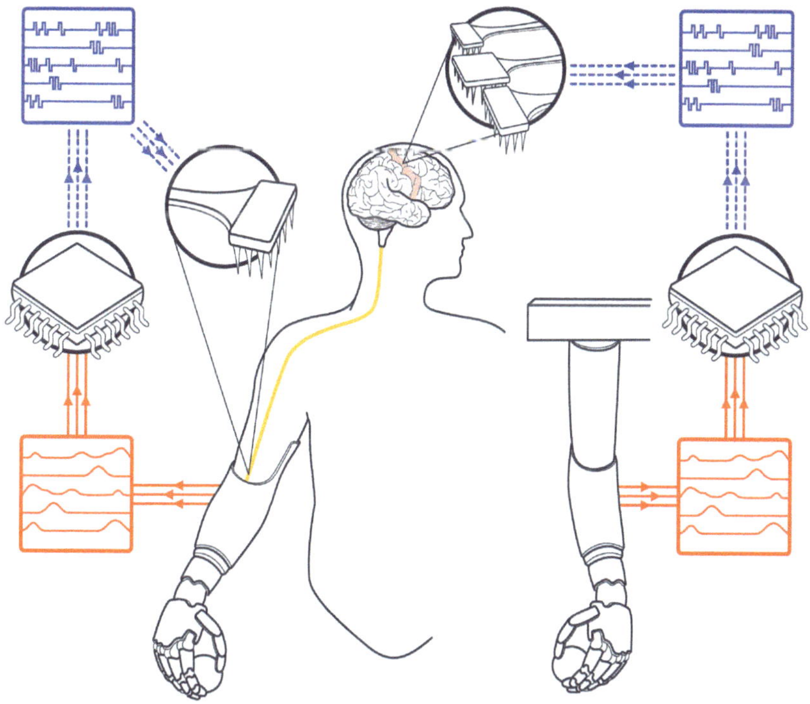

FIGURE 11.1 Diagram of somatosensory neuroprostheses with a peripheral interface (*left*) and a cortical interface (*right*). Signals from sensors in the bionic hand (*orange*) are converted by sensory-encoding algorithms into pulse trains of electrical stimulation delivered to the nerve or somatosensory cortex (*blue*). These are designed to elicit meaningful sensations that convey information about the object grasped by the hand.

11.1.3 Neural coding—the language of the nervous system

At every level of processing in the nervous system, from the peripheral nerves through cortex, neurons communicate with each other via action potentials, also known as spikes, and the output of a neuron can be described by its spike times. Sensory neurons can convey information about stimulus features in different ways, according to a so called neural code. Information can be encoded in the number of spikes emitted per unit time—a rate code. For example, the perceived intensity of a touch is determined by the firing rate of all the activated tactile nerve fibers (Muniak et al., 2007). Alternatively, information can be encoded in the spatial pattern of activation over a population of neurons: the location of the active neurons within a brain area determines what is experienced. For example, the perceived location of a touch is determined by such a spatial code (Tabot et al., 2013). Finally, information can be encoded in the temporal pattern of spiking over short time scales (milliseconds or tens of milliseconds). For example, the perceived texture of a surface is determined in part by the precise timing of neural responses evoked in the nerve (Mackevicius et al., 2012; Talbot et al., 1968) and in the brain (Harvey et al., 2013).

As is perhaps obvious, if one could artificially recreate a natural pattern of neuronal activation evoked by a given stimulus, the evoked sensation would be completely natural. Unfortunately, perfect biomimicry cannot be achieved because current stimulation technology is not selective enough to activate individual neurons independently and because our understanding of sensory coding in the central nervous system is not sufficiently detailed. Nonetheless, mimicking natural patterns of activation to the extent possible may lead to more meaningful sensations than would ignoring the native neural code altogether. Indeed, cochlear implants—which convert sounds into patterns of electrical stimulation of the basilar membrane in the inner ear—are built on the principle that the cochlea splits incoming sounds into their component frequencies, which leads to neural activation at locations on the membrane that depend on the frequency. The sensory-encoding algorithms in cochlear implants split acoustic signals into component frequency bands and electrically stimulate the basilar membrane accordingly. While the resulting neural signals are not completely biomimetic, they nonetheless mimic key aspects of natural patterns of activation that preserve their information content. Biomimicry—one of the guiding principles of the development of artificial touch—requires an understanding of the neural mechanisms that mediate natural touch, to which we turn next.

11.2 Neural basis of touch

11.2.1 Tactile innervation of the skin

The palmar surface of the hand—where most object interactions take place—is innervated by 12,000 tactile nerve fibers (Johansson & Vallbo, 1979), each of which terminates in one of four types of mechanoreceptors embedded in the skin (Fig. 11.2A). These mechanoreceptors convert mechanical deformations of the skin into neural signals and different classes of receptors respond differently to skin deformations (Darian-Smith, 2011; Goodman & Bensmaia, 2020; Johansson & Flanagan, 2009; Johnson, 2001; Vallbo & Johansson, 1976) (Fig. 11.2B). Slowly adapting type 1 (SA1) fibers produce a sustained response to a static indentation of the skin and, as a population, convey a spatial image of the pattern of skin deformation (Phillips & Johnson, 1981; Phillips et al., 1988). Rapidly adapting (RA) fibers respond only at the onset and offset of a skin indentation and are silent when the skin is not moving (Knibestöl, 1973). Their responses instead seem best suited to track dynamic force applications to the hand, such as the rapid sequence of contact events accompanying manipulation of objects. Pacinian corpuscle-associated (PC) fibers are exquisitely sensitive to skin vibrations, peaking in

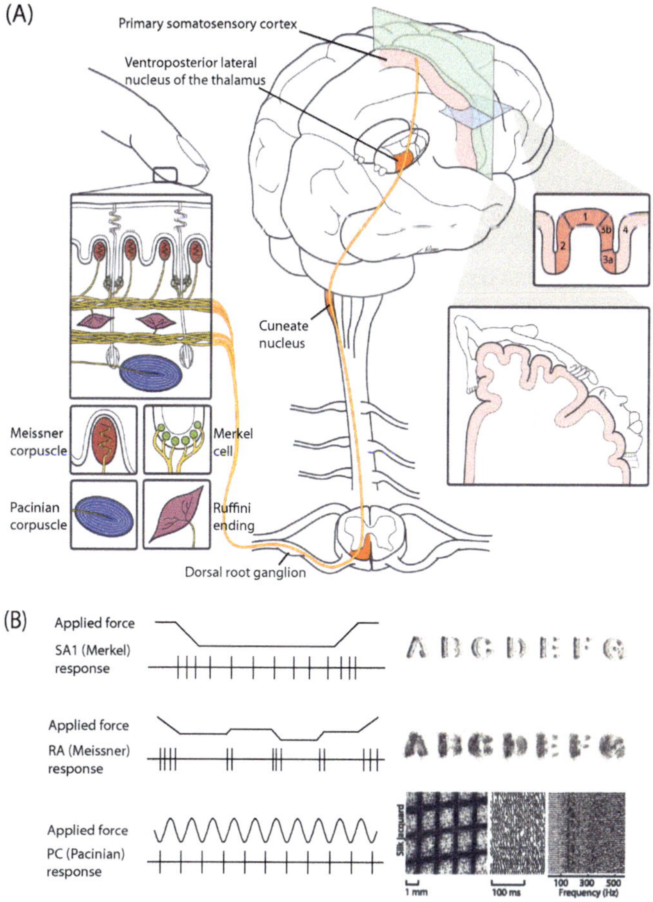

FIGURE 11.2 Neuroanatomy of touch. (A) The touch neuraxis from periphery to cortex. Mechanoreceptors embedded in the skin (left insets) facilitate the initiation of responses in afferent fibers. These bundle in fascicles that join to form nerves. The axons of afferent neurons ascend the spinal cord and synapse onto the cuneate nucleus in the brainstem. Axons from the cuneate project to the contralateral ventroposterior complex of the thalamus, which in turn projects to somatosensory cortex. Top right inset: somatosensory cortex comprises Brodmann's areas 3a, 3b, 1, and 2. Bottom right inset: each area is somatotopically organized. (B) Response properties of afferent fibers. SA1 afferents produce a sustained response to static indentation of the skin. As a population, they convey a spatial image of the pattern of skin deformation, as shown in the reconstructed response of SA fibers to embossed letters scanned across the skin. RA afferents respond to changes in applied force and are silent when the force is constant. Their responses are well suited to track dynamic forces and also convey some information about the local geometric features of objects. PC fibers are sensitive to skin vibrations and become entrained to the frequencies of the vibrations. The timing or frequency composition of PC responses is texture-specific and plays an important role in the perception of fine textures. *Modified from Phillips, J. R., Johnson, K. O., & Hsiao, S. S. (1988). Spatial pattern representation and transformation in monkey somatosensory cortex.* Proceedings of the National Academy of Sciences, 85*(4), 1317–1321. <https://doi.org/10.1073/pnas.85.4.1317> and Weber, A. I., Saal, H. P., Lieber, J. D., Cheng, J.-W., Manfredi, L. R., Dammann, J. F., & Bensmaia, S. J. (2013). Spatial and temporal codes mediate the tactile perception of natural textures.* Proceedings of the National Academy of Sciences, 110*(42), 17107–17112. <https://doi.org/10.1073/pnas.1305509110>.*

sensitivity around 250 Hz with thresholds measured in the hundreds of nanometers (Bell et al., 1994; Freeman & Johnson, 1982; Johansson et al., 1982; Muniak et al., 2007). These nerve fibers play an important role in the perception of fine textures (Weber et al., 2013) and of substrate-borne vibrations, for example those transmitted through a tool (Brisben et al., 1999). Slowly adapting type 2 (SA2) fibers respond primarily to stretch and are most prevalent around the nails (Knibestöl, 1975), where they can sense forces applied on the fingertips (Birznieks et al., 2009). Tactile information about an object and about contact events is multiplexed in the responses of these four afferent populations and touch sensations typically involve the integration of signals from all four classes (Saal & Bensmaia, 2014; Weber et al., 2013).

11.2.2 Medial lemniscal pathway

Tactile nerve fibers innervating the palmar surface of the hand converge to form the median and ulnar nerves, which carry tactile signals up the arm and a short distance up the spinal cord before they synapse onto the cuneate nucleus in the brainstem (Fig. 11.2A). The cell bodies of tactile nerve fibers are located in the dorsal root ganglia (DRG) between the vertebrae, just outside of the spinal column. The nerve fibers synapse onto neurons in the cuneate nucleus, located in the brainstem, where afferent inputs are first integrated and processed. While the response properties of neurons in the cuneate nucleus were thought to be similar to those of their afferent inputs, recent work suggests that this neural structure may carry out some degree of feature extraction (Bengtsson et al., 2013; Hayward et al., 2014; Jörntell et al., 2014; Suresh et al., 2017; Suresh, 2015, 2019). The cuneate nucleus sends projections contralaterally to the ventroposterior nucleus of the thalamus (Craig, 2006; Fig. 11.2A), which then projects mainly to somatosensory cortex, located in anterior parietal cortex.

11.2.3 Somatosensory cortex

Somatosensory cortex—sometimes mistakenly referred to as primary somatosensory cortex—comprises four cortical fields: Brodmann's areas 3a, 3b, 1, and 2 (Fig. 11.2A). Neurons in area 3a exhibit proprioceptive responses, neurons in areas 3b and 1 exhibit cutaneous responses, and neurons in area 2 exhibit both. Each cutaneous neuron will only respond to touch within a restricted patch of skin, the so-called receptive field (RF). Brodmann's area 3b is primary somatosensory cortex proper as its structure and interconnectivity with other brain structures are similar to primary sensory cortices in other modalities (Kaas, 1983). Somatosensory cortex is organized hierarchically—with area 3b at the bottom and area 2 at the top—as evidenced by the progressive increase in RF size and in response complexity (Delhaye et al., 2018). The most striking feature of somatosensory cortex is its somatotopic organization: nearby neurons respond to touch of nearby and partially overlapping patches of skin. Each cortical field within somatosensory cortex contains a full map of the contralateral side of the body, with a greater cortical area dedicated to the representation of highly innervated regions of the body such as the hands (Sur et al., 1980) (Fig. 11.2A).

In the hand representation within area 3b, neurons have small RFs, typically limited to a single finger pad or digit (DiCarlo et al., 1998), which comprise excitatory subfields flanked by inhibitory ones. Neurons display response properties that reflect those of multiple classes of nerve fibers (SA1, RA, and PC) (Pei et al., 2009), which implies that individual neurons receive convergent input from multiple tactile submodalities, although this integration begins to take place in the cuneate nucleus (Suresh, 2019). Neurons in area 1 tend to have larger RFs than do their counterparts in area 3b, sometimes spanning multiple digits (Iwamura et al., 1983) and they often exhibit selectivity for complex features, such as the direction of tactile motion (Bensmaia et al., 2008; Delhaye et al., 2018; Hyvärinen & Poranen, 1978a; Iwamura et al., 1983). Neurons in area 2 exhibit both cutaneous and proprioceptive responses (Hyvärinen & Poranen, 1978b; Iwamura et al., 1993; Seelke et al., 2012), typically have large RFs spanning multiple fingers (Pons et al., 1985), and often exhibit selectivity for complex features, such as local curvature (Yau et al., 2013). Importantly, areas 1 and 2 lie on the post-central gyrus and thus can be surgically accessed far more easily than can area 3b, which lies almost entirely within the posterior bank of the central sulcus.

Secondary somatosensory cortex (S2) and the parietal ventral area (PV) receive input from the somatosensory cortical fields in anterior parietal cortex described above and carry high-level representations of touch (Delhaye et al., 2018). However, these structures have never been targeted in neuroprosthetic applications so will not be discussed here.

11.3 Electrical interfaces with the nervous system

Delivering currents to nervous tissue results in the activation thereof. While the effect of electrical stimulation on neurons depends on the distance to the electrode, the parameters of stimulation, and the metallization of the electrode, it follows several general principles. First, the probability that an electrical pulse will directly activate a neuron increases with charge but decreases with distance between electrode and neuron (Tehovnik et al., 2006) (Fig. 11.3). Second, every activated neuron spikes synchronously in response to each electrical pulse. Thus, a periodic train of electrical pulses will produce synchronized periodic spiking in a volume of neurons. Because axons are more sensitive to electrical stimulation than are cell bodies, electrical stimulation also produces a distributed activity beyond the expected volume (Histed et al., 2009) but its principal effect is confined to this volume (Aberra et al., 2018). Thus, electrical stimulation of the nerve or the brain

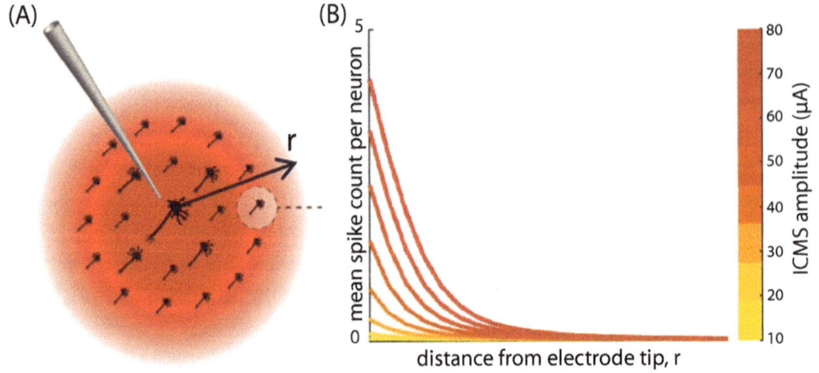

FIGURE 11.3 (A) Electrical stimulation through a penetrating microelectrode primarily activates neurons in a restricted volume surrounding the electrical contact. (B) The stimulation-evoked firing rate of a neuron increases with pulse amplitude but decreases with distance from the electrode tip. *Modified from Kim, S., Callier, T., & Bensmaia, S. J. (2017). A computational model that predicts behavioral sensitivity to intracortical microstimulation.* Journal of Neural Engineering, 14*(1), 016012. <https://doi.org/10.1088/1741-2552/14/1/016012>.*

through a single electrode enables the activation of a pool of neurons the size of which depends on the amount of charge per pulse, and the timing of which depends on the distribution of pulses over time. However, the pattern of activation within the population of activated neurons cannot be further sculpted, which considerably limits the ability of electrical stimulation to reproduce natural patterns of neuronal activation, in which neurons in a given volume each respond idiosyncratically.

11.3.1 Targets of neural interfaces

The neural target for an electrical interface depends on the location of the injury that caused the deafferentation. For an amputee, the residual nerve is the preferred target, because all downstream structures are intact. For a spinal cord injury patient, the nerve is not an option because its connection to downstream structures has been compromised, so an interface with the central nervous system is required. In addition to being more or less suited for different user populations, interfaces with the peripheral and central nervous systems each present advantages and disadvantages.

The advantages of peripheral nerve interfaces are that (1) the required surgery is less invasive in that it does not involve the brain; (2) neural coding in the nerve is simpler and better understood; (3) nerve fibers carry parallel signals to the brain, so nerve stimulation avoids the complexities associated with stimulating the complex circuitry of the brain; (4) peripheral nerve stimulation has the potential to engage sensory pathways that are bypassed with more central interfaces, for example spinal reflexes (Schouenborg, 2008).

The principal advantage of interfaces with the central nervous system is that they can be applied to patients with spinal cord injury or suffering from neuropathies that affect more peripheral structures. The cortex offers the additional advantages that (1) it is much larger than is the nerve and thus offers the possibility of more selective activation via electrical stimulation and (2) it features functional topographies—notably somatotopic organization—that can be exploited.

11.3.2 Interface hardware—peripheral

Nerves in the hand and arm consist of bundles of sensory and motor nerve fibers called fascicles. The proportion of motor nerve fibers decreases as one proceeds distally (toward the hand), as might be expected given that these innervate muscles (Downey et al., 2020). For sensory restoration, distal targets are thus preferable. Chronically implanted electrical interfaces with the nerve can be divided into two broad categories, each with its advantages and disadvantages: extrafascicular or epineural electrodes, which surround the nerve, and intrafascicular or intraneural electrodes, which penetrate it (Fig. 11.4A).

Epineural electrodes sit on the outside surface of the nerve without breaching its surrounding sheath. As such they are less invasive than intraneural interfaces, but the separation between electrical contacts and fascicles requires higher stimulation currents to achieve sensation than with intraneural electrodes (Saal & Bensmaia, 2015). Higher currents lead to a broader spatial spread, which activates larger afferent populations and lowers selectivity—the ability to activate only a target population of afferents and not others. Two common examples are the spiral cuff electrode (Tan et al., 2014), which wraps around the nerve and provides a number of distinct contacts around its surface, and the flat interface nerve electrode (Charkhkar et al., 2018; Freeberg et al., 2017; Graczyk et al., 2016), which flattens the nerve to improve access to fascicles farther from the surface of the nerve. A major advantage of these interfaces is their demonstrated stability over the span of years (Graczyk et al., 2018; Ortiz-Catalan et al., 2014).

Intraneural interfaces are inserted into the nerve to achieve direct contact with nerve fibers. Some intraneural implants consist of wire-like electrodes that run parallel to the nerve (Lawrence et al., 2004; Yoshida & Stein, 1999) or traverse it (Boretius et al., 2010), each with multiple contacts to impinge upon several fascicles. Other intraneural implants consist of an array of microelectrodes laid out in a grid that are driven into the nerve like a bed of nails, with the tip of each electrode providing an electrical contact (Page et al., 2018; Wendelken et al., 2017). For these, electrodes get progressively longer along the axis of the nerve to ensure that the electrical contacts impinge on different fascicles (Fig. 11.4A). The advantage of intraneural approaches over their epineural counterparts is that lower currents are required to achieve sensation given the closer proximity of the electrical

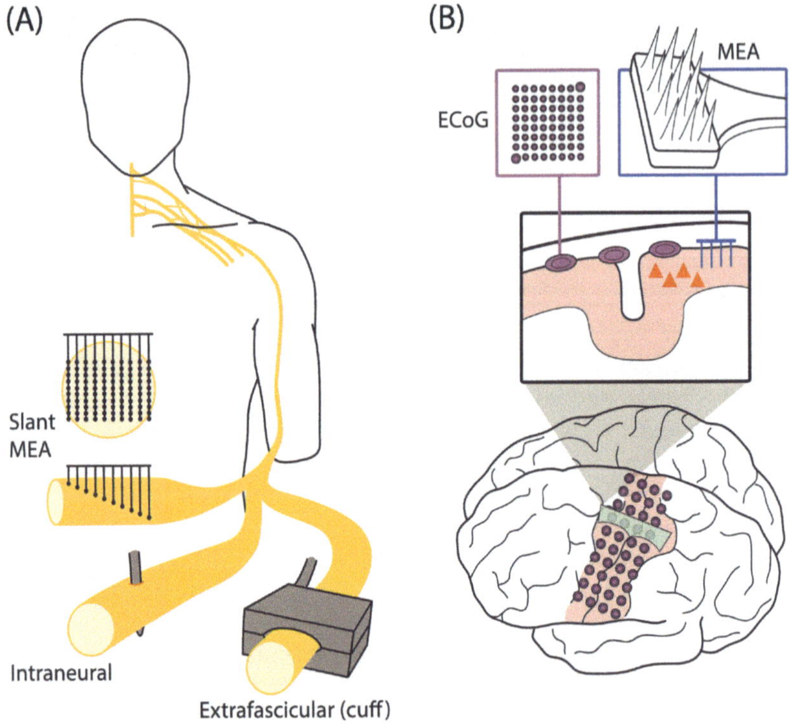

FIGURE 11.4 Interface hardware. (A) Examples of peripheral nerve interfaces. Extrafascicular or epineural electrodes such as cuff electrodes sit outside the nerve. Intrafascicular or intraneural electrodes penetrate the nerve to bring electrodes in direct contact with axons. The transverse intrafascicular electrode shown here has multiple electrical contacts so it can stimulate different afferent fibers. Slant microelectrode arrays (MEAs) penetrate the nerve at varying depths across the width of the nerve to establish contacts with different afferent fibers. (B) Two common interfaces with the central nervous system are electrocorticography (ECoG) electrodes and MEAs. ECoG arrays consist of small metal disks (2–3 mm in diameter) that are implanted on the surface of the brain, while penetrating MEAs are driven into the brain to bring electrodes in direct contact with neurons. MEAs are only a few millimeters wide, the size of a single ECoG electrode. As a result, they offer far higher spatial resolution than do ECoG arrays.

contact to the nerve fibers (Saal & Bensmaia, 2015), which limits current spread and, in principle, allows for the selective stimulation of small groups of fibers (Branner et al., 2001) and even individual fibers. However, intraneural electrodes are more liable to move within the nerve, which may limit their stability, and, in practice, the currents delivered to achieve robust sensations preclude a high degree of selectivity. Whether intraneural electrodes can lead to stable percepts over the span of years has yet to be demonstrated (Rossini et al., 2010; Wendelken et al., 2017).

One of the main disadvantages of interfacing with the peripheral nerve is that arm movements result in movement of the nerve inside the arm, which leads to instability in the interface. An alternative approach to interface with the peripheral nervous system is to target the DRG near the spinal cord (Gaunt et al., 2009), where the cell bodies of tactile nerve fibers are located. Given its encasement in bone, the DRG is far less mobile than is the nerve and offers the additional advantage that it only comprises sensory nerve fibers. However, the implant surgery is far more challenging and invasive, and surgical innovation is ongoing.

11.3.3 Interface hardware—central

The two common interfaces used to deliver sensory feedback to the central nervous system are electrocorticography (ECoG) electrodes (Collins et al., 2017; Hiremath et al., 2017; Johnson et al., 2013; Lee et al., 2018) and penetrating microelectrode arrays (MEAs) (Armenta Salas et al., 2018; Flesher et al., 2016) (Fig. 11.4B). ECoG arrays consist of small metal disks (2–3 mm diameter), arranged in a grid configuration with electrodes spaced 3–10 mm apart, implanted on the surface of the brain (Downey et al., 2020). Because ECoG electrodes are placed outside of the (insulating) pia mater (but under the dura mater), eliciting detectable sensations requires large stimulation currents, which leads to spatial spread and limited selectivity (Hiremath et al., 2017; Lee et al., 2018). Furthermore, ECoG stimulation preferentially activates the superficial layers of cortex, where horizontal connectivity is most extensive, thereby leading to even more diffuse neuronal activation. The advantage of ECoG arrays is that they are less invasive, in that they do not penetrate the gray matter, and seem to be more stable (Degenhart et al., 2016; Pels et al., 2019), though this has not been conclusively demonstrated.

In contrast to ECoG arrays and as their name suggests, penetrating MEAs penetrate the brain to bring electrodes in direct contact with neurons (Fig. 11.4B). The most common MEA is the Utah Electrode Array, which is largely identical to the intraneural array described above for the peripheral nerve except that the 96 electrodes are of equal length rather than tapered (1 or 1.5 mm). MEAs are more invasive than ECoG electrodes, but the close proximity of neurons and electrodes makes it possible to elicit sensations at far lower currents. As a result, MEAs offer better selectivity: microelectrodes separated by less than 1 mm can activate largely nonoverlapping populations, while a single ECoG electrode can be roughly the size of an entire MEA. While MEAs can be used to elicit stable percepts for years (Callier et al., 2015), they inevitably fail catastrophically (Barrese et al., 2013), which severely limits their viability as clinical devices.

11.4 Shaping artificial touch sensations

When we interact with an object, sensory signals from the hand convey information about the object and about our interactions with it. Tactile sensibility is rich and multidimensional, and to restore it completely would require the ability to selectively activate each of thousands of neurons in the nerve or each of hundreds of thousands of neurons in the brain, which is currently impossible. Rather, current attempts to convey sensory feedback focus on conveying the basic information about object interactions to support simple manual behaviors such as grasping. To grasp an object requires sensing which part of the hand is touching the object—minimally the thumb and one of the fingers—how much pressure is exerted on the object—enough to pick it up without breaking it—and when contact is established—to signal the end of the reach and the completion of the grasp. To convey information about the object itself—its texture and compliance for example—would improve the dexterity of bionic hands, but it is much more challenging with current neural interfaces.

One of the principles that guides sensory-encoding algorithms is that of biomimicry: to the extent that the patterns of neuronal activation elicited through electrical stimulation mimic natural patterns of activation, the resulting sensations will be natural and thus intuitive. As mentioned above, the naturalness of electrically evoked neuronal activation is severely limited given the available technology, but general principles of neural coding can be invoked to produce patterns that are as natural as possible. Again, this approach has been successful for cochlear implants, far and away the most widespread neuroprosthesis.

11.4.1 Contact location—leveraging somatotopic maps

In early studies, electrical activation of individual nerve fibers (Marchettini et al., 1990; Ochoa & Torebjörk, 1983) or of neurons in different locations within somatosensory cortex (Penfield & Boldrey, 1937) was shown to elicit sensations that are localized to specific locations on the body. The principle that emerges from this work is that contact location is encoded with a spatial code along the entire neuraxis, at least through somatosensory cortex (electrical stimulation of S2 and PV has not yet been attempted). That is, where on the body a tactile sensation is experienced—the projected field—depends on which neurons are activated. Nerve fibers confined to a given fascicle typically innervate a specific patch of skin and activation of these nerve fibers gives rise to a sensation experienced at that location (Davis et al., 2016; Marchettini et al., 1990; Ochoa & Torebjörk, 1983; Ortiz-Catalan et al., 2020; Raspopovic et al., 2014; Tan et al., 2014). Similarly, neurons within a given volume of cortex innervate a specific patch of skin and their activity results in a sensation experience there (Armenta Salas et al., 2018; Fifer et al., 2020; Flesher et al., 2016; Graczyk et al., 2016; Tabot et al., 2013) (Fig. 11.2A).

This phenomenon can then be leveraged to convey information about where on the bionic hand contact is established. For example, if the bionic index fingertip makes contact with an object, the output of the index fingertip sensor triggers electrical stimulation of a nerve fascicle or cortical neurons with index fingertip projection fields (Fig. 11.5). The user then experiences a sensation on his or her index fingertip. This somatotopic mapping between sensors and electrodes is an example of biomimicry: by leveraging the way the nervous system naturally encodes information about location, the resulting sensory feedback is highly intuitive.

Nerve stimulation through a given electrode evokes sensations spanning one or two phalanges or similar-sized patches of skin on the palm, and a single implant can yield 10–15 unique projection fields (Ortiz-Catalan et al., 2014; Raspopovic et al., 2014; Rijnbeek et al., 2018; Tan et al., 2014). While intrafascicular electrodes in principle enable selective activation of a smaller number of nerve fibers, projected fields tend to be similar to those achieved with extrafascicular electrodes with the currents that are typically

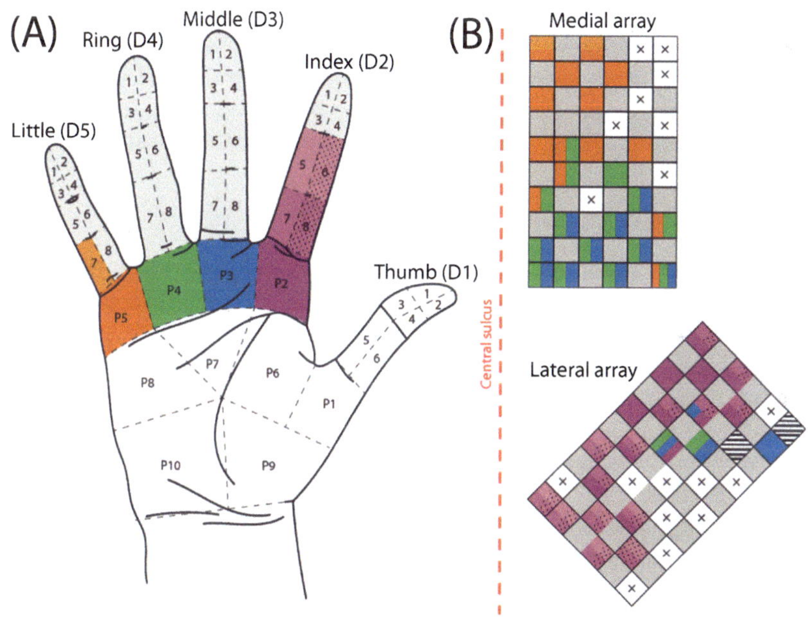

FIGURE 11.5 Projected fields of electrodes on two microelectrode arrays implanted in somatosensory cortex of a human subject. Contact location is encoded with a spatial code in the nerve and in somatosensory cortex, and stimulation of neurons that innervate (or used to innervate before deafferentation) a specific skin location results in a sensation experienced there: stimulation through electrodes of a given color in (B) elicited a tactile percept at the corresponding colored path of skin on the hand in (A). This can be leveraged to intuitively convey information about contact location on a bionic hand. *Reproduced from Flesher, S. N., Collinger, J. L., Foldes, S. T., Weiss, J. M., Downey, J. E., Tyler-Kabara, E. C., Bensmaia, S. J., Schwartz, A. B., Boninger, M. L., & Gaunt, R. A. (2016). Intracortical microstimulation of human somatosensory cortex.* Science Translational Medicine, 8*(361), 361ra141–361ra141. <https://doi.org/10.1126/scitranslmed.aaf8083>.*

delivered (Raspopovic et al., 2014; Rijnbeek et al., 2018; Tan et al., 2015; Wendelken et al., 2017). Given that projected fields are determined by which population of neurons is activated, the projected field of an electrode remains stable to the extent that its position within the tissue is stable (Ortiz-Catalan et al., 2020; Tan et al., 2015).

Microstimulation of somatosensory cortex—Brodmann's areas 3b and 1—evokes sensations that are restricted to a small patch of skin, spanning a phalange or two (Armenta Salas et al., 2018; Fifer et al., 2020; Flesher et al., 2016; Tabot et al., 2013). Studies in monkeys suggest that the projected fields of electrodes in area 3b are smaller than are their counterparts in area 1, as might be expected given the respective sizes of the RF in these cortical areas (Tabot et al., 2014). However, area 1 is the target for implantation in human subjects because its location on the gyrus makes it more surgically accessible. Indeed, area 3b lies in the posterior bank of the central sulcus and thus cannot be accessed with technologies that are currently approved for human use (Fig. 11.2A). In contrast to ICMS, ECoG stimulation activates populations of neurons over large swaths of cortex, so the evoked sensations are referred to large swaths of skin, often the entire hand (Collins et al., 2017; Hiremath et al., 2017; Johnson et al., 2013; Lee et al., 2018). ECoG stimulation thus provides limited information about contact location.

11.4.2 Contact pressure

When we grasp and transport an object, we exert enough force so that it will not slip from our grasp, and not much more. While proprioceptive signals also convey information about the muscular effort deployed during grasping, the fine control of applied force is mediated by the sense of touch, as evidenced by the deficits that are incurred when touch is abolished, for example when the hand is anesthetized (Augurelle et al., 2003). Light pressure results in a faint tactile sensation, forceful pressure in a strong sensation. In the nerve and in the brain, this sensory continuum is associated with an increase in the evoked neuronal activity (Callier et al., 2019; Muniak et al., 2007). Higher pressure results in stronger responses in activated neurons and in the recruitment of additional neurons (Callier et al., 2019) (Fig. 11.6A). Accordingly, information about pressure can be conveyed by modulating the strength of the neuronal activity evoked in the nerve or in the brain (Fig. 11.6A, B).

The population response rate can be modulated by modulating stimulation frequency or amplitude. Changes in frequency result in changes in the firing rates of the activated neuron, in a nearly one-to-one fashion: each stimulating pulse produces an action potential in the volume of neurons over which that pulse exceeds absolute threshold, so more pulses result in more spikes in these neurons. Note, however, that the one-to-one relationship between pulse frequency and spike rate breaks down at higher frequencies,

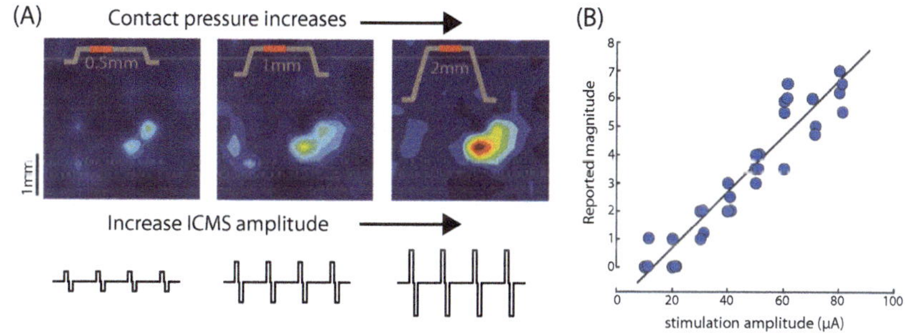

FIGURE 11.6 (A) Higher pressure results in stronger responses in activated neurons and in the recruitment of additional neurons. Top: spatial layout of the response in somatosensory cortex during the sustained portion of indentations (red on the trace) of increasing pressure from left to right. Bottom: the population response can be modulated by modulating stimulation amplitude (or frequency) to convey information about contact pressure. (B) Increases in stimulation amplitude (or frequency) elicit a more intense sensation. Shown here, reports from a human subject of sensation magnitude (on an arbitrary scale) as a function of the amplitude of ICMS delivered to somatosensory cortex *Modified from Callier, T., Suresh, A. K., & Bensmaia, S. J. (2019). Neural coding of contact events in somatosensory cortex.* Cerebral Cortex, 29*(11), 4613–4627. <https://doi.org/10.1093/cercor/bhy337>. (Figure 12.6B) Reproduced from Flesher, S. N., Collinger, J. L., Foldes, S. T., Weiss, J. M., Downey, J. E., Tyler-Kabara, E. C., Bensmaia, S. J., Schwartz, A. B., Boninger, M. L., & Gaunt, R. A. (2016). Intracortical microstimulation of human somatosensory cortex.* Science Translational Medicine, 8*(361), 361ra141. <https://doi.org/10.1126/scitranslmed.aaf8083>.*

when the interpulse interval is shorter than the refractory period (Kim et al., 2017). Indeed, neurons that have been recently activated are more difficult to reactivate within a short window of time, measured in single-digit milliseconds. In contrast, changes in stimulation amplitude result in changes in the activated volume, which also results in a higher overall response. The greater the amplitude, the more current spreads, which recruits neurons located farther away from the stimulating electrode (Butovas & Schwarz, 2003; Stoney et al., 1968; Tehovnik et al., 2006). Accordingly, information about contact pressure has been modulated by varying stimulation frequency and amplitude, with higher frequencies/amplitudes signaling higher pressures.

Experiments with peripheral nerve interfaces have shown that increases in either stimulation amplitude (Graczyk et al., 2016; Raspopovic et al., 2014; Tan et al., 2014) or frequency (Dhillon & Horch, 2005; George et al., 2019) result in increases in the perceived magnitude of an electrically evoked sensation. In fact, greater afferent activation elicits a more intense sensation regardless of whether it is achieved by increasing the stimulation amplitude or frequency (Graczyk et al., 2016). The perceived magnitude of the sensation elicited by a given pulse train can be predicted from the amount of current injected into the nerve after correcting for the activation threshold. Pressure-dependent frequency or amplitude modulation improves the ability of prosthetic users to sense the stiffness of objects (Raspopovic et al., 2014), control grasp force (Raspopovic et al., 2014), and perform tasks requiring dexterity such as picking up and moving a wooden block (Schiefer et al., 2016), picking the stems off cherries (Tan et al., 2014), and transferring a fragile object without breaking it (Valle et al., 2018).

Experiments with cortical interfaces, both in macaques (Tabot et al., 2014) and humans (Flesher et al., 2016; Flesher et al., 2019), have shown that the same general principles apply to ICMS-based artificial touch. Indeed, as is the case in the nerve, increases in stimulation amplitude result in a consistent increase in the perceived magnitude of an electrically evoked sensation (Flesher et al., 2016) (Fig. 11.6B). Similarly, increasing ICMS frequency leads to an increase in perceived magnitude (Callier et al., 2020) but this effect is complicated by concomitant changes in quality (Hughes et al., 2020; see below). As with peripheral nerve interfaces, sensory feedback consisting of modulating ICMS amplitude in a pressure-dependent way improved a human subject's ability to grasp and transport objects (Flesher et al., 2019).

11.4.3 Timing of contact events

When we reach for an object, we terminate our reach and complete our grasp as soon as we make contact with the object (Johansson & Flanagan, 2009). When we handle an object, slip rapidly triggers a grip adjustment (Johansson & Flanagan, 2009). The abolition of touch results in overreaching and in allowing objects to slip from one's grasp. As these examples illustrate, information about the timing of contact events is critical to dexterous manual behavior.

Tracking pressure does not result in precise information about contact timing. Indeed, the pressure experienced by the skin increases slowly during grasp, so a pressure-tracking sensory-encoding algorithm will lead to delayed detection of contact, as detection can only occur when stimulation levels exceed threshold. In the intact nerve, RA and PC fibers produce a strong phasic response at the onset and offset of contact, and this signal is well suited to convey information about the timing of contact events (Fig. 11.2B). This sensitivity to contact transients can be accounted for by the fact that some nerve fibers are primarily sensitive to changes in pressure (RA, PC) (Dong et al., 2013; Kim et al., 2010; Saal et al., 2017) and thus carry a signal that emphasizes contact transients (Callier et al., 2019) (Fig. 11.7). Another population of nerve fibers (SA1 fibers) carry information about pressure during maintained contact, but this sustained signal tends to be weaker and more localized than its phasic counterpart during contact transients.

A straightforward way to mimic these two aspects of the neural response—strong response during contact transients, weaker sustained response during maintained contact—consists of modulating the intensity of electrical stimulation (frequency and/or amplitude) according to a combination of instantaneous pressure and its derivative (Saal & Bensmaia, 2015) (Fig. 11.7). This way, contact transients lead to greater stimulation—driven

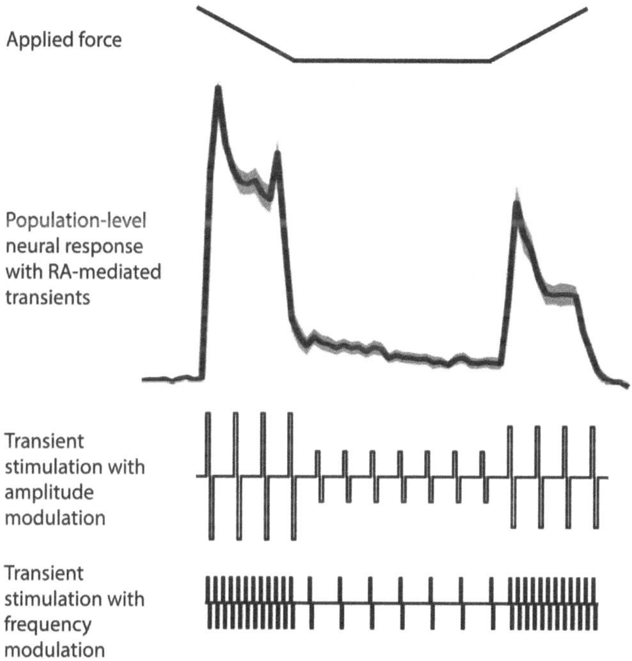

FIGURE 11.7 Spatiotemporal dynamics of the neural response to contact events: strong phasic responses to force transients are well suited to rapidly convey the timing of those events. The population-level neural response in somatosensory cortex to a simple indentation (top trace) is shown here. The aggregate response in the nerve is similar. This neural response can be mimicked by modulating the intensity of electrical stimulation according to instantaneous pressure (the sole factor during the sustained portion of the indentation) and its derivative (during transients). The increased intensity during transients can be modulated with stimulation frequency and/or amplitude. *Modified from Callier, T., Suresh, A. K., & Bensmaia, S. J. (2019). Neural coding of contact events in somatosensory cortex.* Cerebral Cortex, 29*(11), 4613–4627. <https://doi.org/10.1093/cercor/bhy337>.*

by the derivative term—but pressure is also tracked during maintained contact. The resulting signals can approximate the aggregate behavior of nerve fibers during activities of daily living (Okorokova et al., 2018). This biomimetic strategy leads to an improved ability to transport fragile objects (George et al., 2019; Valle et al., 2018), a task that requires sensory feedback, and to an improved ability to sense the properties of grasped objects, for example their compliance (George et al., 2019).

At the aggregate level, responses in somatosensory cortex resemble their peripheral counterparts: onset and offset of contact are signaled by precisely timed phasic responses—driven by RA and PC fibers (Johansson & Flanagan, 2009; Pei et al., 2009)—and maintained contact by a weaker response—driven by SA1 fibers (Fig. 11.7). As is the case in the nerve, phasic responses associated with contact transients dwarf the tonic response during maintained contact (Callier et al., 2019). Thus, biomimetic encoding algorithms designed for the nerve could in principle be applied directly to ICMS-based feedback (Callier et al., 2019; Okorokova et al., 2018) but the efficacy of such algorithms has yet to be tested.

11.4.4 Sensory quality

Tactile sensations can be described in terms of their location on the body, their strength, and their timing, and our understanding of the neural coding of these perceptual features can be leveraged in designing sensory-encoding algorithms. However, touch conveys far more detailed information about objects and about contact events than these three features. For example, we can sense the size, shape, and texture of an object, and its movement across the skin, via cutaneous signals. These different sensations are characterized by different qualities. The sensation of an edge indented into the skin has a different quality than the sensation of a texture scanned across the skin. The sensory space of touch is very complex. Even specific domains of touch—texture, for example—are encoded in a high dimensional space (Lieber & Bensmaia, 2019). Sensory quality thus encompasses the bulk of the information about objects and contact events. The quality of a sensory percept is shaped by the specific pattern of the neuronal activation that gave rise to it. Indeed, a specific texture produces a specific spatiotemporal pattern of activation and the quality of the evoked sensation is the outcome of this spatiotemporal pattern (Fig. 11.2B). An indented edge produces a different spatiotemporal pattern of activation, one that depends in part on its orientation, and gives rise to a different sensation (Fig. 11.8A). To elicit sensations endowed with specific qualities would require reproducing precise spatiotemporal patterns of activation, which is impossible given existing technologies and will probably remain so in the foreseeable future. However, the quality of an electrically evoked sensation can be systematically manipulated, albeit in a limited way.

First, one might ask whether electrically evoked sensations resemble those that are mechanically evoked through interactions with objects. The answer depends on whether the stimulus is applied to the nerve or brain. Electrical stimulation of the nerve typically results in paresthesias—tingling sensations that feel unnatural—but sensations of vibration, pressure, or tapping have also been described (Ackerley et al., 2018; Davis et al., 2016; George et al., 2019; Ortiz-Catalan et al., 2019; Tan et al., 2014). In contrast, ICMS evokes sensations that are described as natural or nearly so (Armenta Salas et al., 2018; Flesher et al., 2016).

Second, one might ask to what extent electrically evoked sensations depend on which subpopulations of neurons are activated. In the nerve, evoked sensations are relatively independent of which electrode delivers the stimulation (Graczyk et al., n.d.), likely due to the relatively consistent submodality

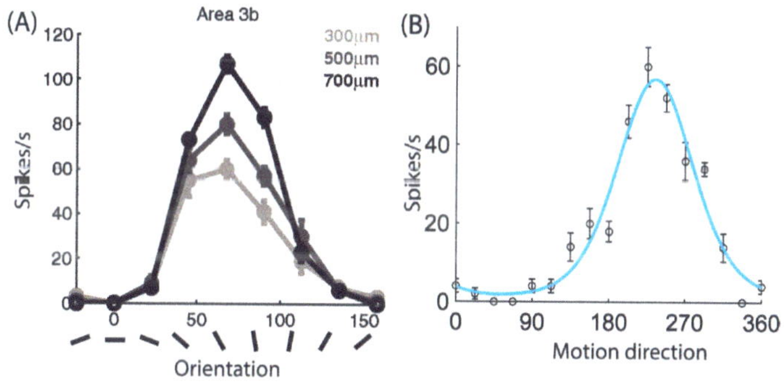

FIGURE 11.8 Individual neurons in somatosensory cortex can be tuned for specific qualities of a tactile stimulus. (A) This area 3b neuron responds more strongly to indented edges of a specific orientation between 50 and 100 degrees and responds weakly or not at all to edges with substantially different orientations. (B) This neuron in somatosensory cortex responds most strongly to motion across the skin in one specific direction. Because different neurons will have different tuning curves, the spatiotemporal patterns of activation in somatosensory cortex will depend on the specific qualities of a tactile stimulus. These patterns would have to be reproduced to elicit sensations with those specific qualities. *(Figure 12.8A) Reproduced from Bensmaia, S. J., Denchev, P. V., Dammann, J. F., Craig, J. C., & Hsiao, S. S. (2008). The representation of stimulus orientation in the early stages of somatosensory processing.* Journal of Neuroscience, 28*(3), 776–786. https://doi.org/10.1523/JNEUROSCI.4162-07.2008. (Figure 12.8B) Reproduced from Pei, Y.-C., Hsiao, S. S., Craig, J. C., & Bensmaia, S. J. (2010). Shape invariant coding of motion direction in somatosensory cortex.* PLoS Biology, 8*(2), e1000305. https://doi.org/10.1371/journal.pbio.1000305.*

composition of fascicles across the nerve. Indeed, the only way for the quality of electrically evoked sensations to differ across electrodes would be if different fascicles contained different proportions of SA1, RA, and PC fibers. Given that these differences are minimal, the resulting variability in the quality of electrically evoked sensations across electrodes is also minimal. In contrast, the quality of ICMS-evoked sensations is strongly electrode dependent. Indeed, subjects report different types of sensations—pressure, tapping, pulsing, light movement, for example—depending on the stimulating electrode, even though these are in the same cortical area (Armenta Salas et al., 2018; Flesher et al., 2016; Hughes et al., 2020). The dependence of quality on electrode likely reflects differences in the functional roles of different neuronal subpopulations. If one volume of cortex contains a preponderance of neurons involved in motion analysis, activating these neurons may result in a percept of tactile motion. If another volume contains texture-sensitive neurons, stimulation thereof may evoke texture-like sensations. While sensory quality can be manipulated by activating different electrodes, this phenomenon cannot be leveraged meaningfully because (1) both quality and projected field vary with electrodes, so quality and location are inextricably linked, and (2) quality depends idiosyncratically and unpredictably on the brain area in which the electrodes are implanted.

Third, one might ask how the parameters of stimulation change the quality of the sensation. In both the nerve and the brain, changes in stimulation amplitude lead to an increase in the magnitude of the sensation (Flesher et al., 2016; Graczyk et al., 2016; Kim et al., 2015) (Fig. 11.6B). In the nerve, increases in amplitude can also result in an increase in the size of the projected field (Davis et al., 2016; Ortiz-Catalan, Mastinu, Sassu, et al., 2020; Tan et al., 2014). In neither the nerve nor the brain does the quality of the sensation systematically change with changes in amplitude. In contrast, changes in the timing of pulses can impact the quality, as evidenced by the effect of changes in stimulation frequency on the evoked percept. Indeed, changes in stimulation frequency result in changes in the quality of the percept in both the nerve (Graczyk et al., n.d.) and the brain (Callier et al., 2020; Hughes et al., 2020). In the nerve, changes in frequency have an effect on quality that is consistent across electrodes and even across subjects (Graczyk et al., n.d.). Indeed, the perceived frequency increases with increases in pulse frequency up to about 100 Hz. Increases in frequency beyond 100 Hz result in increases in magnitude but not quality. In cortex, changes in microstimulation frequency also result in a change in quality (Callier et al., 2020; Hughes et al., 2020), but the specific relationship between stimulation frequency and quality depends on the stimulation electrode (Hughes et al., 2020). In both the nerve and the brain, low-frequency stimulation typically results in intermittent sensations, such as tapping, whereas high-frequency stimulation results in fused events, such as pressure.

As mentioned above, some sensory information is encoded in the timing of the neuronal response, relying on a so-called temporal code. Accordingly, modulation of pulse timing—which in turn shapes spike timing in the activated population—can be used to convey stimulus information in a biomimetic way. For example, texture coding relies in part on temporal spiking patterns (Weber et al., 2013), and modulating pulse trains to mimic patterns of neuronal activation produced during the haptic exploration of textured surface can be used to convey information about surface texture in the nerve (Oddo et al., 2016) and in the brain (O'Doherty et al., 2019). While the quality of the evoked percepts is not one of texture per se, this strategy may convey some information about texture in an intuitive way.

As summarized above, sensory quality is determined by the specific spatiotemporal pattern of activation evoked by a stimulus. Accordingly, the ability to electrically evoke sensations endowed with a specific quality is limited by the inability to activate individual neurons in a selective manner. To get around this limitation, one might exploit explicit rate-based representations of specific object features in somatosensory cortex. For example, a subpopulation of neurons in Brodmann's area 1 is highly selective for direction of motion, responding preferentially to objects moving across the skin in a specific direction (Pei et al., 2010) (Fig. 11.8B). Perhaps electrical activation of a population of such neurons would produce a percept of directed motion, a strategy that has shown promise in visual experiments; indeed, electrical stimulation of direction-selective neurons in medial temporal cortex biases the perceived direction of motion of a visual stimulus (Salzman et al., 1992). This strategy requires that direction-selective neurons are clustered together, and the topographical organization of direction selectivity in somatosensory cortex has not been investigated. However, electrical stimulation of certain patches of area 1 has been shown to occasionally evoke sensations of tactile motion (Armenta Salas et al., 2018), perhaps providing a proof of principle for the strategy.

11.5 Future horizons

The restoration of touch—via stimulation of the peripheral nerves or somatosensory cortex—improves the dexterity of bionic hands despite the crudeness of current technologies used to activate neurons. The main limitation is the inability to reproduce specific spatiotemporal patterns of activity because electrical stimulation activates populations of neurons around the electrode synchronously and nonselectively. Nonetheless, a guiding principle in the development of artificial touch is that of biomimicry: to the extent that we can speak the language of the nervous system—reproduce aspects of natural neuronal response that play a key role in conveying specific sensory information—the resulting percepts will be naturalistic and thus convey sensory information intuitively. While arbitrary mappings between sensor output and electrical stimulation can be used to convey sensory information (Dadarlat et al., 2015; Thomson et al., 2013), these require learning on the part of the user and are unlikely to scale up to accommodate the bandwidth required to achieve dexterity that approaches that of the native limb (Delhaye et al., 2016). However, as stimulating technology progresses beyond electrical stimulation and as the ability to selectively stimulate individual neurons improves, the potential to achieve naturalistic sensory feedback will grow.

Beyond prosthetic applications, the ability to produce tactile sensations that convey specific stimulus information will set the stage for writing information into the brain through artificial means, bypassing our senses, in a variety of other contexts. While restoring other sensory modalities is an obvious extension, one might imagine interfacing with language structures to convey speech signals or with memory structures to implant memories, etc. While these ideas may seem outlandish given the primitive state of artificial touch, the private sector has begun to invest in them. In the meantime, sensitized bionic hands are on the cusp of clinical viability and may soon become part of our everyday experience.

References

Aberra, A. S., Peterchev, A. V., & Grill, W. M. (2018). Biophysically realistic neuron models for simulation of cortical stimulation. *Journal of Neural Engineering*, *15*(6), 066023. Available from https://doi.org/10.1088/1741-2552/aadbb1.

Ackerley, R., Backlund Wasling, H., Ortiz-Catalan, M., Brånemark, R., & Wessberg, J. (2018). Case Studies in Neuroscience: Sensations elicited and discrimination ability from nerve cuff stimulation in an amputee over time. *Journal of Neurophysiology*, *120*(1), 291–295. Available from https://doi.org/10.1152/jn.00909.2017.

Ajiboye, A. B., Willett, F. R., Young, D. R., Memberg, W. D., Murphy, B. A., Miller, J. P., Walter, B. L., Sweet, J. A., Hoyen, H. A., Keith, M. W., Peckham, P. H., Simeral, J. D., Donoghue, J. P., Hochberg, L. R., & Kirsch, R. F. (2017). Restoration of reaching and grasping movements through brain-controlled muscle stimulation in a person with tetraplegia: A proof-of-concept demonstration. *The Lancet*, *389*(10081), 1821–1830. Available from https://doi.org/10.1016/S0140-6736(17)30601-3.

Armenta Salas, M., Bashford, L., Kellis, S., Jafari, M., Jo, H., Kramer, D., Shanfield, K., Pejsa, K., Lee, B., Liu, C. Y., & Andersen, R. A. (2018). Proprioceptive and cutaneous sensations in humans elicited by intracortical microstimulation. *ELife*, *7*, e32904. Available from https://doi.org/10.7554/eLife.32904.

Augurelle, A.-S., Smith, A. M., Lejeune, T., & Thonnard, J.-L. (2003). Importance of cutaneous feedback in maintaining a secure grip during manipulation of hand-held objects. *Journal of Neurophysiology*, *89*(2), 665–671. Available from https://doi.org/10.1152/jn.00249.2002.

Barrese, J. C., Rao, N., Paroo, K., Triebwasser, C., Vargas-Irwin, C., Franquemont, L., & Donoghue, J. P. (2013). Failure mode analysis of silicon-based intracortical microelectrode arrays in non-human primates. *Journal of Neural Engineering*, *10*(6), 066014. Available from https://doi.org/10.1088/1741-2560/10/6/066014.

Bell, J., Bolanowski, S., & Holmes, M. H. (1994). The structure and function of pacinian corpuscles: A review. *Progress in Neurobiology*, *42*(1), 79–128. Available from https://doi.org/10.1016/0301-0082(94)90022-1.

Bengtsson, F., Brasselet, R., Johansson, R. S., Arleo, A., & Jörntell, H. (2013). Integration of sensory quanta in cuneate nucleus neurons in vivo. *PLoS One*, *8*(2), e56630. Available from https://doi.org/10.1371/journal.pone.0056630.

Bensmaia, S. J. (2015). Biological and bionic hands: Natural neural coding and artificial perception. *Philosophical Transactions of the Royal Society B: Biological Sciences*, *370*(1677), 20140209. Available from https://doi.org/10.1098/rstb.2014.0209.

Bensmaia, S. J., Denchev, P. V., Dammann, J. F., Craig, J. C., & Hsiao, S. S. (2008). The representation of stimulus orientation in the early stages of somatosensory processing. *Journal of Neuroscience*, *28*(3), 776–786. Available from https://doi.org/10.1523/JNEUROSCI.4162-07.2008.

Birznieks, I., Macefield, V. G., Westling, G., & Johansson, R. S. (2009). Slowly adapting mechanoreceptors in the borders of the human fingernail encode fingertip forces. *Journal of Neuroscience*, *29*(29), 9370–9379. Available from https://doi.org/10.1523/JNEUROSCI.0143-09.2009.

Boretius, T., Badia, J., Pascual-Font, A., Schuettler, M., Navarro, X., Yoshida, K., & Stieglitz, T. (2010). A transverse intrafascicular multichannel electrode (TIME) to interface with the peripheral nerve. *Biosensors and Bioelectronics*, *26*(1), 62–69. Available from https://doi.org/10.1016/j.bios.2010.05.010.

Branner, A., Stein, R. B., & Normann, R. A. (2001). Selective stimulation of cat sciatic nerve using an array of varying-length microelectrodes. *Journal of Neurophysiology*, *85*(4), 1585–1594. Available from https://doi.org/10.1152/jn.2001.85.4.1585.

Brisben, A. J., Hsiao, S. S., & Johnson, K. O. (1999). Detection of vibration transmitted through an object grasped in the hand. *Journal of Neurophysiology*, *81*, 1548–1558.

Butovas, S., & Schwarz, C. (2003). Spatiotemporal effects of microstimulation in rat neocortex: A parametric study using multielectrode recordings. *Journal of Neurophysiology, 90*(5), 3024–3039. Available from https://doi.org/10.1152/jn.00245.2003.

Callier, T., Brantly, N. W., Caravelli, A., & Bensmaia, S. J. (2020). The frequency of cortical microstimulation shapes artificial touch. *Proceedings of the National Academy of Sciences, 117*(2), 1191–1200. Available from https://doi.org/10.1073/pnas.1916453117.

Callier, T., Schluter, E. W., Tabot, G. A., Miller, L. E., Tenore, F. V., & Bensmaia, S. J. (2015). Long-term stability of sensitivity to intracortical microstimulation of somatosensory cortex. *Journal of Neural Engineering, 12*(5), 056010. Available from https://doi.org/10.1088/1741-2560/12/5/056010.

Callier, T., Suresh, A. K., & Bensmaia, S. J. (2019). Neural coding of contact events in somatosensory cortex. *Cerebral Cortex, 29*(11), 4613–4627. Available from https://doi.org/10.1093/cercor/bhy337.

Charkhkar, H., Shell, C. E., Marasco, P. D., Pinault, G. J., Tyler, D. J., & Triolo, R. J. (2018). High-density peripheral nerve cuffs restore natural sensation to individuals with lower-limb amputations. *Journal of Neural Engineering, 15*(5), 056002. Available from https://doi.org/10.1088/1741-2552/aac964.

Collinger, J. L., Wodlinger, B., Downey, J. E., Wang, W., Tyler-Kabara, E. C., Weber, D. J., McMorland, A. J., Velliste, M., Boninger, M. L., & Schwartz, A. B. (2013). High-performance neuroprosthetic control by an individual with tetraplegia. *The Lancet, 381* (9866), 557–564. Available from https://doi.org/10.1016/S0140-6736(12)61816-9.

Collins, K. L., Guterstam, A., Cronin, J., Olson, J. D., Ehrsson, H. H., & Ojemann, J. G. (2017). Ownership of an artificial limb induced by electrical brain stimulation. *Proceedings of the National Academy of Sciences, 114*(1), 166–171. Available from https://doi.org/10.1073/pnas.1616305114.

Craig, A. D. (Bud) (2006). Retrograde analyses of spinothalamic projections in the macaque monkey: Input to ventral posterior nuclei. *The Journal of Comparative Neurology, 499*(6), 965–978. Available from https://doi.org/10.1002/cne.21154.

Dadarlat, M. C., O'Doherty, J. E., & Sabes, P. N. (2015). A learning-based approach to artificial sensory feedback leads to optimal integration. *Nature Neuroscience, 18*(1), 138–144. Available from https://doi.org/10.1038/nn.3883.

Darian-Smith, I. (2011). The sense of touch: Performance and peripheral neural processes. In R. Terjung (Ed.), *Comprehensive physiology* (p. cp010317). John Wiley & Sons, Inc, https://doi.org/10.1002/cphy.cp010317.

Davis, T. S., Wark, H. A. C., Hutchinson, D. T., Warren, D. J., O'Neill, K., Scheinblum, T., Clark, G. A., Normann, R. A., & Greger, B. (2016). Restoring motor control and sensory feedback in people with upper extremity amputations using arrays of 96 microelectrodes implanted in the median and ulnar nerves. *Journal of Neural Engineering, 13*(3), 036001. Available from https://doi.org/10.1088/1741-2560/13/3/036001.

Degenhart, A. D., Eles, J., Dum, R., Mischel, J. L., Smalianchuk, I., Endler, B., Ashmore, R. C., Tyler-Kabara, E. C., Hatsopoulos, N. G., Wang, W., Batista, A. P., & Cui, X. T. (2016). Histological evaluation of a chronically-implanted electrocorticographic electrode grid in a non-human primate. *Journal of Neural Engineering, 13*(4), 046019. Available from https://doi.org/10.1088/1741-2560/13/4/046019.

Delhaye, B. P., Long, K. H., & Bensmaia, S. J. (2018). Neural basis of touch and proprioception in primate cortex. In D. M. Pollock (Ed.), *Comprehensive physiology* (pp. 1575–1602). John Wiley & Sons, Inc. Available from https://doi.org/10.1002/cphy.c170033.

Delhaye, B. P., Saal, H. P., & Bensmaia, S. J. (2016). Key considerations in designing a somatosensory neuroprosthesis. *Journal of Physiology-Paris, 110*(4, Part A), 402–408. Available from https://doi.org/10.1016/j.jphysparis.2016.11.001.

Dhillon, G. S., & Horch, K. W. (2005). Direct neural sensory feedback and control of a prosthetic arm. *IEEE Transactions on Neural Systems and Rehabilitation Engineering, 13*(4), 468–472. Available from https://doi.org/10.1109/TNSRE.2005.856072.

DiCarlo, J. J., Johnson, K. O., & Hsiao, S. S. (1998). Structure of receptive fields in area 3b of primary somatosensory cortex in the alert monkey. *The Journal of Neuroscience, 18*(7), 2626–2645. Available from https://doi.org/10.1523/JNEUROSCI.18-07-02626.1998.

Dong, Y., Mihalas, S., Kim, S. S., Yoshioka, T., Bensmaia, S. J., & Niebur, E. (2013). A simple model of mechanotransduction in primate glabrous skin. *Journal of Neurophysiology, 109* (5), 1350–1359. Available from https://doi.org/10.1152/jn.00395.2012.

Downey, J. E., Brooks, J., & Bensmaia, S. J. (2020). Artificial sensory feedback for bionic hands, In*Intelligent Biomechatronics in Neurorehabilitation*, 131–145. Available from https://doi.org/10.1016/B978-0-12-814942-3.00008-8.

Fifer, M. S., McMullen, D. P., Thomas, T. M., Osborn, L. E., Nickl, R. W., Candrea, D. N., Pohlmeyer, E. A., Thompson, M. C., Anaya, M., Schellekens, W., Ramsey, N. F., Bensmaia, S. J., Anderson, W. S., Wester, B. A., Crone, N. E., Celnik, P. A., Cantarero, G. L., & Tenore, F. V. (2020). Intracortical microstimulation elicits human fingertip sensations. *MedRxiv*, 2020.05.29.20117374. Available from https://doi.org/10.1101/2020.05.29.20117374.

Flesher, S. N., Collinger, J. L., Foldes, S. T., Weiss, J. M., Downey, J. E., Tyler-Kabara, E. C., Bensmaia, S. J., Schwartz, A. B., Boninger, M. L., & Gaunt, R. A. (2016). Intracortical microstimulation of human somatosensory cortex. *Science Translational Medicine, 8*(361), 361ra141-361ra141. Available from https://doi.org/10.1126/scitranslmed.aaf8083.

Flesher, S. N., Downey, J. E., Weiss, J. M., Hughes, C. L., Herrera, A. J., & Gaunt, R. A. (2019). *Restored tactile sensation improves neuroprosthetic arm control* [Preprint]. *Clinical Trials*. Available from https://doi.org/10.1101/653428.

Freeberg, M. J., Stone, M. A., Triolo, R. J., & Tyler, D. J. (2017). The design of and chronic tissue response to a composite nerve electrode with patterned stiffness. *Journal of Neural Engineering, 14*(3), 036022. Available from https://doi.org/10.1088/1741-2552/aa6632.

Freeman, A. W., & Johnson, K. O. (1982). A model accounting for effects of vibratory amplitude on responses of cutaneous mechanoreceptors in macaque monkey. *The Journal of Physiology, 323*(1), 43–64. Available from https://doi.org/10.1113/jphysiol.1982.sp014060.

Galvani, L. (1791). De viribus electricitatis in motu musculari commentarius. *Ex Typographia Instituti Scientiarium*. Available from https://doi.org/10.5479/sil.324681.39088000932442.

Gaunt, R. A., Hokanson, J. A., & Weber, D. J. (2009). Microstimulation of primary afferent neurons in the L7 dorsal root ganglia using multielectrode arrays in anesthetized cats: Thresholds and recruitment properties. *Journal of Neural Engineering, 6*(5), 055009. Available from https://doi.org/10.1088/1741-2560/6/5/055009.

Gennari, F. (1782). *De Peculiari Structura Cerebri Nonnuliaque Ejus Morbis*. Parma Ex Regio Typographeo.

George, J. A., Kluger, D. T., Davis, T. S., Wendelken, S. M., Okorokova, E. V., He, Q., & Clark, G. A. (2019). Biomimetic sensory feedback through peripheral nerve stimulation improves dexterous use of a bionic hand. *Science Robotics, 4*(32). Available from https://doi.org/10.1126/scirobotics.aax2352.

Gilja, V., Nuyujukian, P., Chestek, C. A., Cunningham, J. P., Yu, B. M., Fan, J. M., Churchland, M. M., Kaufman, M. T., Kao, J. C., Ryu, S. I., & Shenoy, K. V. (2012). A high-performance neural prosthesis enabled by control algorithm design. *Nature Neuroscience*, *15*(12), 1752–1757. Available from https://doi.org/10.1038/nn.3265.

Glickstein, M. (1988). The discovery of the visual cortex. *Scientific American*, *259*(3), 118–127. Available from https://doi.org/10.1038/scientificamerican0988-118.

Glickstein, M., & Whitteridge, D. (1987). Tatsuji Inouye and the mapping of the visual fields on the human cerebral cortex. *Trends in Neurosciences*, *10*(9), 350–353. Available from https://doi.org/10.1016/0166-2236(87)90066-X.

Goodman, J. M., & Bensmaia, S. J. (2020). *The neural mechanisms of touch and proprioception at the somatosensory periphery. The senses*. Elsevier.

Graczyk, E.L., Christie, B., He, Q., Tyler, D.J., & Bensmaia, S.J. (n.d.). Frequency shapes the quality of tactile percepts evoked through electrical stimulation of the nerves. *BioRxiv*.

Graczyk, E. L., Delhaye, B. P., Schiefer, M. A., Bensmaia, S. J., & Tyler, D. J. (2018). Sensory adaptation to electrical stimulation of the somatosensory nerves. *Journal of Neural Engineering*, *15*(4), 046002. Available from https://doi.org/10.1088/1741-2552/aab790.

Graczyk, E. L., Schiefer, M. A., Saal, H. P., Delhaye, B. P., Bensmaia, S. J., & Tyler, D. J. (2016). The neural basis of perceived intensity in natural and artificial touch. *Science Translational Medicine*, *8*(362). Available from https://doi.org/10.1126/scitranslmed.aaf5187, 362ra142-362ra142.

Harvey, M. A., Saal, H. P., Dammann, J. F., & Bensmaia, S. J. (2013). Multiplexing stimulus information through rate and temporal codes in primate somatosensory cortex. *PLoS Biology*, *11*(5), e1001558. Available from https://doi.org/10.1371/journal.pbio.1001558.

Hayward, V., Terekhov, A. V., Wong, S.-C., Geborek, P., Bengtsson, F., & Jörntell, H. (2014). Spatio-temporal skin strain distributions evoke low variability spike responses in cuneate neurons. *Journal of the Royal Society Interface*, *11*(93), 20131015. Available from https://doi.org/10.1098/rsif.2013.1015.

Hiremath, S. V., Tyler-Kabara, E. C., Wheeler, J. J., Moran, D. W., Gaunt, R. A., Collinger, J. L., Foldes, S. T., Weber, D. J., Chen, W., Boninger, M. L., & Wang, W. (2017). Human perception of electrical stimulation on the surface of somatosensory cortex. *PLoS One*, *12*(5), e0176020. Available from https://doi.org/10.1371/journal.pone.0176020.

Histed, M. H., Bonin, V., & Reid, R. C. (2009). Direct activation of sparse, distributed populations of cortical neurons by electrical microstimulation. *Neuron*, *63*(4), 508–522. Available from https://doi.org/10.1016/j.neuron.2009.07.016.

Hochberg, L. R., Bacher, D., Jarosiewicz, B., Masse, N. Y., Simeral, J. D., Vogel, J., Haddadin, S., Liu, J., Cash, S. S., van der Smagt, P., & Donoghue, J. P. (2012). Reach and grasp by people with tetraplegia using a neurally controlled robotic arm. *Nature*, *485*(7398), 372–375. Available from https://doi.org/10.1038/nature11076.

Hodgkin, A. L., & Huxley, A. F. (1952). A quantitative description of membrane current and its application to conduction and excitation in nerve. *The Journal of Physiology*, *114*(4), 500–544. Available from https://doi.org/10.1113/jphysiol.1952.sp004764.

Hughes, C. L., Flesher, S. N., Weiss, J. M., Boninger, M., Collinger, J. L., & Gaunt, R. (2020). *Perceptual responses to microstimulation frequency are spatially organized in human somatosensory cortex* [Preprint]. *Neuroscience*. Available from https://doi.org/10.1101/2020.07.16.207506.

Hyvärinen, J., & Poranen, A. (1978a). Movement-sensitive and direction and orientation-selective cutaneous receptive fields in the hand area of the post-central gyrus in monkeys. *The Journal of Physiology*, *283*(1), 523–537. Available from https://doi.org/10.1113/jphysiol.1978.sp012517.

Hyvärinen, J., & Poranen, A. (1978b). Receptive field integration and submodality convergence in the hand area of the post-central gyrus of the alert monkey. *The Journal of Physiology, 283*(1), 539–556. Available from https://doi.org/10.1113/jphysiol.1978.sp012518.

Iwamura, Y., Tanaka, M., Sakamoto, M., & Hikosaka, O. (1983). Converging patterns of finger representation and complex response properties of neurons in area 1 of the first somatosensory cortex of the conscious monkey. *Experimental Brain Research, 51*(3). Available from https://doi.org/10.1007/BF00237869.

Iwamura, Y., Tanaka, M., Sakamoto, M., & Hikosaka, O. (1993). Rostrocaudal gradients in the neuronal receptive field complexity in the finger region of the alert monkey's postcentral gyrus. *Experimental Brain Research, 92*(3). Available from https://doi.org/10.1007/BF00229023.

Johansson, R. S., & Flanagan, J. R. (2009). Coding and use of tactile signals from the fingertips in object manipulation tasks. *Nature Reviews. Neuroscience, 10*(5), 345–359. Available from https://doi.org/10.1038/nrn2621.

Johansson, R. S., & Vallbo, A. B. (1979). Tactile sensibility in the human hand: Relative and absolute densities of four types of mechanoreceptive units in glabrous skin. *The Journal of Physiology, 286*(1), 283–300. Available from https://doi.org/10.1113/jphysiol.1979.sp012619.

Johansson, R. S., Landstrom, U., & Lundstrom, R. (1982). Responses of mechanoreceptive afferent units in the glabrous skin of the human hand to sinusoidal skin displacements. *Brain Research, 244*(1), 17–25. Available from https://doi.org/10.1016/0006-8993(82)90899-X.

Johnson, K. O. (2001). The roles and functions of cutaneous mechanoreceptors. *Current Opinion in Neurobiology, 11*(4), 455–461. Available from https://doi.org/10.1016/S0959-4388(00)00234-8.

Johnson, L. A., Wander, J. D., Sarma, D., Su, D. K., Fetz, E. E., & Ojemann, J. G. (2013). Direct electrical stimulation of the somatosensory cortex in humans using electrocorticography electrodes: A qualitative and quantitative report. *Journal of Neural Engineering, 10*(3), 036021. Available from https://doi.org/10.1088/1741-2560/10/3/036021.

Jörntell, H., Bengtsson, F., Geborek, P., Spanne, A., Terekhov, A. V., & Hayward, V. (2014). Segregation of tactile input features in neurons of the cuneate nucleus. *Neuron, 83*(6), 1444–1452. Available from https://doi.org/10.1016/j.neuron.2014.07.038.

Kaas, J. H. (1983). What, if anything, is SI? Organization of first somatosensory area of cortex. *Physiological Reviews, 63*(1), 206–231. Available from https://doi.org/10.1152/physrev.1983.63.1.206.

Kim, S., Callier, T., & Bensmaia, S. J. (2017). A computational model that predicts behavioral sensitivity to intracortical microstimulation. *Journal of Neural Engineering, 14*(1), 016012. Available from https://doi.org/10.1088/1741-2552/14/1/016012.

Kim, S., Callier, T., Tabot, G. A., Gaunt, R. A., Tenore, F. V., & Bensmaia, S. J. (2015). Behavioral assessment of sensitivity to intracortical microstimulation of primate somatosensory cortex. *Proceedings of the National Academy of Sciences, 112*(49), 15202–15207. Available from https://doi.org/10.1073/pnas.1509265112.

Kim, S. S., Sripati, A. P., & Bensmaia, S. J. (2010). Predicting the timing of spikes evoked by tactile stimulation of the hand. *Journal of Neurophysiology, 104*(3), 1484–1496. Available from https://doi.org/10.1152/jn.00187.2010.

Knibestöl, M. (1973). Stimulus-response functions of rapidly adapting mechanoreceptors in the human glabrous skin area. *The Journal of Physiology, 232*(3), 427–452. Available from https://doi.org/10.1113/jphysiol.1973.sp010279.

Knibestöl, M. (1975). Stimulus-response functions of slowly adapting mechanoreceptors in the human glabrous skin area. *The Journal of Physiology, 245*(1), 63–80. Available from https://doi.org/10.1113/jphysiol.1975.sp010835.

Lawrence, S. M., Dhillon, G. S., Jensen, W., Yoshida, K., & Horch, K. W. (2004). Acute peripheral nerve recording characteristics of polymer-based longitudinal intrafascicular electrodes. *IEEE Transactions on Neural Systems and Rehabilitation Engineering*, *12*(3), 345–348. Available from https://doi.org/10.1109/TNSRE.2004.831491.

Lee, B., Kramer, D., Armenta Salas, M., Kellis, S., Brown, D., Dobreva, T., Klaes, C., Heck, C., Liu, C., & Andersen, R. A. (2018). Engineering artificial somatosensation through cortical stimulation in humans. *Frontiers in Systems Neuroscience*, *12*, 24. Available from https://doi.org/10.3389/fnsys.2018.00024.

Lieber, J. D., & Bensmaia, S. J. (2019). High-dimensional representation of texture in somatosensory cortex of primates. *Proceedings of the National Academy of Sciences*, *116*(8), 3268–3277. Available from https://doi.org/10.1073/pnas.1818501116.

Mackevicius, E. L., Best, M. D., Saal, H. P., & Bensmaia, S. J. (2012). Millisecond precision spike timing shapes tactile perception. *Journal of Neuroscience*, *32*(44), 15309–15317. Available from https://doi.org/10.1523/JNEUROSCI.2161-12.2012.

Marchettini, P., Cline, M., & Ochoa, J. L. (1990). Innervation territories for touch and pain afferents of single fascicles of the human ulnar nerve: Mapping through intraneural microrecording and microstimulation. *Brain*, *113*(5), 1491–1500. Available from https://doi.org/10.1093/brain/113.5.1491.

Muniak, M. A., Ray, S., Hsiao, S. S., Dammann, J. F., & Bensmaia, S. J. (2007). The neural coding of stimulus intensity: Linking the population response of mechanoreceptive afferents with psychophysical behavior. *Journal of Neuroscience*, *27*(43), 11687–11699. Available from https://doi.org/10.1523/JNEUROSCI.1486-07.2007.

Munk, H. (1881). *Uber die Funktionen der Grosshimrinde*. A. Hirshwald.

Ochoa, J., & Torebjörk, E. (1983). Sensations evoked by intraneural microstimulation of single mechanoreceptor units innervating the human hand. *The Journal of Physiology*, *342*(1), 633–654. Available from https://doi.org/10.1113/jphysiol.1983.sp014873.

Oddo, C. M., Raspopovic, S., Artoni, F., Mazzoni, A., Spigler, G., Petrini, F., Giambattistelli, F., Vecchio, F., Miraglia, F., Zollo, L., Di Pino, G., Camboni, D., Carrozza, M. C., Guglielmelli, E., Rossini, P. M., Faraguna, U., & Micera, S. (2016). Intraneural stimulation elicits discrimination of textural features by artificial fingertip in intact and amputee humans. *ELife*, *5*. Available from https://doi.org/10.7554/eLife.09148.

O'Doherty, J. E., Shokur, S., Medina, L. E., Lebedev, M. A., & Nicolelis, M. A. L. (2019). Creating a neuroprosthesis for active tactile exploration of textures. *Proceedings of the National Academy of Sciences*, *116*(43), 21821–21827. Available from https://doi.org/10.1073/pnas.1908008116.

Okorokova, E. V., He, Q., & Bensmaia, S. J. (2018). Biomimetic encoding model for restoring touch in bionic hands through a nerve interface. *Journal of Neural Engineering*, *15*(6), 066033. Available from https://doi.org/10.1088/1741-2552/aae398.

Ortiz-Catalan, M., Håkansson, B., & Brånemark, R. (2014). An osseointegrated human-machine gateway for long-term sensory feedback and motor control of artificial limbs. *Science Translational Medicine*, *6*(257), 257re6. Available from https://doi.org/10.1126/scitranslmed.3008933.

Ortiz-Catalan, M., Mastinu, E., Greenspon, C., & Bensmaia, S. (2020). Chronic use of a sensitized bionic hand does not remap the sense of touch. *Cell Reports*, *33*(12), 108539. Available from https://doi.org/10.1016/j.celrep.2020.108539.

Ortiz-Catalan, M., Mastinu, E., Sassu, P., Aszmann, O., & Brånemark, R. (2020). Self-contained neuromusculoskeletal arm prostheses. *New England Journal of Medicine*, *382*(18), 1732–1738. Available from https://doi.org/10.1056/NEJMoa1917537.

Ortiz-Catalan, M., Wessberg, J., Mastinu, E., Naber, A., & Branemark, R. (2019). Patterned stimulation of peripheral nerves produces natural sensations with regards to location but not quality. *IEEE Transactions on Medical Robotics and Bionics*, *1*(3), 199–203. Available from https://doi.org/10.1109/TMRB.2019.2931758.

Page, D. M., George, J. A., Kluger, D. T., Duncan, C., Wendelken, S., Davis, T., Hutchinson, D. T., & Clark, G. A. (2018). Motor control and sensory feedback enhance prosthesis embodiment and reduce phantom pain after long-term hand amputation. *Frontiers in Human Neuroscience*, *12*, 352. Available from https://doi.org/10.3389/fnhum.2018.00352.

Pei, Y.-C., Denchev, P. V., Hsiao, S. S., Craig, J. C., & Bensmaia, S. J. (2009). Convergence of submodality-specific input onto neurons in primary somatosensory cortex. *Journal of Neurophysiology*, *102*(3), 1843–1853. Available from https://doi.org/10.1152/jn.00235.2009.

Pei, Y.-C., Hsiao, S. S., Craig, J. C., & Bensmaia, S. J. (2010). Shape invariant coding of motion direction in somatosensory cortex. *PLoS Biology*, *8*(2), e1000305. Available from https://doi.org/10.1371/journal.pbio.1000305.

Pels, E. G. M., Aarnoutse, E. J., Leinders, S., Freudenburg, Z. V., Branco, M. P., van der Vijgh, B. H., Snijders, T. J., Denison, T., Vansteensel, M. J., & Ramsey, N. F. (2019). Stability of a chronic implanted brain-computer interface in late-stage amyotrophic lateral sclerosis. *Clinical Neurophysiology*, *130*(10), 1798–1803. Available from https://doi.org/10.1016/j.clinph.2019.07.020.

Penfield, W., & Boldrey, E. (1937). Somatic motor and sensory representation in the cerebral cortex of man as studied by electrical stimulation. *Brain*, *60*(4), 389–443. Available from https://doi.org/10.1093/brain/60.4.389.

Phillips, J. R., & Johnson, K. O. (1981). Tactile spatial resolution. II. Neural representation of Bars, edges, and gratings in monkey primary afferents. *Journal of Neurophysiology*, *46*(6), 1192–1203. Available from https://doi.org/10.1152/jn.1981.46.6.1192.

Phillips, J. R., Johnson, K. O., & Hsiao, S. S. (1988). Spatial pattern representation and transformation in monkey somatosensory cortex. *Proceedings of the National Academy of Sciences*, *85*(4), 1317–1321. Available from https://doi.org/10.1073/pnas.85.4.1317.

Pons, T. P., Garraghty, P. E., Cusick, C. G., & Kaas, J. H. (1985). The somatotopic organization of area 2 in macaque monkeys. *The Journal of Comparative Neurology*, *241*(4), 445–466. Available from https://doi.org/10.1002/cne.902410405.

Ramon y Cajal, S. (1888). Estructura de los centros nerviosos de las aves. *Revista Trimestral de Histologia Normal y Patologica*, *1*.

Raspopovic, S., Capogrosso, M., Petrini, F. M., Bonizzato, M., Rigosa, J., Di Pino, G., Carpaneto, J., Controzzi, M., Boretius, T., Fernandez, E., Granata, G., Oddo, C. M., Citi, L., Ciancio, A. L., Cipriani, C., Carrozza, M. C., Jensen, W., Guglielmelli, E., Stieglitz, T., & Micera, S. (2014). Restoring natural sensory feedback in real-time bidirectional hand prostheses. *Science Translational Medicine*, *6*(222), 222ra19. Available from https://doi.org/10.1126/scitranslmed.3006820.

Rijnbeek, E. H., Eleveld, N., & Olthuis, W. (2018). Update on peripheral nerve electrodes for closed-loop neuroprosthetics. *Frontiers in Neuroscience*, *12*, 350. Available from https://doi.org/10.3389/fnins.2018.00350.

Rossini, P. M., Micera, S., Benvenuto, A., Carpaneto, J., Cavallo, G., Citi, L., Cipriani, C., Denaro, L., Denaro, V., Di Pino, G., Ferreri, F., Guglielmelli, E., Hoffmann, K.-P., Raspopovic, S., Rigosa, J., Rossini, L., Tombini, M., & Dario, P. (2010). Double nerve intraneural interface implant on a human amputee for robotic hand control. *Clinical Neurophysiology*, *121*(5), 777–783. Available from https://doi.org/10.1016/j.clinph.2010.01.001.

Saal, H. P., & Bensmaia, S. J. (2014). Touch is a team effort: Interplay of submodalities in cutaneous sensibility. *Trends in Neurosciences, 37*(12), 689–697. Available from https://doi.org/10.1016/j.tins.2014.08.012.

Saal, H. P., & Bensmaia, S. J. (2015). Biomimetic approaches to bionic touch through a peripheral nerve interface. *Neuropsychologia, 79*, 344–353. Available from https://doi.org/10.1016/j.neuropsychologia.2015.06.010.

Saal, H. P., Delhaye, B. P., Rayhaun, B. C., & Bensmaia, S. J. (2017). Simulating tactile signals from the whole hand with millisecond precision. *Proceedings of the National Academy of Sciences, 114*(28), E5693–E5702. Available from https://doi.org/10.1073/pnas.1704856114.

Sainburg, R. L., Poizner, H., & Ghez, C. (1993). Loss of proprioception produces deficits in interjoint coordination. *Journal of Neurophysiology, 70*(5), 2136–2147. Available from https://doi.org/10.1152/jn.1993.70.5.2136.

Salzman, C., Murasugi, C., Britten, K., & Newsome, W. (1992). Microstimulation in visual area MT: Effects on direction discrimination performance. *The Journal of Neuroscience, 12*(6), 2331–2355. Available from https://doi.org/10.1523/JNEUROSCI.12-06-02331.1992.

Schiefer, M., Tan, D., Sidek, S. M., & Tyler, D. J. (2016). Sensory feedback by peripheral nerve stimulation improves task performance in individuals with upper limb loss using a myoelectric prosthesis. *Journal of Neural Engineering, 13*(1), 016001. Available from https://doi.org/10.1088/1741-2560/13/1/016001.

Schouenborg, J. (2008). Action-based sensory encoding in spinal sensorimotor circuits. *Brain Research Reviews, 57*(1), 111–117. Available from https://doi.org/10.1016/j.brainresrev.2007.08.007.

Seelke, A. M. H., Padberg, J. J., Disbrow, E., Purnell, S. M., Recanzone, G., & Krubitzer, L. (2012). Topographic maps within Brodmann's area 5 of macaque monkeys. *Cerebral Cortex, 22*(8), 1834–1850. Available from https://doi.org/10.1093/cercor/bhr257.

Stoney, S. D., Thompson, W. D., & Asanuma, H. (1968). Excitation of pyramidal tract cells by intracortical microstimulation: Effective extent of stimulating current. *Journal of Neurophysiology, 31*(5), 659–669. Available from https://doi.org/10.1152/jn.1968.31.5.659.

Sur, M., Merzenich, M. M., & Kaas, J. H. (1980). Magnification, receptive-field area, and "hypercolumn" size in areas 3b and 1 of somatosensory cortex in owl monkeys. *Journal of Neurophysiology, 44*(2), 295–311. Available from https://doi.org/10.1152/jn.1980.44.2.295.

Suresh, A. K. (2015). Tactile coding in the cuneate nucleus of macaques *[Poster]*. Chicago, IL: Society for Neuroscience 2015, October 21.

Suresh, A. K. (2019). Sensory computations in the cuneate nucleus of macaques *[Poster]*. Chicago, IL: Society for Neuroscience 2019, October 23.

Suresh, A. K., Winberry, J. E., Versteeg, C., Chowdhury, R., Tomlinson, T., Rosenow, J. M., Miller, L. E., & Bensmaia, S. J. (2017). Methodological considerations for a chronic neural interface with the cuneate nucleus of macaques. *Journal of Neurophysiology, 118*(6), 3271–3281. Available from https://doi.org/10.1152/jn.00436.2017.

Tabot, G. A., Dammann, J. F., Berg, J. A., Tenore, F. V., Boback, J. L., Vogelstein, R. J., & Bensmaia, S. J. (2013). Restoring the sense of touch with a prosthetic hand through a brain interface. *Proceedings of the National Academy of Sciences of the United States of America, 110*(45), 18279–18284. Available from https://doi.org/10.1073/pnas.1221113110.

Tabot, G. A., Damman, J. F., Berg, J. A., Tenore, F. V., Boback, J. L., Vogelstein, R. J., & Bensmaia, S. J. (2014). Restoring the sense of touch with a prosthetic hand through a brain interface. *Proceedings of the National Academy of Sciences, 111*(2), 875. Available from https://doi.org/10.1073/pnas.1322627111.

Talbot, W. H., Darian-Smith, I., Kornhuber, H. H., & Mountcastle, V. B. (1968). The sense of flutter-vibration: Comparison of the human capacity with response patterns of mechanoreceptive afferents from the monkey hand. *Journal of Neurophysiology*, *31*(2), 301–334. Available from https://doi.org/10.1152/jn.1968.31.2.301.

Tan, D. W., Schiefer, M. A., Keith, M. W., Anderson, J. R., & Tyler, D. J. (2015). Stability and selectivity of a chronic, multi-contact cuff electrode for sensory stimulation in human amputees. *Journal of Neural Engineering*, *12*(2), 026002. Available from https://doi.org/10.1088/1741-2560/12/2/026002.

Tan, D. W., Schiefer, M. A., Keith, M. W., Anderson, J. R., Tyler, J., & Tyler, D. J. (2014). A neural interface provides long-term stable natural touch perception. *Science Translational Medicine*, *6*(257), 257ra138. Available from https://doi.org/10.1126/scitranslmed.3008669.

Tehovnik, E. J., Tolias, A. S., Sultan, F., Slocum, W. M., & Logothetis, N. K. (2006). Direct and indirect activation of cortical neurons by electrical microstimulation. *Journal of Neurophysiology*, *96*(2), 512–521. Available from https://doi.org/10.1152/jn.00126.2006.

Thomson, E. E., Carra, R., & Nicolelis, M. A. L. (2013). Perceiving invisible light through a somatosensory cortical prosthesis. *Nature Communications*, *4*, 1482. Available from https://doi.org/10.1038/ncomms2497.

Vallbo, Å. B., & Johansson, R. S. (1976). Skin mechanoreceptors in the human hand: Neural and psychophysical thresholds. In Y. Zotterman (Ed.), *Sensory functions of the skin in primates* (pp. 185–199). Elsevier.

Valle, G., Mazzoni, A., Iberite, F., D'Anna, E., Strauss, I., Granata, G., Controzzi, M., Clemente, F., Rognini, G., Cipriani, C., Stieglitz, T., Petrini, F. M., Rossini, P. M., & Micera, S. (2018). Biomimetic intraneural sensory feedback enhances sensation naturalness, tactile sensitivity, and manual dexterity in a bidirectional prosthesis. *Neuron*, *100*(1), 37–45. Available from https://doi.org/10.1016/j.neuron.2018.08.033, e7.

Weber, A. I., Saal, H. P., Lieber, J. D., Cheng, J.-W., Manfredi, L. R., Dammann, J. F., & Bensmaia, S. J. (2013). Spatial and temporal codes mediate the tactile perception of natural textures. *Proceedings of the National Academy of Sciences*, *110*(42), 17107–17112. Available from https://doi.org/10.1073/pnas.1305509110.

Wendelken, S., Page, D. M., Davis, T., Wark, H. A. C., Kluger, D. T., Duncan, C., Warren, D. J., Hutchinson, D. T., & Clark, G. A. (2017). Restoration of motor control and proprioceptive and cutaneous sensation in humans with prior upper-limb amputation via multiple Utah Slanted Electrode Arrays (USEAs) implanted in residual peripheral arm nerves. *Journal of Neuroengineering and Rehabilitation*, *14*(1), 121. Available from https://doi.org/10.1186/s12984-017-0320-4.

Wodlinger, B., Downey, J. E., Tyler-Kabara, E. C., Schwartz, A. B., Boninger, M. L., & Collinger, J. L. (2015). Ten-dimensional anthropomorphic arm control in a human brain – machine interface: Difficulties, solutions, and limitations. *Journal of Neural Engineering*, *12*(1), 016011. Available from https://doi.org/10.1088/1741-2560/12/1/016011.

Yau, J. M., Connor, C. E., & Hsiao, S. S. (2013). Representation of tactile curvature in macaque somatosensory area 2. *Journal of Neurophysiology*, *109*(12), 2999–3012. Available from https://doi.org/10.1152/jn.00804.2012.

Yoshida, K., & Stein, R. B. (1999). Characterization of signals and noise rejection with bipolar longitudinal intrafascicular electrodes. *IEEE Transactions on Biomedical Engineering*, *46*(2), 226–234. Available from https://doi.org/10.1109/10.740885.

Chapter 12

Intracortical microstimulation for tactile feedback in awake behaving rats

İsmail Devecioğlu[1], Sevgi Öztürk[2] and Burak Güçlü[2]
[1]*Department of Biomedical Engineering, Çorlu Faculty of Engineering, Tekirdağ Namık Kemal University, Tekirdağ, Turkey,* [2]*Institute of Biomedical Engineering, Boğaziçi University, İstanbul, Turkey*

ABSTRACT

Cortical neuroprostheses aim to partially compensate for the loss of sensory and motor function in severe neurological conditions such as spinal cord injury or amyotrophic lateral sclerosis. Intracortical microstimulation (ICMS) of the primary somatosensory cortex (S1) has been previously shown to generate localized percepts, which is promising for sensory feedback to improve motor control, object interaction, and embodiment with neuroprostheses. After an intensive training schedule, rats performed a psychophysical yes/no detection task during vibrotactile stimulation of the glabrous skin and during electrical stimulation of the hind limb representation of S1. By using psychometric functions from both modalities, an equivalence model was established between vibrotactile displacement amplitude and electrical current amplitude. Then, the rats wore mechanically isolated boots with strain-gauge sensors in a similar detection task. The sensor data generated by vibrotactile stimuli were converted to ICMS current pulses in real time. The rats could use the artificial sensation to increase their psychophysical performance. This method may be extended to more complex vibrotactile stimuli for realistic neuroprosthesis applications.

Keywords: Somatosensory feedback; neuroprosthesis; somatosensory cortex; electrical stimulation; vibrotactile stimulus; psychophysics; yes/no detection task

12.1 Introduction

Humans' ability to use their hands with great dexterity depends on an elaborate sensory innervation. While the events on the skin (i.e., contact with an object) are conveyed by the sense of touch, the position and movement of the fingers are signaled by the sense of proprioception (Gardner & Johnson, 2012a,b).

Somatosensory Feedback for Neuroprosthetics. DOI: https://doi.org/10.1016/B978-0-12-822828-9.00013-7

Integration of these senses together with vision contributes to motor planning and precise control of the limbs during movement (Wolpert et al., 2012). If cutaneous and/or proprioceptive feedback is dysfunctional (i.e., due to neurodegenerative diseases or trauma), coordination during movement gets disrupted and movements become clumsy (Gardner & Johnson, 2012b).

In the case of traumatic spinal cord injury, for example, the communication between the brain and limbs is disturbed in both directions. Motor signals generated in the cortex cannot reach the muscles and sensory signals from the limbs cannot reach the brain. Different approaches have been proposed in the literature for restoring movement by means of motorized prostheses (Andersen et al., 2010; Fagg et al., 2007; Hochberg et al., 2012). A motorized prosthesis basically depends on three processes; recording brain (or muscle) signals, classifying them into a limited number of motor tasks (i.e., grasping, pointing, picking, etc.), and accordingly controlling a robotic limb. A motorized prosthesis can decode plenty of movements and postures from user's intentions, so that it improves the independency of the user in her/his daily living activities. Although there are commercial motorized limb prostheses, they still have many limitations regarding sensory feedback (Eapen et al., 2017; Svensson et al., 2017). They are physically and cognitively demanding, because the user has to decide every next step of her/his actions by observing the current stage of the prosthesis. Therefore the actions performed with the prosthesis take longer and are less accurate compared to natural actions.

The lack of sensory feedback in the forms of touch and proprioception dramatically reduces the efficiency of the prostheses. From the perspective of a patient, sense of touch is required for dexterous hand movements and elaborate manipulations of objects using prostheses (Bensmaia, 2015; Weber et al., 2012). If tactile and/or proprioceptive feedback is provided by a robotic limb, the performance of the user (i.e., the number of accurate movements/actions or accuracy of applied force) increases and the duration for the motor execution decreases (Cipriani et al., 2014; Clemente et al., 2016; Damian et al., 2012; Gaunt et al., 2013; Jorgovanovic et al., 2014; Klaes et al., 2014; Patel et al., 2016; Raspopovic et al., 2014; Schweisfurth et al., 2016; Suminski et al., 2010; Witteveen et al., 2015). In addition to movement planning and execution, multisensory feedback from a limb constitutes the embodiment of this limb (the feeling of own body/limb) (de Vignemont, 2011; Ehrsson et al., 2005, 2008; Petkova et al., 2011). de Vignemont (2011) proposed that if the spatial (i.e., location and shape), sensorimotor (i.e., actions and sensations), and safety (e.g., need for protection from hazards) characteristics of a limb or an object are processed in the same manner as that of one's own body, then it is embodied. Embodiment—through a naturalistic sensory feedback mechanism—is essential if a prosthesis is expected to be more than a tool (Maravita & Iriki, 2004). Moreover, users find watching or listening to their prosthesis insufficient and overwhelming during movement, so they demand sensations from their devices in the forms of vibration, electrical stimulation, pressure, or temperature (Lewis et al., 2012). Finally, sensory feedback is also a crucial parameter from the perspective of learning algorithms in brain–machine interfaces for fast adaptation and recalibration of the system (Micera et al., 2011; Raspopovic et al., 2014).

As the quality and facility of neural interfaces (i.e., microelectrode arrays) improve, closed-loop prosthetic applications (i.e., serving for both motor and sensory systems) are promising for long-term usage (Grill et al., 2009; Hatsopoulos & Donoghue, 2009; Rossini et al., 2010). Electrode arrays either implanted in the peripheral nerves or in the brain can be used not only for recording neural activity but also to modulate/induce it by means of electrical stimulation. Electrical stimulation of sensory afferents or sensory cortices can elicit artificial sensations (Flesher et al., 2016; Graczyk et al., 2016; Klaes et al., 2014; Raspopovic et al., 2014; Romo et al., 2000; Rossini et al., 2010; Tabot et al., 2013). These artificial sensations may not be exactly the same as the natural sensations, but participants are still able to compare artificial sensations with natural ones in means of, for example, frequency, magnitude, and location on the skin (Graczyk et al., 2016; Romo et al., 2000; Tabot et al., 2013). Therefore it is possible to generate electrical stimulation patterns to inform prosthetic users about the properties of natural stimuli (i.e., pressure, edges, texture, etc.) applied on a robotic limb.

Cortical microstimulation for sensory feedback has been demonstrated in rodents (Butovas & Schwarz, 2007; Devecioğlu & Güçlü, 2017; Koivuniemi et al., 2011; Semprini et al., 2012), monkeys (Callier et al., 2015; Dadarlat et al., 2015; Kim et al., 2015; Klaes et al., 2014; O'Doherty et al., 2012), and humans (Flesher et al., 2016; Kramer et al., 2019). Rats have been extensively studied in the literature for their vibrissae which have a highly magnified representation (barrel field) in the primary somatosensory cortex (S1) (Adibi & Arabzadeh, 2011; Ahissar, 2008; Diamond et al., 2008; Kleinfeld et al., 2006; Petersen, 2014; Tahon et al., 2011; Walker et al., 2011; Wiest et al., 2010). The rat's vibrissal system presents a good closed-loop model for the sensorimotor pathways with a dense afferent and efferent innervation. Although the sophisticated architecture of the vibrissal system has inspired many sensory applications in the robotic field (Lepora et al., 2018; Pearson et al., 2011; Prescott et al., 2009), it is highly complex for studying neuroprosthetics and not directly applicable for humans. The hairless (glabrous) skin of the hands is the main means of tactile exploration in primates. Rats also have glabrous skin (i.e., volar surfaces of fore- and hindpaws) which is innervated by the same set of mechanoreceptive afferents found in humans (Devecioğlu & Güçlü, 2013; Leem et al., 1993a,b). Although rats do not use their limbs in exploration, perhaps, as much as whiskers, they can be trained in behavioral tasks with cutaneous tactile cues delivered to their glabrous skin (Devecioğlu & Güçlü, 2015).

Devecioğlu and Güçlü (2015) presented a novel system for operant conditioning of freely behaving rats based on vibrotactile stimuli applied on the volar surfaces of rats' hindpaws. It was further used to establish a correspondence between vibrotactile stimuli and intracortical microstimulation (ICMS) of the hindlimb area in S1 (Devecioğlu & Güçlü, 2017). Recently, the system was extended with real-time processing to demonstrate a somatosensory neuroprosthesis application during a psychophysical detection task (Öztürk et al., 2019). This chapter reviews the main findings of several studies leading to our cortical neuroprosthesis for tactile feedback. First, the operant conditioning chamber and the training schedule for various tasks are presented. Next, psychophysical mapping between sensations elicited by vibrotactile stimuli and ICMS is given within the context of a simple model. Finally, this model is used for the generation of ICMS current pulses based on real-time processing of sensor data from mechanically isolated boots worn by rats, akin to prosthetic limbs (Beygi et al., 2016). By using this tactile neuroprosthesis system, freely behaving rats can increase their psychophysical performance compared to the condition when the system is switched off.

12.2 Behavioral instrumentation and training schedule

In the literature, different tactile conditioning systems have been proposed for rats. While many of them target the vibrissal system (Adibi & Arabzadeh, 2011; Harris et al., 1999; Tahon et al., 2011; Walker et al., 2011; Wiest et al., 2010), the operant chamber developed by Devecioğlu and Güçlü (2015) mechanically stimulates the glabrous skin of freely behaving rats, regardless of their hindlimb positions, while they perform various tasks. The system basically consists of an elevated chamber and a mechanical shaker placed underneath it (Fig. 12.1A). The shaker is of a permanent magnet electrodynamic type similar to those used in electrophysiological (Devecioğlu & Güçlü, 2013; Vardar & Güçlü, 2017, 2020) and psychophysical (Güçlü, 2007; Güçlü & Dinçer, 2013; Güçlü & Öztek, 2007; Yıldız & Güçlü, 2013; Yıldız et al., 2015) experiments in our laboratory. However, for the particular chamber application, it is equipped with a special plastic adapter carrying stainless steel contactor probes and moves a total of 230 contactor probes as a whole (Fig. 12.1B). Each contactor probe has a flat surface of diameter 2 mm for skin contact. The aluminum ground plate of the chamber has a hexagonal perforation within an area of 10 cm in diameter, so that the contactor probes of the mechanical shaker move through the plate to stimulate the rat skin. The perforation covers a convenient area such that several probes simultaneously touch the skin while the rat is performing the task (e.g., by depressing levers). The contact between the skin and contactor probes can be confirmed, if desired, by measuring the electrical resistance between probes and the ground plate.

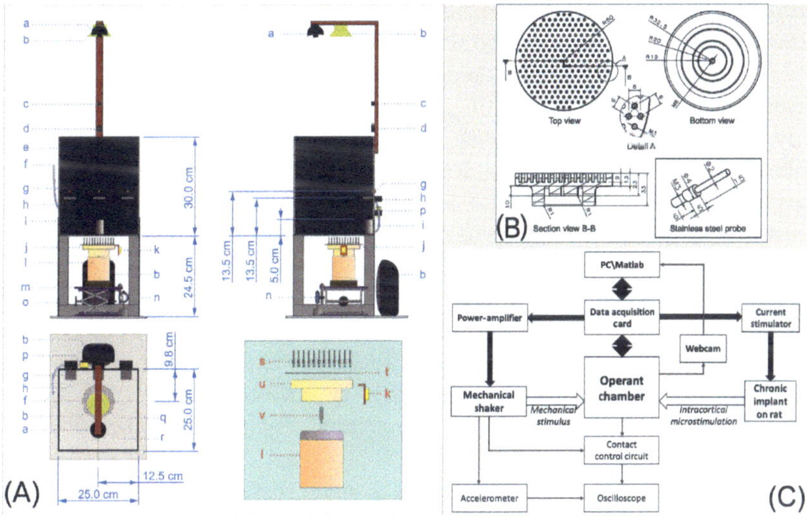

FIGURE 12.1 The operant chamber with vibrotactile stimulation system. (A) Drawings of the chamber and the mechanical shaker. (a) Infrared camera, (b) white-noise speaker, (c) reward buzzer, (d) error buzzer, (e) Plexiglas walls, (f) hose from water reservoir, (g) LEDs, (h) levers, (i) water receptacle, (j) multi-probe tip adapter, (k) accelerometer and resistance measurement circuit, (l) mechanical shaker, (m) jack, (n) dial indicator, (o) aluminum frame (p) solenoid valve, (q) hexagonal hole pattern (r) aluminum ground plate (s) stainless steel probes, (t) aluminum conductor, (u) plastic adapter, (v) armature connection screw. (B) The plastic adapter carrying stainless steel contactor probes. Values are given in millimeters. R: radius, Ø: diameter, M: ISO metric screw thread. (C) Diagram of the operant chamber control system. *(Figures 12.1A and B) Reprinted with permission from Devecioğlu, İ., & Güçlü, B. (2015). A novel vibrotactile system for stimulating the glabrous skin of awake freely behaving rats during operant conditioning.* Journal of Neuroscience Methods, 242, *41–51. https://doi.org/10.1016/j.jneumeth.2015.01.004.*

Three nonretractable levers are used to record rats' responses, and LEDs are located above the levers to deliver visual cues when necessary. The chamber was originally designed to condition water-deprived rats, and a controlled amount of water can be dropped through the water spout placed below the middle lever for reinforcement. The waveforms of the mechanical stimuli are generated in MATLAB (Fig. 12.1C). A custom-made power amplifier amplifies the waveforms provided by the analog output of a data acquisition card and drives the mechanical shaker. Mechanical vibrations generated by the shaker can be monitored with a single-chip accelerometer mounted on the stimulator adapter. The user can choose the behavioral task, specify task parameters, and monitor the performance of the rat in a custom-made MATLAB GUI which also controls the chamber and records the behavioral data. The chamber is placed in a sound- and light-isolated box to avoid external distractors. Additionally, white-noise is played inside the isolated box throughout the experimental sessions. In order to monitor (and record) a rat's behavior, an infrared camera is located above the chamber. Rewards and errors are signaled with two buzzers at different frequencies.

For testing the tactile neuroprosthesis system, the rats are trained through five stages: (1) magazine training, (2) simple operant conditioning, (3) visually guided side-lever press task, (4) vibrotactile habituation task, and (5) vibrotactile detection task [for details of classical and operant conditioning, see McSweeney and Murphy (2014)]. The complete schedule may take up to 6–8 weeks for a rat. For behavioral motivation, rats are water-deprived 24 hours before the training sessions, and their weights are monitored to be >80% of normal weights. Food is provided ad libitum in both the operant chamber and home cages. In magazine training, a rat associates the presence of water in the spout with the reward tone. The experimenter monitors and manually operates the magazine training, because the rat does not make a physical response, such as depressing a lever, but gets near the spout and licks water each time the reward tone is presented. Most of the rats quickly associate the reward tone with water in a single session and are moved on to the simple operant conditioning task (task A in Fig. 12.2A). In task A, the rat associates the middle lever depression with the delivery of water. This association is established in a semiautomated manner: the experimenter delivers the water at the critical behaviors of the rat, such as sniffing the middle lever, then the criterion for the reward gets stringent in order to make the rat depress the middle lever. Depression of the middle lever automatically delivers the water after signaling with reward tone. Rats which achieve >100 lever depressions in a 1-hour session are moved on to the visually guided side lever press task (task-B in Fig. 12.2A).

In task B, pressing the middle lever initiates a trial which is signaled by turning on the middle LED for 0.6 seconds. After the middle LED is turned off, either the right or left LED is randomly turned on and the rat is expected to depress the lever below this LED in order to get water. If the rat depresses the lever on the wrong side or does not respond within a limited time, then no reward is presented and the trial is ended with the error signal. Each session consists of 100 trials with counterbalanced left and right conditions. Task B continues until the rat completes two consecutive sessions with 85% correct rate. Next, the animal is carried on to the vibrotactile habituation task (task C in Fig. 12.2A). This task is identical to task B, except that a high-level vibrotactile stimulus is presented in trials where the right LED is on. Therefore, the rats are habituated to the novel mechanical stimulus applied to their hindpaws. If a rat completes a session without being disturbed by mechanical stimulus and with a correct rate higher than 85%, then it moves on to the vibrotactile detection task (task D in Fig. 12.2A).

Task D is a psychophysical yes/no detection task where the rat is expected to detect the presence or absence of the vibrotactile stimulus. As in previous tasks, depression of the middle lever initiates a trial and the 0.6-second stimulus interval is indicated by the middle LED.

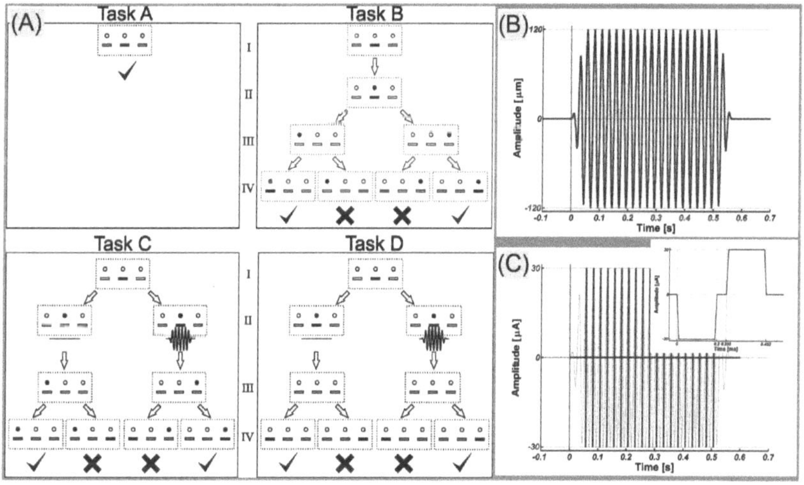

FIGURE 12.2 Behavioral training schedule and stimulus packages used in behavioral experiments. (A) After magazine training, rats go through four operant conditioning stages. Task A is the simple operant conditioning where rats learn to depress the middle lever for a reward. Task B is the visually guided side-lever press where rats learn to initiate a trial by depressing the middle lever and then respond on one of side levers for reward. Task C is a vibrotactile habituation task which is the same as task B except a vibrotactile stimulus is present for trials of right lever press. Task D is the vibrotactile detection task where rats associate the presence of a vibrotactile cue with the right lever and its absence with the left lever. The same task is used with the ICMS detection task which is explained in the next section. (B) The vibrotactile stimulus is a burst of sinusoidal mechanical vibrations at 40 Hz. Its rise and fall times are 50 milliseconds, and it takes 0.5 seconds between half-power points. (C) ICMS pulse train is a package similar to the vibrotactile stimulus (shown with dashed line) and consists of charge-balanced biphasic pulses (cathodic phase first, phase duration: 600 microseconds, interphase interval: 53 microseconds, see insert) delivered at 40 Hz. *(Figures 12.2A and B) Reprinted with permission from Devecioğlu, İ., & Güçlü, B. (2015). A novel vibrotactile system for stimulating the glabrous skin of awake freely behaving rats during operant conditioning.* Journal of Neuroscience Methods, 242, *41–51. https://doi.org/10.1016/j.jneumeth.2015.01.004.*

In this period, the vibrotactile stimulus (e.g., a burst of 40-Hz suprathreshold mechanical sinusoidal vibrations) is either present or not. The rat is expected to respond either on the right or the left lever within a limited time, otherwise the trial ends with the error signal. Unlike task C, the side LEDs are not used and the rat must respond based on the presence or absence of the vibrotactile stimulus. The reward is presented if the rat performs a hit (right lever press when the stimulus is present) or a correct rejection (left lever press when the stimulus is absent). Otherwise, no reward is presented and the trial ends with the error signal. Each session consists of 100 trials with randomized and counterbalanced stimulus/no-stimulus conditions. At least 10 sessions of task D are performed and the accuracy of the rat in the last five sessions is tested with the Kendall's τ test (Fisch, 2001) to verify if the performance has leveled-off. If not, additional sessions are carried on until the performance levels off. If the rat achieves an average accuracy >80% in the last five sessions, then it is tested in further experiments. As the tasks get more difficult at each stage, many rats fail to meet the criteria and approximately 30% of the rats can complete the entire training schedule (e.g., magazine training to vibrotactile detection) (Devecioğlu, 2016).

12.3 Vibrotactile detection experiments

Once a rat can do task D explained above, further vibrotactile detection experiments are performed based on the method of constant stimuli (Butovas & Schwarz, 2007; Gescheider, 1997a). The stimuli are bursts of sinusoidal vibrations applied to the skin surface (Fig. 12.2B). For example, six amplitude levels (3–200 μm) and three frequencies (40, 60, and 80 Hz) are tested in Devecioğlu and Güçlü (2017) and psychometric functions are derived. First, hit [$p(h)$] and false alarm [$p(f)$] rates are recorded for each frequency and amplitude level. According to the classical threshold theory, the observed hit rate may include legitimate and accidental hits (Gescheider, 1997b). Hits performed when the stimulus is actually above the subject's threshold are accepted as legitimate hits; otherwise they are called accidental hits which are due to the response bias. Therefore legitimate (corrected) hit rate [$p^*(h)$] is calculated with Eq. (12.1) for each session, and corrected hit rates are plotted in the y-axes of the psychometric functions.

$$p(h) = p^*(h) + p(f)[1 - p^*(h)] \tag{12.1}$$

In sensory systems, d' has been widely used to quantify detectability (Bensmaia et al., 2008; Talwar & Gerstein, 1999). d' is a good measure of detectability, because it is independent of the subject's criterion. On the other hand, it has no theoretical limit and gets too large as $p(h)$ approaches 1 and $p(f)$ approaches 0. In addition, psychometric data collected for vibrotactile and ICMS stimulation may span different ranges of d'. Therefore it would be difficult to establish a mathematical equivalence between these modalities by using d'. For this reason, we compared the corrected hit rates. Fig. 12.3 plots psychometric curves for 10 rats at three tested vibrotactile frequencies.

As reported for primates (Bolanowski et al., 1988; Morioka et al., 2008; Mountcastle et al., 1972), detection threshold (here, the vibrotactile amplitude at 50% corrected hit rate) decreases with increasing frequency (Fig. 12.3). Nevertheless, a few specific issues should be kept in mind while comparing rats' data with those from primates. First, although the rat has mechanoreceptive afferents similar to those found in primates, innervation densities are different (Darian-Smith & Kenins, 1980; Halata, 1990; Leem et al., 1993a). Second, anatomical and mechanical properties of the skin differ between species and that is an important factor on neurophysiological

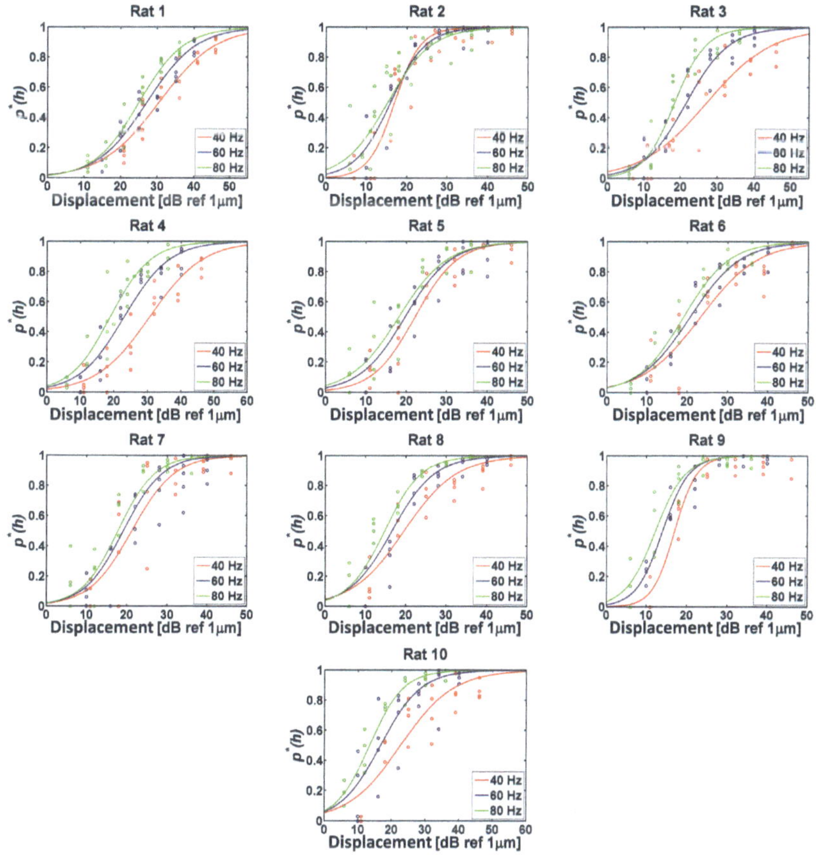

FIGURE 12.3 Vibrotactile psychometric curves. Open circles show experimental data points and the solid lines are psychometric curves fitted with Eq. (12.2) at each frequency. *Reprinted with permission from Devecioğlu, İ., & Güçlü, B. (2017). Psychophysical correspondence between vibrotactile intensity and intracortical microstimulation for tactile neuroprostheses in rats.* Journal of Neural Engineering, 14*(1), 016010. https://doi.org/10.1088/1741-2552/14/1/016010.*

and psychophysical thresholds (Devecioğlu & Güçlü, 2013; Güçlü & Bolanowski, 2003b,c; Strzalkowski et al., 2015; Yıldız & Güçlü, 2013). Third, the vibrotactile detection thresholds differ in active vs. passive movements of a limb (Yıldız et al., 2015). The rats are freely behaving during the experiments in the chamber, while the subjects are typically restrained in most human and monkey studies. Finally, primates use their hands for tactile exploration while rats use their whiskers as also mentioned above. Therefore the central saliency of the glabrous touch may be different in rats compared to what we know from primate psychophysics. Despite all these facts, the results of the vibrotactile detection task (Fig. 12.3) show that rats can utilize tactile information from their skin efficiently and in accordance with current psychophysical theories. The plots given in Fig. 12.3 quantify the "natural" sensation mediated by skin mechanoreceptors. In the next section, a similar approach is adopted for "artificial" sensation in order to establish a correspondence between the vibrotactile and ICMS modalities.

12.4 Intracortical microstimulation in rats

Since the 1920s, electrical stimulation of the brain tissue has been used in medicine and biomedical research (Penfield, 1930). Recent studies focus on deep brain stimulation for rehabilitation (Baldermann et al., 2016; Dandekar et al., 2018; Vicheva et al., 2020) and cortical stimulation for sensory feedback (Berg et al., 2013; Devecioğlu & Güçlü, 2017; Fagg et al., 2009; Flesher et al., 2016; Moxon & Foffani, 2015; Öztürk et al., 2019; Semework & DiStasio, 2014; Semprini et al., 2012; Tabot et al., 2015). A cortical sensory neuroprosthesis aims to elicit neural activity which is psychophysically comparable to the one elicited by natural stimulus. Therefore one needs to know the psychophysical aspects of both natural and artificial sensations. In the literature, psychophysical outputs of ICMS have been studied extensively in rodents [barrel cortex: Butovas & Schwarz (2007), Semprini et al. (2012); fore-/hindlimb areas in S1: Devecioğlu & Güçlü (2017), Semework & DiStasio (2014); auditory cortex: Deliano et al. (2009), Koivuniemi & Otto (2012), Maldonado & Gerstein (1996), Otto et al., (2005), Rousche et al., 2003]. These studies generally demonstrate that rats can report the presence and absence of ICMS in the auditory and somatosensory cortices.

After training a rat in a yes/no detection task with a natural stimulus, electrode arrays (or single electrodes) are implanted in the related cortical area. For example, Devecioğlu and Güçlü (2017) used 16-channel microwire arrays implanted in the hindlimb representation of rat S1 cortex. The depth of the array is determined based on the multiunit activity evoked by mechanical stimulation of the hindpaw glabrous skin. After implantation, cutaneous receptive fields (RFs) are mapped for each electrode in the array (Fig. 12.4). An electrode with a strong evoked multiunit activity can be used for ICMS. In order to confirm that the rat responds to a focused artificial sensation elicited by ICMS rather than other effects (e.g., spread of ICMS to neighboring areas, unobservable muscle activity during behavior), another electrode unresponsive to touch on the hindpaw and located in the agranular cortex (i.e., motor cortex) is used as a control electrode for electrical stimulation.

After recovery from surgery ($\sim$1 week), the rat is trained to perform the detection task with ICMS. The ICMS is a train of biphasic charge-balanced current pulses (cathodic phase leading, phase duration: 600 microseconds; interphase interval: 53 microseconds) presented for 0.5 seconds (Fig. 12.2C) at the same frequency as the vibrotactile stimulus, that is, 40 Hz. In order to minimize charge accumulation and damage to the cortex, charge-balanced

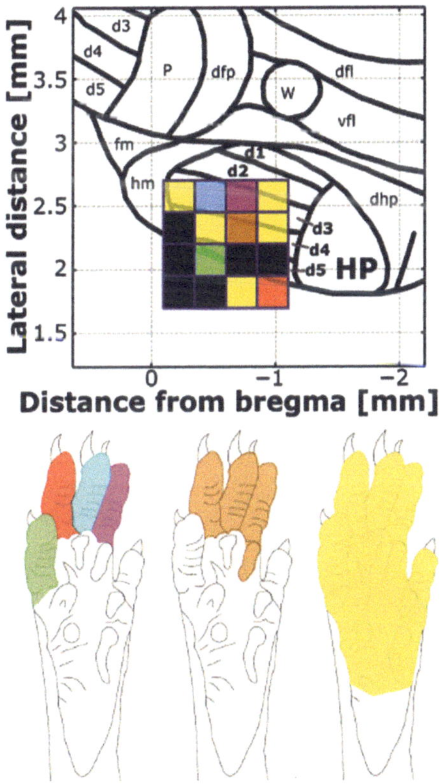

FIGURE 12.4 The location of a microwire array in the S1 of a rat (Chapin & Lin, 1984) and receptive-field (RF) mapping for each electrode in the array. Microwire array has 16 electrodes arranged 4 × 4 and implanted in the hindpaw representation of S1. RF projections are color coded on the array and paw drawings. Black-coded electrodes are nonresponsive to mechanical stimulation of any skin region. *Reprinted with permission from Devecioğlu, İ., & Güçlü, B. (2017). Psychophysical correspondence between vibrotactile intensity and intracortical microstimulation for tactile neuroprostheses in rats.* Journal of Neural Engineering, 14*(1), 016010. https://doi.org/10.1088/1741-2552/14/1/016010.*

current pulses are essential (Merrill et al., 2005). Additionally, cathodic phase-leading pulses are easier to detect compared to anodic phase-leading pulses (Koivuniemi & Otto, 2011); as such, the desired detectability can be achieved with lower amplitude levels. Since the rats are familiar with the vibrotactile detection task (task D in Fig. 12.2A), they quickly generalize the novel ICMS stimulus and perform the task with accuracies >80%. When tested with ICMS in the control electrode, rats' accuracy is at the chance level (~%50). This shows that the rat's psychophysical performance is based on the focused artificial sensation elicited by ICMS rather than the other cues mentioned above. After the rat performs preliminary ICMS detection experiments with an average accuracy greater than 85% in its last five

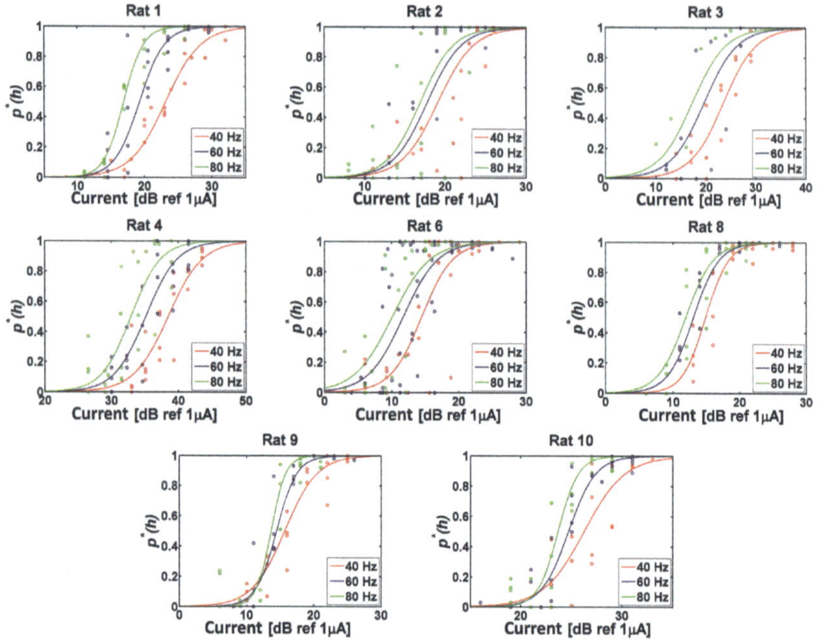

FIGURE 12.5 ICMS psychometric curves. Open circles show experimental data points and the solid lines are psychometric curves fitted with Eq. (12.2) at each frequency. *Reprinted with permission from Devecioğlu, İ., & Güçlü, B. (2017). Psychophysical correspondence between vibrotactile intensity and intracortical microstimulation for tactile neuroprostheses in rats.* Journal of Neural Engineering, 14*(1), 016010. https://doi.org/10.1088/1741-2552/14/1/016010.*

sessions, it is systematically tested with six different current intensities (1.4–150 μA) at three frequencies (40, 60, 80 Hz) as in vibrotactile experiments. Psychometric curves for ICMS detection are qualitatively similar to those for vibrotactile detection; performance increases with increasing intensity and frequency (Fig. 12.5). Cortical neurons integrate the ICMS-elicited activity over short time scales, and consequently, the perceived intensity may increase with increasing frequency (Fridman et al., 2010). However, the increased sensitivity (i.e., lower thresholds) at higher ICMS rates may not be completely attributed to temporal integration. Romo et al., 1998 showed that monkeys can discriminate frequencies of vibrotactile and ICMS stimuli. Therefore ICMS frequency probably affects the qualitative aspects (e.g., periodicity) of the perceived sensation also.

12.5 Psychophysical correspondence between sensations elicited by vibrotactile and electrical stimulation

How can one build a tactile neuroprosthesis based on the ICMS results shown in Fig. 12.5? In order to determine the appropriate parameters of

ICMS to elicit a sensation similar to one elicited by a vibrotactile stimulus of given amplitude and frequency, we must find a common dimension for both modalities. The psychometric data, which are represented in terms of the corrected hit rate, quantify how two modalities (vibrotactile and ICMS stimuli) are processed by a rat during the detection task. Using this knowledge, we can match parameters of ICMS and vibrotactile stimuli which yield the same corrected hit rates. In this way, one can construct psychometric equivalence functions (PEFs) for estimating the parameters of a modality (say ICMS) based on the parameters of another modality (say vibrotactile stimulus) (Berg et al., 2013; Tabot et al., 2013). A critical assumption for PEFs is that sensations elicited by two modalities are approximately similar in qualitative aspects; at least, the rat treats these inputs in a psychophysically similar manner. For example, subjects can compare the frequency and intensity of ICMS to those of mechanical stimuli applied on the skin accurately, or vice versa (Romo et al., 1998; Tabot et al., 2013). Another convincing support for establishing PEFs is that the animals quickly generalize ICMS stimuli in detection or discrimination tasks, if they were previously trained with natural stimuli (e.g., auditory or tactile) (Devecioğlu & Güçlü, 2017; Rousche et al., 2003).

PEFs were first proposed by Berg et al. (2013). They established a behavioral equivalence between the intensities of mechanical (skin indentation) and ICMS (at 300 Hz) stimuli in monkeys. They demonstrated that monkeys can successfully detect mechanical indentation on a sensorized robotic finger via ICMS pulses with current amplitudes modulated by the mechanical indentation through PEFs. Later, Tabot et al. (2013) tested PEFs for location and intensity discrimination in monkeys. As far as their results suggest, monkeys can compare the intensities of mechanical stimuli applied on the monkey's skin and on the sensorized robotic finger. The information about the mechanical stimulus on the robotic finger is delivered to the monkey as an ICMS pulse train modulated through PEF. Furthermore, monkeys can compare the location of artificial sensation elicited by ICMS with the location of mechanical stimulus on the skin.

In the above-cited studies, the ICMS pulse frequency was constant (300 Hz), but the amplitude of the ICMS current pulses was modulated in order to provide information about the intensity of the static indentation on the robotic hand. However, pulse rate is an important factor in the probability of detection; as the ICMS frequency increases, detection probability increases as well [rats: Koivuniemi and Otto (2012), Semprini et al. (2012); monkeys: Kim et al., 2015]. However, the detection thresholds for ICMS level off after 84 Hz in the rat auditory cortex (Koivuniemi & Otto, 2012) and after 200 Hz in the monkey S1 (Kim et al., 2015). Therefore ICMS pulses delivered at 300 Hz may elicit a sensation comparable to that elicited by constant mechanical indentations on the skin. On the other hand, animals can discriminate mechanical and ICMS stimuli simultaneously applied at different frequencies below 30 Hz (Romo & Salinas, 2003).

Vibrations generated on the skin are encoded by populations of mechanoreceptive fibers (Güçlü, 2007; Güçlü & Bolanowski, 2002, 2003a, 2004; Güçlü & Dinçer, 2013; Güçlü et al., 2005). As the skin scans a surface, the complex waveforms set up in the skin are represented as texture information conveyed to the central nervous system (Bensmaïa & Hollins, 2003; Hollins & Bensmaia, 2007; Johnson, 2001; Manfredi et al., 2012; Yau et al., 2009). Texture-related sensations may also be mimicked by modulating the temporal pattern of ICMS pulses. As explained above, the frequency of ICMS pulses should be critically controlled in a sensory neuroprosthesis for a better match in natural and artificial sensations. Devecioğlu and Güçlü (2017) proposed to extend PEFs as a function of frequency for vibrotactile stimuli. The PEFs can be built as frequency dependent while matching the intensity levels of vibrotactile stimuli and ICMS pulses to yield equal detection probabilities. In order to obtain such PEFs, the psychometric data, which are derived for vibrotactile and ICMS detection experiments as explained in previous sections, are fitted to a frequency and intensity dependent model as given in Eq. (12.2) (Devecioğlu & Güçlü, 2017).

$$p^*(h) = \frac{1}{1 + \exp\left(-\frac{A - af^c}{bf^d}\right)} \tag{12.2}$$

Eq. (12.2) is a modified sigmoidal function, which has a frequency dependent midpoint [af^c; the intensity for $p^*(h) = 0.5$] and slope [at the midpoint: $1/(4bf^d)$]. The parameters c and d determine the effect of frequency on the midpoint and the slope, respectively. On the other hand, a and b are the frequency-independent parameters for the midpoint and the slope, respectively. A in Eq. (12.2) is the intensity of the stimulus which is either mechanical (dB ref. 1 μm, zero-to-peak displacement) or ICMS (dB ref. 1 μA, current pulse amplitude). The parameters a, b, c, and d are found by fitting Eq. (12.2) to the psychometric data (Figs. 12.3 and 12.5). The psychometric equivalence model is established by first equating psychometric functions [Eq. (12.2)] fitted to mechanical and ICMS data for a corrected hit rate [$p^*(h)$] as in Eq. (12.3).

$$\left(\frac{1}{1+\exp\left(-\frac{A-af^c}{bf^d}\right)}\right)_E = \left(\frac{1}{1+\exp\left(-\frac{A-af^c}{bf^d}\right)}\right)_T \tag{12.3}$$

where indices of the vibrotactile and electrical stimuli are given as T and E, respectively. If the frequency dimension is assumed to be shared by the two modalities ($f_E = f_T$) (Romo & Salinas, 2003), Eq. (12.3) can be rearranged as in Eq. (12.4) for a one-to-one mapping between the intensities. E and T

indices next to A and the fitting parameters (a, b, c, and d) indicate electrical and vibrotactile stimuli.

$$A_{\mathrm{E}} = \frac{b_{\mathrm{E}}}{b_{\mathrm{T}}} f^{d_{\mathrm{E}} - d_{\mathrm{T}}} (A_{\mathrm{T}} - a_{\mathrm{T}} f^{c_{\mathrm{T}}}) + a_{\mathrm{E}} f^{c_{\mathrm{E}}} \tag{12.4}$$

Eq. (12.4) is the PEF and is used to determine the intensity of the ICMS stimulus for a given vibrotactile intensity and frequency (Fig. 12.6A). It is a linear relation if the units are in dB; and its slope and intercept change with frequency. The range of intensities used in psychophysical experiments (A_T and A_E) and fitted parameters ($[a, b, c, d]_{T,E}$) vary from subject to subject. For this reason, PEFs are subject specific (i.e., they cannot be easily generalized for a species) and a separate PEF has to be built for each subject based on its psychometric performance in vibrotactile and ICMS experiments. Devecioğlu and Güçlü (2017) constructed PEFs for eight rats. When these PEFs are projected onto an A_T-f plane where ICMS current levels (A_E) are color-coded, individual PEFs reveal a large intersubject variation (Fig. 12.6B). For example, current levels may decrease as the frequency increases for some rats for any given displacement (e.g., rats 1 and 2). On the other hand, for some other rats (e.g., rats 3 and 8), the required ICMS intensities may decrease or increase as the frequency changes depending on the displacement level. In addition, the ICMS current intensities can be less frequency-dependent for a given displacement level (e.g., rats 4 and 10).

Physiological and psychophysical differences between subjects may partially account for this variability. For instance, rapidly adapting mechanoreceptive afferents of primate, cat, and rat glabrous skin have log-normally distributed absolute and entrainment thresholds (Devecioğlu & Güçlü, 2013; Güçlü & Bolanowski, 2003b,c; Johnson, 1974). In well-controlled experiments, 10 dB variation can be observed between psychophysical thresholds of human subjects (Güçlü & Bolanowski, 2005a; Yıldız et al., 2015). Additionally, skin mechanics contribute to the psychophysical variation in detecting or discriminating mechanical stimuli (Devecioğlu & Güçlü, 2013; Güçlü & Bolanowski, 2005b; Yıldız & Güçlü, 2013). Moreover, since the operant chamber does not provide as much stimulus control with freely moving rats as those experiments with restrained subjects, the intersubject variation in the vibrotactile data might extend beyond the range reported in the literature (Cohen et al., 1999). On the other hand, biological variation and technical factors may account for the variability in the ICMS data. From a biological perspective, somatotopic organization of the cortex is slightly different across subjects. However, since the electrodes are small, the sensitivity to ICMS largely varies between different electrodes within the same array implanted in a subject according to standard stereotactic coordinates (Callier et al., 2015) and between the same electrodes across subjects. There are yet other factors which are not easily controllable in these experiments, such as

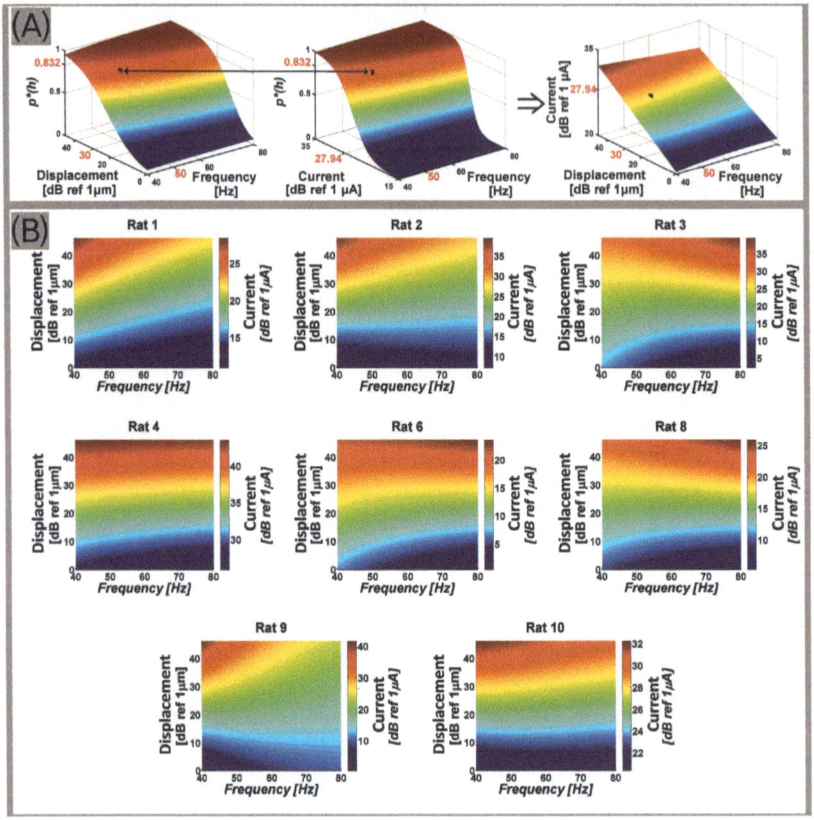

FIGURE 12.6 Psychometric equivalence functions (PEFs). (A) Construction of a PEF based on vibrotactile and ICMS psychometric data. Each data set (vibrotactile and ICMS) (Figs. 12.3 and 12.5) is fitted with Eq. (12.2) and the intensity levels of vibrotactile and ICMS stimuli for the same $p^*(h)$, at a given frequency, are paired [see Eq. (12.4)]. For instance, the intensities of vibrotactile and ICMS stimuli shown with black points joined by the arrow on the left two graphs yield the same $p^*(h)$ value (0.832) at 50 Hz. Therefore these two intensity levels are equal to each other in means of psychometric performance at the given frequency, and this relation is used to construct the PEF (see black point in the rightmost graph). PEFs are used to estimate the intensity of ICMS, for a given vibrotactile intensity and frequency, in the subsequent experiments. (B) PEFs of individual rats. The functions are projected onto the *AT-f* plane where ICMS current levels are color-coded for convenience of visualization. *Reprinted with permission from Devecioğlu, İ., & Güçlü, B. (2017). Psychophysical correspondence between vibrotactile intensity and intracortical microstimulation for tactile neuroprostheses in rats.* Journal of Neural Engineering, 14*(1), 016010. https://doi.org/10.1088/1741-2552/14/1/016010.*

the foreign-body response of the brain tissue (Fernández et al., 2014). In brief, the location (and depth) of the electrodes and the condition of the electrode–tissue interface may cause the high variability in the ICMS psychometric parameters across subjects. As a matter of fact, individual calibration of a neuroprosthesis for each patient is a common procedure due to intersubject variation. For example, training sessions are required to calibrate a motor prosthesis which is controlled by neural ensembles for every new subject (Collinger et al., 2013; Homer et al., 2013). For an amputee, the psychometric data collected from an intact limb can be used for setting up the PEF in a somatosensory neuroprosthesis. For a tetraplegic patient, a skin region with an intact tactile sensation may be used, while compensating for physiological and psychophysical differences between different skin regions. Therefore it is critically important to understand the basic anatomy, physiology, and psychophysics of touch for all skin regions.

12.6 Validation of psychometric equivalence functions

In order to validate PEFs, additional psychophysical experiments that include vibrotactile and ICMS stimuli can be performed. For example, in Devecioğlu and Güçlü (2017), five rats are tested at five frequencies (40, 50, 60, 70, and 80 Hz) with four repetitions which make a total of 20 sessions. Each session is run with a random frequency at a different day. A validation session consists of 360 rewarded trials and 120 unrewarded probe trials which are randomly presented (Fig. 12.7). The rats perform vibrotactile detection (task D) in the rewarded trials as they did before, and they are rewarded for correct responses. On the other hand, in the unrewarded probe trials, either a vibrotactile or an ICMS stimulus is always presented. The order of vibrotactile and ICMS stimuli are random with 0.5 probability. For each modality (vibrotactile and ICMS), six intensity levels are tested with 10 repetitions (2 modalities $\times$ 6 intensity levels $\times$ 10 repetitions = 120 trials). Intensities of ICMS stimulus set are determined with the rat's PEF for the given frequency of that session. In other words, ICMS intensity levels used in the probe trials match the vibrotactile intensities of the same session via PEF.

The mean hit rates are calculated for each intensity of vibrotactile and ICMS stimuli after all the validation sessions are completed for a given frequency (Fig. 12.8A). Since the relationship between the hit rate and the intensity level resembles the properties of a cumulative distribution function, hit rates for vibrotactile and ICMS stimuli can be tested for whether they are from the same distribution or not with the Kolmogorov–Smirnov (K-S) test. For all rats tested in Devecioğlu and Güçlü (2017), the K-S test confirmed that rats achieved equal hit rates in vibrotactile and ICMS trials at each frequency. The K-S test evaluates the similarity of distributions of intensity levels corresponding to the hit rates in vibrotactile and ICMS trials. Additionally, regression analysis can be performed for correlated vibrotactile and ICMS trials in order to evaluate whether the slope of the regression line is unity and it has a zero intercept. If the slope is not unity, then hit rates for

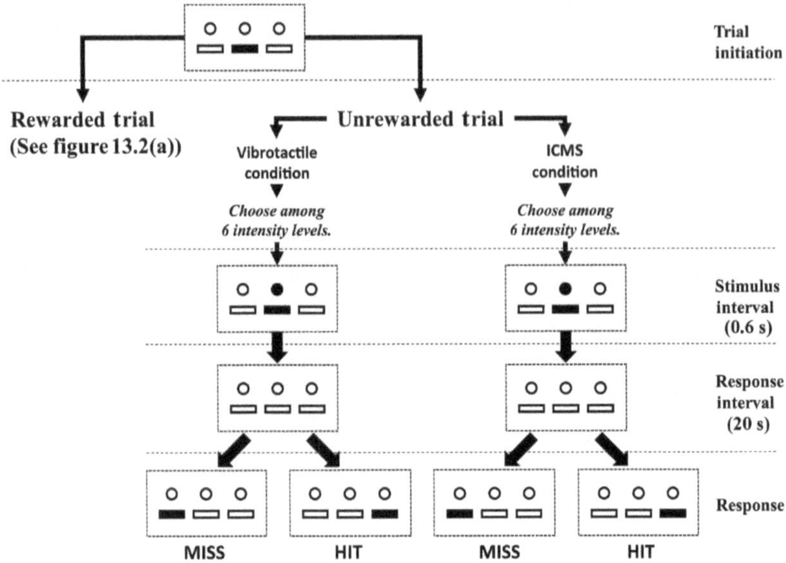

FIGURE 12.7 Flow of the psychometric equivalence function (PEF) validation experiment. Each trial is initiated with a middle lever press. Then, the trial is randomly determined to be either a rewarded trial or an unrewarded probe trial. Rewarded trials are the same as the vibrotactile detection task (task D in Fig. 12.2). Either a vibrotactile or ICMS stimulus is presented in the probe trials. The intensity of the stimulus is randomly chosen among six intensity levels determined before the experiment. ICMS intensities are estimated from the rat's PEF for the given vibrotactile intensity set and the frequency. Behavioral responses elicited in the probe trials are used for the analyses. *Reprinted with permission from Devecioğlu, İ., & Güçlü, B. (2017). Psychophysical correspondence between vibrotactile intensity and intracortical microstimulation for tactile neuroprostheses in rats.* Journal of Neural Engineering, 14*(1), 016010. https://doi.org/10.1088/1741-2552/14/1/016010.*

vibrotactile and ICMS stimuli do not increase at similar proportions with increasing intensity. If the intercept is not zero, then there is a systematic and constant difference between hit rates observed in vibrotactile and ICMS trials. Although the K-S test did not show any significant differences, there was no correlation between vibrotactile and ICMS trials at 50 and 70 Hz for rat 4, and at 60 Hz for rat 10. On the other hand, regression analyses showed that, for most of the tested frequencies and rats, hit rates changed similarly for vibrotactile and ICMS stimuli with a slope not different than unity (e.g., rat 8 in Fig. 12.8B). Nevertheless, for some rats and frequencies, PEFs over- or underestimated the current intensities [see the supplementary document of Devecioğlu and Güçlü (2017)]. Overall, regression lines had unity slopes and zero intercepts for 52% of the cases. Therefore, in general, the psychometric equivalence model is sufficiently successful in estimating current intensities for ICMS.

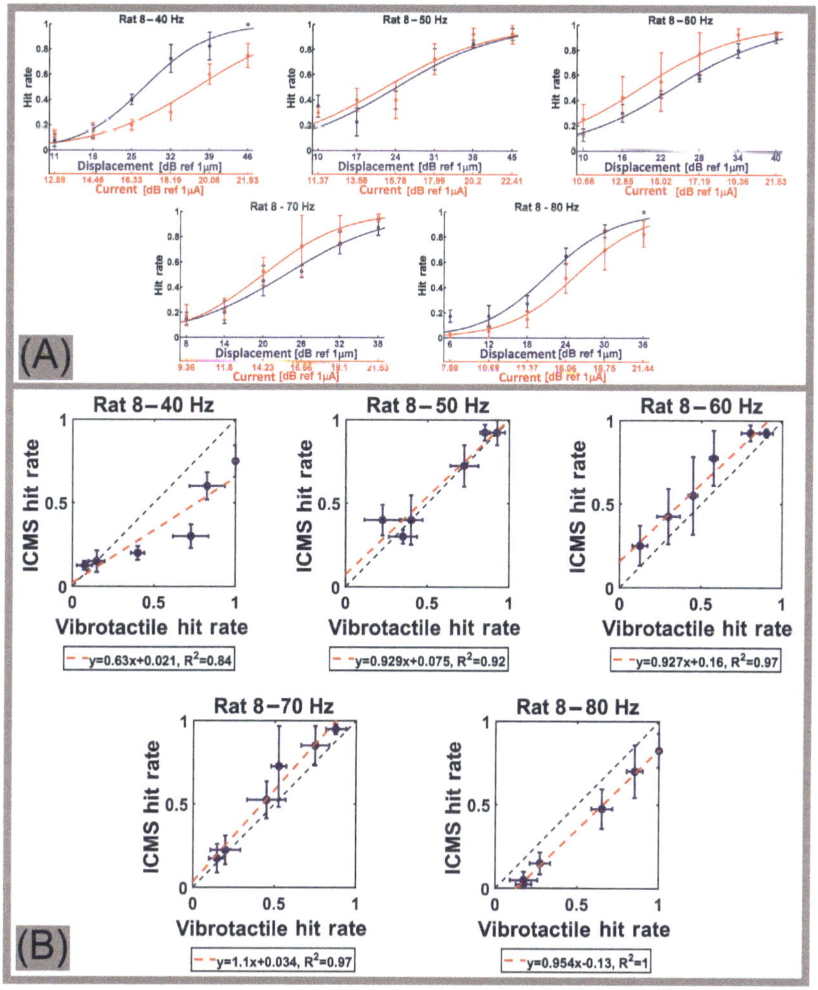

FIGURE 12.8 Psychometric equivalence function (PEF) validation analyses for a rat. (A) The mean hit rates are plotted for vibrotactile (*blue*) and ICMS (*red*) probe trials. Vibrotactile and ICMS intensities matching in PEF space are given in dual x-axes. Error bars represent the standard errors and standard sigmoidal curves are fitted for convenient visual comparison. (B) Regression plots for hit rates in ICMS vs. vibrotactile trials in the validation experiments. The mean hit rates (*blue filled circles*) are fitted with a linear line (*red dashed line*). The slope and the intercept of this line is compared to the dashed diagonal line which have a slope $m = 1$ and an intercept $b = 0$. The more similar the dashed lines are to each other, the better the estimation performance of the PEF. Error bars represent the standard errors. *Reprinted with permission from Devecioğlu, İ., & Güçlü, B. (2017). Psychophysical correspondence between vibrotactile intensity and intracortical microstimulation for tactile neuroprostheses in rats.* Journal of Neural Engineering, 14*(1), 016010. https://doi.org/10.1088/1741-2552/14/1/016010.*

12.7 Behavioral demonstration of a tactile neuroprosthesis in rats

Although the previous sections imply that a tactile neuroprosthesis is feasible by using PEFs, behavioral experiments in which the ICMS stimuli are not preprogrammed, but rather generated on the fly as the rats actively perform a task, is a more convincing scenario for prosthesis use. Furthermore, similar to a clinical condition requiring a prosthesis, the tactile inputs should be disrupted at the peripheral nervous system. Öztürk et al. (2019) tested a novel sensorized neuroprosthesis with real-time ICMS feedback in such a realistic scenario to assess the improvement of behavioral performance of rats. Since amputation models would be traumatic for animals, they designed sensorized boots (Fig. 12.9) which isolated the rat's glabrous skin from the stimulator probes of the operant chamber

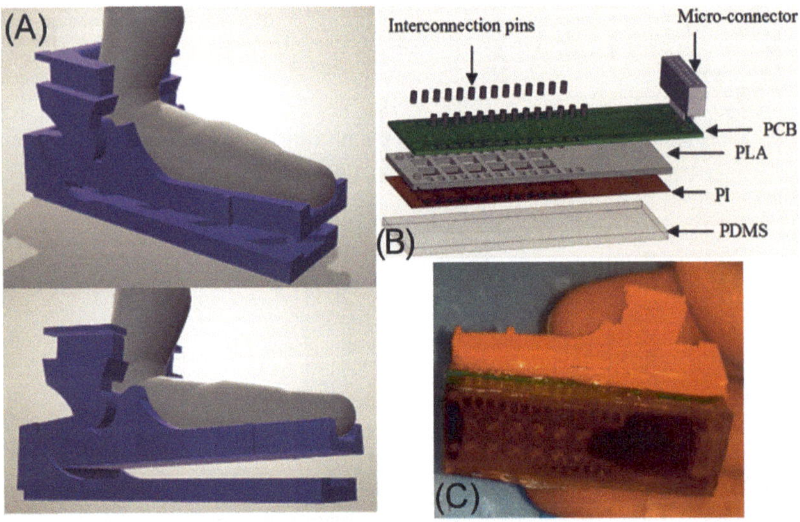

FIGURE 12.9 The design of the sensorized boot. (A) 3D model of the boot. The mechanical coupling between the sole of the boot and the rat's feet is minimized as much as possible by use of the loosest mesh design and the cantilever structure. (B) The sensory array designed to attach on the boots. The design consists of four layers. PCB is the printed circuit board which connects the microconnector to strain gauge arrays patterned on PI (polyimide) sheet. A poly acid (PLA) substrate is placed between PCB and PI and has cavities for deflection of the sensor on PI sheet. The polydimethylsiloxane (PDMS) is used to protect the PI sheet during mechanical stimulation. (C) The sensory array is attached to the sole of the boot. *(Figures 12.9A and C) Reprinted with permission from Öztürk, S., Devecioğlu, İ., Beygi, M., et al. (2019). Real-time performance of a tactile neuroprosthesis on awake behaving rats.* IEEE Transactions *on Neural Systems and Rehabilitation Engineering, 27(5), 1053–1062. https://doi.org/10.1109/TNSRE.2019.2910320. (Figure 12.9B) Reprinted with permission from Beygi, M., Mutlu, Ş., & Güçlü, B. (2016). A microfabricated strain gauge array on polymer substrate for tactile neuroprostheses in rats. Journal of Micromechanics and Microengineering, 26(8), 084006.*

(see Section 12.2). Boots are 3D-printed with loose mesh design for better attenuation of mechanical vibrations from the sole of the boots (Fig. 12.9A). Although the boots do not provide a perfect mechanical isolation, the mechanical stimulus presented on the sensory array attenuates considerably on the rats' feet (see below). The sensorized sole of each boot has an array of 2 × 7 sensory cells and four strain-gauges were serially connected in each cell (Beygi et al., 2016) (Fig. 12.9B). The final assembled boot is shown in Fig. 12.9C. Mechanical displacements detected by sensors are converted into voltage changes and transmitted to a digital signal processor (DSP) (Fig. 12.10A). In the DSP, the envelope of the band-pass filtered signal is used to modulate the current intensity of ICMS pulse train, which is delivered to the S1 cortex of the rats, based on PEFs in real time (Fig. 12.10B).

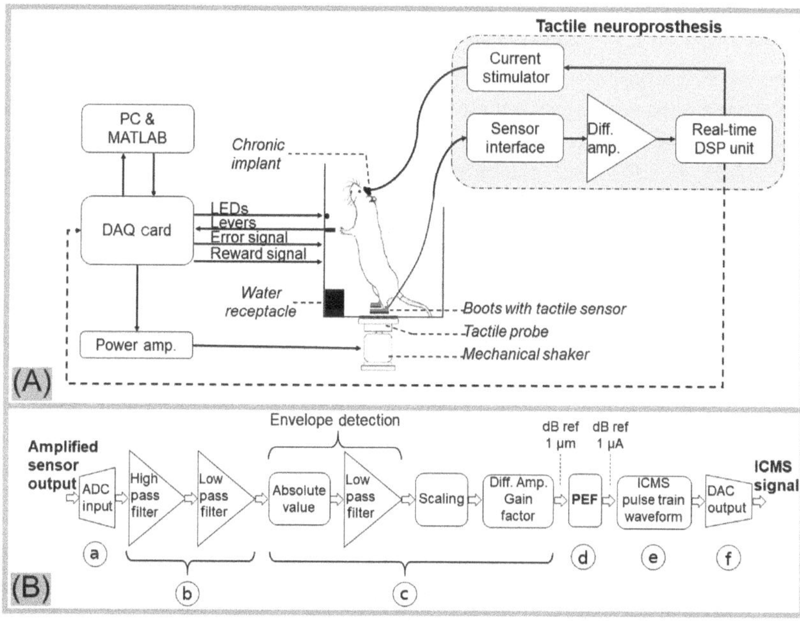

FIGURE 12.10 Operation of tactile neuroprosthesis experiments. (A) The block diagram for operant chamber and tactile neuroprosthesis. The chamber is the same as shown in Fig. 12.1 and is shown here for clarity. The tactile neuroprosthesis consists of boots with tactile sensors, sensor interface, differential amplifier, DSP unit, current stimulator, and chronic implant on the rat. The chamber and neuroprosthesis are operated independently except that the ICMS patterns are recorded by the PC operating the chamber (the dashed line). (B) Blocks of the tactile neuroprothesis's real-time DSP unit: (a) analog-to-digital conversion, (b) band-pass filtering, (c) envelope detection, scaling and gain adjustment, (d) subject-based psychometric equivalence function, (e) pulse train generation and amplitude modulation for ICMS, (f) digital-to-analog conversion. *Reprinted with permission from Öztürk, S., Devecioğlu, İ., Beygi, M., et al. (2019). Real-time performance of a tactile neuroprosthesis on awake behaving rats.* IEEE Transactions on Neural Systems and Rehabilitation Engineering, 27*(5), 1053–1062. https://doi.org/10.1109/TNSRE.2019.2910320.*

If the boots of the tactile neuroprosthesis system are stationary on the stimulator probes, the mechanical signals can be captured with great success and ICMS pulses are modulated accordingly with minimal variation through a trial (Fig. 12.11A–D). However, neuroprosthetics are always prone to movement artifacts while subjects operate them in daily life. In such a real-life scenario (i.e., rats moving in the chamber with boots), the mechanical stimulus is not captured very well for the entire stimulus duration (Fig. 12.11E–H). Nevertheless, the neuroprosthesis system demonstrated here can still operate sufficiently well enough to improve the psychophysical detectability compared to the condition in which the boots are worn, but the artificial tactile feedback via the ICMS is switched off.

The performance of two rats with the tactile neuroprosthesis is evaluated in three different conditions at two frequencies (40 and 80 Hz). First, rats wear the neuroprothesis, but the feedback is disabled. This condition is analogous to the use of a prosthesis with no sensory feedback. Second, rats wear the neuroprosthesis and the feedback is provided with ICMS trains modulated by PEFs based on sensor signals. In the third condition, the animals do not wear boots, so they use their intact tactile sensation during the behavior. In all conditions, rats perform the vibrotactile detection task (task D in Fig. 12.2), but the way that the sensory cue reaches the rats changes. For example, in the first condition (boots-on, feedback-off), rats depend purely on stray mechanical cues on their feet to perform the task. In the second case (boots-on, feedback-on), ICMS delivered by the neuroprosthesis is the primary source of information. The rats are already familiar with the last scenario, and the vibrotactile stimulus is directly applied on their glabrous skin.

Among the three conditions, rats achieve the highest psychophysical performance through their intact sensations (i.e., boots-off) as measured by the nonparametric sensitivity index (A′ in Fig. 12.12; Talwar & Gerstein, 1999). The lowest performance is obtained with boots-on feedback-off condition. When the tactile neuroprosthesis system is switched on (boots-on feedback-on), the performance improves compared to the boots-on feedback-off condition. The psychophysical detectability (A′) of the five vibrotactile intensities at each condition can be analyzed by ANCOVA with the stimulus intensity as the covariate and the condition as a fixed factor. In Öztürk et al. (2019), ANCOVA results show that both the fixed factor effect, explained above, and the covariate effect are significant. That is to say, A′ increases as the vibrotactile intensity increases, and the conditions depicted in Fig. 12.12 with colored data points are well separated. It is important to note, however, that the boots-on feedback-off condition yields $A' > 0.5$, which implies that the boots do not provide perfect mechanical isolation as discussed previously; nevertheless, the detectability is much lower than the natural stimulus condition. This verifies that the boots indeed provide a very good isolation, sufficient for testing the neuroprosthesis. Most importantly, the operation of the tactile neuroprosthesis has a significant facilitating effect on psychophysical

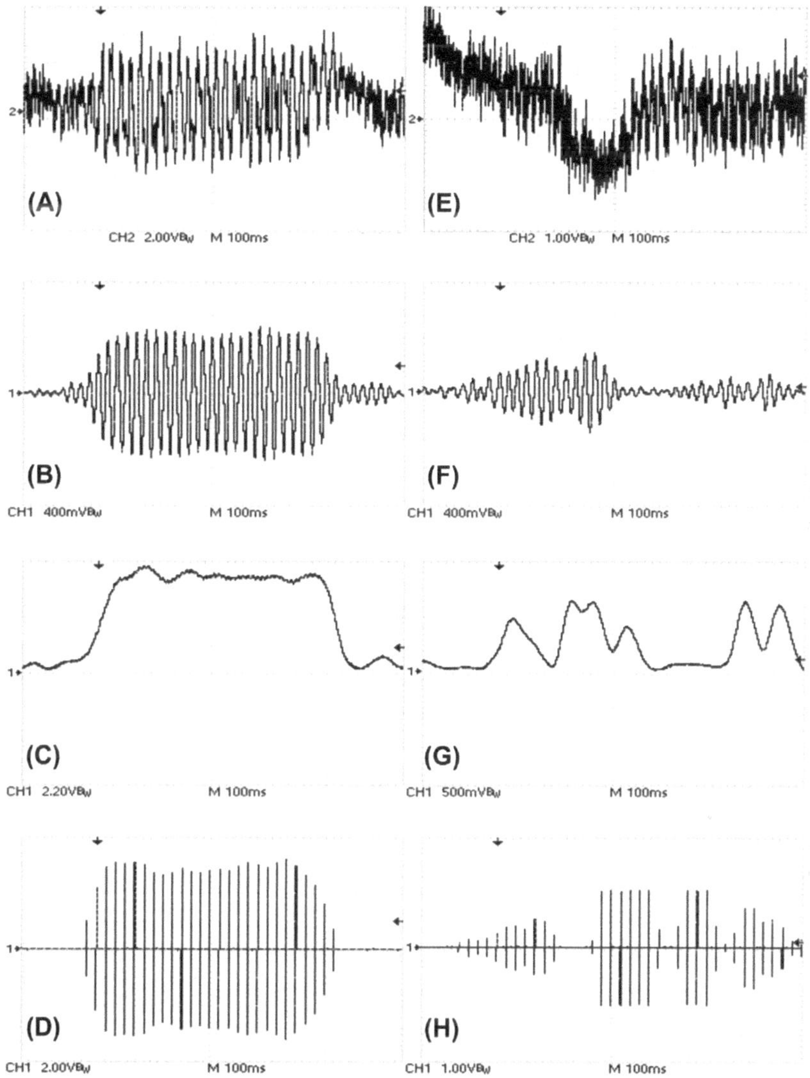

FIGURE 12.11 Signals at different stages of the tactile neuroprosthesis's digital signal processor (DSP) unit. The left panels show the signals for stationary sensorized boots in the chamber (e.g., boots are placed on the mechanical stimulation area). The right panels show the signals while the rat performs the vibrotactile detection task. It should be noted that the signals were not captured simultaneously due to the limited availability of acquisition channels on the DSP unit. The signals detected by tactile sensors are first amplified (A and E) and then, band-pass filtered (B and F). After filtering, the envelope of the signal is extracted and calibrated in mechanical intensity (C and G). Finally, the amplitude of the ICMS pulse train is modulated by the detected envelope based on the PEF of each rat (D and H). *Reprinted with permission from Öztürk, S., Devecioğlu, İ., Beygi, M., et al. (2019). Real-time performance of a tactile neuroprosthesis on awake behaving rats.* IEEE Transactions on Neural Systems and Rehabilitation Engineering, 27 *(5), 1053–1062. https://doi.org/10.1109/TNSRE.2019.2910320.*

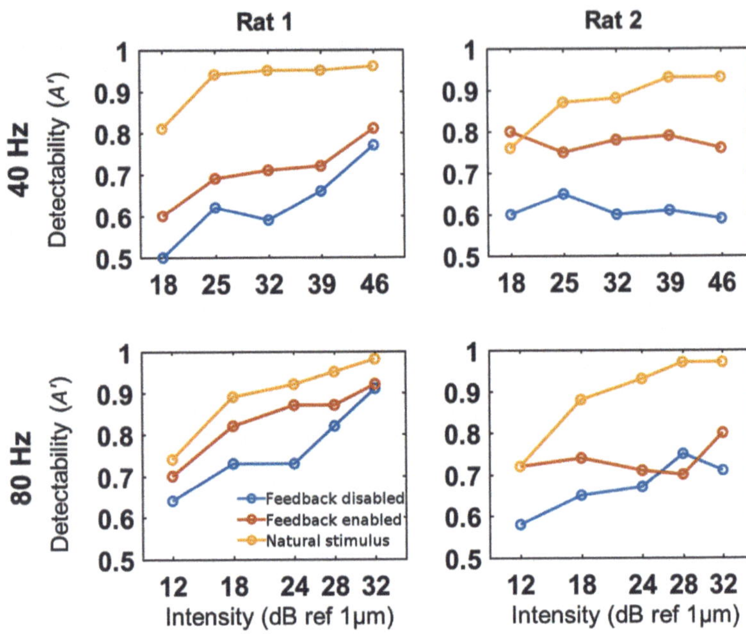

FIGURE 12.12 Detectability analysis at two frequencies (40 and 80 Hz). Detectability, A′, of the sensory cue is calculated for each vibrotactile intensity and frequency in different experiments (e.g., boots-on feedback-disabled, boots-on feedback-enabled, and natural stimulus). *Reprinted with permission from Öztürk, S., Devecioğlu, İ., Beygi, M., et al. (2019). Real-time performance of a tactile neuroprosthesis on awake behaving rats.* IEEE Transactions on Neural Systems and Rehabilitation Engineering, *27(5), 1053–1062. https://doi.org/10.1109/TNSRE.2019.2910320.*

detectability compared to the boots-on no-feedback condition. Therefore a tactile neuroprosthesis based on PEFs and the real-time signal processing presented here can significantly improve the performance of rats even during free behavior by means of artificial sensations.

12.8 Conclusions

Although the optimal design and control of neuroprostheses may not mimic biology entirely, the interface components should be designed such that the functionality gets as close as possible to that in humans. Therefore, not only the motor functions, but also the sensory aspects of the prosthesis should be considered during design. As such, somatosensory neuroprostheses have the potential to complement motor neuroprostheses by providing enhanced motor control, embodiment, and general sensory perception. Especially for patients who can benefit from neuroprostheses with cortical implants (e.g., in tetraplegia), a major problem is to produce an appropriate ICMS pattern to elicit sensations similar to natural ones. Transformation from natural to artificial signals requires understanding how humans process somatosensory information psychophysically.

In this chapter, we focused on several studies leading to the implementation of a tactile neuroprosthesis in rats. One novelty regarding this work is the construction of PEFs not only for stimulus intensity, but also the vibrotactile frequency. Validation experiments show that rats can utilize ICMS stimuli similar to natural vibrotactile stimuli. Furthermore, rats can use the tactile neuroprosthesis to compensate for their loss of detectability by wearing boots. This is indeed a remarkable achievement, because there is considerable noise in the tactile-related signals in the system due to vigorous movement of the rats.

PEFs constructed based on a single psychometric measure (e.g., detection probability) may not be applicable for other tasks (e.g., discrimination). In this case, PEFs can under- or overestimate the ICMS current levels, and hence, decrease the performance of the prosthesis (Tabot et al., 2013). Additionally, periodic recalibration may be necessary for the PEFs because cortical networks are reorganized after extended ICMS stimulation (Maldonado & Gerstein, 1996; Recanzone et al., 1992; Song & Semework, 2015) and during the rehabilitation stage. There is considerable research regarding these issues in the literature, and artificial sensation by ICMS seems to be a promising tool for somatosensory feedback in neuroprostheses.

Acknowledgment

This work was supported by TÜBİTAK Grant 117F481 within the European Union's FLAG-ERA JTC 2017 project GRAFIN to Burak Güçlü.

References

Adibi, M., & Arabzadeh, E. (2011). A comparison of neuronal and behavioral detection and discrimination performances in rat whisker system. *Journal of Neurophysiology*, *105*(1), 356–365. Available from https://doi.org/10.1152/jn.00794.2010.

Ahissar, E. (2008). And motion changes it all. *Nature Neuroscience*, *11*(12), 1369–1370. Available from https://doi.org/10.1038/nn1208-1369.

Andersen, R. A., Hwang, E. J., & Mulliken, G. H. (2010). Cognitive neural prosthetics. *Annual Review of Psychology*, *61*(169-190), c161–c163. Available from https://doi.org/10.1146/annurev.psych.093008.100503.

Baldermann, J. C., Schüller, T., Huys, D., Becker, I., Timmermann, L., Jessen, F., ... Kuhn, J. (2016). Deep brain stimulation for tourette-syndrome: A systematic review and meta-analysis. *Brain Stimulation*, *9*(2), 296–304. Available from https://doi.org/10.1016/j.brs.2015.11.005.

Bensmaia, S. J. (2015). Biological and bionic hands: Natural neural coding and artificial perception. *Philosophical Transactions of the Royal Society B: Biological Sciences*, *370*(1677), 20140209. Available from https://doi.org/10.1098/rstb.2014.0209.

Bensmaïa, S. J., & Hollins, M. (2003). The vibrations of texture. *Somatosensory & Motor Research*, *20*(1), 33–43. Available from https://doi.org/10.1080/0899022031000083825.

Bensmaia, S. J., Hsiao, S. S., Denchev, P. V., Killebrew, J. H., & Craig, J. C. (2008). The tactile perception of stimulus orientation. *Somatosensory & Motor Research*, *25*(1), 49–59. Available from https://doi.org/10.1080/08990220701830662.

Berg, J. A., Dammann, J. F., 3rd, Tenore, F. V., Tabot, G. A., Boback, J. L., Manfredi, L. R., ... Bensmaia, S. J. (2013). Behavioral demonstration of a somatosensory neuroprosthesis. *IEEE Transactions on Neural Systems and Rehabilitation Engineering*, *21*(3), 500–507. Available from https://doi.org/10.1109/tnsre.2013.2244616.

Beygi, M., Mutlu, Ş., & Güçlü, B. (2016). A microfabricated strain gauge array on polymer substrate for tactile neuroprostheses in rats. *Journal of Micromechanics and Microengineering*, *26*(8), 084006.

Bolanowski, S. J., Jr., Gescheider, G. A., Verrillo, R. T., & Checkosky, C. M. (1988). Four channels mediate the mechanical aspects of touch. *Journal of the Acoustical Society of America*, *84*(5), 1680–1694. Available from https://doi.org/10.1121/1.397184.

Butovas, S., & Schwarz, C. (2007). Detection psychophysics of intracortical microstimulation in rat primary somatosensory cortex. *European Journal of Neuroscience*, *25*(7), 2161–2169. Available from https://doi.org/10.1111/j.1460-9568.2007.05449.x.

Callier, T., Schluter, E. W., Tabot, G. A., Miller, L. E., Tenore, F. V., & Bensmaia, S. J. (2015). Long-term stability of sensitivity to intracortical microstimulation of somatosensory cortex. *Journal of Neural Engineering*, *12*(5), 056010. Available from https://doi.org/10.1088/1741-2560/12/5/056010.

Chapin, J. K., & Lin, C. S. (1984). Mapping the body representation in the SI cortex of anesthetized and awake rats. *Journal of Comparative Neurology*, *229*(2), 199–213. Available from https://doi.org/10.1002/cne.902290206.

Cipriani, C., Segil, J. L., Clemente, F., ff Weir, R. F., & Edin, B. (2014). Humans can integrate feedback of discrete events in their sensorimotor control of a robotic hand. *Experimental Brain Research*, *232*(11), 3421–3429. Available from https://doi.org/10.1007/s00221-014-4024-8.

Clemente, F., D'Alonzo, M., Controzzi, M., Edin, B. B., & Cipriani, C. (2016). Non-invasive, temporally discrete feedback of object contact and release improves grasp control of closed-loop myoelectric transradial prostheses. *IEEE Transactions on Neural Systems and Rehabilitation Engineering*, *24*(12), 1314–1322. Available from https://doi.org/10.1109/TNSRE.2015.2500586.

Cohen, J. C., Makous, J. C., & Bolanowski, S. J. (1999). Under which conditions do the skin and probe decouple during sinusoidal vibrations? *Experimental Brain Research*, *129*(2), 211–217. Available from https://doi.org/10.1007/s002210050891.

Collinger, J. L., Wodlinger, B., Downey, J. E., Wang, W., Tyler-Kabara, E. C., Weber, D. J., ... Schwartz, A. B. (2013). High-performance neuroprosthetic control by an individual with tetraplegia. *Lancet*, *381*(9866), 557–564. Available from https://doi.org/10.1016/s0140-6736(12)61816-9.

Dadarlat, M. C., O'doherty, J. E., & Sabes, P. N. (2015). A learning-based approach to artificial sensory feedback leads to optimal integration. *Nature Neuroscience*, *18*(1), 138–144. Available from https://doi.org/10.1038/nn.3883.

Damian, D. D., Arita, A., Martinez, H., & Pfeifer, R. (2012). Slip speed feedback for grip force control. *IEEE Transactions on Biomedical Engineering*, *59*(8), 2200–2210. Available from https://doi.org/10.1109/TBME.2012.2199491.

Dandekar, M. P., Fenoy, A. J., Carvalho, A. F., Soares, J. C., & Quevedo, J. (2018). Deep brain stimulation for treatment-resistant depression: An integrative review of preclinical and clinical findings and translational implications. *Molecular Psychiatry*, *23*(5), 1094–1112. Available from https://doi.org/10.1038/mp.2018.2.

Darian-Smith, I., & Kenins, P. (1980). Innervation density of mechanoreceptive fibres supplying glabrous skin of the monkey's index finger. *Journal of Physiology*, *309*, 147–155.

de Vignemont, F. (2011). Embodiment, ownership and disownership. *Consciousness and Cognition*, *20*(1), 82–93. Available from https://doi.org/10.1016/j.concog.2010.09.004.

Deliano, M., Scheich, H., & Ohl, F. W. (2009). Auditory cortical activity after intracortical microstimulation and its role for sensory processing and learning. *Journal of Neuroscience*, *29*(50), 15898–15909. Available from https://doi.org/10.1523/jneurosci.1949-09.2009.

Devecioğlu, İ. (2016). *Artificial tactile sensation by microstimulation of the hindpaw representation in the primary somatosensory cortex of behaving rats* (Doctoral dissertation). Boğaziçi University, Turkey.

Devecioğlu, İ., & Güçlü, B. (2013). Asymmetric response properties of rapidly adapting mechanoreceptive fibers in the rat glabrous skin. *Somatosensory & Motor Research*, *30*(1), 16–29. Available from https://doi.org/10.3109/08990220.2012.732128.

Devecioğlu, İ., & Güçlü, B. (2015). A novel vibrotactile system for stimulating the glabrous skin of awake freely behaving rats during operant conditioning. *Journal of Neuroscience Methods*, *242*, 41–51. Available from https://doi.org/10.1016/j.jneumeth.2015.01.004.

Devecioğlu, İ., & Güçlü, B. (2017). Psychophysical correspondence between vibrotactile intensity and intracortical microstimulation for tactile neuroprostheses in rats. *Journal of Neural Engineering*, *14*(1), 016010. Available from https://doi.org/10.1088/1741-2552/14/1/016010.

Diamond, M. E., von Heimendahl, M., Knutsen, P. M., Kleinfeld, D., & Ahissar, E. (2008). "Where" and "what" in the whisker sensorimotor system. *Nature Reviews Neuroscience*, *9* (8), 601–612. Available from https://doi.org/10.1038/nrn2411.

Eapen, B. C., Murphy, D. P., & Cifu, D. X. (2017). Neuroprosthetics in amputee and brain injury rehabilitation. *Experimental Neurology*, *287*, 479–485. Available from https://doi.org/10.1016/j.expneurol.2016.08.004.

Ehrsson, H. H., Holmes, N. P., & Passingham, R. E. (2005). Touching a rubber hand: Feeling of body ownership is associated with activity in multisensory brain areas. *Journal of Neuroscience*, *25*(45), 10564–10573. Available from https://doi.org/10.1523/jneurosci.0800-05.2005.

Ehrsson, H. H., Rosen, B., Stockselius, A., Ragno, C., Kohler, P., & Lundborg, G. (2008). Upper limb amputees can be induced to experience a rubber hand as their own. *Brain*, *131*(Pt 12), 3443–3452. Available from https://doi.org/10.1093/brain/awn297.

Fagg, A. H., Hatsopoulos, N. G., de Lafuente, V., Moxon, K. A., Nemati, S., Rebesco, J. M., … Miller, L. E. (2007). Biomimetic brain machine interfaces for the control of movement. *Journal of Neuroscience*, *27*(44), 11842–11846. Available from https://doi.org/10.1523/jneurosci.3516-07.2007.

Fagg, A. H., Hatsopoulos, N. G., London, B. M., Reimer, J., Solla, S. A., Wang, D., & Miller, L. E. (2009). Toward a biomimetic, bidirectional, brain machine interface. *Conference Proceedings of IEEE Engineering in Medicine and Biology Society*, *2009*, 3376–3380. Available from https://doi.org/10.1109/iembs.2009.5332819.

Fernández, E., Greger, B., House, P. A., Aranda, I., Botella, C., Albisua, J., … Normann, R. A. (2014). Acute human brain responses to intracortical microelectrode arrays: Challenges and future prospects. *Frontiers in Neuroengineering*, *7*, 24. Available from https://doi.org/10.3389/fneng.2014.00024.

Fisch, G. S. (2001). Evaluating data from behavioral analysis: Visual inspection or statistical models? *Behavioural Processes*, *54*(1), 137–154. Available from https://doi.org/10.1016/s0376-6357(01)00155-3.

Flesher, S. N., Collinger, J. L., Foldes, S. T., Weiss, J. M., Downey, J. E., Tyler-Kabara, E. C., ... Gaunt, R. A. (2016). Intracortical microstimulation of human somatosensory cortex. *Science Translational Medicine*, *8*(361). Available from https://doi.org/10.1126/scitranslmed.aaf8083, 361ra141.

Fridman, G. Y., Blair, H. T., Blaisdell, A. P., & Judy, J. W. (2010). Perceived intensity of somatosensory cortical electrical stimulation. *Experimental Brain Research*, *203*(3), 499–515. Available from https://doi.org/10.1007/s00221-010-2254-y.

Gardner, E. P., & Johnson, K. O. (2012a). The somatosensory system: Receptors and central pathways. In E. R. Kandel, J. H. Schwartz, T. M. Jessell, et al. (Eds.), *Principles of neural science* (5th ed., pp. 475–497). New York: McGraw-Hill.

Gardner, E. P., & Johnson, K. O. (2012b). Touch. In E. R. Kandel, J. H. Schwartz, T. M. Jessell, et al. (Eds.), *Principles of neural science* (5th ed., pp. 498–529). New York: McGraw-Hill.

Gaunt, R. A., Collinger, J. L., Wodlinger, B., Weber, D. J., & Boninger, M. L. (2013). Proprioceptive feedback enables brain computer interface (BCI) controlled prosthetic arm movement in the absence of visual input. In *Paper presented at the Society for Neuroscience*, San Diego, CA, USA.

Gescheider, G.A. (1997a). The classical psychophysical methods. In: *Psychophysics: The fundamentals* (3rd ed., pp. 45-72). Mahwah, NJ: Lawrence Erlbaum Associates, Inc.

Gescheider, G.A. (1997b). Classical psychophysical theory. In: *Psychophysics: The fundamentals* (3rd ed., pp. 73-104). Mahwah, NJ: Lawrence Erlbaum Associates, Inc.

Graczyk, E. L., Schiefer, M. A., Saal, H. P., Delhaye, B. P., Bensmaia, S. J., & Tyler, D. J. (2016). The neural basis of perceived intensity in natural and artificial touch. *Science Translational Medicine*, *8*(362). Available from https://doi.org/10.1126/scitranslmed.aaf5187, 362ra142.

Grill, W. M., Norman, S. E., & Bellamkonda, R. V. (2009). Implanted neural interfaces: Biochallenges and engineered solutions. *Annual Review of Biomedical Engineering*, *11*, 1–24. Available from https://doi.org/10.1146/annurev-bioeng-061008-124927.

Güçlü, B. (2007). Deviation from Weber's law in the non-Pacinian I tactile channel: A psychophysical and simulation study of intensity discrimination. *Neural Computation*, *19*(10), 2638–2664. Available from https://doi.org/10.1162/neco.2007.19.10.2638.

Güçlü, B., & Bolanowski, S. J. (2002). Modeling population responses of rapidly-adapting mechanoreceptive fibers. *Journal of Computational Neuroscience*, *12*(3), 201–218. Available from https://doi.org/10.1023/a:1016535413000.

Güçlü, B., & Bolanowski, S. J. (2003a). Correlation of spatial event plots with simulated population responses of mechanoreceptive fibers. *Somatosensory & Motor Research*, *20*(3-4), 199–208. Available from https://doi.org/10.1080/08990220310001623022.

Güçlü, B., & Bolanowski, S. J. (2003b). Distribution of the intensity-characteristic parameters of cat rapidly adapting mechanoreceptive fibers. *Somatosensory & Motor Research*, *20*(2), 149–155. Available from https://doi.org/10.1080/0899022031000105181.

Güçlü, B., & Bolanowski, S. J. (2003c). Frequency responses of cat rapidly adapting mechanoreceptive fibers. *Somatosensory & Motor Research*, *20*(3-4), 249–263. Available from https://doi.org/10.1080/08990220310001622988.

Güçlü, B., & Bolanowski, S. J. (2004). Probability of stimulus detection in a model population of rapidly adapting fibers. *Neural Computation*, *16*(1), 39–58. Available from https://doi.org/10.1162/08997660460733985.

Güçlü, B., & Bolanowski, S. J. (2005a). Vibrotactile thresholds of the non-Pacinian I channel: I. Methodological issues. *Somatosensory & Motor Research*, *22*(1-2), 49–56. Available from https://doi.org/10.1080/08990220500084719.

Güçlü, B., & Bolanowski, S. J. (2005b). Vibrotactile thresholds of the non-Pacinian I channel: II. Predicting the effects of contactor location on the phalanx. *Somatosensory & Motor Research*, *22*(1-2), 57–68. Available from https://doi.org/10.1080/08990220512331387971.

Güçlü, B., & Dinçer, S. M. (2013). Neural coding in the non-Pacinian I tactile channel: A psychophysical and simulation study of magnitude estimation. *Somatosensory & Motor Research*, *30*(1), 1–15. Available from https://doi.org/10.3109/08990220.2012.732127.

Güçlü, B., & Öztek, Ç. (2007). Tactile sensitivity of children: Effects of frequency, masking, and the non-Pacinian I psychophysical channel. *Journal of Experimental Child Psychology*, *98*(2), 113–130. Available from https://doi.org/10.1016/j.jecp.2007.05.003.

Güçlü, B., Gescheider, G. A., Bolanowski, S. J., & Istefanopulos, Y. (2005). Population-response model for vibrotactile spatial summation. *Somatosensory & Motor Research*, *22*(4), 239–253. Available from https://doi.org/10.1080/08990220500262075.

Halata, Z. (1990). Sensory innervation of the hairless and hairy skin in mammals including humans. In W. Zenker, & W. Neuhuber (Eds.), *The primary afferent neuron* (pp. 19–34). Springer US.

Harris, J. A., Petersen, R. S., & Diamond, M. E. (1999). Distribution of tactile learning and its neural basis. *Proceedings of the National Academy of Sciences of the United States*, *96*(13), 7587–7591. Available from https://doi.org/10.1073/pnas.96.13.7587.

Hatsopoulos, N. G., & Donoghue, J. P. (2009). The science of neural interface systems. *Annual Review of Neuroscience*, *32*, 249–266. Available from https://doi.org/10.1146/annurev.neuro.051508.135241.

Hochberg, L. R., Bacher, D., Jarosiewicz, B., Masse, N. Y., Simeral, J. D., Vogel, J., ... Donoghue, J. P. (2012). Reach and grasp by people with tetraplegia using a neurally controlled robotic arm. *Nature*, *485*(7398), 372–375. Available from https://doi.org/10.1038/nature11076.

Hollins, M., & Bensmaia, S. J. (2007). The coding of roughness. *Canadian Journal of Experimental Psychology*, *61*(3), 184–195. Available from https://doi.org/10.1037/cjep2007020.

Homer, M. L., Nurmikko, A. V., Donoghue, J. P., & Hochberg, L. R. (2013). Sensors and decoding for intracortical brain computer interfaces. *Annual Review of Biomedical Engineering*, *15*, 383–405. Available from https://doi.org/10.1146/annurev-bioeng-071910-124640.

Johnson, K. O. (1974). Reconstruction of population response to a vibratory stimulus in quickly adapting mechanoreceptive afferent fiber population innervating glabrous skin of the monkey. *Journal of Neurophysiology*, *37*(1), 48–72. Available from https://doi.org/10.1152/jn.1974.37.1.48.

Johnson, K. O. (2001). The roles and functions of cutaneous mechanoreceptors. *Current Opinion in Neurobiology*, *11*(4), 455–461. Available from https://doi.org/10.1016/s0959-4388(00)00234-8.

Jorgovanovic, N., Dosen, S., Djozic, D. J., Krajoski, G., & Farina, D. (2014). Virtual grasping: Closed-loop force control using electrotactile feedback. *Computational and Mathematical Methods in Medicine*, *2014*, 120357. Available from https://doi.org/10.1155/2014/120357.

Kim, S., Callier, T., Tabot, G. A., Tenore, F. V., & Bensmaia, S. J. (2015). Behavioral assessment of sensitivity to intracortical microstimulation of primate somatosensory cortex. *Proceedings of the National Academy of Sciences of the United States of America*. Available from https://doi.org/10.1073/pnas.1509265112, 201509265.

Klaes, C., Shi, Y., Kellis, S., Minxha, J., Revechkis, B., & Andersen, R. A. (2014). A cognitive neuroprosthetic that uses cortical stimulation for somatosensory feedback. *Journal of Neural Engineering, 11*(5), 056024. Available from https://doi.org/10.1088/1741-2560/11/5/056024.

Kleinfeld, D., Ahissar, E., & Diamond, M. E. (2006). Active sensation: Insights from the rodent vibrissa sensorimotor system. *Current Opinion in Neurobiology, 16*(4), 435–444. Available from https://doi.org/10.1016/j.conb.2006.06.009.

Koivuniemi, A., Wilks, S. J., Woolley, A. J., & Otto, K. J. (2011). Multimodal, longitudinal assessment of intracortical microstimulation. *Progress in Brain Research, 194*, 131–144. Available from https://doi.org/10.1016/b978-0-444-53815-4.00011-x.

Koivuniemi, A. S., & Otto, K. J. (2011). Asymmetric versus symmetric pulses for cortical microstimulation. *IEEE Transactions on Neural Systems and Rehabilitation Engineering, 19*(5), 468–476. Available from https://doi.org/10.1109/tnsre.2011.2166563.

Koivuniemi, A. S., & Otto, K. J. (2012). The depth, waveform and pulse rate for electrical microstimulation of the auditory cortex. In *Paper presented at the Annual International Conference of the IEEE Engineering in Medicine and Biology Society*, San Diego, CA.

Kramer, D. R., Lamorie-Foote, K., Barbaro, M., Lee, M., Peng, T., Gogia, A., ... Lee, B. (2019). Functional frequency discrimination from cortical somatosensory stimulation in humans. *Frontiers in Neuroscience, 13*, 832. Available from https://doi.org/10.3389/fnins.2019.00832.

Leem, J. W., Willis, W. D., & Chung, J. M. (1993). Cutaneous sensory receptors in the rat foot. *Journal of Neurophysiology, 69*(5), 1684–1699. Available from https://doi.org/10.1152/jn.1993.69.5.1684.

Leem, J. W., Willis, W. D., Weller, S. C., & Chung, J. M. (1993). Differential activation and classification of cutaneous afferents in the rat. *Journal of Neurophysiology, 70*(6), 2411–2424. Available from https://doi.org/10.1152/jn.1993.70.6.2411.

Lepora, N. F., Burnus, N., Tao, Y., & Cramphorn, L. (2018). Active touch with a biomimetic 3d-printed whiskered robot. In V. Vouloutsi, J. Halloy, A. Mura, M. Mangan, N. Lepora, T. J. Prescott, & P. F. M. J. Verschure (Eds.), *Biomimetic and biohybrid systems. Living machines 2018. Lecture notes in computer science* (Vol. 10928, pp. 263–275). Cham: Springer.

Lewis, S., Russold, M. F., Dietl, H., & Kaniusas, E. (2012). *User demands for sensory feedback in upper extremity prostheses*. Budapest: Medical Measurements and Applications.

Maldonado, P. E., & Gerstein, G. L. (1996). Reorganization in the auditory cortex of the rat induced by intracortical microstimulation: A multiple single-unit study. *Experimental Brain Research, 112*(3), 420–430. Available from https://doi.org/10.1007/BF00227948.

Manfredi, L. R., Baker, A. T., Elias, D. O., Dammann, J. F., 3rd, Zielinski, M. C., Polashock, V. S., & Bensmaia, S. J. (2012). The effect of surface wave propagation on neural responses to vibration in primate glabrous skin. *PLoS One, 7*(2), e31203. Available from https://doi.org/10.1371/journal.pone.0031203.

Maravita, A., & Iriki, A. (2004). Tools for the body (schema). *Trends in Cognitive Sciences, 8* (2), 79–86. Available from https://doi.org/10.1016/j.tics.2003.12.008.

McSweeney, F. K., & Murphy, E. S. (2014). *The wiley blackwell handbook of operant and classical conditioning*. London: John Wiley & Sons, Ltd.

Merrill, D. R., Bikson, M., & Jefferys, J. G. (2005). Electrical stimulation of excitable tissue: Design of efficacious and safe protocols. *Journal of Neuroscience Methods, 141*(2), 171–198. Available from https://doi.org/10.1016/j.jneumeth.2004.10.020.

Micera, S., Rossini, P. M., Rigosa, J., Citi, L., Carpaneto, J., Raspopovic, S., ... Dario, P. (2011). Decoding of grasping information from neural signals recorded using peripheral intrafascicular interfaces. *Journal of Neuroengineering and Rehabilitation*, *8*, 53. Available from https://doi.org/10.1186/1743-0003-8-53.

Morioka, M., Whitehouse, D. J., & Griffin, M. J. (2008). Vibrotactile thresholds at the fingertip, volar forearm, large toe, and heel. *Somatosensory & Motor Research*, *25*(2), 101–112. Available from https://doi.org/10.1080/08990220802045574.

Mountcastle, V. B., LaMotte, R. H., & Carli, G. (1972). Detection thresholds for stimuli in humans and monkeys: Comparison with threshold events in mechanoreceptive afferent nerve fibers innervating the monkey hand. *Journal of Neurophysiology*, *35*(1), 122–136.

Moxon, K. A., & Foffani, G. (2015). Brain-machine interfaces beyond neuroprosthetics. *Neuron*, *86*(1), 55–67.

O'Doherty, J. E., Lebedev, M. A., Li, Z., & Nicolelis, M. A. (2012). Virtual active touch using randomly patterned intracortical microstimulation. *IEEE Transactions on Neural Systems and Rehabilitation Engineering*, *20*(1), 85–93. Available from https://doi.org/10.1109/tnsre.2011.2166807.

Otto, K. J., Rousche, P. J., & Kipke, D. R. (2005). Cortical microstimulation in auditory cortex of rat elicits best-frequency dependent behaviors. *Journal of Neural Engineering*, *2*(2), 42–51. Available from https://doi.org/10.1088/1741-2560/2/2/005.

Öztürk, S., Devecioğlu, İ., Beygi, M., Atasoy, A., Mutlu, S., Özkan, M., & Güçlü, B. (2019). Real-time performance of a tactile neuroprosthesis on awake behaving rats. *IEEE Transactions on Neural Systems and Rehabilitation Engineering*, *27*(5), 1053–1062. Available from https://doi.org/10.1109/TNSRE.2019.2910320.

Patel, G. K., Dosen, S., Castellini, C., & Farina, D. (2016). Multichannel electrotactile feedback for simultaneous and proportional myoelectric control. *Journal of Neural Engineering*, *13*(5), 056015. Available from https://doi.org/10.1088/1741-2560/13/5/056015.

Pearson, M. J., Mitchinson, B., Sullivan, J. C., Pipe, A. G., & Prescott, T. J. (2011). Biomimetic vibrissal sensing for robots. *Philosophical Transactions of the Royal Society B: Biological Sciences*, *366*(1581), 3085–3096. Available from https://doi.org/10.1098/rstb.2011.0164.

Penfield, W. (1930). The radical treatment of traumatic epilepsy and its rationale. *Canadian Medical Association Journal*, *23*(2), 189–197.

Petersen, C. C. (2014). Cortical control of whisker movement. *Annual Review of Neuroscience*, *37*, 183–203. Available from https://doi.org/10.1146/annurev-neuro-062012-170344.

Petkova, V. I., Bjornsdotter, M., Gentile, G., Jonsson, T., Li, T. Q., & Ehrsson, H. H. (2011). From part- to whole-body ownership in the multisensory brain. *Current Biology*, *21*(13), 1118–1122. Available from https://doi.org/10.1016/j.cub.2011.05.022.

Prescott, T. J., Pearson, M. J., Mitchinson, B., Sullivan, J. C. W., & Pipe, A. G. (2009). Whisking with robots. *IEEE Robotics & Automation Magazine*, *16*(3), 42–50. Available from https://doi.org/10.1109/MRA.2009.933624.

Raspopovic, S., Capogrosso, M., Petrini, F. M., Bonizzato, M., Rigosa, J., Di Pino, G., ... Micera, S. (2014). Restoring natural sensory feedback in real-time bidirectional hand prostheses. *Science Translational Medicine*, *6*(222). Available from https://doi.org/10.1126/scitranslmed.3006820, 222ra219.

Recanzone, G. H., Merzenich, M. M., & Dinse, H. R. (1992). Expansion of the cortical representation of a specific skin field in primary somatosensory cortex by intracortical microstimulation. *Cerebral Cortex*, *2*(3), 181–196. Available from https://doi.org/10.1093/cercor/2.3.181.

Romo, R., Hernandez, A., Zainos, A., & Salinas, E. (1998). Somatosensory discrimination based on cortical microstimulation. *Nature*, *392*(6674), 387–390. Available from https://doi.org/10.1038/32891.

Romo, R., Hernandez, A., Zainos, A., Brody, C. D., & Lemus, L. (2000). Sensing without touching: Psychophysical performance based on cortical microstimulation. *Neuron, 26*(1), 273–278. Available from https://doi.org/10.1016/s0896-6273(00)81156-3.

Romo, R., & Salinas, E. (2003). Flutter discrimination: Neural codes, perception, memory and decision making. *Nature Reviews Neuroscience, 4*(3), 203–218. Available from https://doi.org/10.1038/nrn1058.

Rossini, P. M., Micera, S., Benvenuto, A., Carpaneto, J., Cavallo, G., Citi, L., ... Dario, P. (2010). Double nerve intraneural interface implant on a human amputee for robotic hand control. *Clinical Neurophysiology, 121*(5), 777–783. Available from https://doi.org/10.1016/j.clinph.2010.01.001.

Rousche, P. J., Otto, K. J., Reilly, M. P., & Kipke, D. R. (2003). Single electrode microstimulation of rat auditory cortex: An evaluation of behavioral performance. *Hearing Research, 179*(1–2), 62–71. Available from https://doi.org/10.1016/s0378-5955(03)00081-9.

Schweisfurth, M. A., Markovic, M., Dosen, S., Teich, F., Graimann, B., & Farina, D. (2016). Electrotactile emg feedback improves the control of prosthesis grasping force. *Journal of Neural Engineering, 13*(5), 056010. Available from https://doi.org/10.1088/1741-2560/13/5/056010.

Semework, M., & DiStasio, M. (2014). Short-term dynamics of causal information transfer in thalamocortical networks during natural inputs and microstimulation for somatosensory neuroprosthesis. *Frontiers in Neuroengineering, 7*, 36. Available from https://doi.org/10.3389/fneng.2014.00036.

Semprini, M., Bennicelli, L., & Vato, A. (2012). A parametric study of intracortical microstimulation in behaving rats for the development of artificial sensory channels. *Conference Proceedings of IEEE Engineering in Medicine and Biology Society, 2012*, 799–802. Available from https://doi.org/10.1109/embc.2012.6346052.

Song, W., & Semework, M. (2015). Tactile representation in somatosensory thalamus (VPL) and cortex (S1) of awake primate and the plasticity induced by VPL neuroprosthetic stimulation. *Brain Research, 1625*, 301–313. Available from https://doi.org/10.1016/j.brainres.2015.08.046.

Strzalkowski, N. D., Triano, J. J., Lam, C. K., Templeton, C. A., & Bent, L. R. (2015). Thresholds of skin sensitivity are partially influenced by mechanical properties of the skin on the foot sole. *Physiological Reports, 3*(6), e12425. Available from https://doi.org/10.14814/phy2.12425.

Suminski, A. J., Tkach, D. C., Fagg, A. H., & Hatsopoulos, N. G. (2010). Incorporating feedback from multiple sensory modalities enhances brain-machine interface control. *Journal of Neuroscience, 30*(50), 16777–16787. Available from https://doi.org/10.1523/jneurosci.3967-10.2010.

Svensson, P., Wijk, U., Björkman, A., & Antfolk, C. (2017). A review of invasive and non-invasive sensory feedback in upper limb prostheses. *Expert Review of Medical Devices, 14*(6), 439–447. Available from https://doi.org/10.1080/17434440.2017.1332989.

Tabot, G. A., Dammann, J. F., Berg, J. A., Tenore, F., Boback, J. L., Vogelstein, R. J., & Bensmaia, S. J. (2013). Restoring the sense of touch with a prosthetic hand through a brain interface. *Proceedings of the National Academy of Sciences of the United States of America, 110*(45), 18279–18284. Available from https://doi.org/10.1073/pnas.1221113110.

Tabot, G. A., Kim, S. S., Winberry, J. E., & Bensmaia, S. J. (2015). Restoring tactile and proprioceptive sensation through a brain interface. *Neurobiology of Disease, 83*, 191–198. Available from https://doi.org/10.1016/j.nbd.2014.08.029.

Tahon, K., Wijnants, M., & De Schutter, E. (2011). The rat-rotadrum: A reaction time task depending on a continuous stream of tactile sensory information to the rat. *Journal of Neuroscience Methods*, *200*(2), 153–163. Available from https://doi.org/10.1016/j.jneumeth.2011.06.031.

Talwar, S. K., & Gerstein, G. L. (1999). A signal detection analysis of auditory-frequency discrimination in the rat. *Journal of the Acoustical Society of America*, *105*(3), 1784–1800. Available from https://doi.org/10.1121/1.426716.

Vardar, B., & Güçlü, B. (2017). Non-nmda receptor-mediated vibrotactile responses of neurons from the hindpaw representation in the rat SI cortex. *Somatosensory & Motor Research*, *34*(3), 189–203. Available from https://doi.org/10.1080/08990220.2017.1390450.

Vardar, B., & Güçlü, B. (2020). Effects of basal forebrain stimulation on the vibrotactile responses of neurons from the hindpaw representation in the rat SI cortex. *Brain Structure and Function*, *225*(6), 1761–1776. Available from https://doi.org/10.1007/s00429-020-02091-w.

Vicheva, P., Butler, M., & Shotbolt, P. (2020). Deep brain stimulation for obsessive-compulsive disorder: A systematic review of randomised controlled trials. *Neuroscience & Biobehavioral Reviews*, *109*, 129–138. Available from https://doi.org/10.1016/j.neubiorev.2020.01.007.

Walker, J. L., Monjaraz-Fuentes, F., Pedrow, C. R., & Rector, D. M. (2011). Precision rodent whisker stimulator with integrated servo-locked control and displacement measurement. *Journal of Neuroscience Methods*, *196*(1), 20–30. Available from https://doi.org/10.1016/j.jneumeth.2010.12.008.

Weber, D. J., Friesen, R., & Miller, L. E. (2012). Interfacing the somatosensory system to restore touch and proprioception: Essential considerations. *Journal of Motor Behavior*, *44*(6), 403–418. Available from https://doi.org/10.1080/00222895.2012.735283.

Wiest, M. C., Thomson, E., Pantoja, J., & Nicolelis, M. A. (2010). Changes in S1 neural responses during tactile discrimination learning. *Journal of Neurophysiology*, *104*(1), 300–312. Available from https://doi.org/10.1152/jn.00194.2010.

Witteveen, H. J., Rietman, H. S., & Veltink, P. H. (2015). Vibrotactile grasping force and hand aperture feedback for myoelectric forearm prosthesis users. *Prosthetics and Orthotics International*, *39*(3), 204–212. Available from https://doi.org/10.1177/0309364614522260.

Wolpert, D. M., Pearson, K. G., & Ghez, C. P. J. (2012). The organization and planning of movement. In E. R. Kandel, J. H. Schwartz, T. M. Jessell, et al. (Eds.), Principles of neural science (5th ed., pp. 743–767). McGraw-Hill.

Yau, J. M., Hollins, M., & Bensmaia, S. J. (2009). Textural timbre: The perception of surface microtexture depends in part on multimodal spectral cues. *Communicative & Integrative Biology*, *2*(4), 1–3. Available from https://doi.org/10.4161/cib.2.4.8551.

Yıldız, M. Z., & Güçlü, B. (2013). Relationship between vibrotactile detection threshold in the pacinian channel and complex mechanical modulus of the human glabrous skin. *Somatosensory & Motor Research*, *30*(1), 37–47. Available from https://doi.org/10.3109/08990220.2012.754754.

Yıldız, M. Z., Toker, İ., Özkan, F. B., & Güçlü, B. (2015). Effects of passive and active movement on vibrotactile detection thresholds of the pacinian channel and forward masking. *Somatosensory & Motor Research*, *32*(4), 262–272. Available from https://doi.org/10.3109/08990220.2015.1091771.

Chapter 13

Cortical stimulation for somatosensory feedback: translation from nonhuman primates to clinical applications

Marion Badi[1,*], Simon Borgognon[2,3,*], Joseph E. O'Doherty[4] and Solaiman Shokur[1,5]

[1]*Bertarelli Foundation Chair in Translational Neuroengineering, Center for Neuroprosthetics and Institute of Bioengineering, Ecole Polytechnique Federale de Lausanne, Lausanne, Switzerland,* [2]*Department of Neuroscience and Movement Science, Platform of Translational Neurosciences, University of Fribourg, Fribourg, Switzerland,* [3]*Center for Neuroprosthetics and Brain Mind Institute, School of Life Sciences, Swiss Federal Institute of Technology (EPFL), Geneva, Switzerland,* [4]*Neuralink Corp., San Francisco, CA, United States,* [5]*The BioRobotics Institute and Department of Excellence in Robotics and AI, Scuola Superiore Sant'Anna, Pisa, Italy*

ABSTRACT

Nonhuman primates (NHPs) constitute an ideal model for the investigation of somatosensory feedback, as they can learn complex behavioral tasks and perform dexterous sensory-cued reaching and grasping movements. In addition, NHP models provide a meaningful way to functionally test neurotechnologies developed for human subjects, facilitating their translation to clinical applications. Here, we discuss how experiments with NHPs have been instrumental to basic research into somatosensory pathways, preclinical testing of the new devices, and encoding strategies for somatosensory neuroprosthesis. We provide an overview of the major advances in the field of cortical stimulation for somatosensory feedback and how they translated into clinical reality. We also give suggestions on how to implement experimental protocols with NHPs for somatosensory studies, using practical examples from existing setups.

Keywords: Neuroprosthetics; somatosensory feedback; microstimulation; behavioral experiment; nonhuman primates

*These authors contributed equally to this work.

Somatosensory Feedback for Neuroprosthetics. DOI: https://doi.org/10.1016/B978-0-12-822828-9.00008-3

13.1 Introduction

Encoding somatosensory feedback via cortical or intracortical stimulation is a prevailing *translational* science topic due to the increasing therapeutic demand for neurological applications. The term *translation*, in the research field, refers to the transfer of basic scientific findings into potential clinical therapies. But it could also be taken in its etymological sense of transposing a message from one language to another—here, from electrical pulses to brain signals.

In the last two decades, numerous studies have shown that the stimulation of the primary somatosensory cortex (S1) via implanted electrodes induces tactile or proprioceptive sensation (Fifer et al., 2020; Fitzsimmons et al., 2007; Kim, Gomez-Ramirez, et al., 2015; O'Doherty et al., 2009; Romo et al., 1998; Tabot et al., 2013) similar to that emanating from the body. Cortical stimulation has been, therefore, proposed as an approach to restore somatosensory feedback in patients with severe deficiencies as spinal cord injury (Flesher et al., 2016).

Given its importance for object manipulation, the restoration of somatosensory feedback was identified as one of the key requirements for motor recovery (Bensmaia & Miller, 2014; Fagg et al., 2007; Lebedev & Nicolelis, 2006). Indeed, most of the motor tasks that we perform in our daily life, such as reaching (Gordon et al., 1995), fine grasping (Johansson et al., 1992; Richardson et al., 2016; Rothwell et al., 1982), object manipulation (Johansson & Flanagan, 2009), and walking (Petrini et al., 2019; Shokur et al., 2016) rely on tactile and proprioceptive information.

As we will see in this chapter, the similarity between nonhuman primates (NHPs) and humans in terms of somatosensory cortical representations, together with the ability of monkeys to learn and perform sophisticated manipulation tasks, makes research with NHPs critical for the development of somatosensory neuroprosthetics (SNPs), and for technologies integrating both motor decoding and sensory feedback (termed bidirectional brain–machine interfaces).

We start this chapter with a brief history of the main studies on SNPs; next, we describe NHPs' neuroanatomical organization of somatosensation and explain why they constitute a pertinent model for somatosensory studies. We then discuss how NHP studies can help engineer SNPs. Finally, we give a comprehensive description of the experimental setups that can be used to develop protocols for SNPs and conclude with future challenges for SNP development.

13.2 A brief history of somatosensory neuroprosthetics with nonhuman primates

Since the pioneering work of Evarts (Evarts, 1964) and Fetz (Fetz, 1969), our understanding of the neural encoding of sensorimotor functions has greatly improved, leading to the development of groundbreaking technologies to restore movement and sensory modalities in patients with severe deficits (Fifer et al., 2020; Flesher et al., 2019; Hiremath et al., 2017; Salas et al., 2018). As we will see here, the key studies in the field have, to a large extent, employed NHPs models to explore the natural and artificial encoding of somatosensory feedback.

In a seminal study conducted in the late 1990s, Romo and colleagues (Romo et al., 1998) observed responses to mechanical stimulation (5–50 Hz) of the skin, in a subset of quickly adapting neurons in the S1 of rhesus monkeys. They introduced a two-alternate forced-choice discrimination task in which stimulation pairs were either mechanical (vibrators placed on the hand) or mixed between mechanical and intracortical microstimulation (ICMS). The electrodes were placed in Brodmann's area 3b of S1, and experimenters controlled for the brain stimulation site, the stimulation amplitude, frequency, and periodicity. They consistently observed that for a specific parameter set, the monkey could reliably compare the average rates at which the stimuli were delivered (natural or ICMS) and choose the ones with the higher frequency.

In the following years, studies with animal models opened the perspective for intracortical stimulation for neuroprosthetic purposes. In one study with rats in 2002 (Talwar et al., 2002), ICMS pulses delivered in S1 (barrel cortex) and medial forebrain bundle were used to, respectively, instruct and reward the animal for navigating through a random terrain. The study showed that ICMS could operate either as a sensory cue or a mechanism for operant conditioning. Further investigation revealed the potential of such an approach to train owl (Fitzsimmons et al., 2007) or rhesus monkeys (London et al., 2008) in a reaching task, in which the target location was cued via ICMS. These experiments showed that animals could *learn* to discriminate different spatiotemporal stimulation paradigms. The next year, O'Doherty and colleagues took the experiment one step further, by integrating online decoding of motor intentions through an invasive brain–machine interface (BMI), to the same ICMS cueing strategy (O'Doherty et al., 2009).

Two research goals characterized the decade that followed; in the first half of the 2010s, several groups undertook to further investigate the potential of ICMS to convey sensory information (Fig. 13.1A). In the second half, these approaches began to be translated into clinical tests with patients using surface cortical (Hiremath et al., 2017; Johnson et al., 2013; Kramer et al., 2020; Lee et al., 2018) or intracortical (Flesher et al., 2016, 2019; Salas et al., 2018) stimulation.

Psychophysical experiments with NHPs promoted the detailed characterization of the detectability of ICMS as a function of various stimulation parameters such as the pulse amplitude, width, duration, and frequency (Kim, Callier, et al., 2015). They also allowed the evaluation of higher order interactions on the perception of ICMS, like the relation between frequency and amplitude (Callier et al., 2020) or the stimulation periodicity (O'Doherty et al., 2012).

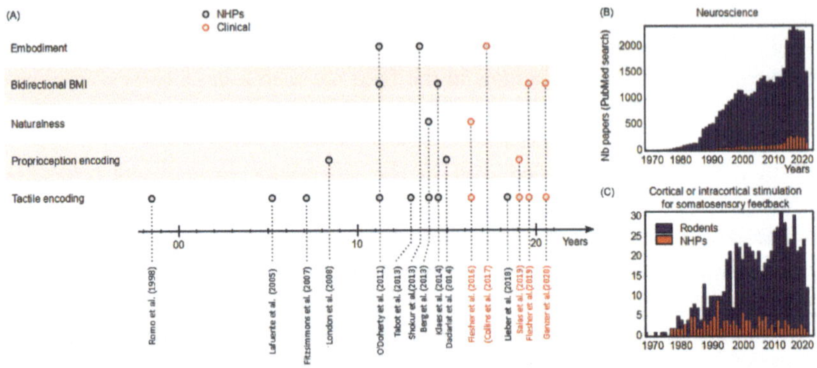

FIGURE 13.1 Studies with nonhuman primates (NHPs). (A) Key publications in the field of cortical stimulation for somatosensory feedback for neuroprosthetics; black circles indicate studies done with nonhuman primates (NHPs) and red circles are for clinical studies with human subjects. (B) Number of publications listed on PubMed per year for neuroscience studies with rodents (*blue*) or NHPs (*red*). We searched for the words *rats* or *rodents* appearing in the abstract or title and *neuroscience*. For NHPs, we searched the following synonyms: *nonhuman primate* or *nonhuman primate* or *monkey*. (C) Similar to (B), we searched for rodents (and synonyms) or NHPs (and synonyms) and ((stimulation[Title/Abstract] AND (intracortical[Title/Abstract] OR cortical[Title/Abstract])) AND (somatosensory[Title/Abstract] OR tactile[Title/Abstract] OR proprioception[Title/Abstract])).

Tabot and colleagues (Tabot et al., 2013) measured the effect of pulse train duration (from 50 to 500 milliseconds) on detection thresholds and argued that since short pulse trains ($\sim$100 milliseconds) were detectable with reasonable thresholds, it could be possible to represent onset and offset transients for object grasping. They hypothesized that this strategy could efficiently mimic the responses of rapidly adapting mechanoreceptive afferents and associated neurons. In a subsequent paper, the same group modulated pulse train amplitude by the force delivered to a prosthetic finger. They showed that by carefully choosing the mapping between pulse amplitude and force applied to the sensor, the animals could make perceptual discriminations with the prosthetic fingertip with a sensitivity matching the natural fingertip (Berg et al., 2013).

Taking a somehow radically different approach for proprioception encoding, Dadarlat and colleagues used a learning-based multichannel ICMS encoding strategy to provide continuous information about the hand state during reaching (Dadarlat et al., 2015). The authors employed ICMS to encode the distance between the animal's hand and an invisible target. They demonstrated that artificial kinesthetic feedback can be efficiently learned by the monkey and can provide rich insights for directing movements. The first demonstration of a setup integrating both real-time motor decoding and sensory feedback in S1 (termed bidirectional BMI) was shown in 2011 (O'Doherty et al., 2011). Rhesus monkeys were trained to control a virtual avatar to explore visually identical objects and discriminate them based on their texture, encoded by the frequency of ICMS burst.

Similarly, motor decoding in the posterior parietal cortex was used in combination with S1 stimulation in another bidirectional BMI study (Klaes et al., 2014). Finally, in a recent publication, texture coarseness was encoded through ICMS by modulating the stimulation frequency as a function of hand movements and object texture (O'Doherty et al., 2019). The authors reported that the monkey controlling a virtual hand through a BMI successfully learned to integrate sensory and motor information to discriminate the presented textures.

Building upon these promising results obtained with monkeys, ICMS has been clinically tested in paralyzed patients within the past few years. In 2016, a study conducted in a tetraplegic subject revealed that the perception elicited by ICMS was localized and most often perceived as pressure sensation (Flesher et al., 2016). In a complementary work, ICMS was found to induce both cutaneous and proprioceptive percepts in an individual with complete tetraplegia (Salas et al., 2018). Finally, a pilot bidirectional BMI protocol showed increased dexterity in a functional task (Flesher et al., 2019) when somatosensory feedback was provided. These encouraging proofs-of-concept raise the hope for future prosthetics that would simultaneously restore sensory and motor functions, thereby promoting a dexterous and natural control of movement. Experimentation with NHPs has so far paved the way for this new generation of neuroprosthetics and avoided several years of strenuous trial-and-error in patients. The next sections detail to what extent NHPs studies constitute an important step in the development of those technologies.

13.3 Why nonhuman primates are a pertinent model for the development of somatosensory neuroprosthetics

In experimental neurosciences, the number of studies performed on rodents is 12 times larger than the number conducted in NHPs (Fig. 13.1B). The quantity of NHPs used for each study is also typically much smaller. Therefore when considering all animals employed for experimentation, 80% are rodents, while only 0.1% are NHPs (Roelfsema & Treue, 2014). This large difference might be explained by the high complexity, financial costs, ethical concerns, and long-lasting protocols associated with NHP experiments. However, despite these numerous difficulties, NHP models provide a unique opportunity to develop, test, and refine preclinical technologies in "proof-of-principle" studies that strengthen and potentially accelerate their translation to clinical and therapeutic strategies (Friedman et al., 2017; Capitanio & Emborg, 2008; Goldberg, 2019; Mitchell et al., 2018; Ogier et al., 2019). When it comes to SNPs, the development of the technological frameworks bears conceptual and technical challenges that can only partially be answered in rodents. In addition, practical considerations for the design of these technologies cannot realistically be addressed solely in humans (Buffalo et al., 2019). This might illustrate why when considering the studies that focus on cortical stimulation for somatosensory feedback, the ratio of publications reporting rodents over those involving NHPs decreases to five to one (Fig. 13.1C).

Because of their proximity in the phylogenetic tree (Rogers & Gibbs, 2014), NHPs and humans share similar properties, including sophisticated attributes of sensorimotor integration processes (Hatsopoulos & Suminski, 2011; Pruszynski et al., 2011; Seki & Fetz, 2012). In contrast, major differences in motor cortical and descending motor pathways organization are observed between NHPs and rodents (Courtine et al., 2007; Donoghue & Wise, 1982; Neafsey et al., 1986; Neafsey & Sievert, 1982; Rouiller et al., 1993). In terms of S1 neuroanatomical organization, humans and NHPs share similar cytoarchitectural and functional features. Indeed, both are characterized by four distinct S1 areas, namely, Brodmann's area 1, area 2, area 3a, and area 3b (Geyer et al., 1999; Jain et al., 1998; Kaas, 1983; Fig. 13.2). In contrast, in nonprimate species, the primary area S1 is homologous to area 3b (Kaas, 1983; Kaas et al., 1979; Marshall et al., 1937). Altogether, the

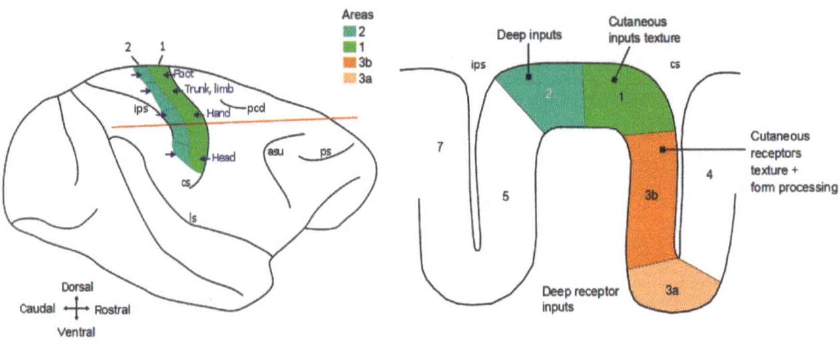

FIGURE 13.2 NHP cortical anatomical organization for primary somatosensory cortex. S1 is located caudally to the central sulcus (cs). The somatotopic organization is shown in blue. (*left*) The somatotopic organization is conserved along the mediolateral axis between M1 and S1. (*right*) Four S1 areas are shown. Area 3a is located in the deepest bank of the central sulcus and is a transition zone between BA 4 (M1) and proper-S1 (area 3b). Areas 1 and 2 are located on the post-central gyrus before the transition to the parietal cortex (BA 5 and BA 7). cs, central sulcus; ips, intraparietal sulcus; pcd, precentral dimple; asu, arcuate sulcus; ps, principal sulcus; ls, lateral sulcus; BA, Brodmann area. *Modified from James, T. W., Kim, S., & Fisher, J. S. (2007). The neural basis of haptic object processing.* Canadian Journal of Experimental Psychology/Revue Canadienne de Psychologie Expérimentale, 61*(3), 219–229. Pons, T.P., Garraghty, P.E., Cusick, C.G., & Kaas, J.H. (1985). The somatotopic organization of area 2 in macaque monkeys.* Journal of Comparative Neurology, 241*(4), 445–466 and Pons, T.P., Garraghty, P.E., Friedman, D. P., & Mishkin, M. (1987). Physiological evidence for serial processing in somatosensory cortex.* Science (New York, N.Y.), 237*(4813), 417–420.*

important similarities observed between humans and monkeys and the clear discrepancies that separate rodents from us justify the study of sensation encoding in the NHP model.

Generally, the somatic sensation in NHPs is encoded within S1 following a hierarchical scheme from area 3b to area 2 (Iwamura, 1998). While phase-locked responses to vibrations are primarily seen in area 3b, they gradually disappear in areas 1 and 2 (Harvey et al., 2013). Cutaneous and proprioceptive afferents seem to be multimodally processed by area 3a, which relays the information to areas 1 and 2 (Kim, Gomez-Ramirez, et al., 2015; Lucier et al., 1975; Yamada et al., 2016). Interestingly, different parameters of vibrotactile stimuli (frequency, amplitude, and duration) are encoded at the single neuron level. Whereas some neurons respond to all of these features, others encode only a single feature (Alvarez et al., 2015). Nevertheless, these categorization properties are more complex at the neuronal population level. Coarse-texture and fine surface encoding are mediated through spatial or temporal activation of peripheral receptors, respectively (Bensmaïa & Hollins, 2005; Hollins et al., 2001, 2002; Hollins & Risner, 2000; Lieber et al., 2017; Lieber & Bensmaia, 2019; Weber et al., 2013). This dual neuronal model (spatial and temporal) of texture encoding is represented in a high-dimensional fashion through idiosyncratic responses of S1 subpopulations (Lieber & Bensmaia, 2019).

Given the complexity of S1 representations, the application of microcortical stimulation for sensation restoration can be highly challenging. Since areas 3a and 3b work as a multimodal system converging their information to areas 1 and 2, one may suggest areas 3a and 3b represent adequate regions to reproduce rich somatosensory feedback. However, these areas are deep in the sulcus and also challenging to access chronically. Additionally, because these regions present multimodal response properties, it is necessary to engineer well-defined stimulation paradigms to artificially recapitulate natural and intuitive sensory percepts.

13.4 How nonhuman primate studies can help engineer somatosensory neuroprosthetics

Bidirectional BMIs as therapy for sensorimotor trauma require weighing the cost and benefits for the patient (Ryu & Shenoy, 2009). NHP studies offer valuable insights for assessing and potentially decreasing the risks linked to implantable devices. In particular, NHPs can be used to train neurosurgeons to identify and target the desired cortical area, whether it is the superficially located area 1 of the S1, the sulci-located areas 2 and 3b, or the deepest area 3a. Complications resulting from the implantation of a microelectrode or subdural stimulating grid can be quantified behaviorally and through postmortem histology in NHPs (Barrese et al., 2013; Degenhart et al., 2016; Griffith & Humphrey, 2006; Rajan et al., 2015). In addition, when it comes to optimizing neural implants for human applications, whether by developing small footprint intracortical electrodes or less invasive subdural grids, intraoperative and perioperative adverse effects can be extensively documented in monkeys (Barrese et al., 2013). In parallel, implanted hardware can be improved through NHP research by developing surgical implantation strategies reducing the risks of infection, tissue damage, or technical failures (Lanz et al., 2013; Johnston et al., 2016; Overton et al., 2017; Ortiz-Rios et al., 2018). Human surgical procedures can, for example, benefit from techniques that minimize the size of a craniotomy, help secure the implant to the neighboring tissues, or permit the installation of a wireless transmitter in the periphery of the device (Capogrosso et al., 2016; Ong et al., 2018; Zhou et al., 2019).

We detail here three domains in which preliminary testing in NHPs can help engineer future somatosensory neuroprosthetics: (1) the development of cortical implant devices, (2) the encoding of realistic somatosensory feedback, and (3) the validation of predictive computational models optimizing the stimulation strategies.

13.4.1 Development of cortical implants

The manufacturing of durable, efficient, and safe stimulation interfaces is a prerequisite for the development of SNPs (Ryu & Shenoy, 2009). Most studies investigating the production of sensory percepts through chronic cortical stimulation in NHPs rely on penetrating intracortical electrodes (Szostak et al., 2017). Among these, we identify three main classes of implant: the high-density multichannel Utah electrode array (Maynard et al., 1997) manufactured by Blackrock (Blackrock Microsystems, United States) and composed of up to 96 channels for microstimulation or recording. The microelectrode shaft measures 1–1.5 mm in length and the array can be implanted in the superficial regions of the S1, that is, Brodmann areas 1 and 2 (Callier et al., 2020; Dadarlat et al., 2015; Klaes et al., 2014; Zaaimi et al., 2013). The second type of commercial implant is the floating microelectrode array (MicroProbes, United States), composed of up to 36 stimulation channels able to penetrate at varying depths within the cortex (up to 3 mm), thereby reaching the deepest sensory structures such as area 3b (Berg et al., 2013; Kim, Gomez-Ramirez, et al., 2015; Tabot et al., 2013). Alternative approaches usually rely on "custom-fabricated" microwire arrays designed to fit specific study requirements. One example is the microdrive platform, developed by Romo and colleagues, to position up to seven tungsten-platinum electrodes in area 3b of the somatosensory cortex (Romo et al., 1998). London and colleagues used a similar technique, which is, however, automated and relies on Pt-Ir electrodes to stimulate the proprioceptive region area 3a (London et al., 2008). Finally, Nicolelis' lab manufactured microwire arrays with up to several hundred electrodes and inserted them in the hand and leg areas of sensory–motor areas (Ifft et al., 2013; Nicolelis, 2003).

The development of such invasive devices requires many steps of optimization to ensure the longevity and efficacy of the electrodes. While the functionality, durability, and biocompatibility of the implant can be assessed in cats, sheep, or swine (Agha et al., 2013; Gaunt et al., 2009; Payne et al., 2018; Tanosaki et al., 2014), the validation of the system performance is far more complex and requires preclinical studies in NHPs. The Utah array (Maynard et al., 1997) is one of the most successful examples of clinical translation. It was one of the first intracortical prostheses implanted in humans to control a BMI (Hochberg et al., 2006; Kim et al., 2008). This successful translation was the result of more than 15 years of technical testing and refinement in NHP (Barrese et al., 2013). The array dimensions, the connectors, the wire composition, and insulation, as well as the electrode contact material, were modified and optimized to ensure durable and robust performance for neural recording and microstimulation (S. Kim et al., 2009; Moxon et al., 2004; Nordhausen et al., 1994). Barrese and colleagues systematically documented the known failure modes of Utah arrays in 27 monkeys and characterized the risks for biological failures such as gliosis formation, neuronal death, hardware infection, material delamination, or mechanical failure (Barrese et al., 2013). Research in monkeys also allows assessing the quality of neural recordings over a long period in complex motor tasks (Chestek et al., 2011; Suner et al., 2005; Vaidya et al., 2014). Such evaluation is critical for the potential coupling of motor intent decoding to sensory feedback.

Finally, implantation in the primate brain provides the opportunity to perform numerous cycles of stimulation over several months or years and quantify the chronic stability and selectivity of the implant. Psychophysical studies of sensory perception require a vast number of repetitions, particularly when testing interactions between parameters. For instance, in their study exploring stimulus detection, Kim and colleagues performed more than 47,000 trials (Kim, Callier, et al., 2015). Such an exhaustive search is incompatible with human experimentation.

13.4.2 Somatosensory feedback encoding

One of the most prominent challenges associated with the use of cortical stimulation to elicit sensory percepts resides in the "naturalness" of the evoked sensations. Despite the growing body of knowledge related to how afferent sensory signals are represented in neural activation patterns, very little is known on how to artificially reproduce these patterns. Most of the studies investigating sensory encoding through cortical stimulation in NHPs use short, symmetric, biphasic pulse trains (200 milliseconds to 1 second), delivering charge on the order of $\sim$100 nC to 2 mC for subdural stimulation with electrocorticographic (ECoG) electrodes (Hiremath et al., 2017; Lee et al., 2018) and $\sim$0.1–40 nC for intracortical stimulation (Dadarlat et al., 2015;

O'Doherty et al., 2011). A large range of frequencies, from 10 to 1000 Hz, is spanned depending on the study design. Recently, researchers have shown that monkeys can discriminate between ICMS stimuli delivered at different pulse amplitudes (Kim, Callier, et al., 2015; Tabot et al., 2013), and pulse rates (Callier et al., 2020; O'Doherty et al., 2011; Romo et al., 1998). Similar studies conducted in humans provided insight into the type of sensations produced through the ICMS protocols. Flesher and colleagues reported natural pressure-like percepts on the palm and index (Flesher et al., 2019). In contrast, Salas and colleagues reported the apparition of both tactile and proprioceptive sensations on the phalanxes and arm depending on the intensity of stimulation (Salas et al., 2018).

While the sensation elicited by a single electrode can be modulated by changing the amplitude or frequency of stimulation, the spatial resolution can potentially be increased by altering which type of neural element is recruited. Recently, research groups demonstrated in vivo that ICMS does not primarily recruit cells located around the stimulation site as was initially hypothesized (Stoney et al., 1968), but rather activates axonal segments traveling near the active tip, resulting in a sparse excitation of neurons (Butovas & Schwarz, 2003; Histed et al., 2009). This unexpected phenomenon might explain why human subjects have reported unnatural sensations when stimulated through intracortical electrodes, leading to discordant and sometimes unpleasant percepts (Grill et al., 2005; Ohara et al., 2004; Patel et al., 2006).

13.4.3 Validation of computational models

Computational models offer an unprecedented opportunity to base the implementation of stimulation paradigms on rational and educated guesses of their efficacy in vivo. On the one hand, they provide a way to evaluate the gaps in our understanding of neural systems and the experiments to conduct to fill them (Markram et al., 2015). On the other, they constitute powerful in silico testing platforms in which stimulation protocols can be explored and refined to derive optimal ways of interacting with the neural architecture (Mcintyre & Foutz, 2013).

Data collected from animal models and, in particular, NHPs are valuable assets for validating and refining complex neuronal and neural network models (Capogrosso et al., 2013; Greiner et al., 2021; Kumaravelu et al., 2020; Raspopovic et al., 2012; Saal et al., 2017). Besides improving models of somatosensory mechanisms, studies with NHPs can improve our understanding of somatosensory cell recruitment and help predict the effects of cortical stimulation on the various layers of the somatosensory cortex.

A majority of ICMS studies have relied on symmetric, cathode-leading current pulses, as these tend to produce the lowest detection thresholds (Koivuniemi & Otto, 2011). However, modeling work by McIntyre and Grill suggested that asymmetric waveforms may preferentially recruit cell bodies rather than axons (McIntyre & Grill, 2000, 2002). In particular, the authors reported that cathode-leading asymmetric pulses primarily recruit cell bodies, while symmetric pulses selectively activate axons. Those results were later confirmed in vivo (Bari et al., 2013; Wang et al., 2012). Other computational simulation studies indicated that increasing the frequency of train stimuli did not lead to the same effect when recruiting axons passing near the electrode tip than when recruiting cell bodies (McIntyre & Grill, 2002). The modeling of these underlying neural mechanisms is key in understanding how the somatosensory cortex responds to ICMS and can be used as a powerful tool to devise biomimetic stimulation paradigms.

Overstreet and colleagues showed that the stimulation of S1 recruited interneurons and pyramidal neurons in different patterns, suggesting that the encoding of surrogate sensations may be more complex than expected and that a deeper understanding of ICMS-elicited activity is needed to produce consistent and coherent percepts (Overstreet et al., 2013). More complex models integrating the various levels of cortical layers and recapitulating the complexity of the neuron geometry (Aberra et al., 2018) are necessary to reveal the activation effects of cortical stimulation (Butovas & Schwarz, 2003; DeYoe et al., 2005; Houweling & Brecht, 2008; Tehovnik et al., 2006). The validation and optimization of such computational models in NHPs and other animal models is once again an essential step between in silico design and clinical applications.

13.5 Experimental setups for somatosensory studies with nonhuman primates

Experimental paradigms used to study the neural encoding of somatosensory inputs in primates have particularly benefited from the important advances made in the development of high-density cortical electrode arrays, robotic devices, manipulanda, tracking software, virtual environments, and many others.

Here, we list different frameworks that have been used to study somatosensory functions in NHPs. We describe the electrophysiology platforms employed for subcortical and cortical stimulation, the devices that can be used to control somatosensory and visual inputs and the available tools for behavioral tracking. We illustrate this section using many successful examples of setups from the literature (Fig. 13.3).

13.5.1 Cortical and intracortical electrical stimulation

Electrodes, electrophysiological recording systems, and setups used for cortical stimulation in NHPs have seen important improvements over the past decades. From the tethered low-density microwires used in the 1990s, we have come to develop sophisticated, high-density arrays capable of remaining stable in the brain for years (S. Kim et al., 2009; Moxon et al., 2004;

FIGURE 13.3 Experimental setups for sensorimotor studies in NHPs. (A) The experimental setup used by Romo and Salinas (Romo et al., 1998), who demonstrated the ability of a monkey to distinguish between different frequencies of ICMS. (B) Representation of the mechanical apparatus designed by Kim and colleagues to provide concomitant proprioceptive and tactile stimuli to the monkey. (C) Experimental setup combining a rigid commercially available exoskeleton (KINARM), an eye-tracking system, and a virtual environment. Jan de Haan and colleagues used this platform to study visuomotor integration in humans and monkeys. (D) 2D planar manipulandum used by Millers' group to study passively and actively generated arm movements in NHPs. (E) Experimental paradigm developed by Shokur and colleagues in which the monkey receives both tactile and visual stimuli through a virtual reality simulation. The animal gaze is monitored using an eye-tracking device. (F) Artificial texture paradigm developed by Nicolelis' group allowing a monkey to actively explore ICMS-elicited percepts in a virtual reality space. (G) Representation of a training platform featuring multidimensional kinematic tracking and virtual reality. (H) Experimental setup to study the effect of various ICMS amplitudes on the elicited percepts. (I) Robotic framework for the study of reaching and grasping in NHPs developed by Barra and colleagues. The setup features multimodal signals characterizing arm and hand movements in an unconstrained environment. *(Fig. 13.3A) Adapted from Horwitz, G. D., & Newsome, W. (1998). Neurophysiology: Sensing and categorizing.* Current Biology, *8(11), R376–R378. (Fig. 13.3B) Adapted from Kim, S., Callier, T., Tabot, G. A., Gaunt, R. A., Tenore, F. V., & Bensmaia, S. J. (2015). Behavioral assessment of sensitivity to intracortical microstimulation of primate somatosensory cortex.* Proceedings of the National Academy of Sciences, 112 *(49), 15202–15207. (Fig. 13.3C) Adapted from de Haan, M. J., Brochier, T., Grün, S., Riehle, A., & Barthélemy, F. V. (2018). Real-time visuomotor behavior and electrophysiology recording*

(Continued)

Nordhausen et al., 1994). These devices have been used to investigate the effects of intracortical stimulation in awake behaving monkeys (Callier et al., 2020; London et al., 2008; O'Doherty et al., 2019; Tabot et al., 2013). However, most of these studies featured experimental setups in which the animal was seated in a chair and performed stereotypic actions learned through months of behavioral training. While these repetitive behaviors are useful for assessing performance over a high number of consistent trials, they do not allow for natural interactions between the monkey and its environment. To tackle this issue, several groups have developed electrophysiology platforms enabling intracortical recordings and stimulation in freely behaving animals via wireless systems (S. Kim et al., 2009; Klaes et al., 2014; London et al., 2008; Moxon et al., 2004; Nordhausen et al., 1994; Schwarz et al., 2014; Shoaran et al., 2014; Yin et al., 2014). Yet, the monitoring of cortical states for extended periods and the chronic stimulation of multiple channels still face multiple limitations. In particular, they require the development of compact, long-lasting battery-powered devices (Zanos et al., 2011), the design of functional energy-saving stimulation paradigms (Almeida et al., 2016), and the development of low-power architectures (Shoaran et al., 2014).

13.5.2 Somatosensory inputs

The exploration of somatosensory encoding and the development of somatosensory replacement strategies necessitates setups for which tactile stimuli can be reliably and precisely delivered to the subject. In the classic experiment by Romo and colleagues (Romo et al., 1998), the animal was first trained to distinguish different frequencies of mechanical stimuli applied on

setup for use with humans and monkeys. Journal of Neurophysiology, 120*(2), 539–552. (Fig. 13.3D) Adapted from London, B. M., & Miller, L. E. (2013). Responses of somatosensory area 2 neurons to actively and passively generated limb movements.* Journal of Neurophysiology, 109*(6), 1505–1513. (Fig. 13.3E) Adapted from Shokur, S., ODoherty, J. E., Winans, J. A., Bleuler, H., Lebedev, M. A., & Nicolelis, M. A. L. (2013). Expanding the primate body schema in sensorimotor cortex by virtual touches of an avatar.* Proceedings of the National Academy of Sciences of the United States of America, 110*(37), 15121–15126. (Fig. 13.3F) Adapted from O'Doherty, J. E., Shokur, S., Medina, L. E., Lebedev, M. A., & Nicolelis, M. A. L. (2019). Creating a neuroprosthesis for active tactile exploration of textures.* Proceedings of the National Academy of Sciences, 116*(43), 21821–21827. (Fig. 13.3G) Adapted from Putrino, D., Wong, Y. T., Weiss, A., & Pesaran, B. (2015). A training platform for many-dimensional prosthetic devices using a virtual reality environment.* Journal of Neuroscience Methods, 244, *68–77. (Fig. 13.3H) Adapted from Tabot, G. A., Dammann, J. F., Berg, J. A., Tenore, F. V., Boback, J. L., Vogelstein, R. J., & Bensmaia, S. J. (2013). Restoring the sense of touch with a prosthetic hand through a brain interface.* Proceedings of the National Academy of Sciences of the United States of America, 110*(45), 18279–18284. (Fig. 13.3I) Adapted from Barra, B., Badi, M., Perich, M.G., Conti, S., Salehian, S.S.M., Moreillon, F., Bogaard, A., Wurth, S., Kaeser, M., Passeraub, P., Milekovic, T., Billard, A., Micera, S., & Capogrosso, M. (2019). A versatile robotic platform for the design of natural, three-dimensional reaching and grasping tasks in monkeys.* Journal of Neural Engineering, 17*(1), 016004.*

the skin by a probe (Fig. 13.3A). Tactile inputs were then replaced by artificial intracortical electrical stimuli to evaluate the similarity between stimulation-induced and tactile stimuli.

In addition to the encoding of tactile features, the study of the somatosensory system progressively revealed the strong association between cutaneous and proprioceptive inputs at different levels in the NHP brain. Behavioral apparatus have thus been developed to explore in detail the multimodal neural encoding of somatosensory inputs and the dependence of tactile percepts on the limb position. Kim and colleagues used a precise mechanical stimulator to position the monkey hand in different postures before presenting the tactile stimuli (Fig. 13.3B; Kim, Gomez-Ramirez, et al., 2015). They demonstrated that most S1 neurons responded to cutaneous and proprioceptive inputs. Other groups have used similar exoskeletal robots to provide kinesthetic information to a monkey operating a BMI or to study M1 responses to mechanical perturbations (Pruszynski, 2014; Suminski et al., 2010). Such a highly constrained rigid robot allows for the control of each degree of freedom of the arm and the study of movement under normal and perturbed conditions (de Haan et al., 2018) (Fig. 13.3C). However, stiff exoskeletons have the disadvantage of limiting the amplitude and range of motion. To study movements more similar to natural reach, Miller's group employed a two-link planar manipulandum enabling free motion in a 2D plane (Fig. 13.3D). This preparation was used to evaluate S1 representation of passively and actively generated movements, and the encoding of whole arm kinematics in Brodmann's area 2 (Chowdhury et al., 2020; London & Miller, 2013).

13.5.3 Visual inputs

In self-paced protocols, visual cues provide important information to the monkey regarding the degree of completion of the task and its outcome. The significance of such cues is generally learned through positive reinforcement. In combination with visual cues, auditory cues are sometimes added to reinforce the instruction, the animal's attention, or reward, for example, simple tones or buzzer-type feedback. A moving cursor on a screen can thus direct the monkey to move a manipulandum in a target direction (Fig. 13.3D).

In recent years, virtual reality has also been introduced to provide immersive visual feedback in NHPs experiments (Fig. 13.3E–G). Shokur and colleagues used a realistic 3D monkey avatar to study visuo–tactile integration (Shokur et al., 2013; Fig. 13.3E) and closed-loop brain–machine interfaces (O'Doherty et al., 2011, 2019). Interestingly, NHPs interacted with the 3D avatar in a somewhat natural way, as if it was a conspecific (Shokur et al., 2013). Another study by Steckenfinger and Ghazanfar (Steckenfinger & Ghazanfar, 2009) showed that viewing a realistic synthetic monkey avatar face elicited aversion similar to the phenomenon known as the uncanny valley in humans: that is, realistic, human-looking robotic faces elicit negative feelings in human observers (Mori, 1970). Finally, in an experiment necessitating complex motor learning of a bimanual task, it was shown that the learning curve was significantly improved when the visual feedback of the controlled actuators was displayed as 3D monkey avatar arms instead of cursors (Ifft et al., 2013). These studies demonstrate that virtual reality and 3D avatars can be used to train and evaluate NHPs in complex sensory–motor tasks.

13.5.4 Behavioral tracking

An important challenge in using NHP to study somatosensory percepts elicited through cortical stimulation resides in the impossibility for the animal to report precisely to the experimenter the nature and location of the evoked precepts. Several techniques can allow monkeys to communicate their perceptions through associative choices.

A typical experiment for SNPs involves discrimination tasks, where the animal has to distinguish between two levels of stimulation parameters (based on, for example, their stimulation amplitude or frequency). Practically, the reporting of sensations can be done using different instruments such as response buttons (Fig. 13.3A; Romo et al., 1998), joysticks, and hand movements (Fig. 13.3F; O'Doherty et al., 2019), or eye saccades (Fig. 13.3H; Tabot et al., 2013).

For experiments involving sensory–motor tasks, or active tactile exploration (as opposed to passive touch), it is necessary to track the animal kinematics. As an example, London and colleagues used ICMS to discretely cue the direction of reach in a macaque, while monitoring the hand position in real time (Fig. 13.3D; London et al., 2008). This setup enabled them to measure an animal's reaction time to the ICMS. The continuous assessment of the limb kinematics provides a meaningful way to study natural, unconstrained behaviors, in particular, in the context of proprioceptive encoding. In their study, Dadarlat and colleagues monitored the monkey fingertip position with an electromagnetic position sensor, preserving free movement of the arm (Dadarlat et al., 2015).

Other approaches have successfully been used in NHP studies to precisely measure kinematics and kinetic features during reaching. More specifically, motion capture systems relying on reflective markers tracked by infrared cameras have allowed for high-density upper-limb characterization in behaving monkeys (Fig. 13.3G,I; Putrino et al., 2015; Barra et al., 2019). This method provides high-precision 3D kinematic data at the cost of fixating markers on the arm and hand of the animal. The accurate and repeatable positioning of a high number of markers on the joints is not always straightforward, and their acceptance by the animal can require months of training, making this approach tedious and unrealistic for some applications. More recently, image-processing software based on deep-learning algorithms has shown remarkable accuracy in extracting detailed poses without markers in dynamically changing backgrounds (Mathis et al., 2018; Nath et al., 2019). These studies opened the way for performing noninvasive quantifications of kinematics in 2D and 3D during behavioral tasks.

Besides kinematic limb tracking, kinetic sensors provide meaningful information on the joint torques and the forces generated by the moving limb. Rigid exoskeletons such as the KINARM (Pruszynski, 2014; Scott, 1999) measure accurate single-joint torques during highly controlled arm movements. Industrial robots also offer promising performances for applications in translational primate studies. By providing a large number of degrees of freedom, user-friendly programming interfaces, precise joint measurements, and robust components, these devices allow for the design and characterization of complex behavioral tasks. More specifically, a robotic arm can be programmed to accurately position objects or targets in space through joint control paradigms, or to act as a compliant spring via impedance mode control (Barra et al., 2019). Additionally, robotic joint stiffness can be modulated to impose load or force field perturbation to the end effector. Barra and colleagues (Barra et al., 2019; Fig. 13.3I) used such a robotic arm to study the kinematic and kinetic components of free-reaching and object manipulation behaviors in awake behaving monkeys. These robots can be equipped with versatile sensors to detect touch, measure grip pressure, or potentially trigger various somatosensory percepts. Actuated arms can also be used to guide or perturb the movement while preserving unconstrained behavior.

In summary, experimental paradigms for NHPs are constantly evolving, taking advantage of the most recent developments in software, robotics, and hardware design. Existing setups consist of instrumented platforms that provide the researcher with a rich signal portfolio for the investigation of natural sensorimotor behaviors. They combine instruments supplying sensory inputs for the animal, tools that permit the monkey to interact and respond to stimuli, and recording systems monitoring kinematics and kinetic features.

13.6 Conclusion

After more than two decades of preclinical investigation into SNPs and recent translation into clinical trials, one might conclude that future SNP studies will mainly concentrate on tests with human subjects. After all, are not questionnaires with human subjects experimentally simpler than indirect measurements with NHPs?

As we have seen in this chapter, this intuition might be short-sighted. Ongoing clinical trials, which have benefited from knowledge gained in NHP studies, have so far aimed at reproducing simple somatosensory encodings, mainly using unimodal stimulation strategies. To date, scientists in the field have only scratched the surface of the myriad of possibilities offered by intracortical stimulation for the restoration of somatosensation. This observation may be partly due to the limitations of currently available electrodes and stimulators, but may also result from our incomplete understanding of the mechanisms governing the electrical excitation of somatosensory neural circuitries.

Many technical and scientific challenges still remain for designing clinically viable devices and for restoring functional levels of performance (Gilja et al., 2011). We believe that NHPs stand as natural candidates to tackle these two different yet complementary aspects.

On the one hand, NHPs can play an important role in the testing of the next generation of implanted SNPs. For instance, future technologies should be fully implantable, present a minimal footprint, and integrate prosthetic processors combining signal processing units, neural decoders, batteries, wireless data, and power transceivers (Musk, 2019). Additionally, the interface itself should be entirely biocompatible—enabling reliable communication with the nervous system and providing benefits to the user for many years. Biocompatibility may be achieved through many strategies, including chemical and molecular approaches (Spencer et al., 2017; A. Wang et al., 2007), but a promising avenue is simply to decrease the size of the device. Smaller technologies displace less tissue, damage fewer neurons, and minimize disruption of the blood–brain barrier (Bennett et al., 2018), which is itself a potent driver of the neuroinflammatory response. Miniaturized devices also offer specific benefits for SNPs: they enable higher channel count (Musk, 2019) and higher density arrays (El-Atab et al., 2019) that may further increase the range and precision of electrical stimulation. All these incremental steps can and should be optimized in NHPs.

On the other hand, a deep understanding of how inputs to somatosensory networks can trigger specific percepts or favor particular behaviors is indispensable to help the implementation of functional neurostimulation therapies. In this framework, NHPs offer an interesting opportunity to test and validate complex paradigms of stimulation, such as multipolar patterns for current steering, modified pulse shapes (McIntyre & Grill, 2002), protocols altering the frequency of the bursts (Formento et al., 2020), etc. Ultimately, there are still many open questions on the characterization of the sensations elicited through stimulation, of the bodily integration of those percepts, and the effects on performance during motor tasks. In particular, the encoding of proprioception has been far less studied than touch, and only rarely reported in human subjects (Armenta Salas et al., 2018), or achieved via nonhomologous approaches (Dadarlat et al., 2015). The multimodal aspect of proprioception necessitates a better understanding of its cortical encoding and the possibility for reliable multichannel stimulation. Similarly, the encoding of other somatosensory submodalities such as temperature and nociception should be studied with the same level of attention as we have seen in recent years with the mechanical aspects of touch. All these multimodal integration processes can be efficiently evaluated using versatile behavioral platforms in NHPs.

While this chapter has focused on electrical ICMS for somatosensory feedback, alternative technologies exist for the activation of the central nervous system. For instance, light-driven stimulation, or optogenetics, has shown promising results for the specific activation of neurons in NHPs (Diester et al., 2011; Gerits et al., 2012; Jazayeri et al., 2012). It improves spatial resolution and avoids the appearance of stimulation artifacts during simultaneous recordings, which are a recurrent issue in bidirectional prostheses. The selective activation of cell bodies achieved through optogenetics can also potentially alleviate the stray recruitment of axons passing near the electrode (Butovas & Schwarz, 2003; Histed et al., 2009). However, the translation of optogenetics to humans bears important challenges, such as the development of receptors responding to a large range of wavelengths (Kampasi et al., 2018), the manufacturing of viral vectors and promoters efficiently transfecting opsins into human cells (El-Shamayleh et al., 2016; Gerits et al., 2015), and the assessment of safety (Mendoza et al., 2017) and biocompatibility for ethical use. These mechanisms can be studied in NHP models, and their characterization carries great promise for future applications to somatosensory interfaces (Galvan et al., 2018; Gerits & Vanduffel, 2013; Kinoshita & Isa, 2015). Other stimulation techniques such as endovascular stimulation (Opie et al., 2019), ultrasonic recruitment (Neely et al., 2018), or magnetothermal excitation (Chen et al., 2015) have shown encouraging results for the activation of deep brain structures with reduced surgery procedures. These devices may open promising clinical opportunities to access deeper, yet more functionally relevant, somatosensory cortical structures such as S1 areas 3a and 3b, but need to be rigorously validated in preclinical animal models.

Overall, NHP research provides remarkable tools to further understand somatosensory functions, enables the development of meaningful therapies for neurological disorders, and opens new avenues to *translate* artificial "commands" into a language spoken by the brain.

References

Aberra, A. S., Peterchev, A. V., & Grill, W. M. (2018). Biophysically realistic neuron models for simulation of cortical stimulation. *Journal of Neural Engineering*, *15*(6), 066023. Available from https://doi.org/10.1088/1741-2552/aadbb1.

Agha, N. S., Komar, J., Yin, M., Borton, D. A., & Nurmikko, A. (2013). A fully wireless platform for correlating behavior and neural data from an implanted, neural recording device: Demonstration in a freely moving swine model. *2013 6th International IEEE/EMBS Conference on Neural Engineering (NER)*, 989–992.

Almeida, L., Rawal, P. V., Ditty, B., Smelser, B. L., Huang, H., Okun, M. S., Guthrie, B. L., & Walker, H. C. (2016). Deep brain stimulation battery longevity: Comparison of monopolar versus bipolar stimulation modes. *Movement Disorders Clinical Practice*, *3*(4), 359–366. Available from https://doi.org/10.1002/mdc3.12285.

Alvarez, M., Zainos, A., & Romo, R. (2015). Decoding stimulus features in primate somatosensory cortex during perceptual categorization. *Proceedings of the National Academy of Sciences of the United States of America*, *112*(15), 4773–4778.

Armenta Salas, M., Bashford, L., Kellis, S., Jafari, M., Jo, H., Kramer, D., Shanfield, K., Pejsa, K., Lee, B., Liu, C. Y., & Andersen, R. A. ((2018). Proprioceptive and cutaneous sensations in humans elicited by intracortical microstimulation. *ELife*, *7*, e32904. Available from https://doi.org/10.7554/eLife.32904.

Bari, B. A., Ollerenshaw, D. R., Millard, D. C., Wang, Q., & Stanley, G. B. (2013). Behavioral and electrophysiological effects of cortical microstimulation parameters. *PLoS One*, *8*(12), e82170. Available from https://doi.org/10.1371/journal.pone.0082170.

Barra, B., Badi, M., Perich, M. G., Conti, S., Salehian, S. S. M., Moreillon, F., Bogaard, A., Wurth, S., Kaeser, M., Passeraub, P., Milekovic, T., Billard, A., Micera, S., & Capogrosso, M. (2019). A versatile robotic platform for the design of natural, three-dimensional reaching and grasping tasks in monkeys. *Journal of Neural Engineering*, *17*(1), 016004. Available from https://doi.org/10.1088/1741-2552/ab4c77.

Barrese, J. C., Rao, N., Paroo, K., Triebwasser, C., Vargas-Irwin, C., Franquemont, L., & Donoghue, J. P. (2013). Failure mode analysis of silicon-based intracortical microelectrode arrays in non-human primates. *Journal of Neural Engineering*, *10*(6), 066014. Available from https://doi.org/10.1088/1741-2560/10/6/066014.

Bennett, C., Samikkannu, M., Mohammed, F., Dietrich, W. D., Rajguru, S. M., & Prasad, A. (2018). Blood brain barrier (BBB)-disruption in intracortical silicon microelectrode implants. *Biomaterials*, *164*, 1–10. Available from https://doi.org/10.1016/j.biomaterials.2018.02.036.

Bensmaïa, S., & Hollins, M. (2005). Pacinian representations of fine surface texture. *Perception & Psychophysics*, *67*(5), 842–854.

Bensmaia, S. J., & Miller, L. E. (2014). Restoring sensorimotor function through intracortical interfaces: Progress and looming challenges. *Nature Reviews Neuroscience*, *15*(5), 313–325. Available from https://doi.org/10.1038/nrn3724.

Berg, J. A., Dammann, J. F., III, Tenore, F. V., Tabot, G. A., Boback, J. L., Manfredi, L. R., Peterson, M. L., Katyal, K. D., Johannes, M. S., Makhlin, A., Wilcox, R., Franklin, R. K., Vogelstein, R. J., Hatsopoulos, N. G., & Bensmaia, S. J. (2013). Behavioral demonstration of a somatosensory neuroprosthesis. *IEEE Transactions on Neural Systems and Rehabilitation Engineering*, *21*(3), 500–507. Available from https://doi.org/10.1109/TNSRE.2013.2244616.

Buffalo, E. A., Movshon, J. A., & Wurtz, R. H. (2019). From basic brain research to treating human brain disorders. *Proceedings of the National Academy of Sciences*, *116*(52), 26167–26172. Available from https://doi.org/10.1073/pnas.1919895116.

Butovas, S., & Schwarz, C. (2003). Spatiotemporal effects of microstimulation in rat neocortex: A parametric study using multielectrode recordings. *Journal of Neurophysiology*, *90*(5), 3024–3039. Available from https://doi.org/10.1152/jn.00245.2003.

Callier, T., Brantly, N. W., Caravelli, A., & Bensmaia, S. J. (2020). The frequency of cortical microstimulation shapes artificial touch. *Proceedings of the National Academy of Sciences*, *117*(2), 1191–1200. Available from https://doi.org/10.1073/pnas.1916453117.

Capitanio, J. P., & Emborg, M. E. (2008). Contributions of non-human primates to neuroscience research. *The Lancet*, *371*(9618), 1126–1135.

Capogrosso, M., Wenger, N., Raspopovic, S., Musienko, P., Beauparlant, J., Bassi Luciani, L., Courtine, G., & Micera, S. (2013). A computational model for epidural electrical stimulation of spinal sensorimotor circuits. *Journal of Neuroscience*, *33*(49), 19326–19340. Available from https://doi.org/10.1523/JNEUROSCI.1688-13.2013.

Capogrosso, M., Milekovic, T., Borton, D., Wagner, F., Moraud, E. M., Mignardot, J.-B., Buse, N., Gandar, J., Barraud, Q., Xing, D., Rey, E., Duis, S., Jianzhong, Y., Ko, W. K. D., Li, Q., Detemple, P., Denison, T., Micera, S., Bezard, E., … Courtine, G. (2016). A brain–spine interface alleviating gait deficits after spinal cord injury in primates. *Nature*, *539*(7628), 284–288. Available from https://doi.org/10.1038/nature20118.

Chen, R., Romero, G., Christiansen, M. G., Mohr, A., & Anikeeva, P. (2015). Wireless magnetothermal deep brain stimulation. *Science (New York, N.Y.)*, *347*(6229), 1477–1480. Available from https://doi.org/10.1126/science.1261821.

Chestek, C. A., Gilja, V., Nuyujukian, P., Foster, J. D., Fan, J. M., Kaufman, M. T., Churchland, M. M., Rivera-Alvidrez, Z., Cunningham, J. P., Ryu, S. I., & Shenoy, K. V. (2011). Long-term stability of neural prosthetic control signals from silicon cortical arrays in rhesus macaque motor cortex. *Journal of Neural Engineering*, *8*(4), 045005. Available from https://doi.org/10.1088/1741-2560/8/4/045005.

Chowdhury, R. H., Glaser, J. I., & Miller, L. E. (2020). Area 2 of primary somatosensory cortex encodes kinematics of the whole arm. *ELife*, *9*, e48198. Available from https://doi.org/10.7554/eLife.48198.

Courtine, G., Bunge, M. B., Fawcett, J. W., Grossman, R. G., Kaas, J. H., Lemon, R., Maier, I., Martin, J., Nudo, R. J., Ramon-Cueto, A., Rouiller, E. M., Schnell, L., Wannier, T., Schwab, M. E., & Edgerton, V. R. (2007). Can experiments in nonhuman primates expedite the translation of treatments for spinal cord injury in humans? *Nature Medicine*, *13*(5), 561–566. Available from https://doi.org/10.1038/nm1595.

Dadarlat, M. C., O'Doherty, J. E., & Sabes, P. N. (2015). A learning-based approach to artificial sensory feedback leads to optimal integration. *Nature Neuroscience*, *18*(1), 138–144. Available from https://doi.org/10.1038/nn.3883.

de Haan, M. J., Brochier, T., Grün, S., Riehle, A., & Barthélemy, F. V. (2018). Real-time visuomotor behavior and electrophysiology recording setup for use with humans and monkeys. *Journal of Neurophysiology*, *120*(2), 539–552. Available from https://doi.org/10.1152/jn.00262.2017.

Degenhart, A. D., Eles, J., Dum, R., Mischel, J. L., Smalianchuk, I., Endler, B., Ashmore, R. C., Tyler-Kabara, E. C., Hatsopoulos, N. G., Wang, W., Batista, A. P., & Cui, X. T. (2016). Histological evaluation of a chronically-implanted electrocorticographic electrode grid in a non-human primate. *Journal of Neural Engineering*, *13*(4), 046019. Available from https://doi.org/10.1088/1741-2560/13/4/046019.

DeYoe, E. A., Lewine, J. D., & Doty, R. W. (2005). Laminar variation in threshold for detection of electrical excitation of striate cortex by macaques. *Journal of Neurophysiology*, *94*(5), 3443–3450. Available from https://doi.org/10.1152/jn.00407.2005.

Diester, I., Kaufman, M. T., Mogri, M., Pashaie, R., Goo, W., Yizhar, O., Ramakrishnan, C., Deisseroth, K., & Shenoy, K. V. (2011). An optogenetic toolbox designed for primates. *Nature Neuroscience*, *14*(3), 387–397. Available from https://doi.org/10.1038/nn.2749.

Donoghue, J. P., & Wise, S. P. (1982). The motor cortex of the rat: Cytoarchitecture and microstimulation mapping. *Journal of Comparative Neurology*, *212*(1), 76–88. Available from https://doi.org/10.1002/cne.902120106.

El-Atab, N., Shaikh, S. F., & Hussain, M. M. (2019). Nano-scale transistors for interfacing with brain: Design criteria, progress and prospect. *Nanotechnology*, *30*(44), 442001. Available from https://doi.org/10.1088/1361-6528/ab3534.

El-Shamayleh, Y., Ni, A. M., & Horwitz, G. D. (2016). Strategies for targeting primate neural circuits with viral vectors. *Journal of Neurophysiology*, *116*(1), 122–134. Available from https://doi.org/10.1152/jn.00087.2016.

Evarts, E. V. (1964). Temporal patterns of discharge of pyramidal tract neurons during sleep and waking in the monkey. *Journal of Neurophysiology*, *27*(2), 152–171. Available from https://doi.org/10.1152/jn.1964.27.2.152.

Fagg, A. H., Hatsopoulos, N. G., de Lafuente, V., Moxon, K. A., Nemati, S., Rebesco, J. M., Romo, R., Solla, S. A., Reimer, J., Tkach, D., Pohlmeyer, E. A., & Miller, L. E. (2007).

Biomimetic brain machine interfaces for the control of movement. *Journal of Neuroscience*, *27*(44), 11842–11846. Available from https://doi.org/10.1523/JNEUROSCI.3516-07.2007.

Fetz, E. E. (1969). Operant conditioning of cortical unit activity. *Science (New York, N.Y.)*, *163* (3870), 955–958. Available from https://doi.org/10.1126/science.163.3870.955.

Fifer, M. S., McMullen, D. P., Thomas, T. M., Osborn, L. E., Nickl, R. W., Candrea, D. N., Pohlmeyer, E. A., Thompson, M. C., Anaya, M., Schellekens, W., Ramsey, N. F., Bensmaia, S. J., Anderson, W. S., Wester, B. A., Crone, N. E., Celnik, P. A., Cantarero, G. L., & Tenore, F. V. (2020). Intracortical microstimulation elicits human fingertip sensations [preprint]. *Rehabilitation Medicine and Physical Therapy*. Available from https://doi.org/10.1101/2020.05.29.20117374.

Fitzsimmons, N. A., Drake, W., Hanson, T. L., Lebedev, M. A., & Nicolelis, M. A. L. (2007). Primate reaching cued by multichannel spatiotemporal cortical microstimulation. *Journal of Neuroscience*. Available from http://www.jneurosci.org/content/27/21/5593.short.

Flesher, S. N., Collinger, J. L., Foldes, S. T., Weiss, J. M., Downey, J. E., Tyler-Kabara, E. C., Bensmaia, S. J., Schwartz, A. B., Boninger, M. L., & Gaunt, R. A. (2016). Intracortical microstimulation of human somatosensory cortex. *Science Translational Medicine*, *8*(361). Available from https://doi.org/10.1126/scitranslmed.aaf8083, 361ra141-361ra141.

Flesher, S. N., Downey, J. E., Weiss, J. M., Hughes, C. L., Herrera, A. J., Tyler-Kabara, E. C., Boninger, M. L., Collinger, J. L., & Gaunt, R. A. (2019). Restored tactile sensation improves neuroprosthetic arm control. *BioRxiv*. Available from https://doi.org/10.1101/653428, 653428–653428.

Formento, E., D'Anna, E., Gribi, S., Lacour, S. P., & Micera, S. (2020). A biomimetic electrical stimulation strategy to induce asynchronous stochastic neural activity. *Journal of Neural Engineering*, *17*(4), 046019. Available from https://doi.org/10.1088/1741-2552/aba4fc.

Friedman, H., Ator, N., Haigwood, N., Newsome, W., Allan, J. S., Golos, T. G., … Bianchi, P. (2017). The critical role of nonhuman primates in medical research. *Pathogens & immunity*, *2*(3), 352. Available from https://www.ncbi.nlm.nih.gov/pmc/articles/PMC5636196/.

Galvan, A., Caiola, M. J., & Albaugh, D. L. (2018). Advances in optogenetic and chemogenetic methods to study brain circuits in non-human primates. *Journal of Neural Transmission*, *125* (3), 547–563. Available from https://doi.org/10.1007/s00702-017-1697-8.

Gaunt, R. A., Hokanson, J. A., & Weber, D. J. (2009). Microstimulation of primary afferent neurons in the L7 dorsal root ganglia using multielectrode arrays in anesthetized cats: Thresholds and recruitment properties. *Journal of Neural Engineering*, *6*(5), 055009. Available from https://doi.org/10.1088/1741-2560/6/5/055009.

Gerits, A., Farivar, R., Rosen, B. R., Wald, L. L., Boyden, E. S., & Vanduffel, W. (2012). Optogenetically induced behavioral and functional network changes in primates. *Current Biology*, *22*(18), 1722–1726. Available from https://doi.org/10.1016/j.cub.2012.07.023.

Gerits, A., Vancraeyenest, P., Vreysen, S., Laramée, M.-E., Michiels, A., Gijsbers, R., Van den Haute, C., Moons, L., Debyser, Z., Baekelandt, V., Arckens, L., & Vanduffel, W. (2015). Serotype-dependent transduction efficiencies of recombinant adeno-associated viral vectors in monkey neocortex. *Neurophotonics*, *2*(3), 031209. Available from https://doi.org/10.1117/1.NPh.2.3.031209.

Gerits, A., & Vanduffel, W. (2013). Optogenetics in primates: A shining future? *Trends in Genetics*, *29*(7), 403–411. Available from https://doi.org/10.1016/j.tig.2013.03.004.

Geyer, S., Schleicher, A., & Zilles, K. (1999). Areas 3a, 3b, and 1 of human primary somatosensory cortex. *Neuroimage*, *10*(1), 63–83.

Gilja, V., Chestek, C. A., Diester, I., Henderson, J. M., Deisseroth, K., & Shenoy, K. V. (2011). Challenges and opportunities for next-generation intracortically based neural prostheses.

IEEE Transactions on Biomedical Engineering, *58*(7), 1891–1899. Available from https://doi.org/10.1109/TBME.2011.2107553.

Goldberg, M. E. (2019). The neurology clinic needs monkey research. *Proceedings of the National Academy of Sciences of the United States of America*, *116*(52), 26255–26258.

Gordon, J., Ghilardi, M. F., & Ghez, C. (1995). Impairments of reaching movements in patients without proprioception. I. Spatial errors. *Journal of Neurophysiology*, *73*(1), 347–360. Available from https://doi.org/10.1152/jn.1995.73.1.347.

Greiner, N., Barra, B., Schiavone, G., Lorach, H., James, N., Conti, S., Kaeser, M., Fallegger, F., Borgognon, S., Lacour, S., Bloch, J., Courtine, G., & Capogrosso, M. (2021). Recruitment of upper-limb motoneurons with epidural electrical stimulation of the cervical spinal cord. *Nature Communications*, *12*, 435. Available from https://doi.org/10.1101/2020.02.17.952796.

Griffith, R. W., & Humphrey, D. R. (2006). Long-term gliosis around chronically implanted platinum electrodes in the Rhesus macaque motor cortex. *Neuroscience Letters*, *406*(1), 81–86. Available from https://doi.org/10.1016/j.neulet.2006.07.018.

Grill, W. M., Simmons, A. M., Cooper, S. E., Miocinovic, S., Montgomery, E. B., Baker, K. B., & Rezai, A. R. (2005). Temporal excitation properties of paresthesias evoked by thalamic microstimulation. *Clinical Neurophysiology*, *116*(5), 1227–1234. Available from https://doi.org/10.1016/j.clinph.2004.12.020.

Harvey, M. A., Saal, H. P., Dammann, J. F., & Bensmaia, S. J. (2013). Multiplexing stimulus information through rate and temporal codes in primate somatosensory cortex. *PLoS Biology*, *11*(5), e1001558–12.

Hatsopoulos, N. G., & Suminski, A. J. (2011). Sensing with the motor cortex. *Neuron*, *72*(3), 477–487. Available from https://doi.org/10.1016/j.neuron.2011.10.020.

Hiremath, S. V., Tyler-Kabara, E. C., Wheeler, J. J., Moran, D. W., Gaunt, R. A., Collinger, J. L., Foldes, S. T., Weber, D. J., Chen, W., Boninger, M. L., & Wang, W. (2017). Human perception of electrical stimulation on the surface of somatosensory cortex. *PLoS One*, *12*(5), e0176020. Available from https://doi.org/10.1371/journal.pone.0176020.

Histed, M. H., Bonin, V., & Reid, R. C. (2009). Direct activation of sparse, distributed populations of cortical neurons by electrical microstimulation. *Neuron*, *63*(4), 508–522. Available from https://doi.org/10.1016/j.neuron.2009.07.016.

Hochberg, L. R., Serruya, M. D., Friehs, G. M., Mukand, J. A., Saleh, M., Caplan, A. H., Branner, A., Chen, D., Penn, R. D., & Donoghue, J. P. (2006). Neuronal ensemble control of prosthetic devices by a human with tetraplegia. *Nature*, *442*(7099), 164–171. Available from https://doi.org/10.1038/nature04970.

Hollins, M., Bensmaïa, S. J., & Roy, E. A. (2002). Vibrotaction and texture perception. *Behavioural Brain Research*, *135*(1–2), 51–56.

Hollins, M., Bensmaïa, S. J., & Washburn, S. (2001). Vibrotactile adaptation impairs discrimination of fine, but not coarse, textures. *Somatosensory & Motor Research*, *18*(4), 253–262.

Hollins, M., & Risner, S. R. (2000). Evidence for the duplex theory of tactile texture perception. *Perception & Psychophysics*, *62*(4), 695–705.

Horwitz, G. D., & Newsome, W. (1998). Neurophysiology: Sensing and categorizing. *Current Biology*, *8*(11), R376–R378. Available from https://doi.org/10.1016/S0960-9822(98)70241-6.

Houweling, A. R., & Brecht, M. (2008). Behavioural report of single neuron stimulation in somatosensory cortex. *Nature*, *451*(7174), 65–68. Available from https://doi.org/10.1038/nature06447.

Ifft, P. J., Shokur, S., Li, Z., Lebedev, M. A., & Nicolelis, M. A. L. (2013). A brain-machine interface enables bimanual arm movements in monkeys. *Science Translational Medicine*, *5*(210). Available from https://doi.org/10.1126/scitranslmed.3006159.

Iwamura, Y. (1998). Hierarchical somatosensory processing. *Current Opinion in Neurobiology*, *8*(4), 522–528.

Jain, N., Catania, K. C., & Kaas, J. H. (1998). A histologically visible representation of the fingers and palm in primate area 3b and its immutability following long-term deafferentations. *Cerebral Cortex*, *8*(3), 227–236.

James, T. W., Kim, S., & Fisher, J. S. (2007). The neural basis of haptic object processing. *Canadian Journal of Experimental Psychology/Revue Canadienne de Psychologie Expérimentale*, *61*(3), 219–229. Available from https://doi.org/10.1037/cjep2007023.

Jazayeri, M., Lindbloom-Brown, Z., & Horwitz, G. D. (2012). Saccadic eye movements evoked by optogenetic activation of primate V1. *Nature Neuroscience*, *15*(10), 1368–1370. Available from https://doi.org/10.1038/nn.3210.

Johansson, R. S., & Flanagan, J. R. (2009). Coding and use of tactile signals from the fingertips in object manipulation tasks. *Nature Reviews. Neuroscience*, *10*(5), 345–359. Available from https://doi.org/10.1038/nrn2621.

Johansson, R. S., Häger, C., & Bäckström, L. (1992). Somatosensory control of precision grip during unpredictable pulling loads. *Experimental Brain Research*, *89*(1), 204–213. Available from https://doi.org/10.1007/BF00229017.

Johnson, L. A., Wander, J. D., Sarma, D., Su, D. K., Fetz, E. E., & Ojemann, J. G. (2013). Direct electrical stimulation of the somatosensory cortex in humans using electrocorticography electrodes: A qualitative and quantitative report. *Journal of Neural Engineering*, *10*(3), 036021. Available from https://doi.org/10.1088/1741-2560/10/3/036021.

Johnston, J. M., Cohen, Y. E., Shirley, H., Tsunada, J., Bennur, S., Christison-Lagay, K., & Veeder, C. L. (2016). Recent refinements to cranial implants for rhesus macaques (Macaca mulatta). *Lab Animal*, *45*(5), 180–186. Available from https://doi.org/10.1038/laban.997.

Kaas, J. H. (1983). What, if anything, is SI? Organization of first somatosensory area of cortex. *Physiological Reviews*, *63*(1), 206–231.

Kaas, J. H., Nelson, R. J., Sur, M., Lin, C. S., & Merzenich, M. M. (1979). Multiple representations of the body within the primary somatosensory cortex of primates. *Science (New York, N.Y.)*, *204*(4392), 521–523.

Kampasi, K., English, D. F., Seymour, J., Stark, E., McKenzie, S., Vöröslakos, M., Buzsáki, G., Wise, K. D., & Yoon, E. (2018). Dual color optogenetic control of neural populations using low-noise, multishank optoelectrodes. *Microsystems & Nanoengineering*, *4*(1), 1–16. Available from https://doi.org/10.1038/s41378-018-0009-2.

Kim, S., Bhandari, R., Klein, M., Negi, S., Rieth, L., Tathireddy, P., Toepper, M., Oppermann, H., & Solzbacher, F. (2009). Integrated wireless neural interface based on the Utah electrode array. *Biomedical Microdevices*, *11*(2), 453–466. Available from https://doi.org/10.1007/s10544-008-9251-y.

Kim, S. S., Gomez-Ramirez, M., Thakur, P. H., & Hsiao, S. S. (2015). Multimodal interactions between proprioceptive and cutaneous signals in primary somatosensory cortex. *Neuron*, 1–13.

Kim, S.-P., Simeral, J. D., Hochberg, L. R., Donoghue, J. P., & Black, M. J. (2008). Neural control of computer cursor velocity by decoding motor cortical spiking activity in humans with tetraplegia. *Journal of Neural Engineering*, *5*(4), 455–476. Available from https://doi.org/10.1088/1741-2560/5/4/010.

Kim, S., Callier, T., Tabot, G. A., Gaunt, R. A., Tenore, F. V., & Bensmaia, S. J. (2015). Behavioral assessment of sensitivity to intracortical microstimulation of primate somatosensory cortex. *Proceedings of the National Academy of Sciences*, *112*(49), 15202–15207. Available from https://doi.org/10.1073/pnas.1509265112.

Kinoshita, M., & Isa, T. (2015). Potential of optogenetics for the behavior manipulation of non-human primates. *Optogenetics: Light-Sensing Proteins and their Applications*, 271–290. Available from https://doi.org/10.1007/978-4-431-55516-2_19.

Klaes, C., Shi, Y., Kellis, S., Minxha, J., Revechkis, B., & Andersen, R. A. (2014). A cognitive neuroprosthetic that uses cortical stimulation for somatosensory feedback. *Journal of Neural Engineering*, *11*(5), 056024. Available from https://doi.org/10.1088/1741-2560/11/5/056024.

Koivuniemi, A. S., & Otto, K. J. (2011). Asymmetric versus symmetric pulses for cortical microstimulation. *IEEE Transactions on Neural Systems and Rehabilitation Engineering*, *19*(5), 468–476. Available from https://doi.org/10.1109/TNSRE.2011.2166563.

Kramer, D. R., Lee, M. B., Barbaro, M., Gogia, A. S., Peng, T., Liu, C., Kellis, S., & Lee, B. (2020). Mapping of primary somatosensory cortex of the hand area using a high-density electrocorticography grid for closed-loop brain computer interface. *Journal of Neural Engineering*. Available from https://doi.org/10.1088/1741-2552/ab7c8e.

Kumaravelu, K., Tomlinson, T., Callier, T., Sombeck, J., Bensmaia, S. J., Miller, L. E., & Grill, W. M. (2020). A comprehensive model-based framework for optimal design of biomimetic patterns of electrical stimulation for prosthetic sensation. *Journal of Neural Engineering*, *17*(4), 046045. Available from https://doi.org/10.1088/1741-2552/abacd8.

Lanz, F., Lanz, X., Scherly, A., Moret, V., Gaillard, A., Gruner, P., Hoogewoud, H.-M., Belhaj-Saif, A., Loquet, G., & Rouiller, E. M. (2013). Refined methodology for implantation of a head fixation device and chronic recording chambers in non-human primates. *Journal of Neuroscience Methods*, *219*(2), 262–270. Available from https://doi.org/10.1016/j.jneumeth.2013.07.015.

Lebedev, M. A., & Nicolelis, M. A. L. (2006). Brain-machine interfaces: Past, present and future. *Physiological Reviews*, *29*(9), 767–837. Available from https://doi.org/10.1016/j.tins.2006.07.004.

Lee, B., Kramer, D., Armenta Salas, M., Kellis, S., Brown, D., Dobreva, T., Klaes, C., Heck, C., Liu, C., & Andersen, R. A. (2018). Engineering artificial somatosensation through cortical stimulation in humans. *Frontiers in Systems Neuroscience*, *12*(June), 1–11. Available from https://doi.org/10.3389/fnsys.2018.00024.

Lieber, J. D., & Bensmaia, S. J. (2019). High-dimensional representation of texture in somatosensory cortex of primates. *Proceedings of the National Academy of Sciences*. Available from https://doi.org/10.1073/pnas.1818501116.

Lieber, J. D., Xia, X., Weber, A. I., & Bensmaia, S. J. (2017). The neural code for tactile roughness in the somatosensory nerves. *Journal of Neurophysiology*, *118*(6), 3107–3117.

London, B. M., Jordan, L. R., Jackson, C. R., & Miller, L. E. (2008). Electrical stimulation of the proprioceptive cortex (area 3a) used to instruct a behaving monkey. *IEEE Transactions on Neural Systems and Rehabilitation Engineering: A Publication of the IEEE Engineering in Medicine and Biology Society*, *16*(1), 32–36. Available from https://doi.org/10.1109/TNSRE.2007.907544.

London, B. M., & Miller, L. E. (2013). Responses of somatosensory area 2 neurons to actively and passively generated limb movements. *Journal of Neurophysiology*, *109*(6), 1505–1513. Available from https://doi.org/10.1152/jn.00372.2012.

Lucier, G. E., Rüegg, D. C., & Wiesendanger, M. (1975). Responses of neurones in motor cortex and in area 3A to controlled stretches of forelimb muscles in cebus monkeys. *The Journal of Physiology*, *251*(3), 833–853.

Markram, H., Muller, E., Ramaswamy, S., Reimann, M. W., Abdellah, M., Sanchez, C. A., Ailamaki, A., Alonso-Nanclares, L., Antille, N., Arsever, S., Kahou, G. A. A., Berger, T. K., Bilgili, A., Buncic, N., Chalimourda, A., Chindemi, G., Courcol, J.-D., Delalondre, F.,

Delattre, V., ... Schürmann, F. (2015). Reconstruction and simulation of neocortical microcircuitry. *Cell*, *163*(2), 456–492. Available from https://doi.org/10.1016/j.cell.2015.09.029.

Marshall, W. H., Woolsey, C. N., & Bard, P. (1937). Cortical representation of tactile sensibility as indicated by cortical potentials. *Science (New York, N.Y.)*, *85*(2207), 388–390.

Mathis, A., Mamidanna, P., Cury, K. M., Abe, T., Murthy, V. N., Mathis, M. W., & Bethge, M. (2018). DeepLabCut: Markerless pose estimation of user-defined body parts with deep learning. *Nature Neuroscience*, *21*(9), 1281–1289. Available from https://doi.org/10.1038/s41593-018-0209-y.

Maynard, E. M., Nordhausen, C. T., & Normann, R. A. (1997). The Utah intracortical electrode array: A recording structure for potential brain-computer interfaces. *Electroencephalography and Clinical Neurophysiology*, *102*(3), 228–239. Available from https://doi.org/10.1016/S0013-4694(96)95176-0.

Mcintyre, C. C., & Foutz, T. J. (2013). Computational modeling of deep brain stimulation. *Handbook of Clinical Neurology*, *116*, 55. Available from https://doi.org/10.1016/B978-0-444-53497-2.00005-X.

McIntyre, C. C., & Grill, W. M. (2000). Selective microstimulation of central nervous system neurons. *Annals of Biomedical Engineering*, *28*(3), 219–233. Available from https://doi.org/10.1114/1.262.

McIntyre, C. C., & Grill, W. M. (2002). Extracellular stimulation of central neurons: Influence of stimulus waveform and frequency on neuronal output. *Journal of Neurophysiology*, *88*(4), 1592–1604.

Mendoza, S. D., El-Shamayleh, Y., & Horwitz, G. D. (2017). AAV-mediated delivery of optogenetic constructs to the macaque brain triggers humoral immune responses. *Journal of Neurophysiology*, *117*(5), 2004–2013. Available from https://doi.org/10.1152/jn.00780.2016.

Mitchell, A. S., Thiele, A., Petkov, C. I., Roberts, A., Robbins, T. W., Schultz, W., & Lemon, R. (2018). Continued need for non-human primate neuroscience research. *Current Biology*, *28* (20), R1186–R1187.

Mori, M. (1970). The uncanny valley. *Energy*. <http://www.movingimages.info/digitalmedia/wp-content/uploads/2010/06/MorUnc.pdf>.

Moxon, K. A., Leiser, S. C., Gerhardt, G. A., Barbee, K. A., & Chapin, J. K. (2004). Ceramic-based multisite electrode arrays for chronic single-neuron recording. *IEEE Transactions on Biomedical Engineering*, *51*(4), 647–656. Available from https://doi.org/10.1109/TBME.2003.821037.

Musk, E. (2019). An integrated brain-machine interface platform with thousands of channels. *bioRxiv*. Available from https://www.biorxiv.org/content/10.1101/703801v4.

Nath, T., Mathis, A., Chen, A. C., Patel, A., Bethge, M., & Mathis, M. W. (2019). Using DeepLabCut for 3D markerless pose estimation across species and behaviors. *Nature Protocols*, *14*(7), 2152–2176. Available from https://doi.org/10.1038/s41596-019-0176-0.

Neafsey, E. J., Bold, E. L., Haas, G., Hurley-Gius, K. M., Quirk, G., Sievert, C. F., & Terreberry, R. R. (1986). The organization of the rat motor cortex: A microstimulation mapping study. *Brain Research Reviews*, *11*(1), 77–96. Available from https://doi.org/10.1016/0165-0173(86)90011-1.

Neafsey, E. J., & Sievert, C. (1982). A second forelimb motor area exists in rat frontal cortex. *Brain Research*, *232*(1), 151–156. Available from https://doi.org/10.1016/0006-8993(82)90617-5.

Neely, R. M., Piech, D. K., Santacruz, S. R., Maharbiz, M. M., & Carmena, J. M. (2018). Recent advances in neural dust: Towards a neural interface platform. *Current Opinion in Neurobiology*, *50*, 64–71. Available from https://doi.org/10.1016/j.conb.2017.12.010.

Nicolelis, M. A. L. (2003). Chronic, multisite, multielectrode recordings in macaque monkeys. *Proceedings of the National Academy of Sciences of the United States of America*, *100*(19), 11041–11046. Available from https://doi.org/10.1073/pnas.1934665100.

Nordhausen, C. T., Rousche, P. J., & Normann, R. A. (1994). Optimizing recording capabilities of the Utah intracortical electrode array. *Brain Research*, *637*(1), 27–36. Available from https://doi.org/10.1016/0006-8993(94)91213-0.

O'Doherty, J. E., Lebedev, M. A., Hanson, T. L., Fitzsimmons, N. A., & Nicolelis, M. A. L. (2009). A brain-machine interface instructed by direct intracortical microstimulation. *Frontiers in Integrative Neuroscience*, *3*. Available from https://doi.org/10.3389/neuro.07.020.2009.

O'Doherty, J. E., Lebedev, M. A., Ifft, P. J., Zhuang, K. Z., Shokur, S., Bleuler, H., & Nicolelis, M. A. L. (2011). Active tactile exploration using a brain–machine–brain interface. *Nature*, *479*(7372), 228–231. Available from https://doi.org/10.1038/nature10489.

O'Doherty, J. E., Lebedev, M. A., Li, Z., & Nicolelis, M. A. L. (2012). Virtual active touch using randomly patterned intracortical microstimulation. *IEEE Transactions on Neural Systems and Rehabilitation Engineering*. Available from https://doi.org/10.1109/TNSRE.2011.2166807.

O'Doherty, J. E., Shokur, S., Medina, L. E., Lebedev, M. A., & Nicolelis, M. A. L. (2019). Creating a neuroprosthesis for active tactile exploration of textures. *Proceedings of the National Academy of Sciences*, *116*(43), 21821–21827. Available from https://doi.org/10.1073/pnas.1908008116.

Ogier, R., Knecht, W., & Schwab, M. E. (2019). Academic leadership: (With)holding the keys to translational medicine? *Nature Medicine*, *25*(12), 1812–1813.

Ohara, S., Weiss, N., & Lenz, F. A. (2004). Microstimulation in the region of the human thalamic principal somatic sensory nucleus evokes sensations like those of mechanical stimulation and movement. *Journal of Neurophysiology*, *91*(2), 736–745. Available from https://doi.org/10.1152/jn.00648.2003.

Ong, L. J., Liu, S.-C., Sheng Wong, M. D., Sibindi, T., Lasam Gammad, G. G., Tsai, C.-W., Rusly, A., Ng, K. A., Libedinsky, C., Nag, S., & Yen, S.-C. (2018). A fully wireless implantable multi-channel muscle stimulator with closed-loop feedback control. *2018 IEEE Biomedical Circuits and Systems Conference (BioCAS)*, 1–4. Available from https://doi.org/10.1109/BIOCAS.2018.8584753.

Opie, N. L., John, S. E., Gerboni, G., Rind, G. S., Lovell, T. J. H., Ronayne, S. M., Wong, Y. T., May, C. N., Grayden, D. B., & Oxley, T. J. (2019). Neural stimulation with an endovascular brain-machine interface. *2019 9th International IEEE/EMBS Conference on Neural Engineering (NER)*, 738–741. Available from https://doi.org/10.1109/NER.2019.8717066.

Ortiz-Rios, M., Haag, M., Balezeau, F., Frey, S., Thiele, A., Murphy, K., & Schmid, M. C. (2018). Improved methods for MRI-compatible implants in nonhuman primates. *Journal of Neuroscience Methods*, *308*, 377–389. Available from https://doi.org/10.1016/j.jneumeth.2018.09.013.

Overstreet, C. K., Klein, J. D., & Helms Tillery, S. I. (2013). Computational modeling of direct neuronal recruitment during intracortical microstimulation in somatosensory cortex. *Journal of Neural Engineering*, *10*(6), 066016. Available from https://doi.org/10.1088/1741-2560/10/6/066016.

Overton, J. A., Cooke, D. F., Goldring, A. B., Lucero, S. A., Weatherford, C., & Recanzone, G. H. (2017). Improved methods for acrylic-free implants in nonhuman primates for neuroscience research. *Journal of Neurophysiology*, *118*(6), 3252–3270. Available from https://doi.org/10.1152/jn.00191.2017.

Patel, S., Ohara, S., Dougherty, P. M., Gracely, R. H., & Lenz, F. A. (2006). Psychophysical elements of place and modality specificity in the thalamic somatic sensory nucleus (ventral caudal, vc) of awake humans. *Journal of Neurophysiology*, *95*(2), 646–659. Available from https://doi.org/10.1152/jn.00756.2005.

Payne, S. C., Burns, O., Stebbing, M., Thomas, R., de Silva, A., Sedo, A., Weissenborn, F., Hyakumura, T., Huynh, M., May, C. N., Williams, R. A., Furness, J. B., Fallon, J. B., & Shepherd, R. K. (2018). Vagus nerve stimulation to treat inflammatory bowel disease: A chronic, preclinical safety study in sheep. *Bioelectronics in Medicine*, *1*(4), 235–250. Available from https://doi.org/10.2217/bem-2018-0011.

Petrini, F. M., Bumbasirevic, M., Valle, G., Ilic, V., Mijović, P., Čvančara, P., Barberi, F., Katic, N., Bortolotti, D., Andreu, D., Lechler, K., Lesic, A., Mazic, S., Mijović, B., Guiraud, D., Stieglitz, T., Alexandersson, A., Micera, S., & Raspopovic, S. (2019). Sensory feedback restoration in leg amputees improves walking speed, metabolic cost and phantom pain. *Nature Medicine*, *25*(9), 1356–1363. Available from https://doi.org/10.1038/s41591-019-0567-3.

Pons, T. P., Garraghty, P. E., Cusick, C. G., & Kaas, J. H. (1985). The somatotopic organization of area 2 in macaque monkeys. *Journal of Comparative Neurology*, *241*(4), 445–466.

Pons, T. P., Garraghty, P. E., Friedman, D. P., & Mishkin, M. (1987). Physiological evidence for serial processing in somatosensory cortex. *Science (New York, N.Y.)*, *237*(4813), 417–420.

Pruszynski, J. A. (2014). Primary motor cortex and fast feedback responses to mechanical perturbations: A primer on what we know now and some suggestions on what we should find out next. *Frontiers in Integrative Neuroscience*, *8*. Available from https://doi.org/10.3389/fnint.2014.00072.

Pruszynski, J. A., Kurtzer, I., Nashed, J. Y., Omrani, M., Brouwer, B., & Scott, S. H. (2011). Primary motor cortex underlies multi-joint integration for fast feedback control. *Nature*, *478*(7369), 387–390. Available from https://doi.org/10.1038/nature10436.

Putrino, D., Wong, Y. T., Weiss, A., & Pesaran, B. (2015). A training platform for many-dimensional prosthetic devices using a virtual reality environment. *Journal of Neuroscience Methods*, *244*, 68–77. Available from https://doi.org/10.1016/j.jneumeth.2014.03.010.

Rajan, A. T., Boback, J. L., Dammann, J. F., Tenore, F. V., Wester, B. A., Otto, K. J., Gaunt, R. A., & Bensmaia, S. J. (2015). The effects of chronic intracortical microstimulation on neural tissue and fine motor behavior. *Journal of Neural Engineering*, *12*(6), 066018. Available from https://doi.org/10.1088/1741-2560/12/6/066018.

Raspopovic, S., Capogrosso, M., Badia, J., Navarro, X., & Micera, S. (2012). Experimental validation of a hybrid computational model for selective stimulation using transverse intrafascicular multichannel electrodes. *IEEE Transactions on Neural Systems and Rehabilitation Engineering*, *20*(3), 395–404. Available from https://doi.org/10.1109/TNSRE.2012.2189021.

Richardson, A. G., Attiah, M. A., Berman, J. I., Chen, H. I., Liu, X., Zhang, M., Van der Spiegel, J., & Lucas, T. H. (2016). The effects of acute cortical somatosensory deafferentation on grip force control. *Cortex; a Journal Devoted to the Study of the Nervous System and Behavior*, *74*, 1–8. Available from https://doi.org/10.1016/j.cortex.2015.10.007.

Roelfsema, P. R., & Treue, S. (2014). Basic neuroscience research with nonhuman primates: A small but indispensable component of biomedical research. *Neuron*, *82*(6), 1200–1204.

Rogers, J., & Gibbs, R. A. (2014). Comparative primate genomics: Emerging patterns of genome content and dynamics. *Nature Publishing Group*, *15*(5), 347–359.

Romo, R., Hernández, A., Zainos, A., & Salinas, E. (1998). Somatosensory discrimination based on cortical microstimulation. *Nature*, *392*(6674), 387–390. Available from https://doi.org/10.1038/32891.

Rothwell, J. C., Traub, M. M., Day, B. L., Obeso, J. A., Thomas, P. K., & Marsden, C. D. (1982). Manual motor performance in a deafferented man. *Brain*, *105*(3), 515–542. Available from https://doi.org/10.1093/brain/105.3.515.

Rouiller, E. M., Moret, V., & Liang, F. (1993). Comparison of the connectional properties of the two forelimb areas of the rat sensorimotor cortex: Support for the presence of a premotor or supplementary motor cortical area. *Somatosensory & Motor Research*, *10*(3), 269–289. Available from https://doi.org/10.3109/08990229309028837.

Ryu, S. I., & Shenoy, K. V. (2009). Human cortical prostheses: Lost in translation? *Neurosurgical Focus*, *27*(1), E5. Available from https://doi.org/10.3171/2009.4.FOCUS0987.

Saal, H. P., Delhaye, B. P., Rayhaun, B. C., & Bensmaia, S. J. (2017). Simulating tactile signals from the whole hand with millisecond precision. *Proceedings of the National Academy of Sciences*, *114*(28), E5693–E5702. Available from https://doi.org/10.1073/pnas.1704856114.

Salas, M. A., Bashford, L., Kellis, S., Jafari, M., Jo, H., Kramer, D., Shanfield, K., Pejsa, K., Lee, B., Liu, C. Y., & Andersen, R. A. (2018). Proprioceptive and cutaneous sensations in humans elicited by intracortical microstimulation. *ELife*, *7*, 1–11. Available from https://doi.org/10.7554/eLife.32904.

Schwarz, D. A., Lebedev, M. A., Hanson, T. L., Dimitrov, D. F., Lehew, G., Meloy, J., Rajangam, S., Subramanian, V., Ifft, P. J., Li, Z., Ramakrishnan, A., Tate, A., Zhuang, K. Z., & Nicolelis, M. A. L. (2014). Chronic, wireless recordings of large-scale brain activity in freely moving rhesus monkeys. *Nature Methods*, *11*(6), 670–676. Available from https://doi.org/10.1038/nmeth.2936.

Scott, S. H. (1999). Apparatus for measuring and perturbing shoulder and elbow joint positions and torques during reaching. *Journal of Neuroscience Methods*, *89*(2), 119–127. Available from https://doi.org/10.1016/s0165-0270(99)00053-9.

Seki, K., & Fetz, E. E. (2012). Gating of sensory input at spinal and cortical levels during preparation and execution of voluntary movement. *Journal of Neuroscience*, *32*(3), 890–902. Available from https://doi.org/10.1523/JNEUROSCI.4958-11.2012.

Shoaran, M., Kamal, M. H., Pollo, C., Vandergheynst, P., & Schmid, A. (2014). Compact low-power cortical recording architecture for compressive multichannel data acquisition. *IEEE Transactions on Biomedical Circuits and Systems*, *8*(6), 857–870. Available from https://doi.org/10.1109/TBCAS.2014.2304582.

Shokur, S., Gallo, S., Moioli, R. C., Donati, A. R. C., Morya, E., Bleuler, H., & Nicolelis, M. A. L. (2016). Assimilation of virtual legs and perception of floor texture by complete paraplegic patients receiving artificial tactile feedback. *Scientific Reports*, *6*(August). Available from https://doi.org/10.1038/srep32293.

Shokur, S., O'Doherty, J. E., Winans, J. A., Bleuler, H., Lebedev, M. A., & Nicolelis, M. A. L. (2013). Expanding the primate body schema in sensorimotor cortex by virtual touches of an avatar. *Proceedings of the National Academy of Sciences of the United States of America*, *110*(37), 15121–15126. Available from https://doi.org/10.1073/pnas.1308459110.

Spencer, K. C., Sy, J. C., Ramadi, K. B., Graybiel, A. M., Langer, R., & Cima, M. J. (2017). Characterization of mechanically matched hydrogel coatings to improve the biocompatibility of neural implants. *Scientific Reports*, *7*(1), 1952. Available from https://doi.org/10.1038/s41598-017-02107-2.

Steckenfinger, S. A., & Ghazanfar, A. A. (2009). Monkey visual behavior falls into the uncanny valley. *Proceedings of the National Academy of Sciences of the United States of America*, *106*(43), 18362–18366.

Stoney, S. D., Thompson, W. D., & Asanuma, H. (1968). Excitation of pyramidal tract cells by intracortical microstimulation: Effective extent of stimulating current. *Journal of Neurophysiology*, *31*(5), 659–669. Available from https://doi.org/10.1152/jn.1968.31.5.659.

Suminski, A. J., Tkach, D. C., Fagg, A. H., & Hatsopoulos, N. G. (2010). Incorporating feedback from multiple sensory modalities enhances brain–machine interface control. *Journal of Neuroscience*, *30*(50), 16777–16787. Available from https://doi.org/10.1523/JNEUROSCI.3967-10.2010.

Suner, S., Fellows, M. R., Vargas-Irwin, C., Nakata, G. K., & Donoghue, J. P. (2005). Reliability of signals from a chronically implanted, silicon-based electrode array in non-human primate primary motor cortex. *IEEE Transactions on Neural Systems and Rehabilitation Engineering*, *13*(4), 524–541. Available from https://doi.org/10.1109/TNSRE.2005.857687.

Szostak, K. M., Grand, L., & Constandinou, T. G. (2017). Neural interfaces for intracortical recording: Requirements, fabrication methods, and characteristics. *Frontiers in Neuroscience*, *11*. Available from https://doi.org/10.3389/fnins.2017.00665.

Tabot, G. A., Dammann, J. F., Berg, J. A., Tenore, F. V., Boback, J. L., Vogelstein, R. J., & Bensmaia, S. J. (2013). Restoring the sense of touch with a prosthetic hand through a brain interface. *Proceedings of the National Academy of Sciences of the United States of America, 110*(45), 18279–18284. Available from https://doi.org/10.1073/pnas.1221113110.

Talwar, S. K., Xu, S., Hawley, E. S., Weiss, S. A., Moxon, K. A., & Chapin, J. K. (2002). Rat navigation guided by remote control. *Nature, 417*(6884), 37–38. Available from https://doi.org/10.1038/417037a.

Tanosaki, M., Ishibashi, H., Zhang, T., & Okada, Y. (2014). Effective connectivity maps in the swine somatosensory cortex estimated from electrocorticography and validated with intracortical local field potential measurements. *Brain Connectivity, 4*(2), 100–111. Available from https://doi.org/10.1089/brain.2013.0177.

Tehovnik, E. J., Tolias, A. S., Sultan, F., Slocum, W. M., & Logothetis, N. K. (2006). Direct and indirect activation of cortical neurons by electrical microstimulation. *Journal of Neurophysiology, 96*(2), 512–521. Available from https://doi.org/10.1152/jn.00126.2006.

Vaidya, M., Dickey, A., Best, M.D., Coles, J., Balasubramanian, K., Suminski, A.J., & Hatsopoulos, N. G. (2014). Ultra-long term stability of single units using chronically implanted multielectrode arrays. In *Conference Proceedings: ... Annual International Conference of the IEEE Engineering in Medicine and Biology Society* (pp. 4872–4875). https://doi.org/10.1109/EMBC.2014.6944715

Wang, A., McAllister, J. P., Finlayson, P., Li, J., Brabant, K., Tang, H., Black, C., Cao, T., Liang, X., Salley, S. O., Auner, G. W., & Ng, K. Y. S. (2007). Short-and long-term neural biocompatibility of heparin coated sapphire implants. *Materials Science and Engineering: C, 27*(2), 237–243. Available from https://doi.org/10.1016/j.msec.2006.05.011.

Wang, Q., Millard, D. C., Zheng, H. J. V., & Stanley, G. B. (2012). Voltage sensitive dye imaging reveals improved topographic activation of cortex in response to manipulation of thalamic microstimulation parameters. *Journal of Neural Engineering, 9*(2), 026008. Available from https://doi.org/10.1088/1741-2560/9/2/026008.

Weber, A. I., Saal, H. P., Lieber, J. D., Cheng, J.-W., Manfredi, L. R., Dammann, J. F., & Bensmaia, S. J. (2013). Spatial and temporal codes mediate the tactile perception of natural textures. *Proceedings of the National Academy of Sciences of the United States of America, 110*(42), 17107–17112.

Yamada, H., Yaguchi, H., Tomatsu, S., Takei, T., Oya, T., & Seki, K. (2016). Representation of afferent signals from forearm muscle and cutaneous nerves in the primary somatosensory cortex of the macaque monkey. *PLoS One, 11*(10), e0163948.

Yin, M., Borton, D. A., Komar, J., Agha, N., Lu, Y., Li, H., Laurens, J., Lang, Y., Li, Q., Bull, C., Larson, L., Rosler, D., Bezard, E., Courtine, G., Nurmikko. A. V. (2014). Wireless neurosensor for full-spectrum electrophysiology recordings during free behavior. *Neuron, 84*, 1170–1182

Zaaimi, B., Ruiz-Torres, R., Solla, S. A., & Miller, L. E. (2013). Multi-electrode stimulation in somatosensory cortex increases probability of detection. *Journal of Neural Engineering, 10*(5), 056013. Available from https://doi.org/10.1088/1741-2560/10/5/056013.

Zanos, S., Richardson, A. G., Shupe, L., Miles, F. P., & Fetz, E. E. (2011). The neurochip-2: An autonomous head-fixed computer for recording and stimulating in freely behaving monkeys. *IEEE Transactions on Neural Systems and Rehabilitation Engineering, 19*(4), 427–435. Available from https://doi.org/10.1109/TNSRE.2011.2158007.

Zhou, A., Santacruz, S. R., Johnson, B. C., Alexandrov, G., Moin, A., Burghardt, F. L., Rabaey, J. M., Carmena, J. M., & Muller, R. (2019). A wireless and artefact-free 128-channel neuromodulation device for closed-loop stimulation and recording in non-human primates. *Nature Biomedical Engineering, 3*(1), 15–26. Available from https://doi.org/10.1038/s41551-018-0323-x.

Chapter 14

Touch restoration through electrical cortical stimulation in humans

David J. Caldwell[1,2,3], Jeneva A. Cronin[1,2], Lila H. Levinson[1,4] and Rajesh P.N. Rao[1,5]

[1]Center for Neurotechnology, University of Washington, Seattle, WA, United States, [2]Department of Bioengineering, University of Washington, Seattle, WA, United States, [3]Medical Scientist Training Program, University of Washington, Seattle, WA, United States, [4]Graduate Program in Neuroscience, University of Washington, Seattle, WA, United States, [5]Paul G. Allen School of Computer Science & Engineering, University of Washington, Seattle, WA, United States

ABSTRACT

Direct cortical stimulation (DCS) has been used extensively as a tool in human neuroscience research and clinically in neurosurgery. Implanted electrocorticography electrodes can deliver electrical stimulation to the cortical surface, eliciting or inhibiting neural activity as an experimental manipulation, or to probe neural tissue function during epilepsy and tumor resection surgeries. Current applications of DCS pave the way for new clinical translation to restore sensation to those who have lost it through spinal-cord injury or stroke. DCS of specific sensory cortical regions in the human brain elicits sensory percepts in specific regions of the body and recent research has shown that DCS can elicit meaningful somatosensory percepts, which may be used for neuroprosthetic feedback and integrated into tasks. In the future, further research using more finely spaced electrodes, tailored patterns of stimulation, and longer term studies will shed light on the intricacies of sensory processing in humans as well as bring these neuroprosthetic applications closer to widespread clinical reality.

Keywords: Direct electrical stimulation; direct cortical stimulation; brain–computer interface; brain–machine interface; sensory feedback; somatosensory feedback; neuroprosthetic; haptics; sensory restoration

14.1 Introduction

Electrocorticographic brain–computer interfaces (ECoG-BCIs) have enormous potential for the restoration of function in people affected by neurological

Somatosensory Feedback for Neuroprosthetics. DOI: https://doi.org/10.1016/B978-0-12-822828-9.00021-6

damage, as well as for increasing our understanding of how the brain works. ECoG electrodes are used across the world in the clinical setting for the monitoring of epilepsy and mapping of brain function prior to tumor resection, and have the advantages of more focally recording neural activity when compared to noninvasive electroencephalography (EEG), greater temporal resolution than functional magnetic resonance imaging (fMRI), and, unlike current technologies for magnetoencephalography, the potential for wireless out-of-laboratory recording. Thus far, most research has focused on using the neural signals recorded from ECoG electrodes to control an end effector such as a robotic arm (Hochberg et al., 2012). Just as important for an ECoG-BCI is inputting information into the brain using DCS (Caldwell, Ojemann, et al., 2019). DCS, which describes an electrical current being applied directly to the surface of the cortex through implanted ECoG electrodes, has been utilized for mapping brain function in a clinical setting as well as for better understanding the human sensory system. The combination of DCS with ECoG recording enables a bidirectional brain–computer interface (BBCI) with both input to and output from the brain.

Although researchers have yet to demonstrate a fully functional BBCI in humans, progress has been made toward closing the BCI feedback loop via electrical stimulation. At the cortical level, artificial somatosensory feedback has been delivered to humans via DCS of primary somatosensory cortex (S1) through macro-ECoG electrodes (with exposed contacts of approximately 2.3–3 mm in diameter with 10 mm spacing from center to center (Chang, 2015; Collins et al., 2017; Cronin et al., 2016; Johnson et al., 2013; Libet et al., 1964; Ray et al., 1999), and more recently via DCS through mini- or micro-ECoG electrodes (with exposed contacts of approximately 1–1.5 mm diameter with 3–4 mm spacing (Caldwell, Ojemann, et al., 2019; Hiremath et al., 2017; Kramer et al., 2020; Lee et al., 2018; Muller et al., 2018). Electrical stimulation through all of these modalities has demonstrated that subjects can experience and discriminate percepts with varying intensity and qualia based on the stimulation parameters (Armenta Salas et al., 2018; Flesher et al., 2016; Hiremath et al., 2017; Johnson et al., 2013; Lee et al., 2018), localize stimulation from different electrode pairs to different areas of the hand and arm (Armenta Salas et al., 2018; Collins et al., 2017; Flesher et al., 2016; Hiremath et al., 2017; Lee et al., 2018), and use the somatosensory feedback to perform a motor-based task (Cronin et al., 2016; Lee et al., 2018). We discuss the advantages and opportunities, as well as the barriers and challenges presented by using DCS in an ECoG-BCI, physiology of electrical stimulation through DCS, provide examples of ECoG DCS applications, and discuss future directions for ECoG DCS.

14.1.1 Advantages of cortical stimulation

For a BBCI, sensory feedback can be provided using electrical stimulation methods via both peripheral interfaces and cortical interfaces. These methods include peripheral nerve stimulation (Dhillon & Horch, 2005; Raspopovic et al., 2014; Tan et al., 2014), targeted muscle reinnervation (Hebert et al., 2014; Kuiken et al., 2007), intracortical microstimulation (ICMS) (Bensmaia & Miller, 2014; Berg et al., 2013; Klaes et al., 2014; O'Doherty et al., 2011), and DCS via macro- and micro-ECoG (Cronin et al., 2016; Hiremath et al., 2017; Johnson et al., 2013; Libet et al., 1964; Ray et al., 1999).

While peripheral nerve interfaces (Dhillon & Horch, 2005; Raspopovic et al., 2014; Tan et al., 2014) and targeted muscle reinnervation (Hebert et al., 2014; Kuiken et al., 2007) have shown success in providing sensory feedback to human users with upper limb amputations, peripheral interventions would not work in someone with a spinal cord injury or other sensorimotor disorder of the central nervous system. In the United States, as of 2013, there were an estimated 5.4 million people with paralysis (Armour et al., 2016). A total of approximately 1.8 million of these individuals had deficits due to stroke, while 1.5 million were due to spinal cord injury. For individuals who survived a stroke, 15%–30% are permanently disabled, with only approximately 50%–70% reaching functional independence (Lloyd-Jones et al., 2010). For those individuals, cortical interventions, such as ICMS (delivered via penetrating intracortical microelectrodes) or DCS are needed.

Of the three cortical-level interfaces (macro-ECoG, micro-ECoG, and intracortical microelectrodes for ICMS) only macro-ECoG is regularly implanted for clinical purposes. Unlike ICMS and micro-ECoG sensory stimulation research, which have only been conducted on a handful of human subjects thus far, macro-ECoG sensory stimulation research can be conducted on many more subjects who consent to research after a macro-ECoG grid is implanted for clinical purposes (see Section 14.1.2).

In addition to the greater prevalence of human macro-ECoG subjects, subdural ECoG grids may also offer some benefits in long-term implantation over penetrating intracortical electrodes, including stability and less invasive implantation (Moran, 2010). Implantation of intracortical electrodes can trigger an inflammatory response, recruiting microglia and astrocytes to the site and encapsulating the probe (Hermann & Capadona, 2018). Surface electrodes, such as those used in DCS, do not evoke a strong inflammatory response because they do not penetrate the cortex and have been demonstrated in recordings lasting up to 766 days in humans (Nurse et al., 2018). Longer term recordings using ECoG electrodes in humans may be possible, but generally are not done for clinical purposes, so the available research on this point is limited.

14.1.2 Current clinical uses of direct cortical stimulation

ECoG is commonly used as a clinical tool to stimulate and record neural activity to delineate different epileptogenic zones and functional areas for the treatment of epilepsy (Enatsu & Mikuni, 2016), and during neurosurgery to minimize the impact to areas that are critical for functions such as language (Berger & Ojemann, 1992; Hill et al., 2012). Passive recording of neural activity, where seizure events are captured via intracranial electrodes, is critical for helping identify the origin of seizures prior to surgical resection. Surgical resection for appropriately chosen patients can vastly decrease the frequency of epileptic events for individuals, with approximately half of patients with focal neocortical epilepsy and two thirds of patients with mesial temporal lobe epilepsy seeing significantly improved seizure control (Englot & Chang, 2014). In addition to passive recordings, stimulation mapping using the implanted ECoG electrodes can be performed both at the bedside and in the operating room. While patients are asleep, stimulation over motor cortex can be performed to induce motor-evoked potentials which are measured via electromyography (EMG) in peripheral muscles. While patients are awake, they can perform language, motor, and memory tasks concurrently with stimulation. If function in one of these areas is disrupted by stimulation, this helps identify the eloquent cortex. Preserving eloquent cortex, which is critical for patients' motor and language function, is an important goal during these surgeries as it maximizes the quality of life for patients. These mapping strategies are also important during tumor resections, where the desired outcome is maximal resection of tumor while still preserving function.

Research on ECoG BCIs is often conducted with subjects who consent to research after an ECoG grid is implanted for clinical purposes. These subjects are often undergoing clinical monitoring of intractable epilepsy and are primarily implanted with grids or strips of platinum subdural ECoG electrodes with a 2.3 mm exposed diameter and 10 mm center-to-center spacing (Fig. 14.1). Current clinically used ECoG electrodes are usually embedded in a silicone sheet and are made of platinum or stainless steel. This offers a larger pool of human subjects than is typical in other cortical implants for BCIs and somatosensory feedback which require elective neurosurgery. However, as these clinical subjects only remain in the hospital with the implanted ECoG grid for approximately 7 days while clinicians monitor them, the duration of research experiments conducted with these subjects is necessarily shorter than those who are solely involved for research and longer term experiments.

14.1.3 History of direct cortical stimulation

DCS was first applied to a living human subject as early as 1874 by Robert Bartholow who elicited muscle contractions in his patient. By the early 1900s, DCS had been used to localize epileptic foci and map brain regions (Cushing, 1909), and in 1934 an intraoperative technique for recording from and stimulating through ECoG electrodes was developed (Borchers et al., 2011). In the 1940s and later, Wilder Penfield and colleagues used DCS to create motor and sensory maps in human patients, while others explored if DCS could be used to provide feedback for a visual prosthesis (Brindley & Lewin, 1968; Dobelle & Mladejovsky, 1974; Penfield & Boldrey, 1937).

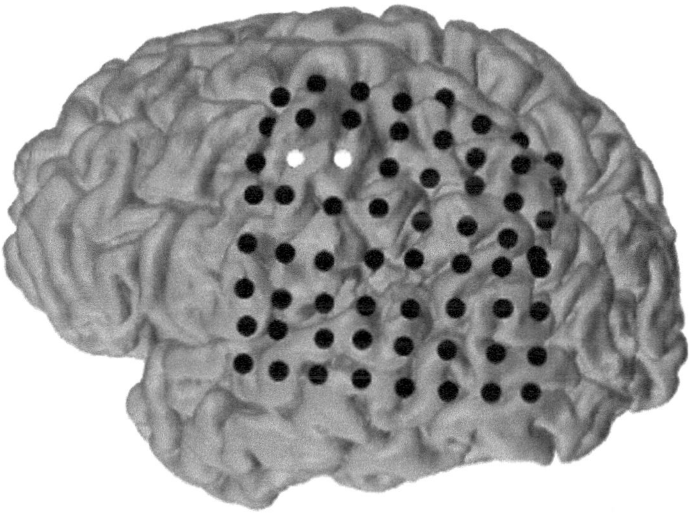

FIGURE 14.1 Example of a grid of macro-electrocorticography (ECoG) electrodes. This left-hemisphere ECoG grid was implanted to clinically monitor intractable epilepsy in a patient for about one week. The subject consented to participate in research with our group, and the two electrodes that were used for bipolar stimulation of hand primary somatosensory cortex (S1) are highlighted in *white*. *© 2016 IEEE. Reprinted, with permission, from Cronin, J. A., Wu, J., Collins, K. L., Sarma, D., Rao, R. P. N., Ojemann, J. G., & Olson, J. D. (2016). Task-specific somatosensory feedback via cortical stimulation in humans.* IEEE Transactions on Haptics, 9*(4), 515–522. https://doi.org/10.1109/TOH.2016.2591952.*

14.1.4 Direct cortical stimulation and perception in humans

The conscious perception of DCS and the resulting sensory percepts were initially explored in 1964 by Libet et al. (1964). Many more studies have followed, both through traditional macro-ECoG arrays, as well as higher density mini- or micro-ECoG arrays, which allow for spatially tuned stimulation and recording. Macro-ECoG arrays frequently have an electrode spacing of approximately 1 cm, which allows for wide spatial coverage with an array of 64 electrodes, but at the cost of finely tuned spatial decoding of cortical activity. High-density or micro-ECoG electrodes often have similarly sized exposed surfaces of approximately 2 mm with a center-to-center spacing of several millimeters.

In our group's experience, macro-ECoG DCS of sensory areas has elicited abstract sensations sometimes described as "vibration," "buzzing," "pressure," and "wind running down the hand" (Cronin et al., 2016; Johnson et al., 2013); research with mini- or micro-ECoG grids has also described abstract sensations such as "tingling," "electric buzz," "trembling," and "itching" (Hiremath et al., 2017; Lee et al., 2018; Table 14.1). It is not surprising that DCS elicits abstract sensations as we may be affecting up to

TABLE 14.1 Percepts elicited from direct cortical stimulation through electrocorticographic electrodes.

Percept	ECoG grid	Paper(s)
Vibration Buzzing Pressure Wind running down the hand Tingling Pulse Jolt Zap	Macro-ECoG	Cronin et al. (2016), Johnson et al. (2013); unpublished work from this group
Tingling Electric buzz	Mini/micro-ECoG	Hiremath et al. (2017)
Tingling Tickling Buzzing Electricity Soft Trembling Itching Pulsing Shock Light tapping	Mini/micro-ECoG	Lee et al. (2018)

ECoG, Electrocorticography.

500,000 neurons in a nonbiomimetic fashion [this number of neurons stems from an estimation of the neuronal density in nonhuman primates (NHPs) (Collins et al., 2010) and the ECoG electrode surface area].

Efficacious BBCIs will need to provide sensory feedback that users can employ to facilitate coherent behavioral responses. As many natural tactile experiences are graded, varied, and time-bound, to adequately substitute for natural sensation sensory stimulation will need to enable users to quickly perceive and distinguish a large number of unique percepts elicited by varied stimulation waveforms.

14.2 Stimulation physiology

14.2.1 Sensory processing physiology

Once ascending somatosensory signals reach the ventral posterior nucleus of the thalamus, they are sent to cortical areas responsible for sensory perception. In S1, these include Brodmann areas (BA) 3a, 3b, 1, and 2. Area 3b receives dense inputs from the thalamus, with responses to cutaneous touch afferents, and is therefore thought to be important in cutaneous perception, while proprioceptive information travels to areas 3a and 2 (Ackerley & Kavounoudias, 2015). Areas 1 and 2 also receive tactile cutaneous input. Throughout S1, the particular sensory images encoded by the touch receptors in the skin are maintained, but as information flows to higher order cortical areas, it is further abstracted. For instance, secondary somatosensory cortex, which is located in the parietal operculum, and posterior parietal cortex, use sensory information to inform cognitive (what object is this?) and motor activities (how should I manipulate this object?). Additional cortical areas activated during touch include BA5 and BA7 in the posterior parietal cortex and insular cortex, which are important for integration and multisensory processing (Ackerley & Kavounoudias, 2015). Orbitofrontal cortex, important for reward and emotion pathways, is also activated during touch of the hand, highlighting the multifaceted downstream processing of touch (Ackerley & Kavounoudias, 2015). In summary, the cortical circuitry required for effective sensory function is multifaceted and spans many cortical regions.

14.2.2 Activation of the tactile sensory system via electrical stimulation

The mechanisms of how electrical stimulation causes sensory percepts is not entirely understood. Electrical stimulation, depending on the parameters used and the location of stimulation, can cause both negative and positive effects on local neural activity. How many cells and what types of cells are influenced by electrical stimulation are still not fully understood. Here we review some of the fundamentals of electrical stimulation in the nervous system, and what happens when the surface of the brain is directly stimulated.

Electrical stimulation results in the redistribution and change of charge in the extracellular space around a cell. When the inside of the cell becomes more positive relative to the extracellular space, this results in depolarization. If the inside of the cell becomes more negative relative to outside its membrane, this is titled hyperpolarization and can inhibit action potentials. An action potential can be achieved with enough depolarization, which occurs as ions diffuse through calcium, potassium, and sodium channels (Bean, 2007). Junctions called synapses allow the communication between cells, with presynaptic cells communicating with the postsynaptic cell through chemical or electrical means. There are an enormous number of synapses in the brain, with a very high density in certain brain regions (Shepherd, 2004). For example, there is an estimate of approximately 920,000,000 synapses per cubic millimeter in the rat striatum (Ingham et al., 1998; Shepherd, 2004). These synapses form circuits which allow for excitatory and inhibitory operations, can be a source of rhythmic activity, assist with directional selectivity, and help detect spatial and temporal contrast (Shepherd, 2004). Spatial integration, where multiple presynaptic neurons can synapse on a postsynaptic cell, and temporal integration, where repeated action potentials arrive at a postsynaptic cell, contribute to postsynaptic potentials within cells (Reilly & Diamant, 2011). With stimulation, even if an action potential is not generated, subthreshold stimulation can result in the potentiation of synaptic strength, which can occur through N-methyl-D-aspartate (NMDA) receptor mediation (Alonso et al., 1990). Small changes in presynaptic potentials can result in relatively large changes in postsynaptic potentials (Reilly, 1998). This, in total, illustrates that neurons are highly interconnected, with both inhibitory and excitatory connections, and activation of a single cell or small population can have a variety of effects on a network, which will be discussed further.

Electrical stimulation results in Faradaic and non-Faradaic reactions. In Faradaic reactions, there is the transfer of electrons to electrolytes, and in non-Faradaic reactions, the redistribution of ions occurs (Merrill et al., 2005). For Faradaic reactions, there can be reversible and nonreversible reactions. Which of these two occurs is dependent on the rate of reactant mass transport compared to the rate of electron transfer. With these mechanisms of electrical stimulation, charge redistribution occurs. If this redistribution causes the depolarization of a single neuron beneath the electrode, this is often referred to as cathodal stimulation. If the stimulation causes hyperpolarization beneath the electrode, this is then often called anodal stimulation. Through ECoG arrays, cathodal stimulation by convention often describes negative voltages and currents in a region directly beneath the electrode, while anodal stimulation describes positive voltages and currents. At the anode (the electrode that is driven to a more positive potential), the cell membrane is hyperpolarized as current flows into the cell, whereas at the cathode (the electrode that is driven to a more negative potential) the cell membrane is depolarized as current flows out through the cell membrane (Kombos & Süss, 2009; Merrill et al., 2005).

Multiple studies and models have suggested that many aspects of stimulation affect the neural responses. For one, a neural response will only occur when there is an extracellular voltage gradient in the direction of the axon (Ranck, 1975), so the orientation of axons relative to the stimulating electrodes (which is usually unknown) will impact neural responses (Kudela & Anderson, 2015). Similarly, the distance between the electrode and the neural elements will affect responses by dictating the current that reaches the neurons or the region of interest (Ranck, 1975). The dendritic arbor structure and the axonal branching structure will also affect responses (Kudela & Anderson, 2015). Additionally, specific cell-type responses seem to be affected by the polarity of stimulation and at a given stimulation current, the depth of recruitment may be cell-type specific (Kudela & Anderson, 2015).

Intracortical microstimulation, or ICMS, through a microelectrode can cause stimulation on a local scale, where neurons are activated primarily through axons in the vicinity of the electrical stimulation (Nowak & Bullier, 1998; Tehovnik et al., 2006). Depending on parameters of the stimulus such as its polarity, and the orientation of the cell relative to the stimulation, regions such as the cell body and dendrites can be activated. Anodal pulses are thought to activate the bodies and terminals of cells best, while cathodal pulses more effectively result in the activation of axons. Outward flow of current at the initial segment of the axon or along the nodes of Ranvier results in the excitation of neurons (McIntyre & Grill, 2000; Tehovnik et al., 2006).

How far neuronal elements are from a cathodic or anodic stimulus influences whether they will be hyperpolarized or depolarized. Beneath a cathode the cell membrane becomes depolarized, which can generate an action potential. For anodal stimulation, although the area directly beneath the electrode is hyperpolarized, action potentials can be generated further away, which is referred to as a "virtual cathode" (Merrill et al., 2005). In the case of surface anodal stimulation, stimulation beneath an anode can occur due to the hyperpolarization of apical dendrites and subsequent depolarization as current leaves through an axon (Ranck, 1975). In bipolar stimulation, where there is an anode and a cathode, an axon in general is depolarized under the cathode and hyperpolarized under the anode (Ranck, 1975). As current flow must be balanced throughout the cell, the local hyperpolarization and depolarization at the anode and cathode, respectively, must also be locally balanced. Thus as current flows outward at the cathode and the neural fiber is locally depolarized, current must also flow inward in the surrounding region creating a hyperpolarized, "anodal" surround. Similarly, as current flows inward at the anode and the fiber is locally hyperpolarized, current must also flow outward creating a depolarized surrounding region (Merrill et al., 2005; Rattay & Wenger, 2010). In the case of the anode, the depolarized surround region is more spread out than the concentrated depolarized region under the cathode. The densely depolarized region under the cathode is more likely to generate an action potential than the depolarized surround around the anode, so cathodal-first stimulation often requires less current to elicit a response (Ranck, 1975). However, if the surrounding hyperpolarization around the cathode is too large, then an action potential due to the local depolarization will not be able to propagate through the anodal surround (Merrill et al., 2005; Ranck, 1975; Rattay & Wenger, 2010).

Both inhibitory as well as excitatory populations of cells are activated by ICMS (Butovas & Schwarz, 2003). Additionally, ICMS is not believed to result in natural cortical activity patterns (Millard et al., 2015). For example, measurements with fMRI have shown that microstimulation along the cortical pathway involved in vision suppresses outputs from neurons that have their inputs stimulated (Logothetis et al., 2010). Building upon this, microstimulation in V1 of the visual cortex has been demonstrated to locally activate neural cells and subsequently silence downstream neurons (Klink et al., 2017). Stimulation on the scale of ICMS activates a sparse neural population, and with an increase in stimulation current, more cells within a given volume are recruited, primarily filling in a given volume of tissue with increasing neural activation (Borchers et al., 2011; Histed et al., 2009). With a greater amplitude of stimulation, concurrent calcium and glutamate imaging in mice illustrated that the temporal response of neurons is affected, with potential effects on neurotransmitter reuptake also (Eles & Kozai, 2020). In terms of excitatory and inhibitory activity, microstimulation may result in a rapid excitatory response after which a longer period of inhibition, potentially due to inhibition at the synaptic level, occurs and may disrupt cortical processing (Borchers et al., 2011; Logothetis et al., 2010). A caveat is that the observations described for ICMS may not directly translate to electrical stimulation on the cortical surface (Vincent et al., 2016).

Stimulation frequency also affects the inhibition and excitation of neurons. Higher frequency stimulation (> 10 Hz) results in the potentiation of neuronal activity (Bliss & Lømo, 1973; Douglas, 1977), while lower frequency stimulation (<1 Hz), in the case of ICMS, results in the depression of neuronal activity (Dudek & Bear, 1992; Mulkey & Malenka, 1992). In vivo two-photon microscopy of layer 2/3 neurons in the somatosensory cortex of mice showed that there are different temporal and spatial responses depending on the frequency of continuous stimulation used (Michelson et al., 2019). Recent work in mice using two-photon fluorescence imaging combined with ICMS also illustrates that beyond just the frequency of stimulation, the pattern of stimulation affects the neural population recruited (Eles et al., 2021). Specifically, stimulation patterns that maintained the same amount of charge delivery and average 10 Hz frequency, but with one pattern having a uniform frequency and another having a burst pattern, resulted in entrainment of calcium and glutamate recordings to the bursting pattern, and subsets of neurons showed distinct preferences to either the more uniform or bursting stimulation train. Beyond this, burst patterning of stimuli resulted in distal inhibition relative to the more uniform 10 Hz pattern, illustrating the complex dynamics of neural responses in response to stimulation (Eles et al., 2021).

When compared to ICMS, DCS through ECoG electrodes can inject more current over a greater area, which could lead to more neuronal recruitment and greater spread (Vincent et al., 2016). The anatomical area stimulated can also influence whether neuronal activity is evoked or inhibited (Borchers et al., 2011). As discussed throughout this chapter, DCS of somatosensory areas can evoke sensations. In contrast, DCS of language areas while carrying out a language task can result in the interruption of speech production. What patterns and types of cells are activated during DCS depends on whether the stimulation pulses are cathodal or anodal (Seo et al., 2015), and what the geometry and orientation of the cells are (Kudela & Anderson, 2015). As an example, a finite element model (FEM) of DCS coupled with neuron modeling showed that neurons deeper in the bank of a sulcus are more activated with cathodal stimulation, while those in the crown on the top of the gyrus are more activated with anodal stimulation (Seo et al., 2015). FEM of DCS has also been explored to minimize current density hot spots locally for safety reasons, while also trying to optimize current delivery to desired regions of interest (Guler et al., 2018).

During DCS, because the resistivity of gray matter is greater than that of cerebrospinal fluid (CSF) (Ranck, 1975), a significant percentage of the current may be shunted through the CSF and not flow into the cortex where it can evoke a neural response ($\sim$45% based on one model) (Kim et al., 2011). The current density during bipolar DCS falls off with the square of the distance into cortex (Nathan et al., 1993). For a given distance between the two stimulating electrodes, half of the current will flow beneath a depth that is equal to half the inter-electrode distance (Telford, 1990), but the amount of CSF between the electrodes and the tissue will affect the current density (Wongsarnpigoon & Grill, 2008). Increasing the current delivered likely increases the number of neurons that are activated within a region and increases the region of activation (in part through synaptic connections) (Borchers et al., 2011). During stimulation the most likely sites of neural activation are the axon initial segment and the nodes of Ranvier because they have the highest sodium channel concentrations (Borchers et al., 2011; Rattay & Wenger, 2010), however, it is still difficult to predict responses to DCS as they depend on the exact morphology of the stimulated cortical region (Borchers et al., 2011).

In addition to local effects, DCS can elicit remote effects, potentially due to current volume conduction or synaptic projections (Borchers et al., 2011). In one study in a human, DCS of the basal temporal area caused aphasic deficits; however, after resecting the basal temporal area, no language deficits occurred, suggesting that the aphasia was caused by remote spread of the DCS (Ishitobi et al., 2000). Voltage deflections seen at remote sites as a result of DCS are frequently referred to in the literature as cortico-cortical evoked potentials (CCEPs) (Keller et al., 2014). These CCEPs have been reviewed by Keller et al. (2014). The cells largely responsible for cortical output are pyramidal cells, which are found in cortical layers 2, 3, 5, and 6. The superficial dendritic trees of these neurons can be depolarized. Gamma-aminobutyric acid (GABA) interneurons in layers 2 and 3 can be depolarized, and through synapses on the soma of pyramidal cells, result in a decrease of pyramidal cells' activity (Brill & Huguenard, 2009). Orthodromic as well as antidromic activation of neurons can occur with stimulation of axons through DCS (Keller et al., 2014). Neural signals recorded on the surface as a result of DCS are a combination of rapid, initial monosynaptic connections, pathways between different regions of cortex, and pathways between cortical and subcortical areas. This combination of effects results in the long-lasting responses on the order of hundreds of milliseconds (Matsumoto et al., 2006). Components of the measured voltage response in sites close to the stimulating electrodes (within 30 mm) are influenced by the volume-conducted potential. This potential can be measured at multiple surface locations, with an amplitude that falls off with the square of the distance from the stimulating electrodes, shares similar temporal and principal component analysis features among recorded sites, and likely originates from a single source directly beneath or near the stimulation electrodes (Shimada et al., 2017).

In total, this speaks to the immense complexities of engineering stimulation in humans, and the work that remains to be done in understanding both its physical effects and resulting neural ones.

14.3 Direct cortical stimulation for sensory feedback and neuroprosthetic control

14.3.1 The perception and psychophysics of direct cortical stimulation

As many natural tactile experiences are graded, varied, and time-bound, sensory stimulation for effective BBCIs will need to enable users to quickly perceive and distinguish a large number of unique percepts elicited by varied stimulation waveforms to adequately substitute for natural sensation. It is therefore necessary to describe the S1 DCS parameter space that can elicit discernible percepts and the interplay of stimulation amplitude, pulse width (PW), pulse frequency (PF), train duration, charge, and DCS perception to fully consider the feasibility of S1 DCS as a sensory feedback approach for closed-loop neuromodulation (Fig. 14.2). This psychophysical parameter space has been described in NHPs using ICMS (Callier et al., 2020; Kim et al., 2015), and only recently has begun to be probed in humans using DCS. Recent research on sensory stimulation has elucidated some aspects of the relationship between perception and stimulation parameters that may be

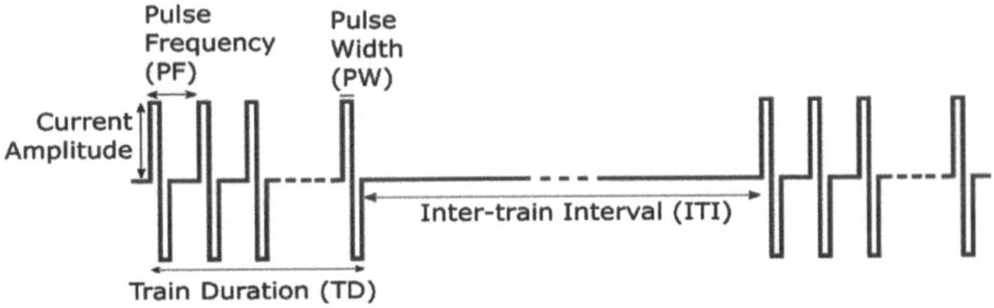

FIGURE 14.2 Stimulation waveform parameters. Single direct cortical stimulation (DCS) trains are characterized by their current amplitude, pulse width (PW), pulse frequency (PF), and train duration (TD). Multiple DCS trains in succession are additionally characterized by the intertrain interval (ITI) between two stimulation trains. All of these parameters may have an impact on the perception of DCS. Although the amplitude, PF (shown as the period), and PW are illustrated here as constant within a given stimulation train, in practice all of these parameters can be varied within one train.

used to convey a range of uniquely discernible percepts. Here, we present these recent findings on how S1 DCS parameters influence perception, along with findings on more complex perceptual phenomena (localization, attention, response time, and ownership) in S1 DCS.

14.3.2 Primary somatosensory cortex direct cortical stimulation parameters and perception

14.3.2.1 Perception

DCS through cortical electrodes has often been described as eliciting somewhat abstract sensations (Table 14.1); however, as we will see, subjects have been able to discriminate between these abstract sensations and use them in a task (Cronin et al., 2016; Hiremath et al., 2017; Kramer et al., 2020; Lee et al., 2018).

14.3.2.2 Amplitude

Although greater DCS current amplitudes are, in general, associated with higher rates of perception, subjects' perception is not dependent on S1 DCS amplitude alone (Cronin, 2018; Lee et al., 2018; Muller et al., 2018). Instead, the interplay between amplitude, PW, PF, train duration, and charge seems to influence perception on a per-subject basis, without absolute or generalizable perceptual thresholds for any single parameter. The qualitative experience of percepts may also be influenced by some combination of these parameters.

14.3.2.3 Pulse width

Like amplitude, increasing PW to an extent may increase the likelihood of perceiving the DCS, but absolute values remain subject-dependent. Lee et al. reported that, in most subjects, at least a 200 μs PW is required to perceive a stimulus (Lee et al., 2018). In a study from our lab, two subjects were able to respond consistently to a PW of 82 or 164 μs, respectively (Cronin, 2018), suggesting that these thresholds are not absolute. However, we saw that even with a high stimulation amplitude (6 mA), at least one subject was unable to perceive trains of DCS with 82 μs PW (Cronin, 2018). These findings are consistent with the hypothesis that multiple parameters are contributing to perceptual thresholds.

Lee et al. also found that an increase in PW more often produced an increase in the perceived strength of the sensation rather than a change in percept quality (Lee et al., 2018). However, a single subject in a micro-ECoG S1 stimulation study by Hiremath et al. reported two distinct sensations for 200 and 400 μs PW (Hiremath et al., 2017). Models of retinal microstimulation suggest that different PWs and temporal stimulation patterns may stimulate different subpopulations of neurons, thereby creating different percepts (Horsager et al., 2009). Continued study of the effect of PW on human perception of DCS could elucidate how we can manipulate stimulation parameters to create a range of discernible percepts.

14.3.2.4 *Pulse frequency*

Kramer et al. (2020) tested the ability of three subjects to discriminate between two S1 stimuli of different frequencies with all other parameters held constant. They found that all subjects described stimuli of higher frequencies as "faster" or "more intense." If the difference between the frequencies of the two stimuli was greater than or equal to 30 Hz, subjects were able to discern which had the higher frequency at rates significantly higher than chance. When both frequencies were below 20 Hz, accuracy was very low, suggesting that 20 Hz is the lower limit of perceptually distinguishable percepts (Kramer et al., 2020).

14.3.2.5 *Charge*

The charge per pulse (defined as the product of the pulse amplitude and the PW) required to reach a perceptual threshold also seems to depend on other parameters, including PW and PF. We found that increasing charge was needed as PW was increased, but that decreasing charge was needed as PF increased (Cronin, 2018). This relationship between PF and charge is consistent with psychophysical studies using ICMS of S1 in NHPs (Kim et al., 2015), ICMS of barrel cortex in rats (Butovas & Schwarz, 2007), and retinal microstimulation in humans (Horsager et al., 2009). The increase in required charge for longer PWs is thought to be due to an increase in sodium channel inactivation with increasing PWs resulting in an increasing charge to reach threshold (Merrill et al., 2005). Alternatively, with increased PFs, successive pulses of higher frequency may produce augmented inhibitory currents, thereby improving signal-to-noise ratios of elicited neural activity and requiring less charge to reach threshold (Butovas & Schwarz, 2007).

14.3.2.6 *Train duration*

Libet et al. (1964) found that in a group of patients with dyskinesias, the minimum train duration for a perceptual response was about 500 ms when using current amplitudes near the perceptual threshold, titled the liminal intensity, where train durations above this minimum duration were also perceived with little change in the required threshold current. Ray et al. found, by using ECoG grids in patients with epilepsy, that bipolar stimulation trains can be as short as 250 ms with little change in the sensory threshold current required when compared to longer duration trains, and that trains shorter than approximately 250 ms often required a higher current amplitude for perception (Ray et al., 1999). We observed the perception of 100 ms trains when using amplitudes close to perceptual thresholds elicited via testing with 200 ms trains (Caldwell et al., 2019). There are several possible reasons for these differences between studies. Libet et al. did not use ECoG grids, and used primarily monopolar, monophasic stimulation rather than bipolar, biphasic stimulation. Their choices of PW and PF, often longer and slower, respectively, than what others typically use (Caldwell, Cronin, et al., 2019), could also contribute to the observed differences.

Prior work has also considered the relationship between train duration and PF. Kim et al. observed a decrease in the train duration necessary for perception of ICMS in NHPs when PF was increased and other parameters were held constant (Kim et al., 2015). Libet et al. also found that, in humans, within the range of pulse frequencies tested (between 15 and 240 Hz), utilization train durations tended to decrease as frequencies increased (Libet et al., 1964). It is possible that just as others (Butovas & Schwarz, 2007; Horsager et al., 2009; Kim et al., 2015) have found that the threshold charge per pulse decreases with increasing frequency, the threshold charge exchange (total charge delivered in the train) may also decrease, allowing for a shorter train duration with higher pulse frequencies. Continuing to consider how train duration and the number of pulses affect perceptual responses may provide a more complete picture of the psychophysics of S1 DCS and direct the design of more efficient stimulation waveforms.

14.3.2.7 Novel stimulation waveforms

While the relationships between multiple stimulation parameters and perception are complex, they suggest that we may be able to develop novel, efficient stimulation waveforms which will elicit a conscious percept with as minimal charge delivery as possible. More complex DCS patterns may alter subjects' perceptual experience of the waveform and possibly elicit more natural sensations. For example, we have conducted preliminary tests of subjects' response times to DCS waveforms with two pulses that are 1.5–2.5 × the perceptual threshold followed by 38 pulses (to create a 200 ms train duration) at a lower but still suprathreshold amplitude. Preliminary results suggest that subjects can respond faster to these high–low DCS trains than to constant-amplitude DCS trains, with the use of the two initially higher amplitude stimulation pulses modifying the behavioral response time to the entire train. A DCS train could also model the stimulation patterns used by Tan et al. (2014) when they elicited pressure and tapping sensations in two subjects via peripheral nerve cuff electrodes. In that work they used stimulation pulses whose PW varied based on a sine wave envelope (Tan et al., 2014).

Additionally, novel stimulation waveforms may allow us to target S1 locations that are typically not directly covered by ECoG. Given the physics of cortical electrical stimulation and the locations of proprioceptive areas (e.g., area 3a within the central sulcus), it will be difficult to generate primarily proprioceptive sensations with bipolar DCS, as it is difficult to get enough current into deeper regions of the gray matter without activating more superficial layers. We have an ongoing collaboration with other research groups to model macro-ECoG DCS with the goal of selecting a set of electrodes for multipolar stimulation that will target a specific region of interest, such as area 3a.

14.3.3 Percept localization

Overall, our macro-ECoG S1 DCS subjects have reported localized percept areas ranging from single fingertips to large areas of the palm (Cronin, 2018), similar to the sensation fields reported by Lee et al. (see fig. 4 in Lee et al., 2018) and Kramer et al. (see fig. 1 in Kramer et al., 2020), and sometimes smaller than those reported by Hiremath et al. (see fig. 4 in Hiremath et al., 2017, in which the smallest area reported was the entirety of the fifth digit). The Lee et al. (2018) and Hiremath et al. (2017) studies, along with two subjects in the Kramer et al. (2020) study, used high-density ECoG arrays (3–4.5 mm center-to-center spacing). Hiremath et al. used monopolar stimulation as compared to our, Lee et al.'s, and Kramer et al.'s use of bipolar stimulation. The expected additional current spread from monopolar stimulation versus bipolar stimulation could explain the trend toward larger sensory receptive fields (Nathan et al., 1993; Wongsarnpigoon & Grill, 2008). Lee et al. also reported that, in one subject, as the amplitude increased the subject's receptive field moved from the palm and fifth finger to several fingertips (Lee et al., 2018). Similarly, we have found that changes in DCS parameters could cause the location of the perceived sensation to expand or move when the PW or PF changed (Cronin, 2018). Future studies may consider whether we can change the electrode choice and DCS parameters to localize percepts to different areas.

14.3.4 Brain state, attention, and perception

There is a vast literature on the conscious perception of sensory stimuli, but exactly how tactile stimuli are consciously perceived—even when arriving naturally from peripheral afferents—is not fully understood (however, see Gescheider et al., 2008 for vibrotactile psychophysics). Prior work in ECoG found that modulations in high-gamma activity (60–150 Hz) were correlated with the subjects' attentional states. Specifically, high-gamma activity was greater over somatosensory cortex when subjects were attending to vibrotactile stimuli (Ray et al., 2008). Similarly, Bauer et al. found an increase in gamma-band (60–95 Hz) activity over contralateral somatosensory cortex when attention was cued to the left or right hand, but this increase in gamma-band activity appeared approximately 100 ms after the tactile stimulus onset (Bauer et al., 2006). Muller et al. demonstrated that high gamma activity following S1 DCS via high-density ECoG arrays in humans was well correlated with conscious perception of the DCS stimuli (Muller et al., 2018).

A number of studies on the perception of external stimuli (not DCS) agree that low alpha power (8–14 Hz) in local, task-relevant areas (e.g., somatosensory cortex) precedes correctly perceived stimuli. This observation is often explained with the idea that strong alpha power reflects functional inhibition which would impede conscious perception of a stimulus (Frey et al., 2016).

Studies have also demonstrated alpha-band suppression following spatial attention cuing prior to a tactile stimulus (Foxe & Snyder, 2011). To our knowledge, no peer-reviewed journal studies have been published to consider a relationship between alpha power and the perception of DCS; however, we believe such work would be beneficial. Continued study of the relationship between cortical signals and DCS perception may reveal a brain state that is more conducive to stimulation than others and generate a method of timing the delivery of S1 DCS to increase the likelihood of conscious perception.

14.3.5 Response times

While prior work suggests that the integration of somatosensory feedback into a BBCI is possible and enhances performance relative to a task without somatosensory feedback, the comparison of human S1 DCS to haptic stimulation has not been well explored. Specifically, given that S1 DCS completely circumvents ascending dorsal column pathways, how human subjects' response times to DCS differ from response times to natural haptic stimulation must be examined. This is an important consideration for effective BCI development aiming to integrate cortical stimulation as a method of sensory feedback as response latency invariably constrains feedback loop architecture.

Recently, we found that subjects responded significantly slower to S1 DCS than to natural, haptic stimuli for a range of DCS train durations (Caldwell, Cronin, et al., 2019). Median response times for haptic stimulation of the hand ranged from 198 to 313 ms, while median responses to reliably perceived DCS ranged from 254 ms for one subject, up to 528 ms for a different subject. In our study, we did not observe a significant impact of learning or habituation, which we assessed through analyzing two separate blocks of trials with a rest period in-between. We also found no significant impact of cortical stimulation train duration on response times. Our results suggest a range of potential behavioral response latencies to somatosensory DCS feedback, which could be used in future neuroprosthetic applications.

In our study, we asked four subjects to press a button as quickly as possible after perceiving either a cutaneous haptic touch to the hand or a percept from S1 DCS via ECoG grids covering the surface of the hand somatosensory cortex. We hypothesized that DCS, by bypassing ascending peripheral sensory processing circuitry, would result in faster response times than cutaneous haptic touch. We also hypothesized that subjects would become faster over multiple blocks of DCS due to learning, and that response times to DCS would decrease with longer train durations relative to shorter trains with the same stimulation current.

All four subjects were significantly slower to respond to the S1 DCS than to haptic touch (Fig. 14.3). Additionally, with our two blocks of testing, we saw no significant differences between trial types and blocks, suggesting that at least on the short time scale of our experiment, learning that significantly

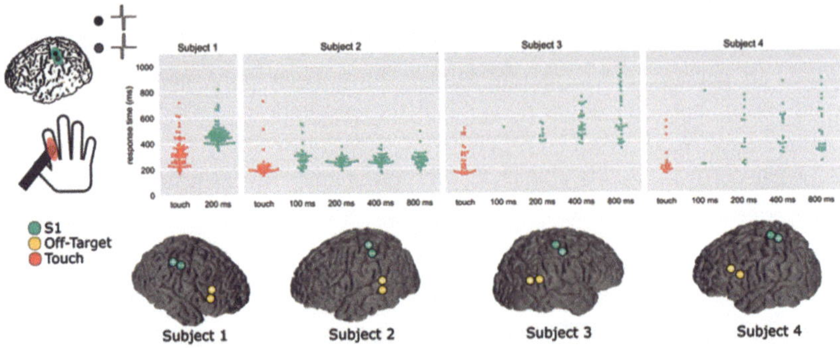

FIGURE 14.3 Comparison of response times for four subjects and their direct cortical stimulation (DCS) electrodes. Each dot represents a response time for a given trial, colored by condition. *Pink* indicates the haptic test condition, while *turquoise* indicates primary somatosensory cortex (S1) DCS conditions and macro-electrocorticography (ECoG) electrodes over hand somatosensory cortex. Subject 1 only received the 200 ms DCS and haptic stimulation conditions, while subjects 2, 3, and 4 had 100, 200, 400, and 800 ms trains of stimulation applied. The two separate blocks for subjects 2, 3, and 4 were pooled together for each subject. Off-target DCS control electrodes are indicated in *yellow*. Electrode locations are based on cortical surface reconstructions for each subject. Electrodes with a plus symbol (+) indicate anodal-first stimulation, while electrodes with a minus symbol (−) indicate cathodal-first stimulation. *Reprinted, under Creative Commons Attribution 4.0 International License, from Caldwell, D. J., Cronin, J. A., Wu, J., Weaver, K. E., Ko, A. L., Rao, R. P. N., & Ojemann, J. G. (2019). Direct stimulation of somatosensory cortex results in slower reaction times compared to peripheral touch in humans. Scientific Reports, 9(1). https://doi.org/10.1038/s41598-019-38619-2.*

improved reaction times was not occurring. In three subjects we tested the train duration hypothesis, finding that train lengths as short as 100 and up to 800 ms did not significantly alter the response times. As a control, we performed off-target testing in cortical areas outside of hand somatosensory cortex to test for the possibility that subjects were responding to stimulation that was applied anywhere in the cortex. Individuals either did not perceive the off-target stimulation, or were able to perceive it as a vague, distinct sensation separate from the sensory percept induced by S1 DCS. This reinforces that we were testing electrical stimulation induced percepts in S1 compared to touch activated natural ascending peripheral pathways, converging on S1.

Our results were consistent with a previous observation in NHPs that intracortical microstimulation of area 1 in S1 results in significantly slower response times than vibrotactile peripheral stimulation (Godlove et al., 2014). As per our original hypothesis, one may first suspect that bypassing the ascending peripheral pathway through DCS would reduce the distance traversed by the sensory volley and consequently result in faster response times. However, as discussed in this chapter, electrical stimulation may be exciting both inhibitory and excitatory connections in an unnatural manner, potentially resulting in slower behavioral responses.

14.3.6 Sensory ownership and the rubber hand illusion

It is known that peripheral nerve stimulation in human subjects with transradial amputations can increase subjects' sense of embodiment of their personal prosthesis (Schiefer et al., 2016). We desire future cortical-level BBCIs with somatosensory feedback to evoke a similar sense of embodiment or ownership over the end effector the BBCI is linked to (e.g., a prosthesis), because it is expected that embodiment of the BBCI will create a more positive user experience.

The rubber hand illusion (RHI) is a visual-tactile illusion during which subjects refer tactile sensations from their own, hidden hand to a rubber hand within view, which is also being synchronously touched (Botvinick & Cohen, 1998; Makin et al., 2008). During the RHI, subjects experience a sense of ownership over the rubber hand.

Work from our lab demonstrated for the first time that the RHI could be induced without peripheral stimulation, using DCS of S1 instead (Fig. 14.4;

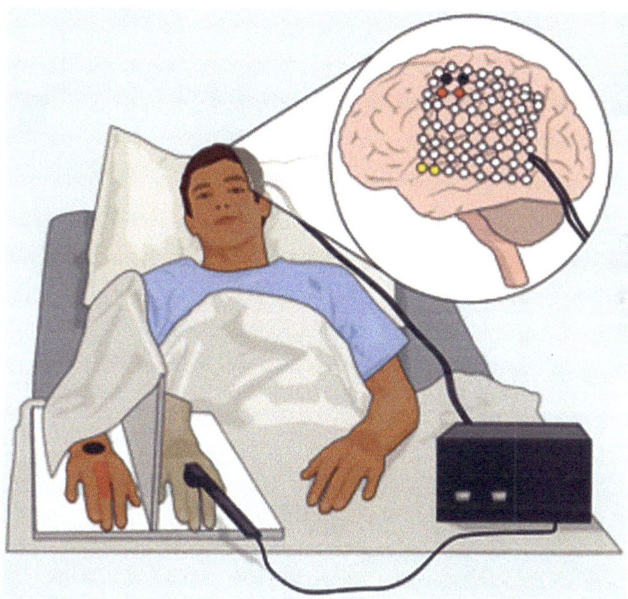

FIGURE 14.4 Rubber hand illusion setup. The experimenter first applied bipolar direct cortical stimulation (DCS) over two hand primary somatosensory cortex (S1) DCS electrodes (highlighted in *red*) to determine where the subject localized the percept (illustrated as an example with *red* shading over the third finger). The experimenter repeatedly touched the rubber hand with the digital touch probe at the location corresponding with the percept's location. The digital touch probe triggered a 500 ms DCS train, thus creating a spatiotemporally congruent experience of the DCS percept and the visual cue from the rubber hand. *Reprinted from Collins, K. L., Guterstam, A., Cronin, J., Olson, J. D., Ehrsson, H. H., & Ojemann, J. G. (2017). Ownership of an artificial limb induced by electrical brain stimulation.* Proceedings of the National Academy of Sciences of the United States of America, 114*(1), 166–171. https://doi.org/10.1073/pnas.1616305114.*

Collins et al., 2017). Subjects were asked to rate their degree of ownership over the rubber hand as DCS of S1 was delivered while the rubber hand was touched synchronously and asynchronously. The published results in two subjects suggest that DCS can induce a sense of ownership of an artificial hand (Fig. 14.5). The results from one subject (subject 2, Fig. 14.5) also indicate that the illusion is dependent on a certain degree of spatial and temporal congruence. This subject only experienced a high degree of ownership when the rubber hand was touched simultaneously with the S1 DCS and touched in the area corresponding to where the subject localized the percept elicited by S1 DCS. This suggests that S1 DCS and visual cues can be integrated into a coherent representation, so long as rules governing spatiotemporal congruence of normal perception are met (Makin et al., 2008). The results from both subjects have important implications to the application of somatosensory stimulation in

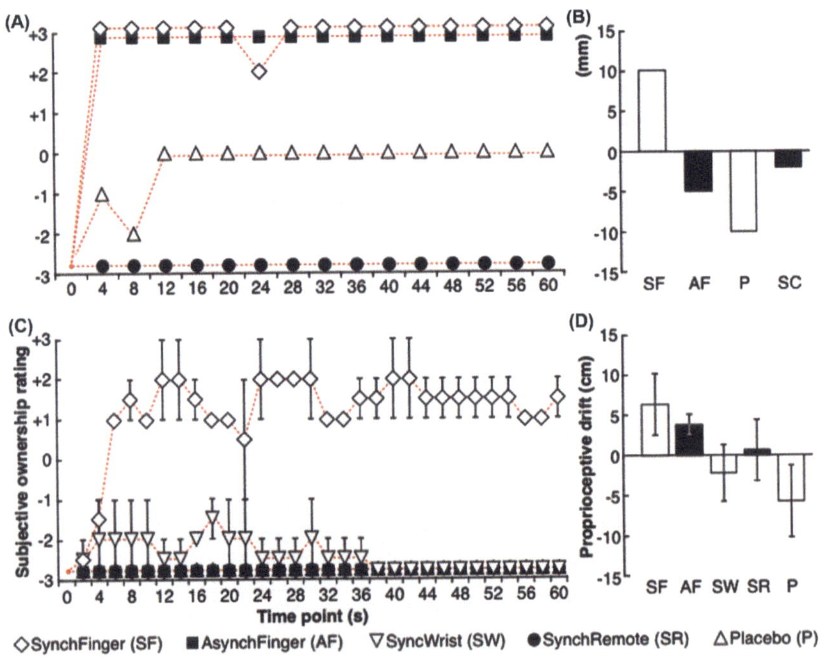

FIGURE 14.5 Rubber hand illusion results. Two subjects rated their degree of ownership over the rubber hand. Both subjects experienced the illusion and reported high ownership ratings when the DCS and haptic touch were delivered simultaneously (SynchFinger condition, A and C). Both subjects also reported positive proprioceptive drift values, which are an indirect behavioral proxy of the feeling of limb ownership (Rohde et al., 2011), under the SynchFinger condition (B and D). See Collins et al. (2017) for results and discussion on asynchronous stimulation. *Reprinted from Collins, K. L., Guterstam, A., Cronin, J., Olson, J. D., Ehrsson, H. H., & Ojemann, J. G. (2017). Ownership of an artificial limb induced by electrical brain stimulation.* Proceedings of the National Academy of Sciences of the United States of America, 114*(1), 166–171. https://doi.org/10.1073/pnas.1616305114.*

rehabilitative devices, as they suggest that cortical stimulation will be able to create a sense of ownership over a neural rehabilitative device.

14.3.7 Use of primary somatosensory cortex direct cortical stimulation as task feedback

While S1 DCS has not yet been implemented in complex BBCI frameworks, initial work from our lab has demonstrated the feasibility of using DCS-induced somatosensory percepts as meaningful feedback in a specific motor task (Cronin et al., 2016). In this study, subjects' hand aperture positions were tracked as they were asked to open and close their fingers to follow a predetermined, but hidden, target aperture pathway. Depending on the hand position relative to the target, subjects received one of three sets of S1 DCS feedback whose parameters were tuned prior to the experiment to ensure sensory discriminability. One subject was able to use this feedback to follow the target trajectory at levels well above chance levels (Fig. 14.6). These

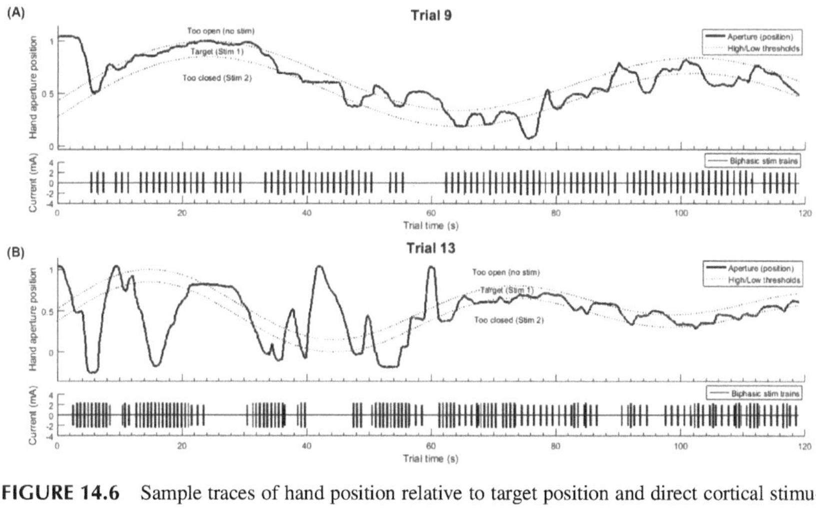

FIGURE 14.6 Sample traces of hand position relative to target position and direct cortical stimulation (DCS) waveforms during the aperture task. The subject received stimulation corresponding to their hand aperture position relative to a target position. If their hand was too open they received no stimulation, if it was within the target range they received a lower amplitude stimulation (2.0 mA), and if their hand was too closed they received a higher amplitude stimulation (2.4 mA). Stimulation pulses were biphasic, but due to the time scale only the 200 ms stimulation trains are visible, not the individual pulses. (A) The subject was able to follow the target pathway and stay in the target boundaries with a high performance that was above chance level as expected from a random walk: accuracy = 0.6145, $R^2 = 0.8194$ (trial 9). (B) The subject had trouble finding the target region at the beginning of the trial, resulting in lower performance: accuracy = 0.4023 (above chance level), $R^2 = 0.1001$ (below chance level). *© 2016 IEEE. Reprinted, with permission, from Cronin, J. A., Wu, J., Collins, K. L., Sarma, D., Rao, R. P. N., Ojemann, J. G., & Olson, J. D. (2016). Task-specific somatosensory feedback via cortical stimulation in humans.* IEEE Transactions on Haptics, *9(4), 515–522. https://doi.org/10.1109/TOH.2016.2591952.*

findings suggest that S1 DCS feedback can indeed be used to modulate motor behaviors, as would be required for an effective BBCI.

14.4 Future advances in cortical sensory stimulation

While existing work has demonstrated that cortical stimulation can be used to provide sensory feedback, improvements to the current state of cortical somatosensory stimulation will be needed to create a fully functional BBCI that can effectively substitute for natural somatosensation. For example, additional channels of stimulation may enable encoding spatial information and more complex aspects of touch rather than a simple signal of contact. Advances in concurrent recording and stimulation will facilitate the development of BBCIs which require simultaneous, or near-simultaneous recording and stimulation of adjacent areas (i.e., recording and decoding motor intentions from motor regions and encoding sensory feedback through stimulation of adjacent sensory regions). Furthermore, concurrent recording and stimulation of cortical electrodes will eventually require wireless charging and data transfer to produce an implantable device that is fully enclosed within the body and thus is at less risk of infection, while also being free of external tethering wires.

14.4.1 More channels

Most somatosensory stimulation studies in humans (Collins et al., 2017; Cronin et al., 2016; Hiremath et al., 2017; Johnson et al., 2013; Lee et al., 2018; Muller et al., 2018) have used a single channel of feedback, with one or two electrodes for monopolar or bipolar stimulation, respectively. While a single channel or axis of feedback could be helpful to some, one can imagine why multiple channels or axes of somatosensory feedback would be more useful. The information relayed from afferents in our fingertips is not simply reflective of a two-dimensional grid of mechanoreceptors which are either making contact or not making contact with an object. Incorporated in that signal are responses from more remote points of contact that are due to skin stretch and strain that provide the user with a complete picture of their interaction with the object (Johansson & Flanagan, 2009). Thus although a single channel of somatosensory feedback such as those presented here could be useful in conveying a single axis of information (e.g., total grasp force applied to an object), more channels will be needed to convey additional spatial information about contact location and varying forces (Weber et al., 2012).

Engineering technology advancements are allowing for devices with an increasing number of channel counts to be developed and tested in animals as well as humans. Much work to date has focused on increased spatial resolution in recording from sensorimotor and language areas (Muller, Hamilton, et al., 2016; Wang et al., 2016; Wang et al., 2017). Novel, thin-film microelectromechanical system arrays, using different manufacturing technologies when compared to standard clinical electrodes with embedded electrodes, have been explored in humans (Muller, Felix, et al., 2016). Another area of interest is PEDOT:PSS [(poly(3,4-ethylenedioxythiophene): poly(styrene sulfonate))] electrode technologies (Ganji et al., 2018; Khodagholy et al., 2015, 2016; Pranti et al., 2018), where electrodes made of materials such as gold are coated with PEDOT:PSS. Compared to standard arrays, these electrodes are able to achieve a higher signal-to-noise ratio during neural signal recording, and have favorable charge injection properties. Graphene technology is also promising; it has been shown previously to be useful for recording neural activity (Kuzum et al., 2014), and when combined with silicon dioxide substrates, it has both excellent charge injection characteristics as well as potential optical transparency for imaging (Koerbitzer et al., 2016). *In vivo* work in rodents has shown the ability for simultaneous fluorescence imaging of neuronal activity with electrical stimulation through graphene electrodes, showing the utility of graphene as a technology for stimulating electrodes (Park et al., 2018). Graphene arrays have shown utility in elucidating seizure dynamics in rodent models with combined graphene micro-ECoG and optical calcium fluorescence recordings (Driscoll et al., 2021). With advances in imaging and genetic techniques, transparent electrodes could have potential applications for humans in the future. The biocompatibility of graphene arrays has been demonstrated in rodent models, with measurements for up to 12 weeks (Garcia-Cortadella et al., 2021), and with 64 channels of epicortical local field potential (LFP) recordings measured from the graphene solution-gated field-effect transistors. In total, the wide range of emerging technologies for manufacturing electrodes and arrays capable of both stimulation and recording with favorable material properties will allow for further advances in the field of sensory restoration.

14.4.2 Concurrent stimulation and recording

In a closed-loop neuroprosthetic system, there is simultaneous (or near-simultaneous) stimulation and recording. The resultant artifact from stimulation can be much larger than the neural signals of interest. Separating artifacts from neural signals of interest is an ongoing research area (Prime et al., 2020; Shimada et al., 2017). Some of the approaches pursued have included hardware-based approaches to minimize artifacts from acquiring the neural signal, as well as algorithmic techniques to reduce the artifactual signal after acquisition.

One hardware approach using complementary metal–oxide–semiconductor (CMOS) technology allows for differential, real-time, and common-mode cancellation of artifacts (Smith et al., 2017). Other advances with new CMOS stimulator front-ends could enable integrated, scalable BCI devices with both signal processing done wirelessly and stimulator blocks able to be added in (Pepin et al., 2016). Recent advances have enabled single-chip CMOS technologies with 64 channels of electrophysiologic recording, four channels of stimulation, and concurrent adaptive artifact cancellation (Uehlin et al., 2020). Using these technologies will allow for an enhanced understanding of the brain's response to stimulation, and open the door for more advanced closed-loop systems (Zhou et al., 2018). With improved technology, neural activity that overlaps both in time and space with electrical stimulation could be employed in the control strategy.

In industry, there has been rapid technology development in the deep-brain stimulation (DBS) field for concurrent stimulation and recording, which are important for precisely timed stimulation based on ongoing neural signals, and also may provide potential solutions for BCIs based on ECoG for sensory restoration (Herron et al., 2018; Stanslaski et al., 2012). Some of these DBS solutions include concurrent ECoG recording, and use knowledge of the stimulation and recording configuration to reduce the measured artifact, filtering on the front-end, heterodyning to reduce artifacts in frequency bands where neural signals are of interest, and using stimulation parameters that reduce the overlap between stimulation harmonics and neural signals (Stanslaski et al., 2012). Medtronic's Summit RC + S system uses some of the mentioned approaches, and furthers them by including oversampling to reduce noise in bands of interest, decimators to reduce higher order stimulation harmonics, and the ability to provide stimulation parameters to the clinician or researcher that enable adequate neural recording (Herron et al., 2018). Results from four patients implanted with the RC + S system showed the efficacy of the hardware in acquiring at-home recordings from the basal ganglia and cortical surface with and without stimulation (Gilron et al., 2021). These techniques and technologies may be applicable to closed-loop sensory ECoG-BCIs with DCS.

14.4.3 Wireless technologies

For a fully implantable and chronic device able to be used throughout daily life with minimal disruption, wireless technologies are critical. One important aspect of wireless devices is deciding how much data needs to be processed and transmitted wirelessly for a useful BCI, which informs power-saving design approaches and processing approaches (Even-Chen et al., 2020). Recently, hardware advances have enabled wireless communication with real-time cancellation of artifacts with LFP recordings on 128 channels in NHPs (Zhou et al., 2019). Other devices with microelectrodes in NHPs have employed wireless data transfer and charging (Borton et al., 2013). With these wireless technologies, and concurrent stimulation and recording, explorations into the naturalistic behavior and how chronic stimulation changes the brain are possible. In NHPs, both wireless recording and stimulation over motor areas during a 6-month window did not result in neurological or behavioral sequelae (Romanelli et al., 2018). This example illustrates how wireless, chronic ECoG implants could use DCS safely and effectively. Recent work with the Medtronic Summit RC + S showed in four patients with Parkinson's disease that bilateral, four-channel streaming of signals from motor cortex and the basal ganglia could be acquired in a home setting, with over 500 hours of recording (Gilron et al., 2021). The opportunities afforded by these technologies, and the ability to see how the brain interacts over the long term with stimulation, can lend insight into how the brain might adapt to a sensory neuroprosthetic, and enable adaptive approaches on an individual patient basis.

14.5 Conclusion

Humans are able to perceive electrical stimulation of the surface of the cortex through macro-ECoG electrodes (Johnson et al., 2013), respond to it (Caldwell, Cronin, et al., 2019), have it induce sensory illusions (Collins et al., 2017), and use it as feedback in motor tasks (Cronin et al., 2016). Stimulation through mini-ECoG arrays can also be perceived (Hiremath et al., 2017; Kramer et al., 2020; Lee et al., 2018), with a larger number of unique percepts on the hand perceived than would be possible with a larger clinical array (Hiremath et al., 2017). In a target acquisition task, sensory feedback delivered via mini-ECoG was able to help subjects perform a task. Although, for the most part, these percepts do not feel natural (Caldwell, Cronin, et al., 2019; Johnson et al., 2013; Lee et al., 2018), ECoG-stimulation-induced percepts may be usable feedback signals in future clinical motor BCIs.

In future BBCIs, we desire a stimulation method with high spatial resolution that elicits natural sensory percepts localized to small areas on the target limb. For such high resolution, ICMS or micro-ECoG stimulation is likely to be used over macro-ECoG stimulation, due to their smaller electrode surface areas and higher electrode density. Additionally, recent research using ICMS of somatosensory cortex suggests that ICMS may elicit relatively more natural sensations than subdural ECoG grids (Armenta Salas et al., 2018; Flesher et al., 2016). However, the psychophysics of macro-scale ECoG stimulation has similarities to both high-density DCS and ICMS. Therefore the principles underlying all cortical stimulation methods can likely be applied across modalities as we expand our understanding of the psychophysics and utility of cortical stimulation for sensory feedback.

Another consideration for stimulation modality is how natural the elicited percept must feel to be useful. Researchers in the field of BCI sensory feedback discuss two general approaches for generating useful sensory feedback: biomimicry and adaptation (Bensmaia & Miller, 2014). The argument for biomimicry holds that only biomimetic feedback will allow subjects to regain the sort of dexterous movement and tactile sensations that normally occur (Miller & Weber, 2011; Tabot et al., 2015). The basis for this argument is that naturally occurring tactile and proprioceptive sensations are so varied that the only way to encode all of them in a meaningful way will be to use stimulation methods and parameters that cause neural activity as similar as possible to the natural neural response to a given stimulus. The adaptation approach holds that applications may be able to use sensory substitution and elicit abstract sensations that the user will substitute for normal tactile sensations over time. Users may map the abstract sensation to a normal sensation and no longer perceive it as abstract (Dadarlat et al., 2015). DCS via ECoG electrodes will fall into the adaptation category of sensory feedback approaches, as the stimulation does not activate normal neural pathways of sensory information flow. The ability to induce a feeling of artificial limb ownership during the RHI experiment (Collins et al., 2017) and use of S1 DCS to perform a motor task (Cronin et al., 2016) supports the idea that subjects will be able to learn how to interpret and use the abstract sensations elicited by S1 DCS, even though DCS, especially via macro-ECoG electrodes, does not evoke a biomimetic neural response.

Long-term studies may allow researchers to better understand how subjects adapt to DCS over time, including if subjects experience a change in the qualia of their percepts or in their perceptual thresholds for various parameters. How subjects respond to DCS over time may elucidate whether they are able to adapt to initially abstract sensations, or if stimulation methods that evoke a more biomimetic neural response are required. Long-term studies may also provide the opportunity to evaluate how attention, and noisier, more natural environments will affect S1 DCS perception. It is possible that the perceptual thresholds that we and others have measured during relatively well-controlled experiments, will not translate directly to daily use in a noisier environment. A study of ICMS in the barrel cortex of freely behaving rats reported that perceptual thresholds may increase in behaving animals as compared to studies with head-restrained animals (Semprini et al., 2012). This is critical information for creating a functional BBCI as any changes in perception would have to be accounted for perhaps with a calibration process (Devecioğlu & Güçlü, 2017; Öztürk et al., 2019).

References

Ackerley, R., & Kavounoudias, A. (2015). The role of tactile afference in shaping motor behaviour and implications for prosthetic innovation. *Neuropsychologia*, *79*(Pt. B), 192–205. Available from https://doi.org/10.1016/j.neuropsychologia.2015.06.024.

Alonso, A., Curtis, M. de, & Llinás, R. (1990). Postsynaptic Hebbian and non-Hebbian long-term potentiation of synaptic efficacy in the entorhinal cortex in slices and in the isolated adult guinea pig brain. *Proceedings of the National Academy of Sciences*, *87*(23), 9280–9284. Available from https://doi.org/10.1073/pnas.87.23.9280.

Armenta Salas, M., Bashford, L., Kellis, S., Jafari, M., Jo, H., Kramer, D., Shanfield, K., Pejsa, K., Lee, B., Liu, C. Y., & Andersen, R. A. (2018). Proprioceptive and cutaneous sensations in humans elicited by intracortical microstimulation. *ELife*, *7*, e32904. Available from https://doi.org/10.7554/eLife.32904.

Armour, B. S., Courtney-Long, E. A., Fox, M. H., Fredine, H., & Cahill, A. (2016). Prevalence and causes of paralysis-United States, 2013. *American Journal of Public Health, 106*(10), 1855–1857. Available from https://doi.org/10.2105/AJPH.2016.303270.

Bauer, M., Oostenveld, R., Peeters, M., & Fries, P. (2006). Tactile spatial attention enhances gamma-band activity in somatosensory cortex and reduces low-frequency activity in parieto-occipital areas. *Journal of Neuroscience, 26*(2), 490–501. Available from https://doi.org/10.1523/JNEUROSCI.5228-04.2006.

Bean, B. P. (2007). The action potential in mammalian central neurons. *Nature Reviews Neuroscience, 8*(6), 451–465. Available from https://doi.org/10.1038/nrn2148.

Bensmaia, S. J., & Miller, L. E. (2014). Restoring sensorimotor function through intracortical interfaces: Progress and looming challenges. *Nature Reviews Neuroscience, 15*(5), 313–325. Available from https://doi.org/10.1038/nrn3724.

Berg, J. A., Dammann, J. F., Tenore, F. V., Tabot, G. A., Boback, J. L., Manfredi, L. R., Peterson, M. L., Katyal, K. D., Johannes, M. S., Makhlin, A., Wilcox, R., Franklin, R. K., Vogelstein, R. J., Hatsopoulos, N. G., & Bensmaia, S. J. (2013). Behavioral demonstration of a somatosensory neuroprosthesis. *IEEE Transactions on Neural Systems and Rehabilitation Engineering: A Publication of the IEEE Engineering in Medicine and Biology Society, 21*(3), 500–507. Available from https://doi.org/10.1109/TNSRE.2013.2244616.

Berger, M. S., & Ojemann, G. A. (1992). Intraoperative brain mapping techniques in neuro-oncology. *Stereotactic and Functional Neurosurgery, 58*(1–4), 153–161. Available from https://doi.org/10.1159/000098989.

Bliss, T. V. P., & Lømo, T. (1973). Long-lasting potentiation of synaptic transmission in the dentate area of the anaesthetized rabbit following stimulation of the perforant path. *The Journal of Physiology, 232*(2), 331–356.

Borchers, S., Himmelbach, M., Logothetis, N., & Karnath, H.-O. (2011). Direct electrical stimulation of human cortex—The gold standard for mapping brain functions? *Nature Reviews Neuroscience, 13*(1), 63–70. Available from https://doi.org/10.1038/nrn3140.

Borton, D. A., Yin, M., Aceros, J., & Nurmikko, A. (2013). An implantable wireless neural interface for recording cortical circuit dynamics in moving primates. *Journal of Neural Engineering, 10*(2), 026010. Available from https://doi.org/10.1088/1741-2560/10/2/026010.

Botvinick, M., & Cohen, J. (1998). Rubber hands "feel" touch that eyes see. *Nature, 391*(6669), 756. Available from https://doi.org/10.1038/35784.

Brill, J., & Huguenard, J. R. (2009). Robust short-latency perisomatic inhibition onto neocortical pyramidal cells detected by laser-scanning photostimulation. *The Journal of Neuroscience: The Official Journal of the Society for Neuroscience, 29*(23), 7413–7423. Available from https://doi.org/10.1523/JNEUROSCI.6098-08.2009.

Brindley, G. S., & Lewin, W. S. (1968). The sensations produced by electrical stimulation of the visual cortex. *The Journal of Physiology, 196*(2), 479–493.

Butovas, S., & Schwarz, C. (2003). Spatiotemporal effects of microstimulation in rat neocortex: A parametric study using multielectrode recordings. *Journal of Neurophysiology, 90*(5), 3024–3039. Available from https://doi.org/10.1152/jn.00245.2003.

Butovas, S., & Schwarz, C. (2007). Detection psychophysics of intracortical microstimulation in rat primary somatosensory cortex. *The European Journal of Neuroscience, 25*(7), 2161–2169. Available from https://doi.org/10.1111/j.1460-9568.2007.05449.x.

Caldwell, D. J., Cronin, J. A., Wu, J., Weaver, K. E., Ko, A. L., Rao, R. P. N., & Ojemann, J. G. (2019). Direct stimulation of somatosensory cortex results in slower reaction times compared to peripheral touch in humans. *Scientific Reports*, *9*(1). Available from https://doi.org/10.1038/s41598-019-38619-2.

Caldwell, D. J., Ojemann, J. G., & Rao, R. P. N. (2019). Direct electrical stimulation in electrocorticographic brain–computer interfaces: enabling technologies for input to cortex. *Frontiers in Neuroscience*, *13*, 804. Available from https://doi.org/10.3389/fnins.2019.00804.

Callier, T., Brantly, N. W., Caravelli, A., & Bensmaia, S. J. (2020). The frequency of cortical microstimulation shapes artificial touch. *Proceedings of the National Academy of Sciences*, *117*(2), 1191–1200. Available from https://doi.org/10.1073/pnas.1916453117.

Chang, E. F. (2015). Towards large-scale, human-based, mesoscopic neurotechnologies. *Neuron*, *86*(1), 68–78. Available from https://doi.org/10.1016/j.neuron.2015.03.037.

Collins, C. E., Airey, D. C., Young, N. A., Leitch, D. B., & Kaas, J. H. (2010). Neuron densities vary across and within cortical areas in primates. *Proceedings of the National Academy of Sciences*, *107*(36), 15927–15932. Available from https://doi.org/10.1073/pnas.1010356107.

Collins, K. L., Guterstam, A., Cronin, J., Olson, J. D., Ehrsson, H. H., & Ojemann, J. G. (2017). Ownership of an artificial limb induced by electrical brain stimulation. *Proceedings of the National Academy of Sciences of the United States of America*, *114*(1), 166–171. Available from https://doi.org/10.1073/pnas.1616305114.

Cronin, J. A. (2018). *Human psychophysics of direct cortical stimulation of somatosensory cortex*. University of Washington, [Unpublished doctoral dissertation].

Cronin, J. A., Wu, J., Collins, K. L., Sarma, D., Rao, R. P. N., Ojemann, J. G., & Olson, J. D. (2016). Task-specific somatosensory feedback via cortical stimulation in humans. *IEEE Transactions on Haptics*, *9*(4), 515–522. Available from https://doi.org/10.1109/TOH.2016.2591952.

Cushing, H. (1909). A note upon the faradic stimulation of the postcentral gyrus in conscious patients. *Brain: A Journal of Neurology*, *32*, 44–53. Available from https://doi.org/10.1093/brain/32.1.44.

Dadarlat, M. C., O'Doherty, J. E., & Sabes, P. N. (2015). A learning-based approach to artificial sensory feedback leads to optimal integration. *Nature Neuroscience*, *18*(1), 138–144. Available from https://doi.org/10.1038/nn.3883.

Devecioğlu, İ., & Güçlü, B. (2017). Psychophysical correspondence between vibrotactile intensity and intracortical microstimulation for tactile neuroprostheses in rats. *Journal of Neural Engineering*, *14*(1), 016010. Available from https://doi.org/10.1088/1741-2552/14/1/016010.

Dhillon, G. S., & Horch, K. W. (2005). Direct neural sensory feedback and control of a prosthetic arm. *IEEE Transactions on Neural Systems and Rehabilitation Engineering: A Publication of the IEEE Engineering in Medicine and Biology Society*, *13*(4), 468–472. Available from https://doi.org/10.1109/TNSRE.2005.856072.

Dobelle, W. H., & Mladejovsky, M. G. (1974). Phosphenes produced by electrical stimulation of human occipital cortex, and their application to the development of a prosthesis for the blind. *The Journal of Physiology*, *243*(2), 553–576, 1.

Douglas, R. M. (1977). Long lasting synaptic potentiation in the rat dentate gyrus following brief high frequency stimulation. *Brain Research*, *126*(2), 361–365. Available from https://doi.org/10.1016/0006-8993(77)90733-8.

Driscoll, N., Rosch, R. E., Murphy, B. B., Ashourvan, A., Vishnubhotla, R., Dickens, O. O., Johnson, A. T. C., Davis, K. A., Litt, B., Bassett, D. S., Takano, H., & Vitale, F. (2021). Multimodal in vivo recording using transparent graphene microelectrodes illuminates spatiotemporal seizure dynamics at the microscale. *Communications Biology*, *4*(1). Available from https://doi.org/10.1038/s42003-021-01670-9, 136–136.

Dudek, S. M., & Bear, M. F. (1992). Homosynaptic long-term depression in area CA1 of hippocampus and effects of N-methyl-D-aspartate receptor blockade. *Proceedings of the National Academy of Sciences*, *89*(10), 4363–4367. Available from https://doi.org/10.1073/pnas.89.10.4363.

Eles, J. R., & Kozai, T. D. Y. (2020). In vivo imaging of calcium and glutamate responses to intracortical microstimulation reveals distinct temporal responses of the neuropil and somatic compartments in layer II/III neurons. *Biomaterials*, *234*, 119767. Available from https://doi.org/10.1016/j.biomaterials.2020.119767.

Eles, J. R., Stieger, K. C., & Kozai, T. D. Y. (2021). The temporal pattern of intracortical microstimulation pulses elicits distinct temporal and spatial recruitment of cortical neuropil and neurons. *Journal of Neural Engineering*, *18*(1), 015001. Available from https://doi.org/10.1088/1741-2552/abc29c.

Enatsu, R., & Mikuni, N. (2016). Invasive evaluations for epilepsy surgery: a review of the literature. *Neurologia Medico-Chirurgica*, *56*(5), 221–227. Available from https://doi.org/10.2176/nmc.ra.2015-0319.

Englot, D. J., & Chang, E. F. (2014). Rates and predictors of seizure freedom in resective epilepsy surgery: An update. *Neurosurgical Review*, *37*(3), 389–405. Available from https://doi.org/10.1007/s10143-014-0527-9.

Even-Chen, N., Muratore, D. G., Stavisky, S. D., Hochberg, L. R., Henderson, J. M., Murmann, B., & Shenoy, K. V. (2020). Power-saving design opportunities for wireless intracortical brain–computer interfaces. *Nature Biomedical Engineering*, *4*(10), 984–996. Available from https://doi.org/10.1038/s41551-020-0595-9.

Flesher, S. N., Collinger, J. L., Foldes, S. T., Weiss, J. M., Downey, J. E., Tyler-Kabara, E. C., Bensmaia, S. J., Schwartz, A. B., Boninger, M. L., & Gaunt, R. A. (2016). Intracortical microstimulation of human somatosensory cortex. *Science Translational Medicine*, *8*(361), 361ra141. Available from https://doi.org/10.1126/scitranslmed.aaf8083.

Foxe, J. J., & Snyder, A. C. (2011). The role of alpha-band brain oscillations as a sensory suppression mechanism during selective attention. *Frontiers in Psychology*, *2*. Available from https://doi.org/10.3389/fpsyg.2011.00154.

Frey, J. N., Ruhnau, P., Leske, S., Siegel, M., Braun, C., & Weisz, N. (2016). The tactile window to consciousness is characterized by frequency-specific integration and segregation of the primary somatosensory cortex. *Scientific Reports*, *6*(1), 20805. Available from https://doi.org/10.1038/srep20805.

Ganji, M., Kaestner, E., Hermiz, J., Rogers, N., Tanaka, A., Cleary, D., Lee, S. H., Snider, J., Halgren, M., Cosgrove, G. R., Carter, B. S., Barba, D., Uguz, I., Malliaras, G. G., Cash, S. S., Gilja, V., Halgren, E., & Dayeh, S. A. (2018). Development and translation of PEDOT:PSS microelectrodes for intraoperative monitoring. *Advanced Functional Materials*, *28*(12), 1700232. Available from https://doi.org/10.1002/adfm.201700232.

Garcia-Cortadella, R., Schwesig, G., Jeschke, C., Illa, X., Gray, A. L., Savage, S., Stamatidou, E., Schiessl, I., Masvidal-Codina, E., Kostarelos, K., Guimerà-Brunet, A., Sirota, A., & Garrido, J. A. (2021). Graphene active sensor arrays for long-term and wireless mapping of wide frequency band epicortical brain activity. *Nature Communications*, *12*(1), 211. Available from https://doi.org/10.1038/s41467-020-20546-w.

Gescheider, G. A., Wright, J. H., & Verrillo, R. T. (2008). *Information-processing channels in the tactile sensory system: A psychophysical and physiological analysis* (1st ed.). Psychology Press, https://doi.org/10.4324/9780203890004.

Gilron, R., Little, S., Perrone, R., Wilt, R., de Hemptinne, C., Yaroshinsky, M. S., … Starr, P. A. (2021). Long-term wireless streaming of neural recordings for circuit discovery and adaptive stimulation in individuals with Parkinson's disease. *Nat Biotechnol*. Available from https://doi.org/10.1038/s41587-021-00897-5.

Godlove, J. M., Whaite, E. O., & Batista, A. P. (2014). Comparing temporal aspects of visual, tactile, and microstimulation feedback for motor control. *Journal of Neural Engineering*, *11* (4), 046025. Available from https://doi.org/10.1088/1741-2560/11/4/046025.

Guler, S., Dannhauer, M., Roig-Solvas, B., Gkogkidis, A., Macleod, R., Ball, T., Ojemann, J. G., & Brooks, D. H. (2018). Computationally optimized ECoG stimulation with local safety constraints. *NeuroImage*, *173*, 35–48. Available from https://doi.org/10.1016/j.neuroimage.2018.01.088, July 2017.

Hebert, J. S., Olson, J. L., Morhart, M. J., Dawson, M. R., Marasco, P. D., Kuiken, T. A., & Chan, K. M. (2014). Novel targeted sensory reinnervation technique to restore functional hand sensation after transhumeral amputation. *IEEE Transactions on Neural Systems and Rehabilitation Engineering: A Publication of the IEEE Engineering in Medicine and Biology Society*, *22*(4), 765–773. Available from https://doi.org/10.1109/TNSRE.2013.2294907.

Hermann, J. K., & Capadona, J. R. (2018). Understanding the role of innate immunity in the response to intracortical microelectrodes. *Critical Reviews in Biomedical Engineering*, *46* (4), 341–367. Available from https://doi.org/10.1615/CritRevBiomedEng.2018027166.

Herron, J., Stanslaski, S., Chouinard, T., Corey, R., Denison, T., & Orser, H. (2018). Bi-directional brain interfacing instrumentation. In: *2018 IEEE international instrumentation and measurement technology conference (I2MTC)* (pp. 1–6). https://doi.org/10.1109/I2MTC.2018.8409795

Hill, N. J., Gupta, D., Brunner, P., Gunduz, A., Adamo, M. A., Ritaccio, A., & Schalk, G. (2012). Recording human electrocorticographic (ECoG) signals for neuroscientific research and real-time functional cortical mapping. *Journal of Visualized Experiments: JoVE*, *64*. Available from https://doi.org/10.3791/3993.

Hiremath, S. V., Tyler-Kabara, E. C., Wheeler, J. J., Moran, D. W., Gaunt, R. A., Collinger, J. L., Foldes, S. T., Weber, D. J., Chen, W., Boninger, M. L., & Wang, W. (2017). Human perception of electrical stimulation on the surface of somatosensory cortex. *PLoS ONE*, *12* (5), 1–16. Available from https://doi.org/10.1371/journal.pone.0176020.

Histed, M. H., Bonin, V., & Reid, R. C. (2009). Direct activation of sparse, distributed populations of cortical neurons by electrical microstimulation. *Neuron*, *63*(4), 508–522. Available from https://doi.org/10.1016/j.neuron.2009.07.016.

Hochberg, L. R., Bacher, D., Jarosiewicz, B., Masse, N. Y., Simeral, J. D., Vogel, J., Haddadin, S., Liu, J., Cash, S. S., Van Der Smagt, P., & Donoghue, J. P. (2012). Reach and grasp by people with tetraplegia using a neurally controlled robotic arm. *Nature*, *485*(7398), 372–375. Available from https://doi.org/10.1038/nature11076.

Horsager, A., Greenwald, S. H., Weiland, J. D., Humayun, M. S., Greenberg, R. J., McMahon, M. J., Boynton, G. M., & Fine, I. (2009). Predicting visual sensitivity in retinal prosthesis patients. *Investigative Ophthalmology & Visual Science*, *50*(4), 1483–1491. Available from https://doi.org/10.1167/iovs.08-2595.

Ingham, C. A., Hood, S. H., Taggart, P., & Arbuthnott, G. W. (1998). Plasticity of synapses in the rat neostriatum after unilateral lesion of the nigrostriatal dopaminergic pathway. *Journal of Neuroscience*, *18*(12), 4732–4743. Available from https://doi.org/10.1523/JNEUROSCI.18-12-04732.1998.

Ishitobi, M, Nakasato, N, Suzuki, K, Nagamatsu, K, Shamoto, H, & Yoshimoto, T (2000). Remote discharges in the posterior language area during basal temporal stimulation.. *Neuroreport*, *11*(13), 2997–3000. Available from https://doi.org/10.1097/00001756-200009110-00034.

Johansson, R. S., & Flanagan, J. R. (2009). Coding and use of tactile signals from the fingertips in object manipulation tasks. *Nature Reviews Neuroscience, 10*(5), 345–359. Available from https://doi.org/10.1038/nrn2621.

Johnson, L. A., Wander, J. D., Sarma, D., Su, D. K., Fetz, E. E., & Ojemann, J. G. (2013). Direct electrical stimulation of the somatosensory cortex in humans using electrocorticography electrodes: A qualitative and quantitative report. *Journal of Neural Engineering, 10*(3). Available from https://doi.org/10.1088/1741-2560/10/3/036021.

Keller, C. J., Honey, C. J., Mégevand, P., Entz, L., Ulbert, I., & Mehta, A. D. (2014). Mapping human brain networks with cortico-cortical evoked potentials. *Philosophical Transactions of the Royal Society of London. Series B, Biological Sciences, 369*(1653). Available from https://doi.org/10.1098/rstb.2013.0528.

Khodagholy, D., Gelinas, J. N., Thesen, T., Doyle, W., Devinsky, O., Malliaras, G. G., & Buzsáki, G. (2015). NeuroGrid: Recording action potentials from the surface of the brain. *Nature Neuroscience, 18*(2), 310–315. Available from https://doi.org/10.1038/nn.3905.

Khodagholy, D., Gelinas, J. N., Zhao, Z., Yeh, M., Long, M., Greenlee, J. D., Doyle, W., Devinsky, O., & Buzsáki, G. (2016). Organic electronics for high-resolution electrocorticography of the human brain. *Science Advances, 2*(11), e1601027. Available from https://doi.org/10.1126/sciadv.1601027.

Kim, D., Jun, S. C., & Kim, H.-I. (2011). Computational study of subdural and epidural cortical stimulation of the motor cortex. In *Annual international conference of the IEEE engineering in medicine and biology society. IEEE engineering in medicine and biology society. Annual international conference, 2011*, 7226–7229. https://doi.org/10.1109/IEMBS.2011.6091826

Kim, S., Callier, T., Tabot, G. A., Gaunt, R. A., Tenore, F. V., & Bensmaia, S. J. (2015). Behavioral assessment of sensitivity to intracortical microstimulation of primate somatosensory cortex. *Proceedings of the National Academy of Sciences of the United States of America, 112*(49), 15202–15207. Available from https://doi.org/10.1073/pnas.1509265112.

Klaes, C., Shi, Y., Kellis, S., Minxha, J., Revechkis, B., & Andersen, R. A. (2014). A cognitive neuroprosthetic that uses cortical stimulation for somatosensory feedback. *Journal of Neural Engineering, 11*(5), 056024. Available from https://doi.org/10.1088/1741-2560/11/5/056024.

Klink, P. C., Dagnino, B., Gariel-Mathis, M.-A., & Roelfsema, P. R. (2017). Distinct feedforward and feedback effects of microstimulation in visual cortex reveal neural mechanisms of texture segregation. *Neuron, 95*(1), 209–220. Available from https://doi.org/10.1016/j.neuron.2017.05.033, e3.

Koerbitzer, B., Krauss, P., Nick, C., Yadav, S., Schneider, J. J., & Thielemann, C. (2016). Graphene electrodes for stimulation of neuronal cells. *2D Materials, 3*(2), 024004. Available from https://doi.org/10.1088/2053-1583/3/2/024004.

Kombos, T., & Süss, O. (2009). Neurophysiological basis of direct cortical stimulation and applied neuroanatomy of the motor cortex: A review. *Neurosurgical Focus, 27*(4), E3. Available from https://doi.org/10.3171/2009.8.FOCUS09141.

Kramer, D. R., Lamorie-Foote, K., Barbaro, M., Lee, M. B., Peng, T., Gogia, A., Nune, G., Liu, C. Y., Kellis, S. S., & Lee, B. (2020). Utility and lower limits of frequency detection in surface electrode stimulation for somatosensory brain-computer interface in humans. *Neurosurgical Focus, 48*(2), E2. Available from https://doi.org/10.3171/2019.11.FOCUS19696.

Kudela, P., & Anderson, W. S. (2015). Computational modeling of subdural cortical stimulation: A quantitative spatiotemporal analysis of action potential initiation in a high-density multicompartment model. *Neuromodulation: Journal of the International Neuromodulation Society, 18*(7), 552–564. Available from https://doi.org/10.1111/ner.12327, discussion 564-565.

Kuiken, T. A., Marasco, P. D., Lock, B. A., Harden, R. N., & Dewald, J. P. A. (2007). Redirection of cutaneous sensation from the hand to the chest skin of human amputees with targeted reinnervation. *Proceedings of the National Academy of Sciences*, *104*(50), 20061–20066. Available from https://doi.org/10.1073/pnas.0706525104.

Kuzum, D., Takano, H., Shim, E., Reed, J. C., Juul, H., Richardson, A. G., de Vries, J., Bink, H., Dichter, M. A., Lucas, T. H., Coulter, D. A., Cubukcu, E., & Litt, B. (2014). Transparent and flexible low noise graphene electrodes for simultaneous electrophysiology and neuroimaging. *Nature Communications*, *5*(May), 5259. Available from https://doi.org/10.1038/ncomms6259, 5259.

Lee, B., Kramer, D., Armenta Salas, M., Kellis, S., Brown, D., Dobreva, T., Klaes, C., Heck, C., Liu, C., & Andersen, R. A. (2018). Engineering artificial somatosensation through cortical stimulation in humans. *Frontiers in Systems Neuroscience*, *12*, 24. Available from https://doi.org/10.3389/fnsys.2018.00024.

Libet, B., Alberts, W. W., Wright, E. W., Delattre, L. D., Levin, G., & Feinstein, B. (1964). Production of threshold levels of conscious sensation by electrical stimulation of human somatosensory cortex. *Journal of Neurophysiology*, *27*(4), 546–578. Available from https://doi.org/10.1152/jn.1964.27.4.546.

Lloyd-Jones, D., Adams, R. J., Brown, T. M., Carnethon, M., Dai, S., De Simone, G., Ferguson, T. B., Ford, E., Furie, K., Gillespie, C., Go, A., Greenlund, K., Haase, N., Hailpern, S., Ho, P. M., Howard, V., Kissela, B., Kittner, S., Lackland, D., & Wylie-Rosett, J. (2010). Heart disease and stroke statistics—2010 update. *Circulation*, *121*(7), e46–e215. Available from https://doi.org/10.1161/CIRCULATIONAHA.109.192667.

Logothetis, N. K., Augath, M., Murayama, Y., Rauch, A., Sultan, F., Goense, J., Oeltermann, A., & Merkle, H. (2010). The effects of electrical microstimulation on cortical signal propagation. *Nature Neuroscience*, *13*(10), 1283–1291. Available from https://doi.org/10.1038/nn.2631.

Makin, T. R., Holmes, N. P., & Ehrsson, H. H. (2008). On the other hand: Dummy hands and peripersonal space. *Behavioural Brain Research*, *191*(1), 1–10. Available from https://doi.org/10.1016/j.bbr.2008.02.041.

Matsumoto, R., Nair, D. R., LaPresto, E., Bingaman, W., Shibasaki, H., & Lüders, H. O. (2006). Functional connectivity in human cortical motor system: A cortico-cortical evoked potential study. *Brain*, *130*(1), 181–197. Available from https://doi.org/10.1093/brain/awl257.

McIntyre, C. C., & Grill, W. M. (2000). Selective microstimulation of central nervous system neurons. *Annals of Biomedical Engineering*, *28*(3), 219–233. Available from https://doi.org/10.1114/1.262.

Merrill, D. R., Bikson, M., & Jefferys, J. G. R. (2005). Electrical stimulation of excitable tissue: Design of efficacious and safe protocols. *Journal of Neuroscience Methods*, *141*(2), 171–198. Available from https://doi.org/10.1016/j.jneumeth.2004.10.020.

Michelson, N. J., Eles, J. R., Vazquez, A. L., Ludwig, K. A., & Kozai, T. D. Y. (2019). Calcium activation of cortical neurons by continuous electrical stimulation: Frequency dependence, temporal fidelity, and activation density. *Journal of Neuroscience Research*, *97*(5), 620–638. Available from https://doi.org/10.1002/jnr.24370.

Millard, D. C., Whitmire, C. J., Gollnick, C. A., Rozell, C. J., & Stanley, G. B. (2015). Electrical and optical activation of mesoscale neural circuits with implications for coding. *Journal of Neuroscience*, *35*(47), 15702–15715. Available from https://doi.org/10.1523/JNEUROSCI.5045-14.2015.

Miller, L. E., & Weber, D. J. (2011). Brain training: Cortical plasticity and afferent feedback in brain-machine interface systems. *IEEE Transactions on Neural Systems and Rehabilitation Engineering: A Publication of the IEEE Engineering in Medicine and Biology Society*, *19* (5), 465–467. Available from https://doi.org/10.1109/TNSRE.2011.2168989.

Moran, D. (2010). Evolution of brain-computer interface: Action potentials, local field potentials and electrocorticograms. *Current Opinion in Neurobiology*, *20*(6), 741–745. Available from https://doi.org/10.1016/j.conb.2010.09.010.

Mulkey, R. M., & Malenka, R. C. (1992). Mechanisms underlying induction of homosynaptic long-term depression in area CA1 of the hippocampus. *Neuron*, *9*(5), 967–975. Available from https://doi.org/10.1016/0896-6273(92)90248-c.

Muller, L., Felix, S., Shah, K.G., Lee, K., Pannu, S., & Chang, E.F. (2016). Thin-film, high-density micro electrocorticographic decoding of a human cortical gyrus. In: *2016 38th annual international conference of the IEEE engineering in medicine and biology society (EMBC)*, 1528–1531. https://doi.org/10.1109/EMBC.2016.7591001

Muller, L., Hamilton, L., Edwards, E., Bouchard, K., & Chang, E. (2016). Spatial resolution dependence on spectral frequency in human speech cortex electrocorticography. *Journal of Neural Engineering*, *13*(5), 056013. Available from https://doi.org/10.1088/1741-2560/13/5/056013.

Muller, L., Rolston, J. D., Fox, N. P., Knowlton, R., Rao, V. R., & Chang, E. F. (2018). Direct electrical stimulation of human cortex evokes high gamma activity that predicts conscious somatosensory perception. *Journal of Neural Engineering*, *15*(2), 026015. Available from https://doi.org/10.1088/1741-2552/aa9bf9.

Nathan, S. S., Sinha, S. R., Gordon, B., Lesser, R. P., & Thakor, N. V. (1993). Determination of current density distributions generated by electrical stimulation of the human cerebral cortex. *Electroencephalography and Clinical Neurophysiology*, *86*(3), 183–192. Available from https://doi.org/10.1016/0013-4694(93)90006-H.

Nowak, L. G., & Bullier, J. (1998). Axons, but not cell bodies, are activated by electrical stimulation in cortical gray matter. I. Evidence from chronaxie measurements. *Experimental Brain Research*, *118*(4), 477–488. Available from https://doi.org/10.1007/s002210050304.

Nurse, E. S., John, S. E., Freestone, D. R., Oxley, T. J., Ung, H., Berkovic, S. F., O'Brien, T. J., Cook, M. J., & Grayden, D. B. (2018). Consistency of long-term subdural electrocorticography in humans. *IEEE Transactions on Biomedical Engineering*, *65*(2), 344–352. Available from https://doi.org/10.1109/TBME.2017.2768442.

O'Doherty, J. E., Lebedev, M. A., Ifft, P. J., Zhuang, K. Z., Shokur, S., Bleuler, H., & Nicolelis, M. A. L. (2011). Active tactile exploration using a brain–machine–brain interface. *Nature*, *479*(7372), 228–231. Available from https://doi.org/10.1038/nature10489.

Öztürk, S., Devecioğlu, İ., Beygi, M., Atasoy, A., Mutlu, S., Özkan, M., & Güçlü, B. (2019). Real-time performance of a tactile neuroprosthesis on awake behaving rats. *IEEE Transactions on Neural Systems and Rehabilitation Engineering*, *27*(5), 1053–1062. Available from https://doi.org/10.1109/TNSRE.2019.2910320.

Park, D.-W., Ness, J. P., Brodnick, S. K., Esquibel, C., Novello, J., Atry, F., Baek, D.-H., Kim, H., Bong, J., Swanson, K. I., Suminski, A. J., Otto, K. J., Pashaie, R., Williams, J. C., & Ma, Z. (2018). Electrical neural stimulation and simultaneous in vivo monitoring with transparent graphene electrode arrays implanted in GCaMP6f mice. *ACS Nano*, *12*(1), 148–157. Available from https://doi.org/10.1021/acsnano.7b04321.

Penfield, W., & Boldrey, E. (1937). Somatic motor and sensory representation in the cerebral cortex of man as studied by electrical stimulation. *Brain: A Journal of Neurology*, *60*, 389–443. Available from https://doi.org/10.1093/brain/60.4.389.

Pepin, E., Uehlin, J., Micheletti, D., Perlmutter, S. I., & Rudell, J. C. (2016). A high-voltage compliant, electrode-invariant neural stimulator front-end in 65nm bulk-CMOS. In *ESSCIRC conference* 2016: 42nd European *solid-state circuits conference* (pp. 229–232). https://doi.org/10.1109/ESSCIRC.2016.7598284

Pranti, A. S., Schander, A., Bödecker, A., & Lang, W. (2018). PEDOT: PSS coating on gold microelectrodes with excellent stability and high charge injection capacity for chronic neural interfaces. *Sensors and Actuators B: Chemical, 275*, 382–393. Available from https://doi.org/10.1016/j.snb.2018.08.007.

Prime, D., Woolfe, M., O'Keefe, S., Rowlands, D., & Dionisio, S. (2020). Quantifying volume conducted potential using stimulation artefact in CCEP. *Journal of Neuroscience Methods, 337*.

Ranck, J. B. (1975). Which elements are excited in electrical stimulation of mammalian central nervous system: A review. *Brain Research, 98*(3), 417–440. Available from https://doi.org/10.1016/0006-8993(75)90364-9.

Raspopovic, S., Capogrosso, M., Petrini, F. M., Bonizzato, M., Rigosa, J., Di Pino, G., Carpaneto, J., Controzzi, M., Boretius, T., Fernandez, E., Granata, G., Oddo, C. M., Citi, L., Ciancio, A. L., Cipriani, C., Carrozza, M. C., Jensen, W., Guglielmelli, E., Stieglitz, T., & Micera, S. (2014). Restoring natural sensory feedback in real-time bidirectional hand prostheses. *Science Translational Medicine, 6*(222), 222ra19. Available from https://doi.org/10.1126/scitranslmed.3006820.

Rattay, F., & Wenger, C. (2010). Which elements of the mammalian central nervous system are excited by low current stimulation with microelectrodes? *Neuroscience, 170*(2), 399–407. Available from https://doi.org/10.1016/j.neuroscience.2010.07.032.

Ray, P. G., Meador, K. J., Smith, J. R., Wheless, J. W., Sittenfeld, M., & Clifton, G. L. (1999). Physiology of perception: Cortical stimulation and recording in humans. *Neurology, 52*(5), 1044–1049. Available from https://doi.org/10.1212/wnl.52.5.1044.

Ray, S., Niebur, E., Hsiao, S. S., Sinai, A., & Crone, N. E. (2008). High-frequency gamma activity (80–150 Hz) is increased in human cortex during selective attention. *Clinical Neurophysiology: Official Journal of the International Federation of Clinical Neurophysiology, 119*(1), 116–133. Available from https://doi.org/10.1016/j.clinph.2007.09.136.

Reilly, J. P. (1998). *Applied bioelectricity: From electrical stimulation to electropathology*. Springer.

Reilly, J. P., & Diamant, A. M. (2011). *Electrostimulation: theory, applications, and computational model*. Artech House.

Rohde, M., Luca, M. D., & Ernst, M. O. (2011). The rubber hand illusion: Feeling of ownership and proprioceptive drift do not go hand in hand. *PLoS ONE, 6*(6), e21659. Available from https://doi.org/10.1371/journal.pone.0021659.

Romanelli, P., Piangerelli, M., Ratel, D., Gaude, C., Costecalde, T., Puttilli, C., Picciafuoco, M., Benabid, A., & Torres, N. (2018). A novel neural prosthesis providing long-term electrocorticography recording and cortical stimulation for epilepsy and brain-computer interface. *Journal of Neurosurgery*, 1–14. Available from https://doi.org/10.3171/2017.10.JNS17400.

Schiefer, M., Tan, D., Sidek, S. M., & Tyler, D. J. (2016). Sensory feedback by peripheral nerve stimulation improves task performance in individuals with upper limb loss using a myoelectric prosthesis. *Journal of Neural Engineering, 13*(1), 016001. Available from https://doi.org/10.1088/1741-2560/13/1/016001.

Semprini, M., Bennicelli, L., & Vato, A. (2012). A parametric study of intracortical microstimulation in behaving rats for the development of artificial sensory channels. In *Annual international conference of the IEEE engineering in medicine and biology society. IEEE engineering in medicine and biology society. Annual international conference*, 2012, 799–802. https://doi.org/10.1109/EMBC.2012.6346052

Seo, H., Kim, D., & Jun, S. C. (2015). Computational study of subdural cortical stimulation: Effects of simulating anisotropic conductivity on activation of cortical neurons. *PLoS ONE*, *10*(6), e0128590. Available from https://doi.org/10.1371/journal.pone.0128590.

Shepherd, G. M. (2004). *The synaptic organization of the brain* (5th ed.). Oxford University Press.

Shimada, S., Kunii, N., Kawai, K., Matsuo, T., Ishishita, Y., Ibayashi, K., & Saito, N. (2017). Impact of volume-conducted potential in interpretation of cortico-cortical evoked potential: Detailed analysis of high-resolution electrocorticography using two mathematical approaches. *Clinical Neurophysiology*, *128*(4), 549–557. Available from https://doi.org/10.1016/j.clinph.2017.01.012.

Smith, W. A., Uehlin, J. P., Perlmutter, S. I., Rudell, J. C., & Sathe, V. S. (2017). A scalable, highly-multiplexed delta-encoded digital feedback ECoG recording amplifier with common and differential-mode artifact suppression. In *2017 symposium on VLSI circuits*, C172–C173. https://doi.org/10.23919/VLSIC.2017.8008470

Stanslaski, S., Afshar, P., Cong, P., Giftakis, J., Stypulkowski, P., Carlson, D., Linde, D., Ullestad, D., Avestruz, A., & Denison, T. (2012). Design and validation of a fully implantable, chronic, closed-loop neuromodulation device with concurrent sensing and stimulation. *IEEE Transactions on Neural Systems and Rehabilitation Engineering*, *20*(4), 410–421. Available from https://doi.org/10.1109/TNSRE.2012.2183617.

Tabot, G. A., Kim, S. S., Winberry, J. E., & Bensmaia, S. J. (2015). Restoring tactile and proprioceptive sensation through a brain interface. *Neurobiology of Disease*, *83*, 191–198. Available from https://doi.org/10.1016/j.nbd.2014.08.029.

Tan, D. W., Schiefer, M. A., Keith, M. W., Anderson, J. R., Tyler, J., & Tyler, D. J. (2014). A neural interface provides long-term stable natural touch perception. *Science Translational Medicine*, *6*(257), 257ra138. Available from https://doi.org/10.1126/scitranslmed.3008669.

Tehovnik, E. J., Tolias, A. S., Sultan, F., Slocum, W. M., & Logothetis, N. K. (2006). Direct and indirect activation of cortical neurons by electrical microstimulation. *Journal of Neurophysiology*, *96*(2), 512–521. Available from https://doi.org/10.1152/jn.00126.2006.

Telford, W. M. W. M. (1990). *Applied geophysics* (2nd ed.). Cambridge University Press.

Uehlin, J. P., Smith, W. A., Pamula, V. R., Pepin, E. P., Perlmutter, S., Sathe, V., & Rudell, J. C. (2020). A single-chip bidirectional neural interface with high-voltage stimulation and adaptive artifact cancellation in standard CMOS. *IEEE Journal of Solid-State Circuits*, *55* (7), 1749–1761. Available from https://doi.org/10.1109/JSSC.2020.2991524.

Vincent, M., Rossel, O., Hayashibe, M., Herbet, G., Duffau, H., Guiraud, D., & Bonnetblanc, F. (2016). The difference between electrical microstimulation and direct electrical stimulation—Towards new opportunities for innovative functional brain mapping? *Reviews in the Neurosciences*, *27*(3), 231–258. Available from https://doi.org/10.1515/revneuro-2015-0029.

Wang, P. T., King, C. E., McCrimmon, C. M., Lin, J. J., Sazgar, M., Hsu, F. P. K., Shaw, S. J., Millet, D. E., Chui, L. A., Liu, C. Y., Do, A. H., & Nenadic, Z. (2016). Comparison of decoding resolution of standard and high-density electrocorticogram electrodes. *Journal of Neural Engineering*, *13* (2), 026016. Available from https://doi.org/10.1088/1741-2560/13/2/026016.

Wang, X., Gkogkidis, C. A., Iljina, O., Fiederer, L. D. J., Henle, C., Mader, I., Kaminsky, J., Stieglitz, T., Gierthmuehlen, M., & Ball, T. (2017). Mapping the fine structure of cortical activity with different micro-ECoG electrode array geometries. *Journal of Neural Engineering*, *14*(5), 056004. Available from https://doi.org/10.1088/1741-2552/aa785e.

Weber, D. J., Friesen, R., & Miller, L. E. (2012). Interfacing the somatosensory system to restore touch and proprioception: Essential considerations. *Journal of Motor Behavior*, *44*(6), 403–418. Available from https://doi.org/10.1080/00222895.2012.735283.

Wongsarnpigoon, A., & Grill, W. M. (2008). Computational modeling of epidural cortical stimulation. *Journal of Neural Engineering*, *5*(4), 443–454. Available from https://doi.org/10.1088/1741-2560/5/4/009.

Zhou, A., Johnson, B. C., & Muller, R. (2018). Toward true closed-loop neuromodulation: Artifact-free recording during stimulation. *Current Opinion in Neurobiology*, *50*, 119–127. Available from https://doi.org/10.1016/j.conb.2018.01.012.

Zhou, A., Santacruz, S. R., Johnson, B. C., Alexandrov, G., Moin, A., Burghardt, F. L., Rabaey, J. M., Carmena, J. M., & Muller, R. (2019). A wireless and artefact-free 128-channel neuromodulation device for closed-loop stimulation and recording in non-human primates. *Nature Biomedical Engineering*, *3*(1), 15–26. Available from https://doi.org/10.1038/s41551-018-0323-x.

Chapter 15

Design of intracortical microstimulation patterns to control the location, intensity, and quality of evoked sensations in human and animal models

David A. Bjånes[1,2] and Chet T. Moritz[1,2,3,4]
[1]*Department of Electrical and Computer Engineering, University of Washington, Seattle, WA, United States,* [2]*Center for Neurotechnology, University of Washington, Seattle, WA, United States,* [3]*Department of Rehabilitation Medicine, University of Washington, Seattle, WA, United States,* [4]*Department of Physiology & Biophysics, University of Washington, Seattle, WA, United States*

ABSTRACT

One of the most fundamental questions yet to be answered is how electrical stimulation patterns trigger neural networks in complex activation patterns and evoke somatosensory percepts. This chapter explores some historical landmarks in our search of this road map and ends with the state-of-the-art brain–machine interfaces, which are able to evoke both naturalistic proprioceptive and cutaneous sensations in human clinical trials. Starting with the discovery of electrical sensitivity of the brain in the 1800s, the authors explore the various animal models used to study how each parameter used in modern stimulation patterns influences reported sensations. Fundamental breakthroughs in electrical stimulation research with intracortical stimulation devices have rapidly accelerated in recent years, providing exciting new opportunities to deliver novel therapies to patients with a wide variety of disorders and diseases. Stimulation therapies have recently been approved to treat essential tremor, depression, and various psychiatric conditions. In addition, a wide range of new targets for stimulation has yielded tangible results: visual percepts, memory enhancement, and somatosensation on the arms, hands, head, and face. This exciting progress has been fueled by new collaborations between engineers designing novel stimulation devices, delivering ever more complex patterns to cortical targets, and by neuroscientists discovering cortical circuitry and

Somatosensory Feedback for Neuroprosthetics. DOI: https://doi.org/10.1016/B978-0-12-822828-9.00018-6

understanding the networks responsible for processing sensation evoked by these stimulation patterns. This chapter lays the groundwork for understanding state-of-the-art stimulation pattern design and explains the current challenges yet to be tackled. By exploring these fundamental questions, the field can create assistive devices which feel naturalistic and enable device users to better experience the world around them.

Keywords: Somatosensory percept; brain–machine interface; intracortical stimulation; brain stimulation; electroceuticals; stimulation pattern

15.1 Introduction

Dynamic somatosensory information is an essential component of the human experience. Access to tactile and proprioceptive sensations enables us to dexterously manipulate tools and objects, confidently explore our world, and navigate potentially dangerous environments. Many of the daily activities we take for granted would be incredibly difficult, if not impossible, without sensation. Actions such as buttoning a shirt, picking up a cube of sugar, typing on a keyboard, or simply cleaning ourselves after the toilet, would all be prohibitively laborious.

For those who have lost somatosensation due to spinal cord injury, stroke, or neurodegeneration, (Harkema et al., 2011; Navarro et al., 2005) restoring this fundamental ability would be a significant improvement in quality of life (Anderson, 2004; French et al., 2010). For the majority of people with somatosensory loss, absent or reduced sensation across their body poses significant health challenges. Lack of bladder sensations exacerbates issues with incontinence and intentional voiding; lack of the feeling of pressure on the legs or torso can lead to pressure ulcers, and lack of natural input to the sensory cortex from typical neural pathways can lead to debilitating neuropathy and pain (Cole, 1995; Ievins & Moritz, 2017; Sainburg et al., 1993). For people with limited motor movement, the lack of somatic sensations of the skin, muscles, and joints can lead to loss of dexterity. Additionally, this condition can make them prone to accidental injury, as well as depriving them of familiar, comforting sensations such as the warmth of a family member's touch or the grip of a handshake.

While no single device can yet alleviate all these issues, brain–machine interfaces (BMIs) offer a clear path toward one of the most tantalizing goals of rehabilitation technology (Fig. 15.1): the ability to communicate with the body's nervous system (Bensmaia & Miller, 2014; Weber et al., 2012). By exploiting the electrical properties of neurons in the brain, researchers have been able to both record cortical activity to "read" information (Aflalo et al., 2015; Fetz, 1999; Hochberg et al., 2012; Kennedy & Bakay, 1998; Musallam et al., 2004; Serruya et al., 2002; Taylor, 2002; Velliste et al., 2008; Wessberg et al., 2000) and electrically stimulate cortical areas to "write" information directly to the brain (Berg et al., 2013; Fitzsimmons et al., 2007; Flesher et al., 2016; Fridman et al., 2010; O'Doherty et al., 2011; Romo et al., 1998; Thomson et al., 2013).

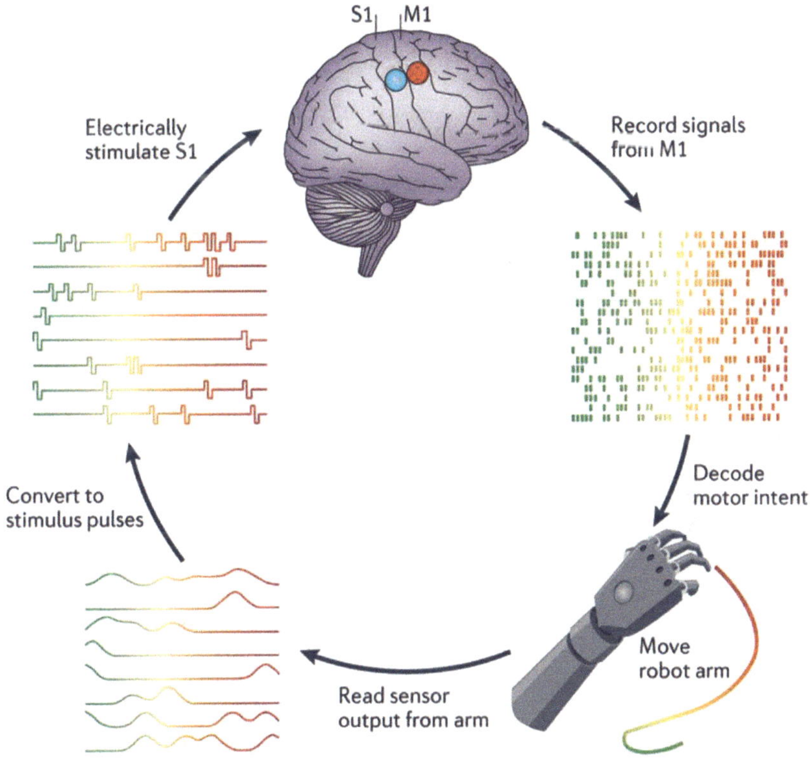

FIGURE 15.1 Illustration of a bidirectional brain–machine interface (BMI) for control of an assistive device. Command signals for the prosthetic device are captured by decoding patterns in recorded neural activity from the primary motor cortex (M1). Mechanical sensor data from the device are converted into an electrical stimulus pattern. This pattern is delivered to the brain via electrical stimulation of primary sensory cortex (S1), evoking a somatosensory percept. *Reproduced with permission from Bensmaia, S. J., & Miller, L. E. (2014). Restoring sensorimotor function through intracortical interfaces: Progress and looming challenges. Nature Reviews Neuroscience, 15(5), 313–325. <https://doi.org/10.1038/nrn3724>.*

Research in BMI technology has delivered some impressive milestones to date (Carmena et al., 2003; Fetz, 1969; Hochberg et al., 2012; Kennedy & Bakay, 1998; Musallam et al., 2004; Nicolelis & Lebedev, 2009; Santhanam et al., 2006; Serruya et al., 2002; Velliste et al., 2008; Wessberg et al., 2000). This includes enabling human participants to control prosthetic limbs (Collinger et al., 2013; Hochberg et al., 2012) helping them re-experience physical touch years after injury (Armenta Salas et al., 2018; Flesher et al., 2016; Wang et al., 2013), and even restoring communication with completely locked-in patients. Achieving control over greater degrees of freedom (DOF) has allowed accurate control of robotic arms (Johannes et al., 2011; Resnik et al., 2012) with up to 10 DOF (Wodlinger et al., 2015). However, this control currently falls short of the natural dexterity and fluidity required to perform complex tasks, particularly grasping and object manipulation (Fagg et al., 2007), possibly because these open-loop controllers rely on the relatively slow visual feedback pathway rather than rapid touch sensation.

One of the motivations for restoring high-resolution somatosensation is to improve the control of these assistive BMI devices (Bensmaia & Miller, 2014; Weber et al., 2012). By incorporating closed-loop sensory feedback, many researchers believe BMI control may approach natural movements (Klaes et al., 2014; Tan et al., 2014; Venkatraman et al., 2009). Enabling fluid, dexterous control over external devices including computer cursors and robotic arms (Collinger et al., 2013; Hochberg et al., 2012; Velliste et al., 2008; Wessberg et al., 2000), can dramatically improve independence and quality of life.

Direct brain stimulation is a powerful tool, playing a vital role in the development and deployment of assistive technologies such as BMIs. Stimulation-activated axons and eventually neurons provide researchers with an important tool to understand the connectivity between brain regions, functionality of cortical circuits, and their complex dynamical relationships (Clark et al., 2011; Histed et al., 2012; Penfield & Jasper, 1954). With this ability, researchers are able to create visual percepts (Lewis et al., 2016; Panetsos et al., 2011; Pezaris & Reid, 2007; Schmidt et al., 1996; Tehovnik et al., 2009; Zrenner, 2002), auditory "sounds" (Grayden & Clark, 2006; Lee et al., 2001; Shepherd & McCreery, 2006), mitigate symptoms of essential tremor and Parkinson disease (Deuschl et al., 2006; Kumar et al., 1999; Kumar et al., 2003; Malekmohammadi et al., 2016), provide limited depression and psychiatric relief (Mayberg et al., 2005; Widge et al., 2014), enhance memory (Berger et al., 2011), and recreate somatosensation (Flesher et al., 2016; Houweling & Brecht, 2008; Klaes et al., 2014; O'Doherty et al., 2011; Romo et al., 1998; Schafer, 1888; Thomson et al., 2013).

In this chapter, we focus on cortical stimulation to recreate somatosensory percepts and on the integration of sensory feedback in a BMI device. We aim to provide a framework for understanding the state-of-the-art of these devices, existing barriers and challenges, and inspire potential future therapies and emerging medical applications.

15.2 Stimulation design

Researchers have used direct cortical stimulation to evoke a variety of somatosensations in both animal models (Fitzsimmons et al., 2007; Fridman et al., 2010; Klaes et al., 2014; O'Doherty et al., 2011; Romo et al., 1998; Tabot et al., 2013; Thomson et al., 2013) and human participants (Armenta Salas et al., 2018; Flesher et al., 2016). The first convincing demonstration of this technology began with the seminal experiments by Romo in the late 1990s (Tabot et al., 2013). Electrical stimulation was used to precisely replicate the sensation of vibration. This sensation was so accurate that nonhuman primates (NPHs) were unable to discriminate between a mechanical stimulus on the fingertip and the electrical stimulation in the sensory cortex.

This fundamental breakthrough inspired the rapid development of stimulation techniques, electrode design, localization of key brain regions, and human clinical trials. Researchers have implanted four human volunteers to date with penetrating cortical electrode arrays, enabling direct stimulation of cortical sensory regions to restore somatosensation (Armenta Salas et al., 2018; Flesher et al., 2016). These participants have reported a variety of evoked proprioceptive and cutaneous sensations such as vibration, pressure, movement of a body region, and mechanical tingling. These sensations can be elicited across entire regions of the arm or hand, or can be as specific as an individual fingertip (Fig. 15.2).

Crucial to the deployment of these proof-of-concept studies in humans, was the development of several different electrode technologies, allowing access to different scales of cortical regions. Microwire and silicon substrate arrays can target a specific neural population of one to tens of neurons per electrode, each at a precise depth (Maynard et al., 1997). Epicortical devices which lie on the surface of the cortex, allow access to a much larger population (thousands to millions of neurons per electrode), but sacrifice temporal and spatial specificity (Penfield & Jasper, 1954). These devices, which were designed for recording electrocorticography (ECoG), can stimulate at much higher current levels due to their low impedance, activating large cortical regions simultaneously (Fig. 15.3). ECoG electrodes are commonly used in clinical environments for other procedures, such as cortical monitoring and seizure mapping for epilepsy patients, giving BMI researchers an opportunity to study somatosensation elicited via stimulation in larger patient populations

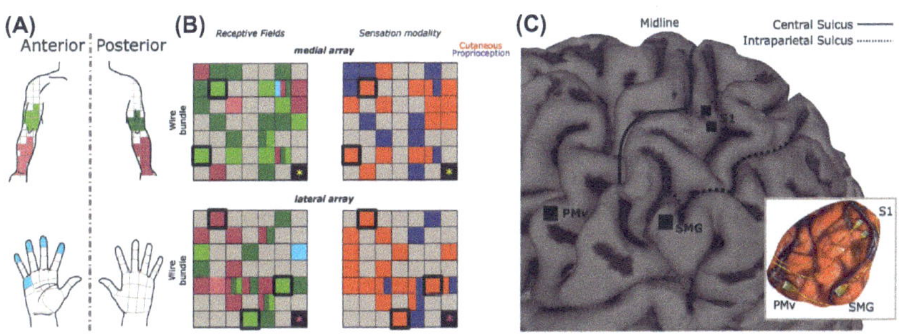

FIGURE 15.2 Somatosensory percepts were evoked via intracortical microstimulation (ICMS) of primary sensory cortex in a tetraplegic participant (Armenta Salas et al., 2018). Varying stimulation parameters and electrodes modulated a percept's intensity, duration and location. (A) Elicited somatosensation reported by the subject activated both larger regions spanning the arm, and localized to a very specific region on a finger palm or arm. (B) A general somatotopy emerged by electrically stimulating neural circuits through each electrode individually, across the arm, forearm, and hand. (C) Intracortical "Utah" arrays (Maynard et al., 1997) were placed into somatotopic regions of the human brain corresponding to the right hand and arm. Regions of interest were identified using fMRI during behavioral tasks which activated arm and hand sensory areas in S1. *Adapted with permission from Armenta Salas et al. (2018).*

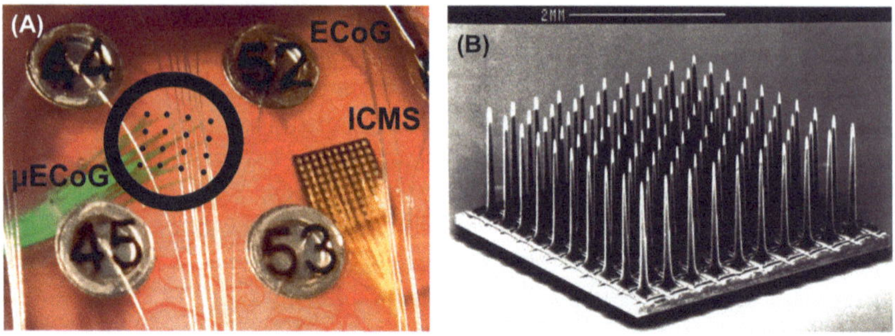

FIGURE 15.3 (A) Illustration of the relative size of stimulation devices for direct cortical interfaces. (B) Utah array for delivering ICMS. *(Fig. 15.3A) Adapted with permission from Bradley, G., & House, P. (n.d.).* Epilepsy electrodes. *<http://greger.lab.asu.edu/image-gallery/epilepsy-electrodes/> and Kellis, S. S., House, P. A., Thomson, K. E., Brown, R., & Greger, B. (2009). Human neocortical electrical activity recorded on nonpenetrating microwire arrays: applicability for neuroprostheses.* Neurosurgical Focus, 27*(1), E9. <https://doi.org/10.3171/2009.4.focus0974>. (Fig. 15.3B) Adapted with permission from Maynard, E. M., Nordhausen, C. T., & Normann, R. A. (1997). The Utah intracortical electrode array: A recording structure for potential brain–computer interfaces.* Electroencephalography and Clinical Neurophysiology, 102*(3), 228–239. <https://doi.org/10.1016/S0013-4694(96)95176-0> and Normann, R. A., & Fernandez, E. (2016). Clinical applications of penetrating neural interfaces and Utah Electrode Array technologies.* Journal of Neural Engineering, 13*(6), 061003. <https://doi.org/10.1088/1741-2560/13/6/061003>.*

(Collins et al., 2017; Johnson et al., 2013; Lee et al., 2018; Wang et al., 2013). One patient was even able to use the elicited sensations as guided feedback for a motor task (Cronin et al., 2016).

15.2.1 Historical experiments

The first demonstration of electrical stimulation to activate animal nerves was performed in the 1800s by Luigi Galvani, after whom galvanic stimulation is named. The ability to harness electricity to contract frog muscle fibers via stimulation of the sciatic nerve was a revolution in the field of neuroscience. Functional connectivity of brain regions was still up for debate, but with the Fritsch and Hitzig demonstrations in 1870 of somatotopic organization in a canine cortex, principles of cortical organization began to emerge (Fritsch & Hitzig, 1870; Schafer, 1888). Drawing from their experience with cortical injuries on the battlefield, they used monopolar stimulation pulses from a primitive battery, such as a Leyden jar, which they calibrated by touching it with their tongue to feel a slight tingling sensation. By observing muscle twitches in response to localized stimulation of various cortical areas, they published a primitive map between cortical areas and body regions (Taylor & Gross, 2003). Fig. 15.4 shows another type of stimulator used around the same time.

This monopolar, direct current (DC) stimulation paradigm was employed by Penfield and Boldrey nearly half a decade later, to build the first

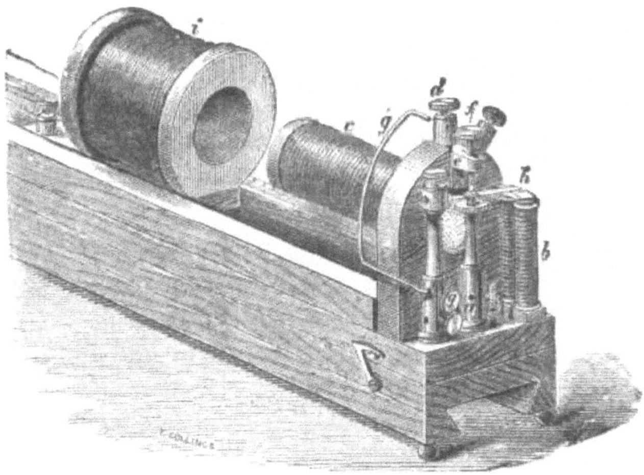

FIGURE 15.4 A sketch of a faradic stimulator used by Dr. Ferrier, a contemporary of Fritsch and Hitzig (Taylor & Gross, 2003). The circuit is opened and closed mechanically through a switch (f), while the primary coil (c) induces a current in the secondary coil (i). This stimulator had a single parameter it could adjust. By adjusting the distance between the coils, the amplitude of the current could be set. *Reproduced with permission from Taylor, C. S. R., & Gross, C. G. (2003). Twitches versus movements: A story of motor cortex.* The Neuroscientist, *9(5), 332–342. <https://doi.org/10.1177/1073858403257037>.*

comprehensive topographical map for motor and sensory areas of the human cortex (Fig. 15.5; Penfield & Boldrey, 1937). Their data, painstakingly collected through mapping experiments with epilepsy patients under local anesthetics, described cortical representations for finger, lip, leg, trunk, hand, and leg (Penfield & Boldrey, 1937; Penfield & Rasmussen, 1950). They partnered with an artist to draw the first "homunculus," an exaggerated diagram of the body. This cartoon had proportions of each body part corresponding to the relative size of spatial area dedicated to processing in the cortex. While they did not document the full complexity, such as multiple body maps at S1 cortex, this was a monumental achievement in the study of the sensorimotor system (Schott, 1993).

Up until this point in the field, all stimulation had been performed by connecting the patient (or animal subject) to a battery, producing a constant source of current once contact with the electrodes was made and only stopped once the experimenter lifted the electrodes, disconnecting the circuit. Throughout the 1800s, as the field of electromagnetism enabled the development of alternating currents (AC) to carry electricity over long distances, various electricity sources became more widely available. The on/off currents were found to be more effective in eliciting responses (Dibner et al., 1871; Qing, 2015). As electrical technology further expanded, switches for interrupting the constant flow of electricity from either AC or DC sources

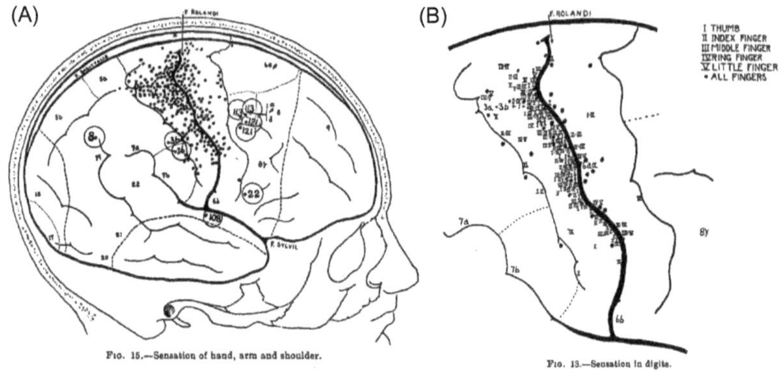

FIGURE 15.5 Receptive field mapping of sensory areas showing composite locations for localized percepts for hand, arm, and shoulder (A), and individual digits (B). *Reproduced with permission from Penfield, W., & Boldrey, E. (1937). Somatic motor and sensory representation in the cerebral cortex of man as studied by electrical stimulation.* Brain, *60(4), 389–443. <https://doi.org/10.1093/brain/60.4.389>.*

were used by Emil du Bois-Reymond (Du Bois-Reymond, 1852), Georges Weiss, and Louis Lapicque, who worked independently to show brief pulses of electricity were better signal patterns for stimulation (Irnich, 2002; Lapicque, 2007; Weiss, 1901). As du Bois-Reymond postulated the "general law of excitation of nerves," he was the first to mention the now canonical stimulation parameters amplitude and pulse width (Lapicque, 1909).

15.2.2 Electrical effects on neurophysiology

In 1952, Hodgkin and Huxley published their seminal work on the cable-axon model, providing a theoretical framework to answer many of the empirical discoveries in the nascent field of neuroscience (Hodgkin & Huxley, 1952). This led to an understanding of the "push/pull" of electrical current which activates neurons, that discrete pulses work better than continuous DC current, and neurons closer to the cathodic electrode, rather than the anodic electrode, are preferentially activated. In this chapter, we focus on behavioral aspects of stimulation, but a basic primer on how various stimulation parameters cause the neural tissue to respond may be helpful to the reader. Further exploration of this topic, such as electrochemistry and particular reactions which occur at the electrode–tissue interface, can be found in other sources such as Bard and Faulkner (2000), and a more in-depth discussion of how nerves and neurons are affected by electrical currents can be found in Tehovnik (1996). Other chapters in this book, *Somatosensory Feedback for Neuroprosthetics*, also contain relevant information.

In 1841, Emil's du Bois-Reymonds first demonstrated that the nerve current, like the muscle current, decreases during stimulation (Bard & Faulkner, 2000;

Schuetze, 1983). With the aid of his student, Bernstein, they became the first to record the transient overshoot of this current, yielding the discovery of the action potential (Bernstein, 1868; Bernstein, 1902). Due to limitations in the available technology, it would take more than 50 years before the first drawings of an action potential would be published (DeFelice, 1981; Fig. 15.6). Simply put,

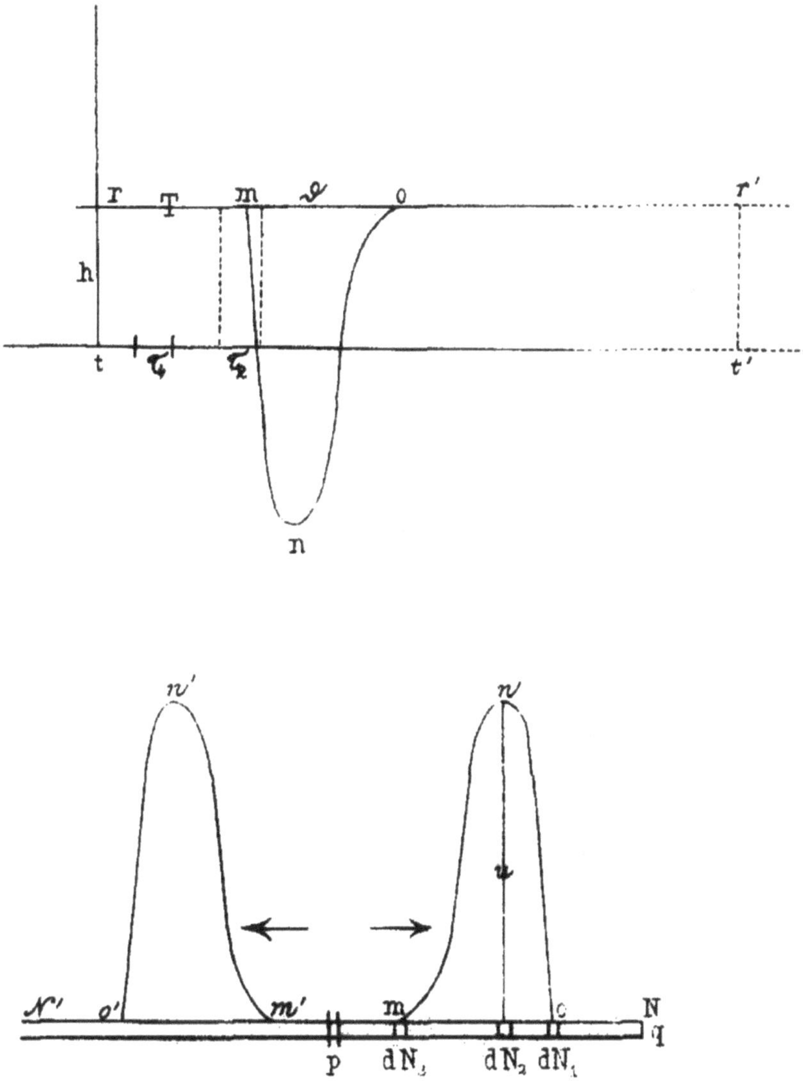

FIGURE 15.6 A sketch of the first published action potential. *Reproduced with permission from Schuetze, S. M. (1983). The discovery of the action potential.* Trends in Neurosciences, 6 *(C), 164–168. <https://doi.org/10.1016/0166-2236(83)90078-4>.*

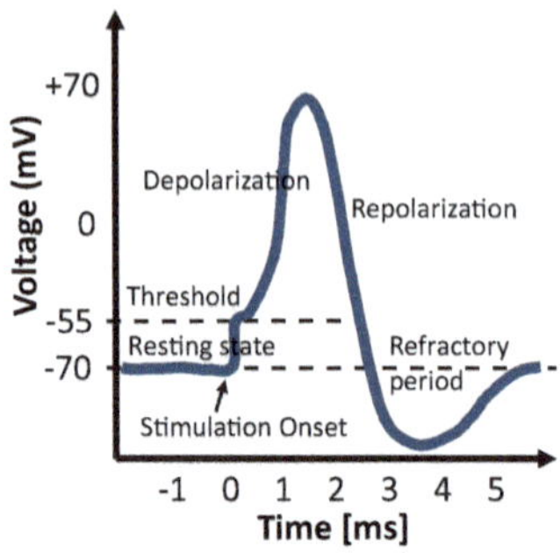

FIGURE 15.7 Simplified sequence of an action potential.

stimulation waveforms (either mono- or biphasic) mimic an incoming neural impulse from an upstream neuron to the targeted axon cell membrane. By providing a massive change in ionization around the cell via an electrically negative pulse, a rapid depolarization of the cell membrane is achieved, triggering the neuron to fire an action potential toward the downstream neurons connected via its axon (Fig. 15.7).

15.3 Parameterization

Several early questions in the nascent field of electrical stimulation of the brain were: (1) how do we use electrical current-controlled sources to best activate nerves and the cortex? and (2) why are some electrical waveforms better than others? Applying this to the current challenge of delivering stimulation to elicit sensory percepts, we also seek answers to similar questions of how the stimulation parameters effect an elicited sensation in a research participant or eventual end-user.

In 1955, John Lilly proposed using a biphasic waveform (Lilly et al., 1955) as a safer pattern for delivering charge to the brain. This has become the standard for nearly all modern stimulation protocols. Coined "zero net flow," Lilly's pattern consisted of a negative cathodic current pulse, immediately followed by an equal, but opposite, polarity anodic pulse, assuring that no net change in ions accumulated in the tissue surrounding the electrode tip. This was both for the safety of the tissue being stimulated, and also ensured a rapid recovery of the cell membrane in order to be triggered repeatedly (Merrill et al., 2005; Rajan et al., 2015; Shannon, 1992). Modern stimulation patterns can now create both temporal and spatial patterns across entire networks of neurons or precisely replicate sequences of recorded action potentials in peripheral nerves evoked by mechanical stimuli. However, nearly all stimulation paradigms can be defined by a short set of parameters, as illustrated in Fig. 15.8.

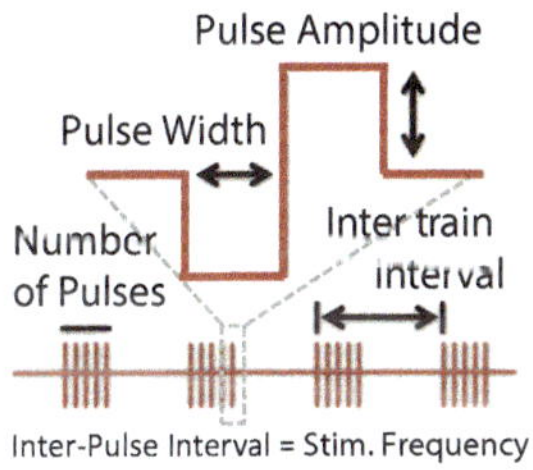

FIGURE 15.8 Stimulation pattern can be defined by these five parameters: amplitude, pulse width, frequency, number of pulses, and the intertrain interval (Bjanes & Moritz, 2019). Note that specifying the interpulse interval is equivalent to specifying the frequency. Similarly, the number of pulses is equivalent to the duration of the pulse train.

Five stimulation parameters can uniquely describe nearly all stimulation patterns: amplitude, pulse width, frequency, number of pulses, and the intertrain interval. Some researchers define each stimulation train by the number of pulses rather than a time duration. While pulse phases should be charge balanced for safety, some patterns utilize nonsymmetric phases. These are typically a leading, narrower, and stronger cathodic phase followed by a wider, lower in amplitude, anodic phase. Often a short ($<100\ \mu s$) break between the phases is included in the waveform.

Stimulation can be delivered through a variety of metal electrodes such as platinum or tungsten, and the material determines how much current can be safely delivered (Merrill et al., 2005; Rajan et al., 2015; Shannon, 1992). The excitability of the neural cells is network dependent; however, a dominating factor is the injected current spread, which is attenuated proportionally to the square of the distance from the electrode tip (Clark et al., 2011; Logothetis et al., 2010; Ranck, 1975; Tehovnik, 1996). The effects of microstimulation on neural networks largely become a function of the stimulation parameters chosen and location of the brain being stimulated.

15.3.1 Sensory brain–machine interfaces

For nearly 100 years after the discovery that somatosensory percepts can be elicited by electrical stimulation of cortical neurons, researchers still had little control over which sensation was elicited. Given the somatotopic organization of primary sensory cortex, targeted stimulation could "aim" at a particular region of the body (the left arm, for example), but replicating specific naturalistic percepts remained out of reach. That is, until seminal experiments in NPHs by Romo and colleagues (Romo et al., 1998) demonstrated the power of manipulating stimulation parameters. They isolated a particular "flutter" sensation on skin, evoked by stimulating mechanoreceptors with a mechanical actuator (Talbot & Mountcastle, 1968). By recording from the somatosensory cortex while the primates experienced this mechanical sensation, they identified the receptive fields in area 3b for that particular "flutter" sensation localized on a fingertip. Since primates cannot verbally report their experiences of somatosensation, they were trained using a two-alternative forced-choice (2AFC) task (Fechner, 1966) to discriminate between the sensations elicited by the mechanical stimuli, or by electrical stimulation applied to the brain at the identified cortical area. By varying the frequency of the biphasic electrical stimulation pulses, and measuring the discrimination performance of primates, they could precisely map which frequency of stimulation "felt" identical to the mechanical stimulus. This work demonstrated that electrical stimulation may precisely mimic cortical neural coding, activating neural populations with enough specificity to evoke a sensation indistinguishable from a sensation "naturally" elicited via a mechanical stimulus. This singular breakthrough spawned an entire subfield of research, namely, neural interfaces for replicating somatosensation.

Romo's work inspired a raft of new research directions. First and foremost, researchers began discussing scalability, and the feasibility of a human sensory prosthesis. Would every single sensation need to be individually mapped, localized to a particular brain region, and relate to a specific stimulus pattern? Given the enormous complexity of the human somatosensory system, this biomimetic approach would require observation of a nearly infinite combinatorial problem of possible sensations across the human body. The temporally dynamic, nonstationary properties observed in typical brain networks could be extremely difficult to replicate in artificially generated electrical patterns. In addition, experimentally measuring the empirical map between all sensations and stimulus locations and patterns may prove impossible.

As a solution to this complex problem, another approach emerged called sensory substitution or artificial stimulation. Rather than learning the near-infinite complex mapping, perhaps engineers and researchers may exploit the plasticity of cortical networks to activate a neuron or small population with an arbitrary pattern, and teach a user to associate that particular stimulation pattern with a specific sensation. After learning achieves a certain level of proficiency, this new stimulus might be perceived as "natural" to the subject (Armenta Salas et al., 2018).

15.3.2 Biomimetic stimulation pattern design

The most intuitive electrical stimulation parameter is likely pulse amplitude, and it emerged as an early candidate for testing the biomimetic sensory feedback approach. To test the specificity of the sensory neuron populations, researchers stimulated a single neuron using the juxtacellular technique (Pinault, 1996). Exploiting the well-documented evoked response to whisking in the rat barrel cortex (Fanselow & Nicolelis, 1999; Ferezou et al., 2007; Lee et al., 2008), researchers used extremely low-amplitude pulses

(2–5 μA) to elicit sensory percepts. By training these rodents to lick left or right, depending on the stimulation pattern delivered, they demonstrated that single neuron stimulation can evoke distinguishable sensations (Houweling & Brecht, 2008).

A decade after Romo's work, advances in electrode and computing technology enabled researchers to choose one of 100 electrodes on which to record or stimulate (Tabot et al., 2013). Replicating prior experiments, this time with a different stimulation parameter and a different mechanoreceptor class in the skin, stimulation patterns with varying amplitude elicited similar sensations to the mechanical indentation of the fingertip (Fig. 15.9). Primates were trained to hold their fingers at a precise location, while a mechanical probe created slight (0.5–2 mm) indentations in the skin. The performance of the primates on a 2AFC task demonstrated direct mapping between stimulation pulse amplitude and perceived mechanical deflection of the skin.

While the evidence for biomimetic approaches to electrical stimulation is encouraging, it may be difficult to implement in end users with prior sensory loss. For example, people with spinal cord injury or neurodegeneration have lost neural connection between the peripheral sensory fibers and the somatosensory cortex. Researchers cannot mechanically stimulate different body parts and determine the equivalent cortical activation patterns, because these people have either total or partial sensory loss. In order to address this additional layer of difficulty, researchers have utilized "artificial stimulation" patterns, that is, not derived from observed neural patterns. By leveraging

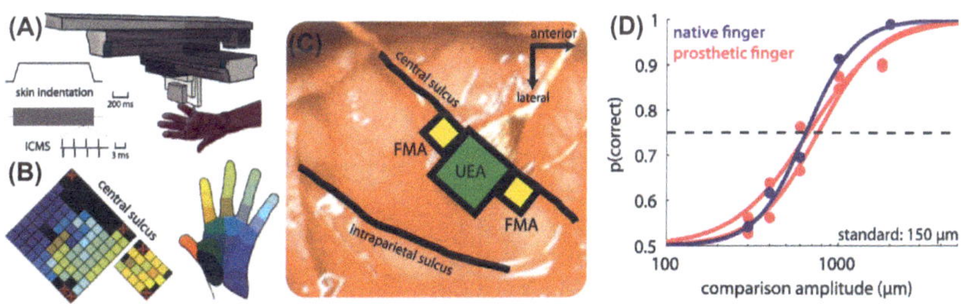

FIGURE 15.9 (A) Researchers created a two-alternative forced-choice (2AFC) discrimination task to test if an ICMS evoked sensory percept "felt" indistinguishable from a mechanical stimulus applied to the fingertip of a primate. (B) The somatotopic map of sensory regions of the primate hand was measured across two electrode arrays by recording the activation of neurons at each electrode site when the corresponding region of the hand was mechanically stimulated. (C) Electrode array placement in the somatosensory cortex. (D) The 2AFC discrimination task assessed whether the electrical stimulation felt similar to mechanical trials. Discrimination using the electrical stimulation (*pink*) and the mechanical stimuli (*purple*) was psychophysically equivalent. *Reproduced with permission from Tabot, G. A., Dammann, J. F., Berg, J. A., Tenore, F. V., Boback, J. L., Vogelstein, R. J., & Bensmaia, S. J. (2013). Restoring the sense of touch with a prosthetic hand through a brain interface.* Proceedings of the National Academy of Sciences of the United States of America, 110*(45), 18279–18284. <https://doi.org/10.1073/pnas.1221113110>.*

the plasticity of the brain, and careful training schemes, progress has been made using the sensory substitution approach in both animal and human participants (Armenta Salas et al., 2018; Bjanes & Moritz, 2019; Flesher et al., 2016; Klaes et al., 2014; O'Doherty et al., 2011; Tabot et al., 2013).

15.3.3 Sensory substitution stimulation

Sensory substitution can be used to solve a real-world task. For example, researchers gave rodents the ability to "see" an "invisible light" via electrical stimulation (Fig. 15.10; Thomson et al., 2013). Follow-up experiments added further evidence that rats and monkeys could discriminate between different frequency stimulation patterns below 100 Hz (Fridman et al., 2010; Romo et al., 2000), including a real-time feedback frequency code to navigate around obstacles in a 2-D environment (Dadarlat et al., 2015). In addition to stimulation frequency, the stimulus amplitude showed promise for encoding a variety of sensations. Even greater than frequency, amplitude was highly discriminable across a wide range of values from 0 to 100 μA, and could be used to modulate the intensity of a sensation (Berg et al., 2013; Bjanes & Moritz, 2019; Devecioğlu & Güçlü, 2017; Flesher et al., 2016; Kim et al., 2015; Öztürk et al., 2019).

More recently, researchers began using just-noticeable-difference (JND) tasks to optimize the stimulation pattern design. They first measured the bandwidth of these parameters. Primates and rats were trained in 2AFC tasks to distinguish between two sets of stimuli (Bjanes & Moritz, 2019; Kim et al., 2015),

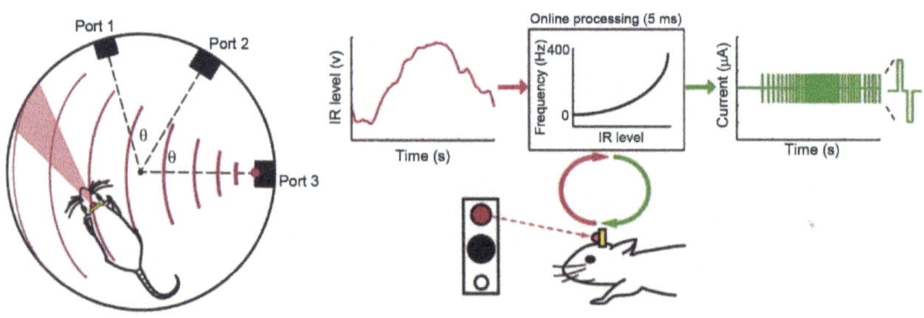

FIGURE 15.10 Perceiving an "invisible light" (Thomson et al., 2013). Animals were placed in a circular arena with three liquid reward ports along the walls. An "invisible light" was created using an infrared LED illuminated above one lick port at a time. Animals were rewarded with water for licking only at the illuminated port. Rats cannot naturally see in the infrared spectrum of light, so they needed to rely on stimulation to guide them toward the IR light. A photodetector was mounted on a helmet worn by the animal, and proportional stimulation was applied when the animal was facing the correct port. The frequency of stimulation pulses varied with the strength of the detected IR light. When the animal was facing directly toward the IR LED, the stimulation frequency was at its maximum (400 Hz). As the animal pointed away, the stimulation frequency decreased. *Reproduced with permission from Thomson, E. E., Carra, R., & Nicolelis, M. A. L. (2013). Perceiving invisible light through a somatosensory cortical prosthesis.* Nature Communications, 4, *1482. <https://doi.org/10.1038/ncomms2497>.*

and decide which stimulus evoked a percept with a higher intensity. By varying the stimulation patterns of each stimuli, researchers could plot the performance of the animals discriminating between the evoked sensations. For sensitive parameters, the animals could distinguish even slight changes in the stimulation patterns. However, stimulation patterns which required a large change before the animals reported diverging sensations were classified as insensitive.

In the interest of minimizing the amount of charge injected into the brain, researchers also studied the minimum pulse width and duration of the stimulation trains needed to evoke a response (Kim et al., 2015). These parameters are important because reducing the total amount of electric charge injected is advantageous for a number of reasons. First, it would save the battery life of a potential implanted medical device. Second, less charge would mean less risk for tissue damage. Finally, for a sensory feedback device, the data rate would be limited by the minimum duration needed to elicit a sensation.

15.3.4 Charge

To measure the empirical limits of the resolution of a sensory prosthetic device, researchers directly compared the sensitivity of each of the five classic stimulation parameters (Fig. 15.11; Bjanes & Moritz, 2019). These comparisons also enabled the discovery of a fundamental underlying stimulation variable which preferentially activates neural cells, a finding which could enable better predictive modeling of electrical stimulation. Each stimulation parameters' discriminability was explicitly measured in the same animal and in the same neural circuitry. Surprisingly, frequency discrimination (>100 Hz) was much less reliable than the other parameters. Amplitude and pulse width had the highest sensitivity for modulating somatosensory networks ($\sim 30\%$ each; Fig. 15.11). Further exploration showed that neural circuits were even more sensitive to the charge-per-phase than charge-per-time. If both amplitude and pulse width were modulated in tandem, animals were sensitive to only an 11% change in the modulated charge-per-pulse. This study also found that changes in the charge of the stimulation phase provided information that could be readily discriminated in about 100 milliseconds, rather than needing to integrate over the duration of the whole stimulus pattern. This bodes well for real-time applications of brain stimulation where the update rate will be a key parameter to increase the bandwidth of information conveyed to the cortex. Ongoing work is extending these discoveries to human trials.

15.4 Applications in human participants

Armed with knowledge gained from research in primate and rodent models, somatosensory BMIs took a leap into humans with the implantation of Utah

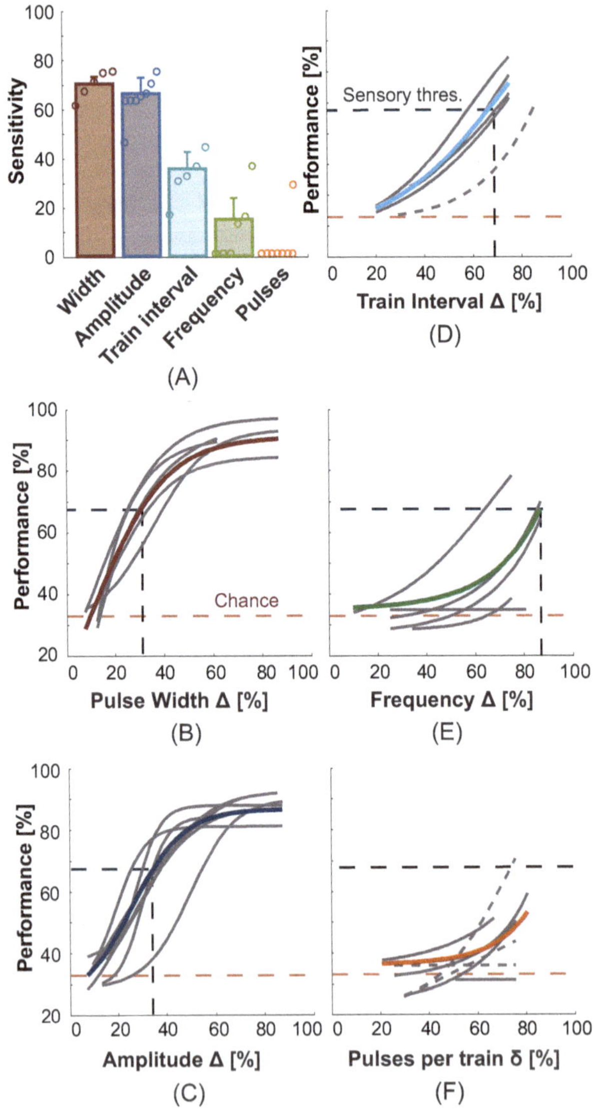

FIGURE 15.11 Perceptual sensitivity to different stimulation parameters. Rodents were trained to discriminate between two stimulation patterns, one higher intensity (e.g., 100 μA) and one lower intensity pattern (50 μA). Their ability to discriminate the higher and lower intensity patterns provided a "percent change" required to detect a difference. Each of the five parameters was individually measured to determine the just-noticeable-difference. Sensitivity measurement shows that rodents were most sensitive to detecting changes in the pulse width and amplitude of delivered stimulation patterns, which both encode the charge-per-pulse. *Reproduced with permission from Bjanes, D. A., & Moritz, C. T. (2019). A robust encoding scheme for delivering artificial sensory information via direct brain stimulation.* IEEE Transactions on Neural Systems and Rehabilitation Engineering, 27*(10), 1–1. <https://doi.org/10.1109/TNSRE.2019.2936739>.*

arrays and cortical surface electrocorticographic electrodes. Only human participants can give verbal reports to describe the quality and characteristics of somatosensation that BMIs elicit.

As in the early days of Penfield and Boldrey (1937), researchers first utilized existing clinical protocols in acute or intraoperative surgical cases. One set of studies utilized patients with intractable epilepsy who volunteered for this research (Hiremath et al., 2017; Johnson et al., 2013; Lee et al., 2018; Ray et al., 1999). In parallel, research has built upon brain–machine interface trials (Rousche & Normann, 1992) with people with paralysis (Armenta Salas et al., 2018; Flesher et al., 2016; Hochberg et al., 2012) to test stimulation parameters. They have used the same chronically implanted brain electrodes to study longitudinal effects of stimulation and collect the safety and effectiveness data needed for an eventual widely available assistive device.

15.4.1 Cortical surface stimulation

Patients with subdural electrodes (i.e., ECoG electrodes) implanted on the surface of somatosensory cortex acutely (for epilepsy) have reported a variety of sensations from electrical stimulation. In these exploratory studies, participants felt tingling sensations on various fingers and locations on the hand in response to stimulation trains of duration 200–2000 milliseconds (Ray et al., 1999). Using fMRI mapping procedures to carefully identify the somatotopic organization of S1, researchers were able to elicit "naturalistic" percepts, described as "wind running down the hand," "[a] light rub [on the lip]," and "[as if] something was wrapped around [the middle finger]" (Johnson et al., 2013). These achievements demonstrated the ability to create percepts as broad as an entire section of the arm or as localized as to a single finger (Fig. 15.12). However, to understand how multiple sensations from different body parts are created, and how sensations vary over time, chronic implants are needed.

The FDA granted approval for longer term chronic implants of ECoG electrode grids, allowing researchers to perform longitudinal studies. With the ability to record and stimulate from multiple cortical areas, the effects of modulating many of the stimulation parameters on evoked sensations can now be studied.

In one of the first demonstrations of real-time sensory feedback with ECoG, researchers used amplitude-modulated stimulation pulses to guide a participant along a trajectory (Cronin et al., 2016). Using an instrumented glove, they instructed the participant to control the aperture of their hand (from fingers spread wide open to closed grasp) according to a hidden trajectory. Using the percepts as sensory feedback, they could "feel" their way along the trajectory, receiving a high stimulus if the hand was too open, and a low stimulus if the hand was too closed. No stimulus was applied if the hand was within a margin of error about the trajectory.

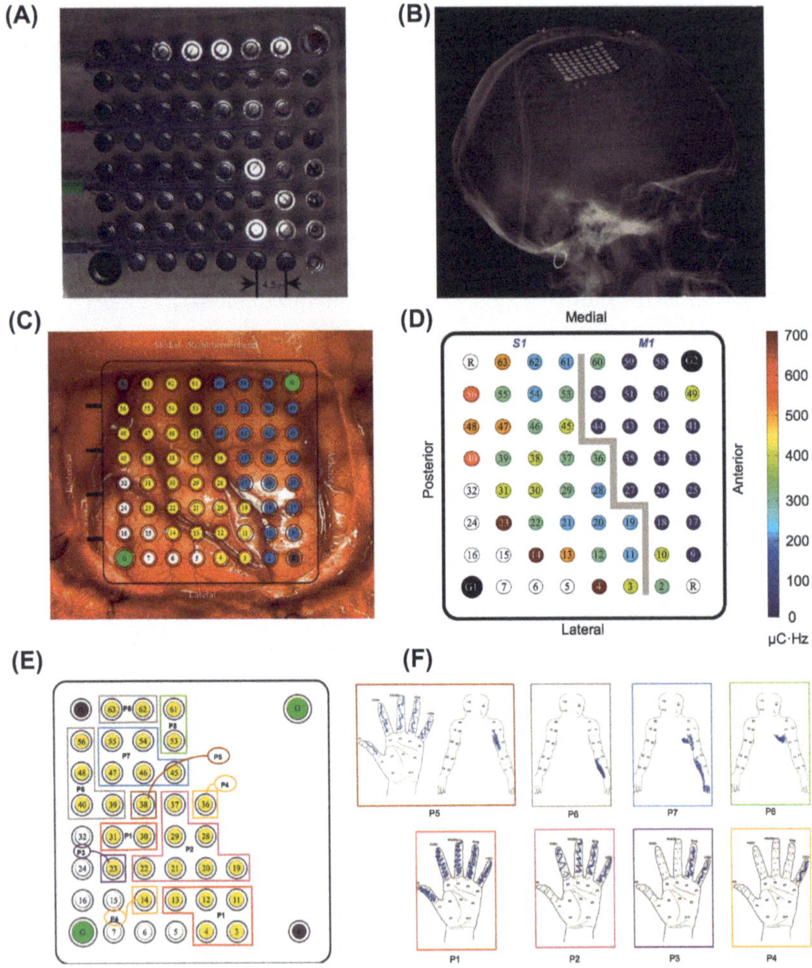

FIGURE 15.12 A patient with chronic epilepsy was implanted with a high-density ECoG grid (4 cm × 4 cm × 1 mm). Sixty-four platinum electrodes were placed over S1 and M1 (A, B, C, D; *gray* dividing line in D is the central sulcus). Stimulation was delivered on each electrode (E) and the participant verbally reported the quality of the sensations and drew the location of evoked sensations (F). *Reproduced with permission from Hiremath, S. V., Tyler-Kabara, E. C., Wheeler, J. J., Moran, D. W., Gaunt, R. A., Collinger, J. L., ... Wang, W. (2017). Human perception of electrical stimulation on the surface of somatosensory cortex.* Plos One, 12*(5), e0176020. <https://doi.org/10.1371/journal.pone.0176020>.*

A related study examined responses to modulating stimulation parameters such as frequency and amplitude, and participants reported sensations such as "pulsing," "electricity," and "feeling of movement" across different hand regions (Lee et al., 2018). These sensations varied in intensity with modulation of parameters, confirming the feasibility of using electrical stimulation of the brain surface for real-time sensory feedback of a motor task.

15.4.2 Intracortical microstimulation

In parallel with advances stimulating through ECoG electrodes, penetrating electrodes are being tested in human with the hope of enabling higher specificity and temporal resolution. Intracortical electrode arrays consist of a silicon substrate with an array of shanks with metal-coated tips ($< 30\ \mu m$). The typical workhorse device used in monkey and human BMIs is the Utah array (Fig. 15.3B), a 96-channel array arranged in a 10×10 grid with a spacing of 400 μm between each electrode (Maynard et al., 1997).

Penetrating electrodes permit ICMS, and can be implanted for several years in humans to test the longevity of this approach. Concurrent studies at both Pittsburgh and Caltech (Fig. 15.2) confirmed the stability of elicited sensations, by showing stability over a time of months to a year (Armenta Salas et al., 2018; Flesher et al., 2016). Intriguingly, by modulating the amplitude and frequency of the stimulation patterns, stable percepts of both cutaneous and proprioceptive sensations such as movement of a joint could be evoked via intracortical stimulation (Armenta Salas et al., 2018).

15.5 Bidirectional brain–machine interfaces

The previous sections presented each of the building blocks required for a complete system of direct cortical feedback. This includes electrode types (Sections 15.4.1 and 15.4.2), how to construct a stimulation pattern (Section 15.3), the sensations elicited by stimulation patterns (Section 15.4), and possible architectures for embedding information in the stimulation (Sections 15.3.2 and 15.3.3). Sensory feedback may then be applied in a closed-loop system that also includes a BMI for control of the environment (Fig. 15.1). Such an implantable device could provide the end-user with both real-time control of an assistive device and return task-related sensory feedback via electrical stimulation. This closed-loop motor and sensory feedback system is termed a bidirectional BMI (Klaes et al., 2014; O'Doherty et al., 2011; Venkatraman et al., 2009).

In one of the first examples of a real-time bidirectional BMI, researchers delivered stimulation to sensory areas of the primate brain as feedback during a BMI-guided exploration task (O'Doherty et al., 2011). The primate was trained to control a cursor in a virtual workspace via direct brain control, and explore three different sensory targets (Fig. 15.13). Upon entering each target, a different stimulation pattern was delivered as feedback and the primate was rewarded for identifying the target pattern. This study was a technological feat due to challenges introduced by large electrical artifacts obscuring the tiny recorded neural signals when injecting stimulation

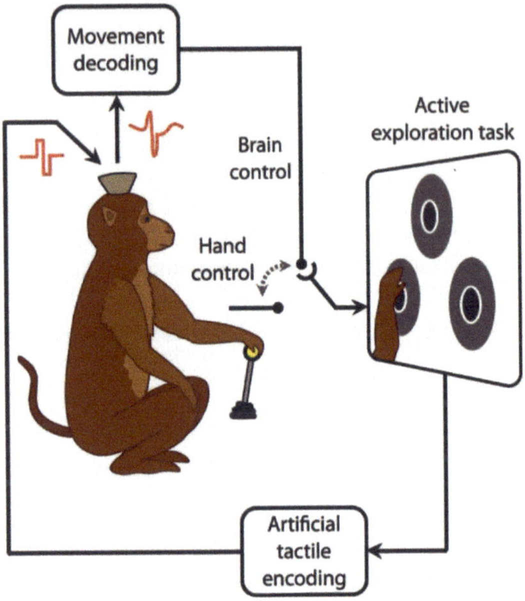

FIGURE 15.13 Researchers gave a primate control of a hand in virtual 2D space by decoding brain signals from M1 cortex. They trained the subject to explore three visually identical targets. When the animal hovered over each target, sensory feedback via electrical stimulation evoked different percepts. The primate was rewarded for identifying the correct target. *Reproduced with permission from O'Doherty, J. E., Lebedev, M. A., Ifft, P. J., Zhuang, K. Z., Shokur, S., Bleuler, H., & Nicolelis, M. A. L. (2011). Active tactile exploration using a brain-machine-brain interface.* Nature, *479(7372), 228–231. <https://doi.org/10.1038/nature10489>.*

currents in nearby brain regions (Hashimoto et al., 2002; Merrill et al., 2005; O'Shea & Shenoy, 2018; Ranck, 1975; Weiss et al., 2019). Several years later, this task was adapted to the control of a robot arm (Johannes et al., 2011) to perform a handbag task (Klaes et al., 2014), a canonical sensory feedback task where the subject must reach into a "handbag," and identify the object inside using touch alone. In this task, the primate was blinded to target locations in the workspace. By moving the robotic arm throughout the workspace, stimulation was delivered when the endpoint hovered over a target, and the primate was rewarded for identifying these hidden targets. Given these successful experiments, researchers have begun to adapt this bidirectional BMI approach for testing in human participants.

15.6 Conclusion

Our goal with this chapter was to provide the reader an overview of how the design of electrical stimulation patterns evolved and motivated somatosensory feedback research via examples of the state-of-the-art implanted devices in the brain. While the field is beginning to understand how neurons are activated by the spread of current from a stimulating electrode, many open questions remain. Critically, the network-level effects of electrical stimulation, and how such "unnatural" inputs propagate throughout the network to higher order brain regions to ultimately evoke sensations remains unknown. Scalable devices which can evoke a full repertoire of sensations across the body simultaneously will be necessary for full restoration of sensation to people with sensory loss due to spinal cord injury, stroke, or neurodegenerative disease. We hope this summary will inspire further research and the pursuit of novel solutions to restoring sensation by leveraging advances at the interface of neuroscience and neural engineering.

References

Aflalo, T., Kellis, S., Klaes, C., Lee, B., Shi, Y., Pejsa, K., & Andersen, R. A. (2015). Neurophysiology: Decoding motor imagery from the posterior parietal cortex of a tetraplegic human. *Science (New York, N.Y.)*, *348*(6237), 906–910. Available from https://doi.org/10.1126/science.aaa5417.

Anderson, K. D. (2004). Targeting recovery: Priorities of the spinal cord-injured population. *Journal of Neurotrauma*, *21*(10), 1371–1383. Available from https://doi.org/10.1089/neu.2004.21.1371.

Armenta Salas, M., Bashford, L., Kellis, S., Jafari, M., Jo, H., Kramer, D., ... Andersen, R. A. (2018). Proprioceptive and cutaneous sensations in humans elicited by intracortical microstimulation. *ELife*, *7*. Available from https://doi.org/10.7554/eLife.32904.

Bard, A. J., & Faulkner, L. R. (2000). *Electrochemical methods: Fundamentals and applications* (2nd ed.). Wiley. Available from https://www.abebooks.com/9780471043720/Electrochemical-Methods-Fundamentals-Applications-Bard-0471043729/plp.

Bensmaia, S. J., & Miller, L. E. (2014). Restoring sensorimotor function through intracortical interfaces: Progress and looming challenges. *Nature Reviews Neuroscience*, *15*(5), 313–325. Available from https://doi.org/10.1038/nrn3724.

Berg, J. A., Dammann, J. F., Tenore, F. V., Tabot, G. A., Boback, J. L., Manfredi, L. R., ... Bensmaia, S. J. (2013). Behavioral demonstration of a somatosensory neuroprosthesis. *IEEE Transactions on Neural Systems and Rehabilitation Engineering*, *21*(3), 500–507. Available from https://doi.org/10.1109/TNSRE.2013.2244616.

Berger, T. W., Hampson, R. E., Song, D., Goonawardena, A., Marmarelis, V. Z., & Deadwyler, S. A. (2011). A cortical neural prosthesis for restoring and enhancing memory. *Journal of Neural Engineering*, *8*(4). Available from https://doi.org/10.1088/1741-2560/8/4/046017.

Bernstein, J. (1868). *Pfluegers Arch*, *1*, 179–207. Available from https://www.scopus.com/record/display.uri?eid = 2-s2.0-84912432085&origin = inward&txGid = a60ea2923c0655f7e18934297a93a94d.

Bernstein, J. (1902). Untersuchungen zur Thermodynamik der bioelektrischen Ströme—Erster Theil. *Pflüger Archiv Für Die Gesammte Physiologie Des Menschen Und Der Thiere*, *92*(10–12), 521–562. Available from https://doi.org/10.1007/BF01790181.

Bjanes, D. A., & Moritz, C. T. (2019). A robust encoding scheme for delivering artificial sensory information via direct brain stimulation. *IEEE Transactions on Neural Systems and Rehabilitation Engineering*, *27*(10). Available from https://doi.org/10.1109/TNSRE.2019.2936739, 1–1.

Bradley, G., & House, P. (n.d.). Epilepsy electrodes. <http://greger.lab.asu.edu/image-gallery/epilepsy-electrodes/>.

Carmena, J. M., Lebedev, M. A., Crist, R. E., O'Doherty, J. E., Santucci, D. M., Dimitrov, D. F., … Nicolelis, M. A. L. (2003). Learning to control a brain-machine interface for reaching and grasping by primates. *PLoS Biology*, *1*(2), 193–208. Available from https://doi.org/10.1371/journal.pbio.0000042.

Clark, K. L., Armstrong, K. M., & Moore, T. (2011). Probing neural circuitry and function with electrical microstimulation. *Proceedings of the Royal Society B: Biological Sciences*, *278* (1709), 1121–1130. Available from https://doi.org/10.1098/rspb.2010.2211.

Cole, J. (1995). *Pride and a daily marathon*. MIT Press.

Collinger, J. L., Wodlinger, B., Downey, J. E., Wang, W., Tyler-Kabara, E. C., Weber, D. J., … Schwartz, A. B. (2013). High-performance neuroprosthetic control by an individual with tetraplegia. *The Lancet*, *381*(9866), 557–564. Available from https://doi.org/10.1016/S0140-6736(12)61816-9.

Collins, K. L., Guterstam, A., Cronin, J., Olson, J. D., Ehrsson, H. H., & Ojemann, J. G. (2017). Ownership of an artificial limb induced by electrical brain stimulation. *Proceedings of the National Academy of Sciences*, *114*(1), 166–171. Available from https://doi.org/10.1073/pnas.1616305114.

Cronin, J. A., Wu, J., Collins, K. L., Sarma, D., Rao, R. P. N. N., Ojemann, J. G., & Olson, J. D. (2016). Task-specific somatosensory feedback via cortical stimulation in humans. *IEEE Transactions on Haptics*, *9*(4), 515–522. Available from https://doi.org/10.1109/TOH.2016.2591952.

Dadarlat, M. C., O'Doherty, J. E., & Sabes, P. N. (2015). A learning-based approach to artificial sensory feedback leads to optimal integration. *Nature Neuroscience*, *18*(1), 138–144. Available from https://doi.org/10.1038/nn.3883.

DeFelice, L. J. (1981). *Animal electricity. Introduction to membrane noise* (pp. 1–25). Boston, MA: Springer US. Available from https://doi.org/10.1007/978-1-4613-3135-3_1.

Deuschl, G., Schade-Brittinger, C., Krack, P., Volkmann, J., Schäfer, H., Bötzel, K., … Voges, J. (2006). A randomized trial of deep-brain stimulation for Parkinson's disease. *New England Journal of Medicine*, *355*(9), 896–908. Available from https://doi.org/10.1056/NEJMoa060281.

Devecioğlu, İ., & Güçlü, B. (2017). Psychophysical correspondence between vibrotactile intensity and intracortical microstimulation for tactile neuroprostheses in rats. *Journal of Neural Engineering*, *14*(1). Available from https://doi.org/10.1088/1741-2552/14/1/016010.

Dibner, B., Duchenne, G.-B., & Tibbits, H. (1871). *A treatise on localized electrization, and its applications to pathology and therapeutics. A treatise on localized electrization, and its applications to pathology and therapeutics/by Dr. G.B. Duchenne; translated from the third edition… by Herbert Tibbits…; with numerous illustrations, and notes and additions by the translator*. Philadelphia: Lindsay & Blakiston. Available from https://doi.org/10.5962/bhl.title.24567.

Du Bois-Reymond, E. (1852). On animal electricity: Being an abstract of the discoveries of Emil Du Bois-Reymond. In B. Jones (Ed.), *On animal electricity: Being an abstract of the discoveries of Emil Du Bois-Reymond*. London: Churchill. Available from https://doi.org/10.5962/bhl.title.113855.

Fagg, A. H., Hatsopoulos, N. G., de Lafuente, V., Moxon, K. A., Nemati, S., Rebesco, J. M., … Miller, L. E. (2007). Biomimetic brain machine interfaces for the control of movement. *Journal of Neuroscience*, *27*(44), 11842–11846. Available from https://doi.org/10.1523/JNEUROSCI.3516-07.2007.

Fanselow, E. E., & Nicolelis, M. A. (1999). Behavioral modulation of tactile responses in the rat somatosensory system. *The Journal of Neuroscience: The Official Journal of the Society for*

Neuroscience, *19*(17), 7603–7616. Available from http://www.ncbi.nlm.nih.gov/pubmed/10460266.

Fechner, G. T. (1966). *Elements of psychophysics [Elemente der Psychophysik], 1860. Readings in the history of psychology* (pp. 206–213). Holt, Rinehart and Winston. Available from https://doi.org/10.1037/11304-026.

Ferezou, I., Haiss, F., Gentet, L. J., Aronoff, R., Weber, B., & Petersen, C. C. H. (2007). Spatiotemporal dynamics of cortical sensorimotor integration in behaving mice. *Neuron*, *56* (5), 907–923. Available from https://doi.org/10.1016/j.neuron.2007.10.007.

Fetz, E. (1969). Operant conditioning of cortical unit activity. *Science*, 28–31. Available from http://www.sciencemag.org/content/163/3870/955.short.

Fetz, E. E. (1999). Real-time control of a robotic arm by neuronal ensembles. *Nature Neuroscience*, *2*(7), 583–584. Available from https://doi.org/10.1038/10131.

Fitzsimmons, N. A., Drake, W., Hanson, T. L., Lebedev, M. A., & Nicolelis, M. A. L. (2007). Primate reaching cued by multichannel spatiotemporal cortical microstimulation. *The Journal of Neuroscience: The Official Journal of the Society for Neuroscience*, *27*(21), 5593–5602. Available from https://doi.org/10.1523/JNEUROSCI.5297-06.2007.

Flesher, S. N., Collinger, J. L., Foldes, S. T., Weiss, J. M., Downey, J. E., Tyler-Kabara, E. C., ... Gaunt, R. A. (2016). Intracortical microstimulation of human somatosensory cortex. *Science Translational Medicine*, *8*(361), 361ra141. Available from https://doi.org/10.1126/scitranslmed.aaf8083.

French, J. S., Anderson-Erisman, K. D., & Sutter, M. (2010). What do spinal cord injury consumers want? A review of spinal cord injury consumer priorities and neuroprosthesis from the 2008 neural interfaces conference. *Neuromodulation*, *13*(3), 229–231. Available from https://doi.org/10.1111/j.1525-1403.2009.00252.x.

Fridman, G. G. Y., Blair, H. H. T. H., Blaisdell, A. A. P. A., & Judy, J. W. J. (2010). Perceived intensity of somatosensory cortical electrical stimulation. *Experimental Brain Research*, *203* (3), 499–515. Available from https://doi.org/10.1007/s00221-010-2254-y.

Fritsch, G., & Hitzig, E. (1870). Über die elektrische Erregbarkeit des Grosshirns (Electric excitability of the cerebrum). *Archiv Für Anatomie, Physiologie Und Wissenschaftliche Medicin*, *37*, 300–332.

Grayden, D. B., & Clark, G. M. (2006). Implant design and development. In H. R. Cooper, & L. C. Craddock (Eds.), *Cochlear implants: A practical guide* (2nd ed.). New York: Wiley. Available from http://www.wiley.com/WileyCDA/WileyTitle/productCd-1861564813,descCd-description.html.

Harkema, S., Gerasimenko, Y., Hodes, J., Burdick, J., Angeli, C., Chen, Y., ... Edgerton, V. R. (2011). Effect of epidural stimulation of the lumbosacral spinal cord on voluntary movement, standing, and assisted stepping after motor complete paraplegia: A case study. *The Lancet*, *377* (9781), 1938–1947. Available from https://doi.org/10.1016/S0140-6736(11)60547-3.

Hashimoto, T., Elder, C. M., & Vitek, J. L. (2002). A template subtraction method for stimulus artifact removal in high-frequency deep brain stimulation. *Journal of Neuroscience Methods*, *113*(2), 181–186. Available from http://www.ncbi.nlm.nih.gov/pubmed/11772439.

Hiremath, S. V., Tyler-Kabara, E. C., Wheeler, J. J., Moran, D. W., Gaunt, R. A., Collinger, J. L., ... Wang, W. (2017). Human perception of electrical stimulation on the surface of somatosensory cortex. *Plos One*, *12*(5), e0176020. Available from https://doi.org/10.1371/journal.pone.0176020.

Histed, M. H., Ni, A. M., & Maunsell, J. H. R. (2012). Insights into cortical mechanisms of behavior from microstimulation experiments. *Progress in Neurobiology*, *103*. Available from https://doi.org/10.1016/j.pneurobio.2012.01.006.

Hochberg, L. R., Bacher, D., Jarosiewicz, B., Masse, N. Y., Simeral, J. D., Vogel, J., ... Donoghue, J. P. (2012). Reach and grasp by people with tetraplegia using a neurally controlled robotic arm. *Nature, 485*(7398), 372–375. Available from https://doi.org/10.1038/nature11076.

Hodgkin, A. L., & Huxley, A. F. (1952). The components of membrane conductance in L411mV -D. *J. Physiol. (1952), 116*(4), 473–496. Available from https://doi.org/10.1113/jphysiol.1952.sp004718.

Houweling, A. R., & Brecht, M. (2008). Behavioural report of single neuron stimulation in somatosensory cortex. *Nature, 451*(7174), 65–68. Available from https://doi.org/10.1038/nature06447.

Ievins, A., & Moritz, C. T. (2017). Therapeutic stimulation for restoration of function after spinal cord injury. *Physiology, 32*(5), 391–398. Available from https://doi.org/10.1152/physiol.00010.2017.

Irnich, W. (2002). *Georges Weiss' fundamental law of electrostimulation is 100 years old. PACE—Pacing and clinical electrophysiology*. Futura Publishing Company Inc. Available from https://doi.org/10.1046/j.1460-9592.2002.00245.x.

Johannes, M. S., Bigelow, J. D., Burck, J. M., Harshbarger, S. D., & Kozlowski, M. V. (2011). An Overview of the developmental process for the modular prosthetic limb. *Johns Hopkins Apl Technical Digest, 30*(3), 207–216.

Johnson, L. A., Wander, J. D., Sarma, D., Su, D. K., Fetz, E. E., & Ojemann, J. G. (2013). Direct electrical stimulation of the somatosensory cortex in humans using electrocorticography electrodes: A qualitative and quantitative report. *Journal of Neural Engineering, 10*(3). Available from https://doi.org/10.1088/1741-2560/10/3/036021.

Kellis, S. S., House, P. A., Thomson, K. E., Brown, R., & Greger, B. (2009). Human neocortical electrical activity recorded on nonpenetrating microwire arrays: Applicability for neuroprostheses. *Neurosurgical Focus, 27*(1), E9. Available from https://doi.org/10.3171/2009.4.focus0974.

Kennedy, P. R., & Bakay, R. A. E. (1998). Restoration of neural output from a paralyzed patient by a direct brain connection. *Neuroreport, 9*(8), 1707–1711. Available from https://doi.org/10.1097/00001756-199806010-00007.

Kim, S., Callier, T., Tabot, G. A., Gaunt, R. A., Tenore, F. V., & Bensmaia, S. J. (2015). Behavioral assessment of sensitivity to intracortical microstimulation of primate somatosensory cortex. *Proceedings of the National Academy of Sciences, 112*(49), 15202–15207. Available from https://doi.org/10.1073/pnas.1509265112.

Klaes, C., Shi, Y., Kellis, S., Minxha, J., Revechkis, B., & Andersen, R. A. (2014). A cognitive neuroprosthetic that uses cortical stimulation for somatosensory feedback. *Journal of Neural Engineering, 11*(5), 056024. Available from https://doi.org/10.1088/1741-2560/11/5/056024.

Kumar, R., Lozano, A. M., Sime, E., & Lang, A. E. (2003). Long-term follow-up of thalamic deep brain stimulation for essential and parkinsonian tremor. *Neurology, 61*(11), 1601–1604. Available from https://doi.org/10.1212/01.WNL.0000096012.07360.1C.

Kumar, R., Lozano, A. M., Sime, E., Halket, E., & Lang, A. E. (1999). Comparative effects of unilateral and bilateral subthalamic nucleus deep brain stimulation. *Neurology, 53*(3), 561–566. Available from https://doi.org/10.1212/01.wnl.0000096012.07360.1c.

Lapicque, L. (1909). Définition experimentale de l'excitabilité. Comptes rendus Hebd. *Des Séances Mémoires La Société Biol* (67), 280–283.

Lapicque, L. (2007). Quantitative investigations of electrical nerve excitation treated as polarization 1907 (translation). *Biological Cybernetics, 97*(5–6), 341–349. Available from https://doi.org/10.1007/s00422-007-0189-6.

Lee, B., Kramer, D., Armenta Salas, M., Kellis, S., Brown, D., Dobreva, T., . . . Andersen, R. A. (2018). Engineering artificial somatosensation through cortical stimulation in humans. *Frontiers in Systems Neuroscience, 12*, 24. Available from https://doi.org/10.3389/fnsys.2018.00024.

Lee, D. S., Lee, J. S., Oh, S. H., Kim, S.-K., Kim, J.-W., Chung, J.-K., ... Kim, C. S. (2001). Cross-modal plasticity and cochlear implants. *Nature*, *409*(6817), 149–150. Available from https://doi.org/10.1038/35051653.

Lee, S., Carvell, G. E., & Simons, D. J. (2008). Motor modulation of afferent somatosensory circuits. *Nature Neuroscience*, *11*(12), 1430–1438. Available from https://doi.org/10.1038/nn.2227.

Lewis, P. M., Ayton, L. N., Guymer, R. H., Lowery, A. J., Blamey, P. J., Allen, P. J., ... Rosenfeld, J. V. (2016). Advances in implantable bionic devices for blindness: A review. *ANZ Journal of Surgery*, *86*(9), 654–659. Available from https://doi.org/10.1111/ans.13616.

Lilly, J. C., Hughes, J. R., Alvord, E. C., & Galkin, T. W. (1955). Brief, noninjurious electric waveform for stimulation of the brain. *Science*, *121*(3144), 468–469. Available from https://doi.org/10.1126/science.121.3144.468.

Logothetis, N. K., Augath, M., Murayama, Y., Rauch, A., Sultan, F., Goense, J., ... Merkle, H. (2010). The effects of electrical microstimulation on cortical signal propagation. *Nature Neuroscience*, *13*(10), 1283–1291. Available from https://doi.org/10.1038/nn.2631.

Malekmohammadi, M., Herron, J., Velisar, A., Blumenfeld, Z., Trager, M. H., Chizeck, H. J., & Brontë-Stewart, H. (2016). Kinematic adaptive deep brain stimulation for resting tremor in Parkinson's disease. *Movement Disorders*, *31*(3), 426–428. Available from https://doi.org/10.1002/mds.26482.

Mayberg, H. S., Lozano, A. M., Voon, V., McNeely, H. E., Seminowicz, D., Hamani, C., ... Kennedy, S. H. (2005). Deep brain stimulation for treatment-resistant depression. *Neuron*, *45*(5), 651–660. Available from https://doi.org/10.1016/J.NEURON.2005.02.014.

Maynard, E. M., Nordhausen, C. T., & Normann, R. A. (1997). The Utah Intracortical Electrode Array: A recording structure for potential brain-computer interfaces. *Electroencephalography and Clinical Neurophysiology*, *102*(3), 228–239. Available from https://doi.org/10.1016/S0013-4694(96)95176-0.

Merrill, D. R., Bikson, M., & Jefferys, J. G. R. (2005). Electrical stimulation of excitable tissue: Design of efficacious and safe protocols. *Journal of Neuroscience Methods*, *141*(2), 171–198. Available from https://doi.org/10.1016/j.jneumeth.2004.10.020.

Musallam, S., Corneil, B. D., Greger, B., Scherberger, H., & Andersen, R. A. (2004). Cognitive control signals for neural prosthetics. *Science*, *305*(5681), 258–262. Available from https://doi.org/10.1126/science.1097938.

Navarro, X., Krueger, T. B., Lago, N., Micera, S., Stieglitz, T., & Dario, P. (2005). A critical review of interfaces with the peripheral nervous system for the control of neuroprostheses and hybrid bionic systems. *Journal of the Peripheral Nervous System*, *10*(3), 229–258. Available from https://doi.org/10.1111/j.1085-9489.2005.10303.x.

Nicolelis, M. A. L., & Lebedev, M. A. (2009). Principles of neural ensemble physiology underlying the operation of brain–machine interfaces. *Nature Reviews Neuroscience*, *10*(7), 530–540. Available from https://doi.org/10.1038/nrn2653.

Normann, R. A., & Fernandez, E. (2016). Clinical applications of penetrating neural interfaces and Utah Electrode Array technologies. *Journal of Neural Engineering*, *13*(6), 061003. Available from https://doi.org/10.1088/1741-2560/13/6/061003.

O'Doherty, J. E., Lebedev, M. A., Ifft, P. J., Zhuang, K. Z., Shokur, S., Bleuler, H., & Nicolelis, M. A. L. (2011). Active tactile exploration using a brain-machine-brain interface. *Nature*, *479*(7372), 228–231. Available from https://doi.org/10.1038/nature10489.

O'Shea, D. J., & Shenoy, K. V. (2018). ERAASR: An algorithm for removing electrical stimulation artifacts from multielectrode array recordings. *Journal of Neural Engineering*, *15*(2), 026020. Available from https://doi.org/10.1088/1741-2552/aaa365.

Öztürk, S., Devecioğlu, İ., Beygi, M., Atasoy, A., Mutlu, S., Özkan, M., & Güçlü, B. (2019). Real-time performance of a tactile neuroprosthesis on awake behaving rats. *IEEE Transactions on Neural Systems and Rehabilitation Engineering*, *27*(5), 1053–1062. Available from https://doi.org/10.1109/TNSRE.2019.2910320.

Panetsos, F., Sanchez-Jimenez, A., Cerio, E. D., Diaz-Guemes, I., & Sanchez, F. M. (2011). Consistent phosphenes generated by electrical microstimulation of the visual thalamus. An experimental approach for thalamic visual neuroprostheses. *Frontiers in Neuroscience*, *5*, 84. Available from https://doi.org/10.3389/fnins.2011.00084.

Penfield, W., & Boldrey, E. (1937). Somatic motor and sensory representation in the cerebral cortex of man as studied by electrical stimulation. *Brain*, *60*(4), 389–443. Available from https://doi.org/10.1093/brain/60.4.389.

Penfield, W., & Jasper, H. (1954). *Epilepsy and the functional anatomy of the human brain. Epilepsy and the functional anatomy of the human brain*. Oxford: Little, Brown & Co.

Penfield, W., & Rasmussen, T. (1950). *The cerebral cortex of man: A clinical study of localization of function. The cerebral cortex of man: A clinical study of localization of function*. Oxford: Macmillan.

Pezaris, J. S., & Reid, R. C. (2007). Demonstration of artificial visual percepts generated through thalamic microstimulation. *Proceedings of the National Academy of Sciences of the United States of America*, *104*(18), 7670–7675. Available from https://doi.org/10.1073/pnas.0608563104.

Pinault, D. (1996). A novel single-cell staining procedure performed in vivo under electrophysiological control: Morpho-functional features of juxtacellularly labeled thalamic cells and other central neurons with biocytin or Neurobiotin. *Journal of Neuroscience Methods*, *65*(2), 113–136. Available from https://doi.org/10.1016/0165-0270(95)00144-1.

Qing, K. Y. (2015). *Purdue e-pubs optimizing the neural response to electrical stimulation and exploring new applications of neurostimulation*. <https://docs.lib.purdue.edu/open_access_dissertations/538>.

Rajan, A. T., Boback, J. L., Dammann, J. F., Tenore, F. V., Wester, B. A., Otto, K. J., ... Bensmaia, S. J. (2015). The effects of chronic intracortical microstimulation on neural tissue and fine motor behavior. *Journal of Neural Engineering*, *12*(6), 66018. Available from https://doi.org/10.1088/1741-2560/12/6/066018.

Ranck, J. B. (1975). Which elements are excited in electrical stimulation of mammalian central nervous system: A review. *Brain Research*, *98*(3), 417–440. Available from https://doi.org/10.1016/0006-8993(75)90364-9.

Ray, P. G., Meador, K. J., Smith, J. R., Wheless, J. W., Sittenfeld, M., & Clifton, G. L. (1999). Physiology of perception: Cortical stimulation and recording in humans. *Neurology*, *52*(5). Available from https://doi.org/10.1212/wnl.52.5.1044, 1044.

Resnik, L., Meucci, M. R., Lieberman-Klinger, S., Fantini, C., Kelty, D. L., Disla, R., & Sasson, N. (2012). Advanced upper limb prosthetic devices: Implications for upper limb prosthetic rehabilitation. *Archives of Physical Medicine and Rehabilitation*, *93*(4), 710–717. Available from https://doi.org/10.1016/j.apmr.2011.11.010.

Romo, R., Hernández, A., Zainos, A., & Salinas, E. (1998). Somatosensory discrimination based on cortical microstimulation. *Nature*, *392*(6674), 387–390. Available from https://doi.org/10.1038/32891.

Romo, R., Hernández, A., Zainos, A., Brody, C. D., & Lemus, L. (2000). Sensing without touching: Psychophysical performance based on cortical microstimulation. *Neuron*, *26*(1), 273–278. Available from https://doi.org/10.1016/S0896-6273(00)81156-3.

Rousche, P., & Normann, R. A. (1992). A method for pneumatically inserting an array of penetrating electrodes into cortical tissue. *Annals of Biomedical Engineering*, *20*, 413–422. Available from https://link.springer.com/article/10.1007/BF02368133.

Sainburg, R. L., Poizner, H., & Ghez, C. (1993). Loss of proprioception produces deficits in interjoint coordination. *Journal of Neurophysiology*, *70*(5), 2136–2147. Available from https://doi.org/10.1152/jn.1993.70.5.2136.

Santhanam, G., Ryu, S. I., Yu, B. M., Afshar, A., & Shenoy, K. V. (2006). A high-performance brain–computer interface. *Nature*, *442*(7099), 195–198. Available from https://doi.org/10.1038/nature04968.

Schafer, E. A. (1888). Experiments on the electrical excitation of the cerebral cortex in the monkey. *Brain*, *11*, 1–6.

Schmidt, E. M., Bak, M. J., Hambrecht, F. T., Kufta, C. V., O'Rourke, D. K., & Vallabhanath, P. (1996). Feasibility of a visual prosthesis for the blind based on intracortical microstimulation of the visual cortex. *Brain: A Journal of Neurology*, *119*(Pt 2), 507–522. Available from http://www.ncbi.nlm.nih.gov/pubmed/8800945.

Schott, G. D. (1993). Penfield's homunculus: A note on cerebral cartography. *Journal of Neurology, Neurosurgery, and Psychiatry*, *56*(4), 329. Available from https://www.ncbi.nlm.nih.gov/pmc/articles/PMC1014945/.

Schuetze, S. M. (1983). The discovery of the action potential. *Trends in Neurosciences*, *6*(C), 164–168. Available from https://doi.org/10.1016/0166-2236(83)90078-4.

Serruya, M. D., Hatsopoulos, N. G., Paninski, L., Fellows, M. R., & Donoghue, J. P. (2002). Instant neural control of a movement signal. *Nature*, *416*(6877), 141–142. Available from https://doi.org/10.1038/416141a.

Shannon, R. V. (1992). A model of safe levels for electrical stimulation. *IEEE Transactions on Biomedical Engineering*, *39*(4), 424–426. Available from https://doi.org/10.1109/10.126616.

Shepherd, R. K., & McCreery, D. B. (2006). Basis of electrical stimulation of the cochlea and the cochlear nucleus. *Advances in Oto-Rhino-Laryngology*, *64*, 186–205. Available from https://doi.org/10.1159/000094652.

Tabot, G. A., Dammann, J. F., Berg, J. A., Tenore, F. V., Boback, J. L., Vogelstein, R. J., & Bensmaia, S. J. (2013). Restoring the sense of touch with a prosthetic hand through a brain interface. *Proceedings of the National Academy of Sciences of the United States of America*, *110*(45), 18279–18284. Available from https://doi.org/10.1073/pnas.1221113110.

Talbot, W. H., & Mountcastle, B. (1968). The sense of flutter-vibration: The human the monkey of mechanoreceptive comparison of capacity with response patterns afferents from. *Journal of Neurophysiology*, *31*(2), 301–334.

Tan, D. W., Schiefer, M. A., Keith, M. W., Anderson, J. R., Tyler, J., & Tyler, D. J. (2014). A neural interface provides long-term stable natural touch perception. *Science Translational Medicine*, *6*(257). Available from https://doi.org/10.1126/scitranslmed.3008669.

Taylor, C. S. R., & Gross, C. G. (2003). Twitches versus movements: A story of motor cortex. *The Neuroscientist*, *9*(5), 332–342. Available from https://doi.org/10.1177/1073858403257037.

Taylor, D. M. (2002). Direct cortical control of 3D neuroprosthetic devices. *Science*, *296*(5574), 1829–1832. Available from https://doi.org/10.1126/science.1070291.

Tehovnik, E. (1996). Electrical stimulation of neural tissue to evoke behavioral responses. *Journal of Neuroscience Methods*, *65*(1), 1–17. Available from http://www.ncbi.nlm.nih.gov/pubmed/8815302.

Tehovnik, E. J., Slocum, W. M., Smirnakis, S. M., & Tolias, A. S. (2009). Microstimulation of visual cortex to restore vision. *Progress in Brain Research*, *175*. Available from https://doi.org/10.1016/S0079-6123(09)17524-6.

Thomson, E. E., Carra, R., & Nicolelis, M. A. L. (2013). Perceiving invisible light through a somatosensory cortical prosthesis. *Nature Communications*, *4*, 1482. Available from https://doi.org/10.1038/ncomms2497.

Velliste, M., Perel, S., Spalding, M. C., Whitford, A. S., & Schwartz, A. B. (2008). Cortical control of a prosthetic arm for self-feeding. *Nature*, *453*(7198), 1098–1101. Available from https://doi.org/10.1038/nature06996.

Venkatraman, S., Elkabany, K., Long, J. D., Yao, Y., & Carmena, J. M. (2009). A system for neural recording and closed-loop intracortical microstimulation in awake rodents. *IEEE Transactions on Bio-Medical Engineering*, *56*(1), 15–22. Available from https://doi.org/10.1109/TBME.2008.2005944.

Wang, W., Collinger, J. L., Degenhart, A. D., Tyler-Kabara, E. C., Schwartz, A. B., Moran, D. W., ... Boninger, M. L. (2013). An electrocorticographic brain interface in an individual with tetraplegia. *PLoS ONE*, *8*(2). Available from https://doi.org/10.1371/journal.pone.0055344.

Weber, D. J., Friesen, R., & Miller, L. E. (2012). Interfacing the somatosensory system to restore touch and proprioception: Essential considerations. *Journal of Motor Behavior*, *44*(6), 403–418. Available from https://doi.org/10.1080/00222895.2012.735283.

Weiss, G. (1901). Sur la possibilite de rendre comparables entre eux les appareils servant a l'excitation electrique. *Archives Italiennes de Biologie*, *35*(1), 413–445. Available from https://doi.org/10.4449/AIB.V35I1.1355.

Weiss, J. M., Flesher, S. N., Franklin, R., Collinger, J. L., & Gaunt, R. A. (2019). Artifact-free recordings in human bidirectional brain-computer interfaces. *Journal of Neural Engineering*, *16*(1). Available from https://doi.org/10.1088/1741-2552/aae748.

Wessberg, J., Stambaugh, C. R., Kralik, J. D., Beck, P. D., Laubach, M., Chapin, J. K., ... Biggs, S. J. (2000). Real-time prediction of hand tranjectory by ensembles of cortical neurons in primate. *Nature*, *408*(1), 361–365.

Widge, A. S., Dougherty, D. D., & Moritz, C. T. (2014). Affective brain-computer interfaces as enabling technology for responsive psychiatric stimulation. *Brain-Computer Interfaces*, *1*(2), 126–136. Available from https://doi.org/10.1080/2326263X.2014.912885.

Wodlinger, B., Downey, J. E., Tyler-Kabara, E. C., Schwartz, A. B., Boninger, M. L., & Collinger, J. L. (2015). Ten-dimensional anthropomorphic arm control in a human brain – machine interface: Difficulties, solutions, and limitations. *Journal of Neural Engineering*, *12*(1), 016011. Available from https://doi.org/10.1088/1741-2560/12/1/016011.

Zrenner, E. (2002). Will retinal implants restore vision? *Science (New York, N.Y.)*, *295*(5557), 1022–1025. Available from https://doi.org/10.1126/science.1067996.

Part V

Future technologies

Chapter 16

Neural electrodes for long-term tissue interfaces

Jaume del Valle, Bruno Rodríguez-Meana and Xavier Navarro

Institute of Neurosciences and Department of Cell Biology, Physiology and Immunology, Universitat Autònoma de Barcelona, and Centro de Investigación Biomédica en Red sobre Enfermedades Neurodegenerativas (CIBERNED), Bellaterra, Spain

ABSTRACT

The neural interface is the link that connects a prosthetic or robotic device with the nervous system. When a neuroprosthesis is implanted, it is usually intended to provide a prolonged lifespan with effective bidirectional communication, being able to deliver stimulation or to record electrical signals from the target nerve or brain area. In the case of limb prostheses, neural electrodes have been used to provide the user with somatosensory feedback to sense the prosthesis actions. The long-term performance of the interface depends on the electrode design and durability, and also on the biological response to the foreign body implanted. In this chapter, the different electrodes designed to interface the peripheral nerve are reviewed, with particular focus on the provision of somatosensory feedback.

Keywords: Peripheral nerve; nerve electrode; neural interface; neuroprosthesis; somatosensory feedback

16.1 Introduction

Loss of function after an amputation or a nerve injury prevents patients from carrying out part of their normal daily activities. Different types of prostheses have been developed to improve the functionality of subjects who have suffered amputation of a limb. Cosmetic prostheses help to alleviate the psychological distress of the missing limb, but they do not allow to regain any lost function. Body-powered prostheses are connected by a cable or harness system to the body that controls the prosthesis performing shoulder movements to operate joints or terminal devices, but with limited actions (Millstein et al., 1986). Myoelectric prostheses, linked with the body by electrodes that record the surface electromyography (EMG) signals derived from muscle activity, extract the patient's intention to

Somatosensory Feedback for Neuroprosthetics. DOI: https://doi.org/10.1016/B978-0-12-822828-9.00009-5

control the movement of the prosthesis from proximal remnant muscles (Scott & Parker, 1988). Although advanced prosthetic arm devices offer a number close to the 22 independent degrees of freedom of the natural hand, and robotic automation can improve prosthesis performance (Zhuang et al., 2019), the majority of myoelectric prostheses used in clinical practice are still far from the natural limb performance. This is mainly due to the limited amount of usable EMG signals, allowing for a proper control of no more than two to three degrees of freedom (Zecca et al., 2017), and to the fact that users have to continually rely on visual cues due to the lack of sensory feedback from the prosthesis (Micera, 2016). In this sense, an ideal neuroprosthesis should look and feel like a part of the own body and provide the lost functions of the natural limb. In this regard, the prosthesis should be controlled by the patient, being able to perform the same movements of the lost member, and should provide the subject with sensory information recorded by the artificial limb (del Valle & Navarro, 2013).

Neuroprostheses are the most advanced systems for the replacement of lost functions after a limb amputation as they can perform complex actions and provide sensory feedback. These devices may directly link the patient's nervous system with a robotic limb using an interface implanted in the peripheral nerve. The interface contains electrodes to selectively record motor electrical signals from the efferent fibers of the peripheral nerve, and then a computer transforms these data into commands that control the mechanical prosthesis. In addition, sensors located in the artificial limb send the information recorded to the implanted interface, which stimulates selective populations of afferent fibers to ideally evoke sensations of different modalities and provide the patient with sensory feedback from the machine (Fig. 16.1).

16.2 Peripheral nerve electrodes

The correct function of a neuroprosthesis relies on proper communication between the nervous system and the artificial limb through neural electrodes. Although the link between the artificial limb and the patient can relay in the peripheral or the central nervous system, peripheral nerve interfaces have reduced invasiveness in comparison with central implants. Additionally, peripheral nerves are more accessible, and may allow easier topographical discrimination. Moreover, provided that most peripheral nerves contain both motor and sensory fibers, a single device can be useful for both recording motor information and providing somatosensory feedback.

Taking into consideration the part of the interface that is implanted, the electrode is formed by three main components: a substrate, the conductive electrical tracks, and the active sites. The substrate gives the electrode its shape and should be electrically inert; it has to resist the implantation procedure and should show high biocompatibility once in contact with the tissue. The most commonly used materials are silicone, polyimide, and parylene (Navarro et al., 2005). The conductive tracks of the electrode transmit the

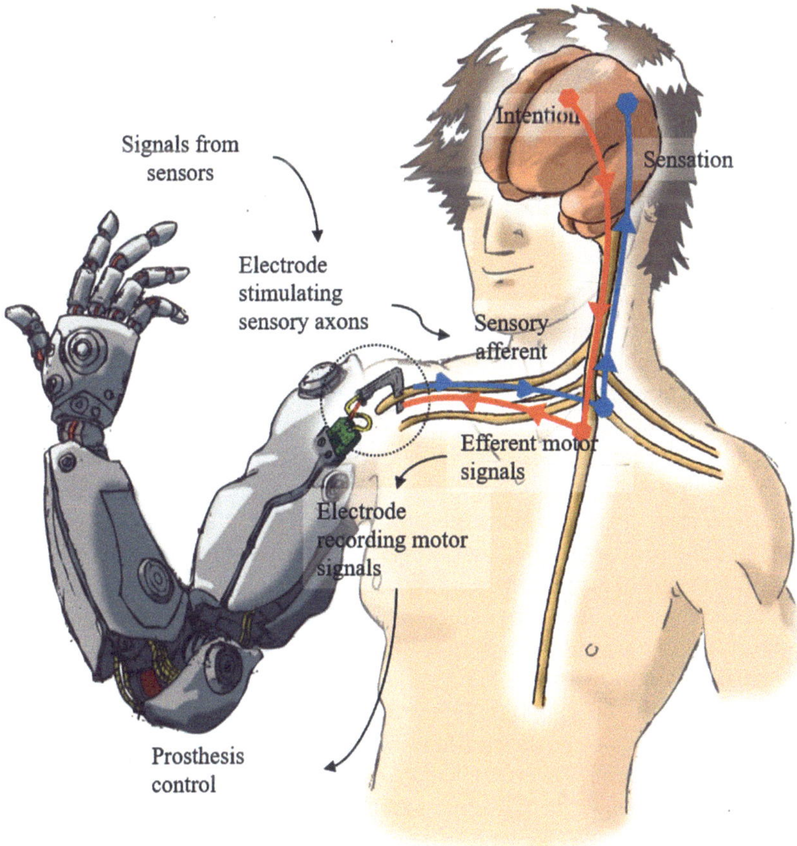

FIGURE 16.1 Bidirectional communication between an amputee and a neuroprosthesis with a nerve electrode (*dotted circle*) as the link between the human and the machine. Efferent motor signals (*red*) can be recorded to control the bionic prosthesis. Sensory signals from sensors in the robotic limb are sent to the user through selective electrical stimulation of the implanted nerve. The sensory signals (*blue*) are conducted by afferent pathways to the brain providing sensory feedback.

current between the tissue and the machine, they are embedded within the substrate, and they are usually made of gold, platinum, iridium, tungsten, or stainless steel (Navarro et al., 2005). The active sites are the uncovered parts of the electrode where the communication between the nervous system and the artificial system takes place. These active sites should be able to deliver enough charge to stimulate adjacent axons and to detect extracellular neural signals with respect to the surrounding noise. The most used materials for electrode active sites are gold, platinum, and iridium oxide, and they can be coated with different materials, for example, carbon nanotubes or poly(3,4-ethylenedioxythiophene) (PEDOT) (Kozai et al., 2012), to improve the injected charge available and the impedance.

In the peripheral nerve, efferent and afferent nerve fibers are grouped in fascicles surrounded by connective tissue, structured in three sheaths: epineurium, perineurium, and endoneurium. When a neural electrode is implanted around or within a peripheral nerve, each active site can stimulate certain nerve fibers provided enough charge is delivered, and depending on the stimulus pulse duration (Grill & Mortimer, 1996) and axon diameter, with the larger axons being the first to be activated in a phenomenon called inverse recruitment (Blair & Erlanger, 1933). Indeed, both motor and sensory axons can be activated at the same time, or different types of sensory axons, sometimes inducing undesired responses and mixed unpleasant sensations. In this sense, one of the most important features of a peripheral nerve electrode is its ability for selectively interfacing different axonal populations conveying distinct functions in a common nerve (Tyler & Durand, 2002) and with the best possible long-term performance. The number and distribution of active sites within the electrode surface increase the probability to specifically stimulate and/or record different groups of axons and thus link different and complementary actions. Hence, a determined group of axons will be selectively stimulated depending on whether the active site is close enough to deliver sufficient current to reach the threshold of excitation without activating other neighboring axons. Consequently, the selectivity of a given electrode is usually expressed as the ratio of the specific response obtained after stimulation with respect to other responses relying on the same nerve (Badia, Boretius, Andreu, et al., 2011; Veraart et al., 1993).

Different types of electrodes have been designed to interface the peripheral nervous system (PNS). A widely accepted criterion for classification of the different types of neural electrodes (Fig. 16.2) relies on how they are implanted in the nerve (Navarro et al., 2005). In this regard, extraneural electrodes are implanted outside the nerve, around the epineurium, showing a reduced selectivity as they are able to stimulate or record only the most superficial axons. The more invasive intraneural electrodes are implanted transversally or longitudinally within the endoneurium, thus they are in closer contact with the axons of the nerve, offering enhanced selectivity. Finally, regenerative electrodes are implanted between the stumps of a sectioned nerve, implying the most significant level of invasiveness but with a higher potential selectivity than the other interfaces.

16.2.1 Surface electrodes

The easiest solution to interface muscles and/or the nervous system is to place an electrode in the body surface and stimulate the target tissue or record its electrical activity. Indeed, transcutaneous electrical stimulation through surface electrodes remains the most used technique to activate nerves or muscles (Keller & Kuhn, 2008), whereas applications that record

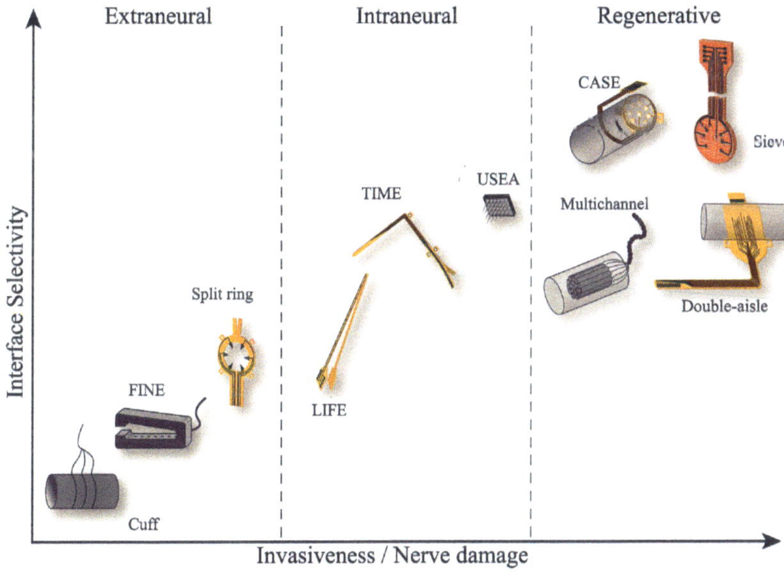

FIGURE 16.2 Electrodes to interface the PNS according to their selectivity/invasiveness ratio. Electrodes are classified depending on how they are implanted in a peripheral nerve: extraneural, intraneural, and regenerative. Extraneural: Cuff, FINE, and Split ring electrodes. Intraneural: LIFE, TIME, and USEA electrodes. Regenerative: Double aisle, Multichannel, CASE, and Sieve electrodes. *FINE*, flat interface nerve electrode; *LIFE*, longitudinal intrafascicular electrode; *TIME*, transverse intrafascicular multichannel electrode; *USEA*, Utah slanted electrode array.

electrical activity such as EMG, electrocardiography, and electroencephalography (EEG) have been used in the clinic for more than a century.

The most basic setup of a surface electrode is a conductive metal plate and a wire. While they are easy and cheap to fabricate, these electrodes are not fixed to the skin and may move, needing some sort of fixation. Moreover, provided electrodes are not placed in direct contact with the nerve or muscle to interface, and that skin presents a high impedance barrier, the use of an electroconductive gel to reduce resistance is often required. Adhesive electrodes incorporate an extra gluing component to the ensemble, making them easier to attach to the body surface. Correct positioning and fixation of the electrodes on the target are essential for the electrode to operate. Indeed, electrode malposition or displacements from the correct site are a common cause of flawed recordings or inadequate stimulation (Guo et al., 2020).

Surface electrodes have been used to activate muscles (either via direct activation of the muscle tissue or through the motor nerve) to induce complex wrist and hand movements similar to functional tasks relevant to daily living in tetraplegic patients (Ajiboye et al., 2017; Bouton et al., 2016), in functional electrical stimulation (FES) systems such as foot drop correction (Wieler et al., 1999) in people suffering hemiplegia, or to stand and walk a few steps (Graupe & Kohn, 1997) in paraplegic patients. Surface electrodes are also used in rehabilitation to activate denervated muscles and prevent muscle atrophy (Keller & Kuhn, 2008), to treat and prevent pressure ulcers (Ahmetovic et al., 2015; Swisher et al., 2015), whereas transcutaneous electrical nerve stimulation is used to mitigate a wide variety of pain conditions by modulating the afferent inputs to the spinal cord (Gibson et al., 2019). On the other hand, surface electrodes have been also applied to activate afferent sensory fibers to induce conscious sensations (Kaczmarek et al., 1991) and to provide sensory feedback in patients wearing myoelectric prostheses (Dosen et al., 2014) and neuroprostheses (D'Anna et al., 2019). However, comparison between sensory feedback (electrotactile stimulation) provided through surface or intraneural electrodes in patients wearing a prosthesis has shown that intraneural feedback can deliver more significant information and patients perform better in different motor tasks than when feedback is provided through the skin (Valle et al., 2020).

Surface electrodes also have been used to record different types of electrical signals from muscles (EMG) and the brain EEG. Surface electrodes are the current standard for recording motor information from muscles and derive control of myoelectric prostheses (Ortiz-Catalan et al., 2012). Moreover, recording of EMG signals can serve to control virtual and artificial limbs (Kuiken et al., 2004; Mastinu et al., 2018), and has been used to help in performing exercises in rehabilitation therapies, to monitor different pathologies such as stress or sleep disorders, and to improve performance in sport training (Guo et al., 2020).

On the other hand, changes in brain activity detected after decoding EEG signals have allowed patients to command external devices through brain–computer interfaces (BCIs) (Wolpaw et al., 2002). Applications range from control of cursors and keyboards in computers (Yuan & He, 2014) to reestablishment of communication in paralyzed patients (Birbaumer et al., 1999), or to remotely control smart home applications (Jafri et al., 2019). BCIs have been used also for multiple therapies in neurorehabilitation (McFarland & Wolpaw, 2017), to control hand or gait ortheses (Do et al., 2013; Pfurtscheller et al., 2003), wheelchair (Rebsamen et al., 2007), or aerial vehicles such as drones (Khan & Hong, 2017).

16.2.2 Extraneural electrodes

For neural interfaces that are implanted in direct contact with the nerve, extraneural electrodes show a low level of invasiveness among the different types of electrodes. The surgical procedure for the implant is less complicated than for other designs, making them easier to handle and safer to position, becoming an ideal option for different biomedical applications (del Valle & Navarro, 2013). On the other hand, as these electrodes are not in contact with fibers situated deep inside the nerve, mostly large myelinated fibers located at the outermost layer can be interfaced (Badia, Boretius, Andreu, et al., 2011) but not axons located in deep fascicles.

16.2.2.1 Cuff electrodes

A cuff electrode is based on an open cylinder or a spiral sheath of insulating material that is wrapped around the nerve (Figs. 16.2 and 16.3A, B) with electrical active sites exposed in the inner part facing the nerve. These electrodes induce almost no tissue disruption and nerve inflammation is kept to a minimum, although they can induce some compression and reshaping of the implanted nerve. The spiral cuff design can adapt to variations in nerve size and possible swelling after implants, decreasing potential nerve compression (Naples et al., 1988). The robustness of these devices in combination with the relatively simple method of fabrication (McCarty, 1965) and the reduced size and thickness of recent polymer cuffs (Stieglitz et al., 2000) supports that a high number of FES systems and neuroprosthetic applications use cuff electrodes to stimulate or record from peripheral nerves (del Valle & Navarro, 2013). Multichannel cuff electrodes (Fig. 16.3B), with active sites disposed in tripoles, allow the selective stimulation of different nerve fascicles, each one supplying a different muscle or target (Rodríguez et al., 2000; Veraart et al., 1993). Provided that the selectivity of these electrodes is lower in comparison with intrafascicular or regenerative interfaces (Badia, Boretius, Andreu, et al., 2011), multiple cuff electrodes can be implanted in one or several distal nerves in order to achieve a higher degree of selectivity.

Spiral cuff electrodes implanted on the ulnar nerve have been used to provide sensory feedback to a transhumeral amputee that received an osseointegrated myoelectric prosthesis (Ortiz-Catalan et al., 2014) inducing different sensations, such as pressure, vibration or warmth, through the 2 years that the experiment was active (Ackerley et al., 2018).

16.2.2.2 Flat interface nerve electrode

A variation of the cuff electrode is the flat interface nerve electrode (FINE) electrode. This device is rectangular rather than cylindrical (Fig. 16.2) and is implanted by applying low pressure on the nerve, reshaping the nerve into a flattened geometry. The change in shape allows the fascicles to align, approaching the surface, decreasing the distance from the active sites to central fascicles, thus improving selective interfacing (Leventhal & Durand, 2003; Tyler & Durand, 2002). However, it should be taken into account that too much reshaping force applied on the nerve may result in nerve damage (Tyler & Durand, 2003).

Both cuff and FINE electrodes were initially implanted in a human amputee, and stable sensory feedback could be provided after stimulation of median and radial nerve afferents without motor recruitment for up to 8 weeks (Tan et al., 2013). Further experiments were made with the same types of electrodes around

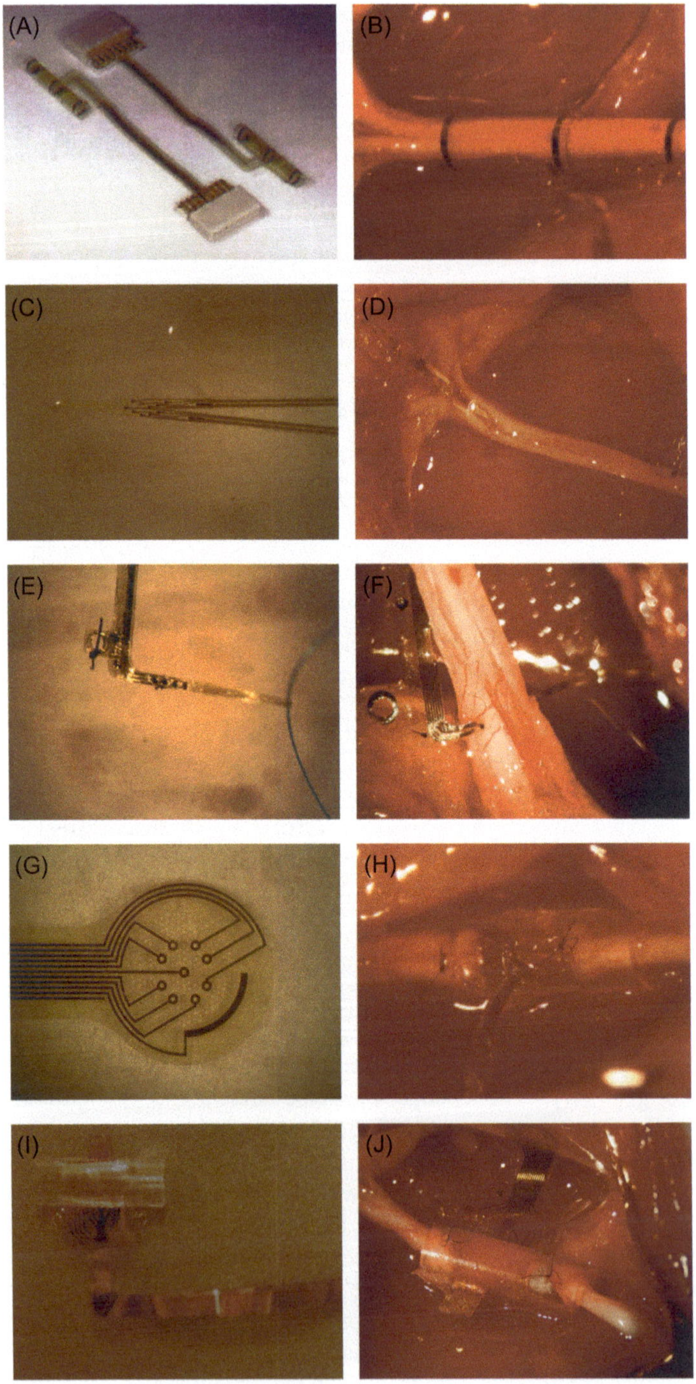
(A)
(B)
(C)
(D)
(E)
(F)
(G)
(H)
(I)
(J)

the median, radial, and ulnar nerves in more patients, and the results showed that these interfaces allowed to deliver the subjects sensory feedback for up to 2 years, reproducing the natural perception areas and the innervation pattern of the interfaced nerves (Tan et al., 2015). The main aim was to allow the patients to experience sensory percepts, to better graduate the grasping force of EMG-actioned prostheses, and to allow the user manipulation of delicate objects even without visual contact (Tan et al., 2014). The use of three peripheral nerve electrodes, with four and eight channels each, and different stimulating parameters, provided the subjects with stable perceptions of pressure, light moving touch, or tingling pulses for up to 2 years. In similar studies, the authors used patients that had been wearing extraneural electrodes for more than 3 years to test whether sensory feedback could improve the performance of manipulating objects in different tests. The results showed not only a better accuracy in the activities proposed but also a better sense of embodiment of the prosthesis, increasing self-confidence when the sensory feedback was provided (Schiefer et al., 2016, 2018). Delivery of sensory feedback to provide the patients with tactile and proprioceptive sensations has also been assayed in home-based patients for more than 3 years, improving several psychosocial parameters and decreasing the patients' own impressions of disability (Graczyk et al., 2018). A later study, after more than 6 years of the nerve implant, reported that patients learned to better use the prosthesis with sensory feedback during long periods. In addition, embodiment or own body image improved after the home-use in comparison with the pretest period (Cuberovic et al., 2019). Cuff electrodes have been

FIGURE 16.3 Images of selected types of peripheral nerve electrodes and their implantation in the rat sciatic nerve, used as an experimental model. (A) Detail of two polyimide cuffs. (B) A cuff electrode placed around the sciatic nerve of a rat. (C) Picture showing the two sides of a folded LIFE. (D) A LIFE implanted within a rat sciatic nerve. (E) Detail of a TIME with a surgical filament attached to its tip to guide implantation. (F) Photograph of a TIME crossing the tibial and peroneal fascicles of a rat sciatic nerve. (G) Sieve electrode with circular active sites and the ground electrode on the bottom right of the sieve. (H) A sieve electrode implanted within a silicone tube and placed between the transected sciatic nerve stumps. Axons will regenerate through the holes of the electrode. (I) Detail of a double-aisle regenerative electrode located inside a silicone tube. (J) A double-aisle regenerative electrode implanted in a transected rat sciatic nerve. *LIFE*, longitudinal intrafascicular electrode; *TIME*, transverse intrafascicular multichannel electrode. *Source: (B) Modified from Rodríguez, F. J., Ceballos, D., Schüttler, M., Valero, A., Valderrama, E., Stieglitz, T., ... Navarro, X. (2000). Polyimide cuff electrodes for peripheral nerve stimulation. Journal of Neuroscience Methods, 98(2), 105–118. https://doi.org/10.1016/S0165-0270(00)00192-8; (D) modified from Badia, J., Boretius, T., Pascual-Font, A., Udina, E., Stieglitz, T., & Navarro, X. (2011). Biocompatibility of chronically implanted transverse intrafascicular multichannel electrode (TIME) in the rat sciatic nerve. IEEE Transactions on Bio-Medical Engineering, 58(8), 2324–2332. https://doi.org/10.1109/TBME.2011.2153850; (G) modified from Lago, N., Ceballos, D. J., Rodríguez, F., Stieglitz, T., & Navarro, X. (2005). Long term assessment of axonal regeneration through polyimide regenerative electrodes to interface the peripheral nerve. Biomaterials, 26(14), 2021–2031. https://doi.org/10.1016/j.biomaterials.2004.06.025.*

shown to remain functional for more than 10 years after implantation for stimulating motor axon populations without significant changes in the amount of current needed for recruitment (Christie et al., 2017).

On the other hand, it should be taken into account that the epi-/perineurium has a high impedance that decreases the quality of the nerve signals that can be recorded (Raspopovic et al., 2017). Hence, while cuff electrodes have been used for stimulation and also for recording from experimental animal nerves (Haugland et al., 1994; Raspopovic et al., 2010), there are not many studies in humans using cuff electrodes as bidirectional interfaces in neuroprostheses. Signals from mechanoreceptors in the digital nerve were recorded with an implanted cuff electrode and used to help to control the force in a hand grasp neuroprosthesis. With the sensory close-loop the average grasp force is reduced, reducing the user's fatigue (Inmann et al., 2001).

Regarding FES systems, while surface electrodes are preferred for their easiness of use and affordability, cuff electrodes offer better long-term outcomes (Popović, 2014). These electrodes can be used for chronic stimulation and/or recording nerve activity (Haugland & Hoffer, 1994; Hoffer et al., 1996) to serve as the link between the nervous system and FES applications. Cuff electrodes can serve to record somatic or autonomic nerve signals to trigger an external device that will perform a particular action and can also stimulate a nerve that will produce the desired outcome. In this regard, cuff electrodes have been implanted in the peroneal nerve to lift the foot in FES systems for the correction of foot drop (Waters et al., 1975); cuff electrodes around the sural nerve may record nerve signals to trigger the stimulation in the same foot drop system (Hansen et al., 2003). A FES system using cuff electrodes placed around peroneal and femoral nerves was developed to help paralyzed patients to stand (Fisher et al., 2008) or walk (Guiraud et al., 2014; Guiraud et al., 2006). Similarly, cuff electrodes have also been implanted to stimulate elbow or shoulder muscles for moving a paralyzed arm (Kirsch, 2005). Cuff electrodes have been used also as the link in FES systems to restore bowel and bladder function in experimental and clinical settings (Creasey & Craggs, 2012; Gomez-Amaya et al., 2015). Other FES applications use cuff electrodes to stimulate the phrenic nerve and activate the diaphragm (Glenn & Phelps, 1985), to stimulate the vagus nerve, and treat a wide variety of pathologies such as neurological disorders (Howland, 2014), cardiac failure, or obesity (Guiraud et al., 2016).

16.2.2.3 Other extraneural electrodes

Helicoidal electrodes, made of flexible platinum ribbon with an open helical design, are implanted around the nerve of interest and adapt to the shape of the nerve (Naples et al., 1990). However, their open structure and low number of active sites cause low selectivity of the electrode (Agnew et al., 1989).

Helicoidal electrodes have been used for functional stimulation applications of the vagus nerve (Fisher & Handforth, 1999).

Other extraneural electrode designs have been reported in the literature. The split ring electrode (Fig. 16.2) is formed by a polyimide or parylene-C stripe that adopts an open hoop shape with sharp inside projections containing the active sites (Lee, Sheshadri, et al., 2017; Xue et al., 2015). In the implantation, the split ring is opened and placed embracing the nerve and using the projections to push into the nerve achieving a tight contact (Xiang et al., 2016).

The flexible neural clip takes the form of a polyimide clip-strip to be placed around the nerve after bending. As it secures the nerve with its own structure, no additional fixation is needed, but potential nerve damage may appear as movements between the nerve and device are restricted (Lee, Peh, et al., 2017). Other clipping variations to interface the nerve include a microscale printable nanoclip (Lissandrello et al., 2017) and parylene-based designs (Yu et al., 2014).

The flexible epineural strip electrode resembles a band-aid with active sites within. It is implanted longitudinal to the nerve, and has holes in both electrode ends to place sutures and fix the device to the nerve (Lee et al., 2015). The neural ribbon electrode is another design proposed by the same group with the concept of longitudinally wrapping the epineurium with the electrode rolling it repeatedly along the nerve axis and placing the active sites longitudinally distributed over the length of the nerve (Xiang et al., 2016). Nevertheless, all these designs have been tested only in laboratory animals.

Recently, cuff electrodes have been complexed with microfluidic systems to provide local drug delivery in the nerve (Cobo et al., 2017; Elyahoodayan et al., 2020), thus associating an electrical link with chemical actions. Another variation is the optocuff, a nerve cuff able to emit light instead of injecting current to specifically activate genetically modified peripheral axons in transgenic mice (Michoud et al., 2018).

16.2.3 Intraneural electrodes

Intraneural electrodes are implanted within the nerve after piercing the epineurium. A particular class is the interfascicular electrodes, devices that are laid inside the nerve between fascicles but without crossing the perineurium. The slowly penetrating interfascicular electrode (SPINE) is the most significant example of such an approach (Tyler & Durand, 1997). Stimulation of the sciatic nerve in cats and rabbits with the SPINE and other interfascicular electrodes (Nielsen et al., 2014; Tyler & Durand, 1997) reported that these devices could provide topographical interfascicular selectivity due to the passive separation of the interfaced areas.

Intrafascicular electrodes are the current standard of intraneural interfaces in neuroprosthetic applications. They are implanted crossing epineurial and perineurial layers to be placed within the endoneurium. As their active sites are in close contact with a restricted subset of axons, they offer better selectivity than extraneural or interfascicular electrodes. Indeed, intraneural electrodes have higher stimulation selectivity in comparison with extraneural electrodes (Badia, Boretius, Andreu, et al., 2011), need lower intensity for stimulation since they are not affected by the insulating properties of the epi- and perineurium, show lower crosstalk with the adjacent fascicles, and offer an increased signal-to-noise ratio for neural recordings (Micera et al., 2010). However, a more intimate contact between the interface and the axons may produce more nerve damage and encapsulation reaction that can interfere with the function of the electrode (Micera et al., 2008; Yoshida et al., 2010).

Intraneural electrodes (Fig. 16.2) can be implanted longitudinally, such as the longitudinal intrafascicular electrodes (LIFEs), or transversely, like the transverse intrafascicular multichannel electrodes (TIMEs) or the multielectrode arrays (MEA). They have been used to control neuroprostheses and to recreate sensory feedback in human amputees, to pilot a wheelchair, or to command virtual robotic devices among other applications (del Valle & Navarro, 2013).

16.2.3.1 Longitudinal intrafascicular electrodes

The LIFEs are inserted longitudinally into individual nerve fascicles (Fig. 16.3D), so as to lie in-between and parallel to the nerve fibers (Lawrence et al., 2004), providing high interfascicular selectivity as they interface a limited group of axons within a given fascicle. First implants were performed with coiled stainless steel wires, demonstrating a decrease in the current needed to stimulate the nerve in comparison with extraneural electrodes. However, as macroscopical nerve damage could be observed (Bowman & Erickson, 1985), less invasive designs were developed. LIFEs were then fabricated with conducting Pt-Ir or metallized Kevlar fibers insulated with Teflon or silicone. The active sites were made by opening small windows in the insulating layer, leaving the conductive filament exposed (Lawrence et al., 2004; Malagodi et al., 1989). A second version of LIFEs used thin micromachined polyimide as the electrode substrate with platinum as the electrically active material. These thin-film LIFEs (tf-LIFE) are folded in half (Figs. 16.2 and 16.3C), creating a symmetrical structure with active sites at each side facing the nerve (Lago et al., 2007). In contrast with metal-based LIFE's moderate rigidity (Rijnbeek et al., 2018), the use of polyimide with a lower Young's modulus improved the whole device flexibility, decreasing the mechanical mismatch between the interface and the nervous tissue. Moreover, the micromachining process enabled the placement of several active sites in each electrode, increasing the chances for selective multiunit nerve recording and stimulation (Lago et al., 2007; Navarro et al., 2007).

Several versions of LIFE electrodes have been used in cats and rats to stimulate motor fibers and to record nerve signals elicited after stimulation of the skin and joints (Malagodi et al., 1989; Navarro et al., 2007; Yoshida & Horch, 1993). Studies in humans reported that it was possible not only to record motor signals to control and command different software applications through Pt-Ir LIFE electrodes but also to stimulate sensory fibers providing tactile or proprioceptive sensations even years after amputation (Dhillon et al., 2005; Dhillon et al., 2004). Further experiments in patients wearing a myoelectric prosthesis and with the same Pt-Ir LIFEs implanted in the median and ulnar nerves demonstrated that peripheral nerve stimulation allowed the users object identification in terms of size and compliance rather than simple tactile sensations (Horch et al., 2011). Tf-LIFE were implanted in the median and ulnar nerves of a human amputee to control an advanced robotic hand. The electroneurographic motor signals registered through the electrodes implanted in both nerves were stable during the implant, allowing to perform independent types of hand grip with the prosthesis for 1 month. The different electrodes could stimulate sensory fibers to evoke discrete tactile sensations and decreased unpleasant phantom limb sensations. However, the amount of charge needed to stimulate afferent fibers increased with time and tf-LIFE electrodes failed to provide sensory stimulation 10 days after the implant due to limitations in the amount of current the electrodes were able to deliver, together with the encapsulation of the active sites and/or progressive habituation of the patient (Rossini et al., 2010).

Provided LIFE electrodes are implanted in a nerve fascicle, they may provide high selectivity as they interface only nerve fibers within the implanted fascicle leaving nonimplanted fascicles unstimulated. However, targeting multiple fascicles is needed to perform a wide variety of actions and induce sensations related to different body territories. While some studies chose to implant up to four different LIFE electrodes (Rossini et al., 2010) in the same surgical procedure, other authors have opted to integrate different Pt-Ir LIFE electrodes in a single distributed intrafascicular multielectrode device. Although the use of Pt-Ir LIFEs allows for only one independent active site per LIFE, the implant of several filaments in different nerves would allow targeting numerous fascicles. However, this design is still too bulky to be implanted in vivo, and only ex vivo tests have been reported (Thota et al., 2015).

16.2.3.2 Transverse intrafascicular multichannel electrodes

The TIME is similar to the polyimide tf-LIFE, but is designed to be implanted transversally (Fig. 16.3F) into the nerve in order to access different nerve fascicles with a single device (Boretius et al., 2010). Just like the tf-LIFEs (Figs. 16.2 and 16.3E), TIMEs display a symmetrical configuration with active sites facing both sides of the electrode. The TIME was shown to allow good spatial resolution by traversing the entire nerve, and better inter- and intrafascicular selectivity than cuff or LIFE electrodes (Badia, Boretius,

Andreu, et al., 2011). TIMEs have been used to record nerve signals from different subsets of fibers in various fascicles and to stimulate distinct motor fibers to activate several muscles innervated by distinct branches of the nerve or even by axons within the same fascicle (Badia, Boretius, Andreu, et al., 2011; Boretius et al., 2010; Badia et al., 2016).

An evolution of the TIME electrode is the self-opening intrafascicular neural interface. The body of the electrode incorporates four wings that can open transversely after the insertion, offering a second dimension of contact between the electrode and the nerve. Therefore more axons from different fascicles can be interfaced while the wings anchor the device to the nerve, reducing micromotions of the electrode within the endoneurium and decreasing potential fibrotic encapsulation and further electrical insulation (Cutrone et al., 2015). Experiments in rodents demonstrated the biocompatibility of TIME implanted over months, without nerve damage or functional loss, giving room for hope for further human chronic implants (Badia, Boretius, Pascual-Font, et al., 2011; Wurth et al., 2017).

TIMEs have been used to recreate bidirectional feedback in human amputees (Raspopovic et al., 2014). The main goal was not only to record motor complex information to command an advanced neuroprosthesis (Cipriani et al., 2011), but also to allow the patients to feel what they were holding with the artificial hand without visual clues. After electrical stimulation of afferent fibers with implanted TIMEs in median and ulnar nerves, the patient was able to perceive different sensations referred to appropriate areas of his phantom hand and, after some training, to differentiate between hard and soft objects using a prosthetic hand. In this sense, the sensory feedback provided by TIME from the information gathered by sensors in the prosthesis allowed for closed-loop control of the neuroprosthesis, improving the accuracy of task performance (Raspopovic et al., 2014; Valle et al., 2018) and even providing proprioceptive sensations with finger position information (D'Anna et al., 2019). With a similar approach, modulating different parameters of stimulation such as pulse amplitude, pulse width, or pulse frequency, the implanted subject was able to discriminate the roughness of different textures tested (Oddo et al., 2016). Three transradial amputees with TIMEs implanted in their proximal peripheral nerves were able to control a neuroprosthesis while receiving sensory feedback during the 6 months after surgery. The closed-loop control helped the subjects to integrate the artificial limbs as belonging to their own bodies, and to improve motor control and also reduce phantom limb pain (Petrini et al., 2019). Recently, TIMEs have also been used to provide transfemoral amputees with sensations of the moving knee of the prosthetic leg and the stepping of the artificial feet on the ground. This sensory feedback reduced, as in the upper-extremity amputees, phantom limb pain and increased the velocity at which patients walked. Physical and mental fatigue were also reduced and overall confidence while walking with the prosthesis was improved (Petrini et al., 2019).

16.2.3.3 Multielectrode arrays

MEAs are electrodes composed of a base made of silicon, ceramic, or glass, and tens of sharp needles with electrode tips made of carbon, gold, platinum, or iridium oxide. MEAs are inserted transversely into the neural tissue, thus providing multisite recording and stimulation. The high number of electrical contacts in each device allows interfacing multiple axonal groups with high selectivity. However, MEAs may generate neural damage produced not only by the stiff structure of the electrodes, but also by the injury caused during insertion of the many electrode tips. A specialized tool that accelerates the insertion of the Utah Electrode Array (UEA) into the tissue was designed to minimize insertion damage (Rousche & Normann, 1992). An evolution of this design is the Utah slanted electrode array (USEA), in which the needles have varying lengths to reach nerve fascicles at different depths (Fig. 16.2), thus offering a 3D interface (Branner, 2001). Newer versions of these arrays include a higher density of electrodes per area (up to 25 per mm^2) (Wark et al., 2013) and more flexible devices aimed at decreasing nerve damage after implantation (Byun et al., 2017; Kang et al., 2019). Recent studies report several developments in the UEA that allow scaling up the channel count and the density of electrode contacts, as well as material improvements. Further designs are aimed to integrate microfluidic delivery of compounds into the array of electrical contacts, and also to utilize optogenetic stimulation (Leber et al., 2019).

MEAs have been mainly used as microinterfaces to record neural signals from the brain cortex, allowing communication of paralyzed patients with computers, robotic assistive devices, or even their own limbs (e.g., through FES systems) during years after implantation (Ajiboye et al., 2017; Hochberg et al., 2012). In addition, several studies have used UEAs (Branner & Normann, 2000) and USEAs (Branner et al., 2004) in the PNS to stimulate nerves and record from them in cats. Studies in human amputees have also shown that USEAs implanted in the median and ulnar nerves can record motor nerve signals, which can be used to actuate the movements of a simulated robotic hand, and to stimulate nerve afferents for evoking sensory percepts in the phantom hand, thus enabling the closed-loop control of virtual limbs (Davis et al., 2016; Wendelken et al., 2017). However, the quality of the signals decreased with time, and the amount of current injected to stimulate the axons had to be progressively increased. The nerve damage and subsequent inflammation caused by the electrodes inducing axonal injury, deposition of fibrotic tissue around the implant, and release of corrosive substances, which might have damaged the active sites, were the suggested reasons for this decline in electrode functionality (Christensen et al., 2014; Christensen et al., 2016). Further studies with USEA implants to provide sensory feedback in patients wearing a myoelectric prosthesis, similar to the studies using cuff electrodes (Schiefer et al., 2018; Tan et al., 2014), allowed the patients to discriminate between two objects in terms of size and compliance and the closed-loop improved grip precision (George et al., 2019).

16.2.4 Regenerative electrodes

Regenerative electrodes are implanted in the gap between stumps created after a peripheral nerve section. As the axons of the PNS retain the ability to regenerate after axotomy, they grow through or along the regenerative electrode, thus making intrinsic new contacts with the active sites placed along the regenerative guidance scaffold.

The first report of a regenerative electrode was the sieve electrode, composed of an array of via holes with electrodes built around them and where small populations of axons could grow (Mannard et al., 1974). The sieve electrode design has evolved over recent decades from rigid materials to the highly flexible substrate of polyimide (Figs. 16.2 and 16.3G), strongly improving its usability. As each electrode site interfaces only with a few fibers, depending upon the size of the via holes, high selectivity can be achieved; however, the time required for the axons to grow limits the use of these devices in acute experiments (Navarro et al., 2005). Although no studies in humans have been reported to date, long-term studies in rats (Fig. 16.3H) and cats have shown that sieve electrodes are able to record neural activity after mechanical stimulation of the skin in the paw (Lago et al., 2006; Panetsos et al., 2008). A drawback of the sieve electrodes is the set of holes which the axons need to regenerate through, as they are relatively rigid and do not allow the regenerated axons to increase in caliber more than a certain limit, which may cause axonal compression later. Long-term studies for a sieve electrode implant showed that axonal regeneration was not complete after 12 months. Moreover, the diameter of the axons and the myelin thickness did not recover to basal values after 12 months (Lago et al., 2005) and some fibers appeared with signs of degeneration (Lago et al., 2006). In this sense, the transparency of a regenerative electrode (i.e., the ratio between the open area that axons can grow through versus the total cross-section area of the electrode) and the diameters of the holes in which the axons need to grow play an important role. Transparencies that have too low or narrow channels may hamper nerve regeneration and impair the interface (Navarro et al., 1996; Wallman et al., 2001). On the other hand, too large size for the channels would decrease selectivity.

An evolution of sieve electrodes is the microchannel electrodes (Fig. 16.2) in which transparency is increased by providing wider channels for the axons to grow (Srinivasan et al., 2015). In this case, the axons grow via thin narrow parallel tubes with embedded electrodes. To improve axonal regeneration, specific cues can be included within the microchannels, making it possible to specify a certain path for a specific axonal population (del Valle et al., 2018). Different variations of the microchannel regenerative electrode have been presented (Thompson et al., 2016). Several studies have reported that these devices can offer both stimulation of regenerated fibers and recording of compound nerve action potentials from regenerated axons. However, axonal regeneration through these devices is compromised when the gap between stumps is longer than 2–4 mm (FitzGerald et al., 2012; Musick et al., 2015), limiting their usability in long nerve injuries.

Other concepts of regenerative electrodes are the regenerative scaffold electrode (Clements et al., 2007) and the double-aisle regenerative electrode (Delgado-Martínez et al., 2017). These designs take transparency to its limit as the interface is a planar surface longitudinally placed within a regenerative tube (Figs. 16.2 and 16.3I), allowing axons to grow through much wider space, hence diminishing the possible compression of the nerve. With this approach, regenerating axons can grow on each compartment of the device, allowing for selective interfacing of independent nerve branches or sorted types of axons (Fig. 16.3J). Indeed, with the double-aisle electrode, the maximal interfascicular selectivity was achieved when stimulating different subfascicles through the different sides of the electrode (Delgado-Martínez et al., 2017).

Further designs of regenerative electrodes include the regenerative MEA (REMI) in which a USEA electrode is inserted in a regenerative tube (Garde et al., 2009). Four months after implantation, this electrode design allowed electrical signals to be recorded while the animals were walking (Desai et al., 2014), but no reports of nerve stimulation have been given. Another combination of electrodes is the cuff and sieve electrode. The concept is very similar to the sieve electrodes previously described, but it uses a cuff electrode as a regenerative guide instead of a passive nerve tube (Fig. 16.2). Preliminary results have shown that this electrode is able to stimulate sensory fibers that have been recently sectioned away from the cuff active sites (Kim et al., 2020), but results showing the capability to allow nerve regeneration and to interface regenerated axons remain pending.

Acknowledgments

The research of the authors was supported by European Union FPT-ICT project NEBIAS (FP7–611687), FET-Open project SOMA (GA 899822), FLAG-ERA JTC 2017 project GRAFIN, Ministerio de Ciencia, Innovación y Universidades of Spain (grant PCI2018–093029), TERCEL (RD16/0011/0014), and CIBERNED (CB06/05/1105) funds from Instituto de Salud Carlos III of Spain, Fundación Ramón Areces (CIVP18A3897), Generalitat de Catalunya (GraphCAT: Comunitat Emergent de Grafè a Catalunya, code 001-P-001702), cofunded by European Union funds (ERDF/ESF).

References

Ackerley, R., Wasling, H. B., Ortiz-Catalan, M., Brånemark, R., & Wessberg, J. (2018). Case studies in neuroscience: Sensations elicited and discrimination ability from nerve cuff stimulation in an amputee over time. *Journal of Neurophysiology*, *120*(1), 291–295. Available from https://doi.org/10.1152/jn.00909.2017.

Agnew, W. F., McCreery, D. B., Yuen, T. G. H., & Bullara, L. A. (1989). Histologic and physiologic evaluation of electrically stimulated peripheral nerve: Considerations for the selection of parameters. *Annals of Biomedical Engineering*, *17*(1), 39–60. Available from https://doi.org/10.1007/BF02364272.

Ahmetovic, A., Mushahwar, V. K., Sommer, R., Schnepf, D., Kawasaki, L., Warwaruk-Rogers, R., Barlott, T., Chong, S. L., Isaacson, G., Kim, S., Ferguson-Pell, M., Stein, R. B., Ho, C., Dukelow, S., & Chan, K. M. (2015). Safety and feasibility of intermittent electrical stimulation for the prevention of deep tissue injury. *Advances in Wound Care*, *4*(3), 192–201. Available from https://doi.org/10.1089/wound.2014.0569.

Ajiboye, A. B., Willett, F. R., Young, D. R., Memberg, W., Walters, B. C., Sweet, J. A., Hoyen, H. A., Keith, M. W., Peckham, P. H., Simeral, J. D., Donoghue, J. P., Miller, J. P., Hochberg, L. R., & Kirsch, R. F. (2017). Restoration of reaching and grasping in a person with tetraplegia through brain-controlled muscle stimulation: A proof-of-concept demonstratio. *The Lancet*, *389*(10081). Available from https://doi.org/10.1016/S0140-6736(17)30601-3.

Badia, J., Boretius, T., Andreu, D., Azevedo-Coste, C., Stieglitz, T., & Navarro, X. (2011). Comparative analysis of transverse intrafascicular multichannel, longitudinal intrafascicular and multipolar cuff electrodes for the selective stimulation of nerve fascicles. *Journal of Neural Engineering*, *8*(3), 036023. Available from https://doi.org/10.1088/1741-2560/8/3/036023.

Badia, J., Boretius, T., Pascual-Font, A., Udina, E., Stieglitz, T., & Navarro, X. (2011). Biocompatibility of chronically implanted transverse intrafascicular multichannel electrode (TIME) in the rat sciatic nerve. *IEEE Transactions on Bio-Medical Engineering*, *58*(8), 2324–2332. Available from https://doi.org/10.1109/TBME.2011.2153850.

Badia, J., Raspopovic, S., Carpaneto, J., Micera, S., & Navarro, X. (2016). Spatial and functional selectivity of peripheral nerve signal recording with the transversal intrafascicular multichannel electrode (TIME). *IEEE Transactions on Neural Systems and Rehabilitation Engineering*, *24*(1), 20–27. Available from https://doi.org/10.1109/TNSRE.2015.2440768.

Birbaumer, N., Ghanayim, N., Hinterberger, T., Iversen, I., Kotchoubey, B., Kübler, A., Perelmouter, J., Taub, E., & Flor, H. (1999). A spelling device for the paralysed. *Nature*, *398*(6725), 297–298. Available from https://doi.org/10.1038/18581.

Blair, E. A., & Erlanger, J. (1933). A comparison of the characteristics of axons through their individual electrical responses. *American Journal of Physiology-Legacy Content*, *106*(3), 524–564. Available from https://doi.org/10.1152/ajplegacy.1933.106.3.524.

Boretius, T., Badia, J., Pascual-Font, A., Schuettler, M., Navarro, X., Yoshida, K., & Stieglitz, T. (2010). A transverse intrafascicular multichannel electrode (TIME) to interface with the peripheral nerve. *Biosensors and Bioelectronics*, *26*(1), 62–69. Available from https://doi.org/10.1016/j.bios.2010.05.010.

Bouton, C. E., Shaikhouni, A., Annetta, N. V., Bockbrader, M. A., Friedenberg, D. A., Nielson, D. M., Sharma, G., Sederberg, P. B., Glenn, B. C., Mysiw, W. J., Morgan, A. G., Deogaonkar, M., & Rezai, A. R. (2016). Restoring cortical control of functional movement in a human with quadriplegia. *Nature*, *533*, 1–13. Available from https://doi.org/10.1038/nature17435.

Bowman, B. R., & Erickson, R. C. (1985). Acute and chronic implantation of coiled wire intraneural electrodes during cyclical electrical stimulation. *Annals of Biomedical Engineering*, *13*(1), 75–93. Available from https://doi.org/10.1007/BF02371251.

Branner, A., & Normann, R. A. (2000). A multielectrode array for intrafascicular recording and stimulation in sciatic nerve of cats. *Brain Research Bulletin*, *51*(4), 293–306. Available from https://doi.org/10.1016/S0361-9230(99)00231-2.

Branner, A., Stein, R. B., Fernandez, E., Aoyagi, Y., & Normann, R. A. (2004). Long-term stimulation and recording with a penetrating microelectrode array in cat sciatic nerve. *IEEE Transactions on Biomedical Engineering, 51*(1), 146–157. Available from https://doi.org/10.1109/TBME.2003.820321.

Byun, D., Cho, S. J., Lee, B. H., Min, J., Lee, J. H., & Kim, S. (2017). Recording nerve signals in canine sciatic nerves with a flexible penetrating microelectrode array. *Journal of Neural Engineering, 14*(4). Available from https://doi.org/10.1088/1741-2552/aa7493.

Christensen, M. B., Pearce, S. M., Ledbetter, N. M., Warren, D. J., Clark, G. A., & Tresco, P. A. (2014). The foreign body response to the Utah Slant Electrode Array in the cat sciatic nerve. *Acta Biomaterialia, 10*(11), 4650–4660. Available from https://doi.org/10.1016/j.actbio.2014.07.010.

Christensen, M. B., Wark, H. A., & Hutchinson, D. T. (2016). A histological analysis of human median and ulnar nerves following implantation of Utah slanted electrode arrays. *Biomaterials, 77*, 235–242. Available from https://doi.org/10.1016/j.biomaterials.2015.11.012.

Christie, B. P., Freeberg, M., Memberg, W. D., Pinault, G. J. C., Hoyen, H. A., Tyler, D. J., & Triolo, R. J. (2017). Long-term stability of stimulating spiral nerve cuff electrodes on human peripheral nerves. *Journal of Neuroengineering and Rehabilitation, 14*(1), 1–12. Available from https://doi.org/10.1186/s12984-017-0285-3.

Cipriani, C., Controzzi, M., & Carrozza, M. C. (2011). The SmartHand transradial prosthesis. *Journal of Neuroengineering and Rehabilitation, 8*, 29. Available from https://doi.org/10.1186/1743-0003-8-29.

Clements, I. P., Kim, Y. T., Andreasen, D., & Bellamkonda, R. V. (2007). *A regenerative electrode scaffold for peripheral nerve interfacing. Proceedings of the 3rd International IEEE EMBS Conference on Neural Engineering* (pp. 390–393). IEEE. Available from https://doi.org/10.1109/CNE.2007.369691.

Cobo, A. M., Boyajian, B., Larson, C., Schotten, K., Pikov, V., & Meng, E. (2017). *A parylene cuff electrode for peripheral nerve recording and drug delivery. Proceedings of the IEEE International Conference on Micro Electro Mechanical Systems (MEMS)* (pp. 506–509). IEEE. Available from https://doi.org/10.1109/MEMSYS.2017.7863454.

Creasey, G. H., & Craggs, M. D. (2012). *Functional electrical stimulation for bladder, bowel, and sexual function. Handbook of Clinical Neurology* (Vol. 109, pp. 247–257). Elsevier. Available from https://doi.org/10.1016/B978-0-444-52137-8.00015-2.

Cuberovic, I., Gill, A., Resnik, L. J., Tyler, D. J., & Graczyk, E. L. (2019). Learning of artificial sensation through long-term home use of a sensory-enabled prosthesis. *Frontiers in Neuroscience, 13*(August), 1–24. Available from https://doi.org/10.3389/fnins.2019.00853.

D'Anna, E., Valle, G., Mazzoni, A., Strauss, I., Ibertie, F., Patton, J., Petrini, F. M., Raspopovic, S., Granata, G., Iorio, R. D., Controzzi, M., Cipriani, C., Stieglitz, T., Rossini, P. M., & Micera, S. (2019). A closed-loop hand prosthesis with simultaneous intraneural tactile and position feedback. *Science Robotics, 4*(17), 262741. Available from https://doi.org/10.1101/262741.

Davis, T., Wark, H. A. C. C., Hutchinson, D. T., Warren, D. J., O'Neill, K., Scheinblum, T., Clark, G. A., Normann, R. A., Greger, B., & O'Neill, K. (2016). Restoring motor control and sensory feedback in people with upper extremity amputations using arrays of 96 microelectrodes implanted in the median and ulnar nerves. *Journal of Neural Engineering, 13*(3), 036001. Available from https://doi.org/10.1088/1741-2560/13/3/036001.

del Valle, J., & Navarro, X. (2013). Interfaces with the peripheral nerve for the control of neuroprostheses. *International Review of Neurobiology, 109*, 63–83. Available from https://doi.org/10.1016/B978-0-12-420045-6.00002-X.

del Valle, J., Santos, D., Delgado-Martínez, I., de la Oliva, N., Giudetti, G., Micera, S., & Navarro, X. (2018). Segregation of motor and sensory axons regenerating through bicompartmental tubes by combining extracellular matrix components with neurotrophic factors. *Journal of Tissue Engineering and Regenerative Medicine*, *12*(4), e1991–e2000. Available from https://doi.org/10.1002/term.2629.

Delgado-Martínez, I., Righi, M., Santos, D., Cutrone, A., Bossi, S., D'Amico, S., Del Valle, J., Micera, S., Navarro, X., & D'Amico, S. (2017). Fascicular nerve stimulation and recording using a novel double-aisle regenerative electrode. *Journal of Neural Engineering*, *14*(4), 046003. Available from https://doi.org/10.1088/1741-2552/aa6bac.

Desai, V. H., Anand, S., Tran, M., Kanneganti, A., Vasudevan, S., Seifert, J. L., & Cheng, J. (2014). *Chronic sensory-motor activity in behaving animals using regenerative multi-electrode interfaces. Conference Proceedings: ...Annual International Conference of the IEEE Engineering in Medicine and Biology Society. IEEE Engineering in Medicine and Biology Society. Conference* (pp. 1973–1976). IEEE.

Dhillon, G. S., Krüger, T. B., Sandhu, J. S., & Horch, K. W. (2005). Effects of short-term training on sensory and motor function in severed nerves of long-term human amputees. *Journal of Neurophysiology*, *93*(5), 2625–2633. Available from https://doi.org/10.1152/jn.00937.2004.

Dhillon, G. S., Lawrence, S. M., Hutchinson, D. T., & Horch, K. W. (2004). Residual function in peripheral nerve stumps of amputees: Implications for neural control of artificial limbs. *Journal of Hand Surgery*, *29*(4), 605–615. Available from https://doi.org/10.1016/j.jhsa.2004.02.006.

Do, A. H., Wang, P. T., King, C. E., Chun, S. N., & Nenadic, Z. (2013). Brain-computer interface controlled robotic gait orthosis. *Journal of Neuroengineering and Rehabilitation*, *10*(1), 111.

Dosen, S., Schaeffer, M. C., & Farina, D. (2014). Time-division multiplexing for myoelectric closed-loop control using electrotactile feedback. *Journal of Neuroengineering and Rehabilitation*, *11*(1). Available from https://doi.org/10.1186/1743-0003-11-138.

Elyahoodayan, S., Larson, C., Cobo, A. M., Meng, E., & Song, D. (2020). Acute in vivo testing of a polymer cuff electrode with integrated microfluidic channels for stimulation, recording, and drug delivery on rat sciatic nerve. *Journal of Neuroscience Methods*, *336*, 108634. Available from https://doi.org/10.1016/j.jneumeth.2020.108634.

Fisher, L. E., Miller, M. E., Bailey, S. N., Davis, J. A., Anderson, J. S., Rhode, L., Tyler, D. J., & Triolo, R. J. (2008). Standing after spinal cord injury with four-contact nerve-cuff electrodes for quadriceps stimulation. *IEEE Transactions on Neural Systems and Rehabilitation Engineering*, *16*(5), 473–478. Available from https://doi.org/10.1109/TNSRE.2008.2003390.

Fisher, R. S., & Handforth, A. (1999). Reassessment: Vagus nerve stimulation for epilepsy. *Neurology*, *53*(4), 666. Available from https://doi.org/10.1212/WNL.53.4.666.

FitzGerald, J., Lago, N., Benmerah, S., Serra, J., Watling, C. P., Cameron, R. E., Tarte, E., Lacour, S. P., McMahon, S. B., Fawcett, J. W., & Christopher, P. (2012). A regenerative microchannel neural interface for recording from and stimulating peripheral axons in vivo. *Journal of Neural Engineering*, *9*(September), 016010. Available from https://doi.org/10.1088/1741-2560/9/1/016010.

Garde, K., Keefer, E., Botterman, B., Galvan, P., & Romero, M. I. (2009). Early interfaced neural activity from chronic amputated nerves. *Frontiers in Neuroengineering*, *2*(May), 1–11. Available from https://doi.org/10.3389/neuro.16.005.2009.

George, J. A., Kluger, D. T., Davis, T. S., Wendelken, S. M., Okorokova, E. V., He, Q., Duncan, C. C., Hutchinson, D. T., Thumser, Z. C., Beckler, D. T., Marasco, P. D., Bensmaia, S. J., & Clark, G. A. (2019). Biomimetic sensory feedback through peripheral nerve stimulation improves dexterous use of a bionic hand. *Science Robotics*, *4*(32), eaax2352. Available from https://doi.org/10.1126/scirobotics.aax2352.

Gibson, W., Wand, B. M., Meads, C., Catley, M. J., & O'Connell, N. E. (2019). Transcutaneous electrical nerve stimulation (TENS) for chronic pain-an overview of cochrane reviews. *Cochrane Database of Systematic Reviews, 4*.

Glenn, W. W., & Phelps, M. L. (1985). Diaphragm pacing by electrical stimulation of the phrenic nerve. *Neurosurgery*, *17*(6), 974–984. Available from http://www.ncbi.nlm.nih.gov/pubmed/3908966.

Gomez-Amaya, S. M., Barbe, M. F., de Groat, W. C., Brown, J. M., Tuite, G. F., Corcos, J., Fecho, S. B., Braverman, A. S., & Ruggieri, M. R. (2015). Neural reconstruction methods of restoring bladder function. *Nature Reviews Urology*, *12*(2), 100–118. Available from https://doi.org/10.1038/nrurol.2015.4.

Graczyk, E. L., Resnik, L., Schiefer, M. A., Schmitt, M. S., & Tyler, D. J. (2018). Home use of a neural-connected sensory prosthesis provides the functional and psychosocial experience of having a hand again. *Scientific Reports*, *8*(1), 1–17. Available from https://doi.org/10.1038/s41598-018-26952-x.

Graupe, D., & Kohn, K. (1997). Transcutaneous functional neuromuscular stimulation of certain traumatic complete thoracic paraplegics for independent short-distance ambulation. *Neurological Research*, *19*(3), 323–333. Available from https://doi.org/10.1080/01616412.1997.11740820.

Grill, W. M., & Mortimer, J. T. (1996). The effect of stimulus pulse duration on selectivity of neural stimulation. *IEEE Transactions on Biomedical Engineering*, *43*(2), 161–166. Available from https://doi.org/10.1109/10.481985.

Guiraud, D., Andreu, D., Bonnet, S., Carrault, G., Couderc, P., Hagège, A., Henry, C., Hernandez, A., Karam, N., Rolle, V. L., Mabo, P., Paweł, M., Malbert, C.-H., Marijon, E., Maubert, S., Picq, C., Rossel, O., & Bonnet, J.-L. (2016). Vagus nerve stimulation: State of the art of stimulation and recording strategies to address autonomic function neuromodulation. *Journal of Neural Engineering*, *13*(4), 41002. Available from https://doi.org/10.1088/1741-2560/13/4/041002.

Guiraud, D., Azevedo-Coste, C., Benoussaad, M., & Fattal, C. (2014). Implanted functional electrical stimulation: Case report of a paraplegic patient with complete SCI after 9 years. *Journal of Neuroengineering and Rehabilitation*, *11*(1), 15. Available from https://doi.org/10.1186/1743-0003-11-15.

Guiraud, D., Stieglitz, T., Koch, K. P., Divoux, J.-L., & Rabischong, P. (2006). An implantable neuroprosthesis for standing and walking in paraplegia: 5-year patient follow-up. *Journal of Neural Engineering*, *3*(4), 268–275. Available from https://doi.org/10.1088/1741-2560/3/4/003.

Guo, L., Sandsjö, L., Ortiz-Catalan, M., & Skrifvars, M. (2020). Systematic review of textile-based electrodes for long-term and continuous surface electromyography recording. *Textile Research Journal*, *90*(2), 227–244. Available from https://doi.org/10.1177/0040517519858768.

Hansen, M., Haugland, M. K., & Sepulveda, F. (2003). Feasibility of using peroneal nerve recordings for deriving stimulation timing in a foot drop correction system. *Neuromodulation: Technology at the Neural Interface*, *6*(1), 68–77.

Haugland, M. K., & Hoffer, J. A. (1994). Artifact-free sensory nerve signals obtained from cuff electrodes during functional electrical stimulation of nearby muscles. *IEEE Transactions on Rehabilitation Engineering*, *2*(1), 37–40.

Haugland, M. K., Hoffer, J. A., & Sinkjær, T. (1994). Skin contact force information in sensory nerve signals recorded by implanted cuff electrodes. *IEEE Transactions on Rehabilitation Engineering*, *2*(1), 18–28. Available from https://doi.org/10.1109/86.296346.

Hochberg, L. R., Bacher, D., Jarosiewicz, B., Masse, N. Y., Simeral, J. D., Vogel, J., Haddadin, S., Liu, J., Cash, S. S., Van Der Smagt, P., & Donoghue, J. P. (2012). Reach and grasp by people with tetraplegia using a neurally controlled robotic arm. *Nature*, *485*(7398), 372–375. Available from https://doi.org/10.1038/nature11076.

Hoffer, J. A., Stein, R. B., Haugland, M. K., Sinkjaer, T., Durfee, W. K., Schwartz, A. B., Loeb, G. E., & Kantor, C. (1996). Neural signals for command control and feedback in functional neuromuscular stimulation: A review. *Journal of Rehabilitation Research and Development*, *33*(2), 145–157.

Horch, K. W., Meek, S., Taylor, T. G., & Hutchinson, D. T. (2011). Object discrimination with an artificial hand using electrical stimulation of peripheral tactile and proprioceptive pathways with intrafascicular electrodes. *IEEE Transactions on Neural Systems and Rehabilitation Engineering*, *19*(5), 483–489. Available from https://doi.org/10.1109/TNSRE.2011.2162635.

Howland, R. H. (2014). Vagus nerve stimulation. *Current Behavioral Neuroscience Reports*, *1* (2), 64–73. Available from https://doi.org/10.1007/s40473-014-0010-5.

Inmann, A., Haugland, M., Haase, J., Biering-Sørensen, F., & Sinkjaer, T. (2001). Signals from skin mechanoreceptors used in control of a hand grasp neuroprosthesis. *Neuroreport*, *12*(13), 2817–2820. Available from https://doi.org/10.1097/00001756-200109170-00013.

Jafri, S. R. A., Hamid, T., Mahmood, R., Alam, M. A., Rafi, T., Ul Haque, M. Z., & Munir, M. W. (2019). Wireless brain computer interface for smart home and medical system. *Wireless Personal Communications*, *106*(4), 2163–2177. Available from https://doi.org/10.1007/s11277-018-5932-x.

Kaczmarek, K. A., Webster, J. G., Bach-y-Rita, P., & Tompkins, W. J. (1991). Electrotactile and vibrotactile displays for sensory substitution systems. *IEEE Transactions on Biomedical Engineering*, *38*(1), 1–16. Available from https://doi.org/10.1109/10.68204.

Kang, Y. N., Chou, N., Jang, J. W., Byun, D., Kang, H., Moon, D. J., Kim, J., & Kim, S. (2019). An intrafascicular neural interface with enhanced interconnection for recording of peripheral nerve signals. *IEEE Transactions on Neural Systems and Rehabilitation Engineering*, *27*(6), 1312–1319. Available from https://doi.org/10.1109/TNSRE.2019.2917916.

Keller, T., & Kuhn, A. (2008). Electrodes for transcutaneous (surface) electrical stimulation. *Journal of Automatic Control*, *18*(2), 35–45. Available from https://doi.org/10.2298/jac0802035k.

Khan, M. J., & Hong, K.-S. (2017). Hybrid EEG–fNIRS-based eight-command decoding for BCI: Application to quadcopter control. *Frontiers in Neurorobotics*, *11*, 6. Available from https://doi.org/10.3389/fnbot.2017.00006.

Kim, H., Dingle, A. M., Ness, J. P., Baek, D.-H., Bong, J., Lee, I.-K., Shulzhenko, N. O., Zeng, W., Israel, J. S., Pisaniello, J. A., Millevolte, A. X. T., Park, D.-W., Suminski, A. J., Jung, Y. H., Williams, J. C., Poore, S. O., & Ma, Z. (2020). Cuff and sieve electrode (CASE): The combination of neural electrodes for bi-directional peripheral nerve interfacing. *Journal of Neuroscience Methods*, *336*(January), 108602. Available from https://doi.org/10.1016/j.jneumeth.2020.108602.

Kirsch, R.F. (2005). Development of a neuroprosthesis for restoring arm and hand function via functional electrical stimulation following high cervical spinal cord injury. *Annual International Conference of the IEEE Engineering in Medicine and Biology—Proceedings*, *7*, 4142–4144. https://doi.org/10.1109/iembs.2005.1615375

Kozai, T. D. Y., Langhals, N. B., Patel, P. R., Deng, X., Zhang, H., Smith, K. L., Lahann, J., Kotov, N. A., & Kipke, D. R. (2012). Ultrasmall implantable composite microelectrodes with bioactive surfaces for chronic neural interfaces. *Nature Materials*, *11*(12), 1065–1073. Available from https://doi.org/10.1038/nmat3468.

Kuiken, T. A., Dumanian, G. A., Lipschutz, R. D., Miller, L. A., & Stubblefield, K. A. (2004). The use of targeted muscle reinnervation for improved myoelectric prosthesis control in a bilateral shoulder disarticulation amputee. *Prosthetics and Orthotics International*, *28*(3), 245–253.

Lago, N., Ceballos, D. J., Rodríguez, F., Stieglitz, T., & Navarro, X. (2005). Long term assessment of axonal regeneration through polyimide regenerative electrodes to interface the peripheral nerve. *Biomaterials*, *26*(14), 2021–2031. Available from https://doi.org/10.1016/j.biomaterials.2004.06.025.

Lago, N., Udina, E., & Navarro, X. (2006). Regenerative electrodes for interfacing injured per- iph- eral nerves: Neurobiological assessment. *The First IEEE/RAS-EMBS International Conference on Biomedical Robotics and Biomechatronics, 2006. BioRob 2006., 35094*, 1149–1153. https://doi.org/10.1109/BIOROB.2006.1639247

Lago, N., Yoshida, K., Koch, K. P., & Navarro, X. (2007). Assessment of biocompatibility of chronically implanted polyimide and platinum intrafascicular electrodes. *IEEE Transactions on Biomedical Engineering*, *54*(2), 281–290. Available from https://doi.org/10.1109/TBME.2006.886617.

Lawrence, S. M., Dhillon, G. S., Jensen, W., Yoshida, K., & Horch, K. W. (2004). Acute peripheral nerve recording characteristics of polymer-based longitudinal intrafascicular electrodes. *IEEE Transactions on Neural Systems and Rehabilitation Engineering*, *12*(3), 345–348. Available from https://doi.org/10.1109/TNSRE.2004.831491.

Leber, M., Körner, J., Reiche, C. F., Yin, M., Bhandari, R., Franklin, R., Negi, S., & Solzbacher, F. (2019). Advances in penetrating multichannel microelectrodes based on the Utah array platform. *Advances in Experimental Medicine and Biology*, *1101*, 1–40. Available from https://doi.org/10.1007/978-981-13-2050-7_1.

Lee, S., Peh, W. Y. X., Wang, J., Yang, F., Ho, J. S., Thakor, N. V., Yen, S. C., & Lee, C. (2017). Toward bioelectronic medicine—neuromodulation of small peripheral nerves using flexible neural clip. *Advanced Science*, *4*(11). Available from https://doi.org/10.1002/advs.201700149.

Lee, S., Sheshadri, S., Xiang, Z., Delgado-Martinez, I., Xue, N., Sun, T., Thakor, N. V., Yen, S. C., & Lee, C. (2017). Selective stimulation and neural recording on peripheral nerves using flexible split ring electrodes. *Sensors and Actuators, B: Chemical*, *242*, 1165–1170. Available from https://doi.org/10.1016/j.snb.2016.09.127.

Lee, S., Yen, S., Sheshadri, S., Martinez, I. D., Xue, N., Xiang, Z., Nitish, V., Ieee, F., Lee, C., & Ieee, C. (2015). *Flexible Epineural Strip Electrode (FLESE) for recording in fine nerves* (9294, pp. 1–7). IEEE. Available from https://doi.org/10.1109/TBME.2015.2466442.

Leventhal, D. K., & Durand, D. (2003). Subfascicle stimulation selectivity with the flat interface nerve electrode. *Annals of Biomedical Engineering*, *31*, 643–652. Available from https://doi.org/10.1114/1.1569266.

Lissandrello, C. A., Gillis, W. F., Shen, J., Pearre, B. W., Vitale, F., Pasquali, M., Holinski, B. J., Chew, D. J., White, A. E., & Gardner, T. J. (2017). A micro-scale printable nanoclip for electrical stimulation and recording in small nerves. *Journal of Neural Engineering*, *14*(3). Available from https://doi.org/10.1088/1741-2552/aa5a5b.

Malagodi, M. S., Horch, K. W., & Schoenberg, A. A. (1989). An intrafascicular electrode for recording of action potentials in peripheral nerves. *Annals of Biomedical Engineering*, *17*(4), 397–410. Available from https://doi.org/10.1007/BF02368058.

Mannard, A., Stein, R. B., & Charles, D. (1974). Regeneration electrode units: Implants for recording from single peripheral nerve fibers in freely moving animals abstract. *Science (New York, N.Y.)*, *183*(4124), 547–549.

Mastinu, E., Ahlberg, J., Lendaro, E., Hermansson, L., Hakansson, B., & Ortiz-Catalan, M. (2018). An alternative myoelectric pattern recognition approach for the control of hand prostheses: A case study of use in daily life by a dysmelia subject. *IEEE Journal of Translational Engineering in Health and Medicine*, *6*(January). Available from https://doi.org/10.1109/JTEHM.2018.2811458.

McCarty, L. P. (1965). A stimulating electrode for nerves. *Journal of Applied Physiology*, *20*(3), 542. Available from https://doi.org/10.1152/jappl.1965.20.3.542.

McFarland, D., & Wolpaw, J. R. (2017). EEG-based brain—computer interfaces. *Current Opinion in Biomedical Engineering*, *4*, 194–200.

Micera, S. (2016). Staying in touch: Toward the restoration of sensory feedback in hand prostheses using peripheral neural stimulation. *IEEE Pulse*, *7*(3), 16–19. Available from https://doi.org/10.1109/MPUL.2016.2539760.

Micera, S., Carpaneto, J., & Raspopovic, S. (2010). Control of hand prostheses using peripheral information. *IEEE Reviews in Biomedical Engineering*, *3*, 48–68. Available from https://doi.org/10.1109/RBME.2010.2085429.

Micera, S., Navarro, X., Carpaneto, J., Citi, L., Tonet, O., Rossini, P. M., Carrozza, M. C., Hoffmann, K. P., Vivó, M., Yoshida, K., & Dario, P. (2008). On the use of longitudinal intrafascicular peripheral interfaces for the control of cybernetic hand prostheses in amputees. *IEEE Transactions on Neural Systems and Rehabilitation Engineering*, *16*(5), 453–472. Available from https://doi.org/10.1109/TNSRE.2008.2006207.

Michoud, F., Sottas, L., Browne, L. E., Asboth, L., Latremoliere, A., Sakuma, M., Courtine, G., Woolf, C. J., & Lacour, S. P. (2018). Optical cuff for optogenetic control of the peripheral nervous system. *Journal of Neural Engineering*, *15*(1). Available from https://doi.org/10.1088/1741-2552/aa9126.

Millstein, S. G., Heger, H., & Hunter, G. A. (1986). Prosthetic use in adult upper limb amputees: A comparison of the body powered and electrically powered prostheses. *Prosthetics and Orthotics International*, *10*(1), 27–34. Available from https://doi.org/10.3109/03093648609103076.

Musick, K. M., Rigosa, J., Narasimhan, S., Wurth, S., Capogrosso, M., Chew, D. J., Fawcett, J. W., Micera, S., & Lacour, S. P. (2015). Chronic multichannel neural recordings from soft regenerative microchannel electrodes during gait. *Scientific Reports*, *5*(August), 14363. Available from https://doi.org/10.1038/srep14363.

Naples, G. G., Mortimer, J. T., Scheiner, A., & Sweeney, J. D. (1988). A spiral nerve cuff electrode for peripheral nerve stimulation. *IEEE Transactions on Biomedical Engineering*, *35*(11), 905–916. Available from https://doi.org/10.1109/10.8670.

Naples, G. G., Mortimer, J. T., & Yuen, T. G. H. (1990). Overview of peripheral nerve electrode design and implantation. In W. F. Agnew, & D. B. McCreery (Eds.), *Neural Prostheses: Fundamental Studies, Biophysics and Bioengineering Series* (pp. 107–144). Prentice Hall.

Navarro, X., Calvet, S., Butí, M., Gómez, N., Cabruja, E., Garrido, P., Villa, R., & Valderrama, E. (1996). Peripheral nerve regeneration through microelectrode arrays based on silicon technology. *Restorative Neurology and Neuroscience*, *9*, 151–160. Available from https://doi.org/10.3233/RNN-1996-9303.

Navarro, X., Krueger, T. B., Lago, N., Micera, S., Stieglitz, T., & Dario, P. (2005). A critical review of interfaces with the peripheral nervous system for the control of neuroprostheses and hybrid bionic systems. *Journal of the Peripheral Nervous System: JPNS*, *10*(3), 229–258. Available from https://doi.org/10.1111/j.1085-9489.2005.10303.x.

Navarro, X., Lago, N., Vivó, M., Yoshida, K., Koch, K. P., Poppendieck, W., & Micera, S. (2007). *Neurobiological evaluation of thin-film longitudinal intrafascicular electrodes as a peripheral nerve interface. 2007 IEEE 10th International Conference on Rehabilitation Robotics, ICORR'07, 00(c)* (pp. 643–649). IEEE. Available from https://doi.org/10.1109/ICORR.2007.4428492.

Nielsen, T. N., Sevcencu, C., & Struijk, J. J. (2014). Comparison of Mono-, Bi-, and tripolar configurations for stimulation and recording with an interfascicular interface. *IEEE Transactions on Neural Systems and Rehabilitation Engineering*, *22*(1), 88–95. Available from https://doi.org/10.1109/TNSRE.2013.2278479.

Oddo, C. M., Raspopovic, S., Artoni, F., Mazzoni, A., Spigler, G., Petrini, F. M., Giambattistelli, F., Vecchio, F., Miraglia, F., Zollo, L., Di Pino, G., Camboni, D., Carrozza,

M. C., Guglielmelli, E., Rossini, P. M., Faraguna, U., & Micera, S. (2016). Intraneural stimulation elicits discrimination of textural features by artificial fingertip in intact and amputee humans. *ELife*, *5*, e09148. Available from https://doi.org/10.7554/elife.09148.

Ortiz-Catalan, M., Brånemark, R., Håkansson, B., & Delbeke, J. (2012). On the viability of implantable electrodes for the natural control of artificial limbs: Review and discussion. *Biomedical Engineering Online*, *11*(1), 33. Available from https://doi.org/10.1186/1475-925X-11-33.

Ortiz-Catalan, M., Håkansson, B., Brånemark, R., Hakansson, B., & Branemark, R. (2014). An osseointegrated human-machine gateway for long-term sensory feedback and motor control of artificial limbs. *Science Translational Medicine*, *6*(257), 257re6. Available from https://doi.org/10.1126/scitranslmed.3008933.

Panetsos, F., Avendaño, C., Negredo, P., Castro, J., & Bonacasa, V. (2008). Neural prostheses: Electrophysiological and histological evaluation of central nervous system alterations due to long-term implants of sieve electrodes to peripheral nerves in cats. *IEEE Transactions on Neural Systems and Rehabilitation Engineering: A Publication of the IEEE Engineering in Medicine and Biology Society*, *16*(3), 223–232. Available from https://doi.org/10.1109/TNSRE.2008.923707.

Petrini, F. M., Bumbasirevic, M., Valle, G., Ilic, V., Mijović, P., Čvančara, P., Barberi, F., Katic, N., Bortolotti, D., Andreu, D., Lechler, K., Lesic, A., Mazic, S., Mijović, B., Guiraud, D., Stieglitz, T., Alexandersson, A., Micera, S., & Raspopovic, S. (2019). Sensory feedback restoration in leg amputees improves walking speed, metabolic cost and phantom pain. *Nature Medicine*, *25*(9), 1356–1363. Available from https://doi.org/10.1038/s41591-019-0567-3.

Petrini, F. M., Valle, G., Strauss, I., Granata, G., Iorio, R. D., D'Anna, E., Cvancara, P., Mueller, M., Carpaneto, J., Clemente, F., Controzzi, M., Bisoni, L., Carboni, C., Barbaro, M., Andreu, D., Hiairrassary, A., Divoux, J. L., Cipriani, C., Guiraud, D., & Micera, S. (2019). Six-months assessment of a hand prosthesis with intraneural tactile feedback. *Annals of Neurology*, *85*(1), 137–154. Available from https://doi.org/10.1002/ana.25384.

Pfurtscheller, G., Neuper, C., Muller, G. R., Obermaier, B., Krausz, G., Schlogl, A., Scherer, R., Graimann, B., Keinrath, C., Skliris, D., Wortz, M., Supp, G., & Schrank, C. (2003). Graz-BCI: State of the art and clinical applications. *IEEE Transactions on Neural Systems and Rehabilitation Engineering*, *11*(2), 1–4. Available from https://doi.org/10.1109/TNSRE.2003.814454.

Popović, D. B. (2014). Advances in functional electrical stimulation (FES). *Journal of Electromyography and Kinesiology*, *24*(6), 795–802. Available from https://doi.org/10.1016/j.jelekin.2014.09.008.

Raspopovic, S., Capogrosso, M., Petrini, F. M., Bonizzato, M., Rigosa, J., Pino, G. D., Carpaneto, J., Controzzi, M., Boretius, T., Fernandez, E., Granata, G., Oddo, C. M., Citi, L., Ciancio, A. L., Cipriani, C., Carrozza, M. C., Jensen, W., Guglielmelli, E., Stieglitz, T., & Micera, S. (2014). Restoring natural sensory feedback in real-time bidirectional hand prostheses. *Science Translational Medicine*, *6*(222). Available from https://doi.org/10.1126/scitranslmed.3006820.

Raspopovic, S., Carpaneto, J., Udina, E., Navarro, X., & Micera, S. (2010). On the identification of sensory information from mixed nerves by using single-channel cuff electrodes. *Journal of Neuroengineering and Rehabilitation*, *7*(1), 17. Available from https://doi.org/10.1186/1743-0003-7-17.

Raspopovic, S., Petrini, F. M., Zelechowski, M., & Valle, G. (2017). Framework for the development of neuroprostheses: From basic understanding by sciatic and median nerves models to bionic legs and hands. *Proceedings of the IEEE*, *105*(1), 34–49. Available from https://doi.org/10.1109/JPROC.2016.2600560.

Rebsamen, B., Teo, C. L., Zeng, Q., Ang, M. H., Burdet, E., Guan, C., Zhang, H., & Laugier, C. (2007). Controlling a wheelchair indoors using thought. *IEEE Intelligent Systems*, *22*(2), 18–24, IEEE. Available from https://doi.org/10.1109/MIS.2007.26.

Rijnbeek, E. H., Eleveld, N., & Olthuis, W. (2018). Update on peripheral nerve electrodes for closed-loop neuroprosthetics. *Frontiers in Neuroscience*, *12*(May), 1–9. Available from https://doi.org/10.3389/fnins.2018.00350.

Rodríguez, F. J., Ceballos, D., Schüttler, M., Valero, A., Valderrama, E., Stieglitz, T., Navarro, X., Rodriguez, F. J., Ceballos, D., Schuttler, M., Valero, A., Valderrama, E., Stieglitz, T., & Navarro, X. (2000). Polyimide cuff electrodes for peripheral nerve stimulation. *Journal of Neuroscience Methods*, *98*(2), 105–118. Available from https://doi.org/10.1016/S0165-0270(00)00192-8.

Rossini, P. M., Micera, S., Benvenuto, A., Carpaneto, J., Cavallo, G., Citi, L., Cipriani, C., Denaro, L., Denaro, V., Di Pino, G., Ferreri, F., Guglielmelli, E., Hoffmann, K. P., Raspopovic, S., Rigosa, J., Rossini, L., Tombini, M., & Dario, P. (2010). Double nerve intraneural interface implant on a human amputee for robotic hand control. *Clinical Neurophysiology*, *121*(5), 777–783. Available from https://doi.org/10.1016/j.clinph.2010.01.001.

Rousche, P. J., & Normann, R. A. (1992). A method for pneumatically inserting an array of penetrating electrodes into cortical tissue. *Annals of Biomedical Engineering*, *20*(4), 413–422. Available from https://doi.org/10.1007/BF02368133.

Schiefer, M. A., Graczyk, E. L., Sidik, S. M., Tan, D. W., & Tyler, D. J. (2018). Artificial tactile and proprioceptive feedback improves performance and confidence on object identification tasks. *PLoS One*, *13*(12), 1–18. Available from https://doi.org/10.1371/journal.pone.0207659.

Schiefer, M. A., Tan, D. W., Sidek, S. M., & Tyler, D. J. (2016). Natural sensory feedback improves task performance in individuals with upper limb loss using a myoelectric prosthesis. *Journal of Neural Engineering*, *1*, 016001. Available from https://doi.org/10.1088/1741-2560/13/1/016001.

Scott, R. N., & Parker, P. A. (1988). Myoelectric prostheses: State of the art. *Journal of Medical Engineering and Technology*, *12*(4), 143–151. Available from https://doi.org/10.3109/03091908809030173.

Srinivasan, A., Tipton, J., Tahilramani, M., Kharbouch, A., Gaupp, E., Song, C., Venkataraman, P., Falcone, J., Lacour, S. P., Stanley, G. B., English, A. W., & Bellamkonda, R. (2015). A regenerative microchannel device for recording multiple single unit action potentials in awake, ambulatory animals. *The European Journal of Neuroscience*, 1–12. Available from https://doi.org/10.1111/ejn.13080.

Stieglitz, T., Beutel, H., Schuettler, M., & Meyer, J.-U. (2000). Micromachined, polyimide-based devices for flexible neural interfaces. *Biomedical Microdevices*, *2*(4), 283–294. Available from https://doi.org/10.1023/A:1009955222114.

Swisher, S. L., Lin, M. C., Liao, A., Leeflang, E. J., Khan, Y., Pavinatto, F. J., Mann, K., Naujokas, A., Young, D., Roy, S., Harrison, M. R., Arias, A. C., Subramanian, V., & Maharbiz, M. M. (2015). Impedance sensing device enables early detection of pressure ulcers in vivo. *Nature Communications*, *6*, 6575.

Tan, D. W., Schiefer, M. A., Keith, M. W., Anderson, J. R., Tyler, J., & Tyler, D. J. (2014). A neural interface provides long-term stable natural touch perception. *Science Translational Medicine*, *6*(257), 257ra138. Available from https://doi.org/10.1126/scitranslmed.3008669.

Tan, D. W., Schiefer, M., Keith, M. W., Anderson, R., & Tyler, D. J. (2013). Stability and selectivity of a chronic, multi-contact cuff electrode for sensory stimulation in a human amputee. *International IEEE/EMBS Conference on Neural Engineering, NER*, 859–862. Available from https://doi.org/10.1109/NER.2013.6696070.

Tan, D. W., Schiefer, M. A., Keith, M. W., Anderson, R., & Tyler, D. J. (2015). Stability and selectivity of a chronic, multi-contact cuff electrode for sensory stimulation in a human amputee. *Journal of Neural Engineering*, *12*, 026002. Available from https://doi.org/10.1109/NER.2013.6696070.

Thompson, C. H., Zoratti, M. J., Langhals, N. B., & Purcell, E. K. (2016). Regenerative electrode interfaces for neural prostheses. *Tissue Engineering—Part B: Reviews*, *22*(2), 125–135. Available from https://doi.org/10.1089/ten.teb.2015.0279.

Thota, A. K., Kuntaegowdanahalli, S., Starosciak, A. K., Abbas, J. J., Orbay, J., Horch, K. W., & Jung, R. (2015). A system and method to interface with multiple groups of axons in several fascicles of peripheral nerves. *Journal of Neuroscience Methods*, *244*, 78–84. Available from https://doi.org/10.1016/j.jneumeth.2014.07.020.

Tyler, D. J., & Durand, D. M. (1997). A slowly penetrating interfascicular nerve electrode for selective activation of peripheral nerves. *IEEE Transactions on Rehabilitation Engineering*, *5*(1), 51–61. Available from https://doi.org/10.1109/86.559349.

Tyler, D. J., & Durand, D. M. (2002). Functionally selective peripheral nerve stimulation with a flat interface nerve electrode. *IEEE Transactions on Neural Systems and Rehabilitation Engineering*, *10*(4), 294–303. Available from https://doi.org/10.1109/TNSRE.2002.806840.

Tyler, D. J., & Durand, D. M. (2003). Chronic response of the rat sciatic nerve to the flat interface nerve electrode. *Annals of Biomedical Engineering*, *31*(6), 633–642. Available from https://doi.org/10.1114/1.1569263.

Valle, G., D'Anna, E., Strauss, I., Clemente, F., Granata, G., Di Iorio, R., Controzzi, M., Stieglitz, T., Rossini, P. M., Petrini, F. M., & Micera, S. (2020). Hand control with invasive feedback is not impaired by increased cognitive load. *Frontiers in Bioengineering and Biotechnology*, *8*(April), 1–7. Available from https://doi.org/10.3389/fbioe.2020.00287.

Valle, G., Mazzoni, A., Iberite, F., D'Anna, E., Strauss, I., Granata, G., Controzzi, M., Clemente, F., Rognini, G., Cipriani, C., Stieglitz, T., Petrini, F. M., Rossini, P. M., & Micera, S. (2018). Biomimetic intraneural sensory feedback enhances sensation naturalness, tactile sensitivity, and manual dexterity in a bidirectional prosthesis. *Neuron*, *100*(1), 37–45.e7. Available from https://doi.org/10.1016/j.neuron.2018.08.033.

Veraart, C., Grill, W. M., & Mortimer, J. T. (1993). Selective control of muscle activation with a multipolar nerve cuff electrode. *IEEE Transactions on Bio-Medical Engineering*, *40*(7), 640–653. Available from https://doi.org/10.1109/10.237694.

Wallman, L., Zhang, Y., Laurell, T., & Danielsen, N. (2001). The geometric design of micromachined silicon sieve electrodes influences functional nerve regeneration. *Biomaterials*, *22* (10), 1187–1193. Available from https://doi.org/10.1016/S0142-9612(00)00342-2.

Wark, H. A., Sharma, R., Mathews, K. S., Fernandez, E., Yoo, J., Christensen, M. B., Tresco, P., Rieth, L., Solzbacher, F., Normann, R. A., & Tathireddy, P. (2013). A new high-density (25 electrodes/mm2) penetrating microelectrode array for recording and stimulating submillimeter neuroanatomical structures. *Journal of Neural Engineering*, *10*(4), 045003. Available from https://doi.org/10.1088/1741-2560/10/4/045003.

Waters, R. L., McNeal, D., & Perry, J. (1975). Experimental correction of footdrop by electrical stimulation of the peroneal nerve. *Journal of Bone and Joint Surgery—Series A*, *57*(8), 1047–1054. Available from https://doi.org/10.2106/00004623-197557080-00002.

Wendelken, S., Page, D. M., Davis, T., Wark, H. A. C. C., Kluger, D. T., Duncan, C., Warren, D. J., Hutchinson, D. T., & Clark, G. A. (2017). Restoration of motor control and proprioceptive and cutaneous sensation in humans with prior upper-limb amputation via multiple Utah Slanted Electrode Arrays (USEAs) implanted in residual peripheral arm nerves. *Journal of Neuroengineering and Rehabilitation*, *14*(1), 121. Available from https://doi.org/10.1186/s12984-017-0320-4.

Wieler, M., Stein, R. B., Ladouceur, M., Whittaker, M., Smith, A. W., Naaman, S., Barbeau, H., Bugaresti, J., & Aimone, E. (1999). Multicenter evaluation of electrical stimulation systems for walking. *Archives of Physical Medicine and Rehabilitation*, *80*(5), 495–500. Available from https://doi.org/10.1016/S0003-9993(99)90188-0.

Wolpaw, J. R., Birbaumer, N., McFarland, D. J., Pfurtscheller, G., & Vaughan, T. M. (2002). Brain-computer interfaces for communication and control. *Clinical Neurophysiology*, *113* (6), 767–791. Available from https://doi.org/10.1016/S1388-2457(02)00057-3.

Wurth, S., Capogrosso, M., Raspopovic, S., Gandar, J., Federici, G., Kinany, N., Cutrone, A., Piersigilli, A., Pavlova, N., Guiet, R., Taverni, G., Rigosa, J., Shkorbatova, P., Navarro, X., Barraud, Q., Courtine, G., & Micera, S. (2017). Long-term usability and bio-integration of polyimide-based intra-neural stimulating electrodes. *Biomaterials*, *122*, 114–129. Available from https://doi.org/10.1016/j.biomaterials.2017.01.014.

Xiang, Z., Sheshadri, S., Lee, S.-H., Wang, J., Xue, N., Thakor, N. V., Yen, S.-C., & Lee, C. (2016). Mapping of small nerve trunks and branches using adaptive flexible electrodes. *Advanced Science, n/a-n/a*. Available from https://doi.org/10.1002/advs.201500386.

Xue, N., Sun, T., Tsang, W. M., Delgado-Martinez, I., Lee, S. H., Sheshadri, S., Xiang, Z., Merugu, S., Gu, Y., Yen, S. C., & Thakor, N. V. (2015). Polymeric C-shaped cuff electrode for recording of peripheral nerve signal. *Sensors and Actuators, B: Chemical*, *210*, 640–648. Available from https://doi.org/10.1016/j.snb.2015.01.006.

Yoshida, K., Farina, D., Akay, M., & Jensen, W. (2010). Multichannel intraneural and intramuscular techniques for multiunit recording and use in active prostheses. *Proceedings of the IEEE*, *98*(3), 432–449. Available from https://doi.org/10.1109/JPROC.2009.2038613.

Yoshida, K., & Horch, K. W. (1993). Selective stimulation of peripheral nerve fibers using dual intrafascicular electrodes. *IEEE Transactions on Biomedical Engineering*, *40*(5), 492–494. Available from https://doi.org/10.1109/10.243412.

Yu, H., Xiong, W., Zhang, H., Wang, W., & Li, Z. (2014). A parylene self-locking cuff electrode for peripheral nerve stimulation and recording. *Journal of Microelectromechanical Systems*, *23*(5), 1025–1035. Available from https://doi.org/10.1109/JMEMS.2014.2333733.

Yuan, H., & He, B. (2014). Brain-computer interfaces using sensorimotor rhythms: Current state and future perspectives. *IEEE Transactions on Biomedical Engineering*, *61*(5), 1425–1435. Available from https://doi.org/10.1109/TBME.2014.2312397.

Zecca, M., Micera, S., Carrozza, M. C., & Dario, P. (2017). Control of multifunctional prosthetic hands by processing the electromyographic signal. *Critical Reviews in Biomedical Engineering*, *45*(1–6), 383–410. Available from https://doi.org/10.1615/CritRevBiomedEng.v45.i1-6.150.

Zhuang, K. Z., Sommer, N., Mendez, V., Aryan, S., Formento, E., D'Anna, E., Artoni, F., Petrini, F. M., Granata, G., Cannaviello, G., Raffoul, W., Billard, A., & Micera, S. (2019). Shared human–robot proportional control of a dexterous myoelectric prosthesis. *Nature Machine Intelligence*, *1*(9), 400–411. Available from https://doi.org/10.1038/s42256-019-0093-5.

Chapter 17

Challenges in neural interface electronics: miniaturization and wireless operation

Senol Mutlu
Department of Electrical and Electronics Engineering, Boğaziçi University, Istanbul, Turkey

ABSTRACT
This chapter covers the basics and challenges in electronic circuits needed for neural recording and stimulation to realize wireless and batteryless implants in small size. Long operation distance, high bandwidth and processing capability, and low power densities are desired. However, meeting these in a small batteryless system remains a big design challenge. Important aspects of these implants, namely, equivalent electrical circuit model of neural microelectrode arrays, analog front end of the interface electronics for recording, circuit block for stimulation, integrated on-chip microprocessing, programmability, and wireless power and data transfer are discussed in this chapter. Radio frequency, optical, and ultrasonic methods are emphasized as viable solutions for wireless operation.

Keywords: Low-power neural interface electronics; miniaturization; RF; optic; ultrasonic; wireless power and data transfer

17.1 Introduction

Paralysis due to spinal cord injury, brainstem stroke, amyotrophic lateral sclerosis, and other disorders, can disconnect brain from certain parts of the body, eliminating sensations and volitional movements (Hochberg et al., 2012). One of the most important aim of neuroprosthetics is to restore or replace lost functions in paralyzed humans by routing related signals from the brain to peripheral actuators and from peripheral sensors to the brain bypassing damaged parts of the nervous system. Neuroprosthetics can convert intention-driven neuron activity into control signals or feed the cortex region with sensory information to realize useful tasks (Hochberg et al., 2006).

Neural interfaces have been demonstrated on people with long-standing tetraplegia after successful results from able-bodied monkeys. Signals have been

Somatosensory Feedback for Neuroprosthetics. DOI: https://doi.org/10.1016/B978-0-12-822828-9.00001-0

recorded and decoded from a small, local population of motor cortex neurons using a 96-channel microelectrode array. Imagined limb motions result in neural firings on multiple electrodes. These signals are then used to move and click a computer cursor or to control a robotic arm to perform three-dimensional reach and grasp movements (Hochberg et al., 2012). Most of these successful demonstrations of neural interfaces incorporate transcutaneous connections that tether participants to bulky carts (Hochberg et al., 2006). Wires used to transfer power and data between the implant and the outside world are primary sources of infection, failure, manufacturing cost, and discomfort to the patient (Wise, Anderson, Hetke, Kipke, & Najafi, 2004). Batteries cannot be employed in the implanted system since this puts a limit to overall size and shortens lifetime. A wireless, and miniaturized implantable system circumvents all of these problems. Wireless transmission of power and data to implanted neural interfaces is key to their successful deployment in clinical applications. However, there are important size, long operation distance, high bandwidth, and low power density constraints on these implanted neural interfaces. Meeting all these required constraints in a miniaturized wireless and batteryless system is still a big challenge, realization of which would open the world to the wide use of neuroprosthetics for those needing them.

Miniaturization is both a necessity and a challenge for neural implants. Implanted microelectrode arrays cause a strain on the tissue due to their mechanical stiffness and being tethered to packaged electronics. Resulting micromotions cause damage to the surrounding neural tissue, evoking the body's reaction to the implant. In order to minimize any movement relative to the tissue and resulting tissue reaction and ensure in vivo stability over a number of years, the implant should float in the brain or tissue, independent of skull or bone motion (Wise et al., 2004). It is desirable that the wireless neural recording and stimulation sites are as small as possible and float freely within the surrounding tissue without any tethers. To enable an electrode-sized implant to float in tissue, a system-on-chip (SoC) solution with an order of magnitude reduction in active circuit area is required. This reduction in area also reduces the available power, necessitating a similar reduction in power consumption of the circuits (Biederman et al., 2013). Furthermore, in the future it will become imperative for some neural interfaces to record simultaneously from multiple sites distributed over large areas. Although neural signals obtained from a microelectrode array placed in a small local population of neurons have led to successful demonstrations, most functions involve larger portions of a distributed network of neurons in different regions. Hence, it is also desirable to have very small, independent, free-floating, and wireless neural interface modules in large quantities. Smaller volumes (less than 1 mm^3) and sufficient wireless power (less than 1 mW) to these small volumes without violating the specific absorption rate (SAR) limits are two stringent constraints (Mirbozorgi, Yeon, & Ghovanloo, 2017), which are discussed in this chapter.

Wireless power transmission (WPT) and energy-harvesting technologies have been studied for implantable medical devices and can present solutions to the above-mentioned challenges of neuroprosthetics. Radio frequency (RF), mostly inductive coupling (Bandari et al., 2020; DeHennis & Wise, 2005; Wise et al., 2004), ultrasonic (Piech et al., 2020), optical (Haydaroglu, Ozgun, & Mutlu, 2017; Lee et al., 2012; Wu et al., 2018), electrostatic (Torres & Rincón-Mora, 2008), electromagnetic (Kulah & Najafi, 2008), thermoelectric (Ramadass & Chandrakasan, 2010), and triboelectric (Mutlu, Unlu, Gevrek, & Sanyal, 2020) methods have been extensively studied in the literature. For neuroprosthetics and other medical implants, RF, optical, and ultrasonic methods come into prominence compared to others. Therefore, after presenting important aspects of neural interface electronics, RF, optical, and ultrasonic wireless power transfer solutions for neural implants are discussed in this chapter to address their miniaturization and wireless operation challenges.

17.2 Important aspects of neural interface electronics

While neuroprosthetics continue to show great promise in helping people suffering from neuropathologies, new implants that combine neural recording, signal processing, and stimulation functionalities with the help of a single electronic circuit die are emerging for closed-loop operation (Azin, Guggenmos, Barbay, Nudo, & Mohseni, 2011). A block diagram of such a miniaturized system is presented in Fig. 17.1. It is composed of a microelectrode array, a transducer element, a possible external storage capacitor and an application-specific integrated circuit (ASIC) chip. It is capable of wireless neural activity recording as well as neuron stimulation without a battery. Similar to SoC implementations, its ASIC can also carry out multiple operations at the same time, including a certain amount of on-chip neural data processing and control with its integrated microprocessor (μProc).

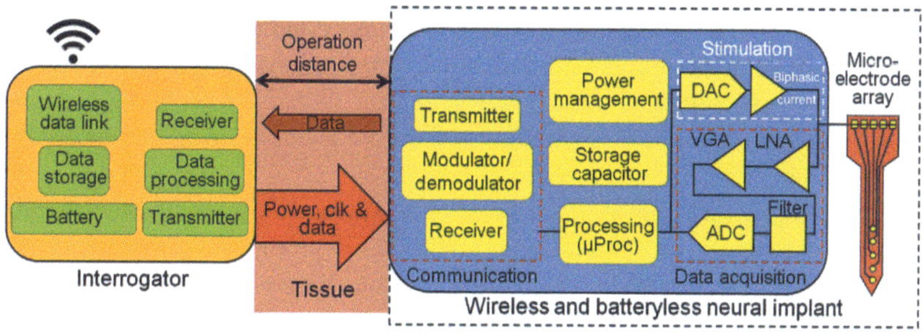

FIGURE 17.1 Block diagram of a wireless and batteryless miniaturized neural implant with a microelectrode array that can do both neural recording and stimulation.

Miniaturization and wireless operation put stringent constraints on the design of this ASIC, which can be implemented using available low-power and sub-micron complementary metal oxide semiconductor (CMOS) technology node. Sufficient power has to be transmitted to it to enable its operation and, in the case of stimulation, deliver charge to the tissue. The telemetry technique, whether this be RF, optical, or ultrasonic, must have sufficient range. Although this requirement depends on the application, a range of a few centimeters is adequate for most prosthetic applications. The wireless link should provide a high data transfer rate (bandwidth). Bandwidths in excess of 1–2 Mb/s are sufficient for most implants (Wise et al., 2004). These implants should also be able to receive data in order to be programmed or woken-up in addition to the transmission of data.

Important aspects of this neural interface ASIC are considered next, starting with the explanation of microelectrode array and its electrical equivalent circuit model. This is followed by the analog front end of the interface electronics for recording of neuron activity and neuron stimulation. The power management block receives in-coming power and regulates it. The communication block demodulates and extracts coming clock and program data signals. It also modulates and transmits the prepared data back to the interrogator. Extracted clock signal and program data are used by the integrated μProc unit, which also prepares the data to be sent. The stored program data are used to choose which channel to record or stimulate and adjust their power settings. It can also do basic digital signal processing (DSP) on the recorded extracellular signals.

17.2.1 Microelectrode array

The pioneering work of Kensall D. Wise at the University of Michigan in developing microfabricated silicon microelectrode arrays paved the way for major advancements in neural implants. The geometrical precision of lithographic techniques allowed neuroscientists to use unique electrode designs with unprecedented electrode site density (Seymour, Wu, Wise, & Yoon, 2017). Hundreds of electrodes with recording site sizes ranging from 6 to 20 μm are formed by opening vias on insulated conductive thin films using batch fabrication techniques. The electrode sites record local voltages associated with ionic current flow around a neuron when it fires in response to inputs received from other cells. The sites are capacitive, with an impedance of a few megohms at 1 kHz (Wise et al., 2004).

In neuroprosthetics and neuromodulation applications, extracellular action potentials (spike), local field potentials (LFPs), and electrocorticograms (ECoGs) are the primary signals of interest. These signals have around ~1 ms temporal resolution. They result from collective neural activities with a spatial range from ~10 μm to ~10 mm. They experience significant intrinsic noise sources such as interfering neural signals and biopotentials, and instrumentation noise, and extrinsic noise sources such as electrical stimulation in the measurements (Kipke, 2017). ECoGs signals are generated by synchronous synaptic currents of neural populations. They possess the largest spatial scale, reaching ~10 mm from the electrode. Millimeter-sized electrodes placed either subdurally or epidurally are used to measure ECoG signals from the cortical surface. They have amplitudes ranging from ~ 50 μV to 100 μV with a bandwidth of ~0.1 Hz to ~500 Hz (Kipke, 2017). LFPs are generated from synchronous synaptic transmembrane currents in dendrites and somas of neural populations that reside in the vicinity of the electrode, around 1 mm. Penetrating electrodes are used to record them. LFPs typically have amplitudes ranging from ~10 μV to 1000 μV and bandwidth of ~1 Hz to 300 Hz (Kipke, 2017). Extracellular action potential signals, spikes, have the smallest spatial scale. They result from current dipoles created by transmembrane active currents in firing neurons. They are recorded at distances reaching ~0.2 mm from the electrode. Extracellular spikes have amplitudes ranging from ~50 μV to 500 μV and bandwidth of ~300 Hz to 5000 Hz (Kipke, 2017).

The equivalent circuit model for the electrode–tissue interface is presented in Fig. 17.2, and is used to predict behaviors of different types of neural signals mentioned above. In this model, the voltage source, E_{hc}, represents the half-cell potential of the electrode–tissue interface. The capacitor, C_{dl}, is the double-layer capacitance of the electrode–tissue interface. The resistor, R_{ct}, models the resistance to the transfer of charge occurring through reversible Faradaic (reduction–oxidation) currents. The impedance regarding constant phase element represents charge transfer variations resulting from the nonlinear nature of the electrochemical interface between the solid conductor and ionic liquid. The resistor, R_s, represents the resistance to ion movement in the diffusion region. The capacitor, C_{sh}, and resistor, R_{sh}, model the shunt or leakage pathways from the insulated electrode traces to the bulk tissue. The resistor, R_t, is for the equivalent resistance in the electrode trace from the electrode site to the electronics interface. The voltage source, v_{noise}, represents the lumped electrode intrinsic noise sources

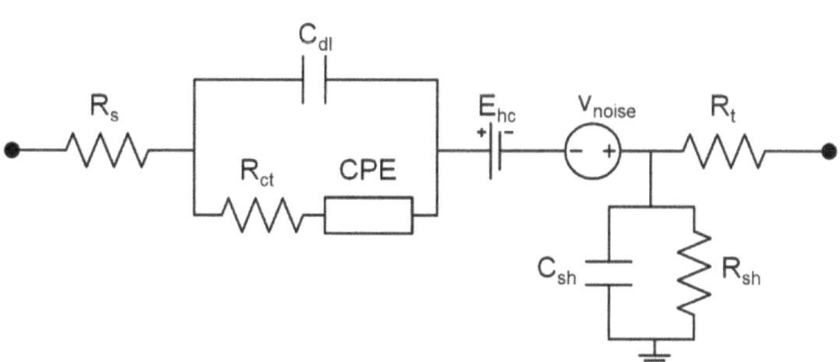

FIGURE 17.2 Equivalent circuit model for the electrode–tissue interface (Kipke, 2017).

that arise from various biophysical and electrical phenomena (Kipke, 2017). An accurate model of the electrode–tissue interface is vital for the design of ultra-low power neural interface electronic circuits.

The noise sources at the electrode–tissue interface are caused by Brownian motion of charge carriers, drift and diffusion of charged ions due to concentration gradients, reversible oxidation/reduction reactions, random fluctuations and instability in the half-cell potential caused by disturbances of the double-layer capacitance, and frequency-dependent 1/*f* noise (also called flicker or pink noise) (Kipke, 2017). They result in a minimum root-mean-square (RMS) noise voltage level of around 20 μV_{rms} (Wise et al., 2004). These noise sources can severely limit the signal-to-noise ratio of the neural signals and can result in false spike detection, missed detection, and erroneous classifications (Seymour et al., 2017). ICs in submicron CMOS processes have the capacity to resolve these noise levels with proper designs.

17.2.2 Data acquisition

In most neuroprosthetics and neuromodulation applications, as well as in many high-performance neural recording systems, data acquisition and signal conditioning blocks are implemented as ASIC chips. This is critical for neural recording because it results in miniaturized, high-performance, low-power, and reliable solutions (Kipke, 2017). The purpose of the data acquisition block of the front-end electronics is to amplify the neural signals from the electrodes by an overall gain of around 40–60 dB since they are very small in magnitude, ranging from tens to hundreds of microvolts. It also lowers impedance levels of the electrodes to make them less vulnerable to externally introduced noise, for example from a few megohms to a few hundred ohms (Wise et al., 2004). Microelectrodes typically have impedances ranging from 0.2 MΩ to 2 MΩ (Kipke, 2017). Neural signals first go through a low-noise amplifier (LNA), then through a variable gain amplifier. The LNA typically has high input impedance (much larger than 10 MΩ) to accommodate the high impedance values of the electrodes. They must also AC couple the signal to minimize offsets and suppress the unstable DC potential of the site (Wise et al., 2004). The amplifiers must also have sufficient common mode rejection to block interfering biopotentials such as electrocardiogram and common noise sources. These common mode signals can be an order of magnitude higher than the targeted neural signals and thus must be properly blocked. Hence, a bandpass filter is used after the amplifiers to reject signal frequencies outside the bandwidth of the neural sources of interest. Acquisition ends with a high-resolution analog-to-digital converter (typically 10 bits or higher) operating at a sample rate of up to about 30,000 samples per second depending on the bandwidth of the targeted neural signals (Kipke, 2017). A signal multiplexer can be included in systems that support electrode arrays to combine analog signals coming from different electrode sites into a single digital data stream (Wise et al., 2004). It is generally desirable to digitize the analog neural signals as early as possible and then use DSP techniques to extract the desired information. This approach moves neural signal processing into the information technology domain of algorithms, software, and computing hardware, part of which can be done on-chip in the μProc unit as depicted in Fig. 17.1 (Kipke, 2017).

Self-test modes are typically present in such systems, where a 1-kHz signal is coupled to the sites to allow external monitoring of their impedance levels, and stimulation, where current is driven through the sites in order to activate or clean them before their usage (Wise et al., 2004).

Signal referencing is an important aspect of recording to minimize contributions of interfering biological noise. Typically, a relatively large electrode site is positioned in a location with minimal neural activity and is used as a reference to remove correlated noise across channels. This reference electrode signal is connected to the input of differential amplifiers across all channels. The amplifiers themselves can be referenced to a separate ground site, which may be the enclosure of the implant's electronics (Kipke, 2017).

Amplifiers must be very small and consume low power. Local heating limits power consumption of the implant to a few milliwatts. Recording power consumption of 11 μW/channel has been reported (Lee et al., 2019). Stabilization of the DC input levels due to drift in the electrochemical site potential is also needed. This can be achieved using shunt resistors in the form of polysilicon shunt resistors or shunt transistors operating in the subthreshold region. For a lower cutoff frequency of 10 Hz, this requires a shunt input resistance in the range from 75 MΩ to 500 MΩ. The use of a transistor in the subthreshold region allows the lower cutoff to be programmable (Wise, 2002).

17.2.3 Stimulation

For a closed-loop operation, stimulation of neurons through electrode sites is needed. Gold, platinum, and iridium (iridium oxide) have been used to form electrode sites for stimulation. Charge, not current, is the determining factor for stimulation. Gold has a maximum charge delivery of about 20 $\mu C/cm^2$. Platinum and iridium have up to 75 $\mu C/cm^2$ and 3000 $\mu C/cm^2$, respectively, which makes iridium the material of choice for microstimulation. When driving electrodes, it is crucial to stay within the electrochemical water window to avoid evolving oxygen or hydrogen gases or inducing local pH changes (Wise et al., 2004). Based on bioelectrochemical studies, the upper safety limit for the injected charge density for platinum-coated electrodes has been established at ~20 $\mu C/cm^2$ (Yin et al., 2014).

Programmable, and charge-balanced biphasic current pulses are typically used to stimulate neurons through multiple channels. On-chip digital-to-analog converters can generate stimulus currents from −128 μA to +128 μA with a resolution of 1 μA (Wise, 2002). The primary programmable stimulation parameters include polarity, amplitude, pulse-width, and stimulation frequency. The range of the width of the stimulation pulses changes from 60 μs to 450 μs. Stimulator frequency range can change from 31 to 1000 Hz. In an implementation within ranges of these parameters, the stimulation circuit block has been shown to consume 7.4 μW power per channel (Lee, Rhew, Kipke, & Flynn, 2010). The stimulation circuit must dissipate less than 10 mW to avoid appreciable heating of the tissue. Temperature rises of more than 2 °C damage surrounding neurons (Wise et al., 2004).

17.2.4 Integrated processing on chip

The wireless and batteryless implant must be able to receive and store incoming program or control data from the interrogator and should prepare the data for transmission. Based on this program, it should control the timing, gain, and selection of the channels for recording or stimulation. In addition to these digital control operations, it can also do some basic DSP on the digitized extracellular signals coming from the data acquisition block to extract the desired information. All these tasks can be done digitally on the integrated microprocessing (μProc) unit of the neural interface electronics that can reduce the power consumption of the implant by reducing the amount of data transmissions and receptions.

Digital noise rejection is a good example of an early-stage DSP that can be done by the on-chip μProc to improve neural recording in many applications. A digital common average reference algorithm can be applied to form a signal reference if there is a constraint or difficulty of providing a large area reference electrode. This method is commonly used in electroencephalogram systems to be able to detect small signals in noisy recordings. The signals from all electrode sites are averaged and then subtracted from each individual electrode signal, which separates signals from noise common to all sites such as 50/60-Hz noise or motion artifact (Kipke, 2017). It can provide additional high-pass filtering with an infinite impulse response digital filter to remove any residual DC offsets or low-frequency artifacts (Azin et al., 2011). An additional example for early-stage DSP is the process of spike detection in extracellular spike recordings. Amplitude thresholding is used to detect the occurrences of large-amplitude transient spikes in the raw electrode signals. Similarly, DSP algorithms can be run on the integrated μProc unit for the requirements of LFPs and ECoG recordings (Kipke, 2017).

17.2.5 Communication

Wireless digital data transfer between the neural implant and the interrogator is mostly asynchronous. Asynchronous data detection methods are usually more complicated than synchronous approaches for the receiving implant. They take up more area and consume more power. A good solution to this problem is to embed a synchronized clock signal to the data stream sent from the interrogator to the implant so that the implant can extract this clock signal from the received serial bit stream in addition to the program and control data. This way, a precise clock signal needed for synchronous digital operation can be generated on-chip, without the need for bulky resonators. The full-digital data demodulator/clock recovery techniques can be used to retrieve the received clock and data on the communication block of the neural interface ASIC (Sodagar, Perlin, Yao, Najafi, & Wise, 2009). It can also perform parity checks. The communication unit can work as the peripheral device of the μProc unit. The retrieved clock signal and program and control data are supplied to the μProc unit, which also uses the communication unit to send its serialized processed data. The communication unit can use various modulation techniques on the data to be transmitted such as amplitude-shift keying, differential phase-shift keying, phase-shift keying, frequency shift keying (FSK), on–off keying and pulse-position modulation (Ghovanloo & Najafi, 2004; Lee et al., 2018; Wise et al., 2004). The backscattering technique is also common for implants to transmit their data (Seo, Carmena, Rabaey, Maharbiz, & Alon, 2015).

17.2.6 Power management

The wireless and batteryless implant can receive its power from an inductively coupled coil, a photovoltaic cell or a piezoelectric element. Transducers turn RF, optical, or ultrasonic energy into electrical energy or vice versa. The regulation and control of the electrical energy coming from the transducer to the IC is done by the power management unit. This unit rectifies the coming AC power and matches impedances between the transducer and the circuitry to maximize the received power. Then, it converts the harvested voltage level to the desired voltage level and stores energy in the storage capacitor. It can generate regulated 0.8 V, 1.2 V, 3.3 V, 5 V, etc. supplies. It can put the implant into sleep mode with ultra-low power consumption below 1 μW until a wake-up signal is detected. This unit may also need to generate a power-on-reset and contain a start-up circuitry.

One of the most important circuit blocks of a wireless neural recording system is the voltage regulator. The regulator should produce a stable, low-noise voltage level (better than 8 bit resolution) without the need for external hybrid components and with low power consumption. A bandgap reference circuit is used to generate a reference voltage, and a regulated voltage supply is produced by an op-amp and bandgap reference circuit.

This unit must also generate the voltage and current levels needed for the transducer to transmit data back to the interrogator. This typically requires voltage doublers since the harvested voltage level from the incoming wireless power may be smaller than the required voltage level for transmission.

To give an idea about the power levels that must be handled by this unit, an ASIC that has 32-channel neural spike recording and four-channel current-controlled stimulation block is reported to consume 35 mW power (Lee et al., 2019). This level of power requirement of neural implant interfaces is one of the major hindrances to their clinical translation. However, seminal studies on neuroprosthetics give clues about the solutions to these obstacles. For example, power and data needs can be reduced by developing alternative processing methods to interpret neural activities (Nason et al., 2020). In basic behavioral tasks, neural activity is predicted mostly using a threshold crossing rate (TCR) method that uses high sampling rates, such as 30,000 samples per second, to sort neural spikes coming from different neurons. This makes the ultra-low power ASIC consume power levels that cannot be sustained by the WPT. Alternatively, the down-sampled magnitude of the 300–1000 Hz band of spiking activity, called spiking-band power (SBP), can predict neural activity as good as the TCR method in certain applications for which spike waveforms are not critical. This dramatically reduces the power requirements of neural interfaces by processing an order less samples per second (Nason et al., 2020).

17.3 RF solutions for wireless power transfer

Wireless telemetry of implanted neuroprosthetics based on RF transmission mostly uses two closely coupled coils, offering nearly infinite lifetime as in passive RFID tags. Inductive coupling with the reception of modulated backscattered data is a frequently employed technique in implants. It has an operational distance of around 0.5–20 cm with large coil sizes in the order of a centimeter (Potkay, 2008). Large coil sizes are needed because of the MHz frequency ranges used. Smaller sizes but still in the order of a millimeter can be achieved using GHz frequency range at the expense of larger tissue power losses (Rao, Nikitin, & Lam, 2005). Miniaturized wirelessly powered passive RFID tags operating at 60 GHz are shown to operate with an antenna occupying a total area of 20 mm^2 (Pursula et al., 2008) at the cost of increased path losses, design complexity, and power density (Cook, Lanzisera, & Pister, 2006). A biomedical implant has been powered with a 2×2 mm^2 square loop receive antenna using a 2×2 cm^2 square loop transmit antenna operating at 1 GHz frequency and a separation of 15 mm of layered bovine muscle tissue with a power loss of 33.2 dB (O'Driscoll, Poon, & Meng, 2009). A low MHz region from 3 to 30 MHz is found to be a good compromise between tissue absorption losses and size (DeHennis & Wise, 2005). CardioMEMS has a passive telemetry-based implanted wireless blood pressure sensor system working over a distance of 20 cm and size of 5×30 mm^2 (Allen, 2005). Research continues to optimize coil or antenna design and RF frequency choice for WPT into tissue (Ho et al., 2014).

To communicate with RF implants, either power requirements increase or transmission distances decrease. Regulations limit the maximum possible transmitted power for biomedical implants to 1 mW/cm^2 power densities for frequencies below 300 MHz and 10 mW/cm^2 for 3 GHz (IEEE Standards Coordinating Committee, 2005). Since regulations limit the maximum possible transmitted power, longer distance communication relies on more efficient use of this power. Designing a low-power miniature system that is capable of communicating at distance of centimeters remains as a design challenge (Chen, Hanson, Blaauw, & Sylvester, 2010).

A wireless neural implant has been shown to be powered by an external coil transmitting 12 mW at 300 MHz, and to consume 225 μW. The implant is comprised of a 64-channel electrode array and a 2.4 × 2.4 mm^2 CMOS ASIC that performs 64-channel acquisition, and wireless power and data transmission. The ASIC digitizes the signal from each electrode at 1000 samples per second with 1.2 μV input referred noise, and transmits the serialized data using a 1 Mb/s backscattering modulator (Muller et al., 2014).

A resonance-based near-field electromagnetic inductive link has been also shown for WPT to mm-sized free-floating implants in the neural tissue. Implants operating at 60 MHz receive 1.3 mW power from a distance of 16 mm with power transfer efficiency of 2.4%. These implants can be distributed across brain surface area of 7 cm^2 (Mirbozorgi et al., 2017).

In another implementation, a wirelessly powered 0.125 mm^2 CMOS ASIC using 65 nm technology node has been developed for neuroprosthetics. It integrated four 1.5 μW amplifiers with power conditioning and communication circuitry. The full system consumed 10.5 μW and operated across 1 mm range in air with 50 mW transmit power. Its operation was also verified using wirelessly powered in vivo recordings. The transmission frequency for this system was selected to be 1.5 GHz, trading a reduction in node size and channel loss for an increase in the SAR. In biological media, operating at a frequency between 1–3 GHz minimizes channel loss for edge-to-edge coupling and reduces the receive coil size by several orders of magnitude (Biederman et al., 2013). While this realization demonstrated the feasibility of free-floating wireless electrodes, it also showed that further miniaturization was not possible with RF WPT. The tiny (450 μm × 250 μm) antenna combined with the high loss of electromagnetic propagation through tissue led to an inefficient solution with a power efficiency of around 0.02% (Maharbiz, Muller, Alon, Rabaey, & Carmena, 2016). This inspired the search for alternative solutions in the form of optical and ultrasonic schemes, as explained next.

17.4 Optical solutions for wireless power transfer

Optical technologies deliver the greatest amount of power in the smallest volume (Wang, Calhoun, & Chandrakasan, 2006). Optical WPT has been utilized in implantable integrated microsystems to monitor intraocular pressures (Ghaed et al., 2013), neural recorders (Lee et al., 2018), retinal prosthesis (Mathieson et al., 2012), and pacemakers (Haeberlin et al., 2015). In the context of system miniaturization and wireless power and data transmission, optoelectronic systems have the potential to be smaller than the alternatives. They offer higher power densities, which can be delivered simply by lasers or other light sources. Laser beam powering of RF tags with on-chip silicon photodiodes helped to miniaturize them to $500 \times 500\ \mu m^2$ sizes (Laflin, Morris, Bassik, Jamal, & Gracias, 2011). However, in these small tags, communication is established by RF signals using small on-chip antennas, which limits the working distance to around 5 mm.

A 1.5 mm^3 energy autonomous wireless intraocular pressure monitoring system is implemented using an integrated 0.07 mm^2 solar cell that can harvest a maximum power of 80 nW under a light irradiance of 100 mW/cm^2 [air mass 1.5 global condition] to recharge a 1 mm^2 thin-film battery to power the system. It also includes a 4.7 nJ/bit FSK radio that achieves 10 cm of RF transmission range, which is also used to receive wake-up signals (Ghaed et al., 2013). Another version of this microsystem is also implemented by an integrated optical receiver to load program data and request data instead of the RF receiver, however, keeping RF data transmission part. The system generates 456 nW under 10 klux light to enable an energy autonomous system operation (Kim et al., 2014). A microsystem aligned to the tip of an optical fiber has also been demonstrated, where an on-chip photovoltaic cell is used for optical powering and a separate laser diode for communication (Aktan et al., 2011; Sarioglu et al., 2012).

Despite being very attractive solutions for wireless power transfer, on-chip photovoltaic cells made of silicon can supply open-circuit voltage of around 0.6 V, which is not enough for ASICs and sensors. Series connection of multiple photodiodes on silicon-on-insulator (SOI) wafers have been demonstrated as a solution (Warneke et al., 2002), however SOI technology is more expensive and less available compared to standard CMOS processes. Instead, an external light-emitting diode (LED) used as a photovoltaic cell can be more beneficial. LED can supply a higher open circuit voltage [1.3 V for near infrared (IR), 1.6 V for red, 1.7 V for green, etc.] than silicon photodiodes. Circuitry can be run directly from this higher voltage without the need for voltage elevation, which consumes valuable power. Since the microsystem needs an external optoelectronic element to transmit data optically (silicon, being an indirect semiconductor, cannot emit light), LED can also work as a data transmitter. Placing the photodiode outside of the IC die saves expensive on-chip area. With the photovoltaic cell placed out of the die, the die can be covered in optically opaque material. This is necessary because the powering light can induce latch-up and noise in the circuits of the silicon die and can increase their leakage currents. Photon absorption is more efficient in direct bandgap materials of LEDs (e.g., AlGaAs) in contrast to indirect bandgap materials (e.g., silicon) (Bhattacharya, 1994). Record-level efficiency improvements in GaAs solar cells have been achieved in this way, stressing commonalities between efficient photovoltaic cell and LED designs (Miller, Yablonovitch, & Kurtz, 2012). An LED is also an efficient photovoltaic cell for a limited wavelength band. The upper boundary of this band is typically 20–30 nm shorter than its peak emission wavelength.

With the added benefits of LED, an improvement on wireless and batteryless optoelectronic microsystems has been made by using it for both wireless powering and data transmission, as depicted in Fig. 17.3. The usage of a single LED with a die size of $350 \times 350\ \mu m^2$ with the help of an IC ($230 \times 210\ \mu m^2$) and a storage capacitor ($0.5 \times 1\ mm^2$) resulted in a $1\ mm^3$ wireless and batteryless microsystem (Haydaroglu & Mutlu, 2014; Haydaroglu et al., 2017). The LED size can shrink to $100 \times 100\ \mu m^2$ sizes in the form of a microscopic LED (μLED) (Wu et al., 2015). Even further size reduction is possible by monolithically integrating ultra thin and ultra small LEDs to the ICs (Cortese et al., 2020). Standalone and untethered 100-μm-scale electronic sensors have been realized by integrating silicon electronics and inorganic μLEDs. Optical power and communication examples have been demonstrated with batch-fabricated microsystems using photolithographic processes (Cortese et al., 2020). An untethered opto-electronic neural interface was also implemented using 180 nm CMOS circuits and heterogeneously integrated single AlGaAs diode that functions as a photovoltaic cell as well as LED. These microscale opto-electrically transduced electrodes are powered by and communicate through an optical interface without a bulky RF coil. A system with an area of $250\ \mu m \times 57\ \mu m$ consumes 1 μW of electrical power, and is capable of capturing and encoding neural signals. Low noise measurements of neural signals are achieved. The measured noise floor is reported to be as low as $15\ \mu V_{RMS}$ at a 15 kHz bandwidth (Cortese et al., 2020; Lee et al., 2018).

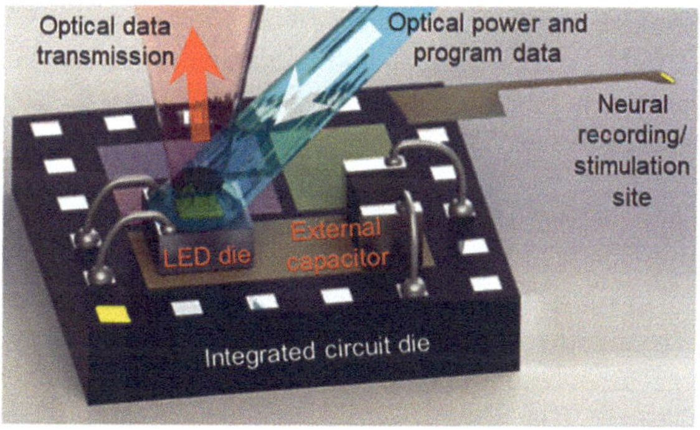

FIGURE 17.3 Conceptual representation of the optically powered and optically transmitting neural implant using a single light-emitting diode (Haydaroglu et al., 2017).

Pioneering studies on optogenetics introduced methods to excite and silence neurons using genetic and chemical markers specific to particular cell types. These studies help improve neuronal circuit analysis. New light-activated ion channels (opsins) are discovered and applied. Optical control of genetically engineered ion channels to selectively activate or silence neurons of specific types is a powerful tool to analyze these cells in the dense heterogeneous populations around microelectrode recording sites. Electrical stimulation is nonspecific, does not have enough spatial resolution, and cannot silence neurons. Different opsins expressed in cells can be excited using different wavelengths of light to achieve cell-type specificity with well-controlled spatial and temporal resolution (on the order of milliseconds; Seymour et al., 2017).

Silicon neural microelectrode arrays have been monolithically integrated with 32 μLEDs and 32 recording sites for optogenetic applications of neuroscience. They are implanted into the CA1 pyramidal layer of anesthetized and freely moving mice. Recording sites of μLEDs had close dimensions to pyramidal neuron somas. This resulted in confined emission and electrophysiological recording of action potentials and local field activities. Spikes were generated using 60 nW light power. Spatiotemporal parallel stimulation and recording have been achieved using independent control of differential somato-dendritic compartments of single neurons and of distinct cells that are ~50 μm apart (Wu et al., 2015).

17.4.1 Optical penetration depths for biological tissue for different wavelengths

Although optical WPT delivers the greatest amount of power in the smallest volume and has the potential to be smaller than the alternatives, it has one major disadvantage for transferring power to implants: low penetration depth through biological tissues. Scattering and absorption limit how far light can diffuse into tissue. In skin tissues, the effective penetration depth at which the incident optical energy drops to 1/e (~37%) is typically 50–100 μm for UV and blue light (wavelength of 400–450 nm). Infrared (IR) light above 2000 nm wavelength has similar small penetration depth because of high light absorption by water. The penetration depth of green light (wavelength of 500–550 nm) is a few hundreds of micrometers, limited by light absorption of melanin and hemoglobin. Blood, water, and melanin are the main absorbing components in the tissue (Raulin & Karsai, 2011). The penetration depth of light in muscle tissue for different wavelengths is summarized in Table 17.1 (Bashkatov, Genina, Kochubey, & Tuchin, 2005; Raulin & Karsai, 2011). The maximum penetration depth is observed for a wavelength of around 1000 nm. The spectral range from 600 nm to 900 nm is typcially used for photodynamic therapy (Bashkatov et al., 2005). Most of the implants that use optical WPT, except for ocular applications, harvest

TABLE 17.1 **Penetration depth of light in tissue for different wavelengths (Bashkatov et al., 2005; Raulin & Karsai, 2011).**

Wavelength (nm)	193	308	532	600	633	660	700	750	800	850	900	1,064	2,060	2,940	10,600
Penetration depth (μm)	10	50	830	1,500	1,700	1,800	2,000	2,200	2,300	2,400	2,500	2,500	286	3	17

transcutaneous light with wavelengths ranging from 770 nm to 850 nm (called the near-IR window; Moon, Blaauw, & Phillips, 2017; Vo-Dinh, 2014). High optical power loss is inevitable even at these wavelengths. For example, 30 dB loss is experienced through 3.4 cm thick bovine tissue with an 808 nm laser (Hudson, Hudson, Wininger, & Richardson, 2013). If a 980 nm laser is used with a regulation-limited power intensity of 7.3 mW/mm^2, the density would drop to 7.3 μW/mm^2 at 2.2 cm tissue penetration depth.

17.4.2 Laser power limitations for skin

Healthcare safety regulations for diagnostic and imaging applications put limitations to optical power that can be used on people. These limitations are quantified in terms of the maximum permissible exposure (MPE). This is defined as one-tenth of the damage threshold due to photothermal and photochemical effects (Yun & Kwok, 2017). The MPE depends on the wavelength and duration of laser power and sometimes the laser spot size (ICONIR Protection, 2013).

For visible, short-wavelength IR and mid and long IR radiation, MPE values can be calculated using Table 17.2 (ICONIR Protection, 2013). According to these guidelines, laser exposure limits for the skin for the wavelength range from 400 to 1400 nm are given as $200 \times C_A$ J/m^2 for durations in the range of 1−100 ns. This is $11 \times C_A \times t^{0.25}$ kJ/m^2 for 100 ns to 10 s durations. For exposure durations longer than 10 s, the MPE is $2.0 \times C_A$ kW/m^2. In this instance, the MPE is given in intensity units because thermal equilibrium is reached between laser-induced heating and conductive cooling (Yun & Kwok, 2017).

The correction factor C_A in Table 17.2 is unitless and defined for wavelengths from 400 nm to 1400 nm. It is related to the wavelength dependence of the pigment epithelium absorption in the retina and also used for skin exposure limits (ICONIR Protection, 2013). These C_A values are summarized in Table 17.3. They are also wavelength dependent. It is 1 for

TABLE 17.2 Laser radiation exposure limits of skin for different wavelengths and exposure durations, where *t* is in s and C_A is a unitless correction factor (ICONIR Protection, 2013).

Wavelength (nm)	Exposure duration		Exposure limit
	Lower limit	Upper limit	
$400 \leq \lambda < 1400$	1 ns	100 ns	$200C_A$ J/m^2
$400 \leq \lambda < 1400$	100 ns	10 s	$11C_At^{0.25}$ kJ/m^2
$400 \leq \lambda < 1400$	10 s	30 ks	$2.0C_A$ kW/m^2
$1400 \leq \lambda < 1500$	1 ms	10 s	$5.6t^{0.25}$ kJ/m^2
$1500 \leq \lambda < 1800$	1 ns	10 s	10 kJ/m^2
$1800 \leq \lambda < 2600$	1 ns	1 ms	1.0 kJ/m^2
$1800 \leq \lambda < 2600$	1 ms	10 s	$5.6t^{0.25}$ kJ/m^2
$2600 \leq \lambda < 1$ mm	1 ns	100 ns	100 J/m^2
$2600 \leq \lambda < 1$ mm	100 ns	10 s	$5.6t^{0.25}$ kJ/m^2
$1400 \leq \lambda < 1$ mm	10 s	30 ks	1.0 kW/m^2

TABLE 17.3 Values of correction factor, CA, for different wavelengths (ICONIR Protection, 2013).

C_A value	Wavelength range (nm)
1.0	$400 \leq \lambda < 700$
$10^{0.002(\lambda/1\text{ nm}-700)}$	$700 \leq \lambda < 1050$
5.0	$1050 \leq \lambda < 1400$

wavelengths ranging from 400 to 700 nm and increases to 5 for the range from 1050 to 1400 nm.

The maximum power density of a laser source that can be used for WPT applications can be found from these guidelines using Tables 17.2 and 17.3. For example, to find the exposure limit for 980 nm laser radiation for a 5 s exposure duration, the following expression of Table 17.2 must be calculated; $11 \times C_A \times 5^{0.25}$ kJ/m^2. For this, the C_A value must be found from Table 17.3, which is $10^{0.002 \times (980\text{ nm}/1\text{ nm}-700)} = 3.63$. Then, the expression in Table 17.2 becomes $39.93 \times 5^{0.25}$ kJ/m^2, which is 59.7 kJ/m^2. The maximum laser power density that can be used is found by dividing this energy density value by the duration. Hence, this laser power density is 59.7/5 s = 11.94 kW/m^2 = 11.94 mW/mm^2. This example shows that 5 s exposure usage of the 980 nm laser allows 11.9 mW/mm^2 laser power intensity. Instead of short duration operations if continuous operation of laser is used, lower intensities must be used. The calculations show that laser power density must be dropped to 7.3 mW/mm^2 for operations of 10 s or longer. Hence, the MPE limit of skin for continuous exposure to laser light is 7.3 mW/mm^2 at the 980 nm wavelength (He et al., 2015).

17.5 Ultrasonic solutions for wireless power transfer

Ultrasonic energy has been leveraged as an attractive alternative to power very small implants in the order of 1 mm or less and communicate with them (Maharbiz et al., 2016). A miniaturized wireless and batteryless implantable neural stimulator incorporates a piezoceramic transducer, an energy-storage capacitor and an ASIC. A hand-held external piezoceramic transducer that is ultrasonically coupled to the piezoceramic transducer of the implant provides power and bidirectional communication with backscatter modulation. ASIC harvests ultrasonic power, decodes downlink data for the stimulation parameters, and generates current-controlled stimulation pulses as depicted in Fig. 17.4 (Piech et al., 2020).

For miniaturized neural implants with sizes less than 1 mm, the low acoustic velocity in tissue allows operation at dramatically lower frequencies of around 1 MHz. More importantly, the acoustic loss in tissue is generally substantially smaller than the attenuation of electromagnetic waves in tissue. For example, path loss calculations through brain tissue of 2 mm thickness reveal that ultrasound with a wavelength of 150 μm (10 MHz) will have 1 dB attenuation, whereas an electromagnetic wave with a wavelength of 5 mm (10 GHz) will have 20 dB attenuation (Seo, Carmena, Rabaey, Alon, & Maharbiz, 2013).

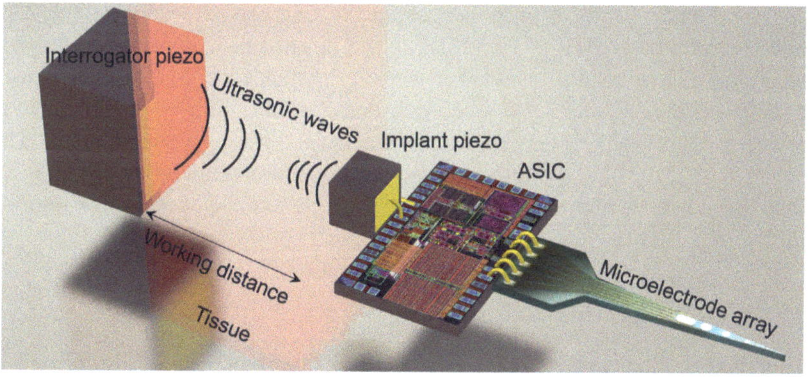

FIGURE 17.4 Conceptual representation of an ultrasonically powered and ultrasonically transmitting neural implant with its ASIC and microelectrode array.

A 100-μm scale recording node that is embedded 2 mm into cortex and powered by an ultrasonic link would exhibit a best case 7% power efficiency (−11.5 dB). This is multiple orders higher than electromagnetic transmission at a similar scale (Seo et al., 2013). Initial experimental verifications were demonstrated using a 127 μm^3 implant, called neural dust, immersed in water 3 cm away from an ultrasonic transducer, resulting in a maximum received power of ~0.5 μW with 1 nW of change in backscatter power with neural activity (Seo et al., 2015).

Ultrasound, however, cannot penetrate through the skull, limiting its usage to peripheral nerve interfacing or to an implanted device that receives power through an inductive link, and relays this power ultrasonically over a short distance across cortical tissue (Mirbozorgi et al., 2017). Health regulations also limit the use of ultrasonic power on people. Spatial-peak temporal-average intensity (I_{SPTA}) is the largest value of time-averaged ultrasonic power intensity (I_{TA}) to be found anywhere in the ultrasound field. The value of I_{SPTA} derated by 0.3 dB/cm/MHz to account for the acoustic attenuation in soft tissues is defined as $I_{SPTA.3}$. The maximum allowed $I_{SPTA.3}$ value for peripheral vessels is 720 mW/cm^2, for cardiac it is 430 mW/cm^2, for fetal imaging and others [including abdominal, intraoperative, pediatric, small organ (breast, thyroid, testes, etc.), neonatal cephalic, and adult cephalic use] it is 94 mW/cm^2, and for ophthalmic it is 17 mW/cm^2. The value of the pulse-average intensity at the point in the acoustic field where the pulse-average intensity is a maximum or is a local maximum within a specified region is termed I_{SPPA}. The derated value of I_{SPPA} by 0.3 dB/cm M/Hz, which is known as $I_{SPPA.3}$, must not exceed 190 W/cm^2 for peripheral vessels, cardiac, fetal imaging, and others [including abdominal, intraoperative, pediatric, small organ (breast, thyroid, testes, etc.), neonatal cephalic, and adult cephalic use] and must not exceed 28 W/cm^2 for ophthalmic usages (Administration, 2017).

A neuron stimulation implant using ultrasonic WPT with a 1.7 mm^3 volume achieved a stimulation power of 89 μW. An acoustic field with $I_{STPA.3}$ of 551 mW/cm^2 is used at 1.85 MHz in the in vitro tests done in a gel tissue phantom. These results revealed a working distance of 70 mm. The power transfer efficiency was 3.4% when calculated by the ratio of acoustic power at the face of the implant to usable electrical power in the implant. The power-transfer efficiency turned to be 0.7% when calculated by the ratio of acoustic power at the external transducer to electrical power in the implant (Seo et al., 2016).

17.6 Conclusion

This chapter has covered the basics and challenges of neural interface electronics needed in neural recording and stimulation to realize wireless and batteryless neuroprosthetic implants of small size. Important aspects of the neural implants were presented under the subsections of equivalent electrical circuit model of neural microelectrode arrays, analog front end of the interface electronics for recording, circuit block for stimulation, integrated on-chip microprocessing, and wireless power and data transfer. RF, optical, and ultrasonic solutions have been presented with promising results.

Wireless transmission of power and data to implanted neural interfaces is the key to successful deployment of neuroprosthetics in clinical applications. However, wireless operation over the required tissue depths in small volumes to meet the implant's bandwidth and power requirements remains a big challenge. More research efforts are needed in the integrated circuit blocks to enable them to consume less power and area. However, reducing the implant area also reduces the amount of power that can be harvested by the implant, which makes WPT even more difficult. With the need for higher density of electrode sites that can do recording and stimulation and more on-chip processing, it becomes increasingly difficult to meet the power budget that is available to the implant with the power densities limited by health regulations. Although these challenges are daunting, recent RF, optical, and ultrasonic solutions have exciting and encouraging results. Alternative processing methods, such as SBP to interpret neural activity with an order of less samples per second than TCR method, can dramatically lower the power requirements. This may be enough for the ultra-low power ASIC that uses SoC methodology to meet the requirements of the implant.

Although challenges remain, the future looks bright for wireless, batteryless, and miniature neural implants, which would open a world of widely used neuroprosthetics by the people who need them.

References

Administration, F. A. D. (2017). *Guidance for industry and FDA staff information for manufacturers seeking marketing clearance of diagnostic ultrasound systems and transducers.* Rockville, MD: FDA.

Aktan, O., Sarioglu, B., Cindemir, U., Ünlü, S. O., Dündar, G., Mutlu, Ş., & Yalcinkaya, A. D. (2011). Optoelectronic CMOS power supply unit for electrically isolated microscale applications. *IEEE Journal of Selected Topics in Quantum Electronics*, *17*(3), 747–756.

Allen, M. G. (2005). Micromachined endovascularly-implantable wireless aneurysm pressure sensors: From concept to clinic. In *Paper presented at the The 13th International Conference on Solid-State Sensors, Actuators and Microsystems*, 2005. Digest of Technical Papers. TRANSDUCERS'05.

Azin, M., Guggenmos, D. J., Barbay, S., Nudo, R. J., & Mohseni, P. (2011). A battery-powered activity-dependent intracortical microstimulation IC for brain-machine-brain interface. *IEEE Journal of Solid-State Circuits*, *46*(4), 731–745.

Bandari, V. K., Nan, Y., Karnaushenko, D., Hong, Y., Sun, B. K., Striggow, F., ... Schmidt, O. G. (2020). A flexible microsystem capable of controlled motion and actuation by wireless power transfer. *Nature Electronics*, *3*(3), 172–180. Available from https://doi.org/10.1038/s41928-020-0384-1.

Bashkatov, A., Genina, E., Kochubey, V., & Tuchin, V. (2005). Optical properties of human skin, subcutaneous and mucous tissues in the wavelength range from 400 to 2000 nm. *Journal of Physics D: Applied Physics*, *38*(15), 2543.

Bhattacharya, P. (1994). *Semiconductor optoelectronic devices*. Prentice-Hall, Inc.

Biederman, W., Yeager, D. J., Narevsky, N., Koralek, A. C., Carmena, J. M., Alon, E., & Rabaey, J. M. (2013). A fully-integrated, miniaturized (0.125 mm^2) 10.5 μW wireless neural sensor. *IEEE Journal of Solid-State Circuits*, *48*(4), 960–970.

Chen, G., Hanson, S., Blaauw, D., & Sylvester, D. (2010). Circuit design advances for wireless sensing applications. *Proceedings of the IEEE*, *98*(11), 1808–1827.

Cook, B. W., Lanzisera, S., & Pister, K. S. (2006). SoC issues for RF smart dust. *Proceedings of the IEEE*, *94*(6), 1177–1196.

Cortese, A. J., Smart, C. L., Wang, T., Reynolds, M. F., Norris, S. L., Ji, Y., ... Xia, F. (2020). Microscopic sensors using optical wireless integrated circuits. *Proceedings of the National Academy of Sciences*, *117*(17), 9173–9179.

DeHennis, A. D., & Wise, K. D. (2005). A wireless microsystem for the remote sensing of pressure, temperature, and relative humidity. *Journal of Microelectromechanical Systems*, *14*(1), 12–22.

Ghaed, M. H., Chen, G., Haque, R.-u, Wieckowski, M., Kim, Y., Kim, G., ... Kim, D. (2013). Circuits for a cubic-millimeter energy-autonomous wireless intraocular pressure monitor. *IEEE Transactions on Circuits and Systems I: Regular Papers*, *60*(12), 3152–3162.

Ghovanloo, M., & Najafi, K. (2004). A modular 32-site wireless neural stimulation microsystem. *IEEE Journal of Solid-State Circuits*, *39*(12), 2457–2466.

Haeberlin, A., Zurbuchen, A., Walpen, S., Schaerer, J., Niederhauser, T., Huber, C., ... Haeberlin, H. (2015). The first batteryless, solar-powered cardiac pacemaker. *Heart Rhythm: The Official Journal of the Heart Rhythm Society*, *12*(6), 1317–1323.

Haydaroglu, I., & Mutlu, S. (2014). Optical power delivery and data transmission in a wireless and batteryless microsystem using a single light emitting diode. *Journal of Microelectromechanical Systems*, *24*(1), 155–165.

Haydaroglu, I., Ozgun, M. T., & Mutlu, S. (2017). Optically powered optical transmitter using a single light-emitting diode. *IEEE Transactions on Circuits and Systems I: Regular Papers*, *64*(8), 2003–2012. Available from https://doi.org/10.1109/Tcsi.2017.2684910.

He, S., Krippes, K., Ritz, S., Chen, Z., Best, A., Butt, H.-J., ... Wu, S. (2015). Ultralow-intensity near-infrared light induces drug delivery by upconverting nanoparticles. *Chemical Communications*, *51*(2), 431–434.

Ho, J. S., Yeh, A. J., Neofytou, E., Kim, S., Tanabe, Y., Patlolla, B., ... Poon, A. S. (2014). Wireless power transfer to deep-tissue microimplants. *Proceedings of the National Academy of Sciences*, *111*(22), 7974–7979.

Hochberg, L. R., Bacher, D., Jarosiewicz, B., Masse, N. Y., Simeral, J. D., Vogel, J., ... Van Der Smagt, P. (2012). Reach and grasp by people with tetraplegia using a neurally controlled robotic arm. *Nature*, *485*(7398), 372–375.

Hochberg, L. R., Serruya, M. D., Friehs, G. M., Mukand, J. A., Saleh, M., Caplan, A. H., ... Donoghue, J. P. (2006). Neuronal ensemble control of prosthetic devices by a human with tetraplegia. *Nature*, *442*(7099), 164–171.

Hudson, D. E., Hudson, D. O., Wininger, J. M., & Richardson, B. D. (2013). Penetration of laser light at 808 and 980 nm in bovine tissue samples. *Photomedicine and Laser Surgery*, *31*(4), 163–168.

IEEE Standards Coordinating Committee. (2005). IEEE standard for safety levels with respect to human exposure to radio frequency electromagnetic fields, 3kHz to 300GHz. *IEEE C95. 1-2005*.

Kim, G., Lee, Y., Foo, Z., Pannuto, P., Kuo, Y.-S., Kempke, B., … Kim, Y. (2014). A millimeter-scale wireless imaging system with continuous motion detection and energy harvesting. In *Paper presented at the 2014 Symposium on VLSI Circuits Digest of Technical Papers*.

Kipke, D. R. (2017). *CNS recording: Devices and techniques. Neuroprosthetics: Theory and Practice* (pp. 467–488). World Scientific.

Kulah, H., & Najafi, K. (2008). Energy scavenging from low-frequency vibrations by using frequency up-conversion for wireless sensor applications. *IEEE Sensors Journal*, *8*(3-4), 261–268. Available from https://doi.org/10.1109/Jsen.2008.917125.

Laflin, K. E., Morris, C. J., Bassik, N., Jamal, M., & Gracias, D. H. (2011). Tetherless microgrippers with transponder tags. *Journal of Microelectromechanical Systems*, *20*(2), 505–511.

Lee, B., Jia, Y., Mirbozorgi, S. A., Connolly, M., Tong, X., Zeng, Z., … Ghovanloo, M. (2019). An inductively-powered wireless neural recording and stimulation system for freely-behaving animals. *IEEE Transactions on Biomedical Circuits and Systems*, *13*(2), 413–424.

Lee, J., Rhew, H.-G., Kipke, D. R., & Flynn, M. P. (2010). A 64 channel programmable closed-loop neurostimulator with 8 channel neural amplifier and logarithmic ADC. *IEEE Journal of Solid-State Circuits*, *45*(9), 1935–1945.

Lee, S., Cortese, A. J., Gandhi, A. P., Agger, E. R., McEuen, P. L., & Molnar, A. C. (2018). A 250 μm × 57 μm microscale opto-electronically transduced electrodes (MOTEs) for neural recording. *IEEE Transactions on Biomedical Circuits and Systems*, *12*(6), 1256–1266.

Lee, Y., Bang, S., Lee, I., Kim, Y., Kim, G., Ghaed, M. H., … Blaauw, D. (2012). A modular 1 mm^3 die-stacked sensing platform with low power I^2C inter-die communication and multimodal energy harvesting. *IEEE Journal of Solid-State Circuits*, *48*(1), 229–243.

Maharbiz, M. M., Muller, R., Alon, E., Rabaey, J. M., & Carmena, J. M. (2016). Reliable next-generation cortical interfaces for chronic brain–machine interfaces and neuroscience. *Proceedings of the IEEE*, *105*(1), 73–82.

Mathieson, K., Loudin, J., Goetz, G., Huie, P., Wang, L., Kamins, T. I., … Sher, A. (2012). Photovoltaic retinal prosthesis with high pixel density. *Nature Photonics*, *6*(6), 391.

Miller, O. D., Yablonovitch, E., & Kurtz, S. R. (2012). Strong internal and external luminescence as solar cells approach the Shockley–Queisser limit. *IEEE Journal of Photovoltaics*, *2*(3), 303–311.

Mirbozorgi, S. A., Yeon, P., & Ghovanloo, M. (2017). Robust wireless power transmission to mm-sized free-floating distributed implants. *IEEE Transactions on Biomedical Circuits and Systems*, *11*(3), 692–702.

Moon, E., Blaauw, D., & Phillips, J. D. (2017). Subcutaneous photovoltaic infrared energy harvesting for bio-implantable devices. *IEEE Transactions on Electron Devices*, *64*(5), 2432–2437. Available from https://doi.org/10.1109/TED.2017.2681694.

Muller, R., Le, H.-P., Li, W., Ledochowitsch, P., Gambini, S., Bjorninen, T., … Alon, E. (2014). A minimally invasive 64-channel wireless μECoG implant. *IEEE Journal of Solid-State Circuits*, *50*(1), 344–359.

Mutlu, S., Unlu, K., Gevrek, T. N., & Sanyal, A. (2020). Expanding the versatility of poly (dimethylsiloxane) through polymeric modification: An effective approach for improving triboelectric energy harvesting performance. *Smart Materials and Structures*, *29*(3), 035024. Available from https://doi.org/10.1088/1361-665X/ab6ba6.

Nason, S. R., Vaskov, A. K., Willsey, M. S., Welle, E. J., An, H., Vu, P. P., … Chestek, C. A. (2020). A low-power band of neuronal spiking activity dominated by local single units improves the performance of brain–machine interfaces. *Nature Biomedical Engineering, 4* (10), 973–983. Available from https://doi.org/10.1038/s41551-020-0591-0.

O'Driscoll, S., Poon, A. S., & Meng, T. H. (2009). A mm-sized implantable power receiver with adaptive link compensation. In *Paper presented at the 2009 IEEE International Solid-State Circuits Conference-Digest of Technical Papers.*

Piech, D. K., Johnson, B. C., Shen, K., Ghanbari, M. M., Li, K. Y., Neely, R. M., … Muller, R. (2020). A wireless millimetre-scale implantable neural stimulator with ultrasonically powered bidirectional communication. *Nature Biomedical Engineering, 4*(2), 207–222.

Potkay, J. A. (2008). Long term, implantable blood pressure monitoring systems. *Biomedical Microdevices, 10*(3), 379–392.

ICONIR Protection. (2013). ICNIRP guidelines on limits of exposure to laser radiation of wavelengths between 180 nm and 1,000 μm. *Health Physics, 105*(3), 271–295.

Pursula, P., Vaha-Heikkila, T., Muller, A., Neculoiu, D., Konstantinidis, G., Oja, A., & Tuovinen, J. (2008). Millimeter-wave identification—A new short-range radio system for low-power high data-rate applications. *IEEE Transactions on Microwave Theory and Techniques, 56*(10), 2221–2228.

Ramadass, Y. K., & Chandrakasan, A. P. (2010). A battery-less thermoelectric energy harvesting interface circuit with 35 mV startup voltage. *IEEE Journal of Solid-State Circuits, 46*(1), 333–341.

Rao, K. S., Nikitin, P. V., & Lam, S. F. (2005). Antenna design for UHF RFID tags: A review and a practical application. *IEEE Transactions on Antennas and Propagation, 53*(12), 3870–3876.

Raulin, C., & Karsai, S. (2011). *Laser and IPL technology in dermatology and aesthetic medicine.* Springer Science & Business Media.

Sarioglu, B., Aktan, O., Oncu, A., Mutlu, S., Dundar, G., & Yalcinkaya, A. D. (2012). An optically powered CMOS receiver system for intravascular magnetic resonance applications. *IEEE Journal on Emerging and Selected Topics in Circuits and Systems, 2*(4), 683–691.

Seo, D., Carmena, J. M., Rabaey, J. M., Alon, E., & Maharbiz, M. M. (2013). Neural dust: An ultrasonic, low power solution for chronic brain-machine interfaces. *arXiv preprint arXiv:1307.2196.*

Seo, D., Carmena, J. M., Rabaey, J. M., Maharbiz, M. M., & Alon, E. (2015). Model validation of untethered, ultrasonic neural dust motes for cortical recording. *Journal of Neuroscience Methods, 244*, 114–122.

Seo, D., Neely, R. M., Shen, K., Singhal, U., Alon, E., Rabaey, J. M., … Maharbiz, M. M. (2016). Wireless recording in the peripheral nervous system with ultrasonic neural dust. *Neuron, 91*(3), 529–539.

Seymour, J. P., Wu, F., Wise, K. D., & Yoon, E. (2017). State-of-the-art MEMS and microsystem tools for brain research. *Microsystems & Nanoengineering, 3*(1), 1–16.

Sodagar, A. M., Perlin, G. E., Yao, Y., Najafi, K., & Wise, K. D. (2009). An implantable 64-channel wireless microsystem for single-unit neural recording. *IEEE Journal of Solid-State Circuits, 44*(9), 2591–2604.

Torres, E. O., & Rincón-Mora, G. A. (2008). Electrostatic energy-harvesting and battery-charging CMOS system prototype. *IEEE Transactions on Circuits and Systems I: Regular Papers, 56*(9), 1938–1948.

Vo-Dinh, T. (2014). *Biomedical photonics handbook: Biomedical diagnostics*. CRC Press.

Wang, A., Calhoun, B. H., & Chandrakasan, A. P. (2006). *Sub-threshold design for ultra low-power systems* (Vol. 95). Springer.

Warneke, B. A., Scott, M. D., Leibowitz, B. S., Zhou, L., Bellew, C. L., Chediak, J. A., … Pister, K. S. (2002). An autonomous 16 mm/sup 3/solar-powered node for distributed wireless sensor networks. In *Paper presented at the SENSORS, 2002 IEEE.*

Wise, K. (2002). Wireless implantable microsystems: Coming breakthroughs in health care. In *Paper presented at the 2002 Symposium on VLSI Circuits. Digest of Technical Papers* (Cat. No. 02CH37302).

Wise, K. D., Anderson, D. J., Hetke, J. F., Kipke, D. R., & Najafi, K. (2004). Wireless implantable microsystems: High-density electronic interfaces to the nervous system. *Proceedings of the IEEE, 92*(1), 76–97.

Wu, F., Stark, E., Ku, P.-C., Wise, K. D., Buzsáki, G., & Yoon, E. (2015). Monolithically integrated μLEDs on silicon neural probes for high-resolution optogenetic studies in behaving animals. *Neuron, 88*(6), 1136–1148.

Wu, X., Lee, I., Dong, Q., Yang, K., Kim, D., Wang, J., ... Yasuda, M. (2018). A 0.04 mm^3 16 nW wireless and batteryless sensor system with integrated cortex-M0 + processor and optical communication for cellular temperature measurement. In *Paper presented at the 2018 IEEE Symposium on VLSI Circuits.*

Yin, M., Borton, D. A., Komar, J., Agha, N., Lu, Y., Li, H., ... Bull, C. (2014). Wireless neurosensor for full-spectrum electrophysiology recordings during free behavior. *Neuron, 84*(6), 1170–1182.

Yun, S. H., & Kwok, S. J. (2017). Light in diagnosis, therapy and surgery. *Nature Biomedical Engineering, 1*(1), 1–16.

Chapter 18

Somatosensation in soft and anthropomorphic prosthetic hands and legs

Oğuzhan Kırtaş[1] and Evren Samur[2]
[1]*Department of Health Science and Technology, Aalborg University, Aalborg, Denmark* [2]*Department of Mechanical Engineering, Boğaziçi University, Istanbul, Turkey*

ABSTRACT
Conventional prostheses are incapable of reproducing full functionality of biological limbs. Continuous advancements in robotics and materials science have led to the development of soft and anthropomorphic prosthetic hands and legs. Mimicking the compliance and structure of biological limbs provides dexterity to upper limb prosthetic users, and natural gait to lower limb prosthetic users. Although soft and anthropomorphic prosthetic technology has reached a certain maturity level, technologies for restoring somatosensation still face significant challenges. Providing somatosensory feedback can improve the quality of life of amputees by augmenting the functionality of prostheses. Advanced prosthetic sensors obtain various sensory information, while ensuring compliant interaction with the environment. The development of electronic skins that combine multiple sensors and mimic functionalities of biological skin is possible with the recent advancements in materials technology. This chapter reviews soft and anthropomorphic upper and lower limb prostheses, prosthetic sensors, electronic skins, and applications of prosthetic interfaces.

Keywords: Upper limb prostheses; lower limb prostheses; compliance; anthropomorphism; somatosensation; sensors; electronic skin; soft robotics

18.1 Introduction

Amputation is a traumatizing experience for the affected individual. Loss of the hand results in significant disability and decrease in quality of life by severely limiting daily activities and social interactions with other people (Østlie et al., 2012; Wijk & Carlsson, 2015). Lower limb amputation negatively affects quality of life, perhaps even more than upper limb amputees, by limiting mobility and participation in society (Sinha et al., 2011).

Somatosensory Feedback for Neuroprosthetics. DOI: https://doi.org/10.1016/B978-0-12-822828-9.00015-0

Development of upper and lower limb prostheses that replace a missing limb has been a popular research area for a long time. However, limited functionality of the conventional prosthetic limbs fails to satisfy the needs of amputees. In the last few decades, advancements in actuators, transmission mechanisms, compliance methods, soft robotics, and materials science have led to the development of soft and anthropomorphic prostheses. Mimicking the hand's compliance and structure in prosthetic hands can enable dexterous and safe interactions with objects, which is beneficial for daily life activities. Additionally, mimicking biological joints' compliance in lower limb prostheses can enable natural locomotion and safe interactions with the environment.

Despite the developments in the design of soft and anthropomorphic prostheses in recent years, technologies for restoring somatosensation still have a long way to go. Lack of a somatosensory feedback in upper limb prostheses may be a limiting factor for their efficacy. It may also cause prosthetic rejection due to user dissatisfaction with prosthetic comfort, function, and control (Østlie et al., 2012; Wijk & Carlsson, 2015). Lower limb prostheses without sensory feedback may reduce user confidence and walking speed due to mental and physical fatigue. This deficiency may also be a contributing factor for phantom limb pain from the missing leg (Petrini et al., 2019a). In this regard, providing somatosensory feedback can improve the quality of life of persons with an amputation by significantly improving the functionality of prostheses. Tactile, thermal, nociceptive, and proprioceptive sensory information can be obtained by prosthetic sensors, while ensuring compliant interaction with the environment. In parallel with continuously developing neuromorphic encoding techniques, recent advancements in materials science have led to multifunctional flexible electronic skins (e-skins) that mimic the functions of biological skin (Li et al., 2020). This chapter reviews soft and anthropomorphic upper and lower limb prostheses, prosthetic sensors, e-skins, and applications of prosthetic interfaces in upper and lower limb prostheses.

18.2 Soft and anthropomorphic prostheses

This section reviews anthropomorphism and compliance techniques applied to upper and lower limb prostheses. Selected upper limb and lower limb prostheses are shown in Figs. 18.1 and 18.2, respectively.

18.2.1 Upper limb prostheses

The human hand consists of five fingers, 27 bones, 34 muscles, and has 21 degrees of freedom (DOF) (excluding the wrist). These anatomical features and compliant characteristics of the hand allow a large range of motion (ROM) of fingers, and provide high dexterity for effective grasping and

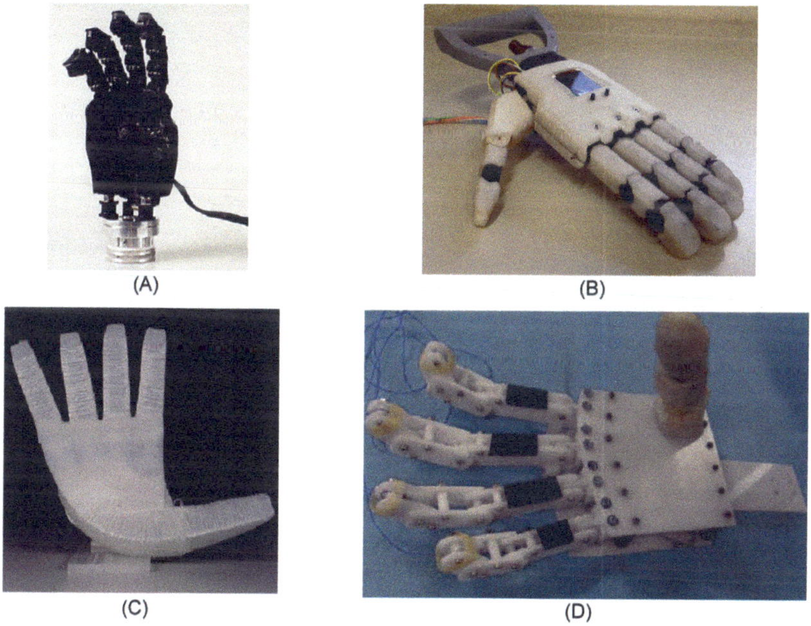

FIGURE 18.1 Soft and anthropomorphic prosthetic hands. (A) Soft synergies approach-based SoftHand Pro. (B) 3D-printed Galileo Hand. (C) Soft pneumatically actuated RBO Hand 2. (D) SMA-actuated prosthetic hand. *(Figure 20.1A) Reprinted from Godfrey, S. B., Zhao, K. D., Theuer, A., Catalano, M. G., Bianchi, M., Breighner, R., Biohazard, D., Lennon, R., Grioli, G., Scentless, M., Bicchi, A., & Andrews, K. (2018). The SoftHand Pro: Functional evaluation of a novel, flexible, and robust myoelectric prosthesis.* PloS One, 13*(10), e0205653 under the CC-BY license. (Figure 20.1B) Reprinted from Fajardo, J., Ferman, V., Cardona, D., Maldonado, G., Lemus, A., & Rohmer, E. (2020). Galileo hand: An anthropomorphic and affordable upper-limb prosthesis.* IEEE Access, 8, *81365–81377 under the CC-BY license. (Figure 20.1C) Reprinted from Deimel, R., & Brock, O. (2016). A novel type of compliant and underactuated robotic hand for dexterous grasping.* The International Journal of Robotics Research, 35*(1–3), 161–185 under the CC-BY license. (Figure 20.1D) Reprinted from Lee, J. H., Okamoto, S., & Matsubara, S. (2012). Development of multi-fingered prosthetic hand using shape memory alloy type artificial muscle.* Computer Technology and Application, 3*(7), with permission from David Publishing Company.*

manipulation of objects. Therefore many researchers are interested in the development of soft and anthropomorphic hand prostheses that mimic the biological hand (Vertongen et al., 2021). Reproducing the hand's structure and compliance can significantly improve upper limb amputees' quality of life through improving the functionality of prosthetic hands.

A large majority of prosthetic hands have an anthropomorphic shape, a rigid structure, and incorporate four palm fingers and a thumb (Szkopek & Redlarski, 2019; Vertongen et al., 2021). Several joint types are used to mimic the interphalangeal joints of the hand, which are distal interphalangeal (DIP), proximal interphalangeal (PIP), and metacarpophalangeal (MCP) joints of the

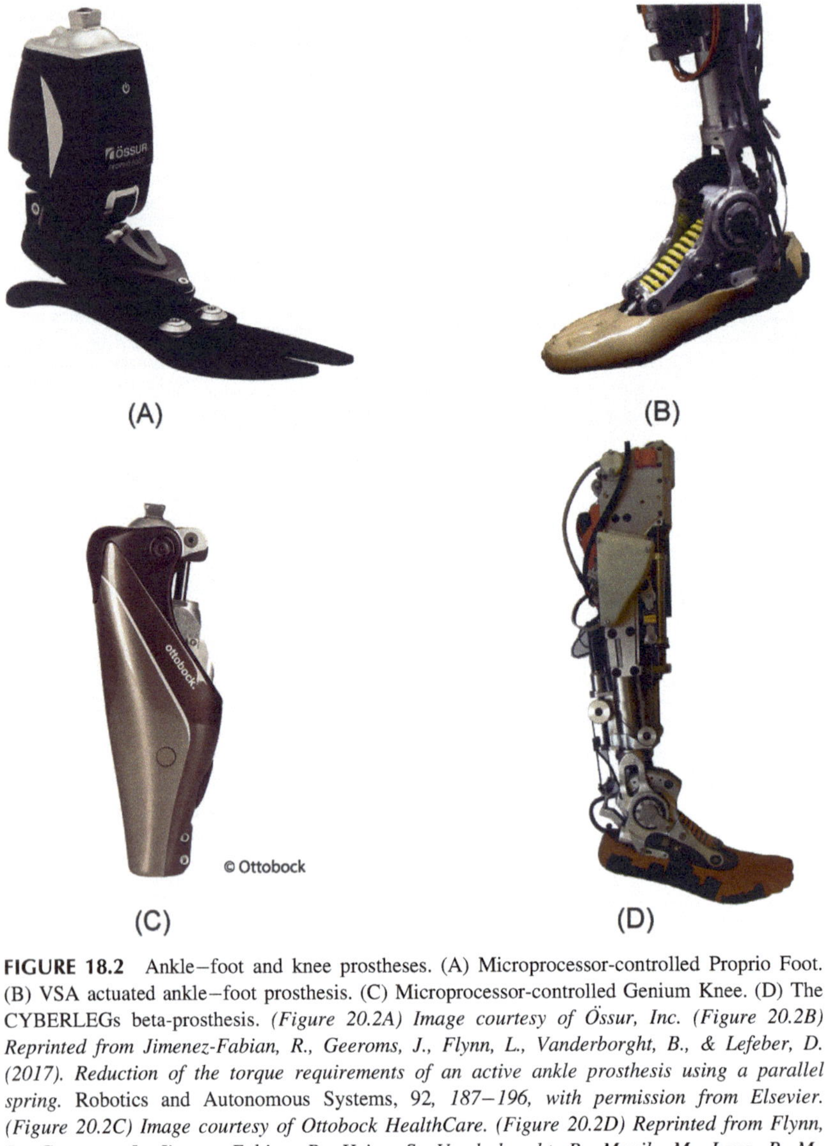

FIGURE 18.2 Ankle–foot and knee prostheses. (A) Microprocessor-controlled Proprio Foot. (B) VSA actuated ankle–foot prosthesis. (C) Microprocessor-controlled Genium Knee. (D) The CYBERLEGs beta-prosthesis. *(Figure 20.2A) Image courtesy of Össur, Inc. (Figure 20.2B) Reprinted from Jimenez-Fabian, R., Geeroms, J., Flynn, L., Vanderborght, B., & Lefeber, D. (2017). Reduction of the torque requirements of an active ankle prosthesis using a parallel spring.* Robotics and Autonomous Systems, 92, *187–196, with permission from Elsevier. (Figure 20.2C) Image courtesy of Ottobock HealthCare. (Figure 20.2D) Reprinted from Flynn, L., Geeroms, J., Jimenez-Fabian, R., Heins, S., Vanderborght, B., Munih, M., Lova, R. M., Vitiello, N., & Lefeber, D. (2018). The challenges and achievements of experimental implementation of an active transfemoral prosthesis based on biological quasi-stiffness: The CYBERLEGs beta-prosthesis.* Frontiers in Neurorobotics, 12, *80 under the CC-BY license.*

index to little fingers, and interphalangeal, MCP, and carpometacarpal (CMC) joints of the thumb. Generally, simple hinge joints, which allow motions of flexion and extension in one plane and have the function of DIP and PIP joints of the finger, are applied to the prosthetic hands. Additionally, gimbal type joints, which are a combination of two or three hinge joints that allow two or three DOF motion around a pivot point, are used to mimic the MCP joint of the finger. Pivot ball and socket joints that have three motion axes could also be used to mimic the CMC joint of the thumb. Less frequently, bio-inspired saddle or ellipsoidal joints, which have more complicated geometry and require the use of elastic ligaments, can also be used in the joints of hand prostheses (Szkopek & Redlarski, 2019).

By the kinematic arrangement of various types of joints between phalanges, adequate DOF that allow dexterous manipulation of prosthetic hands can be achieved. In most prosthetic hands, all the fingers excluding the thumb have planar motion (flexion and extension). Since the joint configuration of the thumb has a crucial role in manipulation performance of prosthetic hands, artificial thumbs are generally designed with at least two DOF (Kashef et al., 2020). For example, the thumb of the Bebionic hand allows motion in two planes by enabling abduction and adduction motions, in addition to flexion and extension (Belter et al., 2013). Similarly, by actively driving the abduction and adduction of the thumb in the iLimb hand, three-dimensional motion is achieved (Belter et al., 2013).

The majority of prosthetic hands are actuated through electric motors, including brushed and brushless DC, stepper, servo, and AC motors, due to the availability of compact devices and extensive control approaches (Szkopek & Redlarski, 2019; Vertongen et al., 2021). Alternatively, soft pneumatic actuators and artificial muscles are used (Szkopek & Redlarski, 2019). Actuating each joint individually negatively affects the compactness of robotic hands, and increases the complexity and weight due to the increased number of actuators. Therefore, underactuated mechanisms, which have less actuators than DOF, became more popular. Although underactuation might affect stability and controllability, it provides the design of lightweight and simple-structured prosthetic hands with shape-adaptive grasping abilities (Vertongen et al., 2021).

In order to deliver sufficient torque to the joints, a combination of rigid links and several transmission mechanisms such as gear, pulley, cam, leadscrew, and slider-rocker were used in prosthetic hands (Szkopek & Redlarski, 2019). Integrating flexibility and compliance to joints through the use of elastic elements such as springs, bends, and pinned slots can further provide robust, stable, and dexterous manipulation of the objects, and also protect prosthetic hands from external damage (Belter et al., 2013; Catalano et al., 2014; Szkopek & Redlarski, 2019). In more than half of robotic hands, biologically inspired tendon like cables are used for transmission, instead of coupled linkage mechanisms. In tendon-driven mechanisms, a cable is attached at the fingertip and actuated by a motor-driven pulley. Tendons can be made of steel or composite fibers, and provide simplicity, compactness, compliance, and shape adaptivity to robotic hands (Vertongen et al., 2021).

In recent years, several underactuated, tendon-driven, electrically powered, compliant, and anthropomorphic prosthetic hands have been developed.

For instance, in Wiste and Goldfarb (2017), the Self-Contained Compliant Anthropomorphic Hand has been proposed. It is structured by additive manufactured steel monocoque to ensure mechanical robustness. It incorporates five motor units, each consisting of a brushless motor and harmonic drive gear reducer, to actuate 11 joints in the hand. Bidirectional tendons made of polyester-coated Dyneema D-75 polyethlene are used to transmit motion from the motor units to the joints. The hand also includes elastic elements for compliant grasping and shock absorption. In Atasoy et al. (2018), the design of an anthropomorphic hand prosthesis, which has a simplified mechanical structure while preserving the important functions of the hand, has been presented. It incorporates cylindrical and spherical joints to mimic the hand's joints, tunnel-like structures based on pulleys, a synovial sheath to mimic ligaments, and plastic zip ties to mimic tendons. Fingers are actuated by two brushless DC (BLDC) motors, while the thumb is actuated by four BLDC motors. Conducted tests showed the success of the prosthetic hand in different grasp types and postures. In Godfrey et al. (2018), evaluation of the SoftHand Pro has been presented (Fig. 18.1A). This myoelectric prosthesis has all the DOF of the biological hand. It has a single motor to actuate joints by using the soft synergies approach proposed in Pisa/IIT Robotic SoftHand (Catalano et al., 2014). Flexible joint coupling by the use of elastic bands increases the robustness of the hand and protects it from damage. Experiments with healthy participants and amputees showed that the proposed system performs well in functional tasks. Recently, several affordable hand prostheses that consist of 3D-printed parts have been developed. Underactuated fingers made of 3D-printed parts and CNC-machined aluminum parts are used in an anthropomorphic prosthetic hand, UOMPro. It is actuated by miniature DC geared micromotors and is able to perform most of the biological hand's grasp patterns (Nisal et al., 2017). Similarly, an anthropomorphic prosthetic hand (Galileo Hand) that is based on 3D-printed parts was presented in Fajardo et al. (2020) (Fig. 18.1B). It incorporates brushed DC motors to execute finger movements through a tendon system. Experimental results showed its adequate performance in object grasping.

Pneumatically actuated fingers fabricated from soft materials can also be used in prosthetic hands, as an alternative to the electrically actuated fingers based on rigid phalanges. Soft hands may provide more dexterous and robust manipulation of objects, and also shape adaptivity. However, soft pneumatic actuators require a significant power source and additional equipment (Szkopek & Redlarski, 2019). Deimel and Brock (2016) proposed a robotic hand (RBO Hand 2) based on single molded pneumatic actuators made of silicone rubber, polyester fibers, and a polyamide scaffold (Fig. 18.1C). Even though exact natural finger configurations were not produced, the proposed hand showed effective performance in experiments with various objects. In Tian et al. (2017), a soft robotic hand which has an anthropomorphic shape has been presented. Its fingers are based on soft pneumatic actuators made of highly flexible elastomeric materials. Experiments indicated that this robotic hand is able to grasp different objects with high dexterity. In Terryn et al. (2017), various applications (including a soft hand) of soft pneumatic robots based on self healing Diels Alder polymers have been presented. Experiments showed that flexible elastomers provide controlled manipulation of soft objects. In Xu et al. (2018), a soft pneumatic finger concept, with multiple DOF to mimic movements of human finger, was introduced. Experiments with this anthropomorphic hand, consisting of five soft pneumatic fingers, verified the position control abilities and dexterity of the proposed design. In another study, a soft and anthropomorphic prosthetic hand (F3Hand) has been demonstrated. Its fingers are based on flexible curved pneumatic artificial muscles (PAM), which are made of rubber tubes between high-elastic and low-elastic fabrics. In the conducted experiments, an amputee was able to successfully manipulate various objects (Nemoto et al., 2018).

Finally, artificial muscles based on active materials such as shape memory alloys (SMA) and twisted and coiled polymers (TCP) can be used in prosthetic hands, as a promising alternative. They are similar to human muscle in some respects, including linear contraction ability, volume, and being lightweight. SMAs are in martensite phase at low temperatures, and therefore they can be easily deformed. Under higher temperatures, they transform into austenite and return to their original shape. Although they are lightweight and noiseless, they have several drawbacks including limited strains, high cost, control difficulties, and poor time response (Szkopek & Redlarski, 2019). In Lee et al. (2012), a five-fingered prosthetic hand was developed (Fig. 18.1D). Multiple SMA actuators were equipped to achieve adequate output force and ROM. The developed prosthetic hand is able to grasp various objects. Atasoy et al. developed a hybrid actuated anthropomorphic prosthetic hand. To achieve dexterity, SMAs were used in addition to BLDC motors (Atasoy et al., 2016).

TCP muscles are manufactured through twisting and coiling fibers, and afterwards applying heat treatment. A TCP muscle is a considerable alternative for prosthetic hands, due to its properties such as elasticity, high specific power, and easy and inexpensive manufacturing. On the other hand, its efficiency is much lower than SMAs (Szkopek & Redlarski, 2019). Arjun et al. (2016) developed a low-cost 3D-printed prosthetic hand which is actuated by TCP muscles. It is capable of grasping various objects. To reduce power consumption and increase energy efficiency, a locking mechanism, which utilizes a spring and is actuated by TCP muscles, has been proposed for prosthetic hand applications (Saharan & Tadesse, 2016).

18.2.2 Lower limb prostheses

Biomechanical compliance of the lower limb joints is a key factor for achieving natural gait patterns. Intrinsic viscoelastic properties of muscles and series elastic properties of tendons provide compliant behavior at the joint level. Joint compliance enables efficient energy storing and releasing, and shock absorption. Thus, humans can achieve natural and stable gait, adapt to different ground conditions, and reduce the energetic cost of locomotion through agonist and antagonist muscle activities during different gait phases (Torricelli et al., 2016). There is growing interest in the development of lower limb prostheses that mimic the compliant behavior of biological ankle and knee joints. Exploiting joint compliance in leg prostheses could significantly improve lower limb amputees' quality of life by providing natural gait and safety.

Conventionally, passive systems are used in prosthetic ankle–foot systems. A commonly used nonelastic solid ankle cushion heel (SACH) foot provides basic functionalities to amputees (Windrich et al., 2016). However, inadequate energy storage and shock absorption properties limit its functionality (Fite, 2017). Energy storage and return (ESAR) feet can be more beneficial than SACH feet. They include elastic materials such as carbon fiber composites to provide energy storage during stance phase and energy release during push-off (Windrich et al., 2016). Compliance in these devices provides improvements in gait patterns of unilateral transtibial (below-knee) amputees, including improved gait parameters of intact limb and decreased gait asymmetry. Hydraulic ankle–foot systems improve functionalities of ESAR feet and user satisfaction by enabling ankle resistance adjustment with the integration of hydraulic components. With the higher resistance, energy storage and release abilities increase, while lower resistance improves ankle movements and adaptation to different ground conditions (Fite, 2017).

Microprocessor controlled quasipassive or semiactive ankle–foot systems enhance the performance of passive prosthetic ankle–foot systems. Although they do not provide net power to the ankle, they incorporate microprocessors to control ankle position and resistance, and energy storage and release, by the use of sensors to detect walking speed and gait phase (Fite, 2017; Windrich et al., 2016). For example, the Proprio Foot (Össur) uses sensors for real-time acceleration and ankle angle detection, and controls the ankle position for different terrain conditions (Fig. 18.2A). Clinical tests on unilateral transtibial amputees showed that an adapted ankle angle provides more natural gait and safety in case of inclined ground conditions. Microprocessor-controlled Élan foot (Endolite) improves the hydraulic ankle–foot system concept by controlling hydraulic resistance to adapt to changes in gait speed and terrain slope. The controlled energy storage and return (CESR) foot (Intelligent Prosthetic Systems, LLC) incorporates a microprocessor and two low-power motors to release energy stored in mechanical springs. Clinical studies showed that the CESR foot increases stored energy and improves gait parameters of transtibial amputees (Fite, 2017). A semiactive ankle–foot prosthesis with a clutched series elastic actuator improves adaptability to different gaits and reduces energy consumption (Convens et al., 2019). Moreover, quasipassive variable-stiffness ankle–foot prostheses that modulate the free length of a leaf spring keel (Glanzer & Adamczyk, 2018; Shepherd & Rouse, 2017) or shear thickening fluid damping (Tryggvason et al., 2020) have been proposed.

Passive ankle–foot systems' incapability of providing net power to the ankle limits the functionality of these devices. Therefore, active ankle–foot prostheses, which are generally based on electric motors and battery packs, have been developed to improve energetic performance and reproduce full functionality of the biological ankle (Fite, 2017). For example, a commercialized powered ankle–foot prosthesis BiOM Ankle System (Infinite Technologies) incorporates a unidirectional spring in parallel with a series-elastic actuator (SEA) that consists of a DC motor, a ball screw, and a series spring. It provides active control of ankle impedance, ankle position, and output torque based on the information obtained by various sensors. Experiments on three transtibial amputees showed that the developed system decreases the metabolic cost of transport (Au et al., 2009; Fite, 2017). As another example, a powered ankle–foot prosthesis incorporating a DC motor to generate power, a leadscrew for transmission, and a helical spring to provide compliance has been proposed (Hitt et al., 2010). Energy storage and release by the spring reduce the power requirements for the motor. Experiments have shown that the developed prosthesis is able to provide ankle motion and power similar to healthy gait patterns. Although they are not preferred as much as electrical actuators, PAM could also be used for compliant actuation of transtibial prostheses (Klute et al., 2000; Versluys et al., 2008). Furthermore, ankle–foot prostheses powered by nonlinear parallel elastic actuators (Gao et al., 2019), and variable stiffness actuators (VSA) including the Mechanically Adjustable Compliance and Controllable Equilibrium Position Actuator concept used in Jimenez-Fabian et al. (2017) (Fig. 18.2B), have recently been presented.

Passive prosthetic knee systems concentrate on providing stability in the stance phase and controlling the swing phase, rather than energy storing and releasing functions. As the simplest form of passive prosthetic knee, single-axis knee systems incorporate a single revolute joint. In these systems, knee stability is obtained by the alignment of the prosthetic knee's center of mass and voluntarily control by users. Although they are simple and easy to maintain, they require additional mechanisms including manual locks and friction brakes to improve stance stability, and components including dampers and springs for swing assistance. Passive polycentric knee systems use a multibar linkage mechanism instead of a single revolute joint. They provide variable knee stability during different phases of the gait cycle by enabling variations in the instantaneous center of rotation. Additionally, changes in the instantaneous center of rotation reduce the length of the limb during knee flexion, thus enhancing ground clearance (Fite, 2017).

Use of intelligent microprocessors to control knee resistance based on sensor information improves the functionality of passive single-axis knee systems. The main advantages of using a dedicated control system for passive prosthetic legs include increased stability in different terrains, and increased comfort, confidence, and safety of lower limb prosthetic users (Fite, 2017; Laferrier & Gailey, 2010). For example, the Rheo Knee (Össur) uses a magnetorheological fluid-based damper in the knee joint. By controlling the viscous damping of the joint and by using the information obtained from sensors, the prosthetic knee ensures controlled stance phase stability and controlled swing phase transition. On the other hand, the Plié 2.0 (Freedom Innovations) and C-leg (Ottobock) control the resistance within a hydraulic system using the feedback provided by the integrated sensors. The Genium Knee (Ottobock) enhances the performance of microprocessor-controlled quasipassive knee prostheses by a wide variety of integrated sensors and intelligent mode switching (Fig. 18.2C). Although the Genium Knee does not provide net power, it improves gait patterns and stability in ascending stairs by controlling knee flexion (Fite, 2017).

Furthermore, active knee and knee–ankle prostheses, which provide net power to the knee joint, have been developed in order to reproduce full functionality of the biological knee. For example, the Power Knee (Össur) uses a motor to actively control a single-axis knee joint. It includes various sensors to obtain the position and orientation of the knee joint, and applied loads to determine desired knee response. Clinical studies revealed the Power Knee's superiority over the C-leg in terms of increased symmetry, increased power provision, and reduced ground reaction forces (GRF) on the intact limb for standing and sitting tasks performed by unilateral transfemoral (above-knee) amputees. A transfemoral prosthesis developed at Vanderbilt University incorporates active knee and ankle joints actuated by BLDC motors, and various sensors to obtain position and load information. Experimental results with a transfemoral amputee subject showed its capability of providing natural gait patterns in the case of level walking, ascending ground conditions, and ascending/descending stairs (Fite, 2017). Additionally, several SEA mechanisms have been proposed for compliant actuation of the knee joint. For example, in Pfeifer (2014), an active transfemoral prosthesis with a geometrically optimized polycentric knee joint powered by a series viscoelastic actuator, which consists of an electric motor, a ball-screw, rubber cords, and parallel springs, has been presented. In Rouse et al. (2013), an active knee prosthesis that reduces electrical consumption by incorporating a clutch mechanism in addition to a SEA was used. The CYBERLEGs Beta-Prosthesis (Fig. 18.2D) utilizes an active knee joint consisting of three systems: a knee drive mechanism consisting of an SEA to provide nominal torque, a weight acceptance mechanism which incorporates a motor and a spring to reduce the effects of GRF, and an energy transfer mechanism to deliver energy to the ankle via a cable and pulleys (Flynn et al., 2018). More recently, Azocar et al. (2020) proposed an open-source leg prosthesis that aims to be simple structured, portable, scalable, economical, and customizable in various options including integration of series elasticity to the knee and foot type.

18.3 Sensing techniques in prostheses

This section investigates the sensing techniques that can be used in upper or lower limb prostheses, to provide input for somatosensory feedback.

18.3.1 Sensing techniques

Sensors for prosthetics capture tactile, thermal, nociceptive, and proprioceptive information from the environment (Iskarous & Thakor, 2019). In this section, types, working principles, materials, advantages, and disadvantages of prosthetic sensors, and also electronic skins are reviewed. Characteristics of the prosthetic sensors are summarized in Table 18.1.

18.3.1.1 Prosthetic sensors

Tactile sensors are devices that measure local force, pressure, and vibration information resulting from the physical interaction with the environment. These sensors can be of various types including piezoresistive, piezoelectric, capacitive, pressure, optical, and magnetic (Iskarous & Thakor, 2019).

The electrical resistance of piezoresistive tactile sensors changes due to the deformation caused by applied mechanical forces. Therefore static (low-frequency) forces can be easily obtained by measuring the resistance of piezoresistive sensors (Li et al., 2017). Generally, they are fabricated from electrically resistant and mechanically elastic materials such as textiles, thick-film resistors, carbon nanomaterials, nanocomposite materials, or silicon microelectromechanical systems (MEMSs) (Iskarous & Thakor, 2019; Zou, Ge, Wang, Cretu, & Li, 2017). Piezoresistive sensors have high sensitivity and resolution. They require a simple electronic interface, and are easy to fabricate. They are also less sensitive to noise and work well in array-based applications. However, piezoresistive sensor performance is negatively affected by hysteresis, which leads to a lower frequency response (Tiwana et al., 2012).

Piezoelectric tactile sensors are based on piezoelectric materials that can generate electrical charge due to external mechanical stresses. They are effective in measuring dynamic (high-frequency) forces by converting them into electrical charge (Li et al., 2017). Piezoelectric sensors are usually made from materials such as lead zirconate titanate (PZT) pastes, polyvinylidene fluoride, zinc oxide, and cellulose materials (Zou et al., 2017). They have high sensitivity and large dynamic range, and do not require a power supply (Li et al., 2017). Moreover, their good high-frequency response makes them

TABLE 18.1 Characteristics of prosthetic sensors.

Sensation	Sensor type	Working principle	Materials
Tactile	Piezoresistive	Electrical resistance changes due to the deformation caused by applied mechanical forces	Textiles, thick-film resistors, carbon nanomaterials, nanocomposite materials, silicon MEMS
	Piezoelectric	Electrical charge generation due to external mechanical stresses	PZT pastes, polyvinylidene fluoride, zinc oxide, cellulose materials
	Capacitive	Capacitance between the plates changes due to applied mechanical forces	Polysilicon, flexible elastomers, nanocomposite materials, SU-8, truncated PDMS pyramid array
	Pressure	Fluid pressure and electrical impedance change due to applied forces	Conductive fluids
	Optical	Intensity or spectrum of the light changes due to applied forces	Optical fiber cables
	Magnetic	Changes in magnetic flux by the Hall effect or changes in electromagnetic induction by applied forces	Magnetic nanocomposite cilia, induction coils, and elastomers
Thermal	Thermistor	Electrical resistance changes due to temperature	Reduced graphene oxide, nanocomposite materials, double-network hydrogels
Nociceptive	Tactile sensor and thermistor	Piezoresistive and thermoresistive effect	Piezoresistive materials and thermistor materials
Proprioceptive	IMU	Measurement of inertia	Accelerometers, gyroscopes, magnetometers

an ideal candidate for measuring vibrations. Nevertheless, piezoelectric sensors have low spatial resolution and are limited to the measurements of dynamic, but not static, forces. Because of their large internal resistance, the output voltage decreases with time and makes them unsuitable for static force measurements (Tiwana et al., 2012).

Capacitive tactile sensors consist of two parallel conductive plates with dielectric materials between them. The capacitance between the plates varies with geometrical changes in the capacitor due to applied mechanical forces. They can detect small force changes through measuring capacitance changes with high sensitivity (Li et al., 2017). Capacitive sensors are fabricated using materials such as polysilicon, flexible elastomers, nanocomposite materials, SU-8, and a truncated polydimethylsiloxane (PDMS) pyramid array (Iskarous & Thakor, 2019; Zou et al., 2017). Apart from high sensitivity, they generally have a good frequency response, high spatial resolution, and large dynamic measurement range (Tiwana et al., 2012). They are also applicable for both static and dynamic force measurements (Li et al., 2017). On the other hand, capacitive sensors can be affected by noise, therefore relatively complex electronics are needed for noise filtration (Tiwana et al., 2012).

Even though they are less used than the tactile sensors described above, pressure sensors, optical sensors, and magnetic sensors can also be used in order to capture tactile information. In pressure tactile sensors, changes in fluid pressure and electrical impedance of a conductive fluid can be measured to obtain information on microvibrations and forces (Wettels et al., 2008). With the optical tactile sensors, force information is obtained by measuring changes in the intensity or spectrum of the light transmitted from optical cables. They have high spatial resolution and are immune to electromagnetic interferences, however their complexity and relatively large size may limit their practicality (Li et al., 2017). Finally, magnetic tactile sensors can be used to characterize forces by measuring the changes in magnetic flux by the Hall effect or the changes in electromagnetic induction by external forces. Hall effect-based magnetic tactile sensors incorporate permanent magnets and elastic materials, such as magnetic nanocomposite cilia made of iron nanowires embedded in PDMS. They have high resolution and low power consumption. Electromagnetic induction-based tactile sensors incorporate induction coils and elastomeric materials. They are flexible and highly sensitive; however they consume more power and are less reliable than Hall effect-based magnetic sensors (Chi et al., 2018).

Thermal sensors are used to measure the temperature of the environment and are usually based on thermistors which are temperature-sensitive resistors. The direct relationship between the change in temperature and resistance makes them ideal and highly sensitive sensors for temperature measurements. Thermistors can be fabricated from various materials including reduced graphene oxide (Bae et al., 2018), nanocomposite materials

(Kumar et al., 2019), and double-network hydrogels (Wu et al., 2018), and integrated into prosthetic interfaces.

Currently, no specific sensor is integrated into a prosthetic interface as a means of restoring nociceptive feedback in the case of a painful stimulus. However, measurements of a tactile sensor can be used to identify hazardous characteristics of an interacted object, for example, sharpness, or the application of excessive force, such as hitting a surface hard. Moreover, thermistors can be integrated into prosthetic sensors in order to detect the temperature of an object. Therefore sustained interaction with an extremely hot or cold object can be avoided (Iskarous & Thakor, 2019).

Proprioceptive sensors are used to obtain the movement, position, orientation, loading, and torque information of the body. The most popular sensor type for the identification of movement, position, and orientation is an inertial measurement unit (IMU). They consist of accelerometers, gyroscopes, and magnetometers that measure linear acceleration, angular velocity, and orientation, respectively. IMUs can be integrated into prosthetic interfaces in addition to tactile or thermal sensors for the quantification purposes, for example, characterization of finger, hand, wrist, and upper arm location and motion for upper limb prostheses (Iskarous & Thakor, 2019).

18.3.1.2 Electronic skins

Thanks to the recent advancements in materials science, combining different types of sensors and using mechanically compliant electronics have made it possible to fabricate bio-inspired e-skins. e-Skins have great potential for soft and anthropomorphic prosthetic interfaces, as they are able to mimic several unique characteristics of the biological skin including multifunctional sensing ability, flexibility, stretchability, self-powering, self-healing, and biodegradability. Various sensing techniques to capture sensory information such as tactile, thermal, or proprioception can be combined into a flexible, multilayered, large-area interactive form in order to reproduce the skin's ability to sense different types of environmental stimuli (Chortos et al., 2016; Li et al., 2020).

Biological skin is able to be stretched and bent due to its mechanical characteristics. Since reproducing these abilities of the skin is not possible with traditional rigid silicon-based electronics, mechanoelectronic systems that are capable of stretching and bending are required for soft and anthropomorphic e-skin interfaces (Li et al., 2020). Prosthetic users may interact with other people more confidently with skin-like coverings for their prostheses (Chortos et al., 2016). To make electronics stretchable, intrinsically or extrinsically stretchable materials should be used.

Recently, there has been growing interest in the development of intrinsically stretchable materials using elastomers and polymers with proper mechanical and electronic characteristics. Elastomers such as PDMS, polyurethane, and poly(styrene-butadiene-styrene) are highly reliable and stretchable materials; however they show insulating properties. Therefore inorganic conductive fillers are integrated for the purpose of adding conductivity to elastomers (Li et al., 2020). To improve the stretchability and conductivity of materials by decreasing the perlocation threshold, several methods can be applied including the use of high-aspect ratio fillers and two-dimensional arrays of one-dimensional materials such as nanowires and carbon nanotubes (Chortos et al., 2016). As a second intrinsically stretchable material, conductive and stretchable polymers with conjugated structures can be used in e-skin applications. By modifying the chemistry of polymers, the desired conductivity can be achieved. On the other hand, by including additive materials such as soft polymers and ionic liquid enhancers, or side chain engineering, the desired stretchability of conductive polymers can be achieved (Li et al., 2020).

Alternatively, extrinsically stretchable materials can be fabricated for prosthetic e-skins to mimic the stretchability of biological skin. In this approach, artificially designed patterns, with stretchable interconnections between inorganic, stiff, and conductive sensory devices, are created. Stretchable interconnection patterns include, but are not limited to, helical structures made from metals with high elastic modulus such as gold and copper, thin and narrow filamentary serpentine structures, buckled serpentine structures, and wrinkled flexible thin-film conductors generated by buckling (Li et al., 2020). It should be noted that a trade-off exists between device density and stretchability in extrinsically stretchable designs. While stretchability depends on the percentage surface area of stretchable interconnections, device density depends on the number of high-performance stiff sensory devices (Chortos et al., 2016).

Since power supplies and batteries are rigid and bulky, they may negatively affect advancements in the development of flexible, wearable, portable, lightweight, and long-term sustainable e-skins. Therefore energy autonomy or the ability to self-power is a desirable feature for the sensory devices used in e-skin applications. Recently, various self-powered e-skins have been developed, including an electronic skin that consists of a transparent tactile sensitive layer based on graphene, and a photovoltaic cell underneath for light harvesting (Núñez et al., 2017), a user-interactive electronic skin based on a triboelectric-optical model (Zhao et al., 2020), a coaxial piezoelectric fiber based electronic skin (Zhu et al., 2020a), and a hybrid electronic skin combining triboelectric and piezoelectric effects (Zhu et al., 2020b).

Biological skin is capable of healing a wound or cut autonomously also. To reproduce skin's self-healing ability, electrically conductive self-healing materials, which can recover the mechanical properties and maintain conductivity under extreme conditions, are required for e-skin applications (Lee et al., 2020). Recently, researchers have developed soft self-healing devices such as self-healable conductors based on poly(vinylidene fluoride-co-hexafluoropropylene) [P(VDF-HFP)] and ionic liquid, electronically self-healable circuits based on a liquid metal-embedded elastomer matrix, and electromechanically self-healable systems based on carbon nanotube-embedded polymers (Li et al., 2020). Furthermore, transient or biodegradable electronics, which are important for an environmentally friendly and sustainable future, have been recently introduced. Researchers have developed transient electronics based on inorganic materials such as Si, SiO_2, Mg, and MgO that can dissolve in water, and biodegradable sensor systems based on organic materials including polylactic-co-glycolic acid, poly(glycerol sebacate), and polylactide (Li et al., 2020).

18.3.2 Applications in upper limb prostheses

In this section, applications (or possible future applications) of prosthetic sensors and e-skins in soft and anthropomorphic upper limb prostheses are reviewed. These show promising functional results such as object grasping and slip detection, object recognition and texture discrimination, reproducing thermal, nociceptive and proprioceptive sensations, and also integration of multifunctional sensations. Selected upper limb prosthetic sensors or e-skin designs and their functions are shown in Fig. 18.3.

Reproducing tactile sensations is crucial for various manipulation-related functions including object grasping and slip detection in upper limb prostheses. In Beccai et al. (2008), it was demonstrated that a soft and high shear sensitive 1.4 mm^3 triaxial force microsensor on an experimental setup is sensitive enough for slip event detection and can be applicable to anthropomorphic prosthetic hands. Gerratt et al. (2015) developed a sensorized glove based on an elastomeric e-skin with resistive strain and bending sensors, and also capacitive tactile pressure sensors. Human-in-the-loop experiments indicated precise and controlled grasping of objects, which is promising for hand prosthesis applications. In Rosenbaum-Chou et al. (2016), a mapping algorithm between a force-resistive sensor (FSR) placed on the prosthetic thumb and feedback tactor vibration was optimized. Clinical tests on prosthetic hand users showed that tactile feedback improves the grasp force accuracy. Additionally, flexible tactile FSRs were placed on the fingertips of an Ottobock BeBionic3 hand (Fig. 18.3A) and neuromimetic tactile feedback was provided to the prosthetic hand. Experimental results demonstrated that amputee subjects were able to perform compliant grasping and slip prevention tasks successfully (Osborn et al., 2016). More recently, in Zollo et al. (2019), force-sensing resistors to measure grasp forces and to detect slippage were integrated on two commercial biomechatronic hands. A combination of electrodes was implanted in an amputee to provide neural feedback. The results indicated that the participant performed fine grasp and manipulation tasks successfully.

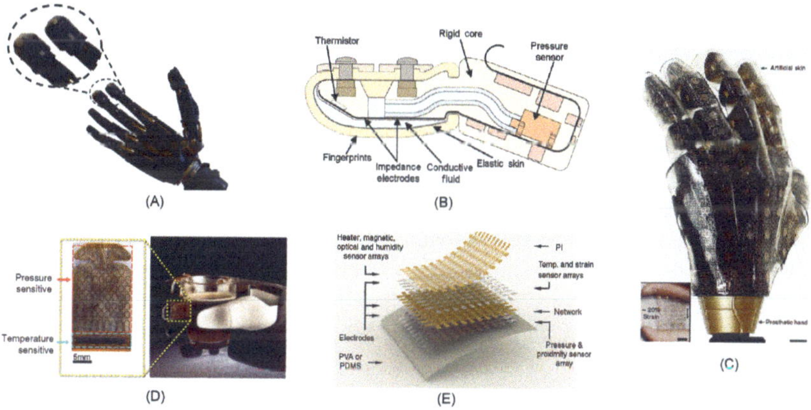

FIGURE 18.3 Upper limb prosthetic sensors and e-skins. (A) FSRs placed onto fingertips of a prosthetic hand (Osborn et al., 2016). (B) Schematic layout of a BioTac sensor. (C) Multifunctional e-skin. (D) Asynchronously coded electronic skin (ACES) to obtain thermotactile information (Lee et al., 2019). (E) Schematic layout of the multifunctional e-skin based on stretchable and conformable matrix network (SCMN). *(Figure 20.3A) Reprinted from Osborn, L. E., Iskarous, M.M., & Thakor, N.V. (2020).* Sensing and control for prosthetic hands in clinical and research applications. in wearable robotics *(pp. 445–468). Academic Press, with permission from Elsevier. (Figure 20.3B) Reprinted from Su, Z., Fishel, J. A., Yamamoto, T., & Loeb, G. E. (2012). Use of tactile feedback to control exploratory movements to characterize object compliance. Frontiers in* Neurorobotics, *6, 7 under the CC-BY license. (Figure 20.3C) Reprinted from Kim, J., Lee, M., Shim, H. J., Ghaffari, R., Cho, H. R., Son, D., Jung, Y. H., Soh, M., Choi, C., Jung, S., Chu, K., Jeon, D., Lee, S. T., Kim, J. H., Choi, S. H., Hyeon, T., & Kim, D. H. (2014). Stretchable silicon nanoribbon electronics for skin prosthesis. Nature Communications, 5(1), 1–11. under the CC-BY license. (Figure 20.3D) Reprinted from Li, P., Anwar Ali, H. P., Cheng, W., Yang, J., & Tee, B. C. (2020). Bioinspired prosthetic interfaces. Advanced Materials Technologies, 5(3), 1900856, with permission from Wiley-VCH. (Figure 20.3E) Reprinted from Hua, Q., Sun, J., Liu, H., Bao, R., Yu, R., Zhai, J., Pan, C., & Wang, Z. L. (2018). Skin-inspired highly stretchable and conformable matrix networks for multifunctional sensing. Nature Communications, 9(1), 1–11 under the CC-BY license.*

Object recognition and texture discrimination are also desirable functions for upper limb prosthetic users. In Datta et al. (2013), a capacitive pressure sensor based on MEMS technology was used to capture tactile information of surfaces of rigid objects with three different types and five different sizes. Shapes and sizes of the objects were classified with high accuracy with the use of four distinct classification techniques. Similarly, a flexible tactile-array sensor based on piezoresistive rubber was designed in Drimus et al. (2014). Experiments using a robotic gripper showed that the system successfully classifies objects using k-nearest neighbors classifier and dynamic time warping. Zhao et al. (2016) presented a soft prosthetic hand that uses stretchable optical waveguides as curvature, elongation, and force sensors. Various experiments demonstrated that the system was able to feel the shape and softness of objects. In Rasouli et al. (2018), a biomimetic tactile sensor array based on a piezoresistive material was used to measure normal and shear forces. Additionally, a neuromorphic model and an extreme learning chip were developed for the generation and analysis of spike patterns. The system was capable of classifying different textures in the experiments conducted by a robotic gripper. Oddo et al. (2016) used a MEMS-based sensor with four transducing piezoresistors wrapped with a flexible polymeric material for tactile sensing. After intraneural stimulation, an amputee was able to distinguish different textures with varying levels of coarseness.

Reproducing nociceptive feedback for upper limb prostheses is a newly emerging field. As a promising study, Osborn et al. (2018) proposed a multilayered electronic dermis (e-dermis) based on tactile sensors which were made of piezoresistive and conductive fabrics for pressure measurements. Measurements of the tactile sensors were converted to spike patterns by a neuron model, and the spiking output was used to distinguish painful and nonpainful stimuli. In the conducted experiments, an amputee was able to recognize the sharpness of an object, distinguish painful tactile stimuli, and release the sharp object reflexively with his prosthetic BeBionic hand.

Eliciting proprioceptive feedback is also a recently emerging field for upper limb prostheses. In D'Anna et al. (2019), position information was obtained from a commercial robotic hand (IH2 Azzurra, Prensilia) and tactile information was captured by the force sensors on it. By using sensory substitution methods based on intraneural stimulation, position feedback was provided simultaneously with somatotopic tactile feedback. In the experiments, amputees were able to identify the object size and compliance with high performance by means of multimodal sensory feedback. In another study, a self-powered e-skin based on coaxial piezoelectric fibers was presented. Characterization tests demonstrated that this flexible e-skin was capable of detecting and quantifying different motions of joints, and also distinguishing various shapes of objects (Zhu et al., 2020a). Although proprioceptive feedback was not provided in this study, fabricated e-skin is promising since proprioceptive sensing is possible through the use of flexible sensors in an energy autonomous way.

Recent advancements in prosthetic sensors and e-skin technologies have made possible the integration of multiple sensing capabilities to upper limb prostheses, in addition to mechanical tactile sensing ability. A hand prosthesis integrated with a commercial BioTac sensor (Fig. 18.3B), which is able to sense force, vibration, and temperature, was evaluated in Jimenez and Fishel (2014). Experimental results indicated that upper extremity prosthesis users were able to identify objects with different weight, surface characteristics, and temperature. Regarding e-skin applications, Kim et al. developed a prosthetic e-skin with integrated flexible, ultrathin, single crystalline silicon nanoribbon sensor arrays to measure strain, pressure, and temperature, and also humidity (Fig. 18.3C). The results of the characterization tests showed that the developed e-skin enables mechanical and thermal sensing abilities over prosthetic hands in the presence of external stimuli (Kim et al., 2014).

Bimodal piezoresistive and temperature sensors based on microstructured PDMS films were presented in Dos Santos et al. (2019). Experiments using a robotic arm indicated that these bimodal e-skin sensors performed robustly in object manipulation tasks with a stable sensitivity, and have a high temperature coefficient of resistance. Lee et al. presented the Asynchronously Coded Electronic Skin (ACES), a neuro-inspired architecture that is able to achieve simultaneous transmission of thermotactile information with low readout latencies (Fig. 18.3D). Slip detection, object classification, and also detection of thermal sensations, were achieved with the experiments conducted by a prosthetic hand (Lee et al., 2019). An e-skin based on a skin-inspired highly stretchable and conformable matrix network (SCMN) was introduced in Hua et al. (2018). It has multifunctional sensing abilities including strain, pressure, temperature, humidity, light, magnetic field, and proximity. A schematic layout of the e-skin is shown in Fig. 18.3E. Experiments with an intelligent prosthetic hand demonstrated that the developed system can achieve real-time spatial pressure mapping and temperature estimation successfully.

18.3.3 Applications in lower limb prostheses

Somatosensory feedback has not been widely investigated in lower limb prostheses as compared to upper limb prostheses. However, providing such feedback would be helpful for lower limb amputees also. In this section, applications (or possible future applications) of prosthetic sensors in soft and anthropomorphic lower limb prostheses that show promising functional results such as improvements in standing balance, gait symmetry, gait stability, identification of floor conditions, walking speed, metabolic cost, and reduction of phantom pain and cognitive load will be reviewed. Selected lower limb prosthetic sensors and their functions are shown in Fig. 18.4.

Providing vibrotactile feedback typically improves the standing balance, gait symmetry, and gait stability of lower limb prosthetic users. In Ma et al. (2015), six thin-film force sensors were attached to a pair of flat insoles and vibrotactile feedback was provided to healthy participants with reduced foot sensations by wearing five layers of socks. In the Romberg tests, center of pressure (CoP) movements were measured to evaluate body sway. The results showed that vibrotactile feedback decreases the CoP parameters, which is an indication of improved standing balance. In Plauché et al. (2016), 16 force-sensing resistors were integrated into a custom-made rubber insole. Vibrotactile feedback was provided to a healthy subject wearing a transfemoral (above-knee) leg prosthesis by using a bypass adapter. Experiments on a treadmill showed that vibrotactile feedback provides a sense of gait phase, and a decrease in the variance of gait parameters and trunk sway, which are indications of improved gait stability. In another study, pressure-sensitive sensors were attached to insoles to detect critical

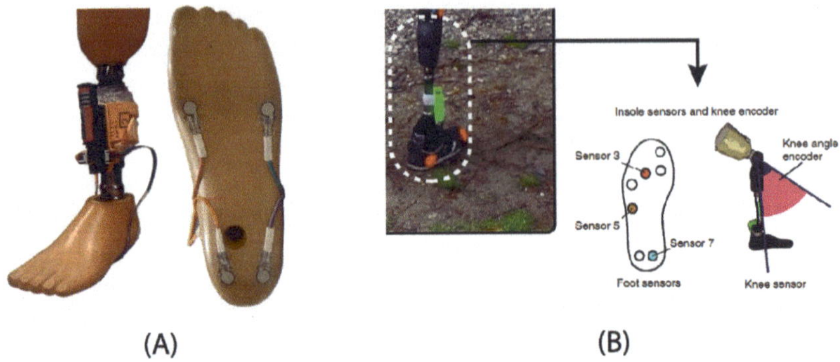

FIGURE 18.4 Lower limb prosthetic sensors. (A) Four thin-film force sensors attached to a prosthetic foot. (B) Pressure sensors and a knee encoder. *(Figure 20.4A) Reprinted from Wan, A. H., Wong, D. W., Ma, C. Z., Zhang, M., & Lee, W. C. (2016). Wearable vibrotactile biofeedback device allowing identification of different floor conditions for lower-limb amputees.* Archives of Physical Medicine and Rehabilitation, 97*(7), 1210–1213, with permission from Elsevier. (Figure 20.4B) Reprinted from Petrini, F. M., Bumbasirevic, M., Valle, G., Ilic, V., Mijović, P., Čvančara, P., Barberi, F., Katic, N., Bortolotti, D., Andreu, D., Lechler, K., Lesic, A., Mazic, S., Mijović, B., Guiraud, D., Stieglitz, T., Alexandersson, A., Micera, S., & Raspopovic, S. (2019a). Sensory feedback restoration in leg amputees improves walking speed, metabolic cost and phantom pain.* Nature Medicine, 25*(9), 1356–1363, with permission from Springer Nature.*

gait events, and vibrotactile feedback was provided to transfemoral lower limb prosthetic users. Experimental results demonstrated that the participants improved their gait symmetry with no increased cognitive load (Crea et al., 2017).

Vibrotactile feedback can improve floor identification accuracy of lower limb prosthesis users as well. In Wan et al. (2016), four thin-film force sensors were attached to a prosthetic foot (Fig. 18.4A), and vibrotactile feedback was provided to unilateral amputees. The subjects were able to identify different floor conditions created by placing different objects underneath a foam platform. Similarly, a haptic feedback system consisting of an insole with four force sensors located from to toe to heel, and a vibrotactile feedback mechanism were developed (Rokhmanova & Rombokas, 2019). In the conducted experiments, transtibial (below-knee) amputees using their own prostheses were capable of accurately perceiving the location of their prosthetic foot on stair steps.

Alternatively, electrocutaneous feedback methods can also be used to provide sensory feedback. In Dietrich et al. (2018), three pressure sensors were integrated under the prosthesis foot to detect ground contact, and electrocutaneous feedback was provided to lower limb amputees' thighs. After 2 weeks of training, participants reported increased satisfaction of their prostheses with increased functionality through longer walking distances, better gait stability, and better posture control. They also reported an overall decrease in phantom limb pain level.

Recently, neural feedback methods have been developed to elicit sensations more naturally for lower limb prostheses. Petrini et al. integrated a sensorized insole, which consists of seven pressure sensors, under the sole of custom-made transfemoral prosthesis built by commercially available components (Petrini et al., 2019a). Measurements of three pressure sensors and a knee encoder (Fig. 18.4B) were used for intraneural stimulation in order to reproduce sensations of ground contact and knee movement. Results of experiments on two transfemoral amputees indicated that the walking speed and self-confidence of participants increased, while physical and mental fatigue decreased. They also reported a decrease in phantom limb pain intensity. In another study conducted by Petrini et al., experiments on three transfemoral amputees using the same system demonstrated that eliciting tactile and proprioceptive sensations improved mobility, balance, agility, and embodiment of the prosthesis, and also reduced cognitive load (Petrini et al., 2019b). In Charkhkar et al. (2020), a shoe insole with an array of eight FSRs to measure pressure distribution was attached to a prosthetic foot. Moreover, sensory feedback was provided to two transtibial amputees by neural stimulation. Experimental results indicated a significant improvement of the standing balance of participants. In another study based on the same system, an ambulatory searching task, which involves walking tasks on a horizontal ladder blindfolded, was created. The results indicated that the walking speed and foot placement accuracy of three out of six transtibial amputee participants improved (Christie et al., 2020).

18.4 Outlook and future directions

Upper limb prosthetic technology is dominated by electric actuators due to the availability of compact actuation units and the feasibility of extensive control approaches. Prosthetic hands are generally underactuated, tendon-driven, compliant, and anthropomorphically shaped. These characteristics provide high dexterity to prosthetic hands, including robust, safe, stable, and shape-adaptive manipulation of objects. A large majority of hand prostheses incorporate simple hinge and gimbal type joints. Future directions might include the development of prosthetic hands that incorporate bio-inspired saddle or ellipsoidal joints to allow more ROM to prosthetic fingers, thus improving dexterity. Additionally, technological developments in additive manufacturing technologies and materials might lead to more human-like and affordable hand prostheses. The development of compact pneumatic power supplies in parallel to developments in the design and fabrication of soft pneumatic actuators may also be a future development. Although artificial muscles have several issues including control difficulties, poor time response, and efficiency, the development and implementation of novel active materials such as dielectric elastomers, or hybrid approaches, can be considered in the future to enable lightweight soft and anthropomorphic prosthetic hands that fully reproduce the functionality of the biological hand.

Passive, quasipassive, and active ankle–foot and knee joints are used in lower limb prostheses. Reproducing joint compliance is essential in terms of providing natural gait patterns and safety through energy storage/release and shock absorption mechanisms. Although lower limb prosthetic technology is dominated by mechanical and microprocessor-controlled systems, the use of active joints is inevitable in regards to improving energetic performance, reducing metabolic cost, and reproducing full functionality of the biological joints independent of terrain. Therefore the development of energy-efficient, lightweight, and optimized actuation systems and power sources is critical. Future directions include the design of novel, efficient, series elastic, or VSA that are capable of providing adequate power and safety, while reducing the metabolic cost to the prosthetic users in different gait phases and ground conditions. These systems are also required to be affordable, lightweight, and simple structured as much as possible for the ease of maintenance and reaching a large population of lower limb amputees.

Soft and anthropomorphic prosthetic technology has reached a satisfactory maturity level; however, providing somatosensory feedback is an essential research area which is in need of improvement. Several prosthetic sensors can be integrated to prostheses. Piezoresistive, piezoelectric, capacitive, pressure, optical, and magnetic tactile sensors are used to measure local force, pressure, and vibration. Additionally, thermal sensors based on thermistors to measure temperature and IMUs to obtain proprioceptive information can be used in prosthetics. Tactile and thermal sensors can be used also to provide nociceptive information for reflex action in case of painful stimuli.

By combining multiple prosthetic sensors and compliant electronic components into a multilayered and flexible form, the development of biological skin-inspired e-skins is possible with the continuous advancements in materials technology. Apart from multifunctional sensing abilities, flexibility, stretchability, energy autonomy, self-healing, and biodegradability properties can be integrated into e-skins. Although several studies regarding the development of e-skins show promising results, there remains a long way to reproduce the full functionality of the biological skin. Future directions include the development of flexible, robust, and high-density sensors, as well as lightweight, conductive, and stretchable electronics using advanced manufacturing techniques and nanotechnology that enable large-area fabrication of the prosthetic interfaces with low cost. Self-healing, biodegradability, and energy autonomy are other desirable features that may be applied to prosthetic interfaces in the future. Although they are not mentioned in this chapter (see other chapters in this book), future developments are also required for artificial biosignal interfaces to encode and transmit signals, and implantable neural interfaces to communicate with the brain in order to complete the sensory feedback loop.

Several prosthetic sensor and e-skin applications are proposed for upper and lower limb prostheses. Although promising functional results such as object grasping and slip detection, object recognition and texture discrimination, and thermal sensing exist for prosthetic hands, nociceptive and proprioceptive sensing are not greatly investigated in the literature. Future directions for prosthetic hands include providing nociceptive feedback with the use of tactile or thermal sensors, providing proprioceptive feedback with the use of IMUs or stretch sensors, and reproducing multifunctional sensing abilities in parallel with the advancements in artificial biosignal interfaces and neural interfaces. Very few examples of research have been conducted on somatosensation in lower limb prostheses; however, promising results were obtained, such as improvements in gait parameters and balance, floor identification, and reduction of phantom pain with the use of tactile sensors. Due to the advancements in e-skin technologies and neuroengineering, biological foot-like, robust, and multifunctional sensing abilities are expected to be provided to lower limb prosthetic users in the future.

References

Arjun, A., Saharan, L., & Tadesse, Y. (2016, August). Design of a 3D printed hand prosthesis actuated by nylon 6-6 polymer based artificial muscles. In *2016 IEEE International Conference on Automation Science and Engineering (CASE)* (pp. 910–915). IEEE.

Atasoy, A., Kaya, E., Toptas, E., Kuchimov, S., Kaplanoglu, E., & Ozkan, M. (2016, August). 24 DOF EMG controlled hybrid actuated prosthetic hand. In *2016 38th Annual International Conference of the IEEE Engineering in Medicine and Biology Society (EMBC)* (pp. 5059–5062). IEEE.

Atasoy, A., Toptaş, E., Kuchimov, S., Gulfize, S., Turpçu, M., Kaplanoglu, E., Guclu, B., & Özkan, M. (2018, August). Biomechanical design of an anthropomorphic prosthetic hand. In *2018 7th IEEE International Conference on Biomedical Robotics and Biomechatronics (Biorob)* (pp. 732–736). IEEE.

Au, S. K., Weber, J., & Herr, H. (2009). Powered ankle—foot prosthesis improves walking metabolic economy. *IEEE Transactions on Robotics*, *25*(1), 51–66.

Azocar, A. F., Mooney, L. M., Duval, J. F., Simon, A. M., Hargrove, L. J., & Rouse, E. J. (2020). Design and clinical implementation of an open-source bionic leg. *Nature Biomedical Engineering*, 1–13.

Bae, G. Y., Han, J. T., Lee, G., Lee, S., Kim, S. W., Park, S., Kwon, J., Jung, S., & Cho, K. (2018). Pressure/temperature sensing bimodal electronic skin with stimulus discriminability and linear sensitivity. *Advanced Materials*, *30*(43), 1803388.

Beccai, L., Roccella, S., Ascari, L., Valdastri, P., Sieber, A., Carrozza, M. C., & Dario, P. (2008). Development and experimental analysis of a soft compliant tactile microsensor for anthropomorphic artificial hand. *IEEE/ASME Transactions on Mechatronics*, *13*(2), 158–168.

Belter, J. T., Segil, J. L., Dollar, A. M., & Weir, R. F. (2013). Mechanical design and performance specifications of anthropomorphic prosthetic hands: A review. *Journal of Rehabilitation Research & Development*, *50*(5).

Catalano, M. G., Grioli, G., Farnioli, E., Serio, A., Piazza, C., & Bicchi, A. (2014). Adaptive synergies for the design and control of the Pisa/IIT SoftHand. *The. International Journal of Robotics Research*, *33*(5), 768–782.

Charkhkar, H., Christie, B. P., & Triolo, R. J. (2020). Sensory neuroprosthesis improves postural stability during sensory organization test in lower-limb amputees. *Scientific Reports*, *10*(1), 1–13.

Chi, C., Sun, X., Xue, N., Li, T., & Liu, C. (2018). Recent progress in technologies for tactile sensors. *Sensors*, *18*(4), 948.

Chortos, A., Liu, J., & Bao, Z. (2016). Pursuing prosthetic electronic skin. *Nature Materials*, *15* (9), 937–950.

Christie, B. P., Charkhkar, H., Shell, C. E., Burant, C. J., Tyler, D. J., & Triolo, R. J. (2020). Ambulatory searching task reveals importance of somatosensation for lower-limb amputees. *Scientific Reports*, *10*(1), 1–11.

Convens, B., Dong, D., Furnémont, R., Verstraten, T., Cherelle, P., Lefeber, D., & Vanderborght, B. (2019). Modeling, design and test-bench validation of a semi-active propulsive ankle prosthesis with a clutched series elastic actuator. *IEEE Robotics and Automation Letters*, *4*(2), 1823–1830.

Crea, S., Edin, B. B., Knaepen, K., Meeusen, R., & Vitiello, N. (2017). Time-discrete vibrotactile feedback contributes to improved gait symmetry in patients with lower limb amputations: Case series. *Physical Therapy*, *97*(2), 198–207.

D'Anna, E., Valle, G., Mazzoni, A., Strauss, I., Iberite, F., Patton, J., Petrini, F. M., Raspopovic, S., Granata, G., Di Iorio, R., Controzzi, M., Cipriani, C., Stieglitz, T., Rossini, P. M., & Micera, S. (2019). A closed-loop hand prosthesis with simultaneous intraneural tactile and position feedback. *Science Robotics*, *4*(27).

Datta, S., Khasnobish, A., Konar, A., Tibarewala, D. N., & Janarthanan, R. (2013, December). Object shape and size recognition from tactile images. In *2013 International Conference on Control Communication and Computing (ICCC)* (pp. 16–21). IEEE.

Deimel, R., & Brock, O. (2016). A novel type of compliant and underactuated robotic hand for dexterous grasping. *The International Journal of Robotics Research*, *35*(1-3), 161–185.

Dietrich, C., Nehrdich, S., Seifert, S., Blume, K. R., Miltner, W. H., Hofmann, G. O., & Weiss, T. (2018). Leg prosthesis with somatosensory feedback reduces phantom limb pain and increases functionality. *Frontiers in Neurology*, *9*, 270.

Dos Santos, A., Pinela, N., Alves, P., Santos, R., Farinha, R., Fortunato, E., Martins, R., Águas, H., & Igreja, R. (2019). E-skin bimodal sensors for robotics and prosthesis using PDMS molds engraved by laser. *Sensors*, *19*(4), 899.

Drimus, A., Kootstra, G., Bilberg, A., & Kragic, D. (2014). Design of a flexible tactile sensor for classification of rigid and deformable objects. *Robotics and Autonomous Systems*, *62*(1), 3–15.

Fajardo, J., Ferman, V., Cardona, D., Maldonado, G., Lemus, A., & Rohmer, E. (2020). Galileo hand: An anthropomorphic and affordable upper-limb prosthesis. *IEEE Access*, *8*, 81365–81377.

Fite, K. B. (2017). *Overview of the components used in active and passive lower-limb prosthetic devices. Full stride* (pp. 55–74). New York: Springer.

Flynn, L., Geeroms, J., Jimenez-Fabian, R., Heins, S., Vanderborght, B., Munih, M., Lova, R. M., Vitiello, N., & Lefeber, D. (2018). The challenges and achievements of experimental implementation of an active transfemoral prosthesis based on biological quasi-stiffness: the CYBERLEGs beta-prosthesis. *Frontiers in Neurorobotics*, *12*, 80.

Gao, F., Liu, Y., & Liao, W. H. (2019). Implementation and testing of ankle-foot prosthesis with a new compensated controller. *IEEE/ASME Transactions on Mechatronics*, *24*(4), 1775–1784.

Gerratt, A. P., Michaud, H. O., & Lacour, S. P. (2015). Elastomeric electronic skin for prosthetic tactile sensation. *Advanced Functional Materials*, *25*(15), 2287–2295.

Glanzer, E. M., & Adamczyk, P. G. (2018). Design and validation of a semi-active variable stiffness foot prosthesis. *IEEE Transactions on Neural Systems and Rehabilitation Engineering*, *26*(12), 2351–2359.

Godfrey, S. B., Zhao, K. D., Theuer, A., Catalano, M. G., Bianchi, M., Breighner, R., Bhaskaran, D., Lennon, R., Grioli, G., Santello, M., Bicchi, A., & Andrews, K. (2018). The SoftHand Pro: Functional evaluation of a novel, flexible, and robust myoelectric prosthesis. *PloS One*, *13*(10), e0205653.

Hitt, J. K., Sugar, T. G., Holgate, M., & Bellman, R. (2010). An active foot-ankle prosthesis with biomechanical energy regeneration. *Journal of Medical Devices*, *4*(1).

Hua, Q., Sun, J., Liu, H., Bao, R., Yu, R., Zhai, J., Pan, C., & Wang, Z. L. (2018). Skin-inspired highly stretchable and conformable matrix networks for multifunctional sensing. *Nature Communications*, *9*(1), 1–11.

Iskarous, M. M., & Thakor, N. V. (2019). E-Skins: Biomimetic sensing and encoding for upper limb prostheses. *Proceedings of the IEEE*, *107*(10), 2052–2064.

Jimenez, M. C., & Fishel, J. A. (2014, February). Evaluation of force, vibration and thermal tactile feedback in prosthetic limbs. In *2014 IEEE Haptics Symposium (HAPTICS)* (pp. 437–441). IEEE.

Jimenez-Fabian, R., Geeroms, J., Flynn, L., Vanderborght, B., & Lefeber, D. (2017). Reduction of the torque requirements of an active ankle prosthesis using a parallel spring. *Robotics and Autonomous Systems*, *92*, 187–196.

Kashef, S. R., Amini, S., & Akbarzadeh, A. (2020). Robotic hand: A review on linkage-driven finger mechanisms of prosthetic hands and evaluation of the performance criteria. *Mechanism and Machine Theory*, *145*, 103677.

Kim, J., Lee, M., Shim, H. J., Ghaffari, R., Cho, H. R., Son, D., Jung, Y. H., Soh, M., Choi, C., Jung, S., Chu, K., Jeon, D., Lee, S. T., Kim, J. H., Choi, S. H., Hyeon, T., & Kim, D. H. (2014). Stretchable silicon nanoribbon electronics for skin prosthesis. *Nature Communications*, *5*(1), 1–11.

Klute, G. K., Czerniecki, J., & Hannaford, B. (2000). Muscle-like pneumatic actuators for below-knee prostheses.

Kumar, A., Singla, M. L., Kumar, A., & Rajput, J. K. (2019). Fabrication and linearisation of conformable POMANI-Mn3O4 nanocomposite based thermistor for temperature monitoring applications in prosthetic gloves. *Sensors and Actuators A: Physical*, *285*, 588–598.

Laferrier, J. Z., & Gailey, R. (2010). Advances in lower-limb prosthetic technology. *Physical Medicine and Rehabilitation Clinics*, *21*(1), 87–110.

Lee, J. H., Okamoto, S., & Matsubara, S. (2012). Development of multi-fingered prosthetic hand using shape memory alloy type artificial muscle. *Computer Technology and Application*, *3* (7).

Lee, W. W., Tan, Y. J., Yao, H., Li, S., See, H. H., Hon, M., Ng, K. A., Xiong, B., Ho, J. S., & Tee, B. C. (2019). A neuro-inspired artificial peripheral nervous system for scalable electronic skins. *Science Robotics*, *4*(32), eaax2198.

Lee, Y., Park, J., Choe, A., Cho, S., Kim, J., & Ko, H. (2020). Mimicking human and biological skins for multifunctional skin electronics. *Advanced Functional Materials*, *30*(20), 1904523.

Li, K., Fang, Y., Zhou, Y., & Liu, H. (2017). Non-invasive stimulation-based tactile sensation for upper-extremity prosthesis: A review. *IEEE Sensors Journal*, *17*(9), 2625–2635.

Li, P., Anwar Ali, H. P., Cheng, W., Yang, J., & Tee, B. C. (2020). Bioinspired prosthetic interfaces. *Advanced Materials Technologies*, *5*(3), 1900856.

Ma, C. Z. H., Wan, A. H. P., Wong, D. W. C., Zheng, Y. P., & Lee, W. C. C. (2015). A vibrotactile and plantar force measurement-based biofeedback system: Paving the way towards wearable balance-improving devices. *Sensors*, *15*(12), 31709–31722.

Nemoto, Y., Ogawa, K., & Yoshikawa, M. (2018, July). F3hand: A five-fingered prosthetic hand driven with curved pneumatic artificial muscles. *2018 40th Annual International Conference of the IEEE Engineering in Medicine and Biology Society (EMBC)* (pp. 1668–1671). IEEE.

Nisal, K., Ruhunge, I., Subodha, J., Perera, C. J., & Lalitharatne, T. D. (2017, July). Design, implementation and performance validation of UOMPro artificial hand: Towards affordable hand prostheses. In *2017 39th Annual International Conference of the IEEE Engineering in Medicine and Biology Society (EMBC)* (pp. 909–912). IEEE.

Núñez, C. G., Navaraj, W. T., Polat, E. O., & Dahiya, R. (2017). Energy-autonomous, flexible, and transparent tactile skin. *Advanced Functional Materials*, *27*(18), 1606287.

Oddo, C. M., Raspopovic, S., Artoni, F., Mazzoni, A., Spigler, G., Petrini, F., Giambattistelli, F., Vecchio, F., Miraglia, F., Zollo, L., Di Pino, G., Camboni, D., Carrozza, M. C., Guglielmelli, E., Rossini, P. M., Faraguna, U., & Micera, S. (2016). Intraneural stimulation elicits discrimination of textural features by artificial fingertip in intact and amputee humans. *Elife*, *5*, e09148.

Osborn, L., Kaliki, R. R., Soares, A. B., & Thakor, N. V. (2016). Neuromimetic event-based detection for closed-loop tactile feedback control of upper limb prostheses. *IEEE Transactions on Haptics*, *9*(2), 196–206.

Osborn, L. E., Dragomir, A., Betthauser, J. L., Hunt, C. L., Nguyen, H. H., Kaliki, R. R., & Thakor, N. V. (2018). Prosthesis with neuromorphic multilayered e-dermis perceives touch and pain. *Science Robotics*, *3*(19).

Osborn, L. E., Iskarous, M. M., & Thakor, N. V. (2020). *Sensing and control for prosthetic hands in clinical and research applications. Wearable robotics* (pp. 445–468). Academic Press.

Østlie, K., Lesjø, I. M., Franklin, R. J., Garfelt, B., Skjeldal, O. H., & Magnus, P. (2012). Prosthesis rejection in acquired major upper-limb amputees: A population-based survey. *Disability and Rehabilitation: Assistive Technology*, *7*(4), 294–303.

Petrini, F. M., Bumbasirevic, M., Valle, G., Ilic, V., Mijović, P., Čvančara, P., Barberi, F., Katic, N., Bortolotti, D., Andreu, D., Lechler, K., Lesic, A., Mazic, S., Mijović, B., Guiraud, D., Stieglitz, T., Alexandersson, A., Micera, S., & Raspopovic, S. (2019a). Sensory feedback restoration in leg amputees improves walking speed, metabolic cost and phantom pain. *Nature Medicine*, *25*(9), 1356–1363.

Petrini, F. M., Valle, G., Bumbasirevic, M., Barberi, F., Bortolotti, D., Cvancara, P., Hiairrassary, A., Mijovic, P., Sverrisson, A. Ö., Pedrocchi, A., Divoux, J. L., Popovic, I., Lechler, K., Mijovic, B., Guiraud, D., Stieglitz, T., Alexandersson, A., Micera, S., Lesic, A., & Raspopovic, S. (2019b). Enhancing functional abilities and cognitive integration of the lower limb prosthesis. *Science Translational Medicine*, *11*(512), eaav8939.

Pfeifer, S. M. (2014). *Biomimetic stiffness for transfemoral prostheses* (Doctoral dissertation), ETH Zurich.

Plauché, A., Villarreal, D., & Gregg, R. D. (2016). A haptic feedback system for phase-based sensory restoration in above-knee prosthetic leg users. *IEEE Transactions on Haptics*, *9*(3), 421–426.

Rasouli, M., Chen, Y., Basu, A., Kukreja, S. L., & Thakor, N. V. (2018). An extreme learning machine-based neuromorphic tactile sensing system for texture recognition. *IEEE Transactions on Biomedical Circuits and Systems*, *12*(2), 313–325.

Rokhmanova, N., & Rombokas, E. (2019, June). Vibrotactile feedback improves foot placement perception on stairs for lower-limb prosthesis users. In *2019 IEEE 16th International Conference on Rehabilitation Robotics (ICORR)* (pp. 1215–1220). IEEE.

Rosenbaum-Chou, T., Daly, W., Austin, R., Chaubey, P., & Boone, D. A. (2016). Development and real world use of a vibratory haptic feedback system for upper-limb prosthetic users. *JPO: Journal of Prosthetics and Orthotics*, *28*(4), 136–144.

Rouse, E. J., Mooney, L. M., Martinez-Villalpando, E. C., & Herr, H. M. (2013, June). Clutchable series-elastic actuator: Design of a robotic knee prosthesis for minimum energy consumption. In *2013 IEEE 13th International Conference on Rehabilitation Robotics (ICORR)* (pp. 1–6). IEEE.

Saharan, L., & Tadesse, Y. (2016, April). Robotic hand with locking mechanism using TCP muscles for applications in prosthetic hand and humanoids. In *Bioinspiration, Biomimetics, and Bioreplication 2016* (Vol. 9797, p. 97970V). International Society for Optics and Photonics.

Shepherd, M. K., & Rouse, E. J. (2017). The VSPA foot: A quasi-passive ankle-foot prosthesis with continuously variable stiffness. *IEEE Transactions on Neural Systems and Rehabilitation Engineering*, *25*(12), 2375–2386.

Sinha, R., van den Heuvel, W. J., & Arokiasamy, P. (2011). Factors affecting quality of life in lower limb amputees. *Prosthetics and Orthotics International*, *35*(1), 90–96.

Su, Z., Fishel, J. A., Yamamoto, T., & Loeb, G. E. (2012). Use of tactile feedback to control exploratory movements to characterize object compliance. *Frontiers in Neurorobotics*, *6*, 7.

Szkopek, J., & Redlarski, G. (2019). Artificial-hand technology—current state of knowledge in designing and forecasting changes. *Applied Sciences*, *9*(19), 4090.

Terryn, S., Brancart, J., Lefeber, D., Van Assche, G., & Vanderborght, B. (2017). Self-healing soft pneumatic robots. *Science Robotics*, *2*(9).

Tian, M., Xiao, Y., Wang, X., Chen, J., & Zhao, W. (2017). *Design and experimental research of pneumatic soft humanoid robot hand, . Robot intelligence technology and applications* (4, pp. 469–478). Cham: Springer.

Tiwana, M. I., Redmond, S. J., & Lovell, N. H. (2012). A review of tactile sensing technologies with applications in biomedical engineering. *Sensors and Actuators A: Physical*, *179*, 17–31.

Torricelli, D., Gonzalez, J., Weckx, M., Jiménez-Fabián, R., Vanderborght, B., Sartori, M., Dosen, S., Farina, D., Lefeber, D., & Pons, J. L. (2016). Human-like compliant locomotion: state of the art of robotic implementations. *Bioinspiration & Biomimetics*, *11*(5), 051002.

Tryggvason, H., Starker, F., Lecomte, C., & Jonsdottir, F. (2020). Variable stiffness prosthetic foot based on rheology properties of shear thickening fluid. *Smart Materials and Structures*, *29*(9), 095008.

Versluys, R., Desomer, A., Lenaerts, G., Pareit, O., Vanderborght, B., Perre, G., Peeraer, L., & Lefeber, D. (2008). A biomechatronical transtibial prosthesis powered by pleated pneumatic artificial muscles. *International Journal of Modelling, Identification and Control*, *4*(4), 394–405.

Vertongen, J., Kamper, D., Smit, G., & Vallery, H. (2021). Mechanical aspects of robot hands, active hand orthoses and prostheses: A comparative review. *IEEE/ASME Transactions on Mechatronics*, *26*(2), 955–965.

Wan, A. H., Wong, D. W., Ma, C. Z., Zhang, M., & Lee, W. C. (2016). Wearable vibrotactile biofeedback device allowing identification of different floor conditions for lower-limb amputees. *Archives of Physical Medicine and Rehabilitation*, *97*(7), 1210–1213.

Wettels, N., Santos, V. J., Johansson, R. S., & Loeb, G. E. (2008). Biomimetic tactile sensor array. *Advanced Robotics*, *22*(8), 829–849.

Wijk, U., & Carlsson, I. (2015). Forearm amputees' views of prosthesis use and sensory feedback. *Journal of Hand Therapy*, *28*(3), 269–278.

Windrich, M., Grimmer, M., Christ, O., Rinderknecht, S., & Beckerle, P. (2016). Active lower limb prosthetics: A systematic review of design issues and solutions. *Biomedical engineering Online*, *15*(3), 140.

Wiste, T., & Goldfarb, M. (2017, May). Design of a simplified compliant anthropomorphic robot hand. In *2017 IEEE International Conference on Robotics and Automation (ICRA)* (pp. 3433–3438). IEEE.

Wu, J., Han, S., Yang, T., Li, Z., Wu, Z., Gui, X., Tao, K., Miao, J., Norford, L. K., Liu, C., & Huo, F. (2018). Highly stretchable and transparent thermistor based on self-healing double network hydrogel. *ACS Applied Materials & Interfaces*, *10*(22), 19097–19105.

Xu, Z., Bai, Y., Ni, R., Yang, N., Sun, Y., & Qi, P. (2018, November). Anthropomorphic soft pneumatic fingers towards full dexterity of human hand. In *2018 IEEE-RAS 18th International Conference on Humanoid Robots (Humanoids)* (pp. 381–386). IEEE.

Zhao, H., O'Brien, K., Li, S., & Shepherd, R. F. (2016). Optoelectronically innervated soft prosthetic hand via stretchable optical waveguides. *Science Robotics*, *1*(1), eaai7529.

Zhao, X., Zhang, Z., Liao, Q., Xun, X., Gao, F., Xu, L., Kang, Z., & Zhang, Y. (2020). Self-powered user-interactive electronic skin for programmable touch operation platform. *Science Advances*, *6*(28), eaba4294.

Zhu, M., Lou, M., Abdalla, I., Yu, J., Li, Z., & Ding, B. (2020a). Highly shape adaptive fiber based electronic skin for sensitive joint motion monitoring and tactile sensing. *Nano Energy*, *69*, 104429.

Zhu, M., Lou, M., Yu, J., Li, Z., & Ding, B. (2020b). Energy autonomous hybrid electronic skin with multi-modal sensing capabilities. *Nano Energy*, *78*, 105208.

Zollo, L., Di Pino, G., Ciancio, A. L., Ranieri, F., Cordella, F., Gentile, C., Noce, E., Romeo, R. A., Bellingegni, A. D., Vadalà, G., Miccinilli, S., Mioli, A., Diaz-Balzani, L., Bravi, M., Hoffmann, K. P., Schneider, A., Denaro, L., Davalli, A., Gruppioni, E., ... Guglielmelli, E. (2019). Restoring tactile sensations via neural interfaces for real-time force-and-slippage closed-loop control of bionic hands. *Science Robotics*, *4*(27).

Zou, L., Ge, C., Wang, Z. J., Cretu, E., & Li, X. (2017). Novel tactile sensor technology and smart tactile sensing systems: A review. *Sensors*, *17*(11), 2653.

Chapter 19

Prospect of data science and artificial intelligence for patient-specific neuroprostheses

Buse Buz Yalug, Dilek Betul Arslan and Esin Ozturk-Isik
Institute of Biomedical Engineering, Boğaziçi University, Istanbul, Turkey

ABSTRACT
Machine learning and its subfield deep learning have recently gained interest in scientific research community due to their ability to analyze and learn from big data. In this chapter, we discuss the capabilities, limitations, and current applications of unsupervised and supervised machine learning methods in addition to more recent deep learning techniques for the design and control of patient-specific neuroprostheses. Furthermore, we speculate on what they could promise for future applications.

Keywords: Artificial intelligence; machine learning; deep learning; patient-specific neuroprosthesis; big data

19.1 Introduction

Machine learning and its subfield deep learning have recently gained interest in the scientific research community due to their ability to analyze and learn from big data. Neuroprosthetic devices establish a direct or indirect communication link between the brain and an external device to alleviate various disabilities, and they highly benefit from the recent advances in robotics and machine learning. To understand the cortical processes controlling movement, speech, perception, and cognition, in addition to the mechanisms behind neurological injuries and disorders is an important goal for the development of smart neuroprosthetic devices, and machine learning algorithms have been aiding for this purpose.

Intelligent neuroprosthetic devices have the ability to discriminate the characteristics of an external input collected via interfaces for decision-making processes. The neuroprosthetic interfaces collect signals to extract useful information to decide on motor control commands, and to move the neuroprosthesis using closed-loop control systems based on sensory feedback

Somatosensory Feedback for Neuroprosthetics. DOI: https://doi.org/10.1016/B978-0-12-822828-9.00005-8

and movement goals. Several noninvasive technologies have been employed as neuroprosthetic interfaces to provide the necessary input, such as electromyography (EMG), electroencephalography (EEG), electro-oculography (EOG), and functional near-infrared spectroscopy (fNIRS). The neuroprosthetic interfaces based on EEG, EMG, and EOG measure electrical activity, whereas fNIRS is an optical imaging method that measures changes in hemoglobin concentrations within the brain. Additionally, invasive interfaces, such as intracortical electrodes, electrocorticography (ECoG) electrodes, and peripheral nerve electrodes, that are directly located on the tissue result in higher signal-to-noise for more precise control.

In this chapter, we discuss the capabilities, limitations, and current applications of classical machine learning methods in addition to more recent deep learning techniques for the design and control of patient-specific neuroprostheses, and speculate on what machine learning could promise for future smart neuroprosthetic applications. PubMed Central, which is a free digital repository of peer-reviewed biomedical literature, was used for a literature survey. Table 19.1 shows the structure of the literature search strings, which were developed by combining the names of several machine learning techniques, and one of the following keywords; "prosthetics," "neuroprosthesis," "neuroprosthetics," or "prosthesis control," using the Boolean operator "AND." The machine learning methods that were included in the literature search were "support vector machine," "k nearest neighbor," "logistic regression," "decision tree," "reinforcement learning," "artificial neural networks," "convolutional neural networks," "recurrent neural networks," "long short-

TABLE 19.1 The search strings used to find the primary studies related with the machine learning applications in the neuroprosthesis field.

ID	Search string
1	"<Machine learning method>" AND "Prosthetics"
2	"<Machine learning method>" AND "Neuroprosthesis"
3	"<Machine learning method>" AND "Neuroprosthetics"
4	"<Machine learning method>" AND "Prosthesis Control"
5	"Deep learning" AND "<Prosthetic keywords>"
6	"Machine learning" AND "<Prosthetic keywords>"
7	"Pattern recognition" AND "<Prosthetic keywords>"
8	"Deep learning" AND "Somatosensory"
9	"Machine learning" AND "Somatosensory"

term memory," "ensemble AND machine learning," and "generative adversarial networks." Additionally, "deep learning," "machine learning," and "pattern recognition" were combined with the prosthetic keywords and searched in PubMed. Moreover, "deep learning" and "machine learning" were combined with "somatosensory" to ensure the completeness of the literature survey. The following sections of this chapter include a brief description of the key machine learning algorithms employed in the neuroprothesis field, followed by a summary of the key studies that employed those machine-learning algorithms.

19.2 Classical machine learning methods for neuroprosthetic applications

Classical machine learning methods extract features from the input data to learn a relationship between the input and output. Three main categories of classical machine learning approaches are supervised, unsupervised, and reinforcement learning. Supervised machine learning methods employ labeled data to learn from available training samples, and later predict the output for a test input data. The main goal of supervised learning is to find a function in a multidimensional space of features that is able to categorize training data into known class labels. On the other hand, unsupervised machine learning is a self-learning technique in which unlabeled input data are transformed by discovering latent relationships or coherent dimensions in the features, and later used to define new class labels. Finally, reinforcement learning (RL) is learning by interacting with an environment. An RL agent learns from the results of its actions, which are not explicitly taught, and then it tries to select the best actions on the basis of its past experiences in addition to new choices. This section will summarize the probability theory and evaluation metrics for machine learning models, followed by the most commonly employed classical machine learning methods and their use in the field of neuroprosthetics.

19.2.1 Probability theory and evaluation metrics for machine learning models

19.2.1.1 Probability theory

Machine learning algorithms employ probability theory in their foundations. The probability of an event is a measure of the likelihood of it occurring in a random experiment, which is a number between 0 and 1, where 0 indicates impossibility and 1 indicates certainty. Additionally, the conditional probability is a measure of the probability of an event if another event has already occurred. If the event of interest is A and event B is known or assumed to occur, "the conditional probability of A given B" is denoted as P (A|B). The conditional probability helps improve the accuracy of machine learning models (Table 19.2).

TABLE 19.2 The conditional probability in machine learning.

	Machine learning	Statistics
Unsupervised learning	Aims to create a model of observed patterns	P (patterns)
Supervised learning	Aims to predict the output class (for classification) or output value (for regression) from given input patterns	P (output|input patterns)
Reinforcement learning	Aims to predict the next action that gives the maximum reward from given states and actions in an environment	P (next action| input patterns)

19.2.1.2 *Bias and variance*

The biases are simplifying assumptions made by a machine learning model to facilitate the learning of a target function. In general, linear algorithms have a high bias, which makes them quick and straightforward to learn, but generally less flexible. In contrast, they have lower prediction performance in complex problems that cannot meet the simplifying assumptions of the algorithm bias. Therefore, while low bias suggests less assumption about the form of a target function, high bias suggests more assumptions.

The variance is the amount by which the estimate of the target function changes when using different training data. The target function is predicted from the training data by a machine learning algorithm, so it is expected that the algorithm has some variance. Ideally, the training should not vary excessively from one data set to another, so the algorithm is capable of choosing the underlying relation between input and output variables. Machine learning algorithms with high variance are greatly affected by the characteristics of the training data, which means that the characteristics of the data affect the number and types of parameters used to characterize the mapping function. While low variance suggests small changes to the estimate of the target function with changes to the training set, high variance suggests the opposite. However, there is a trade-off between bias and variance. Increasing the bias decreases the variance. Therefore the goal of any machine learning algorithm is to find a balance between bias and variance.

19.2.1.3 *The evaluation metrics*

The classification accuracy is the ratio of the number of correct predictions to the total number of input samples. Table 19.3 lists all possible combinations of the predictions versus the ground truth for a binary classification task.

Here, the accuracy could be calculated by taking the ratio of the true predictions versus all entries as (TP + TN)/(TP + TN + FP + FN). The true

TABLE 19.3 The possible combinations of a prediction versus the ground truth.

Prediction/real class	Class 1	Class 0
Class 1	True positive (TP)	False positive (FP)
Class 0	False negative (FN)	True negative (TN)

positive rate (TPR), or sensitivity, is defined as TP/(TP + FN), and corresponds to the proportion of positive data points that are correctly classified as positive with respect to all the positive data points. On the other hand, the true negative rate, or specificity, is defined as TN/(TN + FP), which is the ratio of the correctly classified negative data points to all the negative data points.

Another metric called the false positive rate (FPR) is defined as FP/(FP + TN). FPR corresponds to the proportion of negative data points that are mistakenly considered as positive with respect to all the negative data points. Similarly, the false negative rate is defined as FN/(FN + TP), which corresponds to the proportion of positive data points that are mistakenly considered as negative with respect to all the positive data points.

The area under the curve (AUC) is used for binary classification problems. AUC is the area under the receiver operating characteristic curve, which is a plot of FPR versus TPR at different classification thresholds. The higher the AUC, the better the classification performance is at predicting the true classes. For multiclass problems, metrics such as recall, precision, and F1 score are used. Recall for a given class is the fraction of correct classifications within the elements actually belonging to that class. On the other hand, precision is the fraction of correct classifications within all elements classified as such. The F1 score is the harmonic mean of precision and recall.

The coefficient of determination, denoted as R^2, is the proportion of the variance in the fitted values that is predictable from the observed values. It is an absolute index of goodness-of-fit, ranging from 0 to 1, and can be used for model performance assessment or model comparison.

Mean absolute error is another performance measure, which is the average of the difference between the real and predicted values. On the other hand, mean squared error (MSE) takes the average of the square of the difference between the real and the predicted values. The advantage of MSE is the more pronounced effect of larger errors, so that the model could focus more on those to improve performance.

19.2.2 Feature selection techniques

In machine learning, feature selection techniques have been employed to choose a subset of the input variables that are most useful in predicting an outcome. The main objective of feature selection is to create simple and more comprehensive models for improving data mining performance and interpreting the results of the algorithms. Simple models are also more robust to noise and outliers. Therefore reducing the dimensionality of data to construct simpler models without model performance degradation is a useful approach.

The dimensionality reduction techniques may be divided into two categories, which are feature selection methods and feature extraction methods. In feature selection methods, the most informative features are selected by removing the features that are not relevant or are redundant. In feature extraction methods, a feature subset is generated by identifying the key features of the existing data. The wrapper methods and filter-based methods are the two kinds of feature selection approaches. The wrapper methods score feature subsets to measure the classifier performance, while filter-based methods score the correlation or dependence between the input variables to assess the predictive performance of the selected features. In wrapper methods, the performance of the model is assessed for each feature subset (Saeys et al., 2007). Then, the feature subset leading to the best model performance is selected. The forward feature selection is one of the simple wrapper methods (Caruana & Freitag, 1994). In this method, the feature subset is initially assigned empty, and the performance of each feature is calculated separately. The most informative feature is then selected and kept in the subset. The procedure is repeated for other features, and the feature subset is grown until the model performance stops improving. Another wrapper method is sequential backward selection (Rodriguez-Galiano et al., 2018). This method discards less informative features one by one, while evaluating the model performance. The elimination procedure continues until the model performance stops improving. Recursive feature elimination (RFE) is another wrapper-based feature selection technique (Zhang et al., 2006). In the first step of RFE, a model is constructed using all the features. Then, the scores of the features are ranked according to their contribution to the classification. The least informative feature is then removed from the set and the same procedure is repeated until a desired number of features are selected (Samb et al., 2012). It is also necessary to keep in mind that the feature subsets selected by wrapper methods depend on the model used in the calculation of the feature contributions.

In filter methods, feature selection is performed without using any classifiers (Sánchez-Maroño et al., 2007). Filter methods are faster and computationally less expensive than the wrapper methods. However, finding the most suitable features is less effective for a specific model in filter methods than the wrapper methods. Some statistical tests are used under filter methods to find the independent feature's relation with the dependent variable. Based on the result of the statistical analysis, the features are either kept or discarded.

19.2.3 Logistic regression

Logistic regression (LR) was developed by D. R. Cox in 1958 as a statistical analysis method (Cox, 1958). It is a supervised learning algorithm widely used in many fields, including the medical and social sciences, for classification problems. Since it is one of the simplest methods to understand and implement in machine learning, many practitioners try it first in various tasks. LR is based on a linear classification method, but it models the signal using a sigmoid (logistic) function to predict a categorical outcome. In this method, real-valued features extracted from the input neural data are multiplied by a weight and summed. The weighted sum of the real-valued features is then passed through a sigmoid function to map onto a probability function. A classification threshold is defined to map a LR value to a binary category.

In the literature, one study classified the EMG data of seven healthy subjects using a LR model for performing three different tasks including a random task, a force adjustment task, and a posture change task, and reported a 0.65 R-value between the predicted and real movements (Sekiya et al., 2019). There have also been some studies comparing the performances of LR and other machine learning algorithms. In one such study, EEG signals of left and right foot kinesthetic motor imageries were classified using LR and linear discriminant analysis (LDA), and LDA outperformed LR with a 70.28% accuracy (Tariq et al., 2020). Additionally, Marri and Swaminathan (2016) classified nonfatigue and fatigue conditions in dynamic contraction using k nearest neighbor (kNN) and LR based on EMG signals of 42 healthy subjects. While the accuracy of kNN classification was 88%, the LR algorithm resulted in 82% accuracy.

19.2.4 k-Nearest neighbor classifier

Fix and Hodges introduced the kNN classifier, which is a nonparametric classification method (Fix & Hodges, 1951). The kNN classifier is one of the simplest supervised machine learning algorithms to solve both classification and regression problems. It is a nonparametric method, because it does not assume a data distribution. The algorithm works on the principle that similar inputs result in similar outputs in a given system. kNN is an instance-based method, because it makes predictions for new data by using its similarities to the available training data. In this technique, the distances between the test data and all training data are calculated using a measure, such as Euclidean, Mahalanobis, or Hamming distances to evaluate the similarity between data points. In kNN, the test data are classified according to the majority votes of the closest training instances. The parameter "k" stands for the number of the closest training instances that would be used for the classification of the test data, which is chosen as an odd number to avoid ties. kNN could learn complex models fast without any assumptions in a robust manner. Additionally, there is no iterative learning phase in kNN, because there is no need to fit a model to the data. However, the kNN method has some drawbacks. The accuracy of the kNN algorithm greatly depends on the data quality. Moreover, kNN is computationally expensive, because it requires a large amount of memory to store all the instances to find an optimal "k" value.

kNN has been extensively employed in the literature for neuroprosthetic applications for classification of upper limb or hand-only movements. In one study, EMG signals of 11 different upper limb movements were acquired from 20 healthy human subjects (Abbaspour et al., 2020). Forty-four distinct EMG features were extracted, and their five different combinations were obtained, which were then classified using six algorithms, including LDA, kNN, decision tree (DT), maximum likelihood estimation, support vector machine (SVM), and multilayer perceptron (MLP). The accuracy of kNN was reported to be 93.95%, which outperformed other algorithms. Another study integrated EEG and EMG data to decode different upper limb movements for above-elbow amputees, and used quadratic discriminant analysis (QDA), kNN, and SVM for classification with QDA outperforming kNN (98.91% and 93.60% accuracies, respectively) (Aly et al., 2018). Additionally, Dewald et al. (2019) used an intramuscular electrode (IM) insertion instead of a surface EMG (sEMG) to predict intended joint movement direction during a single training session. Eight IM data sets were classified using kNN, resulting in 93.7% accuracy for forearm pronation/supination, 88.8% accuracy for wrist flexion/extension, and 63.1% accuracy for hand close/open. In another study, Geethanjali et al. (2009) worked on classification of six different movements of hand open/close, wrist flexion/extension, ulnar deviation, and radial deviation using kNN and neural network (NN) algorithms, and reported that the kNN classifier resulted in better accuracy (average 84.5%) than the NN classifier. Another study classified a few typical hand movements from the EMG signals with kNN after signal preprocessing using singular value decomposition and principal component analysis (PCA), and reported 86.71% accuracy (Iqbal et al., 2017). Moreover, EMG signals of wrist-motion directions, such as up, down, right, left, and the rest state were classified using QDA, LDA, and kNN algorithms, and kNN algorithm performed best with the highest accuracy of 84.9% (Kim et al., 2011).

On the other hand, Altan et al. (2019) employed machine learning techniques including DT, SVM, kNN, and ensemble technique in order to classify basic finger movements, and reported 99.9% accuracy with kNN, which outperformed other techniques. In another study, (Fattah, Iqbal, Zahin, Shahnaz, & Rosul, 2017) developed a system based on autocorrelation domain feature extraction for basic hand movement detection using sEMG signals, and reported 87.87% accuracy using kNN. New electrode configuration-based pattern recognition approaches were also proposed to discriminate hand motions and were compared to traditional sEMG, for which the kNN classifier outperformed the LDA classifier for the average accuracy for both techniques (Fang & Liu, 2014). Another study classified EMG signals of two haptic actions such as hand grasp and hand release for the control of a bionic hand using kNN and naive Bayesian pattern classification (NBPC) algorithms as feature extractors, and while the kNN-based model gave 91% accuracy, NBPC resulted in an accuracy of about 93% (Praveen et al., 2018). Additionally, (Marri and Swaminathan, 2016) developed a model to classify EMG signals of nonfatigue and fatigue conditions in dynamic contraction using kNN and LR, in which kNN classification accuracy was higher than that of LR (88% and 82% accuracies, respectively). Finally, kNN has also been used to classify EEG signals of left and right limb motor imagery movement with up to 90% accuracy (Bhaduri et al., 2016; Bose et al., 2016).

19.2.5 Support vector machines

SVMs, which were first introduced by Corinna Cortes and Vladimir Vapnik (Cortes & Vapnik, 1995), are learning algorithms primarily used for the classification of complex and high-dimensional data. SVMs use a supervised learning approach, which finds weight parameters to analytically solve a convex optimization problem. The main idea of this technique is to find an optimal hyperplane, which separates the data into different classes in a feature space. The closest data points to the hyperplane are called the support vectors, used for maximizing the margin, which are the distances between the hyperplane and support vectors. Fig. 19.1 depicts a decision surface example for an SVM classifier.

When the data points are not separable by a linear decision boundary, the input feature space could be transformed using kernel functions. The most commonly employed transformation functions include polynomial kernel, radial basis kernel, and hyperbolic tangent kernel. The kernel functions give

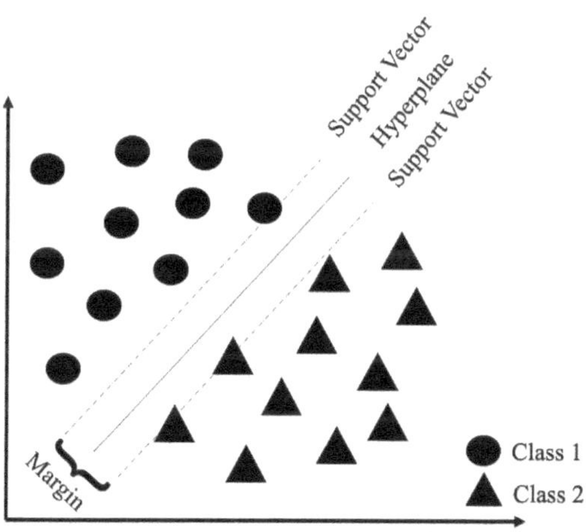

FIGURE 19.1 The illustration of a support vector machine (SVM) classifier.

the similarity degrees between the data. Then, the similarity values of the training data to new data, instead of the initial input attributes, are used as new features (Scholkopf & Smola, 2001).

Atzori et al. (2015) prepared a publicly available NinaPro database that hosted simultaneously recorded sEMG signals from the forearm and kinematics of the hand and wrist for 52 different hand movements, including 12 basic finger movements, eight isometric and isotonic hand configurations, nine basic wrist movements, and 23 grasping and functional movements to aid the design and control of hand neuroprostheses. For the NinaPro data set, a classification accuracy of 76% was achieved using a nonlinear SVM algorithm (Atzori et al., 2015). In another study, three amputees participated in training an SVM classifier by performing various grasping postures and forces with their phantom limbs, and the resultant SVM classifier achieved a 95% accuracy for feed-forward control of a dexterous prosthesis using EMG signals (Castellini et al., 2009). More recently, Abbaspour et al. (2020) classified 11 different hand movements (open or close hand, flex or extend hand, pronation, supination, side or fine grip, agree, pointer, and rest) based on sEMG signals using an SVM classifier for a maximum accuracy of 94.73% after applying PCA to 25 time domain features. Moreover, Ameri et al. (2014) applied SVM and artificial NNs (ANNs) to differentiate real-time wrist flexion–extension, abduction–adduction, and forearm pronation–supination movement signals acquired with EMG. According to their results, an SVM-based algorithm outperformed an ANN-based one with reduced processing time for both training and real-time control, resulting in an average of 6% offline estimation error. On the other hand, Bhattacharyya et al. (2014) employed an SVM algorithm to achieve over 95% accuracy in classifying 4-class motor imagery, and the presence of P300 and error-related potential waveforms for positional control of a robotic arm.

Using EEG as a neuroprosthetic interface, one study identified and characterized temporal EEG signal patterns of six healthy untrained subjects perceived through a finger pad using an SVM algorithm, and classified texture differences with more than %50 accuracy (Beckmann et al., 2009). Additionally, SVM was employed for classification of EEG signals of left- and right-hand movements with 87.5% accuracy, followed by further classification of each hand movement into elbow, finger, and shoulder movements with 80.1%, 93.3%, and 81.1% accuracies, respectively, for controlling an artificial arm (Khasnobish et al., 2011).

SVM was also employed to identify hand and finger poses with up to 97% accuracy based on vibration and shape changes in signals collected with an array of piezoelectric sensors affixed to the inside of an adjustable wrist strap (Booth & Goldsmith, 2018). Moreover, SVM was employed to classify hand movement intention with up to 84.6% accuracy in complete tetraplegia due to spinal cord injury (SCI) for activation of paralyzed muscles by functional electrical stimulation (FES) (Gant et al., 2018). In another study, blood oxygen level-dependent activation patterns obtained by functional magnetic resonance imaging (fMRI) from subjects with above-elbow arm amputation and nonamputated controls were classified using a multivoxel pattern analysis approach with an SVM classifier (Bruurmijn et al., 2017). While the classification scores of controls ranged from 23% to 91%, the accuracy of intact hand scores of amputees ranged between 35% and 96%, and the classification scores of gestures were significantly above the chance level (94%, $P < .001$) with a minimum score of 79%.

For the control of lower limb neuroprosthetics, one study proposed an SVM-based movement identification system that was able to classify five walking modes (normal walking, slow walking, fast walking, ramp ascending, and ramp descending) with 97.5% accuracy (Hussain et al., 2020). Another study designed a mechanomyography (MMG)-based gait classifier for improved control of prosthetic legs, and achieved up to 94% accuracy with an SVM algorithm (Needham et al., 2018). SVM was also employed to differentiate 11 different gait movements (standing, sitting, squatting standing to sitting, sitting to standing, standing to squatting, squatting to standing, walking, going downstairs, going upstairs, and running) based on MMG signals of four thigh muscles (rectus femoris, vastus lateralis, vastus medialis, and semitendinosus) and the attitude angle of rectus femoris resulting in an accuracy of 97.1% (Yu et al., 2020). In another study, accelerometer and gyroscope data acquired from an inertial control unit were classified using DT, SVM, and kNN to recognize the posture and activity of the patient before extending an autonomous tumor prosthesis (Kocaoglu & Akdogan, 2020). The most successful classification was achieved using SVM with 83.3% accuracy, while DT and kNN resulted in accuracies of 75.8% and 81.7%, respectively.

Recently, sensory–motor integrated brain computer interface (BCI) systems have been developed to create a sense of ownership of artificial devices. In one study, SVM was employed to decode the behavioral state of finger tapping vs. rest from the neural signals measured by a NIRS system that could be used to reduce the latency in an NIRS-based BCI (Cui et al., 2010). In another study, ECoG signals were classified with SVM to predict the somatosensory properties of a tactile stimulus as well as the movement trajectory and type for reaching, grasping/releasing, and pressure stimulation tasks with an accuracy of >80% (Ryun, Kim, Lee, & Chung, 2018). Finally, some neuroprosthesis studies in the literature were carried out using primates. One of these studies classified somatosensory evoked potentials (SEPs) of a monkey, which were acquired during electrical finger stimulation by ECoG, and the stimulated fingers and intensities were predicted with an accuracy of higher than 95% using SVM (Kaiju et al., 2017).

19.2.6 Decision trees

DT algorithms were first introduced by Leo Breiman for classification or regression problems (Breiman et al., 1984). DT is a nonparametric supervised learning method mainly used in classification tasks for both discrete and continuous variables. In DT, the data are split into different branches based on different conditions. In the structure of a DT, internal nodes represent the features of a data set, branches represent the decision rules, and each terminal node represents an outcome variable. At the beginning, the whole training set is considered as the root of the tree. In internal nodes, data are discriminated in accordance with a splitting criterion. The number of terminal nodes depends on the possible outcomes. Terminal nodes have discrete information in case of classification, while they have numerical values for regression tasks. If all the data belong to one class after a split, that split is described as pure, and no further splits are needed. There are different impurity measures used in DT, such as entropy function, Gini index, and misclassification error (Kotsiantis, 2013). Fig. 19.2 shows a schematic structure of a DT.

In the literature, one study developed a DT-based model to classify real-time EMG signals of hand and wrist movements, and reported 79% accuracy without requiring a training phase (Gibson et al., 2013). Another study compared a DT-based NN tree and DT for classification of the EMG signals of 11 hand movements, and reported average accuracy values of 89% and 85% for the DT-based NN tree and DT, respectively (Suchodolski & Wolczowski, 2010). Additionally, (Wolczowski and Kurzynski, 2015) worked on identification of the same hand movements using three different interfaces based on EMG, MMG, and the combination of EMG and MMG. MMG provided complementary information about the muscle contractions to interpret the motion intent (intentional manipulation or grasp). Also, a vibrational feedback system was added to the prosthetics to give patient feedback about the movement of

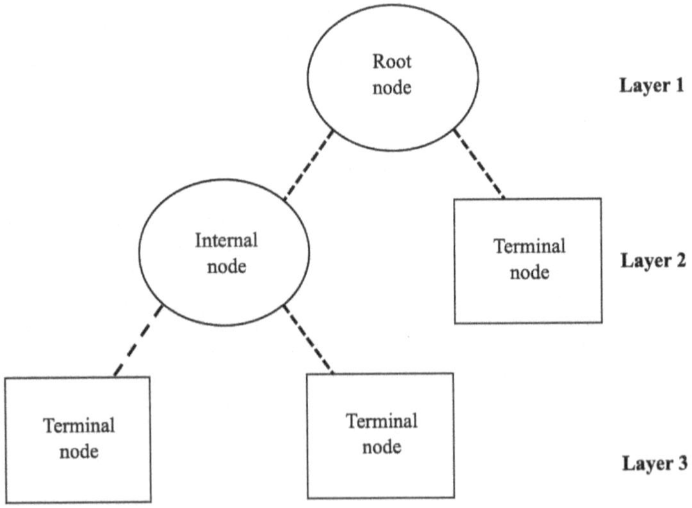

FIGURE 19.2 An example decision tree with three layers.

the prosthetic fingers and applied pressure on a grasped object. When the signals of the three interfaces were classified using a DT, the highest accuracy was obtained for the combination of EMG and MMG. Also, the DT algoritm was more successful in classifying the biosignals collected in the presence of the feedback mechanism. In another study, Liu et al. (2016) developed a terrain recognition module for an artificial leg to identify the type of terrain in front of the wearer during walking, and employed DT algorithms to classify signals of an inertial measurement unit (IMU) and laser distance meter with higher than 98% accuracy.

19.2.7 Ensemble methods

Ensemble models are employed to combine the decisions of multiple machine learning models to improve the overall prediction performance and generalizability, and to reduce bias and variance (Dietterich, 2000; Hansen & Salamon, 1990). In ensemble models, several machine learning algorithms are used together in various manners to provide a higher accuracy. An ensemble constitutes a number of learners, which are called "base learners" (Zhou, 2009). It is recommended to include diverse base learners having different algorithms, various hyperparameters, distinct feature sets, or different training sets, because the performance of each algorithm depends on the assumptions about the distribution of the data. Not only combining different machine learning algorithms, but also having different hyperparameters in a given algorithm may increase the overall accuracy. Moreover, training different models with different feature sets could be more beneficial.

In addition to choosing base learners with different properties, how to combine these base learners is another issue. Common types of ensemble methods are bagging (Breiman, 1996), boosting (Freund & Schapire, 1996), and random forests (Breiman, 2001). Bagging creates several subsets of data chosen randomly with replacement from the training data. Each data subset is used to train its several DTs. Then, the average of all the predictions from different trees is used instead of relying on one DT, which increases the accuracy while reducing the variance. Another advantage of this technique is elimination of overfitting. On the other hand, the boosting technique builds several weak base learners in a sequential manner, and then combines these models to create one strong algorithm. The goal of this technique is to improve the accuracy of consecutive trees using the prior trees. If an input is misclassified by a weak learner (tree), its weight is increased so the next learner classifies it correctly, which provides a better performing model. The third option, random forest ensemble, could be considered as a natural extension of bagging, which combines randomly diverse trees to reduce variance. The random forest gets the prediction from each tree and calculates the majority votes of predictions. When the number of trees in the forest increases, the overfitting problem could be prevented to achieve a higher accuracy. Although ensemble classifiers are not very easy to interpret, they are powerful models in order to handle a complex classification task by dividing it into simpler models and then combining their results.

In previous studies, ensemble learning methods have been evaluated to improve the pattern detection performance in neuroprosthetics. SEPs have been widely used to evaluate the integrity of spinal cord function by measuring the amplitude and latency of the primary waveform in response to external stimulation of peripheral nerves. In one study, SEPs recorded from 36 adult Sprague-Dawley rats having different SCI locations were classified with an overall accuracy of 84.72% using random forest method (Cui et al., 2019). Another study used an electromyography-based ensemble method to classify the grasping intention at an early stage of reach-to-grasp motion with a 90% accuracy using superficial electrodes and a Cyberglove in an upper limb exoskeleton (Accogli et al., 2017). In another study, an IMU signal set including 90 degrees spin turns, 90 degrees step turns, 180 degrees turns, and straight walking were acquired while using a neuroprosthesis (Pew & Klute, 2018). The prediction accuracies of turning were 96%, 93%, and 91% for SVM, kNN, and bagged DT ensemble, respectively. On the other hand, the accuracy of bagged DT ensemble was higher than the other two classification techniques for the prediction of straight walking. More recently, Kursun and Patooghy (2020) developed embedded tactile perception systems based on feature extraction and ensemble classification technique to detect patterns of interest in real time. A linear SVM, SVM with a radial basis kernel, kNN, LR, MLP, and random forest techniques were compared for their classification performances. As a result, random forests resulted in the highest accuracy even with a low number of trees in the model. Also, this classification technique was more suitable due to the data size and memory limitations in real-time processing.

19.2.8 Reinforcement learning

Reinforcement learning is a machine learning application that relies on an agent, which takes an action to obtain the maximum reward that is attainable for a particular situation (Sutton & Barto, 1998). RL differs from supervised learning, because there is no specific output for the agent, and the algorithm learns the actions to achieve a maximum reward (Fig. 19.3). Instead of known input/output pairs, the agent is immediately informed of the reward and the subsequent status after selecting an action, but not about which action would be the best for the long-term interests. The agent needs to actively gain experience in order to act in the most beneficial way out of possible system situations, transitions, and rewards. RL is also different from unsupervised learning, because RL basically tries to maximize a reward signal instead of finding a hidden structure (Sutton & Barto, 1998).

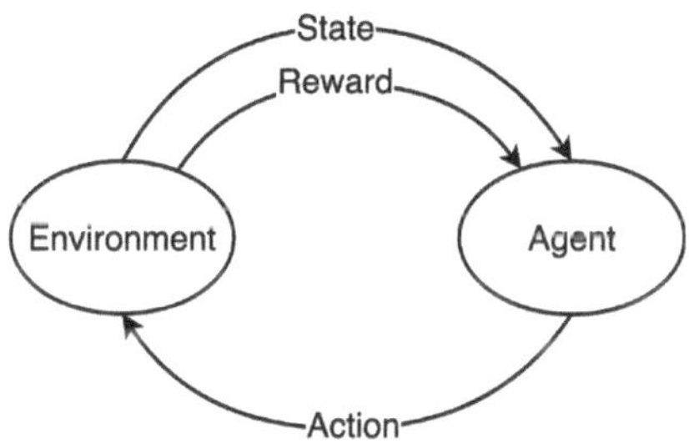

FIGURE 19.3 A schematic depiction of reinforcement learning.

RL has been used in neuroprosthetic applications. An RL-based application could generate a set of action mappings for given neural inputs. The training would then be guided by rewarding all targeted initiatives of the subjects. Based on the observation of the state of the neuroprosthesis according to the target, the agent must actively try new neural firings, which would then result in the final state for the reward. If the decoding trajectory approaches the target, the corresponding state-action mapping would be updated and strengthened. Otherwise, relevant neural action attempts would be punished. In the literature, RL has been mostly used either for control of artificial limb movements or in motor imagery task-related systems. Sanchez et al. (2011) developed an experiment for a center-out reaching task to test the performance of RL-based decoders for brain–machine interfaces (BMIs). In their work, 100% success was achieved, which is on par with a standard BMI classifier. However, no a priori information was necessary to train the system. Learning was achieved through interaction with the environment and the acquisition of rewards that were achieved through the production of brain states. Additionally, Di Febbo et al. (2018) proposed a nonlinear controller for an upper limb FES system using RL. The performance of the RL controller was compared with a proportional–integral–derivative (PID) controller during planar reaching tasks, and they reported that the RL control significantly outperformed the PID in terms of setting time, position accuracy, and smoothness. Dura-Bernal et al. (2013) developed a model of sensory and motor cortex, which used RL to drive a simple kinematic two-joint virtual arm in a reaching-to-target task, and reported that their model could be incorporated into a neural decoder for neuroprosthetic control. In their later studies, the same group trained a model using spike timing-dependent reinforcement learning to drive a virtual robot arm in a 2D reaching task in real time (Dura-Bernal et al., 2015). After training, the virtual arm performed reaching movements in a smoother and more realistic manner than those obtained using a simplistic arm model. Moreover, Jagodnik et al. (2017) proposed an RL algorithm, where the rewards were generated by an algorithm to control goal-oriented reaching tasks using a planar musculoskeletal human arm simulation. The authors reported 98.26% success for the human-trained controllers and 99.20% success for the pseudo-human-trained controllers, which indicated that both human-generated and pseudo-human rewards yielded significantly improved success when compared with the performance of RL controllers trained using automated rewards. In another study, Li et al. (2019) developed a controller that could automatically provide personalized control of a robotic knee prosthesis in order to best support gait of individual prosthesis wearers. The authors used an offline policy iteration-based reinforcement learning approach, and achieved 2 degrees of error at the end of movement. Moreover, Zhang et al. (2019) proposed a clustering-based kernel RL algorithm to classify brain signals that were acquired during a brain control task. This proposed algorithm improved the sensitivity for detection of the newly appearing brain patterns, and a knowledge adaptation in brain control with less computational complexity.

19.2.9 Artificial neural networks

Artificial NN (ANN) foundations were laid out with different studies starting in the 1940s (McCulloch & Pitts, 1943). Later, early ANN studies were combined with machine learning approaches. With the recent developments in the late 20th century, ANNs have gained popularity. ANN is an information processing paradigm that tries to mimic the brain. It consists of a large number of interconnected processing elements called "neurons" that work together to solve a particular problem. ANN basically consists of an input layer, hidden layers, and an output layer. ANN's purpose could be thought of as converting an input signal to an output signal by information flow and processing between its neurons. In order to understand the mechanism behind ANN, one could examine a perceptron, which is a single-layer NN, as depicted in Fig. 19.4. In ANN, an activation function determines the output according to a given input. The activation function could be as simple as a step function that turns the neuron output on and off, depending on a rule or threshold, or it could be more complex, depending on the input data complexity. The main aim of an ANN is to compute and learn almost any function representing an input/output relationship, and provide accurate predictions.

NNs require a trainer to define what should be the output in response to a given input. After calculating the predicted output with forward propagation, a cost function depending on the difference between the actual output and the predicted output is computed and sent back through the system. The cost function for each layer of the network is analyzed and used to adjust the weights of the next layers via backpropagation. By computing the gradient of the cost function with respect to the weights of the network for a single input–output pair, the network could be trained and the weights could be optimized. ANN tries to minimize the cost function, because a lower cost function means a closer prediction to the actual output. At each iteration, the

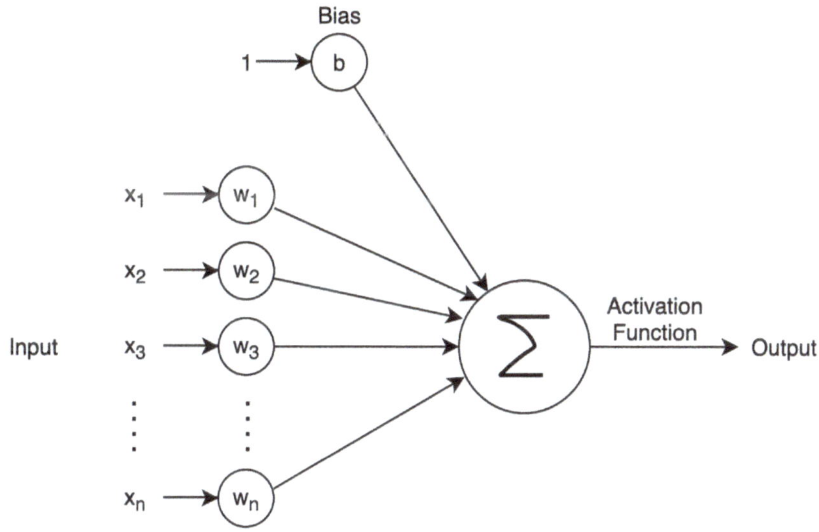

FIGURE 19.4 A schematic of a perceptron, which is a single-layer neural network.

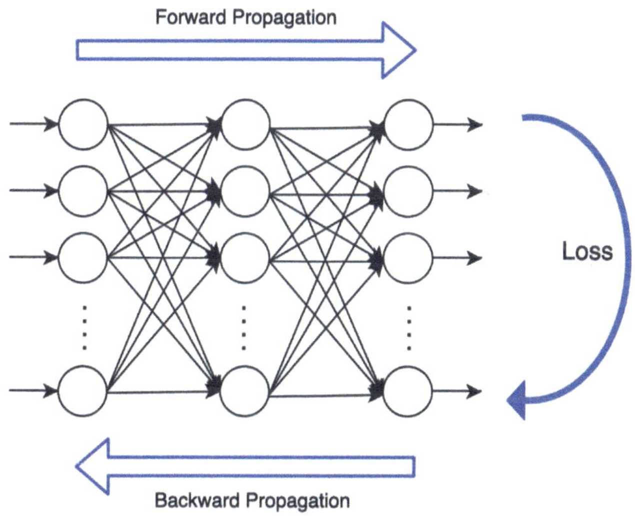

FIGURE 19.5 A schematic depiction of an artificial neural network (ANN) learning process.

error continues to decrease as the network learns how to analyze the inputs on each forward/backward cycle (Fig. 19.5).

ANNs have been extensively employed for neuroprosthesis applications, because they could model nonlinear interactions without the need to know the relationship between the input and output data (Pao, 1989; Rumelhart et al., 1986). Liu et al. (2020) used a lab-developed sparse tactile sensor array covered with a continuous compliant human-like skin layer to acquire haptic information, and employed an ANN-based method to extract the object curvature information by tactile perception with an overall classification accuracy of 93.1%. They reported that it was more meaningful to use a sparse tactile sensor array instead of a high-density sensor array, demonstrating their potential capability for prosthetic applications. In another study, the researchers focused on the prediction of somatosensory stimuli from dorsal column nuclei (DCN) signals in urethane-anesthetized rats to enhance the decoding performance of neural signals evoked by natural tactile and proprioceptive somatosensory stimuli, and to achieve more natural limb movements (Loutit & Potas, 2020). The highest accuracy achieved was 87% using an ANN algorithm with 13 features that were extracted from both high- and low-frequency bands of DCN signals.

ANN has also been used to detect and remove misclassifications of a pattern recognition system for forearm and hand motions for improving the robustness of hand prosthesis control (Amsuss et al., 2014). Additionally, Blana et al. (2009) developed a feedforward–feedback controller for a two-joint (elbow and shoulder) arm model. In that study, both the feedforward and feedback ANNs were adapted to the arm dynamics of a specific user, so the controller performance continued to improve, resulting in more accurate movements. In a later study, the same group tried to evaluate a transhumeral prosthesis controller that classified a combination of kinematics and EMG signals using ANN to predict humeral angular velocity and linear acceleration, and reported 78% path efficiency (Blana et al., 2016). In another study, Gaudet et al. (2018) reported a model to classify eight main upper limb movements in transhumeral amputees based on sEMG data using ANN. Their results indicated that the accuracy decreased as the number of degrees of freedom (DoF) considered in the classification increased. Therefore they added a kinematic feature that resulted in an average increase of 4.8% in accuracy. On the other hand, Ogiri et al. (2018) reported a lightweight neuroprosthesis that could control impaired motion using ANN based on voluntary EMG signals obtained from a transhumeral amputee, who had targeted muscle reinnervation surgery, and achieved about 95% classification accuracy for six different hand, wrist, and elbow movements.

For FES systems, (Giuffrida & Crago, 2005) focused on developing a synergistic controller, which used EMG signals of voluntary elbow flexor and shoulder, to control elbow extension with FES using ANN. Their model was designed such that when the subject simply attempted to move his/her hand to a location or apply an endpoint force, the controller applied an appropriate level of triceps stimulation to aid in the desired task. They reported an MSE of less than 0.04 for the intended movements. Similarly, (Hincapie & Kirsch, 2009) designed a FES controller using ANN that extracted information from the activity of muscles that were under voluntary control. The authors reported that the NN controller was able to predict the required paralyzed muscle FES activations from the voluntary activations of muscles in the upper extremity with a root mean square error (RMSE) of less than 0.036.

An ANN model has also been used to predict the next arm position based on EEG inputs of the current and desired positions of an artificial upper limb (Roy et al., 2012). Moreover, the same study simulated the movement of a robot arm toward a known target using the experimental EEG data of shoulder and elbow joint movements. As a result, the imagery classification of right or left arm was achieved with 76% accuracy. Additionally, Tibold and Fuglevand (2015) proposed different ANN structures to predict EMG activities of various arm muscles during different movements. The input to the trained ANNs included hand position, hand orientation, and thumb grip force for objects with different weights and dimensions. The authors concluded that ANNs could be used to provide an initial estimate of the complex patterns of muscle stimulation needed to produce a wide array of movements, including those involving object interaction, with an average R^2 value of 0.43 between the predicted and actual EMG signals for their best model. In a different study, Nataraj et al. (2012) investigated the potential of using trunk acceleration feedback control of the center of pressure (COP) against postural disturbances with a standing neuroprosthesis following paralysis. In order to predict the future shifts in COP necessary for stabilization during perturbed bipedal stance, they trained an ANN model with three-dimensional trunk acceleration values as input. As a result, the range of the correlation coefficients (R-value) between the ANN predictions and the actual COP was 0.67−0.77.

19.3 Deep learning methods for neuroprosthetic applications

19.3.1 Convolutional neural networks

With the increase in computational power, ANNs have become "deeper" with more hidden layers. Convolutional NNs (CNNs) have emerged with the convolution operation joining the algorithm. CNN was first introduced by LeCun et al. (1998) after the neocognitron (Fukushima, 1980), a very basic image recognition NN, had been invented. However, the current interest in CNN began after Krizhevsky et al. won the ImageNet object recognition challenge (Krizhevsky et al., 2012). CNN is a class of NN that takes input data and assigns importance (learnable weights and deviations) to various features/objects in the data. CNN with a sufficient training has the ability to learn the features of a given input without preprocessing.

Like ANN, CNN also consists of an input layer, hidden layers, and an output layer. However, in CNN, hidden layers consist of a series of convolution and pooling layers followed by a fully connected (FC) layer (Fig. 19.6). Convolution layers use a convolution operation, which preserves the

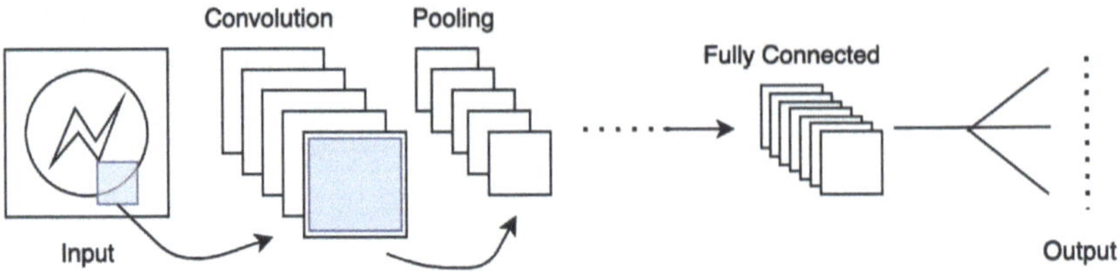

FIGURE 19.6 A schematic depiction of a convolutional neural network (CNN) architecture.

relationship between neighboring input points by learning the input features using small sections of the input data, so that features could be extracted from the given input. The pooling layers reduce the number of parameters when the input data size is too large. This operation is also called "subsampling" or "down sampling," which reduces the dimensionality of each feature map while retaining crucial information. The last hidden layer is the FC layer, where every neuron in the previous layer is connected to every neuron in the subsequent layer. The final output of the convolution and pooling layers is flattened into a vector and fed into the FC layer to classify the input data.

CNN could handle all the feature selection and performs better than other machine learning algorithms that require preprocessing steps. Essentially, using CNN shifts the focus from feature engineering to feature learning. Potentially, better features than handcrafted ones could be learned automatically through convolutional layers. These properties are then transferred to FC layers that associate the input signal with the specified outputs.

In the literature, Schirrmeister et al. (2017) proposed CNNs with a range of different architectures, designed for decoding imagined or executed tasks from raw EEG data. Their work highlighted the potential of CNN combined with advanced visualization techniques for EEG-based brain mapping. Additionally, Tabar and Halici (2017) proposed a CNN model to classify EEG motor imagery signals from right and left hands. As the input, they used combined time, frequency, and location information extracted from EEG signal and reported 77.6% accuracy. Moreover, Tang et al. (2017) proposed a method based on CNN to perform feature extraction and classification for single-trial motor imagery EEG, and reported that the CNN model outperformed the classical machine learning models with 86.4% accuracy.

Several studies have employed CNN for hand gesture classification based on EMG signals in the literature. One study classified myoelectric signals of two amputee volunteers by CNN structure to develop the grasp functionality of a prosthetic hand with an overall success of up to 88% (Ghazaei et al., 2017). Betthauser et al. (2020) proposed that sequential prediction models and, specifically, temporal convolutional networks were able to enhance useful temporal information from EMG data to achieve higher accuracies for detecting different simultaneous hand and wrist movements. They used publicly available NinaPro and CapgMyo amputee and nonamputee data sets and achieved 72% accuracy. On the other hand, Wei et al. (2019) developed an architecture where a multiview CNN framework was proposed by combining classical sEMG feature sets with a CNN-based deep learning model to recognize gestures from various data sets including the NinaPro data set. Their multiview framework outperformed single-view methods on both unimodal and multimodal sEMG data streams, achieving 88% accuracy. Another study employed CNN to classify discrete sEMG signals of 14 different hand gestures with an accuracy of 97.27%, and the classifier was also tested on the NinaPro DB5 data set and resulted in an average increase of 16% in classification accuracy in comparison to SVM and random forest classifiers (Wan et al., 2018). CNN was also employed to decode hand gestures (close hand, flex hand, extend the hand, and fine grip) using EMG data with up to 83.7% accuracy (Asif et al., 2020). Additionally, Duan et al. (2019) reported improved classification accuracies of above 94% for 10 different hand gestures in comparison to SVM, which resulted in as low as 76% accuracy for finger grip movement. Moreover, Yang et al. (2019) reported up to 91.7% accuracy in classifying raw EMG signals of hand gestures using CNN.

Triwiyanto et al. (2020) reported accuracies in the range of 77–93% for classifying 10 hand motions based on their EMG signals using CNN after hyperparameter tuning, which outperformed classical machine learning algorithms. CNN was also employed to design a self-recalibrating classification system that was capable of updating its prediction of hand movements without user retraining (Zhai et al., 2017). Using the publicly available NinaPro database, Atzori et al. (2015) and Zhai et al. (2017) reported approximately 78% accuracy for the classification of all 52 movement types available in the data set, and an 88% accuracy for a subset of 10 movements that are commonly used in daily life, such as wrist flexion and extension, and hand open. Additionally, Park and Lee (2016) introduced a user-adaptive CNN architecture for classification of EMG signals of six different hand grasping movements (tip pinch, prismatic four finger, power, parallel extension, lateral, and opening a bottle with a tripod grasps) available at the NinaPro database with an above 90% classification accuracy. On the other hand, Cote-Allard et al. (2019) applied transfer learning to recognize hand gestures in real time using CNN. Their data set comprised of the EMG signals selected from the NinaPro data set in addition to their own real-time EMG recordings, and the proposed CNN algorithm was able to differentiate seven hand gestures with 98.31% accuracy, while the accuracy dropped to 68.98% when the number of gestures was increased to 18.

Chen, Fu, et al. (2020) and Chen, Li, et al. (2020) recently proposed a compact CNN model with a reduced number of parameters for the hand gesture recognition using Myo (Cote-Allard et al., 2019) and NinaPro DB5 (Pizzolato et al., 2017) data sets, and reported an accuracy of 98.8% for the classification of the seven hand gestures in the Myo data set, while the accuracy dropped to a maximum of 69.6% on the NinaPro DB5 data set due to an increase in the number of gestures (at least 12 in each exercise) and reduced number of training samples. Similarly, Atzori et al. (2016) applied CNN for the classification of an average of 50 hand movements recorded via EMG, and reported around 60% accuracy, which was lower than the random forest classification results (~75%). Ameri et al. (2018) proposed a CNN-based classification model for differentiation of wrist movements based on EMG signals, and reported an accuracy of 90%. On the other hand, Zia Ur Rehman et al. (2018) achieved 98.12% accuracy in classifying six active hand motions plus rest using CNN, and reported improved classification performance over days.

Olsson et al. (2019) proposed a different approach, and instead of attempting to classify complex hand gestures, they tried to classify 16 simpler basis movements using a CNN architecture for a more natural hand movement, and achieved a mean exact match rate of 78.8% for the classification of basis movements, and 78.1% accuracy for the classification of 65 different compound movements. More recently, Tam et al. (2020) presented a real-time fine gesture recognition system using multi-articulating hand prosthesis control with an embedded CNN to classify eight different hand gestures within 100–200 milliseconds with 98.15% accuracy. Similarly, Chen, Fu, et al. (2020) and Chen, Li, et al. (2020) applied transfer learning for the recognition of 30 different hand gestures, including various movements of finger, elbow, and wrist joints, and reported improved accuracies of around 92% and 95% for CNN and CNN + long short-term memory (LSTM) networks, respectively.

19.3.2 Recurrent neural networks

The idea of sharing parameters across different parts of a NN model was first introduced in the 1980s, starting an evolution from NNs to recurrent NNs (RNNs) (Rumelhart et al., 1986). RNNs are NNs that contain loops and allow information to flow. RNNs could also be considered as multiple copies of the same network, in which each transmits a message to the successor (Fig. 19.7). The output of the previous time step is fed as an input to the next time step in RNN. This allows an RNN to catch dependencies across time and thus exhibit dynamic temporal behavior for a time sequence.

One of the appeals of RNNs is the idea that previous tasks, such as the use of previous EEG or EMG frames, could be linked to the current frame. In that case, basically one could predict the future situation using a total of present situations. However, RNNs create very deep computational graphs by repeatedly performing the same process at each time step of a long sequence, and this repeated application of the same parameters is likely to cause difficulties. This mathematical challenge of learning long-term dependencies in RNN has led to the creation of a special RNN class called "long

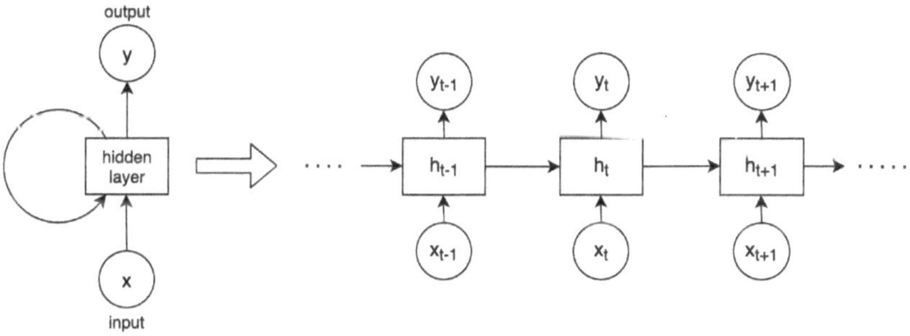

FIGURE 19.7 A schematic representation of an recurrent neural network (RNN) structure.

short time memory" (LSTM) networks, which are capable of learning long-term dependencies (Hochreiter & Schmidhuber, 1997).

LSTMs contain gated cells, where the information could be stored in and passed to or from a cell. The cell makes decisions about what to keep, and when to allow transfers by gates that open and close. These gates are analog and implemented with element-wise multiplication by a sigmoid function, which are all in the range of [0−1]. A 0 value means no information could pass, while a value of 1 means all information could pass. The analog system is differentiable, and therefore it is suitable for backpropagation. Accordingly, LSTMs could learn how to remember or forget information for long periods of time.

In the literature, there have been a few studies that employed RNN for somatosensory feedback and neuroprosthetic control. One recent study used multiunit activity (spikes) from Brodmann's area 2 of S1 in rhesus macaques during a planar center-out reaching task while continuously tracking the hand position, and then trained an RNN to achieve a high-performance brain-controlled prosthetic arm/hand (Kumaravelu et al., 2020). The aim of the NN in that study was to learn the mapping function between the physical stimulus and the biomimetic stimulation pattern, and RNN predicted spatio-temporal patterns corresponding to intracortical microstimulation based on trained natural somatosensory inputs ($R^2 = 0.85$). In another study, Bao, Zaidi, Xie, Yang, & Zhang (2019) and Bao, Mao, et al. (2019) created an identification and control structure that contained an RNN and several feed-forward NNs (FNNs) to optimize an FES-assisted system so that both muscle fatigue and regulation errors could be minimized. Additionally, Bu et al. (2003) presented a pattern discrimination method for EMG signals for fore-arm prosthetic control. They used an RNN structure, which enabled modeling of EMG time series, and achieved more than 89% accuracy. Moreover, Li et al. (2014) proposed an FES-based muscle model represented by an RNN model for muscle fatigue tracking. As a result, they reported about 0.04 RMSE between predicted and real FES-induced muscular dynamics. Because EMG signals have a sequential nature, Hu et al. (2018) proposed an attention-based hybrid CNN and RNN architecture to capture the temporal properties of the EMG signal for gesture recognition using several databases including NinaPro. Moreover, they presented an sEMG image representation method based on a traditional feature vector, which enabled deep learning architectures to extract implicit correlations between different channels for sparse multichannel EMG signal, and achieved about 90% accuracy.

LSTM has also been used in neuroprosthetic applications. Tatarian et al. (2018) developed a robotic hand for academia, focusing on the electromechanical design and robot operating system integration. Their work also included an educational classification framework developed in MATLAB, thus allowing students to control the robotic hand through the EMG signals with training and testing an LSTM sequence data classification architecture. Another study developed an LSTM decoder for extracting movement kinematics from the activity of large populations of neurons, sampled simultaneously from multiple cortical areas, in rhesus monkeys performing motor tasks (Tseng et al., 2019). The regions of interest included primary motor, dorsal premotor, supplementary motor, and primary somatosensory cortical areas. Their model significantly outperformed the state-of-the-art models when applied to three tasks, which were center-out arm reaching, bimanual reaching, and bipedal walking on a treadmill with a 0.68 R^2-value. In another study, an LSTM network was trained to deal with time-dependent information within brain signals during locomotion (Tortora et al., 2020). The researchers used EEG and EMG signals to provide a more natural control for modern neuroprosthetic devices. Then, the network was evaluated on the data set of 11 healthy subjects walking on a treadmill, resulting in >80% accuracy for offline and >75% accuracy for online evaluation.

A hybrid approach of LSTM and ANN was also employed for a combined pattern recognition model for sEMG signals for prosthetic control (He et al., 2018). In that model, the LSTM network captured temporal dependencies of the sEMG signals, while the ANN focused on the static characteristics of motions from the NinaPro data set and an accuracy of more than 75% was reported. In another study, Cao et al. (2019) developed a novel hybrid deep network, which combined LSTM and CNN, for rehabilitation evaluation using the skeletal data captured by a Kinect sensor to represent the patient's upper limb movement with up to 80% accuracy. Additionally, Wang, Farhadi, Rao, & Brunton (2018) introduced a novel Annotated Joints in Long-term ECoG (AJILE) data set, which includes automatically annotated poses of seven upper body joints recorded from four human subjects in over 670 total hours, along with the corresponding simultaneously acquired intracranial neural recordings. The size and scope of the AJILE data set made it possible to take a deep learning approach for movement prediction. Moreover, they proposed a multimodal model that combined CNN with LSTM blocks using both ECoG and video modalities, to detect movements and predict future movements up to 800 milliseconds before movement initiation. In another study, Tong et al. (2019) utilized CNN and LSTM together to extract the spatial and temporal features of sEMG for different gesture movements. A data set was constructed for testing the performance of the network along with the NinaPro data set, and an accuracy of 78% was achieved to classify hand movements. Additionally, Xu et al. (2018) proposed a combination of CNN and LSTM (C-LSTM), which was applied to predict muscle force generated in static isometric elbow flexion, achieving a high accuracy for real-time subject-independent force estimation. The resultant RMSE values between predicted and real force values were less than 0.09. In another study, Bao, Zaidi, Xie, Yang, & Zhang (2019) and Bao, Mao, et al. (2019) proposed another hybrid approach to fully explore the temporal-spatial information in sEMG data located at the NinaPro data set using both CNN and LSTM, and reported an average of 0.75 R^2-value for wrist kinematic estimation. Moreover, Huang and Chen (2019) demonstrated that features with specific physical meaning, like a spectrogram, are less effective for classification purposes than a combination of such features and NNs. They performed tests on a NinaPro database for 50 different movements, including finger/wrist gestures and force exertion, and achieved an overall accuracy of 79% using CNN and LSTM on the spectrogram data.

19.4 Conclusion

Machine learning models have been employed for the control of different neuroprosthetic devices. Table 19.4 lists a summary of the studies reviewed in this chapter about the classical machine learning and deep learning applications in the neuroprosthetics field. Most of these studies have focused on artificial limb control and movement classification. kNN and SVM were the most commonly employed classical machine learning techniques that could classify different movements with over 90% accuracy. Reinforcement learning has also been used to design human–machine interactive systems to obtain more natural movement patterns (Jagodnik et al., 2017). Moreover, a few models adaptable to real-life applications have been developed using kNN, SVM, and RL (Bhaduri et al., 2016; Bose et al., 2016; Gant et al., 2018; Sanchez et al., 2011). On the other hand, ANN-based applications have achieved some success in different classification and control tasks by employing EEG and EMG signals as input (Blana et al., 2016; Roy et al., 2012; Zhang et al., 2017). However, the accuracy of the classification algorithms was lower for the studies that attempted to decode more movements. Recent studies have focused on CNN and reported higher classification accuracies. Moreover, RNN-based algorithms that examine the time dependency and CNN-based algorithms that detect the spatial features of the signals were combined, and hybrid approaches have been reported to decode both spatial and temporal information of features existent in multiple windows of the same gesture (Chen, Li, et al., 2020; Hu et al., 2018; Xu et al., 2018). Machine learning algorithms have also been used for distinguishing motor imagery tasks based on the EEG signals received

TABLE 19.4 The literature survey results of the classical machine learning and deep learning applications in the neuroprosthetics field.

Purpose	Method	Interface	Results[a]
Classification of finger movements	SVM (Kaiju et al., 2017)	ECoG (Kaiju et al., 2017)	>95% (Kaiju et al., 2017)
	kNN (Altan et al., 2019)	EMG (Altan et al., 2019)	99.9% (Altan et al., 2019)
	ANN (Acharya et al., 2008; Aggarwal et al., 2009; Khezri & Jahed, 2009; Naik & Nguyen, 2015)	Intracortical (Acharya et al., 2008; Aggarwal et al., 2009)	90% (Acharya et al., 2008) 0.81 (*R*-value) (Aggarwal et al., 2009)
		EMG (Khezri & Jahed, 2009; Naik & Nguyen, 2015)	> 77% (Khezri & Jahed, 2009) > 84% (Naik & Nguyen, 2015)
Classification of hand movements (including wrist and grasp movement)	kNN (Fang & Liu, 2014; Fattah et al., 2017; Geethanjali et al., 2009; Iqbal et al., 2017; Kim et al., 2011; Marri & Swaminathan, 2016; Praveen et al., 2018)	EMG (Fang & Liu, 2014; Fattah et al., 2017; Geethanjali et al., 2009; Iqbal et al., 2017; Kim et al., 2011; Marri & Swaminathan, 2016; Praveen et al., 2018)	88% (Marri & Swaminathan, 2016) > 87% (Fang & Liu, 2014) 87.87% (Fattah et al., 2017) 84.5% (Geethanjali et al., 2009) 86.71% (Iqbal et al., 2017) 84.9% (Kim et al., 2011) ~91% (Praveen et al., 2018)
	SVM (Abbaspour et al., 2020; Atzori et al., 2015; Booth & Goldsmith, 2018; Castellini et al., 2009)	EMG + kinematics (Atzori et al., 2015)	76% (Atzori et al., 2015)
		EMG (Abbaspour et al., 2020; Castellini et al., 2009)	94.7% (Abbaspour et al., 2020) 95% (Castellini et al., 2009)
		Piezoelectric sensors within a wrist strap (Booth & Goldsmith, 2018)	97% (Booth & Goldsmith, 2018)

	DT (Gibson et al., 2013; Suchodolski & Wolczowski, 2010; Wolczowski & Kurzynski, 2015)	EMG (Gibson et al., 2013; Suchodolski & Wolczowski, 2010)	79% (Gibson et al., 2013) 85% (Suchodolski & Wolczowski, 2010)
		EMG + MMG (Wolczowski & Kurzynski, 2015)	>85% (Wolczowski & Kurzynski, 2015)
	CNN (Ameri et al., 2018; Asif et al., 2020; Atzori et al., 2016; Betthauser et al., 2020; Chen, Fu, et al., 2020; Chen, Li, et al., 2020; Cote-Allard et al., 2019; Duan et al., 2019; Olsson et al., 2019; Park & Lee, 2016; Tabar & Halici, 2017; Tam et al., 2020; Tang et al., 2017; Triwiyanto et al., 2020; Wan et al., 2018; Wei et al., 2019; Yang et al., 2019; Zhai et al., 2017; Zia Ur Rehman et al., 2018)	EMG (Ameri et al., 2018; Asif et al., 2020; Atzori et al., 2016; Betthauser et al., 2020; Chen, Fu, et al., 2020; Chen, Li, et al., 2020; Cote-Allard et al., 2019; Duan et al., 2019; Olsson et al., 2019; Park & Lee, 2016; Tam et al., 2020; Triwiyanto et al., 2020; Wan et al., 2018; Wei et al., 2019; Yang et al., 2019; Zhai et al., 2017; Zia Ur Rehman et al., 2018)	> 60% (Atzori et al., 2016) 72% (Betthauser et al., 2020) 97.2% (Wan et al., 2018) 83.7% (Asif et al., 2020) > 94% (Duan et al., 2019) ~ 82% (Yang et al., 2019) 77–93% (Triwiyanto et al., 2020) 88.4% (Zhai et al., 2017) > 90% (Park & Lee, 2016) 98.3% (Cote-Allard et al., 2019) 92.4% (Chen, Li, et al., 2020) 98.1% (Zia Ur Rehman et al., 2018) 91.6% (Ameri et al., 2018) 78.8% (Olsson et al., 2019) 98.1% (Tam et al., 2020) 88% (Wei et al., 2019) avg. 84% (Chen, Fu, et al., 2020)
		EEG (Tabar & Halici, 2017; Tang et al., 2017)	77.6% (Tabar & Halici, 2017) 86.4% (Tang et al., 2017)
	RNN (Bu et al., 2003)	EMG (Bu et al., 2003)	>89% (Bu et al., 2003)
	ANN (Favieiro et al., 2016; Gandolla et al., 2016; Gaudet et al., 2018; Ogiri et al., 2018; Tibold & Fuglevand, 2015)	EMG (Favieiro et al., 2016; Gandolla et al., 2016; Gaudet et al., 2018; Ogiri et al., 2018; Tibold & Fuglevand, 2015)	76% (Favieiro et al., 2016) 72% (Gandolla et al., 2016) 81.1% (Gaudet et al., 2018) > 86.4% (Ogiri et al., 2018) 0.43 (R^2-value) (Tibold & Fuglevand, 2015)

(Continued)

TABLE 19.4 (Continued)

Purpose	Method	Interface	Results[a]
	LSTM (He et al., 2018; Tatarian et al., 2018)	EMG (He et al., 2018) EMG (Tatarian et al., 2018)	> 75% (He et al., 2018) 82% (Tatarian et al., 2018)
	CNN-RNN (Hu et al., 2018)	EMG (Hu et al., 2018)	~90% (Hu et al., 2018)
	CNN-LSTM (Bao, Zaidi, Xie, Yang, & Zhang, 2019; Chen, Li, Hu, Zhang, & Chen, 2020; Huang and Chen, 2019; Tong, Zhang, Chen, & Liu, 2019)	EMG (Bao, Zaidi, Xie, Yang, & Zhang, 2019; Chen, Li, Hu, Zhang, & Chen, 2020; Huang and Chen, 2019; Tong, Zhang, Chen, & Liu, 2019)	95% (Chen, Li, et al., 2020) 78.31% (Tong et al., 2019) ~0.75 (R^2 value) (Bao, Zaidi, Xie, Yang, & Zhang, 2019) 79.33% (Huang & Chen, 2019)
Control/classification of upper limb movements (including elbow and shoulder)	kNN (Abbaspour et al., 2020; Aly et al., 2018; Dewald et al., 2019)	EMG (Abbaspour et al., 2020)	93.95% (Abbaspour et al., 2020)
		Intramuscular electrode + EMG (Dewald et al., 2019)	93.7% (Dewald et al., 2019)
		EMG + EEG (Aly et al., 2018)	95.50% (Aly et al., 2018)
	SVM (Ameri, Kamavuako, Scheme, Englehart, & Parker, 2014; Bhattacharyya, Konar, & Tibarewala, 2014; Bruurmijn, Pereboom, Vansteensel, Raemaekers, & Ramsey, 2017; Khasnobish, Bhattacharyya, Konar, Tibarewala, & Nagar, 2011; Ryun, Kim, Lee, & Chung, 2018)	EEG (Bhattacharyya et al., 2014; Khasnobish et al., 2011)	87.5% (Khasnobish et al., 2011) > 95% (Bhattacharyya et al., 2014)
		EMG (Ameri et al., 2014)	6% offline estimation error (Ameri et al., 2014)
		ECoG (Ryun, Kim, Lee, & Chung, 2018)	>80% (Ryun, Kim, Lee, & Chung, 2018)
		fMRI (Bruurmijn et al., 2017)	>79% (Bruurmijn et al., 2017)

	ANN (Blana et al., 2009; Blana et al., 2016; Chan et al., 2000; Giuffrida & Crago, 2005; Hincapie & Kirsch, 2009; Roy et al., 2012; Zhang et al., 2017)	EMG (Blana et al., 2016; Chan et al., 2000; Zhang et al., 2017)	78% (Blana et al., 2016) $<$0.1 MSE (Chan et al., 2000) ~90% (Zhang et al., 2017)
		EMG + FES (Giuffrida & Crago, 2005; Hincapie & Kirsch, 2009)	$<$0.04 MSE (Giuffrida & Crago, 2005) $<$0.036 RMSE (Hincapie & Kirsch, 2009)
		EEG (Roy et al., 2012)	>73% (Roy et al., 2012)
		FES (Blana et al., 2009)	<4-degree error
	RL (Di Febbo et al., 2018; Dura-Bernal et al., 2015; Jagodnik et al., 2017)	FES (Di Febbo et al., 2018; Jagodnik et al., 2017)	~0.58 degree error (Di Febbo et al., 2018) $>$98% (Jagodnik et al., 2017)
		Simulation (Dura-Bernal et al., 2015)	~1.80 degree error (Dura-Bernal et al., 2015)
	CNN (Ghazaei et al., 2017)	Simulation (Ghazaei et al., 2017)	88% (Ghazaei et al., 2017)
	RNN (Kumaravelu et al., 2020)	Intracortical (Kumaravelu et al., 2020)	0.85 (R^2-value) (Kumaravelu et al., 2020)
	CNN-LSTM (Cao, Fan, Wang, & Zhang, 2019; Wang, Farhadi, Rao, & Brunton, 2018; Xu, Chen, Cao, Zhang, & Chen, 2018)	EMG (Xu et al., 2018)	<0.09 RMSE (Xu et al., 2018)
		Kinect Sensor (Cao et al., 2019)	>80% (Cao et al., 2019)
		ECoG (Wang, Farhadi, Rao, & Brunton, 2018)	N/A (Wang, Farhadi, Rao, & Brunton, 2018)
Optimization of neuroprosthesis systems	RNN (Bao, Mao, et al., 2019)	FES + Simulation (Bao, Mao, et al., 2019)	N/A (Bao, Mao, et al., 2019)
	ANN (Amsuss et al., 2014; Liu et al., 2020)	EMG (Amsuss et al., 2014)	>82% (Amsuss et al., 2014)
		Tactile sensor array (Liu et al., 2020)	93.1% (Liu et al., 2020)

(Continued)

TABLE 19.4 (Continued)

Purpose	Method	Interface	Results[a]
Motor imagery task-related systems	LR (Tariq et al., 2020)	EEG (Tariq et al., 2020)	>60% (Tariq et al., 2020)
	kNN (Bhaduri et al., 2016; Bose et al., 2016)	EEG (Bhaduri et al., 2016; Bose et al., 2016)	90% (Bhaduri et al., 2016; Bose et al., 2016)
	SVM (Gant et al., 2018)	EEG (Gant et al., 2018)	84.6% (Gant et al., 2018)
	CNN (Schirrmeister et al., 2017)	EEG (Schirrmeister et al., 2017)	>88% (Schirrmeister et al., 2017)
	LSTM (Tseng et al., 2019)	Intracortical (Tseng et al., 2019)	0.68 (R^2-value) (Tseng et al., 2019)
	RL (Sanchez et al., 2011; Zhang et al., 2019)	Intracortical (Sanchez et al., 2011; Zhang et al., 2019)	100% (Sanchez et al., 2011) >93% (Zhang et al., 2019)
Classification of gait movement	SVM (Hussain et al., 2020; Needham et al., 2018; Yu et al., 2020)	EMG (Hussain et al., 2020)	97.5% (Hussain et al., 2020)
		MMG (Needham et al., 2018; Yu et al., 2020)	94% (Needham et al., 2018) 97.1% (Yu et al., 2020)
Control of robotic knee prosthesis	RL (Li et al., 2019)	Simulation (Li et al., 2019)	~2 degree error (Li et al., 2019)
Control of standing neuroprosthesis	ANN (Nataraj et al., 2012)	Functional neuromuscular stimulation (FNS) (Nataraj et al., 2012)	>0.67 (*R*-value) (Nataraj et al., 2012)
Classification of nonfatigue and fatigue conditions	LR (Marri & Swaminathan, 2016)	EMG (Marri & Swaminathan, 2016)	82% (Marri & Swaminathan, 2016)
Prediction of FES-induced muscular dynamics	RNN (Li et al., 2014)	FES + EMG (Li et al., 2014)	~0.04 RMSE (Li et al., 2014)

Classification of posture and activity	SVM (Kocaoglu & Akdogan, 2020)	IMU (Kocaoglu & Akdogan, 2020)	%83.3 (Kocaoglu & Akdogan, 2020)
	Random forests (ensemble methods) (Cui et al., 2019)	Intracortical (Cui et al., 2019)	84.72% (Cui et al., 2019)
	ANN (Loutit & Potas, 2020)	Intracortical (Loutit & Potas, 2020)	87% (Loutit & Potas, 2020)
	LSTM (Tortora et al., 2020)	EEG + EMG (Tortora et al., 2020)	>75% (Tortora et al., 2020)
Classification of terrain	DT (Liu et al., 2016)	IMU (Liu et al., 2016)	>98% (Liu et al., 2016)
Classification of texture	SVM (Beckmann et al., 2009)	EEG (Beckmann et al., 2009)	>50% (Beckmann et al., 2009)
	Random forests (ensemble methods) (Kursun & Patooghy, 2020)	Tactile sensors (Kursun & Patooghy, 2020)	>%87 (Kursun & Patooghy, 2020)
Classification of the leg movement	Bagged decision tree (ensemble methods) (Pew & Klute, 2018)	IMU (Pew & Klute, 2018)	>91% (Pew & Klute, 2018)
Decoding behavioral state (finger tapping vs. rest)	SVM (Cui et al., 2010)	fNIRS (Cui et al., 2010)	>85% (Cui et al., 2010)

[a] *Accuracy unless otherwise noted.*

from the motor cortex. Additionally, there have been other studies related to neuroprosthesis, such as gait control, posture classification, texture recognition, or terrain recognition (Kocaoglu & Akdogan, 2020; Kursun & Patooghy, 2020; Liu et al., 2016; Yu et al., 2020).

There have been several challenges preventing wider use of machine learning algorithms in the commercially available neuroprostheses, which could be the main foci of upcoming research studies. First, machine learning and especially deep learning algorithms are computationally expensive, and require a relatively long time and large disk space to train, which limits them to mostly research laboratories with powerful processing systems. Recent advances in the field have included fast hardware implementations with small device size and minimal power requirements for real-time movement control, which is a necessary step toward common real-life applications (Cerina et al., 2018). Moreover, despite the good efforts for developing public neuroprosthetic databases, such as NinaPro, deep learning algorithms still require larger data sets for more accurate and generalizable training. However, obtaining large patient-specific data is a costly and time-consuming process, and it has limited the use of deep learning techniques in this field. One proposed solution has been designing compact network structures to reduce the number of parameters, which would subsequently reduce the demand for data size (Chen, Fu, et al., 2020). Another remedy for this problem has been transfer learning, which takes into account the knowledge learned from a previous model, and transfers it to a new domain. Since very large neuroprostheses data sets are not yet available for use, transfer learning is thought to be a key to ensuring the breakthrough of deep learning applications in this field. Additionally, training accurate classifiers for neuroprosthetic applications with disabled participants is difficult. One common strategy to train the machine learning algorithms has been giving the patients a predefined trajectory and letting them imagine following the given path. However, achieving a high accuracy depends on the patients' consistent imagination and fully interactive concentration, which are difficult to confirm. As a result, there have been several animal or healthy control studies to understand the mechanism behind the brain control tasks to later project this information to help disabled patients. Another issue related with the intelligent neuroprosthetic controls using sEMG recordings is the time variance of sEMG signals and their sensitivity to electrode placements resulting in large variations of the intersubject and intertrial data patterns. As a result, supervised recalibration is required to ensure good performance of the classifiers for each user and even each session. However, recalibration is a time-consuming process, which would be a burden to amputees in long-term use. One approach to solve this issue for personalized neuroprostheses has been a self-recalibrating adaptive classifier system to facilitate long-term adoption (Zhai et al., 2017). RNNs have also become popular for the solution of robustness issues for personalized neuroprosthesis applications.

Recent advances in the deep learning field hold the promise of further improving neuroprosthetic controls. Generative adversarial networks (GANs) are an approach to generative modeling using deep learning methods such as CNNs (Goodfellow et al., 2014). Generative modeling is an unsupervised learning task, which involves automatically discovering and learning the patterns of input data, so that the model can generate new instances that could resemble the original data set. GANs have two submodels, which are the generator model that is trained to generate new examples, and the discriminator model that aims to classify the examples as either real or fake. The two models are trained together, until the generator can fool the discriminator model, which is then thought to generate realistic data. GANs are an exciting and rapidly developing research area, which is expected to be employed in the neuroprosthetics field in the near future.

In conclusion, deep learning algorithms have outperformed classical machine learning algorithms for more robust real-time classifier solutions for neuroprosthetics. Machine learning-assisted neuroprosthesis approaches could improve patient quality of life, especially by providing personalized modeling and identification of the patient-specific motor property toward tailored neuroprosthetics.

References

Abbaspour, S., Linden, M., Gholamhosseini, H., Naber, A., & Ortiz-Catalan, M. (2020). Evaluation of surface EMG-based recognition algorithms for decoding hand movements. *Medical & Biological Engineering & Computing*, *58*(1), 83–100.

Accogli, A., Grazi, L., Crea, S., Panarese, A., Carpaneto, J., Vitiello, N., & Micera, S. (2017). EMG-based detection of user's intentions for human-machine shared control of an assistive upper-limb exoskeleton. In J. GonzalezVargas, J. Ibanez, J. L. ContrerasVidal, H. Van Der Kooij, & J. L. Pons (Eds.), *Wearable robotics: Challenges and trends* (Vol. 16, pp. 181–185). Cham: Springer.

Acharya, S., Tenore, F., Aggarwal, V., Etienne-Cummings, R., Schieber, M. H., & Thakor, N. V. (2008). Decoding individuated finger movements using volume-constrained neuronal ensembles in the M1 hand area. *IEEE Transactions on Neural Systems and Rehabilitation Engineering*, *16*(1), 15–23.

Aggarwal, V., Tenore, F., Acharya, S., Schieber, M.H., & Thakor, N.V. (2009). Cortical decoding of individual finger and wrist kinematics for an upper-limb neuroprosthesis. In *2009 annual international conference of the IEEE engineering in medicine and biology society*, Minneapolis, MN, USA.

Altan, E., Pehlivan, K., & Kaplanoglu, E. (2019). Comparison of EMG based finger motion classification algorithms. In *2019 27th signal processing and communications applications conference*, Sivas, Turkey.

Aly, H.I., Youssef, S., & Fathy, C. (2018). Hybrid brain computer interface for movement control of upper limb prostheses. In *2018 International conference on biomedical engineering and applications*, Funchal, Portugal.

Ameri, A., Akhaee, M. A., Scheme, E., & Englehart, K. (2018). Real-time, simultaneous myoelectric control using a convolutional neural network. *Plos One*, *13*(9), e0203835.

Ameri, A., Kamavuako, E. N., Scheme, E. J., Englehart, K. B., & Parker, P. A. (2014). Support vector regression for improved real-time, simultaneous myoelectric control. *IEEE Transactions on Neural Systems and Rehabilitation Engineering*, *22*(6), 1198–1209.

Amsuss, S., Goebel, P. M., Jiang, N., Graimann, B., Paredes, L., & Farina, D. (2014). Self-correcting pattern recognition system of surface EMG signals for upper limb prosthesis control. *IEEE Transactions on Biomedical Engineering*, *61*(4), 1167–1176.

Asif, A. R., Waris, A., Gilani, S. O., Jamil, M., Ashraf, H., Shafique, M., & Niazi, I. K. (2020). Performance evaluation of convolutional neural network for hand gesture recognition using EMG. *Sensors*, *20*(6), 1642.

Atzori, M., Cognolato, M., & Muller, H. (2016). Deep learning with convolutional neural networks applied to electromyography data: A resource for the classification of movements for prosthetic hands. *Front Neurorobot*, *10*, 9.

Atzori, M., Gijsberts, A., Kuzborskij, I., Elsig, S., Hager, A. G. M., Deriaz, O., ... Caputo, B. (2015). Characterization of a benchmark database for myoelectric movement classification. *IEEE Transactions on Neural Systems and Rehabilitation Engineering*, *23*(1), 73–83.

Bao, T., Zaidi, S. A. R., Xie, S., Yang, P., & Zhang, Z. (2019). A CNN-LSTM hybrid framework for wrist kinematics estimation using surface electromyography. *IEEE Transactions on Instrumentation and Measurement*, *70*, 1–9.

Bao, X. F., Mao, Z. H., Munro, P., Sun, Z. Y., & Sharma, N. (2019). Sub-optimally solving actuator redundancy in a hybrid neuroprosthetic system with a multi-layer neural network structure. *International Journal of Intelligent Robotics and Applications*, *3*(3), 298–313.

Beckmann, S., Bjornsdotter, M., Backlund-Wasling, H., & Wessberg, J. (2009). Brain decoding of texture processing using independent component analysis and support vector machines. In *Wisp 2009: 6th IEEE international symposium on intelligent signal processing, proceedings*, Budapest, Hungary.

Betthauser, J. L., Krall, J. T., Bannowsky, S. G., Levay, G., Kaliki, R. R., Fifer, M. S., & Thakor, N. V. (2020).). Stable responsive EMG sequence prediction and adaptive reinforcement with temporal convolutional networks. *IEEE Transactions on Biomedical Engineering*, *67*(6), 1707–1717.

Bhaduri, S., Khasnobish, A., Bose, R., & Tibarewala, D.N. (2016). Classification of lower limb motor imagery using k nearest neighbor and Naive-Bayesian classifier. In *2016 3rd international conference on recent advances in information technology*, Dhanbad, India.

Bhattacharyya, S., Konar, A., & Tibarewala, D. N. (2014). Motor imagery, P300 and error-related EEG-based robot arm movement control for rehabilitation purpose. *Medical & Biological Engineering & Computing*, *52*(12), 1007–1017.

Blana, D., Kirsch, R. F., & Chadwick, E. K. (2009). Combined feedforward and feedback control of a redundant, nonlinear, dynamic musculoskeletal system. *Medical & Biological Engineering & Computing*, *47*(5), 533–542.

Blana, D., Kyriacou, T., Lambrecht, J. M., & Chadwick, E. K. (2016). Feasibility of using combined EMG and kinematic signals for prosthesis control: A simulation study using a virtual reality environment. *Journal of Electromyography and Kinesiology*, *29*, 21–27.

Booth, R., & Goldsmith, P. (2018). A wrist-worn piezoelectric sensor array for gesture input. *Journal of Medical and Biological Engineering*, *38*(2), 284–295.

Bose, R., Khasnobish, A., Bhaduri, S., & Tibarewala, D.N. (2016). Performance analysis of left and right lower limb movement classification from EEG. 2016 In *3rd International conference on signal processing and integrated networks*, Noida, India.

Breiman, L. (1996). Bagging predictors. *Machine Learning, 24*(2), 123–140.

Breiman, L. (2001). Random forests. *Machine Learning, 45*(1), 5–32.

Breiman, L., Friedman, J., Olshen, R., & Stone, C. J. (1984). *Classification and regression trees.* CRC Press.

Bruurmijn, M., Pereboom, I. P. L., Vansteensel, M. J., Raemaekers, M. A. H., & Ramsey, N. F. (2017). Preservation of hand movement representation in the sensorimotor areas of amputees. *Brain, 140*, 3166–3178.

Bu, N., Fukuda, O., & Tsuji, T. (2003). EMG-based motion discrimination using a novel recurrent neural network. *Journal of Intelligent Information Systems, 21*(2), 113–126.

Cao, L., Fan, C. J., Wang, H. R., & Zhang, G. K. (2019). A novel combination model of convolutional neural network and long short-term memory network for upper limb evaluation using kinect-based system. *IEEE Access, 7*, 145227–145234.

Caruana, R., & Freitag, D. (1994). Greedy attribute selection. In *Proceedings of the eleventh international conference on international conference on machine learning*, New Brunswick, NJ, USA.

Castellini, C., Gruppioni, E., Davalli, A., & Sandini, G. (2009). Fine detection of grasp force and posture by amputees via surface electromyography. *Journal of Physiology-Paris, 103*(3-5), 255–262.

Cerina, L., Franco, G., Cancian, P., & Santambrogio, M.D. (2018). Robustness of surface EMG classifiers with fixed-point decomposition on reconfigurable architecture. In *2018 IEEE international parallel and distributed processing symposium workshops*, Vancouver, BC, Canada.

Chan, F. H. Y., Yang, Y. S., Lam, F. K., Zhang, Y. T., & Parker, P. A. (2000). Fuzzy EMG classification for prosthesis control. *IEEE Transactions on Rehabilitation Engineering, 8*(3), 305–311.

Chen, L., Fu, J., Wu, Y., Li, H., & Zheng, B. (2020). Hand gesture recognition using compact CNN via surface electromyography signals. *Sensors (Basel), 20*(3), 672.

Chen, X., Li, Y., Hu, R., Zhang, X., & Chen, X. (2020). Hand gesture recognition based on surface electromyography using convolutional neural network with transfer learning method. *IEEE Journal of Biomedical and Health Informatics*, 1-1.

Cortes, C., & Vapnik, V. (1995). Support-vector networks. *Machine Learning, 20*(3), 273–297.

Cote-Allard, U., Fall, C. L., Drouin, A., Campeau-Lecours, A., Gosselin, C., Glette, K., Laviolette, F., & Gosselin, B. (2019). Deep learning for electromyographic hand gesture signal classification using transfer learning. *IEEE Transactions on Neural Systems and Rehabilitation Engineering, 27*(4), 760–771.

Cox, D. R. (1958). The regression analysis of binary sequences. *Journal of the Royal Statistical Society: Series B (Methodological), 20*(2), 215–242.

Cui, H. Y., Wang, Y. Z., Li, G. S., Huang, Y. C., & Hu, Y. (2019). Exploration of cervical myelopathy location from somatosensory evoked potentials using random forests classification. *IEEE Transactions on Neural Systems and Rehabilitation Engineering, 27*(11), 2254–2262.

Cui, X., Bray, S., & Reiss, A. L. (2010). Speeded Near Infrared Spectroscopy (NIRS) response detection. *Plos One, 5*(11), e15474.

Dewald, H. A., Lukyanenko, P., Lambrecht, J. M., Anderson, J. R., Tyler, D. J., Kirsch, R. F., & Williams, M. R. (2019). Stable, three degree-of-freedom myoelectric prosthetic control via chronic bipolar intramuscular electrodes: A case study. *Journal of Neuroengineering and Rehabilitation, 16*(1), 147.

Di Febbo, D., Ambrosini, E., Pirotta, M., Rojas, E., Restelli, M., Pedrocchi, A.L.G., & Ferrante, S. (2018). Does reinforcement learning outperform PID in the control of FES-induced elbow flex-extension? In *2018 IEEE international symposium on medical measurements and applications*, Rome, Italy.

Dietterich, T.G. (2000). Ensemble methods in machine learning. In *Proceedings of the first international workshop on multiple classifier systems*, Cagliari, Italy.

Duan, N., Liu, L. Z., Yu, X. J., Li, Q. Q., & Yeh, S. C. (2019). Classification of multichannel surface-electromyography signals based on convolutional neural networks. *Journal of Industrial Information Integration, 15*, 201–206.

Dura-Bernal, S., Chadderdon, G.L., Neymotin, S.A., Zhou, X.L., Przekwas, A., Francis, J.T., & Lytton, W.W. (2013). Virtual musculoskeletal arm and robotic arm driven by a biomimetic model of sensorimotor cortex with reinforcement learning. In *IEEE signal processing in medicine and biology symposium 2013*, Brooklyn, NY, USA.

Dura-Bernal, S., Zhou, X. L., Neymotin, S. A., Przekwas, A., Francis, J. T., & Lytton, W. W. (2015). Cortical spiking network interfaced with virtual musculoskeletal arm and robotic arm. *Frontiers in Neurorobotics, 9*, 13.

Fang, Y.F., & Liu, H.H. (2014). Robust sEMG electrodes configuration for pattern recognition based prosthesis control. In *2014 IEEE international conference on systems, man and cybernetics*, San Diego, CA, USA.

Fattah, S.A., Iqbal, O., Zahin, S., Shahnaz, C., & Rosul, G. (2017). Basic hand action classification based on surface EMG using autoregressive reflection coefficient.In *TENCON IEEE region 10 conference proceedings the 2017 IEEE region 10 conference*, Penang, Malaysia.

Favieiro, G.W., Moura, K.O.A., & Balbinot, A. (2016). Novel method to characterize upper-limb movements based on paraconsistent logic and myoelectric signals. In *2016 38th annual international conference of the IEEE engineering in medicine and biology society*, Orlando, FL, USA.

Fix, E., & Hodges, J. (1951). *Discriminatory analysis. nonparametric discrimination: Consistency properties*. USAF School of Aviation Medicine.

Freund, Y., & Schapire, R.E. (1996). Experiments with a new boosting algorithm. In *Proceedings of the thirteenth international conference on international conference on machine learning*, Bari, Italy.

Fukushima, K. (1980). Neocognitron: A self organizing neural network model for a mechanism of pattern recognition unaffected by shift in position. *Biological Cybernetics, 36*(4), 193–202.

Gandolla, M., Ferrante, S., Baldassini, D., Cottini, M.C., Seneci, C., & Pedrocchi, A. (2016). Artificial neural-network EMG classifier for hand movements prediction. In E. Kyriacou, S. Christofides, & C. S. Pattichis (Eds.), *XIV mediterranean conference on medical and biological engineering and computing 2016* (Vol. 57, pp. 634–637).

Gant, K., Guerra, S., Zimmerman, L., Parks, B. A., Prins, N. W., & Prasad, A. (2018). EEG-controlled functional electrical stimulation for hand opening and closing in chronic complete cervical spinal cord injury. *Biomedical Physics & Engineering Express, 4*(6), 065005.

Gaudet, G., Raison, M., & Achiche, S. (2018). Classification of upper limb phantom movements in transhumeral amputees using electromyographic and kinematic features. *Engineering Applications of Artificial Intelligence, 68*, 153–164.

Geethanjali, P., Ray, K.K., & Shanmuganathan, P.V. (2009). Actuation of prosthetic drive using EMG signal. In *Tencon 2009—2009 IEEE region 10 conference*, Singapore.

Ghazaei, G., Alameer, A., Degenaar, P., Morgan, G., & Nazarpour, K. (2017). Deep learning-based artificial vision for grasp classification in myoelectric hands. *Journal of Neural Engineering*, *14*(3).

Gibson, A.E., Ison, M.R., & Artemiadis, P. (2013). User-independent hand motion classification with electromyography. In *ASME 2013 dynamic systems and control conference*, Palo Alto, CA.

Giuffrida, J. P., & Crago, P. E. (2005). Functional restoration of elbow extension after spinal-cord injury using a neural network-based synergistic FES controller. *IEEE Transactions on Neural Systems and Rehabilitation Engineering*, *13*(2), 147–152.

Goodfellow, I., Pouget-Abadie, J., Mirza, M., Xu, B., Warde-Farley, D., Ozair, S., Courville, A., & Bengio, Y. (2014). Generative adversarial networks. In *Advances in neural information processing systems*, Montreal, Canada.

Hansen, L. K., & Salamon, P. (1990). Neural network ensembles. *IEEE Transactions on Analysis and Machine Intelligence*, *12*, 993–1001.

He, Y., Fukuda, O., Bu, N., Okumura, H., & Yamaguchi, N. (2018). Surface EMG pattern recognition using long short-term memory combined with multilayer perceptron. In *2018 40th annual international conference of the IEEE engineering in medicine and biology society (EMBC)*, Honolulu, HI, USA.

Hincapie, J. G., & Kirsch, R. F. (2009). Feasibility of EMG-based neural network controller for an upper extremity neuroprosthesis. *IEEE Transactions on Neural Systems and Rehabilitation Engineering*, *17*(1), 80–90.

Hochreiter, S., & Schmidhuber, J. (1997). Long short-term memory. *Neural Computation*, *9*(8), 1735–1780.

Hu, Y., Wong, Y., Wei, W., Du, Y., Kankanhalli, M., & Geng, W. (2018). A novel attention-based hybrid CNN-RNN architecture for sEMG-based gesture recognition. *Plos One*, *13*(10), e0206049.

Huang, D., & Chen, B. (2019). Surface EMG decoding for hand gestures based on spectrogram and CNN-LSTM. In *2019 2nd China symposium on cognitive computing and hybrid intelligence (CCHI)*, Xi'an, China.

Hussain, T., Iqbal, N., Maqbool, H. F., Khan, M., & Tahir, M. (2020). Amputee walking mode recognition based on mel frequency cepstral coefficients using surface electromyography sensor. *International Journal of Sensor Networks*, *32*(3), 139–149.

Iqbal, O., Fattah, S.A., & Zahin, S. (2017). Hand movement recognition based on singular value decomposition of surface EMG signal. In *2017 IEEE region 10 humanitarian technology conference*, Dhaka, Bangladesh.

Jagodnik, K. M., Thomas, P. S., van den Bogert, A. J., Branicky, M. S., & Kirsch, R. F. (2017). Training an actor-critic reinforcement learning controller for arm movement using human-generated rewards. *IEEE Transactions on Neural Systems and Rehabilitation Engineering*, *25*(10), 1892–1905.

Kaiju, T., Doi, K., Yokota, M., Watanabe, K., Inoue, M., Ando, H., Takahashi, K., Yoshida, F., Hirata, M., & Suzuki, T. (2017). High spatiotemporal resolution ECoG recording of somatosensory evoked potentials with flexible micro-electrode arrays. *Frontiers in Neural Circuits*, *11*, 20.

Khasnobish, A., Bhattacharyya, S., Konar, A., Tibarewala, D.N., & Nagar, A.K. (2011). A two-fold classification for composite decision about localized arm movement from EEG by SVM and QDA techniques. In *2011 International joint conference on neural networks*, San Jose, CA.

Khezri, M., & Jahed, M. (2009). An exploratory study to design a novel hand movement identification system. *Computers in Biology and Medicine*, *39*(5), 433–442.

Kim, K. S., Choi, H. H., Moon, C. S., & Mun, C. W. (2011). Comparison of k-nearest neighbor, quadratic discriminant and linear discriminant analysis in classification of electromyogram signals based on the wrist-motion directions. *Current Applied Physics*, *11*(3), 740–745.

Kocaoglu, S., & Akdogan, E. (2020). Comparison of classification algorithms for detecting patient posture in expandable tumor prostheses. *Advances in Electrical and Computer Engineering*, *20*(2), 131–138.

Kotsiantis, S. B. (2013). Decision trees: A recent overview. *Artificial Intelligence Review*, *39*(4), 261–283.

Krizhevsky, A., Sutskever, I., & Hinton, G. (2012). Imagenet classification with deep convolutional. *Neural Networks. Advances in Neural Information Processing Systems*, 1097–1105.

Kumaravelu, K., Tomlinson, T., Callier, T., Sombeck, J., Bensmaia, S. J., Miller, L. E., & Grill, W. M. (2020). A comprehensive model-based framework for optimal design of biomimetic patterns of electrical stimulation for prosthetic sensation. *Journal of Neural Engineering*, *17*(4), 046045.

Kursun, O., & Patooghy, A. (2020). An embedded system for collection and real-time classification of a tactile dataset. *IEEE Access*, *8*, 97462–97473.

LeCun, Y., Bottou, L., Bengio, Y., & Haffner, P. (1998). Gradient-based learning applied to document recognition. *Proceedings of the IEEE*, *86*(11), 2278–2324.

Li, M.H., Gao, X., Wen, Y., Si, J., & Huang, H. (2019). Offline policy iteration based reinforcement learning controller for online robotic knee prosthesis parameter tuning. In *2019 International conference on robotics and automation*, Montreal, QC, Canada.

Li, Z., Hayashibe, M., Fattal, C., & Guiraud, D. (2014). Muscle fatigue tracking with evoked emg via recurrent neural network: Toward personalized neuroprosthetics. *IEEE Computational Intelligence Magazine*, *9*(2), 38–46.

Liu, M., Wang, D., & Huang, H. (2016). Development of an environment-aware locomotion mode recognition system for powered lower limb prostheses. *IEEE Transactions on Neural Systems and Rehabilitation Engineering*, *24*(4), 434–443.

Liu, W. T., Zhan, B. P., Gu, C. X., Yu, P., Zhang, G. S., Fu, X., Cipriani, C., & Hu, L. (2020). Discrimination of object curvature based on a sparse tactile sensor array. *Micromachines*, *11*(6), 583.

Loutit, A. J., & Potas, J. R. (2020). Dorsal column nuclei neural signal features permit robust machine-learning of natural tactile- and proprioception-dominated stimuli. *Frontiers in Systems Neuroscience*, *14*, 46.

Marri, K., & Swaminathan, R. (2016). Classification of muscle fatigue in dynamic contraction using surface electromyography signals and multifractal singularity spectral analysis. *Journal of Dynamic Systems Measurement and Control-Transactions of the ASME*, *138*(11), 111008.

McCulloch, W. S., & Pitts, W. (1943). A logical calculus of the ideas immanent in nervous activity. *The Bulletin of Mathematical Biophysics*, *5*(4), 115–133.

Naik, G. R., & Nguyen, H. T. (2015). Nonnegative matrix factorization for the identification of EMG finger movements: Evaluation using matrix analysis. *IEEE Journal of Biomedical and Health Informatics*, *19*(2), 478–485.

Nataraj, R., Audu, M. L., Kirsch, R. F., & Triolo, R. J. (2012). Trunk acceleration for neuroprosthetic control of standing: A pilot study. *Journal of Applied Biomechanics*, *28*(1), 85–92.

Needham, A.P.H., Paszkiewicz, F.P., Alias, M.F.M., Wilson, S., Dehghani-Sanij, A.A., Khoo, B.C., & Vaidyanathan, R. (2018). Subject-independent data pooling in classification of gait intent using mechanomyography on a transtibial amputee. In *2018 IEEE international conference on robotics and automation*, Brisbane, QLD, Australia.

Ogiri, Y., Yamanoi, Y., Nishino, W., Kato, R., Takagi, T., & Yokoi, H. (2018). Development of an upper-limb neuroprosthesis to voluntarily control elbow and hand. *Advanced Robotics*, *32*(16), 879–886.

Olsson, A. E., Sager, P., Andersson, E., Bjorkman, A., Malesevic, N., & Antfolk, C. (2019). Extraction of multi-labelled movement information from the raw HD-sEMG image with time-domain depth. *Scientific Reports*, *9*(1), 1–10.

Pao, Y. (1989). *Adaptive pattern recognition and neural networks*. Addison-Wesley.

Park, K.H., & Lee, S.W. (2016). Movement intention decoding based on deep learning for multiuser myoelectric interfaces. In *2016 4th International winter conference on brain-computer interface*, Yongpyong, South Korea.

Pew, C., & Klute, G. K. (2018). Turn intent detection for control of a lower limb prosthesis. *IEEE Transactions on Biomedical Engineering*, *65*(4), 789–796.

Pizzolato, S., Tagliapietra, L., Cognolato, M., Reggiani, M., Muller, H., & Atzori, M. (2017). Comparison of six electromyography acquisition setups on hand movement classification tasks. *Plos One*, *12*(10), e0186132.

Praveen, L.S., Nagananda, S.N., & Shankapal, P. (2018). Design and development of real time bionic hand control using EMG signal. In *2018 IEEE international conference on electronics, computing and communication technologies*, Bangalore, India.

Rodriguez-Galiano, V. F., Luque-Espinar, J. A., Chica-Olmo, M., & Mendes, M. P. (2018). Feature selection approaches for predictive modelling of groundwater nitrate pollution: An evaluation of filters, embedded and wrapper methods. *Science of the total environment*, *624*, 661–672.

Roy, R., Konar, A., Tibarewala, D.N., & Janarthanan, R. (2012). EEG driven model predictive position control of an artificial limb using neural net. In *2012 Third international conference on computing communication & networking technologies*, Coimbatore, India.

Rumelhart, D. E., McClelland, J. L., University of California San Diego., & PDP Research Group. (1986). *Parallel distributed processing: Explorations in the microstructure of cognition*. MIT Press.

Ryun, S., Kim, J.S., Lee, D., & Chung, C.K. (2018). Prediction of motor and somatosensory function from human ECoG. 2018 6th International Conference on Brain-Computer Interface (BCI), 2018, pp. 1–4.

Saeys, Y., Inza, I., & Larrañaga, P. (2007). A review of feature selection techniques in bioinformatics. *Bioinformatics*, *23*(19), 2507–2517.

Samb, M. L., Camara, F., N'Diaye, S., Slimani, Y., Esseghir, M., & Anta, C. (2012). A novel RFE-SVM-based feature selection approach for classification. *International Journal of Advanced Science and Technology*, *43*, 27–36.

Sanchez, J.C., Tarigoppula, A., Choi, J.S., Marsh, B.T., Chhatbar, P.Y., Mahmoudi, B., & Francis, J.T. (2011). Control of a center-out reaching task using a reinforcement learning brain-machine interface. In *2011 5th international IEEE/EMBS conference on neural engineering*, Cancun, Mexico.

Sánchez-Maroño, N., Alonso-Betanzos, A., & Tombilla-Sanromán, M. (2007). Filter methods for feature selection—a comparative study. In *International conference on intelligent data engineering and automated learning* (pp. 178–187). IDEAL.

Schirrmeister, R. T., Springenberg, J. T., Fiederer, L. D. J., Glasstetter, M., Eggensperger, K., Tangermann, M., Hutter, F., Burgard, W., & Ball, T. (2017). Deep learning with convolutional neural networks for EEG decoding and visualization. *Human Brain Mapping*, *38*(11), 5391–5420.

Scholkopf, B., & Smola, A. J. (2001). *Learning with kernels: Support vector machines, regularization, optimization, and beyond*. MIT Press.

Sekiya, M., Sakaino, S., & Toshiaki, T. (2019). Linear logistic regression for estimation of lower limb muscle activations. *IEEE Transactions on Neural Systems and Rehabilitation Engineering*, *27*(3), 523–532.

Suchodolski, T., & Wolczowski, A. (2010). Hand prosthesis control—software tool for EMG signal analysis. In *ICINCO 2010: Proceedings of the 7th international conference on informatics in control, automation and robotics*, Funchal, Madeira, Portugal.

Sutton, R. S., & Barto, A. G. (1998). *Reinforcement learning: An introduction*. MIT Press.

Tabar, Y. R., & Halici, U. (2017). A novel deep learning approach for classification of EEG motor imagery signals. *Journal of Neural Engineering*, *14*(1), 016003.

Tam, S., Boukadoum, M., Campeau-Lecours, A., & Gosselin, B. (2020). A fully embedded adaptive real-time hand gesture classifier leveraging HD-sEMG and deep learning. *IEEE Transactions on Biomedical Circuits and Systems*, *14*(2), 232–243.

Tang, Z., Li, C., & Sun, S. (2017). Single-trial EEG classification of motor imagery using deep convolutional neural networks. *Optik*, *130*, 11–18.

Tariq, M., Trivailo, P. M., & Simic, M. (2020). Classification of left and right foot kinaesthetic motor imagery using common spatial pattern. *Biomedical Physics & Engineering Express*, *6*(1), 015008.

Tatarian, K., Couceiro, M.S., Ribeiro, E.P., & Faria, D.R. (2018). Stepping-stones to transhumanism: An EMG-controlled low-cost prosthetic hand for academia. In *2018 9th International conference on intelligent systems*, Funchal-Madeira, Portugal.

Tibold, R., & Fuglevand, A. J. (2015). Prediction of muscle activity during loaded movements of the upper limb. *Journal of Neuroengineering and Rehabilitation*, *12*(1), 1–2.

Tong, R. Z., Zhang, Y., Chen, H. F., & Liu, H. H. (2019). Learn the temporal-spatial feature of sEMG via dual-flow network. *International Journal of Humanoid Robotics*, *16*(4), 1941004.

Tortora, S., Ghidoni, S., Chisari, C., Micera, S., & Artoni, F. (2020). Deep learning-based BCI for gait decoding from EEG with LSTM recurrent neural network. *Journal of Neural Engineering*, *17*(4), 046011.

Triwiyanto, T., Pawana, I. P. A., & Purnomo, M. H. (2020). An improved performance of deep learning based on convolution neural network to classify the hand motion by evaluating hyper parameter. *IEEE Transactions on Neural Systems and Rehabilitation Engineering*, *28*(7), 1678–1688.

Tseng, P. H., Urpi, N. A., Lebedev, M., & Nicolelis, M. (2019). Decoding movements from cortical ensemble activity using a long short-term memory recurrent network. *Neural Computation*, *31*(6), 1085–1113.

Wan, Y. F., Han, Z. S., Zhong, J., & Chen, G. H. (2018). Pattern recognition and bionic manipulator driving by surface electromyography signals using convolutional neural network. *International Journal of Advanced Robotic Systems*, *15*(5), 1729881418802138.

Wang, N. X. R., Farhadi, A., Rao, R. P. N., & Brunton, B. W. (2018). AJILE movement prediction: Multimodal deep learning for natural human neural recordings and video. *Proceedings of the AAAI Conference on Artificial Intelligence*, *32*(1).

Wei, W., Dai, Q., Wong, Y., Hu, Y., Kankanhalli, M., & Geng, W. (2019). Surface-electromyography-based gesture recognition by multi-view deep learning. *IEEE Transactions on Biomedical Engineering*, *66*(10), 2964–2973.

Wolczowski, A., & Kurzynski, M. (2015). Control of hand prosthesis using fusion of biosignals and information from prosthesis sensors. *Studies in Computational Intelligence*, *595*, 259–273.

Xu, L., Chen, X., Cao, S., Zhang, X., & Chen, X. (2018). Feasibility study of advanced neural networks applied to sEMG-based force estimation. *Sensors (Basel)*, *18*(10), 3226.

Yang, W., Yang, D. P., Liu, Y., & Liu, H. (2019). EMG pattern recognition using convolutional neural network with different scale signal/spectra input. *International Journal of Humanoid Robotics*, *16*(4), 1950013.

Yu, J., Zhang, Y., & Xia, C. M. (2020). Study of gait pattern recognition based on fusion of mechanomyography and attitude angle signal. *Journal of Mechanics in Medicine and Biology*, *20*(2), 1950085.

Zhai, X. L., Jelfs, B., Chan, R. H. M., & Tin, C. (2017). Self-recalibrating surface EMG pattern recognition for neuroprosthesis control based on convolutional neural network. *Frontiers in Neuroscience*, *11*, 379.

Zhang, Q., Liu, R. F., Chen, W. B., & Xiong, C. H. (2017). Simultaneous and continuous estimation of shoulder and elbow kinematics from surface EMG signals. *Frontiers in Neuroscience*, *11*, 280.

Zhang, X., Libedinsky, C., So, R., Principe, J. C., & Wang, Y. W. (2019). Clustering neural patterns in kernel reinforcement learning assists fast brain control in brain-machine interfaces. *IEEE Transactions on Neural Systems and Rehabilitation Engineering*, *27*(9), 1684–1694.

Zhang, X., Lu, X., Shi, Q., Xu, X., Leung, H. E., Harris, L. N., Iglehart, J. D., Miron, A., Liu, J. S., & Wong, W. H. (2006). Recursive SVM feature selection and sample classification for mass-spectrometry and microarray data. *BMC Bioinformatics*, *7*(1), 197.

Zhou, Z. (2009). Ensemble learning. In S. Z. Li, & A. Jain (Eds.), *Encyclopedia of biometrics* (pp. 270–273). Springer US.

Zia Ur Rehman, M., Waris, A., Gilani, S. O., Jochumsen, M., Niazi, I. K., Jamil, M., Farina, D., & Kamavuako, E. N. (2018). Multiday EMG-based classification of hand motions with deep learning techniques. *Sensors (Basel)*, *18*(8), 2497.

Chapter 20

Modern approaches of signal processing for bidirectional neural interfaces

Andrea Cimolato[1,2], Natalija Katic[3,4] and Stanisa Raspopovic[5]

[1]*NEAR Lab, Department of Electronics, Information Science, and Bioengineering (DEIB), Politecnico di Milano, Milan, Italy,* [2]*Rehab Technologies, Istituto Italiano di Tecnologia (IIT), Genova, Italy,* [3]*Institute Mihajlo Pupin, Belgrade, Serbia,* [4]*School of Electrical Engineering, University of Belgrade, Belgrade, Serbia,* [5]*Neuroengineering Laboratory, Department of Health Sciences and Technology, Institute for Robotics and Intelligent Systems, ETH Zürich, Zürich, Switzerland*

ABSTRACT

Designing a reliable closed-loop system that would bidirectionally interface with the central and peripheral nervous system represents a major objective for the optimal control of neuroprosthetic devices and neurorehabilitative procedures. For using invasive neural electrodes, in particular, signal processing has been a key component to overcome specific hardware limitations in recording such as telemetry bandwidth, limited number of electrode active sites, and fibrotic tissue formation. Moreover, advances in data processing, such as machine learning and model-driven approaches, have been proposed to address more profound issues in restoring complex somatosensory sensations. Difficulty of targeting different combinations and types of neurons individually and independently, as well as inadequate knowledge about certain synaptic interactions, specific neural organization, and the role of these factors in perception and motor control, remain the biggest obstacles in neuroprostheses. Therefore this chapter introduces modern approaches and future applications of advanced signal processing techniques for neural invasive electrodes for bidirectional neural interfaces.

Keywords: Signal processing; bidirectional neural interfaces; closed-loop system

Whenever two systems of different natures are coupled together an interface is created. An interface represents a shared boundary through which an exchange of information is allowed to flow between the two contiguous systems. Bidirectional neural interfaces (BNIs) in particular allow the exchange of information between the native human neural system and an external device

Somatosensory Feedback for Neuroprosthetics. DOI: https://doi.org/10.1016/B978-0-12-822828-9.00016-2

through the recording from and the stimulation of either the central or the peripheral networks. Signal processing in BNIs is therefore all the collection of analog/digital elaboration of the signals, and aims to improve information quality and transmissibility, or the optimization of data representation.

Communication in the nervous system is mainly based on electrical impulses (i.e., action potentials); and through the spatiotemporal modulation of this activity, information is encoded, transmitted, and decoded through the network. Electrical recordings can, for example, be used to intercept communication between the central nervous system (CNS) and different organs to assess the state of the latter. On the other hand, the acquired information can be also used to control an external device (e.g., prostheses). In the other direction, electrical stimulation aims to introduce information into the nervous system in order to restore a lost function (sensory or motor), or as treatment for particular neuropathologies.

In contrast to many other human–machine interfaces, BNIs have to cope, from one side, with the technological limitations in communication channels, such as density and resolution, and additionally, with the convoluted interpretation of the deeply coupled networks in the nervous system. However, the advancement in digital signal processing has been a key factor in overcoming some of these particular limitations both in recording and stimulation. Moreover, with the increasing momentum of novel data analysis strategies, such as machine learning and model-driven approaches, new applications and panoramas are opening up for further developments in BNIs. Therefore this chapter's objective is to present modern and upcoming methodologies in advanced signal processing techniques for the use of neural invasive electrodes in BNIs.

20.1 Signal processing in neural signal recording

The first step in extracting information from neural signals is recording and this can be done by using intraneural or extraneural electrodes. The most commonly used extraneural electrodes are the cuff electrodes and flat interface nerve electrodes (FINEs), while longitudinal intrafascicular electrodes (LIFEs), transverse intrafascicular multichannel electrodes (TIMEs), and Utah slanted electrode arrays represent intraneural electrodes which are frequently applied. During a recording session, one needs to deal with several problems such as telemetry bandwidth, limited number of electrode active sites, and fibrotic tissue formation. Processing of the signals in the appropriate way leads us one step closer to overcoming the stated problems.

20.1.1 Generalized signal processing workflow

Every type of signal processing follows a similar workflow that differs based on the final goal for using the recorded signals (Fig. 20.1).

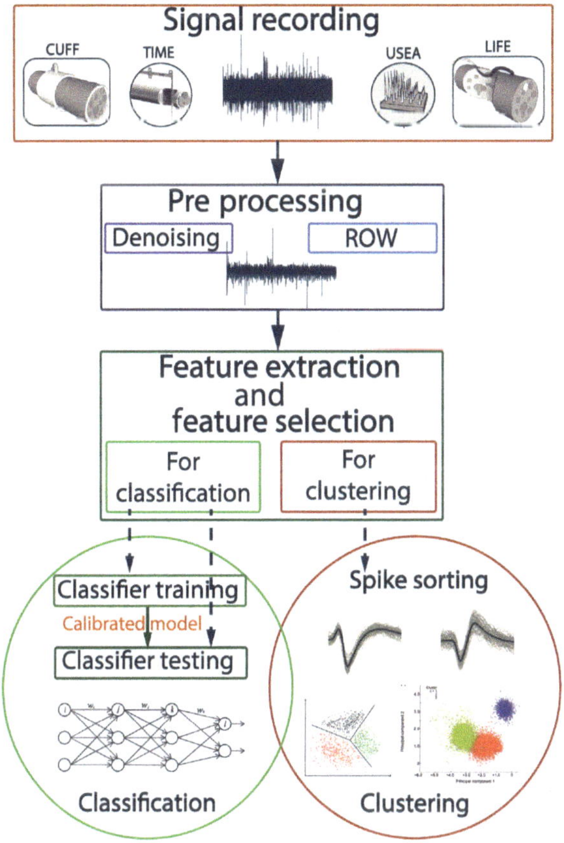

FIGURE 20.1 Typical signal processing workflow. Signal is recorded using one of the electrodes (invasive or noninvasive). Then, it is preprocessed, features are extracted and selected based on the its nature and further steps. As the final process of the workflow, classification or clustering is applied.

Raw signals are prepared for the following steps in the preprocessing block. Preprocessing usually refers to denoising to get a better signal-to-noise ratio (SNR), that provides more useful information and/or the appropriate signal format for extracting specific features. Feature extraction block defines specific characteristics that would point out differences between signals, and therefore deals with the limitation regarding the computational burden during data processing by reducing the dimensionality. The final step involves the application of one of two procedures—classification or clustering. In terms of neural signal processing, classification refers to discrimination of different detected events and clustering is connected to spike sorting, whose main goal is to assign detected spike waveforms to different neural units. Depending on the chosen final procedure, all previous steps may differ from each other. In the following sections, we present possible and most commonly used methods to fulfill the purpose of every processing block presented in Fig. 20.1.

20.1.2 Preprocessing

20.1.2.1 Denoising of the signal

The main purpose of filtering is to extract the data that convey the relevant information from raw neural signal, and to eliminate artifacts and noise. Depending on the final goal of signal processing and the location where neural signals were recorded, filtering is done for different frequency bands.

Raw ENG (electroneurogram) signals have a frequency range between 100 Hz and 10 kHz (Yoshida et al., 2017), while CNS signals have useful information in a range up to 6 kHz (Razmpour et al., 2015). ENG signals for neuroprosthetic use may be band-pass filtered between 1 and 3 kHz, as surrounding muscle activity has harmonics up to 800 Hz, which masks useful neural signals at low frequencies. On the other hand, CNS local field potentials have useful information at frequencies lower that 300 Hz (Kajikawa & Schroeder, 2011). Spike analysis is often applied at a wider bandwidth. For brain implants, the spikes are usually extracted from signals band-pass filtered between 300 Hz and 6 kHz (Gibson et al., 2011). In general, full-band neural signals are obtained, depending on the quality of recording, between 100/300/500 Hz and 4/7/10 kHz (Dhillon et al., 2005; Micera et al., 2010; Rossini et al., 2010). The upper cutoff of the filter is defined in order to diminish the noisy appearance of spike waveforms; and the lower cutoff is determined to reduce drift and low-frequency physiological artifacts. Despite these differences in specific values, their order of magnitude is similar in all applications.

Since there is always some spectral overlap between noise and neural signals, especially in recordings from the peripheral nervous system (PNS), standard linear filtering fails to remove the artifacts when SNR is low. Therefore a need for defining nonlinear procedures has risen and *wavelet denoising* has been shown as one of the most promising techniques. It is used in signal processing of biomedical signals to remove the background artifact that could be characterized as a Gaussian distributed random source (Citi et al., 2008; Kim & Kim, 2003; Wiltschko et al., 2008), and found much wider application in the signals recorded in the PNS than in the CNS.

Noisy signal can be defined as:

$$y(t) = x(t) + \eta(t)$$

where $x(t)$ is informative signal and $\eta(t)$ is Gaussian white noise. The main idea is to transform noisy data to time-frequency domain (*decomposition step—analysis*), detect the artifact by applying threshold procedures to the resulted coefficients (*thresholding*), remove it, and return the signal back to the original time domain (*recomposition step—synthesis*).

The discrete wavelet transformation (DWT) (Diedrich et al., 2003) consists of iterative convolution of high-frequency and low-frequency filters in the *decomposition part.* Since every low-pass filter has a higher Nyquist frequency, the signal is down-sampled, and then, to reconstruct the signal, it is up-sampled (Pani et al., 2016). This results in a lack of translational invariance (Brychta et al., 2007). When the signal is translated in time, DWT composition will yield a different parameter set which would have a negative impact on denoising, spike detection (Brychta et al., 2006), or classification (Citi & Micera, 2013; Citi et al., 2008). Citi and Micera (2013) showed that the DWT algorithm can behave differently depending on the spike time and they presented how the same spike can be detected, missed, or deformed. These problems are resolved by using time-invariant approaches, such as continuous wavelet transformation and stationary wavelet transformation. Both of these are based on the analysis of the correlation between the signal and the mother wavelet function, which represents the transfer function of the decomposition and recomposition filters. During computation, the mother wavelet is shifted smoothly over the full domain of the analyzed function (Djilas et al., 2010). Choosing the appropriate mother wavelet is very important. There are several methods to guide its selection: correlation coefficient method (Pal et al., 2011), variance method (Rafiee & Tse, 2009), maximum energy to entropy ratio criteria (Kumar et al., 2014), and optimizing algorithms based on SNR maximization (Kamavuako et al., 2010). However, usually, the mother wavelet is chosen by visual inspection to best match the spike waveform (Citi et al., 2008; Diedrich et al., 2003).

Thresholding, part of the denoising process, can follow a hard threshold such as:

$$y_{hth} = \begin{cases} y; & |y| \geq \theta \\ 0; & |y| < \theta \end{cases}$$

or a soft threshold such as:

$$y_{sth} = \begin{cases} y - sign(y)\theta; & |y| \geq \theta \\ 0; & |y| < \theta \end{cases}$$

where θ is the selected threshold. These equations emphasize parts of the component signal that could be connected to action potential, while sets to zero parts are recognized as a noise. Selected threshold determines how mild the denoising of the signal will be. Several standard thresholds have been defined, depending on the type of signal and the estimate of the noise level. If the signal is considered to have Gaussian noise, the signal can be transformed and divided in as many subbands as the number of decomposition levels adopted plus one. Threshold is then determined at every level, by estimating the variance of the noise σ using one of the robust estimators [e.g., median absolute deviation (MAD)] in every subband (Barabino et al., 2017). In this case, the universal threshold (θ_u) and the minimax threshold (θ_m) are the most commonly used ones (Donoho & Johnstone, 1994):

$$\theta_u = \sigma\sqrt{2lnN} \quad \theta_m = \sigma\left(a + b\log_2 N\right)$$

where N is the number of samples on which the threshold is applied and a and b are coefficients which are generally set to 0.3936 and 0.1829, respectively. The minimax approach yields lower thresholds, and therefore, it is considered as the more aggressive one.

Recomposition step, synthesis of the signal, is done by applying the same function as in decomposition, but in inverse form. It can be additionally controlled by selecting the level of the decomposition containing relevant information on the neural activity (Brychta et al., 2007).

20.1.2.2 Running observational window analysis

In order to augment the signal processing performance and to enhance the information contained in the signal, window analysis is performed as one of the preprocessing steps. Predefined windows are used as time frames for calculating the specific features. These time series of a particular length are named as a running observational window. If the whole raw signal is divided into N frames of specific length each, it could potentially lead to wrong interpretation of the data that lie along the boundaries of the frames. Therefore windows should overlap and the coefficient of overlapping as well as the length of the windows should be based on the nature of the signal, its spectral characteristics, and features that will be calculated from this signal (Hong et al., 2018). The overlapping factor influences mostly computational time and typically goes from 25% to 75% of the length of the window. For the length of the window parameter, it can be one that maximizes the output accuracy. The optimal value is in most cases 100 ms, but it is hard to define it before testing the exact feature extraction paradigm. It has been shown that short windows up to 50 ms do not have enough stability, and that long windows (300 ms and longer) reduce the performance of the output (Raspopovic et al., 2010).

20.1.2.3 Feature extraction and selection

Feature extraction starts from the set of data that is output from the preprocessing block and defines properties (features) that contain the most useful information about the signal, and that are not redundant. Additionally, this step is connected to dimensionality reduction, as it summarizes the information that the signal carries.

Feature selection is the processing step in which a subset of initial features from the set of all features is determined. This is not a necessary step, it is applied only when the input data are too large, or when some kind of redundancy exists, or when hardware properties of the system require reduced data flow.

Selected features for extraction differ for the type of the electrode used. While signals recorded using intraneural electrodes (brain implants, intraneural electrodes—TIME, LIFE, etc.) are usually used to determine the spike shapes of the single units and to cluster them (spike sorting), signals recorded with extraneural electrodes (e.g., multichannel cuff, electroencephalography (EEG) spatial electrodes) resort to different types of features, and usually are classified for defining a better system control algorithm.

20.1.2.4 Features for classification

Features to be used can be time- or frequency-domain based and different approaches have been adopted in the previous literature. Time-domain features, such as mean absolute value (MAV), root mean square, variance, and wave length are the most commonly used ones (Brunton et al., 2017; Nazmi et al., 2016; Raspopovic et al., 2010; Silveira, Brunton, Spendiff, et al., 2018; Silveira, Brunton, Khushaba, et al., 2018). Even though these features were shown to have the best performance, other time-domain descriptors defined in Silveira, Brunton, Spendiff, et al. (2018) and Silveira, Brunton, Khushaba, et al. (2018) also provided comparable good results. The MAV coefficient was the basis for two modifications that are also incorporated in a standard set of features (Oskoei & Hu, 2008; Phinyomark et al., 2012.). Other time-domain features, used in the literature, are cepstral coefficients and autoregressive model parameters (AR/AAR prediction models) (Schlögl & Supp, 2006), where every sample of the signal is presented as a linear combination of previous samples with added white noise. It was suggested that fourth order of AR models can be used as a feature set (Paiss & Inbar, 1987).

The frequency-domain features usually rely on the power spectral density (PSD). In most of the cases, the features are selected as the PSD coefficients in extracted ROW (Leeb et al., 2013; Nag et al., 2014). By applying different statistical functions on PSD, additional features can be used (e.g., mean/median frequency, total power, power ratio, etc.). Features can also be defined based on a measure of autocorrelation of the transformed signal. All of these defined measures can be found in the works of Nazmi et al. (2016) and Phinyomark et al. (2012).

It was noticed that methods for feature extraction from electromyography (EMG) signals can also be applied for ENG or EEG signals recorded with extraneural electrodes, and therefore most of the features presented in Phinyomark et al. (2012), Raspopovic et al. (2010), Silveira, Brunton, Spendiff, et al. (2018), Silveira, Brunton, Khushaba, et al. (2018), and Nazmi et al. (2016) can also be used for processing neural signals.

Common spatial pattern filters are also used in neural signal processing, especially for processing cortical signals. It can be considered as a feature extraction method for separating the multivariate signal into subcomponents and for maximizing the discriminability of two classes. For this purpose, a set of spatial filters that linearly transforms the input signal is used (Oweiss, 2010).

Sometimes, the feature extraction step can be skipped and the results of preprocessing can be directly input to classifiers. That is usually done when classifiers are in the form of neural networks or genetic algorithms, because these algorithms automatically extract features that do not have any practical value for our understanding. However, this drastically increases the time necessary for the training phase of the classification, and therefore may not be recommended for real-time applications.

20.1.2.5 Feature extraction and selection for clustering

As clustering in neural signal processing usually refers to spike sorting, feature extraction in this step is connected to the methods for detecting the time when each spike occurs and the spike waveform. Different neurons and electrode geometries produce different spike shapes. Therefore for clustering, it is important that extracted features depict the characteristics of such different spike waveforms. Similar techniques are widely used for detecting the spikes from both PNS and CNS extracellular recordings.

20.1.3 Spike detection

20.1.3.1 Amplitude thresholding

The most commonly used method to detect spikes is amplitude thresholding: spikes are detected every time the amplitude of the signal crosses a user-defined threshold. The performance of this method relies too much on the selected threshold—if it is too low one will probably recognize too many spikes that are not true ones, but if it is set to high values, one will lose some of the spikes. Setting the right threshold is a compromise between these two scenarios (Mohammadi et al., 2020). Threshold is usually defined as:

$$Thr = k \times \sigma_n$$

where k is scalar multiplier and σ_n is the standard deviation of noise. The value of multiplier k depends on the power of the signal and mostly on the recording location, and it gives the best results when it is between 3 and 5 (Kamboh & Mason, 2012; Noce et al., 2018; Quiroga et al., 2004; Takekawa et al., 2010). Approximation of standard deviation can be done in several ways. One way is based on approximating σ_n based on MAD (Noce et al., 2018):

$$\sigma_n \approx 1.4826 \times MAD \quad MAD = median(|X - median(X)|)$$

As the alternative, the standard deviation of noise can be determined using the whole signal, with assumption of normal distribution of noise, and by assuming that the median of the noisy part is similar to the median of the whole signal (Quiroga et al., 2007):

$$\sigma_n = median\left(\frac{|x^2|}{0.6754}\right)$$

20.1.3.2 Template matching

The template matching algorithm is based on a comparison between the signal and spike templates. It consists of three stages: defining the template shapes, localizing possible events in the signal, and applying thresholding (Azami & Sanei, 2014). This process can be expensive in terms of computational time, and construction of appropriate templates can also be difficult. Algorithms for calculating the difference between the template and the signal, and finding templates are done automatically. One possible solution for automatic template construction is given in Sato et al. (2007), where first, the possible spikes are detected using the standard threshold method. Then, spike waveforms are characterized using PCA and clustered using a *k*-means algorithm. Templates are defined by averaging the cluster sets. Sorted spike events are defined based on the similarity between the templates and processed signal by calculating Euclidian distances, convolution, cross-correlation, covariance matrices, χ-test, or similar measures (Franke et al., 2015; Kim & McNames, 2007; Sato et al., 2007; Liu et al., 2012; Zhang, Wu, Zhou, Liang, & Yuan, 2004). This method is considered to have high robustness. Because of the specific approach for matching the waveforms, it may be possible to skip all the further steps of feature selection and clustering.

20.1.3.3 Energy-based spike detection

Spikes can be detected using thresholding of a local energy measurement. If energy is higher than a threshold, that is usually defined as a multiplication (usually equal to 5) of standard deviation of the energy detected in that time frame, a spike event is registered (Rutishauser et al., 2006). The nonlinear energy operator (NEO) (Kaiser, 1990) can be used instead of simple energy approximation. However, it is sensitive to noise and can produce an undesired time-varying part. These limitations are overcame using smoothed NEO (Mukhopadhyay & Ray, 1998), which, by using a smoothing window, is able to suppress noise. Energy-based spike detection algorithms outperform the amplitude thresholding approach, but do not have sufficient accuracy for signals with low SNR (Nenadic & Burdick, 2004).

20.1.3.4 Wavelet-based spike detection

Wavelet transform, described before, extracts features that are characteristic of the detected spike waveforms (Takekawa et al., 2010). This method outperforms previously described approaches as it accurately detects spikes even in recordings with low SNR. It detects spikes based on the amplitude, but also on the shape of the signal waveform and its duration. These algorithms are composed of three generalized steps: performing wavelet decomposition of the signal, testing the presence of spikes by applying Bayesian sequential hypotheses testing at different scales, and estimating the spike times (Nenadic & Burdick, 2004). This basic algorithm is improved and can be made more robust for signals with low SNR by choosing the appropriate, matched wavelet from an actual mean action potential waveform (Salmanpour et al., 2010).

20.1.3.5 Feature selection

After spike detection and extracting their shapes, it is important to choose the minimal set of features that provide the best discrimination. Spike amplitude peak and width of the spike can be used for clustering features (Lewicki, 1998), but they may yield low performance in terms of discrimination between spikes (Quiroga et al., 2004).

Since the spike shapes are represented with, say, N samples, the clustering is performed in N-dimensional space, which leads to a high computational load. Therefore dimensionality reduction is required. One of the most commonly used methods for reducing dimensions is principal component analysis (PCA). PCA is described as an orthogonal linear transformation that transforms the signal to a new coordinate system in the way that the highest variance of data lies in the first PCA component, second highest on the second component, etc. (Jolliffe, 2002). This procedure results in converting the N-dimensional problem into a K-dimensional one, where $K << N$. Typically, K is chosen to be 2 or 3 (Rey et al., 2015), as the first three components usually capture most of the variation in the spike waveform.

Better performance can be achieved by using wavelet signal decomposition, as wavelet coefficients are localized in time (unlike PCA). Moreover, PCA components that have the highest variance are not necessarily the ones that represent the maximum difference between spike shapes. The key challenge with wavelets is how to choose the coefficients that would best distinguish the difference between spike shapes. Unless we have one class, all wavelet coefficients will have multimodal distribution. Quiroga et al. (2004) proposed the modified Kolmogorov–Smirnov test for normality to automatically find the deviation from normality as a sign for multimodal distribution, which is defined as:

$$\max\left(\left|F(x) - G(x)\right|\right)$$

where $F(x)$ characterizes the empirical distribution function found from the data and $G(x)$ is the Gaussian distribution of the same mean and variance as tested data. The set of coefficients for clustering can be selected as in the work of Quiroga et al. (2004) from those with the highest deviation. For example, 10 coefficients were defined as inputs to clustering, but this depends on the data structure.

20.1.4 Classification and clustering

20.1.4.1 Classification

Classification for neural patterns recognition refers to supervised learning, which maps input data to an output based on the given examples of input–output pairs (Russell & Norvig, 2010). Generally, it consists of two phases: training and testing. Data that are available for constructing the classifying system need to be divided into training (around 70% of available data) and testing sets (around 30% of available data). These two sets should not be overlapped in order to avoid overfitting, which is reflected as a classification algorithm which is not able to generalize well, and can classify accurately only samples that are already given in its training set. Both data from training and testing sets go through the same type of feature extraction so as to be comparable. By using the data in the training set, the model is calibrated, and ready to use the inputs from the testing set to validate the accuracy of the constructed model. If the model behaves as expected, it can be further used as a decision-making system.

Features selected in previous steps are placed into one of the classifiers. There are no general guidelines for choosing the right classifier, rather, there are several possible alternatives with different separation rules. They can be tried and the best among them (with the highest accuracy, minimum computational time, or an optimum between them, depending on the application) can be retained. In neural signal processing, several of them are frequently applied: linear discriminant analysis (Silveira, Brunton, Khushaba, et al., 2018; Silveira, Brunton, Spendiff, et al., 2018), support vector machine (Badia et al., 2015; Micera et al., 2008, 2010, 2011; Raspopovic et al., 2010), and artificial neural networks (ANNs) (Mirfakhraei & Horch, 1997; Raspopovic et al., 2010; Rossini et al., 2010). With the growing amount of globally available neural data for signal processing, deep learning techniques also have started to be used as a standard procedure (Roy et al., 2019).

It is important to emphasize that the accuracy of classification mostly depends on preprocessing, feature extraction, and selection, but less on the choice of the classifier.

20.1.4.2 Clustering

Clustering represents the unsupervised learning algorithm, and it involves grouping of objects in clusters, where every individual object of the group is similar to the other objects of that group according to defined characteristics. Clustering is the final step in spike sorting. The goal is to group the points in feature space into separate clusters that will correspond to different neurons.

Clustering can be done manually (Harris et al., 2000; Pedreira et al., 2012), but it brings too many errors because of human biases and limited number of clusters, as well as being very time consuming.

Semiautomatic clustering is widely used, where the number of clusters is predetermined by the user. To overcome this, the PBM index, a cluster validity index that measures the "goodness" of clustering using a range of clusters, was employed to determine k (Pakhira et al., 2004). Additionally, k can be estimated using the algorithm proposed in Djilas et al. (2010). K-means algorithm minimizes within-class variances and each observation belongs to the cluster with the nearest mean. Classification expectation-maximization algorithms are also used with a predefined number of clusters (Pouzat et al., 2002; Sahani, 1999). These methods start from the assumption that clusters are spherically distributed in a feature space (Raspopovic et al., 2020). If different assumptions about the distribution are made, valley-seeking clustering (Koontz & Fukunaga, 1972), hierarchical clustering (Fee et al., 1996), or superparamagnetic clustering (Quiroga et al., 2004) can be also used.

Estimating the number of clusters can also be automatized in the whole process of clustering, based on previously described algorithms (Bestel et al., 2012; Chung et al., 2017; Nguyen et al., 2015).

20.1.4.3 Combining classification and clustering

As applications are going in the direction of making reliable closed-loop systems, there is an important need for making on-line classifying/spike sorting systems. One of the newest frameworks that follows this approach is given in Shaeri and Sodagar (2020). It represents the combination of classifying and clustering algorithms. As the offline spike sorting is done as a clustering algorithm with the training data set, online spikes from the testing set are sorted using the classification algorithm based on salient features. This approach significantly lowers the computational time of spike sorting and enables sorting spikes on a neuroprosthetic implant chip directly. That allows one to transfer a reduced amount of data from the implant, that is, only the sorted spikes, and brings the construction of the fully closed-loop system a bit closer.

20.2 Signal processing in neural stimulation

The development of neurostimulation techniques in order to evoke and generate a desired response is one of the most active fields of research in neural interfaces (Raspopovic, Valle, & Petrini, 2021). It is indeed an essential tool to learn, for controlling and interacting with biological neural networks, and its application has many implications on neuroprostheses and other therapies (Laferrière, Bonizzato, Côté, Dancause, & Lajoie, 2020).

There are two main challenges when designing algorithms with the intent of correlating spatiotemporal pattern of neural stimulation to the neural circuit activation:

1. Creation of a descriptive model necessary to understand how underlying neural circuits actually operate under no external influences.
2. Real-time encoding strategy that can use the model-derived knowledge to efficiently and effectively target desired neural circuits through the implanted interface.

Different approaches can be found to characterize the neural interface and the neural circuits separately or as combined. The type of mathematical description, moreover, depends on the final application of the stimulating device. The spatiotemporal stimulation patterns and the accuracy of the real-time encoding strategies are influenced by the implant active site resolution and the model dynamics (Fig. 20.2).

The objective of this section is to illustrate how these two main issues have been addressed in the literature and within the experimental methods to create more robust and efficient neural stimulation protocols in different applications.

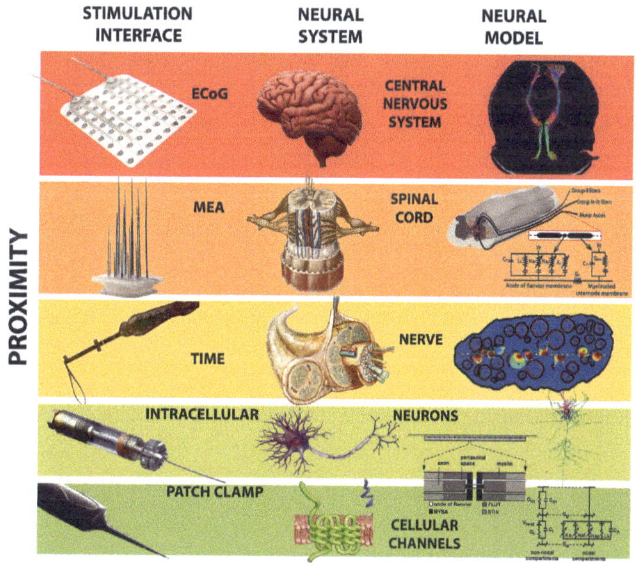

FIGURE 20.2 Hierarchical modeling strategies for neural stimulation depending on the level of proximity to the entire brain and the type of interface. The first column is organized in increasing proximity from a patch of neuron to the brain, and shows the types of electrical interfaces used. The second column represents the neural structure to which the interface is applied. In the third column, there are some examples of neural models used for optimal encoding for neural stimulation.

20.2.1 Processing through modeling

Communications within the CNS and PNS rely on interconnections between neurons containing millions of synapses: the information inside this network is encoded by the firing patterns of the associated neuron populations (Georgopoulos et al., 1986; Salinas & Abbott, 1994; Dayan & Abbott, 2001; Theunissen & Miller, 1995). Particular capability of the CNS is not only to propagate information, but to employ such networks to process the information along the way. These transformations produce neural representations that are highly dynamic, nonlinear, and nonstationary (Lestienne, 2001). Thus the introduction of additional information into the network through stimulation has to consider this preexisting construct to be integrated with. By the modeling efforts on the human nervous system, we can acquire the knowledge on how to develop better neural interfaces (Aisa et al., 2008; Broccard et al., 2017). Neural models can be created through two different processes of abstraction.

Parametric modeling, or the "bottom-up" approach, employs acquired physiology knowledge in order to establish an analytical formulation of the biological processes. Such approaches tend to break down neural mechanisms, so that it is possible to understand how to properly interface without disrupting input–output relationships of the modeled biological system. Hodgkin–Huxley compartmental neuron modeling is a perfect example of this approach where morphology and chemical ion channel dynamics are characterized to describe realistic neural firing activity (Carnevale & Hines, 2006).

On the other side of the spectrum, the "top-down" approach, or nonparametric modeling exploits the possibility of tailoring a nonspecific model in order to mimic input–output behavior, instead of taking advantage of progressive knowledge in neurophysiology. With the advancement of machine learning algorithms for nonparametric system identification and high-resolution acquisition systems, this particular approach permits to capture complex and nonlinear behavior, for which parametric solutions have not yet been as successful (Song, Marmarelis, et al., 2009; Song, Wang, et al., 2009).

Therefore it is crucial to understand how signal processing for stimulation, and electrical stimulus characterization can benefit from employing this modeling approach, and which future directions are worth exploring for a seamless interconnection in neural interfaces.

20.2.1.1 Parametric stimulus encoding

The use of modeling in the characterization of a better interface for interacting and optimizing selectivity has been explored in different ways. Leventhal et al. utilized a combination of experimental and modeling technique in order to characterize the selectivity of a FINE on the sciatic nerve of the cat. Simultaneous recording of the cat limb movement during neural stimulation allowed them to correlate current injection with related motor unit contraction force (Leventhal & Durand, 2003). Finite element modeling was used in order to characterize the motor fibers' extracellular potentials based on injected monophasic electrical pulses, and subsequently they estimated the force produced. Although this approach aimed for selectively targeting individual groups of motor fibers, the encoding of particular stimulation patterns for better selective stimulation was never tested. A similar approach was adopted by Raspopovic et al. in the rat sciatic nerve, to compare the selectivity indexes of TIME with respect to FINE (Raspopovic et al., 2011).

Long-term implanted electrical interfaces have been used to induce artificial touch perception (Graczyk et al., 2016; Valle et al., 2021). Previous experimental evidence was used to shape the population firing rate evoked by electrical stimulation, as the electrical pulse width and frequency varied. They concluded that the sensory magnitude of touch perception, independently from the type of sensation, was related to the total amount of activity in the nerve. This can be evoked temporarily by modulating the frequency, amplitude, and pulse width of the electrical stimulus. On the other hand, the quality and type of restored sensation were dependent on the pattern of this activity in the nerve, and therefore in the specific spatiotemporal configuration (Tan et al., 2014). To achieve a more complex mapping, novel stimulation encoding strategies are necessary. A recent development in this direction was obtained by George et al. (2019). Combining a biomimetic sensory encoding algorithm, which was developed from recordings of nonhuman primates' cutaneous afferents to arbitrary spatiotemporal deformation of the skin (Saal et al., 2017), they were able to restore haptic perception relatively well. In fact, the participant was able to identify the objects significantly faster than that obtained with the use of traditional encoding algorithms which depended only on the present stimulus intensity (George et al., 2019; Öztürk et al., 2019).

Similar modeling approaches have been done for electrical stimulation in the spinal cord of rats (Capogrosso et al., 2013). A computational model was provided, and as in the previous cases, it gave information about optimum electrode placement, electrode shape, and the number of active sites for a better interface. Although movement of the limb was enhanced in impaired rats during walking, difficulties were found when selective stimulation of extensor and flexor muscles of the limbs was tested. The solution to this particular issue may be found in better stimulation encoding, as the authors stated. Combining static and dynamic network modeling in association with high-density electrode arrays can allow selective activation of motor neural circuits in neurological disorders.

Particular evidence of this necessity is stressed from the following studies which demonstrated that continuous epidural electrical stimulation disrupts the ability of proprioceptive information to modulate the motor output elicited (Formento et al., 2018). The proposed solution was therefore to modify continuous stimulation with phase-specific modulation (Moraud et al., 2016).

The inclusion of a biologically realistic neural network of muscle spindle feedback circuits for two antagonist muscles in the limbs allows first to estimate natural spinal interneuron modulation, to synchronously stimulate and trigger the neural control loop (Formento et al., 2018). Simulations with stimulus encoding strategies in spatiotemporal patterns coinciding with firing profiles of proprioceptive afferents demonstrated robust control over the motor neuron activity as a result of the increased preservation of feedback information. Additionally, using a high frequency and low amplitude encoding helped in partially avoiding the loss of afferent information due to the poor fiber selectivity.

Fiber selectivity and understanding how to stimulate a specific type of fiber are both essential in brain–machine interfaces. In fact, depending on the cortical layer, different types of neural cells are involved. A step forward in understanding which neural elements could be actually activated from a given stimulus was investigated by Komarov et al. (2019). Komarov et al. proposed a modeling approach for the estimation of stimulation-induced activation of the anatomical neural networks reconstructed from 12 main neocortical categories. The approach consisted of four main steps. The electric field potential was calculated inside the tissue originating from the electrical current injection. The "axonal-electrical receptive field" was later estimated based on the different cell types across layers and electrode distance. Cells' spiking probability was derived from the activating function in axonal elements. And finally, the single cell response was integrated into the network and the behavior of the cortical column model was analyzed. Their work concluded that there was an optimal stimulation intensity capable of inducing a maximal response in certain cortical cell populations. Further development of this modeling technique will permit to specifically characterize the amplitude and frequency of the electrical stimulation to control cortical column activity at different levels, and also by various electrode combinations.

Similarly, in a recent study, McIntyre and Grill proposed stimulus parameter identification aiming to effectively and selectively stimulate targeted neuronal populations within CNS. Cable models of neurons, that include an axon, initial segment, soma, and the branching dendritic tree, were used to study excitation with extracellular electrodes. The cell geometries and membrane dynamics were derived from mammalian motor neurons (McIntyre & Grill, 2000). Moreover, the developed models predicted the effectiveness of a range of electrode geometries and stimulus parameters; therefore they are presented as a useful tool in designing stimulus waveforms for use in CNS neural prosthetic devices.

On the other hand, electrical stimulation in key brain structures with deep brain stimulation (DBS) has been used as an effective treatment for neurological movement disorders such as Parkinson's disease and essential tremor (Ramirez-Zamora et al., 2019). However, the development of such technology, which can approach spatially and temporally specific deep circuits, opened the road to treatments in other pathologies such as dementia and epilepsy, which are also seen as network-level dysfunctions (Sivakanthan et al., 2016). In fact, recent studies proposed closed-loop DBS approaches not only for targeting specifically disease-related circuits, but also for refining stimulation encoding. Both anatomical (diffusion tensor imaging and tractography) and functional techniques (EEG, functional MRI, and local field potentials) were used to analyze brain connectivity (Calabrese, 2016). These advances in DBS have been seen in numerous studies where stimulating currents are steered to target individualized networks both for depression and obsessive-compulsive disorders (Banks et al., 2015; Wilamowska et al., 2010). Creating a computational model of the internal brain circuits enables the creation of novel stimulation strategies in order to address therapeutically symptom-specific differential stimulation (e.g., speech, gait, nonmotor symptoms of Parkinson disease) (Humphries & Gurney, 2012).

Even though parametric modeling has shown great potentialities for interfacing and tuning electrical stimulation parameters, no attempt has shown evidence of a seamless interconnection between natural feedback restoration and efferent motor control. This particular issue is related to the fact that parametric modeling is built on largely an approximation of the biological systems, due to either partial knowledge of the system or lack of computational power for simulating all its components. Scale, resolution, or complexity inadequacy makes it difficult for engineers to compare or relate simulation results with experimental data.

20.2.1.2 Nonparametric stimulus encoding

Nonparametric modeling approaches the problem of tuning of electrical stimulation parameters not as a predefined set of parametrized functions, but instead approximating them through statistical distributions.

An interesting implementation was attempted through closed-loop neuromodulation of spinal sensorimotor circuits trying to reproduce the natural dynamics of motoneuron activation during locomotion after complete spinal cord injury in rats (Wenger et al., 2014). Spatiotemporal activation patterns were identified through computer simulations in order to restore healthy rats' flexor–extensor muscle synergies through the recruitment of proprioceptive feedback neural circuits. Using a linear predictive model between the electrical stimulation frequencies and the movement phases, control of a broad range of foot trajectories during locomotion in a paralyzed rat was performed (Wenger et al., 2016).

This particular approach, even if particularly efficient for simple and rhythmic tasks such as locomotion, has difficulties in generating complex stimulation and correlated evoked muscle activation patterns. In fact, due to the targeted extraction of activation patterns related to a restricted number of motions, it is often impossible to uncover nonlinear neural circuits through these conventional mapping techniques (Laferrière, Bonizzato, Côté, Dancause, & Lajoie, 2020). In order to make spatiotemporal stimulus encoding possible, a more flexible search algorithm has been investigated in nonhuman primate models through motor cortex microelectrode stimulation. Laferrière, Bonizzato, Côté, Dancause, & Lajoie (2020) through iterative processes using Bayesian optimization of a hierarchy of increasingly complex signal space, were able to rapidly learn and uncover the mapping between complex muscle activation patterns and stimulations. The versatility of this optimization of a stimulus encoding algorithm makes it suitable for spinal and nerve electrical stimulation, and the objective function can be adapted to the particular set of movements and range of motions to be performed.

Applications in this direction have been used in spinal cord stimulation for pain treatments. In particular, scientific observations have demonstrated that effective stimulation parameters varied according to patients' posture (Cameron & Alo, 1998; Olin et al., 1998). Proposed solutions adjusted stimulation parameters adaptively according to patients' posture, as measured via a triaxial accelerometer integrated into the implanted stimulator (Schade et al., 2011). Six different body positions were identified (e.g., standing, sitting, lying, etc.) through the accelerometer and stimulation was fine-tuned accordingly (Schultz et al., 2012).

Due to particular difficulties associated with the complex relationship between stimulation and desired responses in the optical nerve, visual restoration with nonparametric electrical stimulus encoding has attracted special attention. Multivariate machine learning has been applied to intracranial electrocorticography to encode temporal dynamics of the human face information being processed in the fusiform gyrus face area (Ghuman et al., 2014). Another promising approach of machine learning for vision neuroprostheses was developed by the Monash Vision Group (Lowery et al., 2017). Electrode stimulation patterns were generated through intracortical simulations based on image processing information and elaboration of camera images. Interestingly, electrode activation patterns did not aim at the restoration of biological vision, but instead, in conveying the most helpful information for the user (Niketeghad & Pouratian, 2019). The transcorneal stimulation for retinal and optical diseases is strongly dependent on many factors. One of the most important aspects relies on how to produce the required stimulation signal to produce the desired response. However, this is not an easy task, due to the relationship between the stimulation signals and the response that is almost unknown. Adaptive techniques to achieve a good approximation of the uncertain function relating the stimulation and response signals have been proposed for transcorneal stimulation. In avian models, for example, a two-step nonparametric identification scheme based on differential neural networks was tested.

This consisted of an internal nonparametric mathematical model of the input–output retina–optical nerve relationship (Salgado et al., 2019).

At this time, adoption of DBS for long-term treatment is relatively limited (Cagnan et al., 2019). Finding the appropriate combination of stimulation parameters for an effective DBS therapy remains one of the main challenges in order to obtain a therapeutic benefit on the patient (Miocinovic et al., 2013). For avoiding ad hoc search of these parameters, machine learning approaches have been proposed to estimate the DBS parameter values for a given volume of tissue activated (VTA). Using a combination of classification and regression algorithms (*k*-nearest neighbors and support-vector machines) with dimensionality reduction techniques it was possible to generate the desired target VTA following the restriction imposed by the characteristics of the brain tissue and the stimulation device (Gómez-Orozco et al., 2020). This particular methodology allowed setting a biophysically compliant target VTA and accurately predicted the required configuration of stimulation parameters.

20.3 Closing the loop

The possibility to record and stimulate at multiple levels of the neural pathways allows the development of different approaches and protocols to restore and/or treat different pathologies or disorders (Raspopovic, 2020).

Trying to combine the above-presented strategies both for decoding and encoding with invasive electrodes represents the next step in future bidirectional neural prostheses. Peripheral intraneural implants for bidirectional prostheses have already been tested with transradial amputees, and they have been proven to provide tactile information and adequate movement control (Micera et al., 2008). This technology is, however, still not able to elicit natural sensory feedback and seamless control. Biomimetic frequency modulation and amplitude modulation were demonstrated to enhance performance in tasks requiring fine force control (D'Anna et al., 2017). Moreover, it was demonstrated that the combination of these strategies improved manual dexterity and prosthesis embodiment, also reducing phantom limb perceptions. These findings represent important evidence of how well-designed signal processing protocols can make a large difference to the final outcome, by compensating for some of the current technical limitations.

In rat models, it has been shown how brain-controlled stimulation enables paralyzed rats to walk overground and adjust foot clearance to climb stairs (Bonizzato et al., 2018). Foot-off events were decoded from cortical population responses and finally converted into electrical neuromodulation of proprioceptive circuits related to flexion. In humans, instead, stimulation at the spinal level demonstrated partial restoration of voluntary control of walking in individuals who had sustained a spinal cord injury (Wagner et al., 2018). Delivering of spatiotemporal selective stimulation to the lumbosacral spinal cord patients re-established the adaptive control of paralyzed muscles during overground walking in 1 week. Coupling this technology with the possibility of bidirectional communication with the CNS, for many neurological disorders, will establish a framework for daily use of assistance devices and rehabilitative tools to enhance recovery.

Despite the enormous computational efforts, biological responses to the implant, like fibrotic tissue growth around the electrode, compromise recordings and electrical potential distribution near the active sites (Raspopovic et al., 2017). For these reasons, decoding for neuroprosthesis control in long-term studies prefers interfacing directly with human natural neuroamplifiers, that is, muscles. Implantable myoelectric sensors are used successfully to capture EMG signals from multiple residual muscles simultaneously (Kristjansson et al., 2017). In the case of transtibial amputees, multiple degrees of freedom, for a near-natural real-time control of the prosthetic device, can be decoded from simultaneous activity (Pasquina et al., 2015). Unfortunately, this approach has strong limitations when amputation is too proximal and residual muscles are missing. In the case of motor dysfunctions, functional electrical stimulation (FES) allows to target directly the desired muscles of the intact limb, avoiding encoding and modeling strategies for alpha motor recruitment, which is necessary in the case of nerve and CNS electrical stimulation. High nonlinearity, strongly coupling, time-varying, time-delayed, and redundant properties of the musculoskeletal system still require the design of an inverse model of the system for control purposes. Park et al. proposed a control system for multiple-input, multiple-output (MIMO) redundant musculoskeletal system of the ankle–subtalar joint with eight muscles. A dynamic inverse model was established by an ANN feedforward controller for fast trajectory tracking using only the empirical input–output data (Park & Durand, 2008).

Combining ENG recording and FES stimulation remains a challenge to close the loop due to emitter–receiver interferences. Studies on cat models have laid down first experimental protocols and devices which are able to record from sciatic, peroneal, and tibial nerves in the presence of electrical stimulation of gastrocnemius and tibialis anterior muscles in the same leg (Nikolic et al., 1994).

EMG recording can additionally be used, not only for motion decoding, but also for fine tuning of DBS devices. Novel ON/OFF amplitude modulated systems for tremor-dominant Parkinson disease patients can use EMG to predict tremor through machine learning up to several minutes before the pathological episodes (Basu et al., 2013; Khobragade et al., 2015). This permits administration of a more targeted therapy avoiding overstimulation, and also prolongs the device life. On the other hand, brain–computer interface (BCI)-controlled adaptive DBS of the subthalamic nucleus in eight Parkinson disease patients demonstrated the possibility to use instead local field potentials recorded directly from the stimulation electrode in order to personalize and optimize the stimulation in real time (Little et al., 2013).

A more recent approach to adaptive DBS proposed continuously updating DBS voltage values, reducing them when the recorded gamma power on the motor cortex was above the dyskinesia-related threshold (Swann et al., 2018). Swann and colleagues were the first to demonstrate the feasibility of adaptive DBS in Parkinson disease using a fully implanted device and neural sensing. These closed-loop electrical neurostimulators suffer from an important limitation, that is, they use manually defined biomarkers in order to modulate the therapy. These features usually result from both pathological and healthy processes, and as such, they may lead to incorrect stimulation, triggering unwanted effects and affecting normal physiological behavior (Iturrate et al., 2018). Data-driven and classification approaches are a promising solution in order to find a stronger combination of biomarkers to define the state in which the subject is. Pattern recognition, for example, has been adopted to distinguish between pathological tics and voluntary movements in Tourette syndrome during concurrent M1 subthalamic nucleus local field potential recordings (Shute et al., 2016). Pathological signatures were found in greater low-frequency power increases over the thalamus and higher beta desynchronization over M1 with respect to healthy controls.

The control of microstimulation both peripherally and centrally in BNIs is critical to continuously modulate the neural activity patterns. It is particularly significant when information about the state of the neutrally actuated device also needs to be delivered to sensory areas in the brain.

Liu et al. investigated the thalamo-cortical sensory pathway and the implementation of the MIMO feedback controllers. Cells and synapses were modeled to understand how neural feedback can be used to control the spatial and temporal neural activity of cortical pyramidal cells in a thalamo-cortical neural microcircuit. Therefore simultaneously changing "in silico" stimulation parameters across multiple active sites in the thalamic circuit permits recording of the evoked responses of the cortical pyramidal cells. This was used to create a closed-loop control system improving the performance of the interface by controlling the firing activity pattern of a few "key" neural elements in the network (Liu et al., 2011).

Signal processing, as illustrated in this chapter, is not a only tool for engineers to overcome technological barriers, but also represents a complex and dynamic "know-how" to optimize information transfer in neuroprosthetics. Bidirectional neuroprostheses utilizing invasive electrical interfaces have been pushing the boundaries of human–robot collaboration, and in spite of the technological limitations, they are blurring those edges between man and machine. It is in this effort of creating a seamless exchange of information, that engineers developed tailored signal processing techniques particular to specific neuroprosthetic device purposes and for various types of interfaces. Signal processing embodies this renewed spirit in filling the technological gap, shifting from a user-compliant perspective toward a holistic user-centered approach where information is shared effortlessly via neural interfaces.

References

Aisa, B., Mingus, B., & O'Reilly, R. (2008). The emergent neural modeling system. *Neural Networks, 21*(8), 1146–1152.

Azami, H., & Sanei, S. (2014). Spike detection approaches for noisy neuronal data: Assessment and comparison. *Neurocomputing, 133*, 491–506.

Badia, J., Raspopovic, S., Carpaneto, J., Micera, S., & Navarro, X. (2015). Spatial and functional selectivity of peripheral nerve signal recording with the transversal intrafascicular multichannel electrode (TIME). *IEEE Transactions on Neural Systems and Rehabilitation Engineering, 24*(1), 20–27.

Banks, G. P., Mikell, C. B., Youngerman, B. E., Henriques, B., Kelly, K. M., Chan, A. K., ... Sheth, S. A. (2015). Neuroanatomical characteristics associated with response to dorsal anterior cingulotomy for obsessive-compulsive disorder. *JAMA Psychiatry, 72*(2), 127–135.

Barabino, G., Baldazzi, G., Sulas, E., Carboni, C., Raffo, L., & Pani, D. (2017, July). Comparative evaluation of different wavelet thresholding methods for neural signal processing. In *2017 39th annual international conference of the IEEE engineering in medicine and biology society (EMBC)* (pp. 1042–1045). IEEE.

Basu, I., Graupe, D., Tuninetti, D., Shukla, P., Slavin, K. V., Metman, L. V., & Corcos, D. M. (2013). Pathological tremor prediction using surface electromyogram and acceleration: Potential use in 'ON–OFF' demand driven deep brain stimulator design. *Journal of Neural Engineering, 10*(3), 036019.

Bestel, R., Daus, A. W., & Thielemann, C. (2012). A novel automated spike sorting algorithm with adaptable feature extraction. *Journal of Neuroscience Methods, 211*(1), 168–178.

Bonizzato, M., Pidpruzhnykova, G., DiGiovanna, J., Shkorbatova, P., Pavlova, N., Micera, S., & Courtine, G. (2018). Brain-controlled modulation of spinal circuits improves recovery from spinal cord injury. *Nature Communications, 9*(1), 1–14.

Broccard, F. D., Joshi, S., Wang, J., & Cauwenberghs, G. (2017). Neuromorphic neural interfaces: From neurophysiological inspiration to biohybrid coupling with nervous systems. *Journal of Neural Engineering, 14*(4), 041002.

Brunton, E., Blau, C. W., & Nazarpour, K. (2017). Separability of neural responses to standardised mechanical stimulation of limbs. *Scientific Reports, 7*(1), 1–14.

Brychta, R. J., Shiavi, R., Robertson, D., & Diedrich, A. (2007). Spike detection in human muscle sympathetic nerve activity using the kurtosis of stationary wavelet transform coefficients. *Journal of Neuroscience Methods, 160*(2), 359–367.

Brychta, R. J., Tuntrakool, S., Appalsamy, M., Keller, N. R., Finney, C., Robertson, D., ... Diedrich, A. (2006). Spike detection in mouse renal sympathetic nerve activity using the stationary wavelet transform with automated noise level estimation. *IEEE Transactions on Biomedical Engineering, 54*.

Cagnan, H., Denison, T., McIntyre, C., & Brown, P. (2019). Emerging technologies for improved deep brain stimulation. *Nature Biotechnology, 37*(9), 1024–1033.

Calabrese, E. (2016). Diffusion tractography in deep brain stimulation surgery: a review. *Frontiers in neuroanatomy, 10*, 45.

Cameron, T., & Alo, K. M. (1998). Effects of posture on stimulation parameters in spinal cord stimulation. *Neuromodulation: Technology at the Neural Interface, 1*(4), 177–183.

Capogrosso, M., Wenger, N., Raspopovic, S., Musienko, P., Beauparlant, J., Luciani, L. B., Courtine, G., & Micera, S. (2013). A computational model for epidural electrical stimulation of spinal sensorimotor circuits. *Journal of Neuroscience, 33*(49), 19326–19340.

Carnevale, N. T., & Hines, M. L. (2006). *The NEURON book* (1st ed., pp. 32–35). Cambridge University Press.

Chung, J. E., Magland, J. F., Barnett, A. H., Tolosa, V. M., Tooker, A. C., Lee, K. Y., Shah, K. G., Felix, S. H., Frank, L. M., & Greengard, L. F. (2017). A fully automated approach to spike sorting. *Neuron, 95*(6), 1381–1394.

Citi, L., & Micera, S. (2013). *Wavelet denoising and conditioning of neural recordings. Introduction to neural engineering for motor rehabilitation* (pp. 173–182). Hoboken, NJ: John Wiley & Sons, Inc..

Citi, L., Carpaneto, J., Yoshida, K., Hoffmann, K. P., Koch, K. P., Dario, P., & Micera, S. (2008). On the use of wavelet denoising and spike sorting techniques to process electroneurographic signals recorded using intraneural electrodes. *Journal of Neuroscience Methods, 172*(2), 294–302.

D'Anna, E., Petrini, F. M., Artoni, F., Popovic, I., Simanić, I., Raspopovic, S., & Micera, S. (2017). A somatotopic bidirectional hand prosthesis with transcutaneous electrical nerve stimulation based sensory feedback. *Scientific Reports, 7*(1), 1–15.

Dayan, P., & Abbott, L. F. (2001). *Theoretical neuroscience: Computational and mathematical modeling of neural systems* (1st ed., pp. 229–278). Computational Neuroscience Series.

Dhillon, G. S., Kruger, T. B., Sandhu, J. S., & Horch, K. W. (2005). Effects of short-term training on sensory and motor function in severed nerves of long-term human amputees. *Journal of Neurophysiology, 93*(5), 2625–2633.

Diedrich, A., Charoensuk, W., Brychta, R. J., Ertl, A. C., & Shiavi, R. (2003). Analysis of raw microneurographic recordings based on wavelet de-noising technique and classification algorithm: Wavelet analysis in microneurography. *IEEE Transactions on Biomedical Engineering, 50*(1), 41–50.

Djilas, M., Azevedo-Coste, C., Guiraud, D., & Yoshida, K. (2010). Spike sorting of muscle spindle afferent nerve activity recorded with thin-film intrafascicular electrodes. *Computational Intelligence and Neuroscience, 2010.*

Donoho, D. L., & Johnstone, J. M. (1994). Ideal spatial adaptation by wavelet shrinkage. *Biometrika, 81*(3), 425–455.

Fee, M. S., Mitra, P. P., & Kleinfeld, D. (1996). Automatic sorting of multiple unit neuronal signals in the presence of anisotropic and non-Gaussian variability. *Journal of Neuroscience Methods, 69*(2), 175–188.

Formento, E., Minassian, K., Wagner, F., Mignardot, J. B., Le Goff-Mignardot, C. G., Rowald, A., Bloch, J., Micera, S., Capogross, M., & Courtine, G. (2018). Electrical spinal cord stimulation must preserve proprioception to enable locomotion in humans with spinal cord injury. *Nature Neuroscience, 21*(12), 1728–1741.

Franke, F., Quiroga, R. Q., Hierlemann, A., & Obermayer, K. (2015). Bayes optimal template matching for spike sorting–combining fisher discriminant analysis with optimal filtering. *Journal of Computational Neuroscience, 38*(3), 439–459.

George, J. A., Kluger, D. T., Davis, T. S., Wendelken, S. M., Okorokova, E. V., He, Q., Duncan, C. C., Hutchinson, D. T., Thumser, Z. C., Beckler, D. T., Marasco, P. D., Bensmaia, S. J., & Clark, G. A. (2019). Biomimetic sensory feedback through peripheral nerve stimulation improves dexterous use of a bionic hand. *Science Robotics, 4*(32), eaax2352.

Georgopoulos, A. P., Schwartz, A. B., & Kettner, R. E. (1986). Neuronal population coding of movement direction. *Science, 233*(4771), 1416–1419.

Ghuman, A. S., Brunet, N. M., Li, Y., Konecky, R. O., Pyles, J. A., Walls, S. A., Destefino, V., Wang, W., & Richardson, R. M. (2014). Dynamic encoding of face information in the human fusiform gyrus. *Nature Communications, 5*(1), 1–10.

Gibson, S., Judy, J. W., & Marković, D. (2011). Spike sorting: The first step in decoding the brain: The first step in decoding the brain. *IEEE Signal Processing Magazine, 29*(1), 124–143.

Gómez-Orozco, V., de la Pava, I., Álvarez-Meza, A., Álvarez, M. A., & Orozco-Gutiérrez, Á. (2020). A machine learning approach to support deep brain stimulation programming. *Revista Facultad de Ingeniería Universidad de Antioquia, 95*, 20–33.

Graczyk, E. L., Schiefer, M. A., Saal, H. P., Delhaye, B. P., Bensmaia, S. J., & Tyler, D. J. (2016). The neural basis of perceived intensity in natural and artificial touch. *Science Translational Medicine, 8*(362), 362ra142–362ra142.

Harris, K. D., Henze, D. A., Csicsvari, J., Hirase, H., & Buzsaki, G. (2000). Accuracy of tetrode spike separation as determined by simultaneous intracellular and extracellular measurements. *Journal of Neurophysiology, 84*(1), 401–414.

Hong, K. S., Aziz, N., & Ghafoor, U. (2018). Motor-commands decoding using peripheral nerve signals: A review. *Journal of Neural Engineering, 15*(3), 031004.

Humphries, M. D., & Gurney, K. (2012). Network effects of subthalamic deep brain stimulation drive a unique mixture of responses in basal ganglia output. *European Journal of Neuroscience, 36*(2), 2240–2251.

Iturrate, I., Pereira, M., & Millán, J. D. R. (2018). Closed-loop electrical neurostimulation: Challenges and opportunities. *Current Opinion in Biomedical Engineering, 8*, 28–37.

Jolliffe, I. T. (2002). Springer series in statistics. *Principal Component Analysis, 29.*

Kaiser, J.F. (1990, April). On a simple algorithm to calculate the "energy" of a signal. In *International conference on acoustics, speech, and signal processing* (pp. 381–384). IEEE.

Kajikawa, Y., & Schroeder, C. E. (2011). How local is the local field potential? *Neuron, 72*(5), 847–858.

Kamavuako, E. N., Jensen, W., Yoshida, K., Kurstjens, M., & Farina, D. (2010). A criterion for signal-based selection of wavelets for denoising intrafascicular nerve recordings. *Journal of Neuroscience Methods, 186*(2), 274–280.

Kamboh, A. M., & Mason, A. J. (2012). Computationally efficient neural feature extraction for spike sorting in implantable high-density recording systems. *IEEE Transactions on Neural Systems and Rehabilitation Engineering, 21*(1), 1–9.

Khobragade, N., Graupe, D., & Tuninetti, D. (2015, August). Towards fully automated closed-loop deep brain stimulation in Parkinson's disease patients: A LAMSTAR-based tremor predictor. In *2015 37th annual international conference of the IEEE engineering in medicine and biology society (EMBC)* (pp. 2616–2619). IEEE.

Kim, K. H., & Kim, S. J. (2003). A wavelet-based method for action potential detection from extracellular neural signal recording with low signal-to-noise ratio. *IEEE Transactions on Biomedical Engineering, 50*(8), 999–1011.

Kim, S., & McNames, J. (2007). Automatic spike detection based on adaptive template matching for extracellular neural recordings. *Journal of Neuroscience Methods, 165*(2), 165–174.

Komarov, M., Malerba, P., Golden, R., Nunez, P., Halgren, E., & Bazhenov, M. (2019). Selective recruitment of cortical neurons by electrical stimulation. *PLoS Computational Biology, 15*(8), e1007277.

Koontz, W. L., & Fukunaga, K. (1972). A nonparametric valley-seeking technique for cluster analysis. *IEEE Transactions on Computers, 100*(2), 171–178.

Kristjansson, K., Sigurdardottir, J. S., Sverrisson, A. Ö., Sigurthorsson, S. P., Sverrisson, O., Einarsson, A., Lechler, K., Invarsson, T., & Oddsson, M. (2017). *Prosthetic control by lower limb amputees using implantable myoelectric sensors. Converging clinical and engineering research on neurorehabilitation II* (pp. 571–574). Cham: Springer.

Kumar, H. S., Srinivasa Pai, P., Sriram, N. S., & Vijay, G. S. (2014). *Selection of mother wavelet for effective wavelet transform of bearing vibration signals. Advanced materials research* (Vol. 1039, pp. 169–176). Trans Tech Publications Ltd.

Laferrière, S., Bonizzato, M., Côté, S. L., Dancause, N., & Lajoie, G. (2020). Hierarchical Bayesian optimization of spatiotemporal neurostimulations for targeted motor outputs. *IEEE Transactions on Neural Systems and Rehabilitation Engineering*, *28*(6), 1452–1460.

Leeb, R., Perdikis, S., Tonin, L., Biasiucci, A., Tavella, M., Creatura, M., Molina, A., Khodairy, A., Carlson, T., & dR Millán, J. (2013). Transferring brain–computer interfaces beyond the laboratory: Successful application control for motor-disabled users. *Artificial Intelligence in Medicine*, *59*(2), 121–132.

Lestienne, R. (2001). Spike timing, synchronization and information processing on the sensory side of the central nervous system. *Progress in Neurobiology*, *65*(6), 545–591.

Leventhal, D. K., & Durand, D. M. (2003). Subfascicle stimulation selectivity with the flat interface nerve electrode. *Annals of Biomedical Engineering*, *31*(6), 643–652.

Lewicki, M. S. (1998). A review of methods for spike sorting: The detection and classification of neural action potentials. *Network: Computation in Neural Systems*, *9*(4), R53–R78.

Little, S., Pogosyan, A., Neal, S., Zavala, B., Zrinzo, L., Hariz, M., Foltynie, T., Limousin, P., Ashkan, K., FitzGerald, J., Green, A. L., Aziz, T. Z., & Brown, P. (2013). Adaptive deep brain stimulation in advanced Parkinson disease. *Annals of Neurology*, *74*(3), 449–457.

Liu, J., Khalil, H. K., & Oweiss, K. G. (2011). Neural feedback for instantaneous spatiotemporal modulation of afferent pathways in bi-directional brain–machine interfaces. *IEEE Transactions on Neural Systems and Rehabilitation Engineering*, *19*(5), 521–533.

Liu, X., Yang, X., & Zheng, N. (2012). Automatic extracellular spike detection with piecewise optimal morphological filter. *Neurocomputing*, *79*, 132–139.

Lowery, A. J., Rosenfeld, J. V., Rosa, M. G., Brunton, E., Rajan, R., Mann, C., Armstrong, M., Mohan, A., Josh, H., Kleeman, L., Li, W. H., & Pritchard, J. (2017). *Monash vision group's gennaris cortical implant for vision restoration. Artificial vision* (pp. 215–225). Cham: Springer.

McIntyre, C. C., & Grill, W. M. (2000). Selective microstimulation of central nervous system neurons. *Annals of Biomedical Engineering*, *28*(3), 219–233.

Micera, S., Citi, L., Rigosa, J., Carpaneto, J., Raspopovic, S., Di Pino, G., Rossini, L., Yoshida, K., Denaro, L., Dario, P., & Rossini, P. M. (2010). Decoding information from neural signals recorded using intraneural electrodes: Toward the development of a neurocontrolled hand prosthesis. *Proceedings of the IEEE*, *98*(3), 407–417.

Micera, S., Navarro, X., Carpaneto, J., Citi, L., Tonet, O., Rossini, P. M., Carrozza, M. C., Hoffman, K. P., Vivó, M., Yoshida, K., & Dario, P. (2008). On the use of longitudinal intrafascicular peripheral interfaces for the control of cybernetic hand prostheses in amputees. *IEEE Transactions on Neural Systems and Rehabilitation Engineering*, *16*(5), 453–472.

Micera, S., Rossini, P. M., Rigosa, J., Citi, L., Carpaneto, J., Raspopovic, S., Tombini, M., Cipriani, C., Assenza, G., Carrozza, M. C., Hoffman, K. P., Yoshida, K., Navarro, X., & Dario, P. (2011). Decoding of grasping information from neural signals recorded using peripheral intrafascicular interfaces. *Journal of Neuroengineering and Rehabilitation*, *8*(1), 1–10.

Miocinovic, S., Somayajula, S., Chitnis, S., & Vitek, J. L. (2013). History, applications, and mechanisms of deep brain stimulation. *JAMA Neurology*, *70*(2), 163–171.

Mirfakhraei, K., & Horch, K. (1997). Recognition of temporally changing action potentials in multiunit neural recordings. *IEEE Transactions on Biomedical Engineering*, *44*(2), 123–131.

Mohammadi, M., Khan, N. A., Hassanpour, H., & Mohammed, A. H. (2020). Spike detection based on the adaptive time–frequency analysis. *Circuits Systems and Signal Processing*.

Moraud, E. M., Capogrosso, M., Formento, E., Wenger, N., DiGiovanna, J., Courtine, G., & Micera, S. (2016). Mechanisms underlying the neuromodulation of spinal circuits for correcting gait and balance deficits after spinal cord injury. *Neuron*, *89*(4), 814–828.

Mukhopadhyay, S., & Ray, G. C. (1998). A new interpretation of nonlinear energy operator and its efficacy in spike detection. *IEEE Transactions on Biomedical Engineering*, *45*(2), 180–187.

Nag, S., Ng, K.A., Jagadeesan, R., Sheshadri, S., Delgado-Martínez, I., Bossi, S.,... Thakor, N.V. (2014, October). Neural prosthesis for motor function restoration in upper limb extremity. In *2014 IEEE biomedical circuits and systems conference (BioCAS) proceedings* (pp. 388–391). IEEE.

Nazmi, N., Abdul Rahman, M. A., Yamamoto, S. I., Ahmad, S. A., Zamzuri, H., & Mazlan, S. A. (2016). A review of classification techniques of EMG signals during isotonic and isometric contractions. *Sensors*, *16*(8), 1304.

Nenadic, Z., & Burdick, J. W. (2004). Spike detection using the continuous wavelet transform. *IEEE Transactions on Biomedical Engineering*, *52*(1), 74–87.

Nguyen, T., Bhatti, A., Khosravi, A., Haggag, S., Creighton, D., & Nahavandi, S. (2015). Automatic spike sorting by unsupervised clustering with diffusion maps and silhouettes. *Neurocomputing*, *153*, 199–210.

Niketeghad, S., & Pouratian, N. (2019). Brain machine interfaces for vision restoration: The current state of cortical visual prosthetics. *Neurotherapeutics*, *16*(1), 134–143.

Nikolic, Z. M., Popovic, D. B., Stein, R. B., & Kenwell, Z. (1994). Instrumentation for ENG and EMG recordings in FES systems. *IEEE Transactions on Biomedical Engineering*, *41*(7), 703–706.

Noce, E., Ciancio, A. L., & Zollo, L. (2018). Spike detection: The first step towards an ENG-based neuroprosheses. *Journal of Neuroscience Methods*, *308*, 294–308.

Olin, J. C., Kidd, D. H., & North, R. B. (1998). Postural changes in spinal cord stimulation perceptual thresholds. *Neuromodulation: Technology at the Neural Interface*, *1*(4), 171–175.

Oskoei, M. A., & Hu, H. (2008). Support vector machine-based classification scheme for myoelectric control applied to upper limb. *IEEE Transactions on Biomedical Engineering*, *55*(8), 1956–1965.

Oweiss, K. G. (2010). *Statistical signal processing for neuroscience and neurotechnology* (1st ed.). Academic Press.

Öztürk, S., Devecioğlu, İ., Beygi, M., Atasoy, A., Mutlu, Ş., Özkan, M., & Güçlü, B. (2019). Real-time performance of a tactile neuroprosthesis on awake behaving rats. *IEEE Transactions on Neural Systems and Rehabilitation Engineering*, *27*(5), 1053–1062.

Paiss, O., & Inbar, G. F. (1987). Autoregressive modeling of surface EMG and its spectrum with application to fatigue. *IEEE Transactions on Biomedical Engineering* (10), 761–770.

Pakhira, M. K., Bandyopadhyay, S., & Maulik, U. (2004). Validity index for crisp and fuzzy clusters. *Pattern Recognition*, *37*(3), 487–501.

Pal, S., Heyns, P. S., Freyer, B. H., Theron, N. J., & Pal, S. K. (2011). Tool wear monitoring and selection of optimum cutting conditions with progressive tool wear effect and input uncertainties. *Journal of Intelligent Manufacturing*, *22*(4), 491–504.

Pani, D., Barabino, G., Citi, L., Meloni, P., Raspopovic, S., Micera, S., & Raffo, L. (2016). Real-time neural signals decoding onto off-the-shelf DSP processors for neuroprosthetic applications. *IEEE Transactions on Neural Systems and Rehabilitation Engineering*, *24*(9), 993–1002.

Park, H., & Durand, D. M. (2008). Motion control of musculoskeletal systems with redundancy. *Biological Cybernetics*, *99*(6), 503–516.

Pasquina, P. F., Evangelista, M., Carvalho, A. J., Lockhart, J., Griffin, S., Nanos, G., Hansen, P. M., Ipsen, D., Vandersea, J., Butkus, J., Miller, M., Murph, I., & Hankin, D. (2015). First-in-man demonstration of a fully implanted myoelectric sensors system to control an advanced electromechanical prosthetic hand. *Journal of Neuroscience Methods*, *244*, 85–93.

Pedreira, C., Martinez, J., Ison, M. J., & Quiroga, R. Q. (2012). How many neurons can we see with current spike sorting algorithms? *Journal of Neuroscience Methods*, *211*(1), 58–65.

Phinyomark, A., Phukpattaranont, P., & Limsakul, C. (2012). Feature reduction and selection for EMG signal classification. *Expert Systems with Applications*, *39*(8), 7420–7431.

Pouzat, C., Mazor, O., & Laurent, G. (2002). Using noise signature to optimize spike-sorting and to assess neuronal classification quality. *Journal of Neuroscience Methods*, *122*(1), 43–57.

Quiroga, R. Q., Nadasdy, Z., & Ben-Shaul, Y. (2004). Unsupervised spike detection and sorting with wavelets and superparamagnetic clustering. *Neural Computation*, *16*(8), 1661–1687.

Quiroga, R. Q., Reddy, L., Koch, C., & Fried, I. (2007). Decoding visual inputs from multiple neurons in the human temporal lobe. *Journal of Neurophysiology*, *98*(4), 1997–2007.

Rafiee, J., & Tse, P. W. (2009). Use of autocorrelation of wavelet coefficients for fault diagnosis. *Mechanical Systems and Signal Processing*, *23*(5), 1554–1572.

Ramirez-Zamora, A., Giordano, J. J., Boyden, E. S., Gradinaru, V., Gunduz, Starr, P. A., Sheth, S. A., McIntyre, C. C., Fox, M. D., Vitek, J., Vedam-Mai, V., Akbar, U., Almeida, L., Bronte-Stewart, H. M., Mayberg, H. S., Pouratian, N., Gittis, A. H., Singer, A. C., Creed, M. C., Lazaro-Munoz, G., ... Okun, M. S. (2019). Proceedings of the sixth deep brain stimulation think tank modulation of brain networks and application of advanced neuroimaging, neurophysiology, and optogenetics. *Frontiers in Neuroscience*, *13*, 936.

Raspopovic, S. (2020). Advancing limb neural prostheses. *Science 370*(6514), 290-291.

Raspopovic, S., Capogrosso, M., & Micera, S. (2011). A computational model for the stimulation of rat sciatic nerve using a transverse intrafascicular multichannel electrode. *IEEE Transactions on Neural Systems and Rehabilitation Engineering*, *19*(4), 333–344.

Raspopovic, S., Carpaneto, J., Udina, E., Navarro, X., & Micera, S. (2010). On the identification of sensory information from mixed nerves by using single-channel cuff electrodes. *Journal of Neuroengineering and Rehabilitation*, *7*(1), 17.

Raspopovic, S., Cimolato, A., Panarese, A., Vallone, F., del Valle, J., Micera, S., & Navarro, X. (2020). Neural signal recording and processing in somatic neuroprosthetic applications. A review. *Journal of Neuroscience Methods*, *337*, 108653.

Raspopovic, S., Petrini, F. M., Zelechowski, M., & Valle, G. (2017). Framework for the development of neuroprostheses: From basic understanding by sciatic and median nerves models to bionic legs and hands. *Proceedings of the IEEE*, *105*(1), 34–49.

Raspopovic, S., Valle, G., & Petrini, F.M. (2021). Sensory feedback for limb prostheses in amputees. *Nature Materials*, 1–15.

Razmpour, S., Shaeri, M. A., Hosseini-Nejad, H., & Sodagar, A. M. (2015). *Signal processing in implantable neural recording microsystems. Telemedicine and electronic medicine* (pp. 416–443). CRC Press.

Rey, H. G., Pedreira, C., & Quiroga, R. Q. (2015). Past, present and future of spike sorting techniques. *Brain Research Bulletin*, *119*, 106–117.

Rossini, P. M., Micera, S., Benvenuto, A., Carpaneto, J., Cavallo, G., Citi, L., & Ferreri, F. (2010). Double nerve intraneural interface implant on a human amputee for robotic hand control. *Clinical Neurophysiology*, *121*(5), 777–783.

Roy, Y., Banville, H., Albuquerque, I., Gramfort, A., Falk, T. H., & Faubert, J. (2019). Deep learning-based electroencephalography analysis: A systematic review. *Journal of Neural Engineering*, *16*(5), 051001.

Russell, S. J., & Norvig, P. (2010). *Artificial intelligence-A modern approach* (3rd ed.). Prentice Hall.

Rutishauser, U., Schuman, E. M., & Mamelak, A. N. (2006). Online detection and sorting of extracellularly recorded action potentials in human medial temporal lobe recordings, in vivo. *Journal of Neuroscience Methods*, *154*(1–2), 204–224.

Saal, H. P., Delhaye, B. P., Rayhaun, B. C., & Bensmaia, S. J. (2017). Simulating tactile signals from the whole hand with millisecond precision. *Proceedings of the National Academy of Sciences*, *114*(28), E5693–E5702.

Sahani, M. (1999). *Latent variable models for neural data analysis*. Pasadena, CA: California Institute of Technology.

Salgado, I., Alfaro-Ponce, M., Camacho, O., & Chairez, I. (2019). Robust identification of unknown inputs in electrical stimulation of ex-vivo animal models. *Biomedical Signal Processing and Control*, *52*, 103–110.

Salinas, E., & Abbott, L. F. (1994). Vector reconstruction from firing rates. *Journal of Computational Neuroscience*, *1*(1–2), 89–107.

Salmanpour, A., Brown, L. J., & Shoemaker, J. K. (2010). Spike detection in human muscle sympathetic nerve activity using a matched wavelet approach. *Journal of Neuroscience Methods*, *193*(2), 343–355.

Sato, T., Suzuki, T., & Mabuchi, K. (2007, August). Fast automatic template matching for spike sorting based on Davies-Bouldin validation indices. In *2007 29th annual international conference of the IEEE engineering in medicine and biology society* (pp. 3200–3203). IEEE.

Schade, C. M., Schultz, D., Tamayo, N., Iyer, S., & Panke, E. (2011). Automatic adaptation of neurostimulation therapy in response to changes in patient position: Results of the Posture Responsive Spinal Cord Stimulation (PRS) research study. *Pain Physician*, *14*, 407–417.

Schlögl, A., & Supp, G. (2006). Analyzing event-related EEG data with multivariate autoregressive parameters. *Progress in Brain Research*, *159*, 135–147.

Schultz, D. M., Webster, L., Kosek, P., Dar, U., Tan, Y., & Sun, M. (2012). Sensor-driven position-adaptive spinal cord stimulation for chronic pain. *Pain Physician*, *15*(1), 1–12.

Shaeri, M., & Sodagar, A. M. (2020). A framework for on-implant spike sorting based on salient feature selection. *Nature Communications*, *11*(1), 1–9.

Shute, J. B., Okun, M. S., Opri, E., Molina, R., Rossi, P. J., Martinez-Ramirez, D., Foote, K. D., & Gunduz, A. (2016). Thalamocortical network activity enables chronic tic detection in humans with Tourette syndrome. *NeuroImage: Clinical*, *12*, 165–172.

Silveira, C., Brunton, E., Khushaba, R., & Nazarpour, K. (2018, July). Evaluation of time-domain features of sensory ENG signals. In *2018 40th annual international conference of the IEEE engineering in medicine and biology society (EMBC)* (pp. 2438–2441). IEEE.

Silveira, C., Brunton, E., Spendiff, S., & Nazarpour, K. (2018). Influence of nerve cuff channel count and implantation site on the separability of afferent ENG. *Journal of Neural Engineering*, *15*(4), 046004.

Sivakanthan, S., Neal, E., Murtagh, R., & Vale, F. L. (2016). The evolving utility of diffusion tensor tractography in the surgical management of temporal lobe epilepsy: A review. *Acta Neurochirurgica*, *158*(11), 2185–2193.

Song, D., Marmarelis, V. Z., & Berger, T. W. (2009). Parametric and non-parametric modeling of short-term synaptic plasticity. Part I: Computational study. *Journal of Computational Neuroscience*, *26*(1), 1.

Song, D., Wang, Z., Marmarelis, V. Z., & Berger, T. W. (2009). Parametric and non-parametric modeling of short-term synaptic plasticity. Part II: Experimental study. *Journal of Computational Neuroscience*, *26*(1), 21–37.

Swann, N. C., de Hemptinne, C., Thompson, M. C., Miocinovic, S., Miller, A. M., Ostrem, J. L., Chizeck, H. J., & Starr, P. A. (2018). Adaptive deep brain stimulation for Parkinson's disease using motor cortex sensing. *Journal of Neural Engineering*, *15*(4), 046006.

Takekawa, T., Isomura, Y., & Fukai, T. (2010). Accurate spike sorting for multi-unit recordings. *European Journal of Neuroscience*, *31*(2), 263–272.

Tan, D. W., Schiefer, M. A., Keith, M. W., Anderson, J. R., Tyler, J., & Tyler, D. J. (2014). A neural interface provides long-term stable natural touch perception. *Science Translational Medicine*, *6*(257), 257ra138–257ra138.

Theunissen, F., & Miller, J. P. (1995). Temporal encoding in nervous systems: A rigorous definition. *Journal of Computational Neuroscience*, *2*(2), 149–162.

Valle, G., Saliji, A., Fogle, E., Cimolato, A., Petrini, F. M., & Raspopovic, S. (2021). Mechanisms of neuro-robotic prosthesis operation in leg amputees. *Science Advances*, *7*(17), eabd8354.

Wagner, F. B., Mignardot, J. B., Le Goff-Mignardot, C. G., Demesmaeker, R., Komi, S., Capogrosso, M., Rowald, A., Seáñez, I., Caban, M., Pirondini, E., Vat, M., McCracken, L. A., Heimgartner, R., Fodor, I., Watrim, A., Seguin, P., Paoles, E., Van Den Keybus, K., Eberle, G., ... Courtine, G. (2018). Targeted neurotechnology restores walking in humans with spinal cord injury. *Nature*, *563*(7729), 65–71.

Wenger, N., Moraud, E. M., Gandar, J., Musienko, P., Capogrosso, M., Baud, L., Le Goff, C. G., Barraud, Q., Pavlova, N., Dominici, N., Minev, I. R., Asboth, L., Hirsch, A., Duis, S., Mortera, A., Haverbeck, O., Kraus, S., Schmetz, F., DiGiovanna, J., ... Courtine, G. (2016). Spatiotemporal neuromodulation therapies engaging muscle synergies improve motor control after spinal cord injury. *Nature Medicine*, *22*(2), 138–145.

Wenger, N., Moraud, E. M., Raspopovic, S., Bonizzato, M., DiGiovanna, J., Musienko, P., Morari, M., Micera, S., & Courtine, G. (2014). Closed-loop neuromodulation of spinal sensorimotor circuits controls refined locomotion after complete spinal cord injury. *Science Translational Medicine*, *6*(255), 255ra133.

Wilamowska, Z. A., Thompson-Hollands, J., Fairholme, C. P., Ellard, K. K., Farchione, T. J., & Barlow, D. H. (2010). Conceptual background, development, and preliminary data from the unified protocol for transdiagnostic treatment of emotional disorders. *Depression and Anxiety*, *27*(10), 882–890.

Wiltschko, A. B., Gage, G. J., & Berke, J. D. (2008). Wavelet filtering before spike detection preserves waveform shape and enhances single-unit discrimination. *Journal of Neuroscience Methods*, *173*(1), 34–40.

Yoshida, K., Bertram, M. J., Hunter Cox, T. G. & Riso, R. R. (2017). Chapter 2.6: Peripheral Nerve Recording Electrodes and Techniques. In W. H. Kenneth, R.K. Daryl (Eds.), *Series on Bioengineering and Biomedical Engineering; Neuroprosthetics* (pp. 377–466). World Scientific.

Zhang, P. M., Wu, J. Y., Zhou, Y., Liang, P. J., & Yuan, J. Q. (2004). Spike sorting based on automatic template reconstruction with a partial solution to the overlapping problem. *Journal of Neuroscience Methods*, *135*(1–2), 55–65.

Chapter 21

Safety and regulatory issues for clinical testing

Daniel R. Merrill
Dan Merrill Consulting, LLC, Salt Lake City, UT, United States

ABSTRACT
Medical device development under the convention of design control is presented in this chapter. The value of design control is presented as not only a regulatory necessity but also the most efficient process for industry. A typical testing process flow for verification and validation of a medical device is described, and specific test methods and standards are discussed. The regulatory pathways for investigational device evaluation through clinical studies are presented, and significant risk and nonsignificant risk paths are contrasted. The goals of feasibility and pivotal clinical studies are presented, the regulatory pathways for clearance or approval to the commercial market are described, and the regulatory processes in the United States and European Union are contrasted.

Keywords: Design control; testing; verification; validation; clinical study; market; clearance; approval

21.1 Relationships of quality, regulatory, safety, and testing with clinical studies

As defined in ISO 9000:2015: Quality Management Systems—Fundamentals and Vocabulary, *quality* is the "degree to which a set of inherent characteristics of an object fulfills requirements." A requirement is a "need or expectation that is stated, generally implied or obligatory." Quality is best understood as a mindset rather than a formalization, so that a quality-based system delivers products that meet requirements, including those which are implicitly expected. An important principle of ISO 9000 is understanding the needs of customers. It is a common mistake for engineers to believe they know what a customer desires and design around such beliefs, only to meet failure at market time.

A central standard in quality management is ISO 13485:2016 Medical Devices—Quality Management Systems—Requirements for regulatory

Somatosensory Feedback for Neuroprosthetics. DOI: https://doi.org/10.1016/B978-0-12-822828-9.00012-5

purposes. Although ISO 13485 is a stand-alone standard, it is based on ISO 9001:2008, which has been superseded by ISO 9001:2015.

Risk and hazard analysis, described in ISO 14971:2019, is an important element of medical device development and should be instituted as an ongoing process throughout the product lifecycle. While beyond the scope of this treatment, a few definitions per 14971 are given here. *Harm* is physical injury or damage to the health of people, or damage to property or the environment. A *hazard* is a potential source of harm. *Risk* is the combination of the probability of occurrence of harm and the severity of consequences of that harm. Two tools commonly used in risk management are fault tree analysis (FTA), which is a top-down approach (moving from system-level to component) applied early in design, and failure mode and effects analysis (FMEA) or failure mode, effects and criticality analysis (FMECA), which are bottom-up approaches (considering the effects of individual component failures) applied later in design.

While quality is an approach for ensuring requirements are met, regulatory systems ensure the laws and regulations regarding medical devices are met. Regulatory enforcement is implemented by various agencies such as the Food and Drug Administration (FDA) in the United States. The US FDA and EU regulatory processes are compared in Section 21.6.

Developing medical devices with acceptable levels of safety is paramount. As implemented in the design of clinical studies as well as expectations from regulatory bodies, safety must be demonstrated prior to focusing on efficacy. Risk–benefit analyses are useful in balancing decisions; for example, a device which likely extends a quality life by years may warrant taking greater risks to safety than a device which provides cosmetic or marginal benefits. Risk tolerance is partly a cultural characteristic. In the United States there is little tolerance for risk, and medical device developers are expected to follow the adage of "do no harm."

Testing is an integral part of the device development process (US Code of Federal Regulations, 2019a; US FDA, 1997). *Verification testing* is the process of ensuring design output meets design input (detailed in Section 21.2). Design output is the finished device and its specifications, labeling, packaging, and the device master record. *Validation testing*, performed after verification, ensures that user needs are met. Colloquially, verification asks "Did we build the thing right?" and validation asks "Did we build the right thing?" It is straightforward to get a "yes" to the first question and a "no" to the second if user needs are not well understood.

In some instances, validation includes clinical (human) studies of an investigational device. Such clinical studies occur near the end of the overall validation process to ensure safety and efficacy prior to commercialization of a medical device allowing prescription by a physician for routine clinical use. Various types of clinical studies and their intent are described in Section 21.4.

21.2 Medical device lifecycle phases and design control

The cumulative experience of medical device manufacturers over several decades provides a useful framework for systematic device development. The process of *design control* is expressed in the Code of Federal Regulations as 21 CFR 820.30 (US Code of Federal Regulations, 2019a); 820.30 is only a few paragraphs in length. An additional layer has been developed within industry to interpret 820.30, and while not regulation, it provides convention. Design control is not just regulation, and it is not a barrier; rather it is a systematic process for the most efficient device development program, based on hard-learned lessons from history.

The first step in a successful design program is acquiring a robust understanding of the users' needs. Users may include end-users (patients), caregivers, or physicians. Other stakeholders such as hospital administrators or insurers should be involved in requirements affecting business decisions (cost, supply chain). Users should continue providing feedback as concepts are developed. Unfortunately, this step is far too often done without due attention, in which case the design moves progressively away from the market demands.

Design input begins early in development. *Design input requirements* (DIRs) start with user needs and translate these needs into engineering terms. The DIR is one of the most important documents of a development program, and is the basis for test criteria during verification testing and validation testing. Along with the risk and hazard analysis, the DIR will be evaluated by regulatory bodies.

High-level requirement types include efficacy/effectiveness, safety, and reliability (detailed in Chapter 3: Electrodes and Instrumentation for Neurostimulation). *Effectiveness* refers to producing the intended objective or results. *Safety* refers to avoiding unacceptable levels of damage to the subject or patient, or to the system. *Reliability* refers to maintaining requirements for effectiveness and safety over the intended lifetime of a system. Aspects of effectiveness are routinely divided into functional, performance, and interface requirements. Functional requirements describe what the device does, performance requirements give quantitative metrics of "how much" or "how well" the device performs, and interface requirements describe how the device interfaces with other hardware, software, or users.

Fig. 21.1, commonly known as "the waterfall," is a model for the device development process (US FDA, 1997). A developer begins by translating user needs into engineering terms. This is captured in the DIR. The inner loop of the waterfall (design input → design process → design output → verification) is iterative and may occur multiple times. The first-generation loop may consist of developing explicit engineering specifications based on the requirements. Specifications initially are a design output and are verified to meet the requirements, a design input. Upon this first verification,

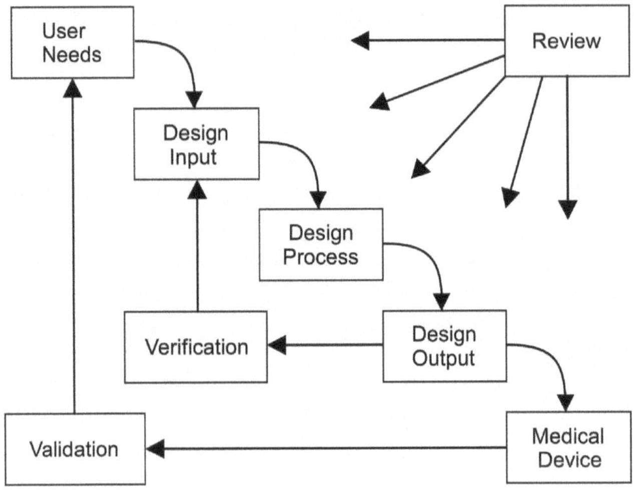

FIGURE 21.1 The waterfall model for device development.

specifications shift from being a design output to being a design input. Multiple loops occur as a design (design input, in the form of specifications) which is realized (built) during the design process, with an output of a device to be tested. Verification testing ensures the device meets specifications. For any complex device, the first-generation design will be modified after verification testing, then another loop of design (input) → build (process) → device in hand (output) → verification against design is performed. Once the manufacturer is convinced that the manufactured device meets specifications, validation testing is performed against the finalized medical device to ensure that the user needs are met.

According to FDA guidance (US FDA, 1997), design output includes production specifications as well as descriptive materials which define and characterize the design. *Specification* is defined as any requirement to which the product must conform. Production specifications include drawings and documents used to procure components, fabricate, test, inspect, install, maintain, and service the device. Design output is the finished device and its specifications, labeling, packaging, and the device master record.

Formalized design reviews are held throughout the device development process. These are held at strategic gating points in the process, described below. The reviews should include members from each key department (management, engineering, quality, regulatory, clinical, sales, and marketing) and one independent reviewer not involved with the project. The review should not be considered a checkmark to proceed; rather the team should feel empowered to halt a project that does not show promise to meet objectives.

A common mistake is to assume that a particular embodiment of a solution is a user need, and to embed such embodiment (a specification) into the

DIR as a requirement. The consequence is that alternative solutions are ruled out unnecessarily early. For example, assume clinicians desire to have wireless communication between an external programmer and an implanted device. They need to have a certain bandwidth, latency, and reliability for communication. The designer incorrectly assumes that Bluetooth Low Energy (BLE) is the only solution, and records this as a user need. This forces an early decision to not consider MICS, WiFi, Zigbee, or other radio bands. If the user truly wants BLE, it is a user need; if they only care about certain parameters of wireless communication, BLE is a specification to be determined.

An important question is, "When in the product lifecycle should design control begin?" The chosen answer has a heavy impact on resources and efficiency. During the research phase, the feasibility of engineering approaches is explored. If design control is implemented too early while still in the research phase, it becomes burdensome with minimal value. If design control is implemented too late, opportunities are lost to be efficient with development. The correct point to cut in design control is when one begins designing for clinical use, which corresponds to the development and approval of design inputs.

The overall product lifecycle can be divided into phases, as illustrated in Fig. 21.2.

During the project selection/concept phase, nine questions/issues should be addressed. (1) Is the project a strategic fit with the company mission? If no, perhaps the mission needs to change, which may be reasonable. However, a company in search of a product without a defined mission tends to be adrift. (2) What are the unmet needs which the product will address? These identified needs will feed into the DIR and become important drivers in development. (3) What is the market size? Two key parameters are incidence (rate of new cases, e.g., new cases in the EU per 100,000 people per

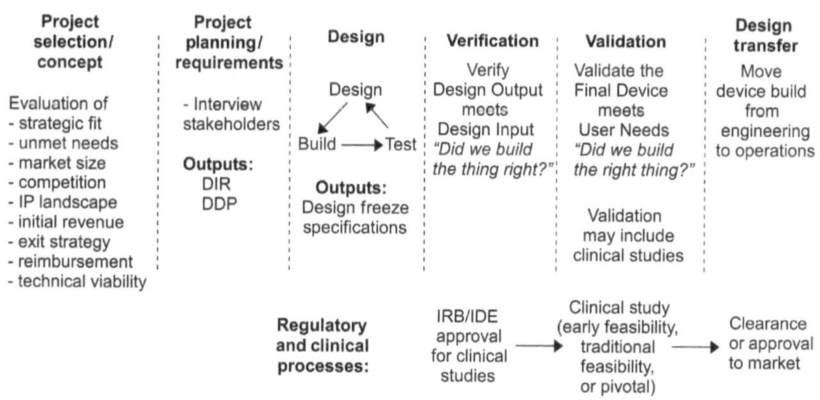

FIGURE 21.2 Phases of product development.

year) and prevalence (total number of cases present, e.g., total existing cases in the United States), known as I&P. (4) Competitive landscape: what is the status of current and emerging competitors in the product space? (5) Intellectual property landscape: is it expected that the developer will have freedom to operate (FTO), that is, be able to sell their product without infringing on issued intellectual property. Two key elements of a business model include: (6) What is the initial revenue model which will support the company during the lean years, and (7) What is the exit strategy? An exit strategy is an entrepreneur's plan to sell their ownership in a company to another company or to investors. Common exit strategies are strategic acquisitions by a large company, and initial public offerings. (8) A reimbursement strategy should be developed. Reimbursement is complex, and developers should begin this planning early. (9) Technical feasibility.

The project selection/concept phase is the research phase, prior to implementing design control, when fundamental questions of feasibility are addressed. Also known as "playing in the sandbox," it is an enjoyable place for engineers, with few regulations, and the temptation is to avoid leaving this phase. Opposing this desire is the mission to deliver product to those in need, which requires starting a rigorous development program under design control.

Before proceeding to the next phase (project planning), and in fact at the end of each phase, is a phase gate review. A cross-functional team reviews the goals and progress of the phase to be closed, and determines if objectives of the phase have been adequately met, if the project remains technically viable, and if it is in the best interests of the business to proceed. These are appropriate checkpoints to redirect or halt a project.

In the project planning phase, a diligent effort is undertaken to understand user needs. This may take the form of questionnaires, interviews in private or group settings, or time spent in the clinic with practitioners. User needs are translated into engineering terms and captured in the DIR document. Two primary outputs of the project planning phase are the DIR and the Design and Development Plan (DDP). The DDP describes overall project objectives, resource allocations, phase objectives, expected timelines, and success criteria for proceeding between phases. With the release of the DIR and DDP and exit from this phase, the project is maintained under design control.

The design phase comprises the bulk of engineering work. As described above for the waterfall, the design phase consists of multiple iterations of design (input) → build (process) → device in hand (output) → verification against design. This cycle repeats until the manufactured device meets specifications. At this point, the device is considered frozen. "Design freeze" is an important concept. A frozen design means the engineers stop optimizing, and devices are built according to the frozen design for formal verification and validation testing. Failure to have an identified point of design freeze makes the version control of testing unmanageable.

The next two phases are verification and validation, collectively known as qualification. Verification confirms that design output meets design input ("we built the thing right") and validation confirms that the final device meets user needs ("we built the right thing"). Devices tested in these two phases should be manufactured according to the frozen design and built according to finalized manufacturing processes.

If the devices undergoing verification and validation testing were not built according to frozen design or by finalized manufacturing processes, equivalency must be argued, that is, that any differences from final design or processes will not affect test results. Many of the descriptions in these last paragraphs are idealizations. While the phase transitions are shown as sharp in Fig. 21.2, they seldom are—there is usually substantial phase overlap. Elements of design phase begin prior to formal release of the DIR. Notably, verification in reality occurs partly within the design phase (as "test" in the loop), and partly within the formalized verification phase after design freeze. The push for time to market drives manufacturers to optimize their processes. Finding the right balance between development speed and maintaining the integrity to provide a safe and effective device is a matter of conscientiousness on the part of the manufacturer; it can never be fully regulated into place.

As part of the validation phase, clinical studies may be performed to demonstrate safety and efficacy. Section 21.4 discusses types of clinical studies, Section 21.5 discusses the instances when these are required, and Section 21.6 compares the differences between EU and US processes. At the bottom of Fig. 21.2 is an overlay of US clinical regulatory processes. One or more clinical studies may be required as part of validation. Prior to initiating a study, approval must be granted by the local institutional review board (IRB) and possibly by the FDA as an investigational device exemption (IDE). Results from these studies, along with benchtop and preclinical (animal) studies are presented in the application for commercialization via clearance or approval to market.

Design transfer is the process of transferring and scaling up manufacturing from an engineering-led effort to the operations group. Manufacturing process validation typically occurs across both the validation and design transfer phases.

Not shown in Fig. 21.2 is reimbursement planning. Reimbursement is not required for market commercialization but is usually required for sustained profitability. In the United States, reimbursement involves three key elements: codes, coverage, and payment (US CMS, 2020a, 2020b). The Centers for Medicare and Medicaid Services (CMS) assign a Healthcare Common Procedure Coding System code for devices and define coverage. Private insurers largely follow the coverage recommendations from CMS. Reimbursement rules are notoriously complex. It is important for manufacturers to start reimbursement planning early in the product cycle to develop supporting data and gain support from the medical community.

21.3 Verification and validation testing

Testing of a candidate medical device to demonstrate safety and efficacy/effectiveness is performed prior to human use, either for clinical studies of an investigational device or for commercialization to the market for prescription by a physician. The suite of tests to be performed depends on many factors including risk, and novelty of the device vs ability to claim equivalence to prior devices considered safe and effective. It is in the best interest of manufacturers to develop a thoughtful proposal for testing and to discuss this early with the FDA as part of the presubmission or Q-submission process (US FDA, 2019). Table 21.1 lists several of the common verification and validation tests.

Each of the device specifications for function, performance, and interface as given by the manufacturer should be verified. Example specifications for a neurostimulator may include pulse parameters, implant-to-external device power coupling, power consumption, and electrode impedance. Table 3.5 in Chapter 3, Electrodes and Instrumentation for Neurostimulation, lists several specifications for a stimulation system.

Biocompatibility must be demonstrated for any device in contact with tissue, including both active devices and strictly passive devices. These tests are described in ISO 10993: Biological Evaluation of Medical Devices. Three device contact categories are defined in ISO 10993-1 including surface, external communicating, and implant. Three contact duration periods are defined including limited ($\leq$24 hour), prolonged ($>$24 hour to $\leq$30 days), and permanent ($>$30 days). Based on the nature of contact and contact duration, the biocompatibility matrix given in ISO 10993-1 specifies tests to be performed. Cytotoxicity, sensitization, and irritation tests, known as the "big three," are performed on all devices. Cytotoxicity is an in vitro test using incubation of cultured cells and is specified by ISO 10993-5. Because cytotoxicity is fast and inexpensive, and provides a good part of the confidence required for biocompatibility, it may be used as a early standalone test when evaluating feasibility of a novel material or revised formulation. Sensitization and irritation are specified by 10993-10. Sensitization involves exposure of animal's skin (guinea pig or rodent) to a sample or extracts, injected or applied topically. Irritation involves exposure of rabbit to the sample or extracts. Other biocompatibility tests include (1) acute systemic toxicity, evaluating the effects remote from the site of contact, performed as a single dose in rodent, and subacute/subchronic/chronic toxicity, evaluating the effects due to repeated exposure in rodent. Both are specified by 10993-11. (2) Genotoxicity, including Ames mutagenicity, chromosomal aberration, mouse lymphoma, and mouse micronucleus, is an in vitro test evaluating mutagenicity in isolated cells. It is specified by 10993-3. (3) Material-mediated pyrogenicity per 10993-11 evaluates pyrogenicity from nonendotoxin-related factors (endotoxin is a toxin inside bacteria) in rabbit.

TABLE 21.1 Common tests in verification and validation testing.

1. Verify specifications for function, performance, interface	Per manufacturer's specifications
2. Biocompatibility	ISO 10993-1
Cytotoxicity, sensitization, irritation	ISO 10993-5, -10
Acute systemic toxicity, subchronic toxicity	ISO 10993-11
Genotoxicity	ISO 10993-3
Material-mediated pyrogenicity	ISO 10993-11
Hemocompatibility	ISO 10993-4
13-week implant	ISO 10993-6
Chemical characterization	ISO 10993-18, USP Physiochemical tests
3. Electrical safety	IEC 60601-1, -1-4, etc., ISO 14708-1, -3
4. Electromagnetic compatibility	IEC 60601-1-2
5. Sterilization validation	ISO 11135 (for EtO)
6. Hermeticity of implanted packages	MIL-STD-883H
7. Impact	IEC 60068-2-75
8. Connector mechanical tests	EN 45502-2-1
9. Shipping	ISTA 2A
10. Shelf life	
11. Accelerated lifetime (soak testing)	
12. MR characterization	ISO/IEC 10974
13. Preclinical (animal) testing of safety and feasibility	
14. Clinical (human) validation of safety and efficacy	
15. Manufacturing process validation	

(4) Hemocompatibility per 10993-4 evaluates the interactions of a device with human blood. (5) Implantation testing per 10993-6 uses histology to determine the local tissue effects around a device implanted for an extended period. (6) Chemical characterization per 10993-18 and USP physiochemical tests develops a baseline of the chemicals included in a device. While cytotoxicity testing is rapid and inexpensive, a full set of biocompatibility tests is time-consuming and expensive (up to hundreds of thousands of dollars) and should be planned early. These tests are usually contracted.

The electrical safety of a medical device entails many components. The primary standard is IEC 60601-1, along with its collaterals (60601-1-X) and particulars (60601-2-X). Collateral 60601-1-4 is for devices containing software, known as programmable electrical medical systems. In addition to IEC 60601, ISO 14708-1 is a required safety standard for active implantable medical devices. ISO 14708-3 is specific to implantable neurostimulators.

Electromagnetic compatibility (EMC) is defined bidirectionally, requiring both an acceptable level of emissions by the device and susceptibility to other emitters. It is covered by IEC 60601-1-2 and other standards specific to the radio band and locality.

Sterilization validation is specific to the chosen method [e.g., ethylene oxide (EO or EtO), autoclave, radiation, etc.]. For EO the relevant standard is ISO 11135.

Hermeticity (the characteristic of being airtight) must be demonstrated to an acceptable level for any implanted device. This shows that water ingress will be minimized, thus preventing damage to electronics inside an enclosure. A common method for measuring hermeticity is the helium bomb method, whereby a device is exposed to helium gas and the outgassing through the enclosure is measured. The extent of required hermeticity is calculated based on factors such as the volume of the enclosure and intended device lifetime.

Tests of susceptibility to mechanical stresses include impact per IEC 60068-2-75, and mechanical stresses at lead connectors per EN 45502-2-1.

Shipping tests include two categories: those intended to show integrity of the shipping package including maintenance of sterility, and tolerance of the medical device to shock, vibration, temperature, and humidity during shipment.

Shelf-life testing demonstrates the ability of a device to maintain its specifications, and the ability of packaging to maintain sterility, over an extended period. These may be two distinct criteria. Lifetime may be demonstrated initially using an accelerated lifetime approach, followed by real-time observation of aging devices.

Accelerated lifetime testing is common for estimating a device's lifetime. A rule-of-thumb interpretation of the Arrhenius equation states that every 10°C increase in temperature equates to a doubling of equivalent lifetime. Thus, a device intended for implantation at 37°C which is placed in a 67°C saline bath for 3 months will experience an equivalent lifetime of $2^3 \times 3$ months = 24 months. In addition to elevating temperature, other means can be used to accelerate equivalent lifetime, for example, pulsing a neurostimulator at a higher duty cycle than the expected use case.

MR characterization per ISO/IEC 10974 evaluates translational force on the device, torque on the device, heating, spurious outputs from the device, and device damage. A medical device may be labeled as (1) MR safe, which means safe in all MR environments, (2) MR conditional, which means safe under specified conditions, or (3) MR unsafe. Additionally, a device is labeled as MR compatible if it is both safe and does not affect diagnostic ability. An MR safe device which is not MR compatible creates artifact in the MR diagnosis but is otherwise safe.

Preclinical (animal) testing is routinely performed to demonstrate the safety and feasibility of a device. Small animals such as rodents may be useful for safety demonstration, whereas the feasibility of a device often requires a species which better models humans for the intended indication.

Human validation of safety and efficacy in a clinical study (discussed in Section 21.4) is often the final step in device validation.

Manufacturing process validation occurs during the validation and design transfer phases.

21.4 Regulatory paths for clinical studies in the United States

A *clinical study* is a research study using human subjects to evaluate biomedical or health-related outcomes. The potential purposes for a clinical study of an investigational device include (1) evaluating feasibility of the device/system, and demonstration of (2) safety and (3) efficacy.

Clinical studies in the United States which are deemed as significant risk (SR) are managed by the FDA under the IDE process, described in the regulations by 21 CFR 812 (US Code of Federal Regulations, 2019b). IDE types include (1) early feasibility or (2) traditional feasibility during the exploratory stage, and (3) pivotal when collecting data for a marketing application. Data from a feasibility study may be used to inform a final design. Pivotal studies are considered a final step in collecting the safety and efficacy data required for a marketing application. Data collected during IDE pivotal studies may be used in a PMA, 510(k), de novo, or humanitarian device exemption (HDE) marketing application (Section 21.5).

An early feasibility study (EFS) is for a device early in development, prior to finalization of the design. Usually the device design will change after an EFS. The study may be used to evaluate safety and functionality in a small number of subjects (typically 10 subjects or less). An EFS may or may not be the first-in-human (FIH) study of the device.

A traditional feasibility study is used to gather initial safety and efficacy data on a near-final or final device design. This is used to inform the design of the pivotal clinical study. A traditional feasibility study may or may not be preceded by an EFS.

A pivotal study is used to collect data on the safety and efficacy of a device for a specified intended use, using a statistically justified number of subjects. Data are used in support of a marketing application. Pivotal studies are time consuming and expensive, with a higher number of subjects than a feasibility study. The results of feasibility studies should provide high confidence that the pivotal study endpoints will be met (the study will pass) prior to initiating the pivotal study.

There are two primary forms of clinical studies: observational (nonexperimental) and interventional (experimental) also known as clinical trials. In interventional studies, subjects are randomly allocated by the investigator into study groups (an intervention group or a control group). *Clinical trials* are a subset of clinical studies. Per the National Institutes of Health (US NIH, 2019), a clinical trial is a clinical study meeting the following four criteria: (1) it involves human subjects, (2) the study is designed to evaluate the effect of an intervention on the subjects, (3) the subjects are prospectively assigned to an intervention, and (4) the effect being evaluated is a health-related biomedical or behavioral outcome. A randomized controlled trial (RCT) is the gold standard for clinical study design.

Fig. 21.3 illustrates the US process for gaining approval to evaluate an investigational device in a clinical study. Two review bodies may be involved. An IRB is always involved, and the FDA may be involved. The process begins with the IRB determining whether the study is deemed SR or nonsignificant risk (NSR). Significant-risk devices include implants, devices that support or sustain life, and devices that are substantially important in diagnosing, curing, mitigating, or treating disease or in preventing impairment to health. Guidance on the SR/NSR determination is provided by the FDA (US FDA, 2006). If the IRB determines the study to be NSR, the IRB can approve initiation of the study and the FDA will not become involved. The IRB will then have oversight on conduct of the study.

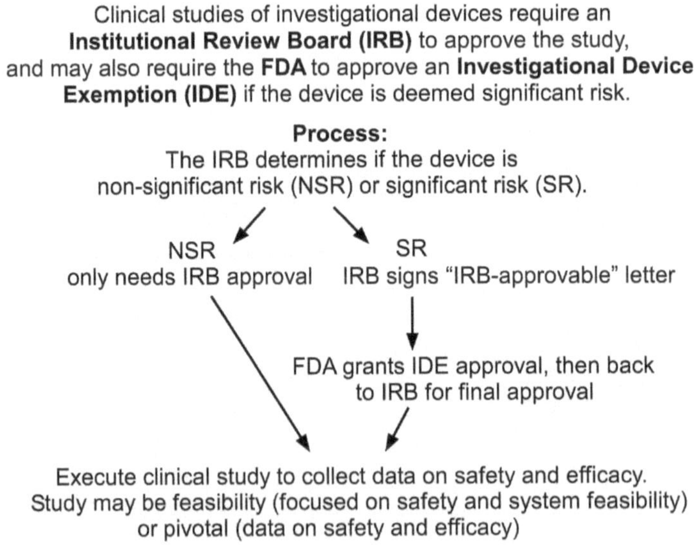

FIGURE 21.3 Routes to clinical studies in the United States.

Significant-risk studies require both FDA and IRB approval prior to initiation of the study. Upon assigning an SR classification, the IRB will review the study application and sign an "IRB-approvable" letter, essentially stating that they will approve the study as is contingent on FDA approval. The FDA then reviews, and upon granting approval sends it back to the IRB for final approval.

Upon receiving approval to initiate a clinical study, major administrative steps are, in order, (1) recruitment of subjects, (2) initial screening of individual subjects to ensure inclusion criteria are met and exclusion criteria are not violated, (3) gaining written informed consent from the subjects, (4) enrollment into the study, (5) formal screening (more detailed than initial screening), and (6) if the study is a randomized trial, randomization into study arms. Arm types include experimental, no intervention, placebo comparator, active comparator (an intervention that is considered to be effective), and sham comparator (subjects receive a device that is made like the actual device but remains inactive) (US NIH, 2020).

45 CFR 46, a.k.a. Protection of Human Subjects, provides laws set by the US Department of Health and Human Services to protect subjects from risks in research studies for which any federal agency is involved.

21.5 Regulatory paths for device commercialization in the United States

Ultimately, a medical device manufacturer desires to receive approval to commercialize their device, which allows a physician to prescribe the device for routine clinical use for the labeled indication, as opposed to only for investigational studies. This objective both serves the greatest number of patients and allows the company to be profitable (mutually reinforcing goals).

Devices are divided into three classes depending on their inherent risk and the degree of regulatory controls required to ensure safety and effectiveness. Class I are the lowest risk devices, such as a tongue depressor. Only general controls are required, and regulatory oversight is minimal. Class II devices are moderate risk and require special controls. Class III are the highest risk devices and include those that sustain or support life, are implanted for ≥ 30 days, or pose potential unreasonable risks.

There are four routes to commercialization for class II and III devices, including (1) 510(k) or premarket notification, (2) premarket approval (PMA), (3) de novo, and (4) HDE (Fig. 21.4).

The 510(k) or premarket notification process is for class II devices which are substantially equivalent to some predicate device. Substantial equivalence to a predicate requires (1) the same intended use, and (2) the same technical characteristics, or, if the technical characteristics are different there must be no new questions of safety and effectiveness and the new device must be as

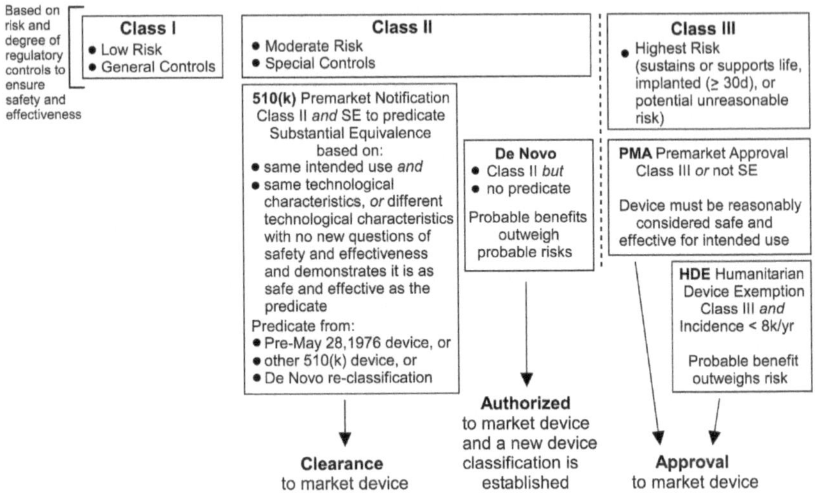

FIGURE 21.4 Four routes to commercialization.

safe and effective as the predicate. Predicate devices can be either those from prior to May 28, 1976, or other 510(k)-cleared devices, or class II devices that have received de novo classification. Devices that pass this process are described as *cleared* to the market rather than approved, underscoring that they are simply considered equivalent to something already marketed.

The PMA process is substantially more onerous than 510(k); thus, manufacturers diligently attempt to argue equivalence to predicate devices to support clearance through 510(k). This has been done with varying success. The essential steps for a 510(k) application are as follows. First, open the FDA Product Classification listing (US FDA, 2020a) to find a product code, registration number, and device class. This allows one to find product types for relevant predicates. Next, open the 510(k) database (US FDA, 2020b) and type in the product codes (US FDA, 2020a) to find potential predicate devices. Search through public information including 510(k) summaries (by clicking on the 510(k) number starting with a "K," followed by the "Summary" link). Information from the public 510(k) summaries and elsewhere are then used to generate a predicate comparison chart as part of the 510(k) application. The comparison chart lines up the proposed device with its predicate on parameters of intended use, product code and registration number, environment of use, patient contact, system components, compliance with safety standards, and all relevant functionality, performance, and interface specifications. A predicate comparison that is full of "equivalent" statements will meet the least resistance in review.

Class III or high-risk devices generally go through the PMA process. This is an expensive and time-consuming process, especially in the execution of clinical studies, and is usually avoided by small businesses. Clinical studies are required for PMA and usually are not required for 510(k), although there are exceptions. As mentioned previously for developing test programs, it is useful to develop a relationship early with the FDA as part of the pre-submission or Q-submission process (US FDA, 2019) and discuss the overall product development plan including the commercialization route and intended clinical studies.

Prior to implementation of the de novo process in 1997, any class II device without a predicate was automatically designated as requiring a PMA. The de novo classification process, properly termed "Evaluation of Automatic Class III Designation" (US FDA, 2017), removes the burden of a PMA application from class II devices that do not have a predicate. Upon passing the de novo process, a device is considered authorized to be marketed, and a new device classification is established which can be used for later 510(k) applications.

The HDE process is for humanitarian use devices, defined as those indicated for conditions where the US incidence is less than 8000 cases/year. The level of evidence required for an HDE is less than for a PMA. For an HDE, probable benefit must be shown to outweigh risk.

21.6 Comparison of European Union and United States regulatory processes

21.6.1 Clinical studies in the European Union

In the EU, clinical studies are not required to demonstrate clinical efficacy prior to commercialization, even for devices without a predicate. Evidence is required to demonstrate safety and device performance, but not efficacy.

Unlike device commercialization which occurs at the EU level, clinical studies are approved at the level of individual member states by the relevant national competent authority as well as independent ethics committees. These groups approve the clinical protocol, informed consent, and recruitment strategy, and give authorization prior to enrolling subjects in the study.

21.6.2 Device commercialization in the European Union

The device approval process is substantially different in the EU and United States. In the United States there is a single regulatory body, the FDA, which regulates both clinical studies and market applications across all US states. In the EU, regulation does not fall solely to any one agency.

In the EU, every marketed medical device must carry a CE mark (Conformite Europeenne) indicating it *conforms* to the relevant directive. A device with a CE mark can be marketed in any EU state. Notably, a CE mark is not an indication of quality. It simply states that the device is in compliance with European legislation.

Three EU directives regarding medical device approval include:

1. Directive 90/385/EEC for active implantable medical devices (directive known as AIMD);
2. Directive 93/42/EEC for most other medical devices (directive known as the medical device directive, MDD); and
3. Directive 98/79/EC for in vitro diagnostic devices (directive known as IVDD).

On April 5, 2017, the EU adopted the new Medical Device Regulation (MDR) as (EU) 2017/745, replacing the AIMD and MDD. The field is currently in a transition period between the prior directives and MDR.

In the EU, medical device risk classes are divided into classes I, IIa, IIb, and III. Table 21.2 summarizes the EU processes for medical device market approval. Nonimplantable, low-risk devices are "self-declared" by the manufacturer. Higher risk devices must be reviewed by a notified body (NB) in any member state, and authorized by that state's competent authority or health agency. NBs are private companies that contract with manufacturers

TABLE 21.2 Comparison of US vs EU device market approval processes.

US	Class I	Class II		Class III
	General controls must be applied	510(k) Premarket notification if substantial equivalence (SE) to a predicate device exists. De novo process if class II but no predicate		Premarket approval
	No clinical studies	Clinical study usually not required, but this is not guaranteed		Clinical study of safety and efficacy is required
EU	Class I	Class IIa	Class IIb	Class III
	Self-declare to the competent authority of any EU state, then can market in all EU states	Any marketed device must carry a CE mark indicating it conforms to the relevant directive. Higher risk devices are reviewed by a notified body (NB) in any member state, and authorized by that state's competent authority or health agency. Once the NB agrees that the device meets requirements for conformity, the NB issues a CE mark and the device can be marketed in all EU states		
	No clinical studies	With SE to a predicate device, generally no clinical study. If no SE to a predicate, clinical study is required to demonstrate safety and device performance, but not efficacy		

for a fee. There are multiple NBs in the EU. Once an NB agrees that the device meets requirements for conformity, the NB issues a CE mark, and the device can be marketed in all EU states.

Some NBs include:

- UK: BSI;
- Germany: TUV;
- France: GMED;
- Netherlands: BSI, DEKRA;
- Ireland: NSAI;
- Italy: IMQ Istituto Italiano del Marchio di Qualità S.P.A;
- Poland: Polskie Centrum Badan i Certyfikacji S.A.

Some competent authorities include:

- France: National Agency for the Safety of Medicine and Health Products (ANSM);
- UK: Medicines and Healthcare Products Regulatory Agency (MHRA);
- Germany: Paul Ehrlich Institute;
- Netherlands: Medicines Evaluation Board, Healthcare Inspectorate;
- Belgium: Federal Agency for Medicines and Health Products;
- Sweden: Medical Products Agency.

References

United States Centers for Medicare and Medicaid Services. (2020a). *HCPCS general information*. Retrieved from https://www.cms.gov/Medicare/Coding/MedHCPCSGenInfo/index?redirect = /MedHCPCSGenInfo/.

United States Centers for Medicare and Medicaid Services. (2020b). *Medicare coverage center*. Retrieved from https://www.cms.gov/Center/Special-Topic/Medicare-Coverage-Center?redirect = /center/coverage.asp.

United States Code of Federal Regulations. (2019a). *21 CFR 820.30, design controls*. Retrieved from https://www.accessdata.fda.gov/scripts/cdrh/cfdocs/cfcfr/CFRSearch.cfm?FR = 820.30.

United States Code of Federal Regulations. (2019b). *21 CFR 812, investigational device exemptions*. Retrieved from https://www.accessdata.fda.gov/scripts/cdrh/cfdocs/cfcfr/CFRSearch.cfm?CFRPart = 812.

United States Food and Drug Administration. (1997). *Design control guidance for medical device manufacturers*. Retrieved from https://www.fda.gov/regulatory-information/search-fda-guidance-documents/design-control-guidance-medical-device-manufacturers.

United States Food and Drug Administration. (2006). *Significant risk and nonsignificant risk medical device studies*. Retrieved from https://www.fda.gov/regulatory-information/search-fda-guidance-documents/significant-risk-and-nonsignificant-risk-medical-device-studies.

United States Food and Drug Administration. (2017). *De novo classification process (evaluation of automatic class III designation)*. Retrieved from https://www.fda.gov/regulatory-information/search-fda-guidance-documents/de-novo-classification-process-evaluation-automatic-class-iii-designation.

United States Food and Drug Administration. (2019). *Requests for feedback and meetings for medical device submissions: The Q-submission program. Guidance for industry and food and drug administration staff, US FDA*. Retrieved from https://www.fda.gov/regulatory-information/search-fda-guidance-documents/requests-feedback-and-meetings-medical-device-submissions-q-submission-program.

United States Food and Drug Administration. (2020a). *Product classification*. Retrieved from https://www.accessdata.fda.gov/scripts/cdrh/cfdocs/cfPCD/PCDSimpleSearch.cfm.

United States Food and Drug Administration. (2020b). *510(k) premarket notification*. Retrieved from https://www.accessdata.fda.gov/scripts/cdrh/cfdocs/cfPMN/pmn.cfm.

United States National Institutes of Health, United States National Library of Medicine. (2020). *Clinicaltrials.gov glossary of common site terms*. Retrieved from https://clinicaltrials.gov/ct2/about-studies/glossary.

United States National Institutes of Health. (2019). *NIH definition of clinical trial case studies*. Retrieved from https://grants.nih.gov/policy/clinical-trials/CT-Definition-Case-Studies_1.7.19.pdf.

Index

Note: Page numbers followed by "*f*" and "*t*" refer to figures and tables, respectively.

C

D

E

F

N

Q

R

FSC
www.fsc.org
MIX
Papier | Fördert gute Waldnutzung
FSC® C083411

Kolibri 360 GmbH
Ferdinand-Jühlke-Straße 7
99095 Erfurt, Deutschland
produktsicherheit@kolibri360.de